T0137140

Springer Optimization and Its Applications

Volume 194

Aims and Scope

Optimization has continued to expand in all directions at an astonishing rate. New algorithmic and theoretical techniques are continually developing and the diffusion into other disciplines is proceeding at a rapid pace, with a spot light on machine learning, artificial intelligence, and quantum computing. Our knowledge of all aspects of the field has grown even more profound. At the same time, one of the most striking trends in optimization is the constantly increasing emphasis on the interdisciplinary nature of the field. Optimization has been a basic tool in areas not limited to applied mathematics, engineering, medicine, economics, computer science, operations research, and other sciences.

The series **Springer Optimization and Its Applications (SOIA)** aims to publish state-of-the-art expository works (monographs, contributed volumes, textbooks, handbooks) that focus on theory, methods, and applications of optimization. Topics covered include, but are not limited to, nonlinear optimization, combinatorial optimization, continuous optimization, stochastic optimization, Bayesian optimization, optimal control, discrete optimization, multi-objective optimization, and more. New to the series portfolio include Works at the intersection of optimization and machine learning, artificial intelligence, and quantum computing.

Volumes from this series are indexed by Web of Science, zbMATH, Mathematical Reviews, and SCOPUS.

Duc A. Tran · My T. Thai ·
Bhaskar Krishnamachari
Editors

Handbook on Blockchain

 Springer

Editors
Duc A. Tran
Department of Computer Science
University of Massachusetts Boston
Boston, MA, USA

My T. Thai
Department of Computer and Information
Science and Engineering
University of Florida
Gainesville, FL, USA

Bhaskar Krishnamachari
Department of Electrical and Computer
Engineering
University of Southern California
Los Angeles, CA, USA

ISSN 1931-6828   ISSN 1931-6836 (electronic)
Springer Optimization and Its Applications
ISBN 978-3-031-07537-7   ISBN 978-3-031-07535-3 (eBook)
https://doi.org/10.1007/978-3-031-07535-3

Mathematics Subject Classification: 68-00, 68-02

This Springer imprint is published by the registered company Springer Nature Switzerland AG
The registered company address is: Gewerbestrasse 11, 6330 Cham, Switzerland

To the memory of my dad

—Duc A. Tran

To the memory of my dearest brother, anh Huan…

—My T. Thai

To Lila, Anuj, Appa, and the memory of my dearest Amma

—Bhaskar Krishnamachari

Preface

Most software applications today are distributed. Research in Distributed Computing started in the 1970s to answer increasing needs for replication of systems on more than one machine to enhance security and robustness, utilization of distributed resources to expand computing power, and support of specific applications that must involve tasks at different places. The computers in a distributed system are however controlled by a central operator. This is perfectly fine for mission-critical applications, but in a system serving consumers or everyday users, they may be unfairly and unverifiably subject to censorship, manipulation, and irregularities.

Decentralized Computing removes the operator role from Distributed Computing. The very first work was in 1979, but developments in Decentralized Computing became popular in the late 1990s when the Peer-to-Peer term was coined. A system can be self-managed by the participants who are seen as equal peers and transact with each other in a peer-to-peer manner. At the time though, the main motivation was to improve scalability due to performance bottleneck at the server/operator. Fast forward to 2008, Satoshi Nakamoto published a white paper titled "Bitcoin: A Peer-to-Peer Electronic Cash System" and later built the Bitcoin network. That was when *true* Decentralized Computing was realized for real-world applications at a large scale. True, because Bitcoin is a breakthrough real-world implementation of a system where admission is completely permissionless, decision-making completely decentralized, participants are truly peers, and transactions directly peer-to-peer without any central roles. That began the era of blockchain computing.

Blockchain is more than just a decentralized computing technology. Interestingly, its best value is not in scalability which is the driver for early P2P networks. Blockchain is about decentralization plus transparency, security, and trust guarantees. These core elements make blockchain a foundational technology to transform in better ways how we live and transact, with freedom, ease, and many barriers, geographical, financial, or political, removed. That said, the field of blockchain is still young. Every new technology needs time to mature. Since the birth of Bitcoin, many developments have followed, including core methods to enhance existing shortcomings or add new vertical features to enable more services, platforms to make the jobs of application developers easier, and practical applications to bring the benefits of blockchain directly to mass adoption.

This book is an effort to be a reliable source of academically produced materials helpful for those who want to venture into the blockchain field. Readers will have a basic understanding of blockchain and know its potentials and limitations. Each chapter, in survey or expository form, is self-contained, making it easy for reading. There are 22 chapters organized into 5 parts:

- Part I (Foundation) walks us through a comprehensive set of essential concepts, protocols, and algorithms that lay the foundation for blockchain. The topics include the peer-to-peer networking layer, consensus mechanisms, incentive designs to achieve security, and how to enable inter-blockchain operation.
- Part II (Scalability) focuses specifically on the most pressing challenge of today's blockchain networks; that is, how to scale a blockchain network to keep up with real-world expectations of transaction processing speed. A tradeoff of making a blockchain network fully secure and decentralized is smaller transaction throughput. The chapters in this part present ways to address this problem in the Ethereum network and in more general contexts.
- Part III (Trust and Security) provides a detailed coverage of the issues of trust, reputation, and security in blockchain. It begins with a description of the interplay of trust, blockchain, and reputation systems, followed by a case study of cyber-physical systems such as supply chains and smart cities. This part also provides an analysis of blockchain security threats, effectively capturing the recent attacks, and a review of security enhancement solutions for blockchain. Formal verification of the correctness of blockchain consensus protocols is also a topic crucial to any serious blockchain development.
- Part IV (Decentralized Finance) is dedicated to a high-impact application of blockchain: finance. Indeed, finance is the area that has been benefiting the most from blockchain and that should continue even more significantly in the future. With blockchain, financial products can be managed and delivered algorithmically, meaning being automated by computer programs. We will see as an example of how decentralized trading can be formulated as a mathematical optimization problem, how digital versions of fiat currency can be created as stablecoins in different ways, and how a government can digitalize their currency, which is referred to as Central Bank Digital Currency, with blockchain technology.
- Part V (Application and Policy) includes several cases where blockchain applies to the real world. One can build a decentralized marketplace to exchange data and a decentralized infrastructure involving multiple stakeholders in supply chain and transportation applications. One can also represent any physical asset as digital tokens such as NFT (non-fungible tokens) that are instantly exchangeable to increase liquidity. The sky is the limit for blockchain applications, but the reality, especially when it comes to financial impacts, depends on regulation per country. This part gives us the current state of crypto regulation, particularly

in Europe. Being a decentralized world, blockchain needs good governance too, which can learn from traditional economic perspectives. This topic is covered in this part.

As the field continues to evolve, we do not intend to cover every possible issue in blockchain, but instead, present areas that we think are the most beneficial to the readers in getting the first overall picture of blockchain technology. More advanced readers can also find deep details about certain topics in the book too.

We would like to extend gratitude to the 57 authors/co-authors who contributed chapters to this book. It would not have been possible without their quality work. We are grateful for their time, efforts, and especially their patience during the process which was somewhat impacted by the COVID-19 pandemic. The guidance from the Springer publishing team, particularly Elizabeth Loew, was awesome. Personally, co-editor Duc A. Tran would like to thank VinUniversity College of Engineering and Computer Science for hosting his sabbatical during which he conducted the book project. Co-editor My T. Thai would like to thank the University of Florida for supporting her adventure into blockchain research. Her work in this project is partially supported by the National Science Foundation under grant CNS-2140477. Co-editor Bhaskar Krishnamachari would like to thank the USC Viterbi Center for Cyberphysical Systems and the Internet of Things and students, staff, and colleagues at USC for supporting his investigations into Blockchain technology and applications.

Boston, MA, USA   Duc A. Tran
Gainesville, FL, USA   My T. Thai
Los Angeles, CA, USA   Bhaskar Krishnamachari
April 2022

Contents

Foundation

Blockchain in a Nutshell

Duc A. Tran and Bhaskar Krishnamachari

Abstract Blockchain enables a digital society where people can contribute, collaborate, and transact without having to second-guess trust and transparency. It is the technology behind the success of Bitcoin, Ethereum, and many disruptive applications and platforms that have positive impact in numerous sectors, including finance, education, health care, environment, transportation, and philanthropy, to name a few. This chapter provides a friendly description of essential concepts, mathematics, and algorithms that lay the foundation for blockchain technology.

1 Introduction

Let us consider the following favorite game of our childhood: Alice and Bob each bet $100 on the outcome of a coin toss, whether it is "head" or "tail". Alice calls the outcome and Bob is the tosser. Alice will win the bet if her guess is correct and Bob will otherwise. It is so easy and a simple game, isn't it? Not really. What if Alice and Bob play this game remotely or separated by a brick wall such that Alice does not see the toss? How can Alice trust that Bob is honest? Bob can easily cheat; knowing Alice's prediction he can say the opposite outcome. Even in the case he is honest, Alice may not be. She can run away not giving Bob the $100 she bet, assuming she sprints so fast that he cannot catch her.

The above game is an example of a big real-world problem we see almost everywhere. That is, how to quickly process transactions for everybody, possibly involving multiple people, in an environment not always honest, where people may not trust one another?

Our society has hundreds, thousands, of years been relying on the intermediaries to solve that problem. If we do not trust each other, let us do the transaction through

D. A. Tran (✉)
University of Massachusetts, 100 Morrissey Blvd, Boston, MA 02125, USA
e-mail: duc.tran@umb.edu

B. Krishnamachari
University of Southern California, 3740 McClintock Avenue, Los Angeles, CA 90089, USA
e-mail: bkrishna@usc.edu

© The Author(s), under exclusive license to Springer Nature Switzerland AG 2022
D. A. Tran et al. (eds.), *Handbook on Blockchain*, Springer Optimization
and Its Applications 194, https://doi.org/10.1007/978-3-031-07535-3_1

a trusted middleman, hence the existence of banks for financial activities, central servers for storage and computation, or, at a larger scale, central governments for maintaining the society. The trust put on the intermediaries is an *assumed* trust: we assume that they will do what they are supposed to do. That is the perfect scenario, which is not the case in practice. Mistakes are made by humans. Machines fail. Hackers are always looking for ways to penetrate into systems. Even in an ideal world where such errors or attacks do not happen, the conventional way of relying on a central authority to store information, process transactions, or manage activities for many people and institutions cannot scale. The authority is the bottleneck. It is increasingly expensive in both money and time when there are more workloads.

This is where blockchain comes in. It is completely decentralized with no intermediary involved. Blockchain overcomes the weaknesses of the centralized intermediary approach in four crucial aspects: trust, security, privacy, and transparency. Blockchain is trustless; there is no need to raise the trust question. Bob and Alice in the aforementioned betting game do not have to worry about the other cheating. Blockchain is secure, while a central server as a single point of contact can be attacked or the data therein stored may maliciously be altered, blockchain as a system always functions correctly 24/7. As identity privacy is of utmost importance today, it can be leaked in a middleman-based system. Blockchain does not allow this to happen as it is designed to hide personal identities. Lastly, about transparency, while today's banks may not disclose to us what they do behind closed doors with our deposited money, blockchain makes all the transactions visible and verifiable. Since there is no concept of personal identity on the blockchain, making transactions visible does not cause loss of privacy.

Blockchain is capable to provide the above desirable properties thanks to its architecture as a decentralized network utilizing many computers owned by people. These computers collectively store and process transactions in a way that although working autonomously they can still achieve consensus in decision-making and be robust against malfunctions, attacks, dishonesty, and self-interests. On the surface, we can think of blockchain as an Internet-like infrastructure for processing transactions. Using the Internet, one can send data from one computer to another without having to worry about how the data finds its way to get delivered or whether the data can be lost; the Internet takes care of all those things so that we can focus on the main business job. Similarly, if people transact on the blockchain, they do not have to worry about many what-ifs, including trust about whether the other side may act as agreed upon or whether money may be lost or data maliciously changed. Blockchain has its name because, as a digital ledger, the transactions are stored in blocks, each new block appended to the previous to form a chain; hence the name *blockchain*. Two consecutive blocks are mathematically linked in such a way that any change in an existing block would violate the mathematics of the link with the next block. The mathematical methods used for this linking are from the field of mathematical cryptography, hence the name *crypto* in "cryptocurrencies" we see trending today.

Trust is the biggest bottleneck in realizing transactions. It is the biggest bottleneck in advancing the society. As a trustless system, blockchain removes that bottleneck. It makes sense that many consider blockchain the next big thing since the birth of

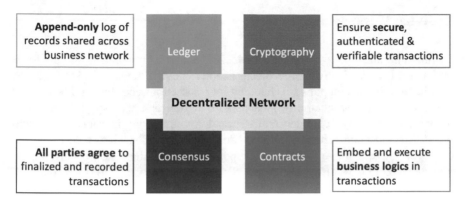

Fig. 1 The five constituent components of blockchain: decentralized network, cryptography, consensus, ledger, and contracts

the Internet. The Internet removes the geographical constraint, moving people closer for communication despite geographical distances. Blockchain, by removing the trust distance, moves people closer for doing actual transactions. Putting blockchain together with AI, a field of great mention today, we can think of AI as the brain of a system whereas Blockchain is the body. AI needs computing resources and training data to realize its promise. Blockchain is no less important because it is the best way to incentivize people to contribute computing power and good data, the only way if we care about trust, security, privacy, and transparency.

Blockchain is still in an early application stage. The space for blockchain-based developments is immense. To consider whether blockchain may apply to your business, at least four out of six following conditions should hold: (1) data is shared by multiple parties, (2) data is updated by multiple parties, (3) verification is required, (4) it is expensive to rely on intermediaries, (5) valid transactions must be eventually executed, and (6) transactions are inter-related. Most applications satisfy this, which are found in almost every sector, including financial services, product manufacturing, energy and utilities, health care, e-government, retail and consumer, entertainment and media, just to name a few.

According to Harvard Business Review [23], one can argue that Blockchain is not only a disruptive technology, but also has the potential to create new foundations for our economic and social systems; it is a foundational technology. A recent PwC report [39] projected that Blockchain by 2030 will potentially add 1.76 trillion USD to the global GDP, create 40 million new jobs, and be used to support 10–20% of global business infrastructures. The 2020 annual global blockchain survey of Deloitte [13] interviewing 1488 business leaders from 14 countries, who had certain knowledge about Blockchain, found that 39% of the businesses applied Blockchain, a 23% increase from 2019, 55% considered Blockchain a top-5 priority, and 82% would hire blockchain staff within 12 months.

2 What is Blockchain

Having introduced the motivation for Blockchain and its potentials, we now focus on what it actually is. To non-technical people, one can define Blockchain based on what it offers: a computing technology for transaction recording and processing that is safe (no loss or mutability of data possible), transparent (easy verification and tracing), and trustless (confidence of transacting without any intermediary). Technically, the most complete definition of Blockchain should see it as a decentralized computing system of five constituent components: decentralized networking, mathematical cryptography, distributed consensus, transaction ledger, and smart contracts, as illustrated in Fig. 1:

- Decentralized networking: For computing, blockchain relies on a decentralized network of computers, called blockchain nodes, that contribute computing resources to help store and process transactions. These computers work autonomously and communicate with each other in a peer-to-peer (P2P) manner. Most blockchain networks including Bitcoin adopt an unstructured P2P topology, i.e., a node chooses its neighbors arbitrarily. Some networks such as Ethereum use a structured one like Kademlia Distributed Hash Table [32] to optimize the P2P communication. Unstructured P2P may be less efficient than structured P2P, but the latter is more difficult to maintain, especially in a permissionless blockchain. Ethereum uses Kademlia but only as an add-on assistance [45]; in other words, it still works with any unstructured P2P topology, albeit less efficient if only so.
- Mathematical cryptography: Cryptographic methods used in blockchain provide mathematical proofs that the blockchain must function as supposed to. Cryptographic hash is used to link data blocks in the chain so that no data alteration is allowed post recording into the blockchain. Each transaction is encrypted with public-key cryptography to ensure that the sender is verifiable using digital signature and only the intended recipient of the transaction can be the receiver. Transaction confidentiality is achieved thanks to the method of Zero Knowledge Proof [4]. The choice of cryptography to use determines the performance and guarantees of the blockchain. For example, Dogecoin blockchain clones Bitcoin but using simpler cryptographic functions to increase transaction throughput; the mining in Dogecoin is based on SCRYPT which is faster and easier to run than SHA256 used in Bitcoin. This, however, results in weaker security, less robust to attacks by dishonest nodes.
- Transaction ledger: As a storage technology, blockchain is a digital ledger that stores the transactions chronologically in blocks which are added in an append-only manner. This is the default data structure of the ledger for almost all blockchain networks. However, some blockchain networks, for example, Hedera [3] and Fantom [36], design the ledger as a directed acyclic graph (DAG) of blocks (or transactions) instead of a chain structure which can only append blocks. A chain is a simple case of DAG because it shares the property of being directed acyclic. The former offers simplicity but the latter is more efficient in transaction processing (for example, searching for a transaction is faster). The ledger struc-

ture, the block structure, and the number of transactions in a block are important considerations when designing the ledger component of the blockchain.

- Distributed consensus: When a decision needs to be made, for example, whether a transaction is valid, there is no central authority to decide. Instead, the decision is made based on consensus reached among the participating nodes. Therefore, a blockchain network must have a consensus protocol to make sure that every transaction or block added to the blockchain is the one and only version of the truth that is agreed upon by all the nodes. Proof-of-work consensus [35], giving more decision power to nodes with more hardware-computing power, is adopted in early blockchain networks (Bitcoin, Litecoin, Ethereum in its original version). Proof-of-stake consensus [17], giving more decision power to nodes with more financial stake, is popular among today's blockchain networks; its first functioning use for cryptocurrency was in Peercoin in 2012 [24]. The choice of consensus protocol is the most critical consideration in designing a blockchain network.
- Smart contracts: A blockchain can be considered a non-conventional kind of computers to perform certain tasks. Instead of being a computer integrating built-in computing processing units (the CPUs), blockchain is a decentralized computer utilizing hundreds or thousands of computers anywhere in the world. Applications that run on the blockchain are implemented as "smart contracts", a term coined by Nick Szabo in the 1990s [43]. A smart contract is nothing but a computer program; the term is used because an application deployed on the blockchain always functions correctly as programmed, like executing the conditions in a legal contract. This contract is smart because of its automated execution without human intervention.

Next, we elaborate further on these components and their importance. We do not attempt to cover every aspect and every detail. Instead, we select certain issues to discuss hoping that the reader can have a quick understanding of what blockchain is and requires. More details will follow later to dig deeper into the technicality of blockchain.

2.1 The Blockchain Computer

We can view blockchain as a computer whose architecture consists of three layers, illustrated in Fig. 2: the P2P networking layer, the consensus layer, and the logic layer. For example, Bitcoin is a blockchain computer that implements all these layers, whereas Ethereum implements the first two layers, leaving the logic layer to application developers. Bitcoin is a purpose-specific blockchain computer that performs only one application: create a digital currency, the Bitcoin cryptocurrency as we all know, and functions for moving this currency between accounts. This application is a built-in logic of the Bitcoin blockchain, and as such smart contract is not a concept of Bitcoin. On the other hand, Ethereum is a universal blockchain computer; it was designed to enable deployment of arbitrarily purposed applications on

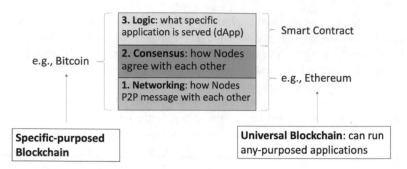

Fig. 2 Architecture of Blockchain as a new kind of computer

the blockchain. Therefore, Ethereum is called a smart-contract blockchain network. In contrast, Bitcoin is an application-specific blockchain, precisely a cryptocurrency blockchain.

Viewing blockchain as a computer is an intuitive observation. Essentially, a computer is a machine that automates processing of applications, and it is thus reasonable that blockchain can be seen as a computer, at least virtually. In early years, with desktop computing, we have applications running on a desktop computer near us, in our home or office; we control this desktop computer. The past decade has seen many businesses moving to cloud computing; the cloud provider controls the "cloud computer" (AWS cloud of Amazon or Azure Cloud of Microsoft). The future, very soon, we argue will be the era of blockchain computing; nobody controls the blockchain computer.

This is a natural evolution in computing. Cloud computing has replaced desktop computing to reduce the cost to maintain the IT system for businesses and at the same time more efficiently utilize computing resources. It is a one-stop shop to satisfy all computing needs so that companies can focus more time on their business logic. Compared to cloud computing, blockchain computing offers the benefit of decentralization and trust guarantees. The cloud provider has the power to manipulate the cloud computer; we have to trust this organization. Blockchain computing is trustless and anyone can be a part owner of it.

2.2 The Blockchain State

To interact with the blockchain, one needs an address or, interchangeably, an account. The blockchain state consists of the set of addresses and information about them. As the state changes from time to time, blockchain can be modeled as a state machine. It starts with a genesis state (when the blockchain is launched) and transitions from one state to the next upon triggering events (when transactions are added to the blockchain). We need to keep track of the blockchain state at any point of time. Depending on how the blockchain is designed, the state's data structure may differ.

It can be transaction based (the state information consists of the list of transactions) or account based (the state information consists of account balances). The data structure to represent transactions can also vary. We compare these models below, assuming for simplicity that each transaction is a transfer of value (asset) between addresses.

Transaction-Based Model

In the transaction-based model, known as Unspent Transaction Output (UTXO) [35] conceived by Bitcoin, each transaction can send value to one or more recipients. It consists of the following information:

- Output field: A list of receiving addresses and the amount of fund to be sent to each, respectively. Each transfer output is called a UTXO transaction.
- Input field: A list of UTXO transactions that will provide the fund for the transaction. These UTXO's previously sent funds to the sender and currently are unspent.

Figure 3 provides an example of Bitcoin transactions. The very first transaction Tx1, called the genesis transaction, sends 25 BTC to Alice. The input field is empty because this is the very first transaction of the blockchain operation, meaning Alice is the first recipient of Bitcoin (somebody has to be the first recipient). This transaction results in creation of a UTXO transaction, Tx1(#1). The second transaction Tx2 is initiated by Alice, sending 17 BTC to Bob and the rest, 8 BTC, to herself. The total fund to send, $17 + 8 = 25$ BTC, comes from the fund that Alice previously received in UTXO Tx1 (#1). Because Tx1 (#1) is unspent, she has enough money for Tx2. After this execution, UTXO Tx1 (#1) is marked as "spent" and new UTXO Tx2 (#1, #2) is created and marked as "unspent". Later, Bob initiates transaction Tx3 to send 8 BTC to Charlie, with the remaining 9 BTC to himself. The total fund to send, $8 + 9 = 17$ BTC, comes from the fund that he previously received in UTXO Tx2 (#2). Because Tx2 (#2) is unspent, he has enough money to execute Tx3. After this execution, UTXO Tx2 (#2) is marked as "spent" and new UTXO Tx3 (#4, #5) is created.

The blockchain state is the set of current UTXO transactions. Each time a UTXO transaction is used as an input in a new transaction, the input UTXO will be marked as "spent" thus no longer usable and each output sending fund out will be created as a new UTXO transaction. The new UTXO transaction(s) may be used later as input providing funds to future transactions. The marking of input UTXO transactions as "spent" is to avoid double spending, which means spending the same UTXO for two different transactions. The UTXO blockchain state does not directly provide account balances. To know how much Alice has in her account, one needs to sum all the funds she received in current UTXO transactions.

In the case that the total input fund has more than the output, the remaining balance can be sent to the sender's own address to avoid losing fund. For example, in transaction Tx4, Alice sends 3 BTC to Dave out of the 8 BTC she has available from UTXO Tx2 (#3), but because UTXO Tx2 (#3) will be marked as "spent", in order

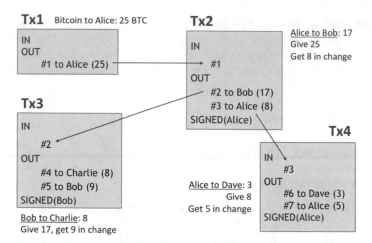

Fig. 3 The Unspent Transaction Output (UTXO) model: the blockchain state at the current time is the list of all unspent transactions

not to lose the $8 - 3 = 5$ BTC she has remaining, she creates a new UTXO Tx4 (#7) to send this 5 BTC to herself. She does not lose any money. In some blockchains, for example, Bitcoin and Ethereum, Alice may not send all of the remaining balance to her address; in this case, the leftover will be sent as reward to the blockchain node that adds this transaction to the blockchain.

Account-Based Model

The account-based model [6] is more intuitive. It is like the account model of a bank. The state consists of the balance information for each address. When there is a transaction, the balances of the sender's and receiver's accounts will be updated immediately and saved in the state. Therefore, when queried the account balance of an address is instantly available without any computation.

A transaction in the account-based model is much simpler than a UTXO transaction. The former consists of only one receiving address and the amount of fund to send. It is much faster to verify if the sender has enough fund, which is done by simply comparing two numbers: whether the sender's balance exceeds the amount to send. In contrast, UTXO requires searching the blockchain state to see if the input UTXOs are indeed unspent. Consequently, the account-based model offers a clear advantage when it comes to enabling "smart contracts" (computer programs to deploy applications on the blockchain). For smart contracts, a transaction can be not only a transfer of value, but also a call to a function of arbitrary logic; it contains code data for executing this function. To process a transaction thus involves execution of the code in the transaction. As smart contracts are computationally expensive, simplicity of computation is important. UTXO creates computational overhead because all spending transactions must be explicitly recorded.

UTXO is suitable for a cryptocurrency blockchain like Bitcoin which serves only one application: transfer of money. Computation is not that complex. Another reason is due to transparency and traceability. Back to Fig. 3, if we want to know how Dave received 3 BTC from Alice in transaction Tx4, we can trace all the way to the beginning how the fund started and flowed. We can find that it started from Alice in Tx1 (#1) to Alice in Tx2 (#3) to Dave in Tx4 (#6). In other words, every transfer has a non-fungible path. With the account-based model, if Dave received 3 BTC from Alice, this fund is fungible; we only know that this 3 BTC came from Alice, not knowing any further where this particular 3 BTC arrived at Alice. In other words, UTXO is more transparent. That said, one could argue that the account-based model offers better privacy.

2.3 The Chain Structure

By default, and adopted in all but a few unpopular blockchain designs, the blockchain ledger follows a chain structure. The data is organized into a chain of data blocks: b_1, b_2, b_3, ...When new transactions need to be saved, they are put in a new block which will be appended to the last block of the existing chain. In an account-based blockchain, e.g., Ethereum, a block also contains the blockchain state information (the balances of all the accounts at the current time).

Besides storing the transaction data, blockchain state if applicable, and necessary header information, the block has two important attributes:

- Block ID $b_i.id$: This is set to the hash value of the block content using a cryptographic hash function H, i.e., $b_i.id = H(b_i)$. This hash function is predefined and publicly known.
- Previous hash $b_i.prev$: This is set to the ID of the previous block b_{i-1} to which b_i is appended, i.e., $b_i.prev = b_{i-1}.id$.

It is noted that the block ID may not necessarily be stored in the block because it can be computed from the block's content.

The previous hash information is critical in maintaining the data integrity of the chain. If any part of any block is changed after it is recorded in the blockchain, this will be detected. This is because for a new block to be added to the blockchain it must pass a procedure called *block validation*. A new block b_{i+1} is valid if and only if

1. Previous hash is consistent: $b_{i+1}.prev = H(b_i)$.
2. All the transactions in b_{i+1} are valid.
3. Previous block b_i is valid.

Let us put Step 2 aside (to be discussed later). The verification in Step 1 requires computing the hash value of b_i and comparing it with $b_{i+1}.prev$. Step 3 requires running the same block validation procedure to verify the validity of block b_i. Consequently, the validation procedure for block b_{i+1} requires checking whether the previous hash value stored in block b_j equals the hash value of its previous block b_{j-1} for all $j \leq i + 1$. If an earlier block, say b_{j-1}, has been changed from its original value, when we compute its hash value, $H(b_{j-1})$, we will find it not identical to the previous hash value $b_j.prev$ stored in block b_j. This is a violation and as a result the new block b_{i+1} is concluded to be invalid and not added to the blockchain.

A consequence of block b_j being changed is that the blockchain will never grow beyond the time of this change. One might say, "that means, the blockchain is useless then, because just one block's modification halts the whole blockchain". This is true if the blockchain network consists of only one computer. In practice, the blockchain network runs many computers, where the blockchain data is replicated on every computer node. For a node to ensure that its blockchain copy is correct (same as the globally correct version), it needs to compare its copy with that of the neighbors and choose to use the longest[1] blockchain as the correct one. Before this comparison takes place, the node needs to check the validity of each neighbor's blockchain copy, which requires validating all the blocks in this copy. Therefore, if a blockchain copy from some node contains a violation, this copy will fail the validation step. As such, the bad copy will not be used by the honest nodes in the network.

2.4 Use of Cryptography

It is now clear that the data immutability of the blockchain is achieved thanks to the previous hash information linking between consecutive blocks in the blockchain. However, in theory, a hash function may have different input values resulting in the same hash output, meaning that block b_{j-1} can be changed from its original value such that its hash value, $H(b_{j-1})$, remains the same as before, which equals the previous hash value $b_j.prev$ stored in block b_j. In this case, the block validation procedure cannot detect the change. The choice of the hash function is therefore critical. We should choose one so that even though such a block alteration without being detected is theoretically possible, realizing it is practically impossible. For this reason, the hash function H used in blockchain must be a *cryptographic* hash function, not any arbitrary hash function.

Recall that a hash function is a one-way function that takes an input of arbitrary length to output a string of constant length, here assuming that values are represented as binary strings. For example, SHA256 is a hash function that outputs a binary string of 256 bits. A cryptographic hash function H is a hash function with three properties:

- Collision-resistant: It is infeasible to find different input messages x and y such that $H(x) = H(y)$.
- Hiding: Given the output $c = H(x)$, it is infeasible to find an input x.

---

[1] Comparing based on blockchain length (the number of blocks in the blockchain) is adopted in most blockchain networks, but other comparison criteria have also been explored, for example, choosing the "heaviest" blockchain copy as the correct one, where "heaviness" is a weighted generalization of the length.

- Puzzle-friendly: If we know the hash value $c = H(r \| x)$ of an input message made by concatenation of r and x, and even if we know part of the input, x, we cannot reconstruct the remaining input r in time complexity faster than 2^n where n is the binary length of output c.

Because of these properties, knowing $b_j.prev = H(b_{j-1})$, it is infeasible to find $b'_{j-1} \neq b_{j-1}$ such that $H(b'_{j-1}) = H(b_{j-1})$. With H being a cryptographic hash function, no one can alter an existing block not to be detected. The blockchain data is tamper-proof.

Cryptographic methods also have many other uses in the operation of a blockchain. Recall the coin bet between Alice and Bob at the start of this chapter, in which a situation is what if Bob cheats. A cryptographic hash function H can solve this cheating problem as follows.

1. Alice: suppose that her prediction is x ("head" or "tail").

   - Generate a secret random number r (of some large binary length n).
   - Compute $c = H(r \| x)$ (called "prediction commitment").
   - Send c to Bob, instead of sending her prediction as raw data.

2. Bob: upon receipt of the prediction commitment c, he will send Alice the honest outcome x^* of the coin toss. Because the hash function H is cryptographic, he does not know the ground-truth prediction x of Alice, and as such he has no reason to cheat.

3. Alice: upon receipt of x^*, if her guess is correct, i.e., $x = x^*$ she will tell Bob that she wins by sending him the secret number r.

4. Bob: upon receipt of number r, he will verify if the commitment c he received earlier from Alice equals $H(r \| x^*)$ and convincingly accept the loss.

This solution is called a commitment scheme in cryptography [12]. It is critical that the secret r generated by Alice must come from a large number space. If the binary length n was small, it would take short time for Bob to exhaustively try all possible values of r and combine with x="head" or x="tail" to see which combination satisfies $H(r \| x) = c$. When that combination is found, he can cheat by telling Alice that the outcome is the opposite value of x found in this combination. When n is large, even though x can take only two possible values, "head" or "tail", Bob cannot reconstruct the secret r thanks to the "puzzle-friendly" property of H as a cryptographic hash function.

The above is a glimpse into how mathematical cryptography helps make a system trustless. Alice and Bob do not need to question each other's honesty thanks to the commitment scheme. However, in the case Alice loses the bet, what if she runs away? Intuitively, a solution is to at least require that they both have to deposit the bet money in a lockbox which when the outcome is announced will be unlocked to transfer all the money to the winner. This is to say that there is a lot more to do and mathematical cryptography is the main tool to realize all that.

2.5   Where is Blockchain Stored

As we explained earlier, the blockchain is a decentralized network of computers contributing computing resources to help with transaction storage and processing. Among these computers, where is the blockchain data stored? Should we distribute the blocks in the blockchain ledger across these nodes so that some blocks are on node 1, some blocks on node 2, etc.? We should not because if node 1 fails, we cannot access the blocks stored there. Hence, some redundancy is needed to guarantee availability, that is, a block should be replicated on more than one node. The next question then is, "how much replication is enough?". In blockchain, the blockchain ledger is replicated fully on every node: each node stores a full copy of the entire blockchain. This is because of the blockchain's vision to provide complete decentralization (no node depending on other nodes to access certain blocks) and complete availability (it is always accessible even in the worst case of failure).

When a new node joins the blockchain network, it must discover existing nodes as neighbors and connect P2P to them. The new node obtains a blockchain copy from these neighbors. The list of blockchain nodes is available publicly. In most blockchain networks, the P2P networking topology can be arbitrary; any existing nodes can be selected at random, not geographically dependent.

Over the time, since nodes work autonomously and independently, their local blockchain copies may disagree. To ensure consistency, they need to frequently, or upon some triggering event such as adding new transactions, send their blockchain copy to the neighbors or pull blockchain copies from the neighbors. When presented with multiple blockchain copies, a node must decide which copy is the globally correct one and uses it. As aforementioned, the default criterion is to choose the longest copy.

2.6   How to Process a Transaction

When someone initiates a transaction with the blockchain, this is usually done in a user-friendly front-end application that can interact with the blockchain network via API calls. This transaction needs to be sent to a blockchain node (in practice, multiple nodes in case one node may fail or behave wrongly) and will be processed as follows:

- Each node X on first receipt of transaction $\texttt{Tx}$:

  - Transaction forwarding: forward transaction $\texttt{Tx}$ to the neighbor nodes of X.
  - Transaction verification: verify that the sender address of transaction $\texttt{Tx}$ has sufficient fund to send. If so $\texttt{Tx}$ is put into a mempool which is a queue of valid transactions waiting to be put in a new block.
  - Blockchain creation: pull pending transactions from the mempool to include in a new block b and append this block to the existing blockchain ledger at node

X. Note that block b must include the previous hash information (the hash value of the last block).

- Block update: send the new block b to the neighbor nodes of X.

- Each node Y on first receipt of block b:

  - Block forwarding: forward block b to the neighbor nodes of Y.
  - Block validation: verify the validity of block b on the existing blockchain ledger of node Y. This validation requires checking on the consistency of previous hash information and the validity of every transaction in block b.
  - Block insertion: append block b to the blockchain ledger if it is valid. Else, ignore b.

To validate a transaction during the Block Validation step may vary from one blockchain design to another. In Bitcoin, we only need to verify that the sender of the transaction has available fund to spend. This verification is successful if the input transactions exist in the blockchain state, meaning they are currently unspent, and the sum of output amounts in these transactions is sufficient. However, in a smart-contract blockchain network like Ethereum, transaction validation may involve more work than just checking the balance sufficiency. If a transaction involves a function call to interact with a smart contract, the verification will need to run this function with the blockchain in the previous blockchain state (recorded in the previous block) and if the resulted blockchain state does not match the blockchain state recorded in the block under validation, the block is considered invalid.

The transaction processing procedure in blockchain is simple and allows for autonomous processing at the blockchain nodes. This simplicity, however, leads to several consistency problems. First, each transaction is broadcast to all the nodes and so the same transaction may be added to different blocks created at different nodes. We need to ensure that each transaction can only be added to the blockchain once. Second, different nodes in parallel create different new blocks to attempt to append to the (same) existing blockchain. We need to ensure that only one of them will be added as the next block. Third, different nodes may have different copies of the blockchain. We need to ensure that they have to agree on a copy as the globally correct version. To resolve these inconsistencies, the nodes have to regularly agree on the current state of the blockchain, and that is what we call consensus achievement. We need a consensus protocol.

2.7 How to Achieve Consensus

Consensus is a research area of computing with more than 30 years of study before blockchain became popular. It started in the 1970s with the NASA sponsored project, "Software Implemented Fault Tolerance (SIFT)" [46], aimed to build a resilient aircraft control system. The challenge was to replicate the system on multiple machines such that the whole system can sustain multi-machine failures. The nominal work by Lamport et al. in 1982 [29] formulated this challenge as the now well-known "Byzantine Generals' Problem" (BGP). It coined the notion of "Byzantine Fault" to model a condition in a distributed system where some nodes are unreliable and may appear arbitrarily normal or malicious and collude with each other such that there is no consistent information for the other nodes to declare their malfunction.

A Byzantine Fault Tolerance (BFT) system must avoid complete failure and for that the nodes must agree on a concerted strategy and live by this consensus, knowing that some nodes may fail or act maliciously. BGP laid the foundation for research in distributed consensus. Companies like Google and Facebook started adopting scientific results in BFT consensus for mission-critical services such as Google Wallet and Facebook Credit. The birth of Bitcoin in 2009 [35] was the first time that consensus is realized in a large-scale practical environment in a permissionless and decentralized manner. The distributed consensus implementation by NASA, Google, or Facebook is not fully decentralized nor permissionless because the participating computers are controlled by these organizations. The Bitcoin network is public, requiring no permission for computers to participate and no centralized authority to make decisions.

To describe BFT formally, consider a broadcast system of nodes where a sender node needs to broadcast a message (value) to all the nodes in a peer-to-peer manner. At the beginning, the sender receives an input value m. The broadcast protocol must result in that at the end each node i will output a value m_i. The sender and receivers may be honest or dishonest. This protocol achieves BFT if it satisfies two requirements:

- Consistency: all honest nodes i and j must output the same value: $m_i = m_j$.
- Validity: if the sender is honest, all honest nodes i must output value $m_i = m$.

A system can be consistent but not valid, when all honest nodes output the same value but this value is not the same as the sender's: $m_i = m_j \neq m$. A system can be valid but not consistent, when the sender is dishonest and some honest nodes output different values: $m_i \neq m_j$ for some i, j. Thus, both requirements are needed.

Blockchain is a BFT system. To address inconsistencies due to the autonomous and independent working of blockchain nodes, the standard solution is for every node to agree on the consensus that the longest blockchain copy, the one with most blocks, is the globally correct version. Because the blockchain copies shorter than the correct blockchain are not used, nodes want to keep their copies as current as possible because otherwise they would waste efforts adding their blocks to a wrong blockchain. As discussed in the previous subsection, nodes frequently update their blockchain copy to make sure its version is the latest (globally correct one). Consequently, even though at times some transaction may be recorded in different blockchain copies at different nodes, different blocks may append to the same last block of the existing blockchain at different nodes, or different nodes may have different blockchain copies, eventually these nodes will have the same blockchain copy.

But that is just theory. If consensus eventuality happens too late, the aforementioned inconsistencies will cause the system to perform incorrectly, for example, double spending can happen. Therefore, we need to (1) minimize the likelihood for inconsistencies to happen and (2) minimize the time it takes to reach blockchain-consensus eventuality. Toward these, different consensus mechanisms have been used for blockchain. Major among them are the methods of Practical Byzantine Fault Tolerance (PBFT) [9], Proof of Work (PoW) [35], and Proof of Stake (PoS) [17].

3 The Bitcoin Network

We present next the actual working of a real-world blockchain network: Bitcoin. It is a blockchain network to build a peer-to-peer digital cash system, where the name of the digital currency is bitcoin (BTC). It has a total supply of 21 million BTC to be minted over time according to a deterministic schedule such that all will have been minted in the year of 2140. Technically, it follows the general blockchain framework described in the previous section. Specifically, it adopts the UTXO model for the blockchain state and the chain structure for the ledger. Newly arriving transactions will be put in a block to be appended to this chain. Any node can create blocks, and in that case it is called a "miner" and the process of creating a block is called "mining". The globally correct blockchain is chosen to be the longest one among all the local copies. We focus below on the key ideas and methods that are characteristic of Bitcoin implementation.

3.1 Addresses

To hold Bitcoin, one needs to create a wallet. Each wallet corresponds to an address (the Bitcoin address). When wallet A is created, it is associated with a pair (K_A^-, K_A^+) of 256-bit private key K_A^- and 256-bit public key K_A^+ generated according to an asymmetric cryptography method called Elliptic Curve Cryptography (ECC) [21, 25, 34]. Only the wallet owner knows the private key. The public key is publicly available. The address of wallet A is a 160-bit hashed version of its public key K_A^+:

$$A = RIPEMD160(SHA256(K_A^+)).$$

This is one-way cryptographic hashing using RIPEDMD160 and SHA256 hash functions. Because only the owner has the private key to unlock the public key, no one else can take ownership of a transaction that sends BTC to her. For ease of human read-ability, Bitcoin addresses are encoded as "Base58Check", which uses 58 characters (a base-58 number system) and a checksum, to produce a string like this example, "*1J7mdg5rbQyUHENYdx39WVWK7fsLpEoXZy*".

3.2 Elliptic Curve Cryptography

The Elliptic Curve Cryptography (ECC) mentioned above is an approach to public-key cryptography based on the algebraic structure of elliptic curves over finite fields. The use of elliptic curves in cryptography was proposed in 1985 by Miller [34] and Koblitz [25] and became popular in 2004. For cryptographic purposes, an elliptic curve is a plane curve over a finite field (rather than the real numbers) with the following equation:

$$y^2 \equiv x^3 + ax + b \pmod{p}$$

. The shape of the curve depends on the values given to a and b. The size of the finite field is given by p, which defines the length of the keys we want to generate. The points on the curve are limited to integer coordinates within the square matrix of size $p \times p$ only. For example, the curve in Fig. 4 is $y^2 = x^3 + 7$ which is used in Bitcoin, and the points in Fig. 5 are integer points of $y^2 \equiv x^3 + 7 \pmod{17}$.

On the elliptic curve, we define an algebraic operator on the points called "point addition". This operator allows to "add" points to obtain a point on the curve, as follows (illustrated in Fig. 4):

- Addition $P + Q$: Draw the line PQ and let R be the point where PQ cuts the curve. Point $P + Q$ is the mirrored point of R over the x-axis.
- Double $2P = P + P$: Draw the line tangent with the curve at point P and let R be the point where this line cuts the curve. Point $2P$ is the mirrored point of R over the x-axis.
- Multiplication $mP = P + P + ... + P$: This is the result of adding P with itself m times.

Despite its simplicity, a nice property of this operation on elliptic curves when applied on a finite field (i.e., all the points must be integer points in a finite square) is the hardness to compute the discrete "logarithm" m such that $mP = Q$ given points P and Q. To date, no algorithm can reconstruct m in time complexity faster than exhaustive search (having to try all possible values for m). On the other hand, if some m is given, it is easy to verify its correctness, that is, to check whether $mP = Q$. For example, if $m = 16$, we need only $\log m = 4$ point additions for verification: $2P$, $4P = 2(2P)$, $8P = 2(4P)$, and $16P = 2(8P)$; in comparison, to find the unknown m in $mP = Q$ would need 16 point additions.

Thanks to this property, ECC uses elliptic curves over finite fields to create a secret that only the private key holder is able to unlock. We can think of Q as the

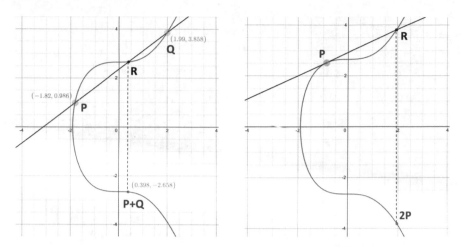

Fig. 4 Point addition on the elliptic curve ($y^2 = x^3 + 7$): (left) adding two different points; (right) adding two identical points

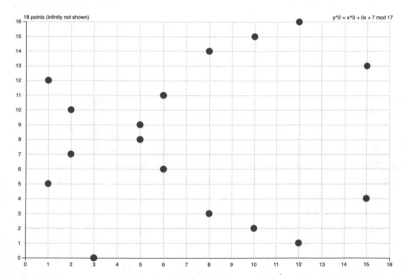

Fig. 5 The integer points of the elliptic curve on a finite field: $y^2 \equiv x^3 + 7 \pmod{17}$

public key and m as the private key. The larger the key size, the larger the curve space, and the harder the problem is to solve. For example, Secp256k1 with equation $y^2 = x^3 + 7$ and $p = 2^{256} - 2^{32} - 2^9 - 2^8 - 2^7 - 2^6 - 2^4 - 1$ is the ECC used by Bitcoin to implement its public-key cryptography. All integer points on this curve are valid Bitcoin public keys.

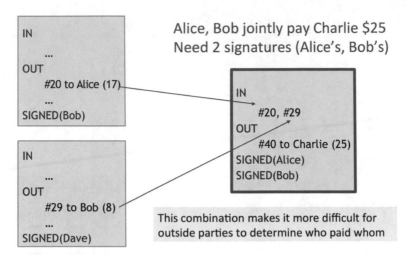

Fig. 6 Bitcoin joint payment: a transaction can use two or more input UTXOs that belong to different payers to collectively provide the fund to send. These multiple payers need to co-sign the transaction

3.3 Transactions

Bitcoin transactions are based on the UTXO model. A transaction by default is a transfer of BTC from a sender to one or more receivers. Every transaction must have a digital signature of the sender who "signs" with her private key. This way, anyone who knows her public key can verify that the signature is valid and the transaction indeed comes from that sender. Each output UTXO is destined for a receiving address. As we described earlier, each Bitcoin address is an encryption of its public key. Only the owner of that address has the corresponding private key to match. Hence, nobody but he can unlock the UTXO to use the fund.

A transaction can also be a joint payment which takes as input multiple UTXO transactions that belong to different addresses. For example, illustrated in Fig. 6, Alice and Bob jointly pay Charlie 25 BTC, where 17 BTC is funded by UTXO #20 of Alice and 8 BTC is funded by UTXO #29 of Bob. This joint transaction needs to be signed by both Alice and Bob. Joint payments make it more difficult for outside parties to determine who paid whom.

Transaction Fee

In a transaction, the input fund amount should be at least the output amount. The leftover is called the "transaction fee" to be sent to the miner who puts this transaction in a new block. Transaction fees are a way to incentivize miners to participate in Bitcoin. Rational miners prefer transactions that offer high transaction fees and so a transaction's sender should choose a generous fee to increase its chance to be

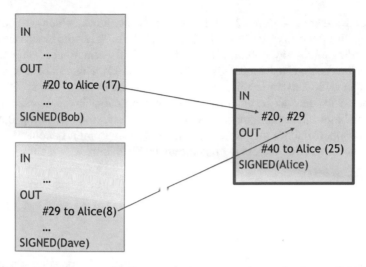

Fig. 7 Bitcoin transaction consolidation: an owner can create a transaction to consolidate the funds from many UTXOs he or she owns

processed earlier. To determine the fee, the sender should consider the transaction size and the network traffic. A block can contain a maximum of 4 MB of data, thus limiting the number of transactions included. A larger transaction will take up more block data. Thus, larger transactions typically pay fees on a per-byte basis.

Transaction Consolidation

A consequence of Bitcoin's being a UTXO ledger is that one address may own many small UTXO transactions. As such, when this address makes a large payment out, it may need to include as input many UTXOs. Not only that the transaction size increases, but also the transaction verification will be more expensive since it involves verifying many input UTXOs. For this large transaction to be included in a block, the sender should pay a high transaction fee. Therefore, it is a good idea for her to consolidate UTXO transactions if she owns too many of them. This can be done easily by creating a new UTXO transaction that consumes these existing UTXO transactions. For example, as illustrated in Fig. 7, Alice has funds in UTXO #20 and UTXO #29 and consolidates them by creating UTXO #40. The decision for transaction consolidation is made at the application level by the wallet owner.

Coinbase Transaction

Transaction fees are not the only incentive for the miners. For each block that is successfully added to the blockchain, the miner who created this block will receive a "block reward". As of March 2022, it is 6.25 BTC per block, which will be halved automatically after every 210,000 new blocks are added. To get the block reward, the corresponding miner, say Bob, inserts into the block a special transaction called the "coinbase transaction" that sends this 6.25 BTC to himself. This coinbase transaction has no input UTXO, meaning this amount will be minted by the network. If the block is validated and added to the blockchain, all the transactions in this block, including Bob's coinbase transaction, are officially recorded, effectively sending the block reward to Bob. Coinbase transactions are the only way to mint bitcoin. Except the genesis bitcoin transfer, bitcoin is minted only by block mining, which is sent to the miners.

3.4 Blocks

Block creation is the main job of the miners. A miner pulls pending transactions from the mempool, typically selecting those with high transaction fees (because these fees will be paid to the miner) and put them into a block. This is called "block mining". The very first block was added to Bitcoin network timestamped at 2009-01-03 13:15, called the genesis block, or block 0. It contains only one transaction, which is a coinbase transaction. This block is hardcoded in the Bitcoin client node software, so that when nodes join Bitcoin, they will always have the information about the genesis block.

Block Structure

A Bitcoin block has the following structure: (1) block size (4 bytes): the size of the whole block in bytes; (2) transaction count (variable size, 1–9 bytes): the number of transactions in the block; (3) transactions (variable size): the list of transactions included in the block; and (4) block header (80 bytes): important information useful for block creation and validation. The block header consists of the following fields:

- Version (4 bytes): the version of the Bitcoin node software.
- Previous hash (32 bytes): the hash (ID) of the previous block.
- Merkle root hash (32 bytes): the hash value of the included transactions according to Merkle tree
- Timestamp (4 bytes): the block creation time in second (Unix epoch).
- Difficulty target (4 bytes): a threshold number that is used for Bitcoin's proof-of-work algorithm.
- Nonce (4 bytes): a counternumber that is used for Bitcoin's proof-of-work algorithm.

In Bitcoin, the ID of a block is a hash of its block header, not the whole block content. It is the value which resulted from hashing the block header twice through

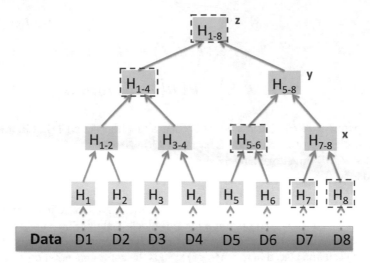

Fig. 8 Merkle tree: a binary tree where each internal node stores the hash value of the children's

the SHA256 algorithm. The block ID is not actually included inside the block's data structure. Anyone can obtain this ID by applying double-SHA256 hashing on the block's header.

Merkle Tree

The transactions are organized in the block as a Merkle Tree [33], a binary tree where each internal node stores the hash value of the children's values. Figure 8 illustrates such a tree for Bitcoin, where there are eight transactions $\{D1, D2, ...,$ $D8\}$, each stored in a leaf node, internal node $H_{1-4} = H(H_{1-2} \| H_{3-4})$, internal node $H_{1-2} = H(H_1 \| H_2)$, internal node $H_1 = H(D1)$, $H_2 = H(D2)$, etc. Bitcoin uses SHA2 for the hash function H.

The value at the tree root, e.g., H_{1-8}, is the Merkle root hash stored in Bitcoin block header. There are two crucial properties. First, any change in the transaction data causes a change in the Merkle root hash. As such, if a block is altered, whether it is in the transaction data or the non-transaction part, the hash of the block will change and be detected. Second, it is fast to verify the existence of a transaction in the block. For example, to prove that transaction $D7$ is in the block, the prover only needs to provide four values as evidence: H_7, H_{1-4}, H_{5-6}, H_8. The verifier will compute the following

$$x = H(H_7 \| H_8)$$

$$y = H\left(x \| H_{5-6}\right) = H\left(H(H_7 \| H_8) \| H_{5-6}\right)$$

$$z = H\left(y \| H_{1-4}\right) = H\left(H\left(H(H_7 \| H_8) \| H_{5-6}\right) \| H_{1-4}\right)$$

and compare z with H_{1-8}. Their equality means that transaction $D7$ is in the block. For a Merkle tree of n transactions, it takes $\mathcal{O}(\log n)$ time for this verification. It would take $\mathcal{O}(n)$ time if we naively store the transactions in a list-like structure.

A node in Bitcoin can be a non-miner node. It is only there to transact with the network, uninterested in creating blocks to receive block reward. Because the block header provides sufficient information for verification and transacting purposes, the node needs the block header only, not the full block content. Since no actual transactions are stored, the storage requirement for such a node is modest (only 80 bytes, which can easily be placed in the memory). So is the communication cost to pull the blockchain copies from the neighbors (only pulling block headers).

Block ID Rule

After choosing transactions to put in a new block and organizing them into a Merkle tree, the miner needs to do a final critical step on the block before submitting it to the network (by block broadcasting as we described in the general blockchain framework in Sect. 2.6). This step is called "mining" or "timestamping". The block needs an ID and recall that the ID is the hash value of the block header. In Bitcoin, the ID is not arbitrary, but follows a rule that it must be less than a target value determined by the *difficulty* target value specified in the block header:

$$\underbrace{SHA256(SHA256(block\_header))}_{ID} < \underbrace{(65535 << 208)/difficulty}_{TARGET}. \qquad (1)$$

The block header is known, except for two values that need to be filled: the *header_nonce* value and the *coinbase_nonce* value. *header_nonce* is the nonce attribute of the block header. *coinbase_nonce* is the value of the coinbase field in the coinbase transaction inserted by the miner to get the block reward. *coinbase_nonce* is set by the miner for flexible purposes. We need to choose these two values such that the resulted block ID satisfies the difficulty target according to Inequality (1).

Proof-of-Work Mining

To find a satisfactory block ID is not easy due to the double-SHA256 hashing. No algorithm is better than brute force, which takes $\mathcal{O}(2^{32})$: scanning all possible combinations of *header_nonce* value and the *coinbase_nonce* value. A pseu-

```
TARGET = (65535 << 208) / DIFFICULTY;
coinbase_nonce = 0;
while (1) {
    header = makeBlockHeader(transactions, coinbase_nonce);
    for (header_nonce = 0; header_nonce < 2^32; header_nonce++) {
        hash_value = SHA256(SHA256(makeBlock(header, header_nonce));
        if (hash_value < TARGET) break; //block found!
    }
    coinbase_nonce++;
}
```

Fig. 9 Proof-of-work mining algorithm in Bitcoin

docode of the mining algorithm is given in Fig. 9. The flexibility for the miner to set *coinbase_nonce* is important here because if we fixed it, we might not be able to find *header_nonce* among all the 2^{32} possible values such that the difficulty target is met. By allowing the miner to freely choose *coinbase_nonce*, eventually the miner will find a satisfactory block ID.

In Inequality (1), if we increase *difficulty*, the value of $TARGET$ will decrease, making it more difficult to meet the inequality $ID < TARGET$. When Bitcoin started with the genesis block, the difficulty was set to *difficulty* = 1, the easiest. Over the time, this target is dynamically adjusted depending on the transaction traffic in the network. The Bitcoin node software updates this difficulty target such that the blockchain grows at a rate of one new block for every 10 min.

The idea of making miners solve the above computationally expensive inequality is to serve several purposes. If blocks are created too easily, since each newly created block is broadcast to the network, the communication cost would be very expensive. Worse, as the miners add blocks autonomously, many blocks simultaneously created by different nodes will be appended to the same last block of the existing blockchain (their local copy). This causes not only severe inconsistency and double-spending vulnerability, but also wasted efforts by most miners due to the fact that only one block can append next to the existing global blockchain. Furthermore, malicious nodes can spam the network by creating many fake blocks and broadcasting them.

Bitcoin resolves these problems by making the task of block creation difficult. A miner must spend some provable effort in order to create a block. This is analogous to real-world miners spending efforts to discover gold; hence the term "Bitcoin miner/mining". The found ID of the block is the proof, hence the term "Proof-of-Work" (PoW) always associated with Bitcoin. The challenge of finding a good ID satisfactory of Inequality (1) is called the PoW problem. The PoW protocol also helps keep Bitcoin, as a currency, from inflation. Its slow minting and finite supply create circulation scarcity, thus making its price valuable.

The concept of PoW is not new. It was proposed by Dwork and Naor [14] in 1992 to prevent email spamming. Every time you send an email, your computer must solve a computational puzzle. The recipient's email program ignores your email if you do not attach the solution to the puzzle; if you do, the solution verification is quick. A similar idea was proposed in HashCash by Adam Back in 1997 and formally documented in 2002 [1] for anti-denial-of-service purposes. Bitcoin extended the PoW idea of HashCash.

3.5 Mining Difficulty

Achieving consensus in a permissionless environment is difficult due to Sybil attacks. As nodes communicate with one another in unauthenticated communication channels, a player can impersonate many others to outnumber the honest players and disrupt the consensus. This does not apply to a permissioned environment where the nodes are known to the authority. The choice of the difficulty target for PoW is critical. The more difficult it is, the more robust Bitcoin is against Sybil attacks, as a bad player must pay a higher cost to harm the network.

Difficulty Setting

In Bitcoin, the PoW difficulty is set such that (1) the network is BFT (Byzantine fault tolerant) as long as more than half of the nodes are honest, say 51%, and (2) on average only one block can be mined in each period of 10 min. To understand how they are related, we present a theoretical method below to find a good value for PoW difficulty (some derivations are similar to [42]).

Let n denote the number of blockchain nodes and p the probability that a given node creates a block in a round (equal to 10 min in Bitcoin). We will set the difficulty target in Inequality (1) to $TARGET = p2^m$ where m is the hash bit length (256 bits for Bitcoin). Hence, p indirectly represents the PoW difficulty. For example, if we set $p = 1$, any given node will 100% certainly create a block, because Inequality (1) is always satisfactory regardless of any block ID. However, that would result in n blocks created, violating the 10 min rule. Our goal is to find a good value for p.

The probability that no honest node creates a block in a round is $(1 - p)^{0.51n}$. The probability to have a block created by some good node, hence a good block, in a round is $1 - (1 - p)^{0.51n}$. Consequently, the number of rounds it takes to mine a good block is

$$\Lambda = \frac{1}{1 - (1 - p)^{0.51n}}.$$

Let Δ be the worse-case network propagation time. It takes this much time for the mined block to reach all the honest nodes to be added to the good blockchain. An honest node would not produce the next block during this Δ period to make sure that the previous block must have reached all the nodes; else, the next block may be invalid (when validated at other network nodes before the previous block arrives there). Therefore, the block-mining efficiency is the ratio between the mining time to the actual time it takes for this block to be added to the blockchain:

$$E = \frac{\Lambda}{\Lambda + \Delta} = \frac{\frac{1}{1-(1-p)^{0.5ln}}}{\frac{1}{1-(1-p)^{0.5ln}} + \Delta} = \frac{1}{1 + \Delta(1 - (1-p)^{0.5ln})}.$$

Let q be the fraction of dishonest mining power, which is the total hashrate of all the dishonest nodes. We need the hashrate of the honest nodes to exceed that of the dishonest, i.e., $(1 - q) > q$. However, due to the efficiency E, the effective hashrate of the honest is $(1 - q)E$, not $(1 - q)$. This block-mining efficiency E does not apply to the dishonest nodes who can do whatever they want, for example, sending block after block without considering the Δ delay. So, we should have $(1 - q)E > q$. To make the blockchain even more secure, we introduce a parameter $\epsilon > 0$ arbitrarily small and make a more stringent requirement:

$$\frac{effective\ hash\ rate\ of\ the\ honest}{hash\ rate\ of\ the\ dishonest} > 1 + \epsilon. \tag{2}$$

The larger ϵ, the more secure the network. The left-hand side is

$$\frac{effective\ hash\ rate\ of\ the\ honest}{hash\ rate\ of\ the\ dishonest} = \frac{(1-q)E}{q} = \frac{1-q}{q(1 + \Delta(1 - (1-p)^{0.5ln}))},$$

and so we require

$$\frac{1-q}{q(1 + \Delta(1 - (1-p)^{0.5ln}))} > 1 + \epsilon$$

$$\Leftrightarrow \qquad \frac{1-q}{q(1+\epsilon)} > 1 + \Delta(1 - (1-p)^{0.5ln})$$

$$\Leftrightarrow \qquad \frac{\frac{1-q}{q(1+\epsilon)} - 1}{\Delta} > 1 - (1-p)^{0.5ln}$$

$$\Leftrightarrow \qquad (1-p)^{0.5ln} > 1 - \frac{\frac{1-q}{q(1+\epsilon)} - 1}{\Delta}$$

$$\Leftrightarrow \qquad 1 - p > \left(1 - \frac{\frac{1-q}{q(1+\epsilon)} - 1}{\Delta}\right)^{1/(0.5ln)},$$

which leads to the following important inequality:

$$p < 1 - \left(1 - \frac{\frac{1-q}{q(1+\epsilon)} - 1}{\Delta}\right)^{1/(0.5ln)}. \tag{3}$$

What this means is that we should choose p to satisfy this inequality and the larger the gap between p and this upper bound, the more secure the blockchain is. We can choose a very small p to make the mining difficult (because we set difficulty target to $TARGET = p2^m$), but doing so will slow down the transaction processing. A good p is a reasonably high value still satisfying Inequality (3).

Inequality (3) also implies that given the same mining difficulty p, the blockchain becomes less secure if the value of the right-hand side upper bound is smaller, because that creates a bigger risk for violating the inequality. The right-hand side will be smaller if the network delay Δ is longer or if the dishonest hashrate q is faster. This explains why the security of a Bitcoin-like blockchain network is weakened in a slow network environment.

Difficulty Adjustment

Because each miner is solving the proof-of-work puzzle in parallel, on average the time taken for the first miner to solve it reduces inversely proportionally with the number of miners. At times when there are a lot of miners active on the network, the time to produce blocks will therefore be lower than when there are fewer miners active on the network. Since the number of miners on the Bitcoin network is going to change over time, unless some measure is employed, the block production time would also vary. In particular, over time if more and more miners joined the network, it would just keep decreasing. This is problematic as it could result in blocks being produced too fast, increasing the bandwidth requirements on the network, and also potentially result in more "forks".

To prevent this, the difficulty level of block production is periodically adjusted in a decentralized manner in such a way so as to ensure that on average a block is produced or mined once every 10 min. Based on this 10 min period, we can calculate that once every 2 weeks, the total number of blocks produced should be 0.1 (block/min) × 60 (min/h) × 24 (h/day) × 7 (days/week) × 2 (weeks) = 2016 blocks. The protocol therefore adjusts the difficulty level after each epoch of 2016 blocks using the following equation: $D(n + 1) = \frac{2D(n)}{T}$, where $D(n)$ is the difficulty at epoch n and T is the time taken in weeks to produce the previous 2016 blocks. If this time is shorter than 2 weeks, then that implies that there must be too many miners on the network, and therefore the difficulty level should be increased, and on the contrary if this time is longer than 2 weeks, then that implies there are too few miners on the network and therefore the difficulty level should be decreased. Each miner independently computes the new difficulty and will only accept blocks that meet the difficulty that they computed. Figure 10 shows how the Bitcoin mining difficulty has changed during the past year (April 2021–April 2022).

3.6 Mining (Un)Fairness

Bitcoin is generally fair in that if a node contributes more computing power, which is referred to as "hashrate", it can solve the PoW problem more quickly, thus having a

Difficulty level

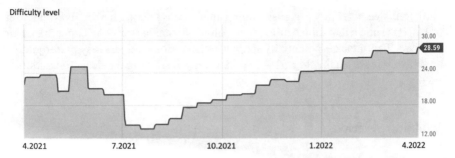

Fig. 10 Bitcoin PoW mining difficulty adjustment over the time as of April 1, 2022. The y-axis is measured in the number of "0" bits in the prefix of the difficulty target

better chance to earn block reward. If a fast node A and a slow node B both want to add a block to the existing blockchain at the same time, the block of A is more likely to be born first (with satisfactory block ID) and will be added to the blockchain copies in the network before the block of B. Naturally, the PoW mechanism encourages nodes to upgrade computing power to receive more block reward. This healthy competition leads to a better Bitcoin network.

However, not all nodes are good citizens. We have honest nodes who follow the protocol precisely and responsibly. There are selfish nodes who follow the protocol but do things to their benefit at the cost of other nodes. The remainder is the nodes who want to harm the network. To be equitably fair, a node contributing a fraction of the network hashrate should have the same fraction of blocks accepted by the network (thus being rewarded). For example, if a node contributes 20% of total network hashrate, it should own 20% of the blocks in the blockchain. In this aspect, Bitcoin can be very unfair. We show below that Bitcoin can encounter a situation where the rate at which a good Bitcoin miner successfully adds a block to the blockchain is much lower than its hashrate contribution.

Selfish Miner Attack

First, let us explain how a selfish miner A can abuse the network. Each time A has created a block a, it does not broadcast the block to the network right away. Instead, A waits on the event of receiving a new block from another miner, say block b of miner B. When this happens, A will immediately broadcast its block a to the network, but ignores block b of miner B (not forwarding it further). If we generalize this strategy such that A represents the pool of selfish miners and B the pool of honest miners, the blocks belonging to the selfish will be faster to reach the blockchain nodes than the honest's blocks. This is because while A honestly forwards every block, including B's, immediately upon its arrival, B ignores A's blocks. As a result, not only B receives more block reward, but also a lot of A's efforts are wasted.

Bitcoin's vulnerability to selfish mining attacks was investigated by Eyal and Sirer in [16]. It is shown that selfish miners can collude to obtain a revenue larger than their fair share. This attack is potentially serious in that rational miners may prefer joining the selfish miners leading to a 51% majority, thus destroying the decentralization of Bitcoin.

Unfairness Severity

Next, let us see how unfair Bitcoin can be from a theoretical view given dishonest behaviors. Consider the formulation in Sect. 3.5. It is shown in [42] that, in an honest node's blockchain, the honest block fraction is approximately $\mu = 1 - \frac{1}{1+\epsilon}$. Assume a blockchain network with zero propagation delay $\Delta = 0$, which is ideal for honest nodes because the block creation efficiency is $E = 100\%$. The honest hashrate is therefore $(1 - q)E = 1 - q$. Recall q as the total hashrate fraction of the dishonest nodes. Inequality (2) becomes

$$\frac{1 - q}{q} > 1 + \epsilon.$$

As a result, we have

$$\mu = 1 - \frac{1}{1 + \epsilon} < 1 - \frac{q}{1 - q} = \frac{1 - 2q}{1 - q},$$

which is smaller than the honest hashrate fraction $(1 - q)$.

What this means is that even in a network ideal for the honest nodes, the rate at which they can add blocks to the blockchain, μ, is lower than the hashrate they contribute to the network, $(1 - q)$. For example, when 51% of the network is honest (i.e., $q = 49\%$), they own a fraction $\mu < \frac{1-2q}{1-q} = 3.9\%$ of the blocks on the blockchain. The dishonest coalition with only 49% hash power owns 96% of the block creation.

So, Bitcoin mining may be unfair when it comes to block reward. Despite this risk in theory, one may argue that it is unlikely or of little impact in practice. Miners have ideological considerations and incentives to keep the network decentralized. If a coalition grows so big to be a concern to the rest of the network, people may leave Bitcoin due to the lack of decentralization; the coalition would not benefit, of course.

3.7 Block Finality

A malicious miner makes a payment, then secretively creates a second conflicting transaction using the same UTXO input in a new block, that allows him to recover the fund. This is an example of the double-spending problem. This is feasible if this miner controls more than 50% hashrate of the whole network, mining faster than the rest of the network combined. Therefore, his local chain is the longest among all local copies and will be accepted by the network as the consensus for the globally correct blockchain.

Even when the bad minor has less than 50% hashrate as in most cases, there is still a non-zero chance that the bad miner can grow the longest blockchain. Although this can only last for a short period of time, double spending is not impossible. To minimize this risk, when somebody pays a merchant to buy something, the merchant should wait some time to make sure the money is in before delivery. In Bitcoin, the wait is for six block confirmations, i.e., six blocks to be added after the block containing the payment transaction.

Why six block confirmations is enough? Consider a miner A with a fraction p of the total hashrate and a miner B with a smaller fraction $q = 1 - p < 1/2$. We are interested in computing the probability that B's blockchain will be longer than A's after A adds k blocks if both nodes start at the same time. This is similar to a race of two players in a Gambler Ruin problem. In this game, block creations form a sequence of independent Bernoulli trials. Each trial is the creation of a block which has two potential outcomes: "success" means that the block is created by A and "failure" if the block is created by B.

We observe this sequence until A has created k blocks (i.e., k successes). The number of blocks B created is a negative binomial random number, $X \sim NB(k, p)$, which has the following probability mass function:

$$P(X = i) = \binom{i + k - 1}{i}(1 - p)^i p^k.$$

During the time that A has added k blocks, the probability that B has created more than k blocks, hence winning the race outright, is

$$P(X > k) = \sum_{i > k} P(X = i) = \sum_{i > k} \binom{i + k - 1}{i} q^i p^k. \tag{4}$$

In the case that B has created less than or equal to k blocks, i.e., $i \le k$, B will be behind by $(k - i)$ blocks and still has a probability $(q/p)^{k-i}$ to catch up with A and thus win. Summing up these probabilities will lead to the probability that B will win the race:

$$P(k) = P(X > k) + \sum_{i=0}^{k}(q/p)^{k-i}P(X = i) \qquad (5)$$

$$= \sum_{i>k}\binom{i+k-1}{i}q^{i}p^{k} + \sum_{i=0}^{k}(q/p)^{k-i}\binom{i+k-1}{i}q^{i}p^{k} \qquad (6)$$

$$= \left(1 - \sum_{i=0}^{k}\binom{i+k-1}{i}q^{i}p^{k}\right) + \sum_{i=0}^{k}\binom{i+k-1}{i}q^{k}p^{i} \qquad (7)$$

$$= 1 - \sum_{i=0}^{k}\binom{i+k-1}{i}(q^{i}p^{k} - q^{k}p^{i}). \qquad (8)$$

This probability converges exponentially to zero as k increases. Grunspan and Perez-Marco [19] provide a closed form for this probability

$$P(k) = I_{4pq}(k, \frac{1}{2}),$$

where $I(.)$ is the regularized incomplete beta function:

$$I_x(a, b) = \frac{\Gamma(a+b)}{\Gamma(a)\Gamma(b)} \int_{0}^{x} t^{a-1}(1-t)^{b-1}dt.$$

For Bitcoin, it is recommended that we wait for $k = 6$ block confirmations before assuming that the transaction is final, which is enough for $P(k)$ to be extremely small. For example, $P(6) = 0.0005914$ for $q = 0.1$. A block becomes "final", hence the blockchain up to this block is considered "finality", if it is followed by this many block confirmations. Thus, Bitcoin finality is not instant. Instead, it is guaranteed asymptotically.

4 Smart-Contract Blockchains

Bitcoin is an example of an application-specific blockchain network where the only application logic is to serve digital payments. Although it allows for some limited programmability, it does not provide arbitrary programmability. As many applications in the real world, not necessarily financial, can benefit from blockchain technology, having a dedicated blockchain network for each individual application is not realistic.

This is the motivation for Ethereum, the first blockchain network created by Buterin et al. [6] to be a universal blockchain computer that can run applications of arbitrary purposes. To develop such applications, developers write computer programs called "smart contracts". Ethereum, therefore, is said to be a smart-contract blockchain. Other public smart-contract blockchain networks include Algorand [10], Tezos [18], and Solana [48].

4.1 Smart Contract

Smart contracts are written using a high-level programming language (e.g., Solidity, Viper, Flint, Bamboo). Solidity is the most popular language for smart-contract networks. It is Turing-complete, meaning that it can simulate any computation. In contrast, Script, the programming language of Bitcoin, is not Turing-complete. Script is thus very light and suitable for Bitcoin. Bitcoin does not need a universal language because digital currency is the only purpose of Bitcoin.

Compared to Bitcoin, a smart-contract blockchain has an additional layer of functioning because of the smart-contract capability. When a smart contract is deployed, it is submitted as a transaction to the blockchain network to run on every node. Each node needs a run-time environment to execute the bytecode of the smart contract. On Ethereum, this is called the Ethereum Virtual Machine (EVM), a powerful sandboxed virtual stack embedded with each full Ethereum node. EVM is where all Ethereum accounts and smart contracts live. It maintains the consensus for the blockchain. While the smart-contract language used in Ethereum is Solidity which is Turing-complete, EVM is a quasi-Turing-complete machine. Quasi, because EVM can theoretically run every smart contract but its execution will stop and be reverted if exceeding the resource allocation limit specified by the deployer.

As an analogy, let us compare running a computer program on a single computer versus the Ethereum blockchain computer. In the former case, all the state of the program is stored on the computer which is the only point of contact for the user and if this computer fails, all the state will be lost. In the blockchain case, the computer program is deployed and runs simultaneously on all computer nodes of the blockchain; these nodes independently and autonomously keep track of the program's state. A user can interact with the program using any node. If a node fails, the program is still running on the other nodes. The consensus mechanism of the blockchain ensures that the states on all the nodes are identical.

The source codes of deployed smart contracts are visible to the public. Therefore, there is nothing to hide in the working of a smart contract and people can be assured that it will work as programmed. That said, in certain cases, a complex smart contract may contain bugs and other security holes that are not easily seen; to fix them is a headache after the application is already deployed with many users; note that the blockchain is immutable. Therefore, a professional project should have its smart contracts certified by reputable smart-contract auditors.

4.2 Token Creation

Bitcoin (BTC) is the only native token (digital currency) of the Bitcoin blockchain network. Its users transact with each other (paying one another) using BTC. A smart-contract network also has a native token (e.g., ETH for Ethereum), whose main use is for the users to deploy smart contracts and interact with them. To deploy a smart

```
 1    interface IERC20 {
 2        function totalSupply() external view returns (uint256);
 3        function balanceOf(address account) external view returns (uint256);
 4        function allowance(address owner, address spender) external view returns (uint256);
 5        function transfer(address recipient, uint256 amount) external returns (bool);
 6        function approve(address spender, uint256 amount) external returns (bool);
 7        function transferFrom(address sender, address recipient, uint256 amount) external returns (bool);
 8        event Transfer(address indexed from, address indexed to, uint256 value);
 9        event Approval(address indexed owner, address indexed spender, uint256 value);
10    }
```

Fig. 11 The IERC20 interface for ERC-20 tokens. ERC-20 tokens must implement these functions

contract on Ethereum, one must pay a certain amount in ETH. This amount depends on the computational complexity of the contract. Besides ETH, many secondary tokens can be created to serve different applications. For example, one can build a loyalty application on top of Ethereum and implement the loyalty point as a token, or a country's government can issue a Central Bank Digital Currency (CBDC) as a token on top of Ethereum.

A token is implemented in the form of a simple smart contract. If the token is meant to be a kind of digital currency, this contract stores the token balance information for each account (blockchain address) and includes essential functions to enable a sender to transfer tokens to a receiver (needed for a real-world payment transaction), a spender to transfer tokens on behalf of their owner to a receiver (useful for a trading exchange or a bank to transfer money from someone's account to a payee, of course, with permission only) or, in many cases, mint new and burn existing tokens (useful for a government to cope with inflation crises). Figure 11 shows the interface for token smart contracts in Ethereum with six required functions. Events can be emitted from a smart contract so that front-end applications can watch and be instantly notified of their happening.

For ease of token creation, several token standards have been defined and template smart contracts created. The first standards were defined for the Ethereum network and their counterparts later followed for other smart-contract networks. For Ethereum, ERC-20 is the standard for fungible tokens, ERC-721 for non-fungible tokens (NFT), and ERC-1155 for generic multi-tokens (one that can represent a fungible token or a non-fungible token or a multiple of them). For example, ERC-20 is used for implementing a cryptocurrency, ERC-721 for digitally representing a

```
1    contract ERC20Basic is IERC20 {
2        string public constant name = "ERC20Basic";
3        string public constant symbol = "ERC";
4        uint8 public constant decimals = 18;
5        mapping(address => uint256) balances;
6        mapping(address => mapping (address => uint256)) allowed;
7        uint256 totalSupply_ = 10 ether;
8        constructor() { balances[msg.sender] = totalSupply_; }
9        function totalSupply() public override view returns (uint256) { return totalSupply_; }
10       function balanceOf(address tokenOwner) public override view returns (uint256) {
11           return balances[tokenOwner];
12       }
13       function transfer(address receiver, uint256 nTokens) public override returns (bool) {
14           require(nTokens <= balances[msg.sender]);
15           balances[msg.sender] = balances[msg.sender]-nTokens;
16           balances[receiver] = balances[receiver]+nTokens;
17           return true;
18       }
19       function approve(address delegate, uint256 nTokens) public override returns (bool) {
20           allowed[msg.sender][delegate] = nTokens;
21           return true;
22       }
23       function allowance(address owner, address delegate) public override view returns (uint) {
24           return allowed[owner][delegate];
25       }
26       function transferFrom(address owner, address buyer, uint256 nTokens) public override returns (bool) {
27           require(nTokens <= balances[owner]);
28           require(nTokens <= allowed[owner][msg.sender]);
29           balances[owner] = balances[owner]-nTokens;
30           allowed[owner][msg.sender] = allowed[owner][msg.sender]+nTokens;
31           balances[buyer] = balances[buyer]+nTokens;
32           return true;
33       }
```

Fig. 12 A basic smart contract of an ERC-20 token implementing IERC20 interface

physical asset uniquely and non-duplicatable as an NFT, and ERC-1155 for digitally representing equity shares of a company. A basic Solidity smart-contract implementation of ERC-20 is shown in Fig. 12.

4.3 Transaction Processing

Since a transaction may involve interacting with a smart contract by calling a function in the smart contract, the processing is not simply a verification such as checking fund availability. Let us explain this for the case of Ethereum. The Ethereum blockchain adopts the account-based state model; its state consists of a set of accounts (blockchain addresses) and their corresponding information. There are two types of accounts:

- Externally owned account: one that is owned by a normal user (like a bitcoin account). The state for an external account is its ETH balance.
- Contract account: one that represents a deployed smart contract. The state for a contract account consists of its ETH balance, contract code, and a storage area to save the run-time state of the contract.

The Ethereum blockchain protocol is essentially the same as that in the blockchain framework we presented in Sect. 2.6, the main difference being in transaction processing and block validation steps.

A transaction is a transfer of asset/value and optional data from a sender to a receiver. Specifically, it has a sender who initiates the transaction, a receiver who receives the transaction, a value (amount of tokens) to be transferred from the sender to the receiver, and a data part if the receiver is a contract account. If the transaction is received by a contract account, the corresponding contract code will be executed, taking as input the data included in the transaction. Ethereum introduces a concept called "gas fee" to represent how much ETH the transaction will pay the miner. A transaction contains two values, $startgas$ and $gasprice$.

- The $startgas$ value represents the maximum number of computational steps the transaction execution is allowed to take. The sender should have an idea as to how complex the transaction is and determines this value properly. If the miner takes more steps than this threshold allows, the transaction will halt and be reverted.
- The $gasprice$ value is the ETH fee the sender will pay the miner per computational step. To expedite the processing, the sender should increase $gasprice$ so that the miner would include the transaction in the next block.

Upon a transaction, the <u>state transition</u> happens as follows:

1. Check if the transaction is well formed and valid. Else, terminate.
2. Set transaction fee to $startgas \times gasprice$. Subtract this fee from the sender's balance. If the balance is not sufficient, terminate.
3. Initialize $GAS = startgas$, minus a certain quantity of gas per byte to pay for the byte count in the transaction.
4. Transfer the value specified in the transaction from the sender's balance to the receiver's balance.

    - If the receiver does not exist, create it.
    - If the receiver exists and is a contract account, run the contract code either to completion or until the execution runs out of GAS.

5. If this transfer fails: revert all state changes except the payment of the fees, and add the fees to the miner's account.
6. Else, refund the remaining GAS to the sender, and send the fees paid for gas consumed to the miner.

4.4 Block Validation

In Ethereum, a block contains a list of transactions and the blockchain state obtained by applying these transactions to the previous state (stored in the previous block). The creation of a new block requires PoW mining similar to Bitcoin (although Ethereum is transitioning to proof-of-stake consensus in version 2.0). The validation of a block

```
1    contract Train {
2        struct Ticket {
3            uint price;
4            address buyer;
5        }
6
7        mapping(uint => Ticket) tickets;
8
9        function booking(uint ticket_id) payable public {
10            require(msg.value == tickets[ticket_id].price);
11            rooms[ticket_id].buyer = msg.sender;
12        }
13        // ... other code ...
14   }
```

Fig. 13 The Train smart contract: this contract will be called later by the Booking smart contract

requires checking the cryptographic link with the previous block as usual, but in addition, it has to replay the running of all the transactions in the block. Specifically, a miner validates a new block as follows:

1. Verify that the previous block referenced exists and is valid.
2. Verify that the timestamp of the new block is greater than that of the previous block and less than 15 min into the future.
3. Verify that the new block's ID, difficulty target, and transaction Merkle root are valid.
4. Starting from the previous blockchain state (stored in the previous block):

   - Run a sequence of state transitions as a result of applying all the transactions in the new block, one by one.
   - If any such transaction replay fails or if the total gas consumed exceeds the limit, terminate.

5. If the Merkle tree root of the final state in the above step equals that stored in the new block, then the new block is valid. Else, invalid.

4.5 Contract Interoperability

In a smart-contract blockchain network, one can call another contract inside a contract. For example, consider a Travel Booking application: allow people to purchase a train ticket and reserve a hotel such that each booking is atomic—either both reservations succeed or neither do. This is referred to as the "train-and-hotel" problem popular as a case study in Ethereum research community.[2]

We have three smart contracts:

---

[2] This problem is described on this page: https://eth.wiki/sharding/Sharding-FAQs.

```
1    contract Hotel {
2        struct Room {
3            uint price;
4            address guest;
5        }
6
7        mapping(uint => Room) rooms;
8
9        function booking(uint room_id) payable public {
10           require(msg.value == rooms[room_id].price);
11           rooms[room_id].guest = msg.sender;
12       }
13       // ... other code ...
14   }
```

Fig. 14 The Hotel smart contract: this contract will be called later by the Booking smart contract

```
1    contract Booking {
2        Train train;
3        Hotel hotel;
4
5        function order(
6            uint ticket_id, uint ticket_price,
7            uint room_id, uint room_price) payable public {
8            require(msg.value == ticket_price + room_price);
9            require(train.call.value(ticket_price).booking(ticket_id));
10           require(hotel.call.value(room_price).booking(room_id));
11       }
12       // ... other code ...
13   }
```

Fig. 15 The Booking smart contract: this contract calls the Train contract and Hotel contract

- Train smart contract (Fig. 13): keep status of all train bookings. A user can book a train ticket by calling the booking() function of this contract.
- Hotel smart contract (Fig. 14): keep status of all hotel bookings. A user can book a hotel room by calling the booking() function of this contract.
- Booking smart contract (Fig. 15): A user can book a trip (a hotel and a train) by calling the order() function of this contract.

The Booking smart contract calls the other two contracts. Intuitively, it is just like calling another computer program inside a computer program. However, the advantage of running this application on the blockchain is that in the case one of the two bookings fails even though the other was order() function call will fail and, as a result, reverting the successful booking. The user will not lose any money. In the traditional deployment of this application on a non-blockchain environment, it would be more complex to the revert the user's successful booking.

The use of a Turing-complete language like solidity for smart-contract networks and the capability for contracts to call one another open limitless creativity when it comes to application. For any computer application in the real world, in theory, we can develop an equivalent version to run on the blockchain. This is why, with the birth of Ethereum and recent smart-contract network alternatives, we have been witnessing many businesses enter the blockchain space, most notably in the finance field with Decentralized Finance (DeFi).

5 Blockchain Scalability

Every transaction is broadcast to the whole network. So is every block. Block validation takes time and efforts too. Due to the chain topology of the blockchain, the fact that only one block can be the next node of the chain leads to many completing blocks being wasted, effectively reducing transaction throughput. On the other hand, we want blockchain to be the universal computer for everyone, for every application, if at all possible. Unfortunately, scalability remains a top challenge of blockchain technology [20].

5.1 The Blockchain Trilemma

Blockchain is aimed at three goals: decentralized, secure, and scalable. They cannot be all perfectly realized, at least according to Ethereum's Founder, Vitalik Buterin, who originated the term "Scalability Trilemma" for blockchain [8]:

- Decentralized: Set to provide trustless computing, blockchain does not rely on a central point of control. It needs to be decentralized such that nodes participate autonomously and equally to each other.
- Secure: The blockchain must operate as expected, robust to malfunctions and attacks. As we discussed in the previous section, blockchain is a Byzantine fault-tolerant system. The more failures of a large threshold of nodes it can sustain, the better security it provides.
- Scalable: Meant to be a "world" computer for all people to run all applications, blockchain should scale with increasingly growing amounts of transactions. This is in terms of both storage and computation demands.

The decentralization goal requires as many nodes as possible to participate in the block validation. Having more validator nodes, however, leads to more difficulty in maintaining consensus, thus security. Both decentralization and security goals are achievable only with small-scale blockchain networks (network size or transaction volume). As such, the scalability goal is not met.

Blockchains are often forced to make tradeoffs in this trilemma. Bitcoin offers excellent decentralization and security, but unattractive scalability. Due to the 10 min block creation rule to guarantee security, the transaction processing is slow, on the order of two to five transactions per second (tps). This is not practical for real-world payment at merchants. Traditional credit card systems such as Visa and Mastercard are three to four orders of magnitude faster.

On the other hand, Solana, a smart-contract network created in 2017 by Anatoly Yakovenko et al. [48], sacrifices decentralization for scalability. It is very fast. Solana does not use PoW consensus which is of course slow. It is based on a unique Proof-of-History (PoH) consensus algorithm, a variation of Proof of Stake (PoS). In theory, Solana's claimed throughput can be as high as 710,000 tps; the practically observed number is about 50,000 tps, still much faster than Bitcoin. However, as pointed out by many, Solana is vulnerable to centralization. In Solana, like other PoS networks, the decision to add a block to the blockchain is made among a small subset of "validator" nodes. Since fewer nodes involve in the consensus decision, it is faster than Bitcoin where all nodes can be miners. The problem with Solana is that the Solana Foundation is the only entity developing core nodes (validators) on the blockchain. This means Solana has a central point of control that reduces the network's overall decentralization. In comparison, several core node developers are building on Ethereum (e.g., Go Ethereum, OpenEthereum, Nevermind, and Besu). As of April 1, 2022, the number of Solana validator nodes is estimated to be around 1,100 nodes; in comparison, the PoS Ethereum network already has more than 200,000 nodes.

Comparing Ethereum (the PoS version) and Bitcoin, both offer excellent decentralization. Ethereum is faster, but the tradeoff is in security where Bitcoin is the superior, mainly due to PoW which has a higher entry barrier for block generation and higher cost to attack. An attacker would need to acquire 51%+ of the computational power in the network, whereas a PoS attacker would need to acquire 51%+ of the money within that system. To get the computational power in PoW, not only the attacker needs money but also physical efforts to acquire hardware. This external and physical factor makes PoW less vulnerable to attacks.

5.2 Layer-2 Scalability

Changing the core design, whether the consensus mechanism, block structure (chain or DAG), or cryptographic methods, at layer-1 of a blockchain network has tradeoffs due to the scalability trilemma. In 2016, at the peak then of high Ethereum gas fee, Joseph Poon and Vitalik Buterin introduced the approach of layer-2 scalability that applies to Ethereum, and, in theory, any layer-1 blockchain. The proposed solution, called Plasma [37], builds a high-throughput blockchain network anchored atop the layer-1 blockchain as follows: (1) layer-2 transaction processing: users transact on the layer-2 blockchain, hence very fast and (2) layer-1 transaction finality: state information records of completed transactions are saved in the layer-1 blockchain, hence assuring security against dishonest transactions.

For example, Polygon[3] is a layer-2 smart-contract blockchain on top of Ethereum network as layer-1. Polygon started with the Plasma approach in the early stage and now is one of the most successful blockchain networks. It is noted that the idea of layer-2 scalability was actually applied in the Lightning Network [38], created by Joseph Poon and Thaddeus Dryja in 2015. The scaling method used is called *State Channels*. This, however, is suitable only to a payment network like Bitcoin (as a fast payment protocol on layer-2), but not to a general smart-contract network. Another scaling method often referenced is *Sidechain* [2], which is a much simpler and less secure version of Plasma.

The invention of Plasma opened a new direction in blockchain scalability, leading to more advanced solutions such as Optimistic Rollups[4] and ZK Rollups[5], which are trending today [41]. To explain the layer-2 scalability's concept and feasibility, let us describe how Plasma works below. We hope that this will be helpful to the reader in understanding recently emerging scalability methods.

Plasma Scaling

There are several variants of Plasma. For example, Plasma Cash [26] is a Plasma solution for non-fungible tokens (NFT). We present a basic version of Plasma—the original proposal [37], which is for fungible assets below.

The Plasma Chain: First, we need to build a separate blockchain network to serve as the layer-2. We refer to this as Plasma Chain and to the layer-1 chain as Root Chain. Any blockchain design can work for Plasma Chain as long as it is fast and scalable, for example, Proof of Stake or Proof of Authority is a better choice than Proof of Work for the consensus mechanism. In the initial Plasma proposal, Plasma Chain adopts the UTXO state model (Bitcoin-like). Although this model is not suitable for enabling smart contracts at layer-2, for simplicity, we assume this model to explain the core idea of layer-2 scalability.

Plasma Chain processes transactions and creates blocks as usual functionalities of the chain. However, there is an additional step for the validator nodes (those that can produce blocks on Plasma Chain) after they have added each block to Plasma Chain: need to save a record of it on Root Chain. This is called an on-chain "block commit" or "checkpoint". By "on-chain" we mean layer-1 activity, whereas "off-chain" we mean layer-2. Adding a block to Plasma Chain provides its finality on Plasma Chain. Committing this block to Root Chain provides its finality on Root Chain, which is the "finalized" finality. The latest checkpoint is the proof that all transactions (and the funds) are permanent up to this point. Blocks are committed on-chain by interacting with a smart contract on Root Chain. There is also an entity, called Plasma Operator, which is watching events from this smart contract and will respond accordingly on Plasma Chain.

---

[3] https://polygon.technology.

[4] https://docs.ethhub.io/ethereum-roadmap/layer-2-scaling/optimistic_rollups/.

[5] https://docs.ethhub.io/ethereum-roadmap/layer-2-scaling/zk-rollups/.

The Root Contract: We need to create a smart contract on Root Chain; let us name it Root Contract. It provides the following functionalities:

- `Block Submission`: Root Contract maintains a list of Plasma block headers, each essentially consisting of the Merkle root of the corresponding original block and the time it is added to the list; transactions are not included. The contract has a public function for inserting such a block into this list. This function is called by Plasma Chain's validator nodes after they have validated a block; alternatively, this can be called by Plasma Operator who watches block insertions on Plasma Chain. It is noted that because Root Contract simply saves the headers of the Plasma blocks, not the actual transactions, it cannot know by itself their validity (honest or malicious purpose).

- `Fund Deposit`: Bob needs some fund in his account before doing any transfer on Plasma Chain. The contract has a public function for anyone like him to deposit this fund. Once this fund is deposited on Root Chain, an event will be emitted to notify Plasma Operator who will mint a new UTXO with the corresponding amount of fund on Plasma Chain for Bob. The amount of fund in circulation on Plasma Chain is the total amount of all deposits (minus withdrawn funds if any).

- `Fund Withdrawal`: Alice can withdraw fund from Root Chain. The contract has a public function to allow so, which asks her to provide the proof for the fund used to withdraw. A withdrawal must correspond to some unspent UTXO on Plasma Chain. The fund proof includes the position of an unspent UTXO belonging to Alice on Plasma Chain and the Merkle proof for this UTXO in its corresponding Plasma block. Because the contract cannot tell instantly whether this UTXO is indeed unspent on Plasma Chain, the withdrawal is not immediate. It has to wait a dispute period, e.g., 7 days, during which anyone can challenge. If the challenge is valid, the contract will revert the withdrawal.

- `Fraud Proofs`: The contract has a public function to allow anyone to challenge the validity of a malicious block committed from Plasma Chain or a withdrawal request within its dispute period. In the case of challenging Alice's withdrawal request above, if Bob observes that the UTXO used in the withdrawal is also spent on the Plasma Chain, he will provide the position of this invalid UTXO on Plasma Chain and the Merkle proof of its existence there as input to the withdrawal-challenge function. The contract will see if this proof matches the corresponding block record in the Plasma block list of the contract. If matching, Alice's withdrawal will be reverted.

Example: Consider a Plasma Chain on top of Ethereum for people to make ETH payments.

1. Alice deposits 10 ETH to Root Contract on Root Chain. As a result, Plasma Operator will mint 10 ETH for her on Plasma Chain (this ETH on Plasma Chain is actually a wrapped version of the Ethereum ETH). At this time, the Plasma blockchain consists of only 1 UTXO:

$$UTXO\ 1 : \emptyset \rightarrow Alice : 10$$

2. On Plasma Chain, Alice transfers 5 ETH to Bob. The new blockchain state is

$$spent : UTXO\ 1 : \emptyset \rightarrow Alice : 10$$
$$UTXO\ 2 : Alice \rightarrow Alice : 5$$
$$UTXO\ 3 : Alice \rightarrow Bob : 5$$

3. Bob then transfers 3 ETH to Charlie. The new blockchain state is

$$spent : UTXO\ 1 : \emptyset \rightarrow Alice : 10$$
$$UTXO\ 2 : Alice \rightarrow Alice : 5$$
$$spent : UTXO\ 3 : Alice \rightarrow Bob : 5$$
$$UTXO\ 4 : Bob \rightarrow Bob : 2$$
$$UTXO\ 5 : Bob \rightarrow Charlie : 3$$

4. Charlie transfers 2 ETH to Alice. The new blockchain state is

$$spent : UTXO\ 1 : \emptyset \rightarrow Alice : 10$$
$$UTXO\ 2 : Alice \rightarrow Alice : 5$$
$$spent : UTXO\ 3 : Alice \rightarrow Bob : 5$$
$$UTXO\ 4 : Bob \rightarrow Bob : 2$$
$$spent : UTXO\ 5 : Bob \rightarrow Charlie : 3$$
$$UTXO\ 6 : Charlie \rightarrow Charlie : 1$$
$$UTXO\ 7 : Charlie \rightarrow Alice : 2$$

5. At this time, on Plasma Chain, Alice has 7 ETH (from UTXO 2 and UTXO 7), Bob has 2 ETH (from UTXO 4), and Charlie has 1 ETH (from UTXO 6). Note that the above transactions were included in Plasma blocks of Plasma Chain and their headers have been saved in Root Contract.
6. Bob requests to withdraw 2 ETH (calling the withdrawal function of Root Contract on Root Chain). He inputs to this function UTXO 4 as the source for the fund. The withdrawal request is pending for 7 days. During these 7 days, no one challenges this request because UTXO 4 is not spent on Plasma Chain during the dispute period. Therefore, Root Contract sends 2 ETH (of Root Chain) to Bob. It is noted that Bob did not have to deposit fund on Ethereum in order to withdraw.
7. Alice requests to withdraw 5 ETH using UTXO 3. During the 7-day dispute period, Charlie who watches Plasma Chain observes that UTXO 3 was spent on Plasma Chain. He will challenge the withdrawal by submitting the Merkle proof of this UTXO 3 to Root Contract. This proof is valid, thus canceling Alice's withdrawal.

It is important that those users who have fund on Plasma Chain should watch the chain frequently to make sure their funds are safe. This requires downloading the chain and verify its correctness. If a user detects or suspects something wrong, the user's wallet (software) will automatically request to withdraw funds.

To avoid spammers and those submitting irresponsible withdrawals while encouraging fraud reporting, one can design Root Contract such that each withdrawal request must include a penalty bond that will be collected to reward the challenger in the case of bad withdrawal. To enable fast withdrawals (7 days are too long), a Plasma solution can involve Liquidity Providers who are incentivized to advance the fund to the withdrawers while taking the risk of bad withdrawals. A solution, e.g., Polygon, can also require that Alice burn the fund on Plasma Chain before requesting to withdraw it on Root Chain; she needs to submit the proof of this burn.

The on-chain block commit in Plasma is the key difference between it and the Sidechain scaling approach [2]; the latter is often mistakenly considered the same as Plasma but it is very different. Sidechain also has a smart contract like Root Contract with functions for deposits and withdrawals, but does not have block commits. It is simpler but a major con is that the sidechain can stop producing blocks and locks everyone's funds up forever. Sidechain is thus much less secure. With Plasma, the block list in the Root Contract is the proof that users have their funds and thus can withdraw them.

Rollups Scaling

The Plasma approach is more suitable for token transfer transactions, but not for smart contracts. The Rollups approach [7] was born to be general purpose. The layer-2 blockchain in Rollups can run smart contracts. For example, one can run an EVM inside the layer-2 chain, allowing existing Ethereum applications to migrate to Rollups without re-writing the smart-contract code.

Rollups can be considered a hybrid Plasma approach. Plasma keeps all the transaction data off-chain and, as such, Root Chain cannot verify Plasma transactions, leaving room for Plasma Chain to do things maliciously. In Rollups, part of transaction data is saved on the Root Chain in addition to block headers. As a result, Root Chain can verify transactions too, thus providing an additional layer to enhance security and decentralization. It is noted that Rollups does not save *all* transactions on Root Chain because doing so makes the Rollups chain meaningless; it does not do any scaling. If Rollups saved none of transaction data, it would become Plasma. To reduce the amount of transaction data saved on Root Chain, it saves only the information necessary to verify transactions. Transaction data involving state storage remains on the Rollups chain.

There are two main Rollups approaches: optimistic Rollups and zero-knowledge (ZK) rollups. The former resembles Plasma in that it also uses fraud proofs to challenge invalid fund withdrawals and invalid layer-2 transactions. ZK Rollups is more disruptive in that it allows instant withdrawals.

- Optimistic Rollups: The name "optimistic" comes from the assumption in this approach that the transaction data submitted to Root Chain is correct. After the Rollups chain commits a batch of transactions to Root Chain, they will be considered permanently finalized if no one submits a fraud proof to challenge any transaction. Whenever a fraud proof is submitted, the suspicious transaction will be re-validated: it will be replayed on Root Chain using the block state and transaction data information already saved in Root Contract. The replay of such transaction is similar to that in the transaction verification procedure of Ethereum. Noticeable implementations of Optimistic Rollups include Optimism[6] and Arbitrum[7].
- ZK Rollups: ZK Rollups leverages a cryptographic method called zk-SNARK (zero-knowledge succinct non-interactive argument of knowledge) [11]. A zk-SNARK is a cryptographic proof that allows one party to prove that it possesses certain information without revealing that information. The verification of the proof is quick and cheap. When a batch of transactions are to be committed on Root Chain, a zk-SNARK proof is computed for this data to prove its validity and sent along to Root Chain. Root Contract verifies this proof on Root Chain when receiving a withdrawal request; if valid, the fund is released immediately. Noticeable implementations of ZK Rollups include dYdX[8], Loopring [44], zkSync [15], and ZKSpace.[9]

To understand ZK Proofs, suppose that Alice wants to prove to Bob her knowing of a value x such that $f(x) = output$ for a given $output$. Can she do that without disclosing value x? For example, can Alice provide a proof that she knows a secret value having a given SHA256 hash without revealing this secret? This is called a zero-knowledge proof. A related example is the well-known Yao's Millionaires' Problem [31]: can two millionaires, Alice and Bob, know who is richer without revealing their actual wealth? Mathematically put, with two numbers a and b, can we determine whether $a \leq b$ without revealing the actual values of a and b?

zk-SNARK is a method for computing ZK proofs. First, assume that we can write a computer program to implement a Boolean function $C(output, x)$ that returns true if and only if $f(x) = output$. For example, if f is SHA256:

```
Boolean function C(output, x) {
    return (SHA256(x) == output);
}
```

A zk-SNARK is a set of three functions, $Generator()$, $Prover()$, and $Verifier()$, defined as follows:

---

[6] https://www.optimism.io.

[7] https://offchainlabs.com/.

[8] https://dydx.exchange.

[9] https://zks.org.

$$Generator(\lambda, C) \rightarrow (pk, vk) \tag{9}$$

$$Prover(pk, output, x) \rightarrow prf \tag{10}$$

$$Verifier(vk, output, prf) \rightarrow \{true, false\} \tag{11}$$

- *Generator()*: This is called the key generator. It takes as input a secret parameter λ and program C and outputs a pair of keys called a "proving key" pk, and a "verification key" vk. These keys are publicly known. It is noted that the secret parameter λ must be known to no one except the generator.
- *Prover()*: This is called the prover: Alice calls this function taking as input the proving key pk, the public value $output$, and her secret value x that she wants to prove that $f(x) = output$. This function will output a value called "proof" prf. Alice will send this proof to Bob.
- *Verifier()*: This is called the verifier: Bob uses this function to take as input the verification key vk, the public value $output$, and the proof prf he received from Alice. This function returns true iff the proof is correct, i.e., the prover knows a value x satisfying $f(x) = output$.

As another example, suppose that Alice wants to transfer tokens of some ERC-20 cryptocurrency to somebody. Using the standard ERC-20 smart contract, the public sees the account balance of Alice, $balance$, and the amount she sends, $value$. In many cases, it is desirable to hide these numbers. For this purpose, we can implement a smart contract that makes public only the following hashes of these numbers, $balanceOld = SHA256(balance)$, $sentValue = SHA256(value)$, and $balanceNew = SHA256(balance - value)$. Knowing these hashed values, the miner (Bob) cannot know the raw values, $balance$ and $value$, but can still verify whether the transfer is valid. In this example, Alice is the prover and miner Bob is the verifier. The corresponding program code to define the logic for this verification, which is input into zk-SNARK, is as follows:

```
Boolean function C(output, x) {
    return (x.balance >= x.value
            && SHA256(x.balance) == output.balanceOld
            && SHA256(x.value) == output.sentValue
            && SHA256(x.balance-x.value) == output.balanceNew);
}
```

Here, $output$ is the object consisting of the three hashed values that Bob observes and x is the secret information about the sender's balance and value sent. With this program code C, the key generator will take it as input, together with a random parameter λ, to generate a proving key pk and a verification key vk. Alice and Bob use these two keys to prove and verify as above.

There is tradeoff between optimistic versus ZK rollups. Due to mathematical complexity, generic constructions for ZK protocols are too expensive to be used in practice. Thus far, it has been suitable for only a few specific applications such as payments and token exchanges, like what is mainly served by Plasma. Optimistic Rollups, on the other hand, thanks to its simplicity, supports layer-2 smart contracts better. It, however, requires more storage in Root Contract (data needed to replay transactions for verification purposes). In contrast, with ZK proofs that can readily verify transactions, ZK Rollups requires less storage for Root Contract.

6 Blockchain Interoperability

Existing blockchain networks are each on their own island isolated from one another. Bitcoin users can only transact with other Bitcoin owners, but not with Ethereum users. Decentralized applications on Ethereum cannot make calls to those on other blockchain networks. Data on one blockchain cannot be shared outside either. This is analogous to the early days of Internet, where different "Internets" (networks) were developed independently to serve their own purposes or groups of users. They adopted different technologies and architectures that do not speak the same language. However, the Internet today is universally interoperable in that even though it consists of many Internet providers' networks, any two computers or applications regardless of where they belong can communicate with each other.

Interoperability between the chains must be a top priority for blockchain. This should be seamless so that one should focus on the logic of the application without having to worry about which underlying blockchain technology stacks to use. Imagine the complexities that would arise for a supply-chain company if it runs the product tracking application on a blockchain and the payment application on another blockchain, and these two blockchains are not compatible.

At the least, we should enable interoperability for digital assets. We should be able to transfer or exchange assets between different networks without intermediaries such as a centralized cryptoexchange. This would allow a Bitcoin user to pay Bitcoin to a merchant that runs its point-of-sale software built on Ethereum. This would benefit immensely decentralized finance (DeFi) applications that would be able to tap into all populations of users who own various types of assets. The next level of interoperability is for cross-chain exchanges of arbitrary data. This would enable smart contracts and applications on different blockchains to communicate and share information. This kind of interoperability is of course much more difficult to achieve.

Efforts to realize blockchain interoperability remain fragmented. Protocols, however, have taken shape into three main approaches: Atomic Swap, Chain Bridge, and Chain Hub.

6.1 Atomic Swap

Atomic Swap [22] is a simple solution for two users to swap assets without involving any third party. They can be on the same chain or different chains. Suppose that Alice wants to transfer some asset X to Bob who in return transfers some asset Y to her. In a naive scenario, Alice will just send X to Bob and expects him to send Y to her. The problem is, in the real world, Bob could just take her asset and run away.

Atomic Swap guarantees that the exchange succeeds or else, nothing happens without either side losing asset. It works as follows. Alice and Bob each need to create a Hash-Time Locked Contract (HTLC) [38] to deposit their respective asset. Specifically, Alice will do:

1. Generate a secret key k_{Alice}. Only she knows it at this time.
2. Compute a crypto-hash value of this key, $m = H(k_{Alice})$. The hash function H is known to Bob.
3. Create a Hash-Time Locked smart contract (HTLC) on her chain to deposit asset X with a lock and an expiration time. This HTLC has a function to unlock X if it is called before expiration and input with a key k such that $H(k) = m$.

   • If asset X is unlocked, it will be transferred to the caller.
   • If X remains locked at expiration time, it will be returned to Alice.

4. Send the hash value m to Bob.

   On his side, Bob will do:

1. Create a Hash-Time Locked Smart Contract (HTLC) on his chain to deposit asset Y with a lock and an expiration time. This HTLC has a function to unlock Y if it is called before expiration and input with a key k such that $H(k) = m$. This value m is the hash value sent from Alice.
2. Wait until the above unlock function is called and succeeds.

   • If asset Y remains locked at expiration time, it will be returned to Bob.
   • Else, the input key k must equal the secret key of Alice, k_{Alice}. Therefore, Bob knows this private key. He will call the HTLC of Alice inputting this key $k = k_{Alice}$ to unlock asset X and have it transferred to him.

Atomic Swap will not do anything if Alice does not claim asset Y on Bob's contract, because if so Bob has no knowledge of her secret key to claim asset X on Alice's contract. If Alice does claim, Bob will know this key and claim his part too. No third party is involved here. On the other hand, Atomic Swap is not instant. It depends on the actions of Alice and Bob. Alice must send the hash value m to Bob for him to set up his smart contract. She must then by herself contact his smart contract and vice versa.

6.2   Chain Bridge

While Atomic Swap is for swapping assets, Chain Bridge enables transfers of assets cross chains. To illustrate its idea and feasibility, suppose that we want to bridge a smart-contract blockchain X (token USDX) with a smart-contract blockchain Y (token USDY). A basic chain bridge solution needs to write two smart contracts, one on X and one on Y. The bridge is owned by an entity called bridge operator, who watches events emitted from these contracts. Bridge operator also has a liquidity pool LP_X of n_X USDX on X and a liquidity pool LP_Y of n_Y UDXY on Y.

Suppose that Alice on chain X wants to transfer 10 USDX to Bob on chain Y. For simplicity, 1 USDX = 1 USDY and so he will receive 10 UDXY. The transfer happens as follows:

- On Chain X: Alice calls the contract on X to deposit 10 USDX to the liquidity pool LP_X on X. The new pool amount will become $n_X := n_X + 10$.
- Bridge Operator: detects this deposit and does the step below.
- On Chain Y: Bridge operator calls the contract Y to transfer to Bob 10 USDY from the liquidity pool LP_Y. The new pool amount will be $n_Y := n_Y - 10$.

Since X and Y are existing chains in which bridge operator has no authority to mint assets, the liquidity pools are needed to provide instant liquidity for the transfer. The reserve amounts n_X and n_Y set the maximum amount one can transfer to X and Y, respectively. Thus, the more reserves, the more transfer volume is allowed. One can be creative by encouraging liquidity providers to contribute to these pools.

In the case that bridge operator owns one of the two chains, say chain X, we do not need liquidity pool LP_X. In place of LP_X, bridge operator simply mints new USDX to the receiver anytime receiving a transfer from chain Y. Similarly, bridge operator burns USDX of the sender when needing to transfer it to chain Y. This is the solution often used when designing a new blockchain network that wants to bridge with an existing blockchain (e.g., Ethereum, so that the new network can host a wrapped version of ETH).

A challenge with chain bridge is how to ensure security given the role of bridge operator [30]. For maximal security, bridge operator needs to be decentralized; ideally, it can itself be a blockchain network. However, that would lead to implementation complexities. In fact, no bridging solution has adopted such a method fully. One can resort to the cryptographic method of secure multi-party communication to partially decentralize the role of bridge operator, as in the multichain framework,[10] but to date weak security remains the biggest concern for chain bridge. Many hacks targeted bridge solutions, most noticeable being the attack on Axie Infinity just this year (March 2022) incurring a loss of 600+ million USD.

---

[10] https://multichain.org/.

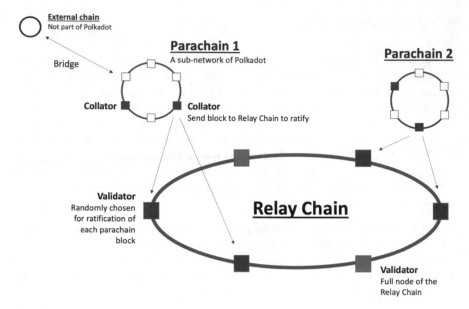

Fig. 16 Polkadot network: Blockchains (parachains) communicate with each other via Relay Chain. Parachain consensus is ensured by the Collators who are the validators of the parachain. Inter-parachain consensus is ensured by the validators who are nodes on Relay Chain

6.3 Chain Hub

Bridging is the interoperability solution to make two blockchains talk to each other. If there are n blockchains, we would need $n(n-1)/2$ bridges to enable any two chains to communicate directly. Chain hub is an approach that builds the Internet of blockchains by providing a "hub" connecting to all the blockchains and dedicated to passing messages between them. More than that, this hub itself is a blockchain network, thus providing maximal decentralization and security. Cosmos [28], Polkadot [5], and Avalanche [40] are major solutions adopting this approach. They call the "hub" by different names (relay in Polkadot and Avalanche, or hub in Cosmos).

For example, consider Polkadot [47], whose architecture is illustrated in Fig. 16. Polkadot is a network of heterogeneous blockchain shards called "parachains". These chains connect to and are secured by a chain called relay chain; this is the hub of Polkadot. Existing blockchains or those not of Polkadot network are called external networks which can talk to any parachain via bridges. There are four main roles for Polkadot keepers: validators, nominators, collators, and fishermen.

- Validators: They must be among the nodes that form Relay Chain. Once new blocks have been validated in their parachains, they must be ratified on Relay Chain. First, a subgroup of validators is chosen randomly to ratify each new parachain block. This results in a new block to add to Relay Chain. This block will be validated on Relay Chain as usual by all the validators.

- Nominators: They are stake-holding parties who risk capital to nominate nodes to become validators. Nominators get earnings if their nominees are chosen as validators. The method to choose validators from nominations is based on Nominated Proof-of-Stake (NPoS) consensus [5]. In some sense, the validators are similar to the mining pools of current PoW blockchains and the nominators are similar to the miners who join these pools.
- Collators: They must be among the parachain nodes. On their parachain, they author new blocks and execute transactions as usual (like miners in PoW blockchains or validators in PoS blockchains). In addition, as collators, they provide validators with valid parachain blocks (and zero-knowledge proofs) as candidate blocks to ratify on Relay Chain. We can think of collators as "local helpers" of validators on each parachain.
- Fishermen: They are "bounty hunters" who monitor Relay Chain and parachains to report irregularities committed by the nodes. They are rewarded by submitting a timely proof showing that at least one bonded party misbehaved. The fishermen are an additional layer for enhancing the network security.

Polkadot can connect a set of independent blockchains while providing pooled security and trust-free cross-chain transactability, which is thanks to Relay Chain with contributions from the above players. However, a Chain Hub solution like Polkadot requires building blockchains from scratch, which must use the same development framework (e.g., Substrate[11] in Polkadot or Tendermint [27] in Cosmos) and abide a shared communication protocol. As such, a blockchain network adopting Chain Hub cannot interface with existing blockchains or those using non-compatible designs. Chain Hub is therefore called a layer-0 blockchain interoperability solution. In the future, one hopes that Chain Hub will be successful and widely adopted. When that happens, we will realize the true vision of blockchain being a universal computer or the next-generation Internet.

7 Conclusions

This chapter has presented how blockchain works fundamentally, together with selective case studies, methods, and challenges, that help the reader understand this technology quickly to be sufficiently ready for further adventures. The coverage includes what blockchain is, its architecture and components, how it works for Bitcoin with proof-of-work consensus, the view of smart-contract blockchains as universal computers, and open challenges in scalability and interoperability, the top-2 priorities for blockchain technology. It should become now clear that there is no limit in potential applications of blockchain and emerging business models that otherwise are not feasible with conventional non-blockchain computing. However, despite its promise, blockchain technology is still in its infancy. Like the evolution of the Internet, it takes time for a new technology to mature and be widely accepted by traditional

---

[11] https://www.parity.io/technologies/substrate/.

businesses. Technically, besides the foremost importance of scalability and interoperability, many other challenges remain to address as we go more deeply into each component of the blockchain architecture: how to optimize the peer-to-peer networking layer; innovate consensus mechanisms to be eco-friendly, incentivize, and evaluate contributions to the security and decentralization of the blockchain; develop smart contracts that are bug-free; enable decentralized finance for everybody; and apply effectively to other meaningful real-world problems. All that makes the research and development of blockchain technology interesting.

Acknowledgements Duc A. Tran's work for this chapter was partially funded by Vingroup Joint Stock Company and supported by Vingroup Innovation Foundation (VINIF) under project code VINIF.2021.DA00128. Bhaskar Krishnamachari's work was supported in part by the USC Viterbi Center for Cyberphysical Systems and the Internet of Things.

References

1. Back, A.: Hashcash-a denial of service counter-measure (2002). http://www.hashcash.org/papers/hashcash.pdf
2. Back, A., Corallo, M., Dashjr, L., Friedenbach, M., Maxwell, G., Miller, A., Poelstra, A., Timón, J., Wuille, P.: Enabling blockchain innovations with pegged sidechains (2014). https://www.peercoin.net/whitepapers/peercoin-paper.pdf
3. Baird, L., Harmon, M., Madsen, P.: Hedera: A public hashgraph network and governing council (2020). https://hedera.com/hh_whitepaper_v2.1-20200815.pdf
4. Blum, M., Feldman, P., Micali, S.: Non-interactive zero-knowledge and its applications (extended abstract). In: J. Simon (ed.) Proceedings of the 20th Annual ACM Symposium on Theory of Computing, May 2–4, 1988, Chicago, Illinois, USA, pp. 103–112. ACM (1988). https://doi.org/10.1145/62212.62222
5. Burdges, J., Cevallos, A., Czaban, P., Habermeier, R., Hosseini, S., Lama, F., Alper, H.K., Luo, X., Shirazi, F., Stewart, A., Wood, G.: Overview of polkadot and its design considerations (2020). CoRR arXiv:2005.13456
6. Buterin, V.: Ethereum: a next-generation smart contract and decentralized application platform (2014). https://ethereum.org/en/whitepaper
7. Buterin, V.: An incomplete guide to rollups (2021). https://vitalik.ca/general/2021/01/05/rollup.html
8. Buterin, V.: Why sharding is great: demystifying the technical properties (2021). https://vitalik.ca/general/2021/04/07/sharding.html
9. Castro, M., Liskov, B.: Practical byzantine fault tolerance. In: Proceedings of the Third Symposium on Operating Systems Design and Implementation, OSDI'99, pp. 173–186. USENIX Association, USA (1999)
10. Chen, J., Micali, S.: Algorand: A secure and efficient distributed ledger. Theor. Comput. Sci. **777**, 155–183 (2019). https://doi.org/10.1016/j.tcs.2019.02.001
11. Chen, T., Lu, H., Kunpittaya, T., Luo, A.: A review of zk-snarks (2022)
12. Damgård, I.: Commitment schemes and zero-knowledge protocols. In: I. Damgård (ed.) Lectures on Data Security, Modern Cryptology in Theory and Practice, Summer School, Aarhus, Denmark, July 1998. Lecture Notes in Computer Science, vol. 1561, pp. 63–86. Springer (1998). DOI https://doi.org/10.1007/3-540-48969-X_3
13. Deloitte: Deloitte's 2020 global blockchain survey (2020). https://www2.deloitte.com/mt/en/pages/technology/articles/2020-global-blockchain-survey.html

14. Dwork, C., Naor, M.: Pricing via processing or combatting junk mail. In: Proceedings of the 12th Annual International Cryptology Conference on Advances in Cryptology, CRYPTO'92, pp. 139–147. Springer, Berlin, Heidelberg (1992)
15. Ethworks: Zero-knowledge blockchain scalability (2018). https://ethworks.io/assets/download/zero-knowledge-blockchain-scaling-ethworks.pdf
16. Eyal, I., Sirer, E.G.: Majority is not enough: bitcoin mining is vulnerable. Commun. ACM **61**(7), 95–102 (2018). https://doi.org/10.1145/3212998
17. Gilad, Y., Hemo, R., Micali, S., Vlachos, G., Zeldovich, N.: Algorand: scaling byzantine agreements for cryptocurrencies. In: Proceedings of the 26th Symposium on Operating Systems Principles, SOSP'17. Association for Computing Machinery, pp. 51–68. New York, NY, USA (2017). https://doi.org/10.1145/3132747.3132757
18. Goodman, L.M.: Tezos: a self-amending crypto-ledger (white paper) (2014). https://tezos.com/whitepaper.pdf
19. Grunspan, C., Pérez-Marco, R.: The mathematics of Bitcoin (2020). CoRR arXiv:2003.00001
20. Hafid, A., Hafid, A.S., Samih, M.: Scaling blockchains: a comprehensive survey. IEEE Access **8**, 125244–125262 (2020). https://doi.org/10.1109/ACCESS.2020.3007251
21. Hankerson, D., Menezes, A.: Elliptic Curve Cryptography, pp. 397. Springer US, Boston, MA (2011)
22. Herlihy, M.: Atomic cross-chain swaps. In: Proceedings of the 2018 ACM Symposium on Principles of Distributed Computing, PODC'18. Association for Computing Machinery, pp. 245–254. New York, NY, USA (2018). https://doi.org/10.1145/3212734.3212736
23. Iansiti, M., Lakhani, K.: The truth about blockchain. Harv. Bus. Rev. **95**, 118–127 (2017)
24. King, S., Nadal, S.: Ppcoin: Peer-to-peer crypto-currency with proof-of-stake (2012). https://www.peercoin.net/whitepapers/peercoin-paper.pdf
25. Koblitz, N.: Elliptic curve cryptosystems. Math. Comput. **48**(177), 203–209 (1987)
26. Konstantopoulos, G.: Plasma cash: towards more efficient plasma constructions (2019). https://doi.org/10.48550/ARXIV.1911.12095. arXiv:1911.12095
27. Kwon, J.: Tendermint: consensus without mining (2014). https://tendermint.com/static/docs/tendermint.pdf
28. Kwon, J., Buchman, E.: A network of distributed ledgers (2016). https://v1.cosmos.network/resources/whitepaper
29. Lamport, L., Shostak, R., Pease, M.: The byzantine generals problem. ACM Trans. Program. Lang. Syst. **4**(3), 382–401 (1982). https://doi.org/10.1145/357172.357176
30. Lan, R., Upadhyaya, G., Tse, S., Zamani, M.: Horizon: a gas-efficient, trustless bridge for cross-chain transactions (2021). https://doi.org/10.48550/ARXIV.2101.06000. arXiv:2101.06000
31. Lin, H.Y., Tzeng, W.G.: An efficient solution to the millionaires' problem based on homomorphic encryption. In: Proceedings of the Third International Conference on Applied Cryptography and Network Security, ACNS'05, pp. 456–466. Springer, Berlin, Heidelberg (2005)
32. Maymounkov, P., Mazieres, D.: Kademlia: A peer-to-peer information system based on the xor metric. Peer-to-Peer Systems, pp. 53–65 (2002)
33. Merkle, R.C.: A digital signature based on a conventional encryption function. CRYPTO'87, pp. 369–378. Springer, Berlin, Heidelberg (1987)
34. Miller, V.S.: Use of elliptic curves in cryptography. In: Williams, H.C. (ed.) Advances in Cryptology–CRYPTO'85 Proceedings, pp. 417–426. Springer, Berlin Heidelberg (1986)
35. Nakamoto, S.: Bitcoin: a peer-to-peer electronic cash system (2008). https://bitcoin.org/bitcoin.pdf
36. Nguyen, Q., Cronje, A., Kong, M., Lysenko, E., Guzev, A.: Lachesis: Scalable asynchronous bft on dag streams (2021). https://doi.org/10.48550/ARXIV.2108.01900. arXiv:2108.01900
37. Poon, J., Buterin, V.: Plasma: scalable autonomous smart contracts (2017). https://plasma.io/plasma.pdf
38. Poon, J., Dryja, T.: The Bitcoin lightning network: Scalable off-chain instant payments (2017). https://lightning.network/lightning-network-paper.pdf
39. PwC: Time for trust: The trillion-dollar reason to rethink blockchain pwc projected (2020). https://www.pwc.com/timefortrust

40. Rocket, T., Yin, M., Sekniqi, K., van Renesse, R., Sirer, E.G.: Scalable and probabilistic lead-erless BFT consensus through metastability (2019). CoRR arXiv:1906.08936
41. Sguanci, C., Spatafora, R., Vergani, A.: Layer 2 blockchain scaling: a survey (2021). arXiv:2107.10881
42. Shi, E.: Foundations of distributed consensus and blockchains (book manuscript) (2020). https://www.distributedconsensus.net
43. Szabo, N.: The idea of smart contracts (1997). https://nakamotoinstitute.org/the-idea-of-smart-contracts
44. Wang, D., Zhou, J., Wang, A.: Loopring: A decentralized token exchange protocol (2018). https://loopring.org/resources/en_whitepaper.pdf
45. Wang, T., Zhao, C., Yang, Q., Zhang, S., Liew, S.C.: Ethna: Analyzing the underlying peer-to-peer network of ethereum blockchain. IEEE Trans. Netw. Sci. Eng. **8**(3), 2131–2146 (2021). https://doi.org/10.1109/TNSE.2021.3078181
46. Wensley, J.H.: Sift: software implemented fault tolerance. In: Fall Joint Computer Conference, Part I, AFIPS'72 (Fall, part I). Association for Computing Machinery, pp. 243–253. New York, NY, USA (1972). https://doi.org/10.1145/1479992.1480025
47. Wood, G.: Polkadot white paper (2016). https://polkadot.network/PolkaDotPaper.pdf
48. Yakovenko, A.: Solana: A new architecture for a high performance blockchain v0.8.13 (2017). https://solana.com/solana-whitepaper.pdf

Blockchain Peer-to-Peer Network: Performance and Security

Phuc D. Thai, Minh Doan, Wei Liu, Tianming Liu, Sheng Li,
Hong-sheng Zhou, and Thang N. Dinh

Abstract Mistrusting nodes in a blockchain can reach consensus without the need
of a trusted central entity. Instead, the nodes reach consensus through exchanging information on a peer-to-peer (P2P) network, without pre-established identities.
Serving as the foundation of the blockchain, the P2P network plays critical roles
in all performance and security aspects of the blockchain system. While P2P networks had been previously examined for many applications domains, including the
file sharing systems, there is relatively less understanding on blockchain P2P networks that differs substantially from traditional P2P systems. In this chapter, we will
cover different aspects of blockchain P2P networks from topology, peer discovery,
known attacks, and defenses to improvement proposals to increase the throughput
and reduce the latency in blockchain. Finally, we investigate theoretical limit on the
throughput of blockchain systems in which nodes have heterogeneous capacities.
We provide insights and discussion on how to construct a network to achieve the
maximum theoretical limit in throughput.

P. D. Thai (✉) · H. Zhou · T. N. Dinh
Virginia Commonwealth University, Richmond, VA, USA
e-mail: thaipd@vcu.edu

H. Zhou
e-mail: hszhou@vcu.edu

T. N. Dinh
e-mail: tndinh@vcu.edu

M. Doan
Harmony.one, Pittsburgh, PA, USA
e-mail: minh@harmony.one

W. Liu
Mayo Clinic, Jacksonville, FL, USA
e-mail: Liu.Wei@mayo.edu

T. Liu · S. Li
University of Georgia, Athens, GA, USA
e-mail: tliu@uga.edu

S. Li
e-mail: sheng.li@uga.edu

© The Author(s), under exclusive license to Springer Nature Switzerland AG 2022
D. A. Tran et al. (eds.), *Handbook on Blockchain*, Springer Optimization
and Its Applications 194, https://doi.org/10.1007/978-3-031-07535-3_2

1 Introduction

In a blockchain system, all nodes share and agree on a common ledger, consisting of transactions, data, and timestamps. The information is organized using a block data structure. Every block is linked to a previous one via a cryptographic hash link, forming a chain (hence the name). Blockchain has been positioned as a disruptive force to decentralize economy and society structures. It can remove the need for trust in a single party and eliminate single point of failures in centralized systems, making those systems more resilient against censorship [1]. Major applications of blockchain include cryptocurrencies such as Bitcoin [2] and Ethereum [3], decentralized finance [4], decentralized AI [5], the Internet of Things [6, 7], and Digital Health [8, 9].

The operation of the blockchain relies on its network foundation, a peer-to-peer (P2P) network in which equally privileged peers contribute their resources, such as computing power and network bandwidth to jointly maintain the consensus. Nodes in the P2P network relay blocks, transactions, or other information to its neighbors who, in turn, relay the information further to other nodes in the network. The P2P network has been found to be the root cause for the all the unstability [10, 11] and cyberattacks [12] of blockchain systems. Thus, it is crucial to study the security and performance of the blockchain P2P networks for the purpose of designing a secure P2P network for high-throughput blockchain systems, matching the throughput of existing financial networks such as Visa and MasterCard.

In this chapter, we will provide different aspects of P2P networks. We will present the existing P2P networks in Bitcoin and Ethereum, known security issues, such as eclipse and partitioning attacks, and performance issues, such as low throughput and high latency. In addition, we provide a summary on improvement proposals for known issues in the P2P networks. Finally, we investigate the theoretical limit on the throughput of P2P networks in which nodes have heterogeneous capacities. While Kumar and Ross [13] establish a theoretical limit on the throughput of a P2P, the limit is only for the case when there is only a single source in the network. In contrast, any nodes in a blockchain system can be the source for broadcasting transactions and blocks. To this end, we show that when having *multiple sources*, the throughput is bounded by the bandwidth of the source nodes and the average bandwidth of all nodes. We also provide a distribution scheme to achieve the maximum throughput, proving the tightness of our bound.

Our contributions are summarized as follows:

- We provide an overview on important aspects of blockchain P2P networks including network topology formation (peer discovery, connection, and data forwarding rules), network-level attacks, and performance issues.
- We establish a theoretical limit on the throughput for multiple source broadcasting in P2P network. The limit serves as an upper bound on maximum throughput of a blockchain system with heterogeneous capacities.
- We formulate the problem of designing a P2P network with maximum throughput and show a simple construction to attain the maximum limit.

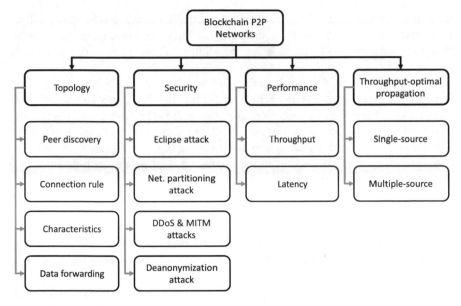

Fig. 1 Paper organization

Organization. A summary of our paper is given in Fig. 1. In Sect. 2, we provide an overview on blockchain P2P networks. We present the topology formation and data forwarding schemes in existing blockchain systems in Sect. 3. Section 4 presents critical network-level attacks on the P2P network. In Sect. 5, we study the performance (throughput and latency) of blockchain systems. Finally, in Sect. 6, we formalize an optimization problem to maximize the throughput in a blockchain system and present a solution to achieve the maximum throughput.

1.1 Related Work

Before the arrival of Bitcoin [2], the problem of maximizing the throughput in a P2P network has been investigated for several applications.

SplitStream [14] constructs a transmission schedule from a single-source node by constructing multiple multicast trees in which an interior node in one tree is a leaf node in all other trees. This ensures that the forwarding load can be well balanced among all nodes, achieving both low latency and sparsity. However, the approach does not consider heterogeneous bandwidth and provides no theoretical guarantees on the throughput and latency in those cases.

In [15], the single-source broadcast problem was solved by dividing nodes into clusters of size $\log(n)$, in which the total capacity of each cluster is roughly the same. Nodes in each cluster are fully connected to each other in order to obtain optimal throughput in each cluster. Clusters are then abstracted as supernodes, and multiple broadcast trees are built over the supernodes. Since the capacity of each cluster is almost the same, they can be achieved within a factor $(1 - \epsilon)$ of the optimal throughput.

Designing an optimal network for data distribution, e.g., video streaming, from a *single source* was investigated in Kumar and Ross [13]. The paper presents a transmission schedule in which a single-source node sends packets to a set of target nodes, optimizing both throughput and latency. In their model, the bandwidth bottlenecks are assumed to be in the uploading and downloading link rates instead of the Internet core, and every node participates in the system until all nodes finish downloading packets. The paper shows a tight bound on the throughput as a function of the upload bandwidth of the source node, and the upload and download bandwidth of the target nodes. The transmission schedule in [13] can achieve both optimal throughput and latency. However, the constructed network is a complete network with $O(n^2)$ links. Our blockchain network design, studied in this paper, is different in two important aspects: (a) the all-to-all broadcast and (b) the designed network is required to be sparse for practicality.

Many blockchain systems construct their P2P network using Distributed Hash Table (DHT). For example, Ethereum[1] uses Kademlia [16] to construct the overlay network. In a DHT, an overlay network is constructed by assigning nodes with different identifiers. For any message that is assigned with a key, nodes can efficiently choose some other nodes to forward the message based on the identifiers that are stored in the DHT. Most DHTs such as Pastry [17], Chord [18], and Kademlia [16] guarantee that any message can be delivered to any nodes in $O(\log n)$ steps.

2 Overview

In a *blockchain* system, participants, called *miners* or *validators*, follow a consensus protocol to maintain the ledger (records of transactions). The protocol dictates the condition in which a miner can create a new block, a collection of transactions, to append to the blockchain. The block is broadcasted through the P2P network to all nodes, who will verify the validity of the block before appending it to their local copy of the blockchain. The blockchain allows the participants to agree on an ordered list of transactions without the use of a central authority as a trusted intermediary. Once transactions are added to the ledger, they cannot be removed or altered. The guarantee holds despite Byzantine behavior of a fraction of the participants maintaining the blockchain. The first application of blockchain arises in the context of cryptocurrencies, by permitting mutually distrusting participants to engage in financial operations securely.

---

[1] https://github.com/ethereum/wiki.

The nodes exchange information via the P2P network to synchronize the ledger. Having multiple copies of the ledgers at nodes create necessary redundancy to remove single point of failure and centralization in the system. Now, we summarize the important aspects of the blockchain networks.

In a permissionless blockchain system, such as Bitcoin and Ethereum, any node can join the P2P network. The nodes run a *peer discovery* protocol to learn the addresses of other nodes in the network. In addition, the nodes follow some rules to establish connections to other nodes (see more details in Sect. 3). When a miner get the right to generate a new block (or nodes create new transactions), the data is forwarded through a P2P network using a propagation scheme. Upon receiving a new block/transaction, a node verifies the correctness and adds to its memory (the mempool), after checking its validity.

The security of a blockchain system critically depends on its P2P network. As we mentioned in the previous paragraphs, data are forwarded through the P2P network, using a propagation scheme. The blockchain P2P network should be resistant to many types of attacks such as eclipse attacks, network partitioning attack, and denial-of-service (DoS) attacks (see more detail in Sect. 4). If the P2P network is not secure, the attacker can leverage vulnerabilities at the network level to perform consensus-level attacks, such as double-spending and selfish mining attacks. Additionally, the P2P network should provide anonymity and privacy. Any curious adversary should not be able to trace down the originality of data.

The design of P2P network also has a significant influence on the system performance. Especially, as nodes may have heterogeneous bandwidth capacities, a random topology will not maximize the potential throughput in the network. We will discuss more details on the throughput and latency in Sect. 5.

In Sect. 6, we formalize an optimization problem to improve performance (throughput and latency) in a blockchain network and present a solution to achieve optimal performance. We show the physical limit of the throughput based on the bandwidth of nodes in the network and summarize a throughput-optimal scheme in [13] that is designed for a single-source node. To capture the limit on the throughput of a blockchain, we propose a new throughput-optimal scheme that is designed for multiple source nodes.

3 Network Topology

To form the P2P network, nodes first find other nodes' addresses by using a peer discovery process. The process typically relies on fixed information sources and/or Distributed Hash Table (DHT) approaches. After discovering addresses of other peers, nodes select some peers to establish *outbound connections*. In this section, we present the process to construct the P2P network in Bitcoin and Ethereum.

3.1 Bitcoin P2P Network

In the Bitcoin network, all nodes are "equally privileged" unlike the classic server-client network model. However, they may take different roles based on their functionalities. A Bitcoin node may support one or more of the four following main functions [19]: routing, the blockchain database, mining, and wallet services. Depending on the functions that the node support, it may download different types of data. For example, a node that supports the blockchain database function should download all blocks on the blockchain with the transactions included. While a node that only supports wallet services (Simplified Payment Verification or SPV node) only download block headers but not the actual data in the block.

Peer discovery. When a new node boots up, it must discover other Bitcoin nodes on the network in order to establish connections. When a node joins the network for the first time, it discovers other nodes by making DNS queries to DNS seed servers. A *DNS seed server* responds to DNS queries from bitcoin nodes with a list of IP addresses of known peers. The size of the list is limited by constraints on DNS. For example, the maximum number of IP addresses that can be returned by a single DNS query is around 4000 [20]. The DNS seed servers are hardcoded into the Bitcoin core client software. At the time of writing (April 2021), there are currently nine seed addresses listed in the bitcoin software. The DNS seed servers are maintained by Bitcoin community members. For example, some of them provide dynamic DNS seed servers which automatically get IP addresses of active nodes by scanning the network. In this way, the hardcoded DNS seed servers act as the trusted, authoritative source for initial nodes. After that, as the node interacts on the network, it builds up a local list of active nodes.

Once a node establishes a connection to another node, they will exchange the information on the known nodes by sending a GETADDR message [21]. The queried node will reply with an *ADDR message*, containing up to 1000 IP addresses and their timestamps. If more than 1000 addresses are sent in an ADDR message, the peer who sent the message is blacklisted. Nodes accept both solicited and unsolicited ADDR messages. An ADDR message is solicited only upon establishing an outgoing connection with a peer; the node responds with up to three ADDR messages, each containing up to 1000 addresses randomly selected from its local list of addresses.

Furthermore, every day, a node sends its own IP address in an ADDR message to each peer. Also, when a node receives an ADDR message with no more than 10 addresses, it forwards the ADDR message to two randomly selected connected peers. To choose these peers, the node takes the hash of each connected peer's IP address and a secret nonce associated with the day, selects the peers with the lexicographically first and second hash values. Finally, to prevent stale ADDR messages from endlessly propagating, each node keeps a known list of the addresses it has sent to or learned from each of its connected peers, and never sends addresses on the known list to its peer. The known lists are flushed daily.

Each node stores addresses it has already seen in two tables: "tried" and "new". The "tried" table consists of 64 buckets, each of which can store up to 64 unique addresses. The table stores peer addresses that the node has already connected to. The "new" table has 256 buckets, each of which can store up to 64 unique addresses. It stores the nodes' addresses that the node has received from other peers.

Since the tables of each node can only store a bounded number of addresses, an attacker can fill up the tables with its address. In this case, for any connection rule, the node has no choice but to connect to the nodes that are controlled by the attacker. The attacker now can manipulate which data the node can receive. More discussion on this attack will be presented in Sect. 4.1.

Connection rules. Each node selects 8 random addresses in its "tried" and "new" table to establish *outbound connections*. In detail, for the i-th ($i \in [8]$) outbound connection, a node selects a random address in the "tried" table with a probability

$$\frac{\sqrt{\rho}(10 - i)}{i + \sqrt{\rho}(10 - i)}.$$

Furthermore, the node selects a random address from the table, with a bias toward addresses with fresher timestamps, i.e., those that join the table earlier have less chance to be selected. Each node may also accept up to 125 *inbound connections* by default.

There are two types of nodes in the Bitcoin network: *private nodes* that do not accept inbound connections and *public nodes* that do accept inbound connections. However, once they have joined the network, public and private nodes are indistinguishable in their operation: both node types perform transaction and block validation and relay valid transactions and blocks to their peers.

Characteristics. Many works have discovered and explored the characteristics of the Bitcoin network. In [22], by running the peer discovery protocol for 45 days during the time horizon from 2018/12/10 until 2019/01/23 and extracts over 162, 000 nodes and 136, 023 unique IP addresses from the Bitcoin main network (some nodes may use the same IP address). 87, 652 IP addresses are reachable (public nodes). The authors also pointed out that the Bitcoin network shows more community structures compared with what should be expected from a random graph network. The top three largest communities consist of almost 40% of nodes.

In [23], using IP geolocation databases, the authors geolocate the address of nodes in the Bitcoin network by country. Most of the nodes are located in the US (23.7%) and EU-Germany (19%), France (6.8%), Netherlands (4.9%), whereas a smaller share is located in China (6.7%).

Although most Bitcoin nodes are located in EU and US, the mining power is actually concentrated in China. Almost 50% of all BTC blocks are mined by 4 major mining pools in China. Plus, in terms of incoming distribution, 4.5% of all nodes holding more than 85% of all mined coins so far.

3.2 Ethereum's P2P Network

Nodes in the Ethereum use a Distributed Hash Table (DHT) approach (Kademlia [16]) to make connections. In the DHT approach, an overlay network is constructed by assigning nodes with different identifiers. For any message that is assigned with a key, nodes can efficiently choose some other nodes to forward the message based on the identifiers that are stored in the DHT. By using the DHT approach, we can guarantee that any message can be delivered to any nodes in $O(\log n)$ steps.

Peer discovery. The peer discovery in the Ethereum network is quite similar to the one in the Bitcoin network. When a new node joins the Ethereum network for the first time, it connects to some bootstrap nodes that maintain a list of all nodes that are connected to them in a period of time. (At the time of writing, April 2021, there are currently eight bootstrap nodes in the main Ethereum network.) Then, the bootstrap nodes share the lists of peers with the new node. Finally, the new node synchronizes with the peers to obtain the list of all nodes.

Connection rule. Nodes in the Ethereum network use Kademlia [16] to establish connections to the discovered nodes. Each node is assigned with a NodeID in the 160-bit identifier space, and {key,value} pairs are stored on nodes with IDs close to the key. Here, keys are also 160-bit identifiers. A NodeID-based routing algorithm will be used to locate nodes near a destination key.

To locate {key,value} pairs, node relies on the notion of distance between two identifiers. Given two 160-bit identifiers, a and b, it defines the distance between them as their XOR value, i.e., $d(id_u, id_v) = id_u \oplus id_v = d(id_v, id_u)$, for all pair of identifiers id_u, id_v. XOR also offers the triangle inequality property $d(id_u, id_v) + d(id_v, id_x) \geq d(id_u, id_x)$. Furthermore, XOR is unidirectional, i.e., for any given identifier id_u and distance $d > 0$, there is exactly one identifier id_v such that $d(id_u, id_v) = d$ (the identifier may not be used by any node in the network, but it does exist). The unidirectional approach makes sure that all lookups for the same key converge along the same path, regardless of the originating node.

The node in the network stores a list of {IP address, NodeID} tuples for nodes of distance between 2^i and 2^{i+1} from itself. These lists are called k-buckets. Figure 2 shows an example of k-buckets of a NodeID. Each k-bucket is kept sorted by last time seen, i.e., least recently accessed node at the head, most recently accessed at the tail. Nodes in the Kademlia routing protocol can send four types of messages:

- PING check if a peer is active.
- STORE store a {key,value} pair in a node's table.
- FIND_NODE takes a 160-bit ID, and returns {IP address, NodeID} tuples for the k nodes it knows that are closest to the target ID.
- FIND_VALUE is similar to FIND_NODE: it returns {IP address, NodeID} tuples, except in the case when a node receives a STORE for the key, in which case it just returns the stored value.

Each node locates the k closest peers to some given NodeID. This lookup initiator starts by picking some nodes from its closest non-empty k-bucket, and then sends

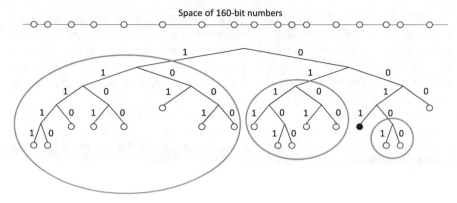

Fig. 2 Kademlia binary tree. The black dot shows the location of the node with NodeID = 0011 ⋯ in the tree. Each gray oval shows a bucket of NodeID as a subtree (reproduced from [16])

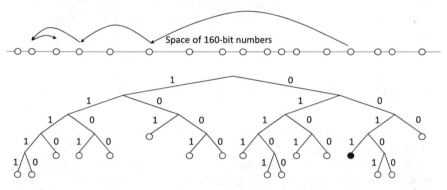

Fig. 3 Node performs a FIND_VALUE lookup to find the k peers. Here, nodes with prefix = 0011 ⋯ finds the nodes with prefix 1110 by successively querying closer and closer nodes. The line segment on top shows how the lookups converge to the target node (reproduced from [16])

FIND_NODE messages to the nodes it has chosen. If FIND_NODE fails to return a node that is any closer than the closest nodes already seen, the initiator resends the FIND_NODE to all of the k closest nodes it has not already queried. It can route for lower latency because it has the flexibility to choose any one of k nodes to forward a request. To find a {key,value} pair, a peer starts by performing a FIND_VALUE lookup to find the k peers with IDs closest to the key. Figure 3 shows an example of a node performing a FIND_VALUE lookup to find the k nodes with NodeIDs closest to the key.

To join the network, a node u inserts the nodes that it found with peer discovery protocol into the appropriate k-bucket. Then, node u lookup for its own NodeID to select which addresses to establish outbound connections.

Note that, since nodes in the Ethereum network establish outbound connections based on the NodeIDs, the attackers can select some carefully selected NodeIDs to ensure that a victim node will establish outbound connections to those addresses with the NodeIDs. Thus, the attacker can isolate a victim node without filling up the address tables. This makes the Ethereum network more vulnerable to this kind of attack. We will discuss more on this issue in Sect. 4.1.

Characteristics. In studies in [24], discover 769, 000 Ethereum active nodes. Most of the nodes only have one IP address. There are 1, 268 nodes changing their communication address once within the day. The node that changes its IP addresses most frequently possesses 514 addresses. Similar to another real-world network, the degree distribution in the Ethereum network follows a power-law distribution. The mean of the indegree (or outdegree) is 118.75, the maximum of the indegree is 986, while the number of outdegree is 586.

Similar to Bitcoin, in Ethereum network [25], a few large communities contain a large number of nodes. 43.2% of the nodes operate in the US and 12.9% in China. Furthermore, most of the nodes in Ethereum network runs on several major cloud service. More than 50% of nodes run on the top eight cloud service providers.

3.3 Data Forwarding

Nodes in the P2P networks forward data to its direct neighbors who will forward the data to all of the neighbors of neighbors, and so on. To prevent malicious nodes from spamming the network with bogus information, nodes use a store-and-forward propagation to ensure that only forward valid data will be forwarded.

Transaction/block propagation. To avoid sending packets to a node that already received them, nodes use a *flooding mechanism* to forward the packets instead of forwarding them directly. As shown in Fig. 4, when a node *u* receives and verifies the validity of a new packet, it advertises the packet to *all* neighbors, except those

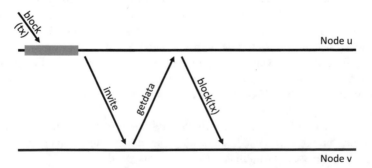

Fig. 4 Data forwarding in Bitcoin's P2P network. Each node *u* verifies the block/transaction before sending the invite message to its neighbor *v*

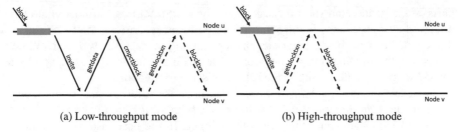

(a) Low-throughput mode   (b) High-throughput mode

Fig. 5 Data forwarding protocol in compact block relay, BIP152

who already advertised the packets, by sending an invite message[2], containing a set of packets' hash values. A node v, receiving an invite message for packets that it does not have, will send a getdata message containing the hashes of the needed packets. After that, the actual transfer of the packet is done via individual block or tx messages.

Compact block relay proposal. Forwarding full blocks (that consists of the transactions) is not efficient. The transactions in the block are already forwarded to most of the nodes in the network. In other words, each transaction is forwarded twice, once as an individual transaction and once in a block. Furthermore, this also causes bandwidth spikes when new blocks are generated. When such spikes occur, the network is swarmed and the propagation process may be delayed.

Matt Corallo proposes an improvement, called *compact block relay* (BIP152 [26]), with the goal of decreasing the bandwidth used during block forwarding. Since nodes in the network may download some transactions in the full blocks, nodes forward lightweight compact blocks, instead of forwarding the full blocks, to receiving nodes. These compact blocks include the following information: the 80-byte header of the new block, shortened transaction identifiers (hash values of transactions), and some full transactions which the nodes predict the receiving nodes have not received yet.

The receiving node then tries to reconstruct the entire block using the received information and the transactions already in its memory pool. If the receiving node is still missing some transactions, it will request those from the node that forwards the compact block.

With compact block relay improvement, in the best case, transactions only need to be forwarded once, when they are originally broadcasted from the users. This provides a large reduction in overall bandwidth consumption. A node can reconstruct a full block of size 1 MB by receiving a compact block of size 9 KB.

There are two different modes in the compact block relay proposal, *low-throughput* mode and *high-throughput* mode (see Fig. 5). In *low-throughput* mode, after verifying the validity of the block, node u sends new block announcements with the usual invite messages. Then the receiving node v requests the block using a getdata message, which will receive a response of the compact block that con-

---

[2] In Ethereum, the invite messages are equivalent to the NewBlockHashes or NewPooledTransactionHashes messages.

sists of the header and short transaction IDs via a cmpctblock. If node u already received all transactions in the compact block, it can reconstruct the block immediately. Otherwise, if some transactions are missing, node v sends a getblocktxn to request the missing transactions from node u. Finally, node u will respond by a blocktxn message that contains the lists of missing transactions.

A node can also enable *high-throughput* mode for a few peers. In this mode, node u can send the compact block to node v with verification or sending invite message. This significantly improves the latency. In the best case, where node v already receives all the transactions in the compact block, it only takes 0.5 round time trip (RTT) to forward data from node u to node v. In the worst, it will take 1.5 RTT which is the same to the legacy data forwarding as shown in Fig. 4. Note that, node v may receive the duplicate compact block from multiple nodes. However, since the size of compact blocks is small, this duplication can be simply ignored.

Note that, although the goal of the compact block relay proposal is not to reduce the latency, it does improve the latency since the size of compact blocks is significantly smaller than the size of the full blocks (see Sect. 5.2 for more details).

4 Attacks on Blockchain P2P Networks

In this section, we discuss the security issues in blockchain P2P networks. The security of blockchain critically depends on the security of the P2P network. The attacker can break the security of a blockchain by attacking its P2P network.

4.1 Eclipse Attacks

In an eclipse attack, the goal of the attacker is to obscure some target node by attempting to have all the target's connections to the attacker-controlled nodes. Upon success, the attacker can eclipse the information from the target's view and conduct further attacks such as double-spending, selfish mining attacks, and 51% attacks.

Flooding address tables attacks. In the Bitcoin network, nodes store the addresses of other nodes in the tried and new tables. In an eclipse attack [27], the attacker fills up the victim's tables with the attacker's addresses. To be precise, the attacker keeps advertising its addresses to the victim. Note that, when an address table of the victim is full, to add a new address to the table, the victim needs to remove an old address. Eventually, all the addresses from honest nodes will be removed from the tables of the victim. Then, when the victim reboot, it can only make connections to the attacker's addresses. The attacker can either actively perform Distributed Denial-of-Service (DDoS) attacks or simply wait out until the victim restart.

Many counter-measures were proposed in [27] with several of them already implemented in Bitcoin software. One obvious counter-measure is to increase the number of connections to improve the chance honest nodes will be connected to each other.

Other counter-measures are to increase the size of the tables or reduce the number of addresses in ADDR messages. This reduces the chance that the addresses of honest nodes get kicked out of the tables. The remaining counter-measures aim to address the vulnerability in adding and removing nodes from the address tables and preventing the adversary from flooding the target's address table with trash addresses. For example, before removing an address from the tried table, the node briefly attempts to connect to the address. If the connection is successful, then the node will not remove the address and add a new address to the table. The proposed counter-measures substantially reduce the chance that the adversary can successfully perform an eclipse attack.

Stubborn mining attacks. In [28], the authors investigate new stubborn mining attacks which combine eclipse attacks with selfish mining [29] attacks. In this work, the authors consider the same model against users who are also eclipsed in the network and show the effect to which eclipsed users help a stubborn mining attacker. Overall, eclipse attacks empower adversarial agents with a larger strategy space to continue running attacks and, when paired with stubborn mining strategies, enable an attacker to improve their relation fraction of block rewards beyond traditional selfish mining strategies.

Eclipse attacks on Ethereum network. The works in [30, 31] show that the Ethereum network is more vulnerable to eclipse attacks. Since nodes in the Ethereum network use Kademlia to select nodes to connect, the attackers can simulate the connection rule of the victims. Thus, it is easier for attackers to perform eclipse attacks.

Newly connected nodes attacks. Due to the block propagation design of Ethereum, a node that newly connects to the network may receive a chain that is longer than the main chain but has a lower total difficulty. To perform this attack [30], the attacker creates a long blockchain starting from the genesis block by decreasing the difficulty for each block. Then, the attacker connects to the victim and advertises a high total difficulty (higher than the difficulty on the main chain). Finally, the victim will request to download the chain from the attacker. The authors point out an implementation bug in Ethereum's difficulty calculation. The attacker can use this bug to present the victim to download the main chain.

Zero-outbound-connections issue. In the Ethereum network, a node may establish outbound connections if it accepts too many inbound connections. Based on this connection rule, when a victim reboots, the attacker can immediately initiate inbound connections to the victim from its addresses. In this case, the victim is eclipsed since all connections of the victim are from the addresses of the attackers. A simple counter-measure to this attack is enforcing an upper limit on the number of inbound connections.

Kademlia-based attacks in the Ethereum network. Even if an upper limit on the number of inbound connections is enforced, the attacker can use a carefully crafted set of node's IDs to repeatedly ping the victim (since a node establishes outbound connections to the nodes with IDs that are closest to the ID of that node). When the victim restarts, the victim establishes all outgoing connections to the attacker's address with high probability. To complete the eclipse, the attacker monopolizes the remaining connection slots by initiating inbound connections to the victim. Several counter-measures was proposed to increase the difficulty to perform eclipse attacks. A simple counter-measure is to make the node ID deterministic to the address. Another counter-measure is to make the lookup process non-public. This prevents the attacker from predicting set of nodes' IDs that the victim will connect to when it reboots.

4.2   Network Partitioning Attacks

A splitting of the P2P network can lead to devastating consequences. Partitions affect the ability of participants to exchange data. Thus, nodes in different parts may maintain different chains. When all nodes are connected, based on the consensus protocol, only one chain will survive, and as a result the mining power on the other chains is wasted. Plus, many transactions are reverted (and potentially double-spent).

The attacker can partition the blockchain network by using *routing attacks*. The attackers control the Border Gateway Protocol (BGP) advertisements to manipulate the connections of nodes. Recent studies [32] show that the attackers can isolate 50% of the Bitcoin mining power by hijacking less than 100 IP prefixes.

SABRE [33] presents a transparent relay network protecting Bitcoin clients from routing attacks by providing them with an extra secure channel for learning and propagating the latest mined block. SABRE is easy to deploy and can run alongside the existing P2P network. The IP addresses of the SABRE relay nodes will be publicly known (e.g., via a website), and that everyone can connect to the relay nodes. SABRE is designed to efficiently handle extremely high load and resistance to denial-of-service attacks. The authors use properties of BGP to predict where would be a good place to host relay nodes—locations that are inherently protected against routing attacks and on paths that are economically preferred by majority of Bitcoin clients. In addition, they provide resiliency through the use of caching, and partially hardware implementation in programmable network devices. This enables SABRE relay nodes to sustain large (D)DoS attackers.

4.3   DDoS Attacks

In a blockchain system, nodes can generate as many transactions as they wish if they are able to pay the transaction fees. An attacker can perform a DDoS attack by generating a large number of transactions and forwarding them to some nodes in the network. Then, nodes will unconditionally forward those transactions to the entire network. The main cost for the attacker is the transaction fees. However, it is possible that the attacker does not need to pay those fees. In fact, if transactions are propagated to the entire network but are not included on the blockchain, the fees are not collected, i.e., the attacker can perform a DDoS attack without any cost. Thus, in Bitcoin, miners only forward transactions that pay a sufficient fee and are likely to be included in a block.

4.4 Man-in-the-Middle Attacks

In [34], the authors study the impact of man-in-the-middle attacks on Ethereum. Based on the properties of the Ethereum public blockchain topology, the authors build a simulated network to mimic the top 10 biggest mining pools of Ethereum. Then, the authors perform BGP hijacking and ARP spoofing to partition the network before issuing a double-spending attack. The results demonstrate the attack will be almost impossible in public blockchains (nodes rely on the Internet for communication). However, consortium blockchains (nodes rely on the multiple organization networks that are connected by the Internet) and private blockchains (nodes rely on a single organization network) are suffering from this attack. The attacker can successfully perform a double-spending attack with a probability of 80% with 12-minute attack duration.

4.5 Deanonymization Attacks

An attacker may wish to identify which node originally generates a transaction. By actively connecting to several nodes, it is possible that the attack will trace back to the source node that generates the transaction. Then, the attack may make an attempt to censor the transactions of a victim node.

In Biryukov et al. [35], a deanonymization method for the Bitcoin network is presented, which allows linking IP addresses of nodes to their pseudonyms. This way, we are able to find out where the transactions are generated. The method explicitly targets nodes behind NAT (network address translation) or firewall. The method also works on nodes that use anonymity services like Tor. Plus, it can distinguish between nodes with the same IP address. The key idea of this deanonymization is that each node can be uniquely identified by a set of nodes he connects to (entry nodes). Each transaction is mapped to a set of entry nodes, which is associated with a node with a similar set. To avoid this deanonymization technique, the authors suggest frequently changing the set of entry nodes.

In [36], Neudecker and Hartenstein evaluate the deanonymization in Bitcoin network in the form of address clustering. The goal of address clustering is to group the addresses into clusters so that the addresses in each cluster are controlled by a single user. The authors compare the blockchain information (e.g., the public keys of users in the transactions)-based clustering approaches and the network-information-based clustering approaches. Majority of nodes have no correlation between network information and the clustering performed on blockchain information. However, a small number of nodes (8%) exhibit correlations that might make them susceptible to network-based deanonymization attacks.

The Dandelion protocol [37] is a transaction relay protocol to improve the anonymity in the presence of honest-but-curious attackers. The Dandelion protocol consists of two phases. In the first phase, each transaction is propagated on a random path, i.e., each node forwards the transaction to exactly one random neighbor for a random number of hops. Then, in the second phase, the transaction will be broadcasted using the same flooding mechanism as in Bitcoin. Dandelion++ [38] extends Dandelion to defend against active attackers that can divert from the protocol. Instead of forwarding transactions through a random path (in the first phase), nodes forward transactions over one of two intertwined paths on a 4-regular graph. The second phase remains the same as in Dandelion.

5   Performance

Another important aspect of a blockchain system is performance. In this work, we focus on the throughput, i.e., the number of transactions per second the system can process, and the latency, i.e., the time it takes for a block to propagate to all nodes in the network.

5.1   Throughput

Scalability remains a thorny issue that prevents the wide adoption of blockchain. Most of existing legacy blockchains suffer from very low throughput. For example, in Bitcoin, for every 10 min (600 s), a new block, which consists of 2,467 transactions on average,[3] is generated. In other words, on average, the Bitcoin system can process 4.1 transactions per second. Ethereum can achieve a better throughput (15 transactions per second). However, comparing with Visa, which does around 1,700 transactions per second on average, the throughput of the blockchain systems is still too low. Thus, it is essential to make blockchain systems more scalable.

One simple solution that may come to mind is to increase the block size or decrease the block generation time. However, this solution comes with consequences. Without sacrificing the security, to increase the block size or decrease the block generation time, nodes in the network require a higher bandwidth capacity which some nodes cannot afford. Presumably, to increase the throughput by this solution, fewer and fewer nodes can participate in the system, leading to the increase of centralization. Thus, to improve the throughput without sacrificing security, we should have a more efficient way to propagate data.

---

[3] According to Bitcoin historical data [39], at the time of this writing (April 2021), the average block size is 1.08 MB and the average transaction size is 459 B.

Reducing redundancy. Although the flooding mechanism is needed for security purposes, it has a bad effect on the performance. Using the flooding mechanism, nodes are required to send many redundant transaction announcements. Indeed, by using the flooding mechanism, each node sends an announcement on each of the links except the one where that announcement originally arrived. In other words, each link sees each announcement once, if no two nodes ever send the same announcement to each other simultaneously, and more than once if they do. Therefore, in Bitcoin, each announcement is sent at least as many times as the number of links. On the other hand, optimally, each node would receive each announcement exactly once, the number of times each announcement is sent should be equal to the number of nodes. As the size of an invite message is 32 B (while the average size of a transaction is 459 B [39]), this redundancy is quite big. According to Naumenko et al. [40], in Bitcoin, nodes use 48% of bandwidth to send invite, getdata messages.

Erlay [40] minimizes the redundant transaction announcements. Instead of announcing every transaction on each link (flooding), a node advertises it to a subset of peers (low-fanout flooding). This low-fanout flooding helps to reduce the number of invite messages for advertising transactions. Note that, to defend against timing attacks (where the attacker can use the timing of transactions arrival to guess whether a transaction originated at its peers), node relay transaction via outbound connections.

Furthermore, to make sure that all transactions reach the entire network, nodes periodically engage in an interactive protocol to discover announcements that were missed and request missing transactions. This can be done using set reconciliation. Each node performs set reconciliation by computing a local set sketch with a predetermined capacity. When the number of elements in the set does not exceed the capacity, it is always possible to recover the entire set from the sketch. Plus, a sketch of the symmetric difference between the two sets can be obtained by XORing the bit representation of sketches of those sets.

Optimizing throughput of the P2P network. Many works have been studied to optimize the throughput for data propagation in the P2P network. We can borrow those techniques to optimize the throughput of the blockchain P2P network. There are two classes of approaches to propagate data via a P2P network, tree-based or mesh-based. Tree-based approaches explicitly construct multiple spanning trees, connecting the source node to all receivers. In mesh-based approaches, nodes exchange packets with several of their neighbors without explicitly constructing the spanning trees.

We now present a tree-based approach in [15] that can achieve near-optimal throughput. Note that, in this construction, we only consider the single-source problem, where a single-source node broadcasts data to other nodes in the network.

The work in [15] solves single-source broadcast problem by constructing multiple broadcast trees over clusters instead of individual nodes (see Fig. 6). Recall that, with the SplitStream construction, we can achieve near-optimal throughput if the capacity of all nodes are roughly the same. Taking this advantage, in [15], nodes are divided into clusters of size $O(\log n)$ such that the total capacity of all nodes in each cluster is roughly the same. Nodes in each cluster are fully connected to each other in order to

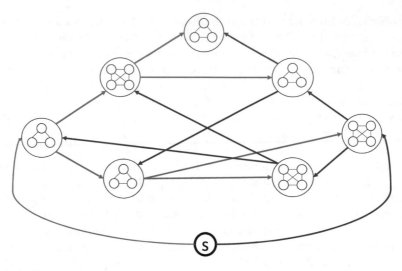

Fig. 6 In this example, the tracker groups the peers into seven clusters. The server s forms two interior-node-disjoint trees, each having tree degree 2, to distribute video to all clusters. Within each cluster, the peers form a full mesh and locally broadcast the video among themselves. There are not enough clusters left to build a third tree with disjoint interior clusters; hence, some clusters, e.g., the cluster on the top, do not get the chance to serve as interior clusters

obtain optimal throughput in each cluster (for example, we can use the construction in [13]). Nodes in each cluster are fully connected to each other in order to obtain optimal throughput in each cluster. Clusters are then abstracted as supernodes and d broadcast trees are then built (using SplitStream [14]) over those supernodes. In detail, we construct d broadcast trees that aim to balance the transmission loads of clusters. Here, a cluster only forwards the data equal to the data that the cluster received. In the context of broadcasting, all clusters in the network forward the same data. To be precise, we form d broadcast trees with the same capacity in which, in each tree, each interior cluster has exact d links to forward data to other clusters. Furthermore, if a cluster is an interior cluster in one tree, there will be a leaf in all other trees.

Since the capacity of the clusters is almost the same, we can achieve within a factor $(1 - \epsilon)$ of the optimal throughput. We can also achieve the latency of $O(\log n)$. Let k be the depth of a broadcast tree. Since each interior cluster has exact d links, the depth of each broadcast tree is $O(\log n)$. Plus, all nodes in each cluster connect to each other. Thus, the length of the propagation path in each cluster is $O(1)$. Thus, we can achieve the latency of $O(\log n)$.

5.2 Latency

The latency of the P2P network affects the security of the systems. The security analysis of Bitcoin [41] is based on the assumption that all packets can be delivered to all nodes with a bounded delay. By reducing the latency (i.e., the bounded delay), we can reduce the confirmation time, i.e., the time it takes for a transaction on the blockchain becomes irreversible.

Even though the goal of the compact block relay proposal (as we presented in the paragraph in Sect. 3.3) is not to reduce the latency, it does improve the latency since the size of compact blocks is significantly smaller than the size of the full blocks. In [42], the authors run simulations of Bitcoin with and without the compact block relay proposal, with 2019 Internet parameters on a network of 9, 000 nodes and the block size is 1 MB. Without compact block relay proposal, the 50, 90% latency (the time it takes for a block to reach 50, 90% of nodes in the network) is 6.4 and 9.4 s, respectively. With the compact block relay proposal, the latency reduces to 1.3 and 2.4 s, respectively. The authors also measure the number of orphan blocks. With the compact block relay proposal, the fraction of orphan blocks reduces from 0.95 to 0.19%.

Relay network is a network that attempts to minimize the latency in the transmission of blocks between miners. The *original Bitcoin relay network* [43] was created by core developer Matt Corallo in 2015 to enable fast synchronization of blocks between miners with very low latency. The network consisted of multiple gateways infrastructure around the world and served to connect the majority of nodes in the network.

The original Bitcoin Relay Network was replaced in 2016 with the introduction of the *Fast Internet Bitcoin Relay Engine* or FIBRE [44], also created by core developer Matt Corallo. FIBRE uses a similar architecture while using UDP-based transmission instead of TCP based. TCP based is designed to provide reliable transmission at reasonable bandwidth across medium-large amounts of data, it is incredibly bad at low-latency relay of small amounts of data. It is generally tuned to send packets (each just under 1500 bytes) once and to only discover that some packets were lost after getting a response from the other side. Only then will the sender retransmit the lost packets, allowing the receiver to (potentially) reconstruct the original transmission. Thus, in order to have minimal latency block transmission, we must avoid the need for retransmissions at all costs. In order to do so, we must transmit enough extra data that the receiving peer can reconstruct the entire block even though some packets were lost on the way. The common solution is UDP-based transmission with some relatively simple linear algebra to send data which can fill in gaps of lost packets efficiently.

Falcon [45] is a relay network that uses cut-through routing instead of store-and-forward propagation model to reduce latency. Here, nodes propagate parts of blocks as soon as they arrive rather than waiting for the entire block to arrive.

BloxRoute [11] is a high-capacity, low-latency blockchain distribution network that is optimized to quickly propagate transactions and blocks for blockchain systems.

Contrary to the relay networks, the BloxRoute propagates data without knowing the content of the data. This prevents BloxRoute from censoring the specific data by intentionally delaying the propagation of the data.

6 Performance Improvement as an Optimization Problem

In this section, we formalize the performance (throughput and latency) improvement problem in a blockchain network as an optimization problem and present a solution to achieve an optimal throughput.

6.1 Optimization Problem

Consider a blockchain system with n participants that are modeled as a set of nodes $V = \{1, 2, \ldots, n\}$. Each node $i \in V$ has an upload bandwidth $c(i)$, or c_i for short. A node i can transmit simultaneously to each neighbor j with rate g_{ij} as long as $\sum_j g_{ij} \leq c(i)$. We refer to this as the *capacity constraint*.

During the execution of the system, some source node i may produce new data (e.g., block) to other nodes. We denote the *arrival rate* λ_i as the average data that is produced by node i. For any node i, the arrival rate cannot exceed the upload throughput of, i.e., $\lambda_i \leq c_i, \forall i \in V$. For example, in Bitcoin, since the size of compact blocks is relatively small (the average size of compact blocks is 9 KB, while the average size of transactions is 459 KB [39]), the arrival rate can be approximated as the transaction generation rate.

Throughput. The throughput is the data nodes can broadcast to all nodes. We say a propagation scheme can achieve a throughput TP iff there exists an arrival rate $\lambda = (\lambda_1, \ldots, \lambda_n)$ such that $\sum_{i \in V} \lambda_i = \mathsf{TP}$, the propagation scheme, which satisfies the capacity constraint, is able to deliver such data to all nodes in the network.

Lemma 1 *Consider the set of node $V = (1, \ldots, n)$ with the upload capacity $C = (c_1, \ldots, c_n)$. Let OPT_{TP} be the optimal throughput the network can achieve. The upper bound on the optimal throughput can be simplified into*

$$\mathsf{OPT}_{TP} \leq \frac{\sum_{i \in V} c_i}{n - 1}. \tag{1}$$

Proof Since each node needs to forward its data to $n - 1$ other nodes, to achieve a throughput of OPT_{TP}, the total data nodes need to forward is $(n - 1)\mathsf{OPT}_{TP}$. The total amount of data nodes that can be forwarded is $\sum_{i \in V} c_i$. Thus, we have,

$$(n - 1)\mathsf{OPT}_{TP} \leq \sum_{i \in V} c_i.$$

Jumping ahead, in Sect. 6.3, we show by construction that we can achieve the optimal throughput $\text{OPT}_{TP} = \frac{\sum_{i \in V} c_i}{n-1}$.

Latency. The latency is the time for any source nodes to transmit the data to all other nodes. In our model, for simplicity, we measure the latency based on the number of hops on the propagation paths. To be precise, the latency equals the length of the longest path that a packet travels from any source node to any target node.

The goal of our problem is to construct a propagation scheme that can support high throughput, while maintaining low latency.

6.2 Throughput-Optimal Propagation Scheme for Single-Source Problem

In [13], Kumar and Ross present a throughput-optimal propagation scheme for single-source problem. In the single-source problem, there is only one source node s which can provide data to other nodes, i.e., for any node $v \in V \backslash \{s\}$ the arrival rate $\lambda_v = 0$.

Note that, since the arrival rate is bounded by the capacity, i.e., $\lambda_s \leq c_s$, the optimal throughput in the single-source problem is also bound by the capacity of the source node s, i.e.,

$$\min \left\{ c_s, \frac{\sum_{i=1}^{n} c_i}{n-1} \right\}.$$

The source node s first forward a small amount of data to all other nodes. We denote p_v as the data node s forwards to node v. Then, each node v forwards $q_v \leq p_v$ data to all node $u \in V \backslash \{s, v\}$. Here, each node needs to receive λ_s data, i.e.,

$$p_v + \sum_{u \subset V \backslash \{v, s\}} q_v = \lambda_s, \forall v \in V \backslash \{s\}.$$

Plus, the capacity constraint is also needed to be satisfied, i.e.,

$$\sum_{v \in V \backslash \{s\}} q_v \leq c_s,$$

$$(n-2)q_v \leq c_v, \forall v \in V \backslash \{s\}.$$

We consider two cases in the bound of optimal throughput.

- *Case 1:* $c_s \leq \frac{\sum_{i \in V} c_i}{n-1}$. In this case, we can achieve a throughput of c_s, i.e., $\lambda_s = c_s$. For all $v \in V \backslash \{s\}$, we set

$$q_v = p_v = \lambda_s \frac{c_v}{\sum_{i \in V \backslash \{s\}} c_i}.$$

- *Case 2:* $c_s > \frac{\sum_{i \in V} c_i}{n-1}$. In this case, we can achieve a throughput of $\frac{\sum_{i \in V} c_i}{n-1}$, i.e., $\lambda_s = \frac{\sum_{i \in V} c_i}{n-1}$. For all $v \in V \backslash \{s\}$, we set

$$q_v = \frac{c_v}{n-2}$$

$$p_v = q_v + \frac{(n-1)c_s - \sum_{i \in V} c_i}{(n-1)(n-2)}.$$

We omit the analysis of this propagation scheme since it can be considered as a subcase of the propagation scheme in Sect. 6.3.

6.3 Throughput-Optimal Propagation Scheme for Blockchain Data Forwarding Problem

We now modify the propagation scheme in [13] to solve the blockchain data forwarding problem where the arrival rate of any node $s \in V$ can be bigger than zero. Here, the bound on the capacity of the source node is removed.

Similar to the single-source problem, each source node s forwards a small amount of data to all other nodes. We denote p_{sv} as the data node s forwards to node v. Then, each node v forwards the same data from node s to all node $u \in V \backslash \{s, v\}$. Each node v needs to receive an λ_s data from s, i.e.,

$$p_{sv} + \sum_{u \in V \backslash \{s,v\}} q_{su} = \lambda_s.$$

For the capacity constraint, when s is the source node, it sends p_{sv} to all other nodes v. Plus, when a node $v \in V \backslash \{s\}$ is the source node, s sends $(n-2)$ packets of size q_{vs} to all nodes $u \in V \backslash \{s, v\}$. Thus, the capacity constraint can be written as

$$\sum_{v \in V \backslash \{s\}} (p_{sv} + q_{vs} (n-2)) \le c_s.$$

In detail, for each source node $s \in V$ and a node $v \in V \backslash \{s\}$, we assign p_{sv} and q_{sv} as in Algorithm 1. We denote $c_v^{(s)}$ as the remaining capacity after nodes forwarded data for the first s source nodes. Note, at the beginning, each node s reserves λ_s of its capacity. This ensure that node s has enough capacity to forward its own data. At lines 1–2, we set $c_v^{(0)} = c_v - \lambda_v$. Then, we iterate through all node in V. For the source node s, we consider two cases (Fig. 7).

- *Case 1:* $\lambda_s \le \frac{\sum_{i \in V \backslash \{s\}} c_i^{(s-1)}}{n-2}$. In this case, for all $v \in V \backslash \{s\}$, we set

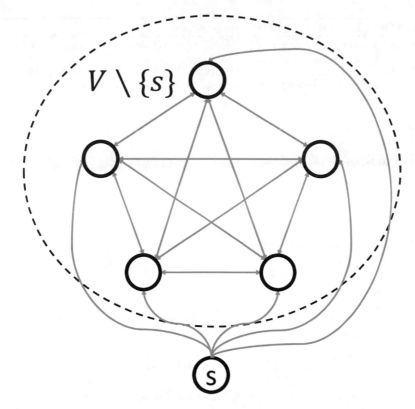

Fig. 7 Each source node first forwards a small amount of data to all other nodes the desire to receive the data. Then, all nodes exchange data of that source node with each other. Note that, data can be forwarded from any source node

$$p_{sv} = q_{sv} = \lambda_s \frac{c_v^{(s-1)}}{\sum_{i \in V \setminus \{s\}} c_i^{(s-1)}}.$$

- *Case 2*: $\lambda_s > \frac{\sum_{i \in V \setminus \{s\}} c_i^{(s-1)}}{n-2}$. In this case, for all $v \in V \setminus \{s\}$, we set

$$q_{sv} = \frac{c_v^{(s-1)}}{n-2}$$

$$p_{sv} = q_{sv} + (\lambda_s - \frac{\sum_{i \in V \setminus \{s\}} c_i^{(s-1)}}{n-2}).$$

Then, we deduct the capacity of each node by the amount of data it uses to forward the data from the source node s. To be precise, each node $v \in V \setminus \{s\}$ forwards an q_{sv} data to $n-2$ nodes. Thus, we set

---

Algorithm 1: Constructing propagation scheme

> **Input** : Given the set of node $V = (1, \ldots, n)$, the upload capacity $C = (c_1, \ldots, c_n)$, and
> the arrival rate $\lambda = (\lambda_1, \lambda_2, \ldots, \lambda_n)$.
> **Output**: For each source node $s \in V$, and node $v \in V \setminus \{s\}$, assign q_{sv} is the data s send to v
> and p_{sv} is the data v forward data from s to all node $u \in V \setminus \{s, v\}$.

1 **for** $v = 1$ *to* n **do**
2    $c_v^{(0)} = c_v - \lambda_v$
3 **for** $s = 1$ *to* n **do**
4   **for** $v = 1$ *to* n **do**
5   **if** $\lambda_s \leq \dfrac{\sum_{i \in V \setminus \{s\}} c_i^{(s-1)}}{n-2}$ **then**
6        $p_{sv} = q_{sv} = \lambda_s \dfrac{c_v^{(s-1)}}{\sum_{i \in V \setminus \{s\}} c_i^{(s-1)}}$
7   **else**
8        $q_{sv} = \dfrac{c_v^{(s-1)}}{n-2}$
9        $p_{sv} = q_{sv} + (\lambda_s - \dfrac{\sum_{i \in V \setminus \{s\}} c_i^{(s-1)}}{n-2})$
10   **for** $v \in V \setminus \{s\}$ **do**
11      $c_v^{(s)} = c_v^{(s-1)} - q_{sv}(n-2)$
12    $c_s^{(s)} = c_s^{(s-1)} + \lambda_s - \sum_{v \in V \setminus \{s\}} p_{sv}$
13 **Return** p, q

$$c_v^{(s)} = c_v^{(s-1)} - q_{sv}(n-2).$$

The source node s sends p_{sv} data to each node v. Thus, we set

$$c_s^{(s)} = c_s^{(s-1)} + \lambda_s - \sum_{v \in V \setminus \{s\}} p_{sv}.$$

Here, we plus λ_s as the reserved data for node s to forward data of its own.

Now, we prove that, by assigning p and q as in Algorithm 1, each node receives λ_s data from each source node s (see Lemma 2), and the data each node s forwarded does not exceed c_v (see Lemma 3).

Lemma 2 *Consider the set of node $V = (1, \ldots, n)$, the upload capacity $C = (c_1, \ldots, c_n)$, and the arrival rate $\lambda = (\lambda_1, \lambda_2, \ldots, \lambda_n)$ such that $\sum_{i \in V} \lambda_i \leq \frac{\sum_{i \in V} c_i}{n-1}$. For each source $s \in V$ and a node $v \in V \setminus \{s\}$, p_{sv} and q_{sv} are assigned as in Algorithm 1. Then, for each source node s, each node $v \in v \setminus s$ receives λ_s data from s, i.e.,*

$$p_{sv} + \sum_{u \in V \setminus \{s, v\}} q_{su} = \lambda_s, \forall s \in V, v \in V \setminus \{s\}.$$

Proof We consider two cases as in Algorithm 1.

- *Case 1*: $\lambda_s \leq \frac{\sum_{i \in V \setminus \{s\}} c_i^{(s-1)}}{n-2}$. In this case, we have

$$p_{sv} + \sum_{u \in V \setminus \{s,v\}} q_{su} = \sum_{u \in V \setminus \{s\}} \left(\lambda_s \frac{c_v^{(s-1)}}{\sum_{i \in V \setminus \{s\}} c_i^{(s-1)}} \right) = \lambda_s$$

- *Case 2:* $\lambda_s > \frac{\sum_{i \in V \setminus \{s\}} c_i^{(s-1)}}{n-2}$. In this case, for all $v \in V \setminus \{s\}$, we set

$$p_{sv} + \sum_{u \in V \setminus \{s,v\}} q_{su} = (\lambda_s - \frac{\sum_{i \in V \setminus \{s\}} c_i^{(s-1)}}{n-2}) + \sum_{u \in V \setminus \{s\}} \frac{c_v^{(s-1)}}{n-2} = \lambda_s.$$

Lemma 3 *Consider the set of node $V = (1, \ldots, n)$, the upload capacity $C = (c_1, \ldots, c_n)$, and the arrival rate $\lambda = (\lambda_1, \lambda_2, \ldots, \lambda_n)$ such that $\sum_{i \in V} \lambda_i \leq \frac{\sum_{i \in V} c_i}{n-1}$.. For each source $s \in V$ and a node $v \in V \setminus \{s\}$, p_{sv} and q_{sv} are assigned as in Algorithm 1. Then, the data each node $s \in V$ forwards does not exceed c_s, i.e.,*

$$\sum_{v \in V \setminus \{s\}} (p_{sv} + q_{vs}(n-2)) \leq c_s, \forall s \in V. \tag{2}$$

Proof Recall from lines 8–10 in Algorithm 1, for each source node s, we deduct the data that each node v uses to forward data from the node s. The capacity constraint in Eq. 2 is violated iff the remaining capacity of any node v after nodes forwarded data from all source nodes is smaller than zero, i.e., $c_v^{(n)} < 0$.

We will prove $c_v^{(n)} \geq 0$, $\forall v \in V$ by showing that

- For all $s \in [n]$, $\lambda_s \leq \frac{\sum_{i \in V} c_i^{(s-1)} + \lambda_s}{n-1}$. From lines 8–10 in Algorithm 1, we have

$$\sum_{i \in [n]} c^{(s)} = \sum_{i \in [n]} c^{(s-1)} - \lambda_s(n-2)$$

$$\Rightarrow \sum_{i \in [n]} c^{(s)} = \sum_{i \in [n]} (c_i - \lambda_i) - \sum_{i \in [s]} (\lambda_i(n-2))$$

$$(\text{Since } c_i^{(0)} = c_i - \lambda_i)).$$

Thus, for any $s \in [0..n-1]$, we have

$$\frac{\sum_{i \in V} c_i^{(s)} + \lambda_s}{n-1} = \frac{\sum_{i \in [n]} (c_i - \lambda_i) - \sum_{i \in [s]} (\lambda_i(n-2)) + \lambda_s}{n-1}$$

$$\geq \frac{(n-2) \sum_{i \in [n]} \lambda_i - (n-2) \sum_{i \in [s]} \lambda_i + \lambda_s}{n-1}$$

$$= \frac{(n-2) \sum_{i \in [s..n]} \lambda_i + \lambda_s}{n-1} \geq \lambda_s.$$

- For all $s \in [n]$, if $\lambda_s \leq \frac{\sum_{i \in V} c_i^{(s-1)} + \lambda_s}{n-1}$, then $c_v^{(s)} \geq 0, \forall v \in V$. We prove by induction as follows. Since $\lambda_v \leq c_v, \forall v \in V$, we have $c_v^{(0)} \geq 0, \forall v \in V$. Now, assuming $c_v^{(s-1)} \geq 0, \forall v \in V$. We consider two cases as in Algorithm 1.

  – *Case 1*: $\lambda_s \leq \frac{\sum_{i \in V \setminus \{s\}} c_i^{(s-1)}}{n-2}$. For the source node s, we have

$$c_s^{(s)} = c_s^{(s-1)} + \lambda_s - \sum_{v \in V \setminus \{s\}} p_{sv}$$

$$= c_s^{(s-1)} + \lambda_s - \sum_{v \in V \setminus \{s\}} \left(\lambda_s \frac{c_v^{(s-1)}}{\sum_{i \in V \setminus \{s\}} c_i^{(s-1)}} \right)$$

$$= c_s^{(s-1)} + \lambda_s - \lambda_s = c_s^{(s-1)} \geq 0.$$

  For each node $v \in V \setminus \{s\}$, we have

$$c_v^{(s)} = c_v^{(s-1)} - q_{sv}(n-2)$$

$$= c_v^{(s-1)} - \lambda_s \frac{(n-2)c_v^{(s-1)}}{\sum_{i \in V \setminus \{s\}} c_i^{(s-1)}}$$

$$\geq c_v^{(s-1)} - c_v^{(s-1)} = 0 \ (\text{since } \lambda_s \leq \frac{\sum_{i \in V \setminus \{s\}} c_i^{(s-1)}}{n-2}).$$

  – *Case 2*: $\lambda_s > \frac{\sum_{i \in V \setminus \{s\}} c_i^{(s-1)}}{n-2}$. For the source node s, we have

$$c_s^{(s)} = c_s^{(s-1)} + \lambda_s - \sum_{v \in V \setminus \{s\}} p_{sv}$$

$$= c_s^{(s-1)} + \lambda_s - \sum_{v \in V \setminus \{s\}} \left(\frac{c_v^{(s-1)}}{n-2} + \lambda_s - \frac{\sum_{i \in V \setminus \{s\}} c_i^{(s-1)}}{n-2} \right)$$

$$= c_s^{(s-1)} + \lambda_s - (n-1)\lambda_s + \frac{n-1}{n-2} \sum_{i \in V \setminus \{s\}} c_i^{(s-1)}$$

$$= \sum_{i \in V} c_i^{(s-1)} - (n-2)\lambda_s + \frac{1}{n-2} \sum_{i \in V \setminus \{s\}} c_i^{(s-1)}$$

$$\geq \frac{1}{n-2} \sum_{i \in V \setminus \{s\}} c_i^{(s-1)} \geq 0 \ \ (\text{Since } \lambda_s \leq \frac{\sum_{i \in V} c_i^{(s-1)} + \lambda_s}{n-1}).$$

For each node $v \in V \setminus \{s\}$, we have

$$
\begin{aligned}
c_v^{(s)} &= c_v^{(s-1)} - q_{sv}(n-2) \\
&= c_v^{(s-1)} - \frac{c_v^{(s-1)}}{n-2}(n-2) \\
&= c_v^{(s-1)} - c_v^{(s-1)} = 0.
\end{aligned}
$$

7 Conclusion

Despite the critical role of P2P network in both security and performance, it remains a relatively less study topic in blockchain. This calls for a new principle approach to design a secure P2P network that can attain (asymptotically) optimal throughput. The problem is even more challenging in the presence of adversarial nodes who not only contribute no bandwidth resource but also carry vandalizing behaviors to put a damper on the system performance. Thus, new network topology, distribution schemes, and incentives in P2P networks are needed which are important open questions in the near future.

Acknowledgements This work was supported in part by NSF under grant CNS 2140411.

References

1. Swan, M.: Blockchain: blueprint for a new economy. O'Reilly Media, Inc. (2015)
2. Nakamoto, S.: Bitcoin: a peer-to-peer electronic cash system (2008)
3. Ethereum, "ethereum/sharding." https://github.com/ethereum/sharding/blob/develop/docs/doc.md
4. Chen, Y., Bellavitis, C.: Blockchain disruption and decentralized finance: the rise of decentralized business models. J. Bus. Ventur. Insights **13**, e00151 (2020)
5. Dinh, T.N., Thai, M.T.: Ai and blockchain: a disruptive integration. Computer **51**(9), 48–53 (2018)
6. Christidis, K., Devetsikiotis, M.: Blockchains and smart contracts for the internet of things. IEEE Access **4**, 2292–2303 (2016)
7. Huckle, S., Bhattacharya, R., White, M., Beloff, N.: Internet of things, blockchain and shared economy applications. Proc. Comput. Sci. **98**, 461–466 (2016)
8. Yue, X., Wang, H., Jin, D., Li, M., Jiang, W.: Healthcare data gateways: found healthcare intelligence on blockchain with novel privacy risk control. J. Med. Syst. **40**(10), 218 (2016)
9. Azaria, A., Ekblaw, A., Vieira, T., Lippman, A.: Medrec: using blockchain for medical data access and permission management. In: International Conference on Open and Big Data (OBD), pp. 25–30. IEEE (2016)
10. Decker, C., Wattenhofer, R.: Information propagation in the bitcoin network. In: 2013 IEEE Thirteenth International Conference on Peer-to-Peer Computing (P2P), pp. 1–10. IEEE (2013)

11. Klarman, U., Basu, S., Kuzmanovic, A., Sirer, E.G.: Bloxroute: a scalable trustless blockchain distribution network whitepaper. IEEE Internet of Things J. (2018)
12. Gervais, A., Karame, G.O., Wüst, K., Glykantzis, V., Ritzdorf, H., Capkun, S.: On the security and performance of proof of work blockchains. In: Proceedings of the 2016 ACM SIGSAC Conference on Computer and Communications Security, pp. 3–16 (2016)
13. Kumar, R., Ross, K.W.: Peer-assisted file distribution: the minimum distribution time. In: 1st IEEE Workshop on Hot Topics in Web Systems and Technologies: HOTWEB'06, vol. 2006, pp. 1–11. IEEE (2006)
14. Castro, M., Druschel, P., Kermarrec, A.-M., Nandi, A., Rowstron, A., Singh, A.: Splitstream: high-bandwidth multicast in cooperative environments. ACM SIGOPS Oper. Syst. Rev. **37**(5), 298–313 (2003)
15. Liu, S., Chen, M., Sengupta, S., Chiang, M., Li, J., Chou, P.A.: P2p streaming capacity under node degree bound. In: 2010 IEEE 30th International Conference on Distributed Computing Systems, pp. 587–598. IEEE (2010)
16. Maymounkov, P., Mazieres, D.: Kademlia: a peer-to-peer information system based on the xor metric. In: International Workshop on Peer-to-Peer Systems, pp. 53–65. Springer (2002)
17. Rowstron, A., Druschel, P.: Pastry: Scalable, decentralized object location, and routing for large-scale peer-to-peer systems. In: IFIP/ACM International Conference on Distributed Systems Platforms and Open Distributed Processing, pp. 329–350. Springer (2001)
18. Stoica, I., Morris, R., Karger, D., Kaashoek, M.F., Balakrishnan, H.: Chord: a scalable peer-to-peer lookup service for internet applications. ACM SIGCOMM Comput. Commun. Rev. **31**(4), 149–160 (2001)
19. Bitcoin book. https://github.com/bitcoinbook/bitcoinbook
20. Heilman, E.: How many ip addresses can a dns query return? https://ethanheilman.tumblr.com/post/110920218915/how-many-ip-addresses-can-dns-query-return
21. Bitcoin protocol documentation. https://en.bitcoin.it/wiki/Protocol_documentation
22. Essaid, M., Park, S., Ju, H.-T.: Bitcoin's dynamic peer-to-peer topology. Int. J. Netw. Manag. **30**(5), e2106 (2020)
23. Mariem, S.B., Casas, P., Romiti, M., Donnet, B., Stütz, R., Haslhofer, B.: All that glitters is not bitcoin–unveiling the centralized nature of the btc (ip) network. In NOMS 2020-2020 IEEE/IFIP Network Operations and Management Symposium, pp. 1–9. IEEE (2020)
24. Gao, Y., Shi, J., Wang, X., Tan, Q., Zhao, C., Yin, Z.: Topology measurement and analysis on ethereum p2p network. In: IEEE Symposium on Computers and Communications (ISCC), vol. 2019, pp. 1–7. IEEE (2019)
25. Kim, S.K., Ma, Z., Murali, S., Mason, J., Miller, A., Bailey, M.: Measuring ethereum network peers. In: Proceedings of the Internet Measurement Conference, vol. 2018, pp. 91–104 (2018)
26. Corallo, M.: Compact block relay. https://github.com/bitcoin/bips/blob/master/bip-0152.mediawiki
27. Heilman, E., Kendler, A., Zohar, A., Goldberg, S.: Eclipse attacks on bitcoin's peer-to-peer network. In: 24th {USENIX} Security Symposium ({USENIX} Security 15), pp. 129–144 (2015)
28. Nayak, K., Kumar, S., Miller, A., Shi, E.: Stubborn mining: Generalizing selfish mining and combining with an eclipse attack. In: IEEE European Symposium on Security and Privacy (EuroS&P), vol. 2016, pp. 305–320. IEEE (2016)
29. Eyal, I., Sirer, E.G.: Majority is not enough: Bitcoin mining is vulnerable. In: International Conference on Financial Cryptography and Data Security, pp. 436–454. Springer (2014)
30. Wüst, K., Gervais, A.: Ethereum eclipse attacks. Technical Report, ETH Zurich (2016)
31. Marcus, Y., Heilman, E., Goldberg, S.: Low-resource eclipse attacks on ethereum's peer-to-peer network. IACR Cryptol. ePrint Arch. **2018**, 236 (2018)
32. Apostolaki, M., Zohar, A., Vanbever, L.: Hijacking bitcoin: routing attacks on cryptocurrencies. In: IEEE Symposium on Security and Privacy (SP), vol. 2017, pp. 375–392. IEEE (2017)
33. Apostolaki, M., Marti, G., Müller, J., Vanbever, L.: Sabre: protecting bitcoin against routing attacks (2018). arXiv:1808.06254

34. Ekparinya, P., Gramoli, V., Jourjon, G.: Impact of man-in-the-middle attacks on ethereum. In: IEEE 37th Symposium on Reliable Distributed Systems (SRDS), vol. 2018, pp. 11–20. IEEE (2018)
35. Biryukov, A., Khovratovich, D., Pustogarov, I.: Deanonymisation of clients in bitcoin p2p network. In: Proceedings of the 2014 ACM SIGSAC Conference on Computer and Communications Security, pp. 15–29 (2014)
36. Neudecker, T., Hartenstein, H.: Could network information facilitate address clustering in bitcoin? In: International Conference on Financial Cryptography and Data Security, pp. 155–169. Springer (2017)
37. Bojja Venkatakrishnan, S., Fanti, G., Viswanath, P.: Dandelion: redesigning the bitcoin network for anonymity. In: Proceedings of the ACM on Measurement and Analysis of Computing Systems, vol. 1, no. 1, p. 22 (2017)
38. Fanti, G., Venkatakrishnan, S.B., Bakshi, S., Denby, B., Bhargava, S., Miller, A., Viswanath, P.: Dandelion++ lightweight cryptocurrency networking with formal anonymity guarantees. Proc. ACM Meas. Anal. Comput. Syst. **2**(2), 1–35 (2018)
39. Bitcoin historical data. https://tradeblock.com/bitcoin/historical/
40. Naumenko, G., Maxwell, G., Wuille, P., Fedorova, S., Beschastnikh, I.: Bandwidth-efficient transaction relay for bitcoin (2019). arXiv:1905.10518
41. Pass, R., Seeman, L., Shelat, A.: Analysis of the blockchain protocol in asynchronous networks. In: Annual International Conference on the Theory and Applications of Cryptographic Techniques, pp. 643–673. Springer (2017)
42. Nagayama, R., Banno, R., Shudo, K.: Identifying impacts of protocol and internet development on the bitcoin network. In: IEEE Symposium on Computers and Communications (ISCC), vol. 2020, pp. 1–6. IEEE (2020)
43. Corallo, M.: Bitcoin relay network
44. Fibre. http://bitcoinfibre.org/
45. Basu, S., Eyal, I., Sirer, E.: Falcon. https://www.falcon-net.org/

Consensus Algorithms for Blockchain

Hyunsoo Kim and Taekyoung Ted Kwon

Abstract A consensus algorithm is an essential component of a blockchain, responsible for reaching an agreement among decentralized nodes. It also determines the performance and characteristics of an application. With more than 2,000 different cryptocurrencies currently in use, we face an ever-growing list of consensus algorithms. Furthermore, the inherent complexity of consensus algorithms and their rapid evolutions make it hard to assess their suitability for blockchain applications. Understanding the pros and cons of a consensus algorithm is crucial in designing new blockchain services and developing more advanced algorithms. We propose a framework with comprehensive criteria to evaluate consensus algorithms in terms of performance, security, and decentralization. In addition, we present the operational mechanisms and analyze the characteristics of mainstream consensus algorithms, namely, proof-based algorithms such as Proof of Work (PoW) and Proof of Stake (PoS), and vote-based algorithms with Byzantine Fault Tolerance (BFT). The algorithms are evaluated based on our proposed framework to provide a better understanding. We hope this article leads us to identify research challenges and opportunities of consensus algorithms.

1   Introduction

Blockchain technologies have received widespread attention across the industry, governments, and academia alike over the past decade. Today's most predominant blockchain applications are cryptocurrencies, for instance, Bitcoin has recently hit $1 trillion in market value [1]. Cryptocurrencies have disrupted the long-established centralized financial system on a global scale. Many developing countries are now

H. Kim · T. T. Kwon (✉)
Department of Computer Science and Engineering, Seoul National University, 1 Gwanak-ro,
Gwanak-gu, Seoul, South Korea
e-mail: tkkwon@snu.ac.kr

H. Kim
e-mail: wayles@snu.ac.kr

D. A. Tran et al. (eds.), *Handbook on Blockchain*, Springer Optimization
and Its Applications 194, https://doi.org/10.1007/978-3-031-07535-3_3

seeing higher rates of cryptocurrency adoption. Take Nigeria, for example; roughly a third of the population owns cryptocurrencies and uses them in everyday lives [2].

Now, we are witnessing the blockchain expands across various industries such as energy, health care, real estate, supply chain, and so on. According to a recently conducted study [3], blockchain technologies have the potential to boost global Gross Domestic Product (GDP) by $1.76 trillion across the industry over the next decade, which is 1.4% of the predicted global GDP.

Blockchain is essentially a decentralized, asynchronous distributed system, often with much more nodes than its traditional counterpart. Making reliable communications between the nodes and maintaining the correct state across the system even in the presence of malicious nodes and network failures are the key issues [4]. This is where a consensus algorithm takes place. At the heart of a blockchain (or its application), the consensus algorithm is responsible for maintaining consistent copies of the current state across all nodes, validating new transactions, and updating the current state while achieving an agreement among the nodes.

The Proof-of-Work (PoW) consensus algorithm used in Bitcoin is the first and most popular consensus algorithm in the blockchain. However, although this consensus algorithm is well-fitted for the application of Bitcoin, it has its shortcomings. Namely, its energy inefficiency of validating and constructing a new block, known as *mining*, and low throughput have been a vexing issue [5]. As a result, researchers and developers have sought to devise new consensus algorithms.

There are over 2,000 different cryptocurrencies, let alone blockchain applications from other industries, which are currently employing diverse consensus algorithms. The list of consensus algorithms is extensive, and even now, newer ones are under way. Researchers and developers must understand the characteristics and limitations of a consensus algorithm since the overlaying blockchain application's performance and usability will highly depend on it. Thus, we believe it is essential to lay out a framework that can be used to analyze and evaluate a consensus algorithm and determine its suitability to a particular application. The contributions of this article are summarized as follows. First, a framework for evaluating consensus algorithms is proposed. The framework consists of comprehensive criteria that can be applied to most consensus algorithms. Second, an in-depth survey of representative consensus algorithms and their characteristics are discussed.

The rest of this article is organized as follows. In Sect. 2, we first review the literature on consensus algorithms with a focus on their evaluation criteria. And then, we present our evaluation framework and discuss each criterion in detail. Section 3 presents major consensus algorithms categorized in PoW, PoS, and vote-based. A thorough evaluation will follow in Sect. 4 based on our proposed framework. Section 5 proposes future research opportunities regarding consensus algorithms and concludes the article.

2 Evaluation Criteria

With the advancement of different blockchain technologies and their applications in multiple domains, a variety of consensus algorithms have been developed. As most of the consensus algorithms have their limitations, there are still ongoing debates on addressing the drawbacks or issues of those consensus algorithms.

This paper aims to identify and provide key criteria for a consensus algorithm from diverse perspectives. In this section, we first review the major characteristics and functionalities of the consensus algorithms in the literature and then present a framework for their evaluation.

2.1 Related Works

In order to define and present an evaluation criteria framework, we first go over a comprehensive review of the prior consensus algorithms and their evaluations.

In [6], the authors focus on the Bitcoin cryptocurrency and list the following criteria: maximum throughput, latency, bootstrap time, cost per confirmed transaction, transaction validation, bandwidth, and storage. Their criteria are mostly related to the performance of the consensus algorithm with a focus on the Proof-of-Work (PoW) algorithm in Bitcoin.

Mingxiao et al. [7] compared the five consensus algorithms: PoW, Proof of Stake (PoS), Delegated Proof of Stake (DPoS), Practical Byzantine Fault Tolerance (PBFT), and Raft. The list of criteria consists of: Byzantine fault tolerance, crash fault tolerance, verification speed, throughput, and scalability. Note that the criteria can be classified into two categories: fault tolerance and performance.

Nguyen and Kim [8] performed a comprehensive survey of consensus algorithms by categorizing them to proof-based consensus algorithms: PoW, PoS, hybrid form of PoW and PoS, and voting-based consensus algorithms: Byzantine fault tolerance and crash fault tolerance. Performance comparison was done between PoW, PoS, and hybrid form of PoW and PoS based on energy efficiency, modern hardware, forking, double-spending attack, block creating speed, and pool mining. Another performance comparison was performed between proof-based consensus algorithms and vote-based consensus algorithms in general. The criteria were agreement making, joining nodes, number of nodes, decentralization, trust, node identity management, security threat, and reward. Although we take the similar categorization of consensus algorithms in this study, the listed criteria are focused on qualitative properties and only applied to each category and not individual algorithms.

Unlike the above studies, [9] did not categorize consensus algorithms for evaluation but viewed the blockchains at multiple levels: consensus level, mining pool, network level, and smart contracts. The authors specifically focused on attack vectors at the consensus level: double spending, Finney attack, Vector76 attack, brute force attack, 51% attack, and nothing-at-stake attack.

Table 1 Comparison of consensus algorithm evaluation criteria of [11, 12]

Ferdous et al. [11]		Bamakan et al. [12]	
Category	Criterion	Category	Criterion
Structural	Node type	Throughput	TPS
	Structure type		Block creation
	Underlying mechanism		Verification time
Block and reward	Genesis date		Block size
	Block reward	Profitability	Mining reward
	Total supply		Power consumption
	Block time		Transaction fees
Security	Authentication		Hardware dependency
	Non-repudiation	Decentralization	Blockchain governance
	Censorship resistance		Permission model
	Adversary tolerance		Trust model
	Sybil protection	Security	Double spending
	DoS resistance		51% attack
Performance	Fault tolerance		Sybil attack
	Throughput		
	Scalability		
	Latency		
	Energy consumption		

Bano et al. [10] performed a survey based on individual applications such as ByzCoin, Ouroboros, Bitcoin, and Spectre which is different from our approach. However, the authors classify the criteria into three categories: committee configuration, safety, and performance. Safety consists of censorship resistance, DoS resistance, and adversary tolerance, while performance consists of throughput (TPS), scalability, and latency.

Ferdous et al. [11] performed an extensive survey of consensus algorithms and classified them into two categories: incentivized consensus and non-incentivized consensus. This is analogous to our proof-based and vote-based categorization. [11, 12] are both noticeable for their structuring evaluation criteria of consensus algorithms, which is found in Table 1.

2.2 Evaluation Framework

Identifying universal criteria that apply to most, if not all, consensus algorithm is key to defining a solid evaluation framework. We also focus on generalized consensus

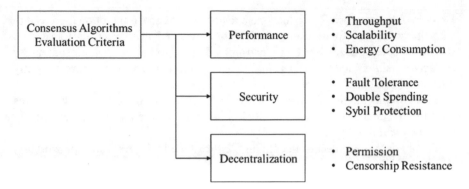

Fig. 1 The criteria of our evaluation framework for consensus algorithms are classified into three categories

algorithms such as PoW and PoS algorithms, not on individual cryptocurrency applications. As a result, structural criteria such as node characteristics and management, or profitability criteria, such as mining reward, mining pools, and transaction fees, are avoided unless necessary.

Based on the above standpoint and our review of the literature that define various criteria for consensus algorithm's performance evaluation, we present a framework to evaluate consensus algorithms in terms of criteria in the three following categories: performance, security, and decentralization, as depicted in Fig. 1. In the following, we will briefly introduce three categories and detail the criteria therein.

Performance Criteria

The performance criteria consist of properties or metrics to measure the quantitative performance of consensus algorithms. In this paper, we consider throughput, scalability, and energy consumption for performance.

Throughput/TPS

The *throughput* of a consensus algorithm is the speed of processing transactions by the participating nodes or members. In other words, the maximum throughput of a blockchain is the maximum rate at which the blockchain can confirm transactions [6]. It is also referred to as *Transactions per Second* (TPS), defined by the number of transactions processed per second. For example, if a particular blockchain processes an average of 600 transactions per minute, the TPS of that blockchain is 10 (10 transactions per second). The higher is the TPS, the faster the transactions will be verified, executed, and confirmed by that blockchain.

Throughput is one of the most essential criteria when discussing the performance of a blockchain. There are several elements that we should also consider when discussing throughput: *latency* and *block size*.

Latency: The *latency* refers to the time it takes from when a transaction is created to when the consensus (for the transaction) has been reached. In between, the transaction will be validated, added to the block, and appended to the chain. *Latency* is sometimes replaced by similar terms such as *block time* or *finality*, in which they have a slightly different meaning.

Block time is the time it takes to make a new block since the last block that was added to the chain. An increase in block time will increase the latency, effectively reducing the throughput.

There are two approaches in defining *finality*. One is *deterministic* (or *absolute*) *finality*, which guarantees that the transaction is verified and immutable as soon as it is added to the chain. The time to deterministic finality is identical to block time and, in most cases, latency as well [13]. Another, *probabilistic finality* is used when a transaction becomes probabilistically immutable as more blocks are added to the chain. This will be explored further in Double-Spending Prevention in Security Criteria (Section "Security Criteria").

Block size: The *block size* refers to the maximum amount of transactions (or bytes) in a block. Larger block size may lead to shorter latency since it can fill the block with more transactions. However, on the contrary, increasing the block size could improve the throughput of the consensus algorithm since more transactions can be included in the block given the same block time.

All in all, the throughput of a consensus algorithm is not determined by a single factor and must consider different variables and their implications. Table 2 presents the TPS, maximum block size, and the minimum and maximum latency of selected cryptocurrencies between March 2018 and February 2021 [14, 15].

Scalability

Scalability refers to the ability to support a growing number of users and nodes. It is considered one of the critical factors in the design of decentralized distributed systems. Throughput is also another aspect of scalability. As the network grows, we can expect the number of transactions to increase proportionally. However, the throughput limitation compared to centralized systems is one of the hindrances to

Table 2 TPS, latency, and max block size of selected cryptocurrencies [14, 15]

Cryptocurrency		Bitcoin	Ethereum	Litecoin	Ripple	Dogecoin	DASH	Monero
TPS		7	15	28	1500	16	56	30
Latency [min]	Min	7.35	0.22	2.12	0	1,03	2.56	1.57
	Max	15.65	0.39	3.48	0	1.05	2.69	10.99
Max block size [MB]		1	Dynamic[a]	1	N/A[b]	1	2	Dynamic[c]

a Variable block size based on gas limit.
b Does not have blocks.
c Variable block size based on last 100 blocks.

blockchain deployment. For example, Bitcoin can handle up to 7 TPS, far from PayPal and VISA's performance, which has approximately 200 TPS and 20 k TPS, respectively [16].

Energy Consumption

Energy consumption is another important performance criterion of a consensus algorithm. It is well known that the total energy consumption for mining in the Bitcoin network can now power a whole country like Portugal, Singapore, and Czech Republic, to name a few [17]. Furthermore, the primary source of electricity that runs the Bitcoin network is from coal-fired power plants in China, which is infamous for its extreme amount of carbon emission. A study conducted in 2018 suggests that the carbon emission related to Bitcoin alone could increase global warming by 2 °C in less than three decades [18].

The high energy consumption is mainly due to the computation of the cryptographic hash functions such as SHA-256 (e.g., Bitcoin), Ethash (e.g., Ethereum), and many other hash functions used by the PoW algorithms. As the difficulty of the PoW algorithm continues to increase, so does the energy consumption, and it is pivotal that future consensus algorithms focus on energy efficiency as a top objective.

Security Criteria

The blockchains have various cybersecurity attack vectors that can threaten any given consensus algorithms. Naturally, one could think of diverse attacks and vulnerabilities when given a particular consensus algorithm. However, in this paper, we present only well-known attacks that can be commonly applied to most of consensus algorithms. Namely, adversary tolerance, double-spending prevention, and Sybil protection resistance will be explored in detail.

Fault (or Adversary) Tolerance

To begin with, fault tolerance typically refers to crash fault tolerance in which nodes or members of a network fail and become offline until they are brought back online. This is different from becoming compromised and sending fraudulent transactions. When f number of nodes or members has crashed, the network requires $2f + 1$ participants or a quorum of $f + 1$ to be crash fault tolerant.

If the nodes are subverted and send fraudulent transactions to the network, this is called Byzantine behavior [19], and their consensus algorithm must be Byzantine fault tolerant. One of the well-known algorithms that can achieve consensus in this attack is Practical Byzantine Fault Tolerant (PBFT) [20], which is capable of handling up to f Byzantine nodes with $3f + 1$ total nodes.

By contrast, when we take proof-based consensus algorithms into account, the term "fault tolerance" refers to the percentage of total network resources that need to assure consensus. When an adversary is able to control more than 50% of a network's computing power, it could maliciously alter or control the consensus process to launch an attack (e.g., double spending). Hence, the term *51% attack* is widely used as it

Fig. 2 A hash rate
distribution of Bitcoin
mining pools in February
2022 is shown

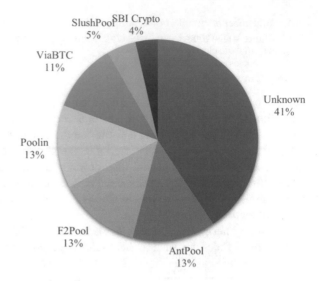

was also discussed in the original Bitcoin paper [21]. A selfish mining strategy shows that an adversary with less than 50% power could withhold block and increase its profitability without affecting the safety or liveness of a blockchain network [22]. It is also worth mentioning that in theory 51% attack is unavoidable, and adversaries with mining pools could always collude with each other. For example, Fig. 2 shows that the top five mining pools of Bitcoin exceed 51% of the total hash rate of the whole network as of early 2022 [23].

Double-Spending Prevention

In a double-spending attack, the adversary creates a typical transaction that reaches consensus, and then the adversary creates a fork with a conflicting transaction to revert the prior transaction. A successful double-spending attack will allow the adversary to spend the same coin more than once, hence double.

However, in order to push the fork with the malformed transaction to the main chain, the adversary must have already broken the adversary tolerance. For example, in the Bitcoin PoW algorithm, the adversary should have at least 51% computing power of the entire network to ensure a successful fork through faster block creation.

Consensus algorithms try to mitigate this attack by introducing a *confirmation count*, which is incremented when a block of the transaction of interest is followed by another block. Subsequent blocks increase the number of confirmations, which in turn increases the probability of the transaction validity. This is also referred to as *Probabilistic finality*. In essence, the probability of an invalid fork that reverts a prior transaction decreases exponentially as more blocks are appended to the chain [13].

Table 3 presents the number of confirmations required and the average validation time of well-known cryptocurrencies. Notice that the number of confirmations multiplied by average latency or block generation time from Table 2 is the average validation time.

Table 3 Number of confirmations and average validation time of selected cryptocurrencies

Cryptocurrency	Number of confirmations	Average validation time [Min]
Bitcoin	6	60
Ethereum	30	6
Litecoin	12	30
Ripple	N/A	N/A
Dogecoin	20	20
DASH	6	15
Monero	15	30

Sybil Protection

A Sybil attack [24] takes place when an adversary attempts to control the network by duplicating fraudulent identities. Within the blockchain, a Sybil attack will be an attempt to create and possess as many nodes or members of the network in order to influence the consensus algorithm. A successful attack may grant the adversary the higher voting power in consensus algorithms that utilize a voting process (e.g., DPoS, PBFT), or enable network layer attacks targeting peer discovery and block broadcasts [25].

To prevent Sybil attacks, consensus algorithms could use combinations of methods from increasing the cost of creating identities, requiring second-channel authentication, or two-step verification for identities [26, 27].

Decentralization Criteria

We identified two factors that can qualitatively evaluate the consensus algorithms decentralization: permission and censorship resistance.

Permission

Depending on blockchains, a node may need a permission to participate in reaching a consensus by validating transactions and creating blocks. A *permissionless* consensus algorithm will allow any anonymous node to participate in the consensus process. While a *permissioned* consensus algorithm will only allow authenticated nodes to participate in the consensus process. This is not to be confused with the concept of *public* and *private* blockchain, where permission refers to the anonymity of miners and validators, while public and private refers to the anonymity of all the nodes participating in the blockchain network. For example, the Bitcoin and the Ethereum network is a public permissionless network. A blockchain-based voting system should be a public but permissioned network, so that the voters remain anonymous while the validators are authenticated to be trustworthy.

The number of nodes in the permissionless network will be large compared to that of a permissioned network, and to mitigate attack vectors such as 51% attacks and Sybil attacks, proof-based consensus algorithms are used. As a result, all validators must spend energy and resources to prove its contribution to the network by participating in the consensus process. In general, permissioned networks are more centralized compared to the permissionless network.

Censorship Resistance

Censorship resistance refers to the network's property that assures any node to freely make transactions as long as they follow the rules of the consensus algorithm and the blockchain network. With traditional finance institutions, some intermediaries would censor transactions that it deemed suspicious or undesirable, justified to prevent financial crimes. Also, if probabilistic finality is achieved, a transaction recorded on the blockchain is technically irreversible, also commonly known as *immutable*, providing further censorship resistance [28].

3 Consensus Algorithms

Performing a consensus algorithm in a blockchain network is a non-trivial process. A newly broadcasted transaction is first verified by a verifying node and added to its candidate block. Similarly, the candidate block will be verified or voted by other nodes in the network before being added to the chain. We would like to emphasize that once a transaction is included in the chain, it is not feasible to modify or delete them due to blockchains' immutability.

The consensus algorithms that will be discussed in this article can be classified into two categories: proof-based and vote-based. A similar distinction between the two categories can be made using the permission criterion. In permissionless blockchains, any nodes are free to join and leave the network, and their behaviors are unpredictable. Therefore, a permissionless blockchain typically relies on a *proving* mechanism that appreciates verifying nodes' contribution toward the network. This usually involves rewards, which incentivizes the nodes to participate in the consensus process. In contrast, a vote-based consensus algorithm does not require contribution or *proof* from the node participating in the consensus process since the participants are *permissioned* in advance, and their participation list is managed. Thus, vote-based consensus algorithms can be adopted in *non-incentivized* blockchain networks and are well-suited for private blockchains and non-cryptocurrency applications.

Note that the voting process does not always go along with vote-based consensus algorithms since it can be used in proof-based consensus algorithms (e.g., DPoS, BFT PoS). The difference between the above two categories lies in whether the verifying node of the consensus algorithm is required to provide a *proof* (e.g., computation, stake) to the network to participate in the consensus process.

Figure 3 shows the classification of consensus algorithms that will be discussed in this section.

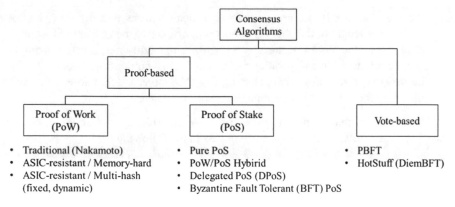

Fig. 3 A classification of consensus algorithms surveyed in this article is summarized

3.1 Proof-Based Consensus Algorithms

The original Proof of Work (PoW) of Bitcoin is the most popular proof-based consensus algorithm to date. As mentioned earlier, the basic concept of a proof-based consensus algorithm is that a participating node of the consensus process performs or provides a sufficient proof to append a new block and is rewarded. Depending on the method of a proof, the proof-based consensus algorithms can be further divided into proof of work (computation), proof of stake (currency locked in escrow), proof of activity (transaction participation), proof of research (Berkeley Open Infrastructure for Network Computing contribution), and so on.

In this article, we will focus on the two major proof-based consensus algorithms, PoW and PoS.

3.2 Proof of Work (PoW)

The Proof-of-Work (PoW) algorithm usually requires a proving node and a verifying node. The proving node performs a resource-intensive computational task to find a solution to a problem of a certain difficulty level. The result is then presented to the verifier who spends a significantly less resource for validation compared to the prover. The asymmetry and the excessive amount of resources required for the prover serves two notable purposes:

1. It mitigates Sybil attacks at the consensus level. Launching a Sybil attack involves the adversary creating multiple fraudulent identities. However, by design, the amount of computational resources (e.g., hash rate) is important for the PoW algorithm, not the number of nodes. Note that Sybil attacks in the network layer are still a vulnerability, but this article focuses on the consensus layer.

2. The workload itself becomes a safeguard against forks and double-spending attacks. The length of the chain is almost proportional to the amount of resources spent mining the blocks. If the adversary wants to modify a transaction from the past, he will first have to acquire more than 51% of computing resource within the network, fork a new chain starting from the target block, and exhaustively mine the blocks until the new chain becomes the longest.

There are two major sub-categories of PoW algorithms: traditional PoW and ASIC-resistant PoW, which are to be explored in the following.

Traditional PoW

Traditional PoW algorithm employs computational tasks that heavily exploit either the CPU or the GPU with little dependency on the system memory size. Another critical characteristic of computation-bound PoW is that the computation can easily be implemented on an ASIC, which leads to mining farms and mining pools, counteracting the notion of decentralized consensus.

The earliest idea of computation-bound PoW algorithm dates to 2002, an anti-spam system called HashCash [29]. This system requires the sender of an email to generate a SHA-1 hash value with at least 20 bits of leading zeros. The list of inputs included the recipient's address and date alongside the random number, called a *nonce*, provided by the sender. The sender should try numerous proof attempts to meet the leading zeros requirement, while the verification by the recipient is relatively trivial.

Bitcoin's PoW algorithm is based on the HashCash's PoW algorithm, but is modified to use SHA-256d (SHA-256 performed twice) instead of SHA-1. Fundamentally, provers of the Bitcoin network are trying to find a 256-bit nonce that, when hashed with the block, will have outcome which is smaller than the *difficulty value*. Recently, Bitcoin's hash rate surpassed 150Exahash/s (Exa $= 10^{18}$) [15], which means in order to append a block, an average of 90×10^{21} nonces is tried. When an appropriate nonce is found and approved by the verifiers, the prover receives a block reward of 6.25 BTC as of March 2021. The process of finding the nonce is known as *mining*, and the proving nodes are called *miners*. Figure 4 shows the process of a miner finding the nonce.

When a miner successfully finds a nonce that satisfies the difficulty, it broadcasts the mined block to the entire network. Other verifying nodes, who might also have been mining a block at the same height, verify the newly broadcasted block by checking whether all the transactions included within the block are valid, whether the previous hash value (*Prev_Hash*) matches with the hash value of the last block from their current chain, and whether the nonce value satisfies the current difficulty. If these conditions are met, the mining (or proving) nodes abandon their current task and append the new block to their chain. And then, they reselect the transactions to be included in the next block by referencing the latest block, form a new transactions list and the next block header, and start over the nonce calculation.

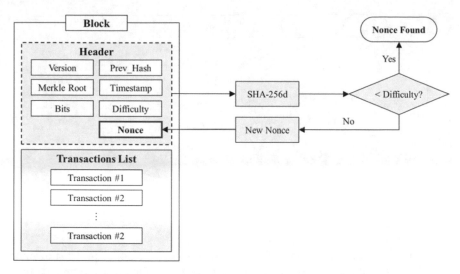

Fig. 4 A process of finding the nonce in Bitcoin PoW is shown

Once in a while, due to the worldwide scale of the Bitcoin network and large pool of miners, there may be more than one new valid blocks broadcasted throughout the network at the same height. As a result, the network is divided based on which one of the legitimate blocks the node received first as shown in Fig. 5. Here, the two black-colored nodes each successfully mined two new blocks N and N' that are appended to the most recent block $N-1$. Due to the size of the network and broadcast latency, the nodes in Group 2 received the block N' faster than the block N. Hence, the nodes in Group 2 appended the block $N-1$ with block N', and the nodes in Group 1 with block N. Now there are two concurrently valid chains in the network and the global consensus is now broken. This situation is called a *fork*.

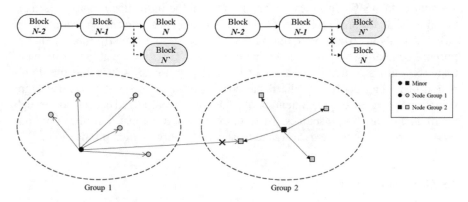

Fig. 5 A fork in blockchain takes place at block height N

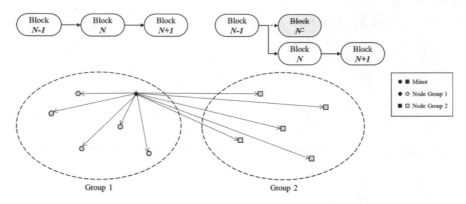

Fig. 6 How a fork is resolved through the longest chain rule is illustrated

To overcome this issue and achieve consensus, first the nodes proceed to mine new blocks based on their version (or branch) of the chain. For example, the next block $N + 1$ is mined by a single miner (highlighted in star shape in Fig. 6). Block $N + 1$ gets broadcasted to all the nodes of the network. The nodes of Group 1 share the same block N as the current block and append it with the newly mined block $N + 1$. However, the nodes of Group 2 do not recognize the previous hash of block $N + 1$, which is N, and are unable to append block $N + 1$ to N'. Now, the two branches of the fork have a length difference of a block.

The chain length indicates the amount of computational resource put into creating the blocks and managing the chain. When given multiple branches of a chain, the branch with the longest chain is allocated with the most computational power, i.e., hash rate. Thus, the nodes of Group 2 abandon the former consensus based on block N' and select $N \rightarrow N + 1$ as their new chain. This is known as the *longest chain rule*, and this helps resolve forks and reachieve global consensus in the network.

As mentioned at the beginning of this subsection, traditional PoW algorithms are subject to ASIC. Specifically, this means that the hash function used in traditional PoW algorithms such as SHA-256 in Bitcoin is easy to implement using an ASIC. A study in 2018 showed that an ASIC-based system outperformed a general-purpose computing system (e.g., PC with high-end GPU) of equal power by more than 1500 times in terms of hash rate [30]. Heavy use of ASICs and the domination of mining pools centralized the consensus layer of traditional PoW algorithms, violating the original intention of achieving decentralized consensus through distributed computing resources. Also, the overheated competition of mining has raised concerns about energy/resource wastes and environmental implications.

ASIC-Resistant PoW

To overcome issues regarding expediting traditional PoW algorithms in ASIC machines, ASIC-resistant PoW algorithms are gaining more and more interests in the community. In this article, we classify ASIC-resistant PoWs into two categories: memory-hard PoWs and multi-hash PoWs.

Memory-hard PoW: While ASICs have an advantage in hash rate, they are also limited by memory access latency, bandwidth, and memory size. Memory-hard PoW algorithms restrict ASICs from having a performance advantage over general-purpose computing systems by requiring the use of random pieces of data from a large dataset. The dataset should be too large to be stored in the on-chip memory of an ASIC chip. Nonetheless, the Ethash of Ethereum [31], an ASIC-resistant memory-hard PoW, was broken in 2018 by Bitmain's Antminer E3 [32], and now there are ASICs available for memory-hard PoW as well. Still, unlike traditional PoW schemes where mining can heavily leverage ASICs, GPU-based general-purpose computing systems are still dominant in Ethereum [33].

The first memory-hard PoW algorithm appeared before cryptocurrencies in the form of password-based Key Derivation Function (KDF) called Scrypt in 2009 [34, 35]. It was later reinstated as a PoW algorithm of Tenebrix (no longer active), Litecoin [36], Dogecoin [37], and few other cryptocurrencies. Scrypt aims to resist against custom hardware such as GPU, FPGA, and ASIC, and hence turns out to be a CPU-friendly PoW. Both memory-hardness and CPU-friendliness are achieved by read-many, write-few memory access patterns of PoW algorithms. However, ASIC engineers reduced the memory size by 1/8 with a $3.5\times$ additional logic calculation. Thus, Scrypt's memory-hardness broke, and ASIC efficiency gains were 300,000 times over a CPU [38].

Ethash, also known as the Dagger-Hashimoto algorithm, is a memory-hard consensus algorithm of Ethereum [31], the second largest cryptocurrency in terms of market capitalization. Noticeably, during the mining process, Ethash requires data from a Directed Acyclic Graph (DAG) dataset that is over 4 GB in size, as shown on the right side of Fig. 7. To construct a DAG, first, the *seed hash* must be hashed N consecutive times using the keccak-256 [39] hashing algorithm, an early version of SHA-3, where N is the current *epoch* of the DAG. Next, the current *seed hash* is used to calculate a *pseudorandom cache* that is again used to generate the actual DAG dataset. After every 30,000 blocks (approximately 5 days), a new *mining season* begins, and the epoch is increased by one, consequently updating the seed hash, pseudorandom cache, and the DAG dataset. By design, the size of the pseudorandom cache and the DAG dataset is configured to increase linearly by each epoch, starting from 16 MB for the cache and 1 GB for the DAG dataset. In more than 20 years, Ethash will reach epoch #2048, where the cache size reaches 285 MB and the DAG dataset reaches 18.2 GB.

The left side of Fig. 7 depicts the mining process using the generated DAG. Like Bitcoin's PoW algorithm, a random nonce is selected, and after several hashes and mixes with input data from various stages, the final output is compared with the difficulty value to determine the result of the nonce. Every *mixing* operation requires a value from a random address of the DAG, involving memory access and enough storage for the DAG. Although the hashing and mixing logic can be built using an ASIC, the memory bandwidth serves as a bottleneck balancing the end performance between an ASIC-based system and a general-purpose computing system.

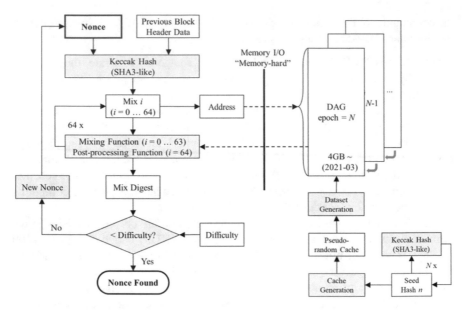

Fig. 7 A process of finding the nonce in Ethash is shown

Multi-hash PoW: Another group of ASIC-resistant PoW algorithms is *multi-hash* PoWs or *chained* PoWs. As their name suggests, these consensus algorithms do not rely on a single hash function but multiple hash functions during the mining process. There are a few ways to do this. First, we can define a sequence of hash functions forming a fixed hash chain. An example of fixed multi-hash PoW algorithm, X11 [39], used in the DASH blockchain, is presented in Fig. 8. Here, the algorithm uses 11 hash functions: Blake, Blue Midnight With (BMW), Grøstl, JH, keccak, Skein, Luffa, CubeHash, SHAVite, SIMD, and Echo. Keccak is the winner of the SHA-3 [40] open competition, and others in the list include those advanced to the second round or the final round. There are also the variants of X11 such as X13, X14, X15, and X17 which utilize more hash functions in their sequence of hashes.

Another way to implement multi-hash PoW algorithms is using a dynamic sequence of hash functions through several methods. One is a fully variable sequence of hash functions where the sequence is permuted by a random value based on block hash or timestamp. Alternatively, a partially variable sequence with randomly selected hash functions in between a fixed sequence is also possible. An example of a fully variable sequence is X16R [41] which is shown in Fig. 9. The X16R algorithm constructs the sequence of hash functions based on the last 8 bytes of the previous block's hash. Each hashing algorithm is mapped with a hexadecimal value, and the last 8 bytes, i.e., 16 hexadecimal values, determine the permutation of hash functions for the current block. Also, note that repeated use of a hash function is possible when the two consecutive hexadecimal values are the same. Nonetheless, we can safely assume that the use of hash functions would be balanced due to the randomness property of cryptographic hash functions.

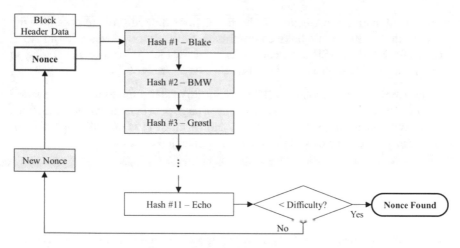

Fig. 8 How to find a nonce in fixed multi-hash PoW, X11, is depicted

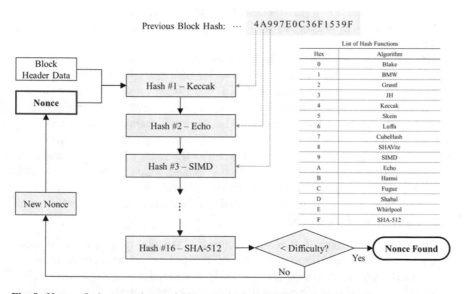

Fig. 9 How to find a nonce in a variable multi-hash PoW, X16R, is shown

Multi-hash PoW algorithms were first thought to be ASIC-resistant since they required implementing multiple hash functions in hardware. Whereas in a general-purpose computing system, it was a simple matter of switching codes in software to execute the different hash functions. This becomes more apparent when we consider variable multi-hash PoW algorithms. It is not profitable to design an ASIC chip that supports all possible sequences of hash functions, for example, 16^{16} in the case of X16R (since it allows repeated usage). However, this is true only if we assume that hash modules in an ASIC must be connected in a fixed pipeline. It is reported that supporting dynamic combinations of hash functions in hardware are possible with some overhead [30]. As to the fixed multi-hash PoWs, it took 2 years for an ASIC targeted for X11 in DASH to be commercially available. We believe developing an ASIC for a variable sequence of hash functions is a matter of time.

Programmatic PoW: Another ASIC-resistant PoW algorithm known as *programmatic* PoW achieves ASIC-resistance by using a large pool of mathematical functions or randomly generated programs [42, 43]. However, there is no PoW algorithm of this category used in an existing blockchain application as of today.

3.3   Proof of Stake (PoS)

To overcome the downside of PoW algorithms, such as mining centralization, energy waste, and low TPS, a new consensus algorithm called Proof of Stake (PoS) is gaining a momentum [44, 45]. A PoS algorithm attempts to validate transactions and achieves consensus in the network without the heavy computational tasks like PoW algorithms. Instead of *working*, a verifying node has to lock its *stake*, a proportion of its wealth on the network.

In the PoS algorithm, stakeholders lock their stakes in escrow and become eligible to participate in transaction validation and block creation. Unlike the PoW algorithm, the stakeholders in PoS do not mine new coins by validating transactions and creating new blocks. Instead, they collect the transaction fees or interests proportional to their stakes as rewards. Here, the term *minting* or *forging* is commonly used instead of *mining,* and the stakeholder who creates a new block is referred to as a *validator, forger,* or *minter,* not a *miner.* If the validator is found to behave maliciously, the escrowed stake and possible reward will be confiscated, which serves as an economic disincentive for adversarial stakeholders in PoS algorithms.

The advantages of PoS algorithms over PoW algorithms are as follows: decentralization, energy efficiency, and high throughput.

1. The high performance of ASICs has centralized the mining of PoW-based applications. Due to the economy of scale, miners can expect an exponential increment of computational resources (e.g., hash power) and the ensuing rewards when investing in PoW-targeted mining equipments. On the contrary, the expected gain in network control in PoS is directly proportional to the amount of additional stake that the validator put in escrow. Although we cannot claim that PoS algorithms are immune to centralization, it will likely be more decentralized compared to PoW-based networks.
2. As discussed in section "Performance Criteria", Bitcoin miners consume more energy than countries like Portugal, Singapore, and the Czech Republic. PoW algorithms are spending natural resources too much to secure the immutability of Bitcoin transactions. PoS algorithms do not require validators to solve resource-intensive computational tasks when forging blocks, thus they can be made highly

energy efficient. Any blockchain applications that are based on PoS will be more sustainable in the long run.

3. PoW algorithms suffer from low TPS and high block generation times. For instance, a payment made by a PoW-based cryptocurrency normally requires at least one block confirmation to be approved. What is worse, if a burst of transactions suddenly floods the network, PoW-based cryptocurrencies may not handle them timely due to the limited TPS. On the other hand, PoS algorithms offer relatively higher TPS and shorter block generation time by using faster agreement mechanisms for accepting a new block to the chain, ensuring higher throughput compared to PoW algorithms.

PoS algorithms can be classified into three major categories: chain-based PoS, Delegated PoS (DPoS), and Byzantine Fault-Tolerant PoS (BFT PoS). These are classified based on their difference in terms of validator selection, block creation, and agreement.

Chain-Based PoS

A chain-based PoS algorithm mimics PoW mechanics by featuring validators competing one another to *mint* (not *mine*) a new block. Instead of going through the resource-intensive process of finding a valid nonce, the validators are pseudo-randomly selected to mint a block based on their stakes. Like PoW algorithms, the chain-based PoS algorithm's finality is not achieved at the time of block creation. Forks could happen when multiple validators mint the new block simultaneously across the network. Thus, it should have a mechanism to converge the branches of the chain.

This article classifies chain-based PoW algorithms into two sub-categories depending on how to select the validator: Pure PoS with randomized probabilistic selection and PoW/PoS Hybrid with coin-age-based selection.

Pure PoS: A chain with a pure PoS algorithm consists of blocks only minted by validators. Recall that in PoW algorithms, the chain's length is almost proportional to the amount of work put into the chain, and thus the network converges to the longest chain if a fork happens. Similarly, in PoS, the chain's length indicates the sum of stakes escrowed by validators, and the nodes are expected to follow the longest chain.

Nxt [46] is a pure PoS-based cryptocurrency that utilizes a randomized probabilistic function in selecting a validator. When a new block is added to the chain, an active Nxt account (or node) i performs a sequence of hashes using its public key and the latest block's signature to generate an *account hit* value H_i. The use of an individual account's public key and the signature value of an independent block provide a pseudo-randomness to H_i. Next, the accounts calculate their target values T_i.

$$T_i = T_b \times t \times B_i$$

T_i: target value of account i,

T_b: base target value,

t: seconds passed since the last block was generated, and

B_i: effective balance (stake) of account i.

As time goes by, the target values of active accounts increase, and the first account with a target value larger than its account hit value, $H_i < T_i$, gains the right to mint the next block. Notice that T_b and t are shared within the network, and B_i determines how fast the target value reaches H_i. The randomness of H_i allows other accounts to be selected as the validator, although the probability diminishes as one's stake gets smaller.

The new block gets broadcasted to the network, and every account updates its account hit and target values to repeat the process from $t = 0$.

PoW/PoS Hybrid: As its name suggests, PoW/PoS hybrid algorithms are a mixture of both PoW and PoS. PeerCoin (PPCoin) [45] is the first variant of this category, but interestingly, it is also the first PoS algorithm to be used in cryptocurrencies.

PeerCoin recognizes two kinds of blocks: PoW blocks and PoS blocks. PoW blocks are similar to the blocks of Bitcoin, mined by miners competing with each other to find a valid block with a nonce satisfying the target difficulty. Their purpose is to increase the net supply of the coins within the network. PoS blocks, on the other hand, are minted by minters (or validators), who are competing against one another based on their stakes for a target difficulty with far less computation. Thus, the minters are in charge of performing the consensus by adding PoS blocks [47]. From a broader view, miners and minters are competing against each other as two groups since the new block could be either a PoW block or a PoS block.

In selecting the minter of the next PoS block, PeerCoin employs the concept of *coin-age*. Coin-age is the product of the amount of token/coin in an account and its holding period. For example, if Alice has held 100 coins in her wallet for 100 days and Bob has just received 500 coins. In terms of coin-age, Alice has accumulated 10,000 coin-age (days) and Bob 500 coin-age (days), making Alice a more prominent stakeholder of the network at the time.

The target difficulty of a valid PoS block gets easier as more coin-age is accumulated, and if a minter successfully mints a block, she burns all the coin-age by including a transaction of paying the stake to herself. This ensures that the stakes used in minting the block are not used until she accumulates a certain amount of coin-age, enabling every participant to mint blocks with fairness long term.

A significant issue of chain-based PoS is its vulnerability to Nothing-at-Stake (NAS) attacks. Whenever there is a fork in the chain, the nodes working on the next block should decide which branch of the fork it will work on. For PoW algorithms, this means that a miner should invest its computational resource (and energy) in finding the nonce. Thus, it is rational for the miner to select only one branch. The PoW algorithm would give a penalty for working on multiple branches. However, for PoS algorithms, it costs little for a minter to work on multiple branches and, in fact, it increases the possibility of minting the next block, which motivates the minter to do so.

One solution to solve NAS, proposed by another chain-based PoW/PoS hybrid algorithm known as Casper the Friendly Finality Gadget (Casper FFG), is penalizing (or *slashing*) the stake of a validator who works on two conflicting blocks.

Delegated PoS (DPoS)

Despite the similarity of terminology, DPoS, is substantially different from the original PoS in the sense that the network allows stakeholders to *delegate* their voting power to a particular participant who can then stake on behalf of her voters [48]. The rewards (i.e., transaction fees and interests) earned by delegates (also called witnesses) can trickle down to voters with a small amount of stakes, allowing them to participate in the consensus process for incentives.

Unfortunately, by design, DPoS centralizes the consensus layer since a small number of selected delegates manage the chain. However, this allows higher throughput and better scalability since not all the nodes have to participate in block creation and chain management.

Figure 10 shows a simplified version of DPoS. First, the stakeholders who possess any amount of stakes give their votes to delegate candidates. The more stake a stakeholder has, the more voting power it can exert. In the end, delegate candidates are ranked based on their stakes, and a predefined number (e.g., EOS: 21, Ark: 51, Lisk: 101) of candidates are elected as the group of delegates. The election is done periodically, and the stakeholders can freely reallocate their voting power to different candidates, and new delegates will be chosen. Also, the system can define a backup pool of delegates so that when a delegate in the main group is unable to participate in the block creation, it will be replaced by a backup delegate.

Within the group of delegates, the witnesses equally take turns (e.g., round-robin) creating and proposing new blocks. Thus, it is essential for the witness to always be online while she is working in the group. When a witness forges a new block, it can be added to the chain directly or go through a voting phase, as shown in Fig. 10. If the proposed block gets more than 2/3 of the votes, it is then appended to the current chain, and the next witness starts creating the next block. The rewards earned from participating as a delegate will be redistributed to the stakeholders who voted for the delegate. (Note that a portion of the rewards can be spent on other causes such as the platform development or charity.) This redistribution plan can also affect the stakeholders in placing their votes on candidates.

Other cryptocurrencies that utilize DPoS include EOS, BitShares, Ark, Lisk, and Tezos.

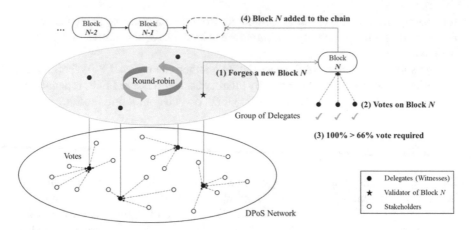

Fig. 10 How to reach consensus in DPoS is illustrated

Byzantine Fault-Tolerant PoS (BFT PoS)

BFT PoS is based on a round-based voting process. Unlike PoW or chain-based PoS algorithms where a consensus is reached through the chain's length, BFT PoS achieves consensus for every voting round whenever a block is forged.

Tendermint [49, 50] is the first BFT PoS that showed BFT consensus could be achieved with PoS. We focus on the Tendermint Core consensus engine that works as a round-based voting mechanism.

Similar to DPoS, the network requires a set of validators who maintain the chain and take turns proposing and committing new blocks for every block height. The validator responsible for the block proposal in a round is called a *proposer*, and other validators validate the proposed block and place votes. The application can define the voting power of the votes cast by the validators. In other words, Tendermint Core allows both even distribution of voting power per validator and weighted voting power based on stakes. In BFT PoS, we assume that the validators are selected by stakes or through delegation like in DPoS, and that their voting power and frequency of being a proposer are also proportional to the stakes involved.

Each round of creating a new block consists of three steps with equal timeouts: *propose, pre-vote, pre-commit*, and two special steps: *commit* and *new height*, as shown in Fig. 11.

In the *propose* step, a new block is forged by the current round's proposer (chosen in round-robin fashion with a weighted probability based on the stake) and broadcasted throughout the network. If the block is valid, every validator goes to the *pre-vote* step and broadcasts a pre-vote. Else, if the block is not valid, or a validator did not receive any block proposal within the defined timeout, it sends pre-vote nil.

During the *pre-vote* step, every validator waits and listens for pre-vote broadcasts and checks whether more than two-third of the voting power has been achieved on

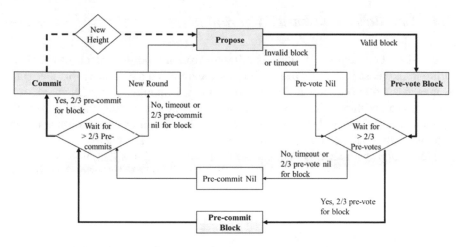

Fig. 11 A simplified state machine of Tendermint [50] is shown

pre-vote. If so, every validator moves on to the *pre-commit* step, broadcasts a pre-commit for that block, and *locks* itself to that block. On the contrary, if the validator did not receive two-third of the voting power on pre-vote until timeout, or received more than two-third of the voting power placed on pre-vote nil, it sends pre-commit nil. Here, we mentioned the concept of locking onto a block. Once a validator is locked on a block, it always sends a pre-vote for that block in future rounds of the same chain height. Also, it can only unlock that block if it receives a newer block of the same chain height with more than two-third of the voting power placed on pre-vote.

If the validator receives more than two-third of the voting power placed on pre-commit during the *pre-commit* step, it enters the *commit* step, where the block is finalized and added to the chain. The chain height is increased by one, and the next proposer starts preparing the next block. However, if two-third of the voting power is not reached during the *pre-commit* step, the next step is the *new round*. A new proposer is selected to propose a block of the same chain height. If the new proposer has locked herself to the block from the previous round, it would propose the same block, and other validators who have also locked themselves to this block would pre-vote this block.

As seen from the above sequence of steps, Tendermint relies heavily on achieving more than two-third of the voting power to achieve consensus. If more than one-third of voting power is somehow offline or showing Byzantine behaviors, the network will fall into an endless loop.

3.4   Vote-Based Consensus Algorithms

A vote-based consensus algorithm is different from a proof-based consensus algorithm in the sense that nodes responsible for creating and maintaining the chain are managed and often controlled. As mentioned in the beginning of Sect. 3, vote-based consensus algorithms mainly target permissioned and non-incentivized blockchain applications. As a result, we can approach vote-based consensus algorithms as we propose traditional methods for fault tolerance in distributed systems.

Nodes of a distributed system can suffer from crash faults (i.e., non-Byzantine faults), where they halt or disconnect from the network caused by hardware failures, broken network, or software issues. Alternatively, the nodes can maliciously forge or tamper with the information, which we refer to as Byzantine faults. Thus, the algorithms of this category can be further classified as being Crash Fault Tolerant (CFT) or Byzantine Fault Tolerant (BFT).

Current mainstream CFT algorithms include Paxos [51] and Raft [52], the latter of which is a simplified and implementation-friendly derivative of the former. CFT algorithms could not guarantee system reliability and resiliency in the presence of Byzantine faults and are thus used in a closed environment like IPFS [53] and IBM Hyperledger Fabric [54]. In this article, we focus on BFT algorithms that can tolerate both crash and Byzantine faults, providing better security and implementation flexibility.

First, we will present PBFT [20], representing a family of protocols for tolerating Byzantine faults. Various adaptations of PBFT include Redundant BFT (RBFT) [55], delegated BFT (dBFT) [56], and BFT-SMaRt [57]. Then, we will present HotStuff [58], a recent variation of PBFT that will be the base consensus algorithm for Facebook's upcoming blockchain payment system *Diem* (formerly known as *Libra*) [59].

Practical Byzantine Fault Tolerance

The original PBFT algorithm is written in the context of the Network File System (NFS), where nodes are addressed as primary and replicas, and a client's request triggers the execution of the state machine. Here, based on IBM's Hyperledger Sawtooth PBFT [60] implementation, we describe the functionality of PBFT as follows.

Normal case operation: The PBFT algorithm reaches a consensus through four phases: *pre-prepare, prepare, commit,* and *finish*, as shown in Fig. 12. In this example, Node 1 is the primary and Node 0 is a crash fault node. Because there are four nodes in this example, the network is both crash fault tolerant ($f = 1$, $4 \geq 2f + 1$) and Byzantine fault tolerant ($4 \geq 3f + 1$).

1. **Pre-prepare**: The primary node creates a new block and publishes it to the network. This is followed by the *pre-prepare* broadcast, which contains the block ID, the block number, view number, and the ID of the primary. Each node will verify the received block and the *pre-prepare* message; if it is valid, it will add those to its log and move on to the prepare phase.

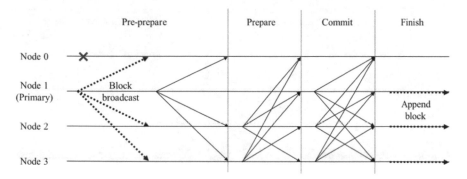

Fig. 12 A normal case operation of sawtooth PBFT [60] is illustrated

2. **Prepare**: In the prepare phase, the nodes will broadcast a *prepare* message that matches the received *pre-prepare* message to the rest of the network. Like the *pre-prepare* message, it contains the block ID and the block number, the node ID, and the view number. After sending the *prepare* message, the node will wait for *prepare* messages from other nodes. If a node has reached $2f + 1$ *prepare* messages (including the *pre-prepare* message sent by the primary) with the same block ID and number, the messages are logged, and it moves on to the commit phase.

3. **Commit**: This phase is similar to the prepare phase; the nodes broadcast a *commit* message to the network, indicating that it received and accepted *prepare* messages from two-third of the nodes or more. And again, the nodes wait until they receive $2f + 1$ matching *commit* messages from other nodes. When they reach the required number of confirmations, they move on to the finish phase. One difference between prepare phase and commit phase for the primary is that the primary is not allowed to broadcast the *prepare* message, whereas it can broadcast the *commit* message to the network after receiving $2f + 1$ *prepare* messages.

4. **Finish**: In the finish phase, the nodes will append the proposed block to their chain, increase the sequence number by one, and get ready to validate the next proposed block.

If the primary node happens to be the faulty node, there must be a way to replace the primary node to guarantee the algorithm's liveness. The primary node is concluded to be faulty in the following cases: no new block or *pre-prepare* message is sent by the primary within the timeout, a commit timeout occurs during the prepare phase, a view-change timeout occurs during the view-change operation (presented below shortly), multiple *pre-prepare* messages are received, or a *prepare* message is sent by the primary.

In any of the above cases, every node broadcasts a *view-change* message. If $f + 1$ *view-change* messages are received from other nodes (this is based on the assumption

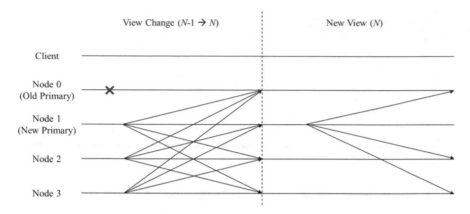

Fig. 13 A view-change operation of sawtooth PBFT is illustrated

that at most f nodes can be faulty, and receiving $f + 1$ messages indicates that regular nodes are joining the view-change), the faulty primary should be replaced by starting a view-change operation.

The view-change operation consists of two phases: view-change and new view, as shown in Fig. 13. Given the current view, $N-1$, if a node decides to start a view-change, it will broadcast a *view-change* message to the network with an updated view, N. If the primary node is indeed faulty, other non-faulty nodes will also broadcast the *view-change* messages. When the primary candidate for the new view, N, receives $2f + 1$ *view-change* messages from other nodes, it will broadcast a *new view* message for N and become the new primary, who can now start publishing blocks and send out *pre-prepare* messages.

The PBFT algorithm tolerates Byzantine faults through the broadcasts of *prepare* and *commit* messages within the network. This ensures that PBFT maintains consistency, availability, and immutability and achieves consensus. However, with the increase in the number of participating nodes, the number of broadcast messages increases quadratically, $O(n^2)$ for normal cases and $O(n^3)$ including view-change. This results in high communication overhead and degradation in performance, which makes PBFT difficult to deploy on a large scale. Many variants of PBFT try to solve this issue and achieve better scalability, one of which is HotStuff we will discuss in the following section.

HotStuff (DiemPBT)

HotStuff is a PBFT algorithm that provides linearity in the communication complexity of $O(n)$ for both normal cases and view-change cases, and view-change responsiveness. This is made possible through a few crucial design choices of HotStuff.

First, HotStuff changes the network topology from a mesh to a star, as shown in Fig. 14. As a result, the nodes no longer broadcast messages to each node; instead, it sends the message directly to the *leader* (primary in PBFT) node, significantly reducing the network's communication complexity.

Second, HotStuff is a leader-based protocol, where the view-change process is not separate but merged into the normal case process since a leader is rotated every round. Now, we will present the normal case process of basic HotStuff, as shown in Fig. 15. Note that the word *basic* is used to differentiate two types of HotStuff, which are *basic* HotStuff and *chained* HotStuff.

The basic HotStuff progresses through a series of views, which is incremented by one for every round. A round consists of the following phases: *prepare, pre-commit, commit,* and *decide*. A different leader exists for each view number, who sends out broadcasts for each phase and receives votes from other nodes. Nodes vote to a proposed branch (or block) by signing the branch with their private keys, and then

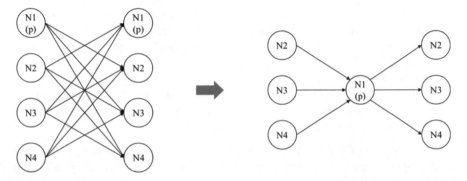

Fig. 14 While PBFT has a mesh network topology, HotStuff has a star topology. Here, p indicates primary node

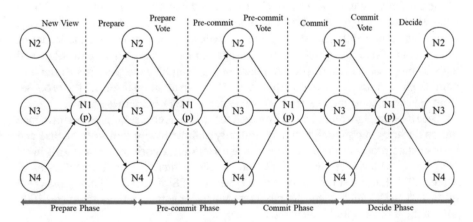

Fig. 15 The phases of basic HotStuff are shown

they send the votes to the leader. The leader collects $2f + 1$ valid votes and combines them into a *QuorumCertificate* (QC). The QC is required for each of the prepare, pre-commit, and commit phases where votes are sent by the nodes.

1. **Prepare**: The leader encapsulates the proposal into the *prepare* message and broadcasts it to the network. Every node verifies the proposal, and if accepted, returns a vote with a partial signature to the leader and moves on to the pre-commit phase.
2. **Pre-commit**: When the leader receives prepare votes from $2f + 1$ nodes for the current proposal, it combines them into a *prepareQC* and then broadcasts the *pre-commit* message to the network. The node receiving the *pre-commit* message returns a vote with a partial signature to the leader and moves on to the commit phase.
3. **Commit**: When the leader receives pre-commit votes from $2f + 1$ nodes for the pre-commit, it assembles them into a *precommitQC* and then broadcasts the *commit* message to the network. The nodes receiving the *commit* message lock their *lockedQC* to *precommitQC* and return votes with partial signatures to the leader and move on to the decide phase. The lock ensures that a consensus can be reached even if this round fails and goes through a view-change (i.e., new round).
4. **Decide**: When the leader receives commit votes from $2f + 1$ nodes for the commit, it merges them into a *commitQC* and broadcasts the *decide* message to the network. After receiving this message, the nodes execute the state transition of the proposal and move on to the next view number.

Note that a similar mechanism is repeated across the four phases; nodes vote (and sign) on a message, and the leader combines them and sends it in the following broadcasted message.

The idea of chained HotStuff is pipelining the basic HotStuff to achieve the high throughput. Specifically, the votes over the *prepare* phase are collected by the leader of the current view v_i into a *genericQC* (which is a generic version of the QCs from basic HotStuff). Then the *genericQC* is relayed to the leader of the next view v_{i+1}, delegating responsibility of starting the *pre-commit* phase. However, the leader of v_{i+1}, instead of making a *pre-commit* message, initiates a new *prepare* message and adds its proposal. This new prepare phase of v_{i+1} simultaneously serves as the pre-commit phase of the sequence started in v_i. After two additional relays, the proposal from v_i is decided as the leader of view v_{i+3} sends the *prepare* message.

The HotStuff algorithm, when compared to the PBFT, ensures higher performance due to linearity in communication complexity and the merging of normal case and view-change process into one. However, the HotStuff algorithm itself is still in its early stage, with limited evaluations done in the *testnet* for about 100 days as of March 2021. Early test results showed an average TPS of 10, with the highest value of 44 TPS [61], which is considerably less when compared to the anticipated 1000 TPS at the time of initial launch [62].

4 Evaluation

This section presents a summary of the various consensus algorithms presented in this article based on our evaluation framework.

Table 4 presents the evaluation of each criterion of the PoW algorithms (traditional, memory-hard, and multi-hash), PoS algorithms (pure PoS, PoW/PoS hybrid, DPoS, and BFT PoS), and vote-based algorithms (PBFT and HotStuff).

Throughput is not a fixed parameter that is unique to a specific category of algorithms. Even if two cryptocurrencies use the same consensus algorithm, the throughput will vary based on their latency, block size configuration, and even network size. For example, DASH started in 2014 with a block size of 1 MB, like Bitcoin. At first, DASH was able to make 28 transactions per second, max. In 2016, DASH decided (by a decentralized vote) to increase the block size to 2 MB. As a result, the current TPS of DASH tops at 56.

Overall, PoW algorithms have a relatively low throughput of one or two digit TPS. This is because they employ resource-intensive computations in reaching a consensus, which also results in their high energy consumptions. In contrast, PoS algorithms and vote-based algorithms have higher throughput of over three-digit TPS and low energy consumption due to their less computations and a fixed set of validators (reducing the accumulated energy consumption of the network).

Scalability can be evaluated in three aspects: the number of regular clients, the number of block validators, and the maximum throughput. First, a regular client who does not participate in the consensus process has minimal to no impact on the communication volume and the throughput. All consensus algorithms scale well to clients [63]. However, the situation changes a little as the number of block validators increases. Non-BFT algorithms offer great scalability due to simple communication protocols with linear communication complexity. However, BFT algorithms have several rounds/phases with a broadcast message from every participating node. This results in high communication overhead and degradation of the performance of BFT algorithms, which limits their scalability significantly. In our survey of BFT-related algorithms, PBFT, which has no predefined group of validators (i.e., primaries), offers low scalability. However, BFT PoS, DPoS, and HotStuff, each of which has a group of delegated validators, offer much better scalability. The last aspect is the throughput. If we take Bitcoin as an example, the block size limit is 1 MB, and each transaction's size is around 256 bytes on average. Thus, approximately 4,000 transactions can be contained on a block which is mined every 10 min. This leads to the maximum throughput of approximately 7 TPS. As noted earlier (Section "Performance Criteria"), reducing the latency or enlarging the block size could increase the throughput. Nevertheless, chain-based algorithms tend to achieve a TPS of under 100. On the other hand, vote-based algorithms have much higher TPS that promises better scalability, provided that their communication complexity is resolved.

In terms of fault (adversary) tolerance, chain-based algorithms all have $2f + 1$ tolerance since an adversary requires 51% of hash rate or stake in the network to

Table 4 Representative consensus algorithms are evaluated in terms of comprehensive criteria

	Consensus algorithm	Performance					Security			Decentralization	
		Throughput			Scalability	Energy consumption	Fault	Sybil protection	Double	Permissionless	Censorship Resistance
		TPS	Latency	Block size			(Adversary) tolerance		Spending prevention		
Proof of Work	Traditional[a] (Nakamoto)	5–7	1 0m	1 MB	High	High	$2f+1$	O	X	O	High
	ASIC-resistant/Memory-hard[b] (Ethash)	13–15 (Max 30)	15 s	Dynamic	High	High	$2f+1$	O	X	O	High
	ASIC-resistant/Multi-hash[c] (X-series)	~56 (Max)	2 m 30 s	2 MB	High	High	$2f+1$	O	X	O	High
Proof of Stake	Chain-based/Pure PoS[d]	100	1m	32 KB	High	Low	$2l+1$	O	X	O	High
	Chain-based/PoW/PoS Hybrid[e]	8	8 m 30 s	1 MB	High	Mid	$2f+1$	O	X	O	High
	DPoS[f]	1000+	0.5 s	Dynamic	High	Low	$3l+1$ $2f+1^s$ >	O	O	O/x	High
	BFT PoS[g]	~14,000	(6 + 0.57?)s R: rounds	N/A	Mid	Low	$3l+1$ $2f+1^s$ 1»	O	O	O/x	High
Vote-based	PBFT[h]	Mid	Low	N/A	Low	Low	$3l+1$	O	O	X	Low
	HotStuff[i]	High	Low	N/A	High	Low	$3l+1$	O	O	X	Low

a Bitcoin
b Ethereum
c DASH
d Nxt
e Peercoin
f EOS
g Tendermint
h 3f + 1 for validators of the voting process, 2f + 1 for delegate selection
i TPS, Latency was qualitatively compared to proof-based algorithms, (e.g., Diem: ~1000 TPS, 10 s latency)

have malicious influence in the mining/minting process. Meanwhile, vote-based algorithms have $3f + 1$ tolerance due to their Byzantine fault-tolerant property.

Interestingly, fault (adversary) tolerance of DPoS and BFT PoS can be approached in two ways. First, if we focus on the election process, the algorithms have $2f + 1$ tolerance since an adversary with more than 51% of stake in the network could take advantage of the election. On the other hand, if we focus on the delegates' voting process, we need $3f + 1$ Byzantine fault tolerance.

As mentioned earlier (Sect. 3.2), proof-based algorithms come with an anti-Sybil attack at their core. Launching Sybil attacks involves multiple fraudulent identities, but proof-based algorithms do not depend on the number of identities; instead, it is the amount of resource/cost that influences the network operations. On the contrary, vote-based algorithms are vulnerable to Sybil attacks if there is no additional PoS or PoW admission mechanisms such as DPoS and BFT PoS [64]. This can be related to decentralization properties, where algorithms that have Sybil protection can be deployed in a permissionless environment with high resistance to censorship. In contrast, those with no Sybil protection should be permissioned, but a closed permissioned system (most likely private or consortium blockchains) may be prone to censorship.

Double-spending prevention can be related to the *finality* mechanism of the algorithm. Chain-based algorithms have *probabilistic finality*, where the immutability of the transaction increases with the number of block confirmations and can be targeted for a double-spending attack. In contrast, the algorithms that utilize the voting process have *deterministic finality*, and a transaction is finalized at the point of block creation, mitigating such attacks.

5 Conclusion

A consensus algorithm lies at the core of a blockchain and offers many research opportunities. The design of a consensus algorithm directly impacts the performance and characteristics of the developed blockchain applications. Thus, it can be argued that a thorough evaluation of a consensus algorithm is mandatory. After reviewing the literature on consensus algorithms and their evaluation criteria, we proposed an evaluation framework with comprehensive criteria for consensus algorithms in terms of performance, security, and decentralization. An in-depth evaluation framework targeted for each group of consensus algorithms (e.g., BFT variants, ASIC-resistant PoWs) is proposed as future work. We also presented the operations and characteristics of representative consensus algorithms and evaluated them with our proposed framework. We infer from our results that no consensus algorithm deems ideal for all situations. Thus, finding a well-balanced consensus algorithm that satisfies given requirements will be crucial for wide-scale adoption of blockchains across various industries and businesses.

References

1. Lam, E.: Bitcoin Hits $1 trillion value as crypto leads other assets. https://www.bloomb erg.com/news/articles/2021-02-19/bitcoin-nears-1-trillion-value-as-crypto-jump-tops-other-assets (2021). Last accessed15 Mar 2021
2. Buchholz, K.: How common is crypto? https://www.statista.com/chart/18345/crypto-currency-adoption/ (2021). Last accessed 15 Mar 2021
3. PwC: Time for trust: how blockchain will transform business and the economy. https://www.pwc.com/hu/en/kiadvanyok/assets/pdf/Time_for_Trust_The%20trillion-dollar_reasons_to_r ethink_blockchain.pdf (2020). Last accessed 15 Mar 2021
4. Zheng, Z., et al.: An overview of blockchain technology: architecture, consensus, and future trends. In: 2017 IEEE international congress on big data, IEEE, pp. 557–564 (2017)
5. Alsunaidi, S.J., Alhaidari, F.A.: A survey of consensus algorithms for blockchain technology. In: 2019 International Conference on Computer and Information Sciences (ICCIS), IEEE, pp. 1–6 (2019)
6. Croman, K., et al.: On scaling decentralized blockchains. In: International conference on financial cryptography and data security, pp. 106–125. Springer, Berlin, Heidelberg (2016)
7. Mingxiao, D., et al.: A review on consensus algorithm of blockchain. In: 2017 IEEE international conference on systems, man, and cybernetics (SMC), IEEE, pp. 2567–2572 (2017)
8. Nguyen, G.-T., Kim, K.: A survey about consensus algorithms used in blockchain. J. Inform. Proc. Syst. **14**(1), 101–128 (2018)
9. Hasanova, H., et al.: A survey on blockchain cybersecurity vulnerabilities and possible countermeasures. Int. J. Network Manage. **29**(2), e2060 (2019)
10. Bano, S., et al.: SoK: Consensus in the age of blockchains. In: Proceedings of the 1st ACM Conference on Advances in Financial Technologies, pp. 183–198 (2019)
11. Ferdous, M.S., et al.: Blockchain consensuses algorithms: a survey (2020). arXiv preprint arXiv:2001.07091
12. Bamakan, S.M.H., Motavali, A., Bondarti, A.B.: A survey of blockchain consensus algorithms performance evaluation criteria. Expert Syst. Appl. 113385 (2020)
13. Chaudhry, N., Yousaf, M.M.: Consensus algorithms in blockchain: Comparative analysis, challenges and opportunities. In: 2018 12th International Conference on Open Source Systems and Technologies (ICOSST), IEEE, pp. 54–63 (2018)
14. Coincheckup. https://coincheckup.com/. Accessed 15 March 2021
15. Bitinfocharts. https://bitinfocharts.com/comparison/confirmationtime-btc-ppc.html. Accessed 15 March 2021
16. Wan, S., et al.: Recent advances in consensus protocols for blockchain: a survey. Wireless Netw. **26**(8), 5579–5593 (2020)
17. Digiconomist: Bitcoin energy consumption index. https://digiconomist.net/bitcoin-energy-con sumption (2021). Last Accessed 15 Mar 2021
18. Mora, C., et al.: Bitcoin emissions alone could push global warming above 2 C. Nat. Clim. Chang. **8**(11), 931–933 (2018)
19. Lamport, L., Shostak, R., Pease, M.: The Byzantine generals problem. In: Concurrency: the Works of Leslie Lamport, pp. 203–226 (2019)
20. Castro, M., Liskov, B.: Practical byzantine fault tolerance. In: OSDI, vol. 99, pp. 173–186. (1999)
21. Nakamoto, S.: Bitcoin: A peer-to-peer electronic cash system. (2008) https://bitcoin.org/bit coin.pdf Last accessed 15 Mar 2021
22. Eyal, I., Sirer, E.G.: Majority is not enough: Bitcoin mining is vulnerable. In: International conference on financial cryptography and data security, pp. 436–454. Springer, Berlin, Heidelberg (2014)
23. Blockchaininfo: https://www.blockchain.com/charts/pools (2022). Last accessed 14 Feb 2022
24. Douceur, J.R.: The sybil attack. In: International workshop on peer-to-peer systems, pp. 251–260. Springer, Berlin, Heidelberg (2002)

25. Neudecker, T., Hartenstein, H.: Network layer aspects of permissionless blockchains. IEEE Communications Surveys & Tutorials **21**(1), 838–857 (2018)
26. Mohaisen, A., Kim, J.: The sybil attacks and defenses: a survey (2013). arXiv preprint arXiv: 1312.6349
27. Zhang, S., Lee, J.-H.: Double-spending with a sybil attack in the bitcoin decentralized network. IEEE Trans. Industr. Inf. **15**(10), 5715–5722 (2019)
28. Coinmarketcap: What is censorship resistance? https://coinmarketcap.com/alexandria/article/what-is-censorship-resistance (2020). Last accessed 15 Mar 2021
29. Back, A.: Hashcash-a denial of service counter-measure. ftp://sunsite.icm.edu.pl/site/replay.old/programs/hashcash/hashcash.pdf (2002). Last accessed 15 Mar 2021
30. Cho, H.: SIC-resistance of multi-hash proof-of-work mechanisms for blockchain consensus protocols. IEEE Access **6**, 66210–66222 (2018)
31. Wood, G.: Ethereum: a secure decentralised generalised transaction ledger. Ethereum project yellow paper. https://ethereum.github.io/yellowpaper/paper.pdf (2021). Last accessed 15 Mar 2021
32. Higgins, S.: Bitmain confirms release of first ethereum ASIC miners. https://www.coindesk.com/bitmain-confirms-release-first-ever-ethereum-asic-miners. (2018) Last accessed 15 Mar 2021
33. O'Neal, S.: ETH miners will have little choice once ethereum 2.0 launches with PoS. https://cointelegraph.com/news/eth-miners-will-have-little-choice-once-ethereum-20-launches-with-pos (2020). Last accessed 15 Mar 2021
34. Percival, C.: Stronger key derivation via sequential memory-hard functions. https://www.tarsnap.com/scrypt/scrypt.pdf (2009). Last accessed 15 Mar 2021
35. Percival, C., Josefsson, S.: The scrypt password-based key derivation function, RFC 7914. https://tools.ietf.org/html/rfc7914 (2016). Last accessed 15 Mar 2021
36. Litecoin: https://litecoin.org/ (2021). Last accessed 15 Mar 2021
37. Dogecoin: https://dogecoin.com/ (2021). Last accessed 15 Mar 2021
38. Medium: What is memory-hard? https://medium.com/Linzhi/what-is-memory-hard-45a363b59dfe (2019). Last accessed 15 Mar 2021
39. DASH: X11 Hash algorithm. https://docs.dash.org/en/stable/introduction/features.html#x11-hash-algorithm (2021). Last accessed 15 Mar 2021
40. Bertoni, G., et al.: Keccak specifications. Submission to NIST (round 2), pp. 320–337. (2009)
41. Black, T., Weight, J.: X16R ASIC resistant by design. https://ravencoin.org/assets/documents/X16R-Whitepaper.pdf (2018). Last accessed 15 Mar 2021
42. Colvin, G., Lanfranchi, A., Carter, M.: EIP-1057: ProgPoW, a programmatic proof-of-work, ethereum improvement proposals, no. 1057. https://eips.ethereum.org/EIPS/eip-1057 (2018). Last accessed 15 Mar 2021
43. Medium: 'Loaded' PoW: a new direction in proof-of-work algorithms. https://jeffreyemanuel.medium.com/loaded-pow-a-new-direction-in-proof-of-work-algorithms-ae15ae2ae66a (2018). Last accessed 15 Mar 2021
44. Bitcointalk: Proof of stake instead of proof of work. https://bitcointalk.org/index.php?topic=27787.0 (2011). Last accessed 15 Mar 2021
45. King, S., Nadal, S.: PPcoin: peer-to-peer crypto-currency with proof-of-stake. https://decred.org/research/king2012.pdf (2012). Last accessed 15 Mar 2021
46. Nxt: Nxt whitepaper. https://nxtdocs.jelurida.com/Nxt_Whitepaper (2021). Last accessed 15 Mar 2021
47. PeerCoin: PeerCoin docs. https://docs.peercoin.net/ (2021). Last accessed 15 Mar 2021
48. Larimer, D.: Delegated proof-of-stake (dpos). Bitshare whitepaper **81**, 85 (2014)
49. Kwon, J.: Tendermint: Consensus without mining. Draft v. 0.6, fall 1(11) (2014)
50. Tendermint: Tendermint core. https://docs.tendermint.com/master/ (2021). Last accessed 15 Mar 2021
51. Lamport, L.: Paxos made simple. ACM Sigact News **32**(4), 18–25 (2001)
52. Ongaro, D., Ousterhout, J.: In search of an understandable consensus algorithm. In: 2014 USENIX Annual Technical Conference, pp. 305–319 (2014)

53. Benet, J.: Ipfs-content addressed, versioned, p2p file system (2014). arXiv preprint arXiv: 1407.3561
54. Hyperledger Fabric: release-2.2. https://hyperledger-fabric.readthedocs.io/en/release-2.2/ (2020). Last accessed 21 Mar 2021
55. Aublin, P.-L., Mokhtar, S.B., Quéma, V.: Rbft: Redundant byzantine fault tolerance. In: 2013 IEEE 33rd International Conference on Distributed Computing Systems, IEEE, pp. 297–306 (2013)
56. NeoReserach: Delegated byzantine fault tolerance: technical details, challenges and perspectives. https://github.com/NeoResearch/yellowpaper/blob/master/sections/08_dBFT. md (2019). Last accessed 21 Mar 2021
57. Bessani, A., Sousa, J., Alchieri, E.E.: State machine replication for the masses with BFT-SMART. In: 2014 44th Annual IEEE/IFIP International Conference on Dependable Systems and Networks, IEEE, pp. 355–362 (2014)
58. Yin, M., et al.: Hotstuff: Bft consensus with linearity and responsiveness. In: Proceedings of the 2019 ACM Symposium on Principles of Distributed Computing, pp. 347–356 (2019)
59. Diem Association: Diem White Paper v2.0. https://www.diem.com/en-us/white-paper/ (2021). Last accessed 21 Mar 2021
60. Hyperledger Sawtooth: Sawtooth PBFT. https://sawtooth.hyperledger.org/docs/pbft/releases/ latest/index.html (2018). Accessed 15 Mar 2021
61. InDiem Blockchain Explorer: https://indiem.info/ (2021). Accessed 21 Mar 2021
62. Amsden, Z., et al.: The libra blockchain. https://mitsloan.mit.edu/shared/ods/documents/?Pub licationDocumentID=5859 (2019). Last accessed 21 Mar 2021
63. Vukolić, M.: The quest for scalable blockchain fabric: Proof-of-work vs. BFT replication. In: International workshop on open problems in network security, Springer, pp. 112–125 (2015)
64. Natoli, C., et al.: Deconstructing blockchains: A comprehensive survey on consensus, membership and structure (2019). arXiv preprint arXiv:1908.08316

Blockchain Incentive Design and Analysis

Jianyu Niu and Chen Feng

Abstract The Bitcoin white paper introduced the blockchain technology to real-ize a decentralized electronic cash system that does not rely on a central authority. A major novelty behind the technology is the incentive design, in which partici-pating nodes obtain rewards by creating blocks in a longest chain. The incentive design is paramount for a secure blockchain system as shown in many recent works. In this chapter, we take a close look at the incentive design of three influential blockchain protocols including Bitcoin, Ethereum, and Bitcoin-NG. For each pro-tocol, we present the potential incentive-based attacks and go through several the-oretical results to characterize the impact of these attacks. We hope that at the end of our journey, our readers can have a deeper understanding of blockchain incentive designs and analysis.

1 Introduction

In 2008, Nakamoto invented Bitcoin in the seminal paper titled "Bitcoin: A Peer-to-Peer Electronic Cash System" [24]. Bitcoin is aimed to realize a decentralized electronic cash system that does not rely on a central authority for currency issuance and transaction processing. One year later, the Bitcoin network was launched based on the open-source code developed by Nakamoto. By now, as the biggest cryptocurrency, Bitcoin has a market value of more than 1 trillion US dollars by the dollar/bitcoin exchange rate. The huge success of Bitcoin is largely owing to two technological innovations: the so-called *Nakamoto consensus* (NC) protocol, which can realize a

J. Niu (✉)
Southern University of Science and Technology, Shenzhen, China
e-mail: niujy@sustech.edu.cn

C. Feng
The University of British Columbia (Okanagan Campus), Kelowna, Canada
e-mail: chen.feng@ubc.ca

© The Author(s), under exclusive license to Springer Nature Switzerland AG 2022   119
D. A. Tran et al. (eds.), *Handbook on Blockchain*, Springer Optimization
and Its Applications 194, https://doi.org/10.1007/978-3-031-07535-3_4

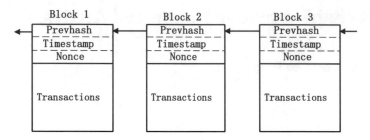

Fig. 1 An illustration of the blockchain data structure. Each block contains a batch of transactions, a hash value of its previous block, and other metadata

public, immutable, and distributed ledger widely known as a blockchain today, and the incentive design, by which nodes obtain rewards from participation. NC protocol has been discussed in other chapters. In this chapter, we will focus on the blockchain incentive design and analysis. This chapter is based closely on our previous work of incentive analysis for Ethereum [12] and Bitcoin-NG [27] as well as the first author's Ph.D. dissertation.

In a blockchain, users can create and exchange transactions to modify their monetary amounts, and new transactions can be batched into blocks for processing. Particularly, each block is linked to its previous block with a cryptographic hash, forming a chain of blocks accepted by the users (which explains the name of blockchain). A simple illustration of the blockchain data structure in Bitcoin is provided in Fig. 1. In addition to transactions and hash reference, each block also contains a timestamp, nonce (i.e., a random string), and other metadata. Here, the nonce is produced by solving computational puzzles, which is often known as Proof of Work (PoW). The searching process for the valid nonce is also called mining, and the participants are called miners. Miners follow the longest chain rule (LCR) to reach an agreement on an increasing sequence of blocks. That is, miners always extend the longest chain that they have received. Informally speaking, LCR together with PoW form the Nakamoto consensus. Despite the simplicity, NC can make blockchains securely work as long as more than half of the computation power is controlled by honest miners who follow the protocol.

When participating in NC-based blockchains, miners have to pay for the computing hardware (e.g., CPU, GPU, or ASIC), electricity, and other fees. The economic cost makes it impractical for any miners to voluntarily support the protocol. To solve this dilemma, a (public) blockchain system often relies on its incentive design, by which miners can receive a block reward, i.e., some amount of self-issued tokens, for every block included in the blockchain. In addition, miners can also get the transaction fees for all the transactions that are contained in the block [24]. These rewards can pay for the mining cost, making mining profitable and incentivizing miners to contribute as much computation power as possible. Here blockchain protocol adopts an implicit assumption that all the miners are individually rational [16, 24]. As stated previously, the security of NC relies on the assumption that the majority of computation power is controlled by honest miners. To guarantee the rationality of being honest, NC should ensure *incentive compatibility*, i.e., miners will suffer from economic loss whenever they deviate from the protocols. In other words, without incentive compatibility, rational miners would deviate from the protocol to obtain higher revenue such that the above assumption does not hold, and, consequently, the system security will be threatened. All of these show the importance of the blockchain incentive design. In this chapter, we take a journey into blockchain incentive design and analysis, with a focus on three influential blockchain protocols: Bitcoin, Ethereum, and Bitcoin-NG. Ethereum and Bitcoin-NG are two variants of Bitcoin; both of them are designed based on NC and the incentive design of Bitcoin. For each protocol, we present the potential incentive-based attacks and go through several theoretical results to show the impact of these attacks. We hope that at the end of our journey, readers can have a deeper understanding of blockchain incentive designs and analysis.

The rest of this chapter is organized as follows. In Sect. 2, we present incentive design and analysis of Bitcoin, particularly on the selfish mining analysis from [11]. We then provide incentive analysis of Ethereum with a focus on comparing its reward design with Bitcoin's in Sect. 3 and conduct a comprehensive incentive analysis of Bitcoin-NG in Sect. 4. We provide further readings on blockchains' incentive design and analysis in Sect. 5 and conclude this chapter in Sect. 6.

2 Incentive Design and Analysis in Bitcoin

In the Bitcoin white paper, miners without coordination are suggested to follow the protocol, i.e., obeying NC to mine blocks and immediately publishing their blocks [24]. In this way, all miners are believed to be fairly rewarded; they obtain revenue in proportion to their computation power. This also encourages miners to devote more computation power to protocol. However, in reality, miners join in mining pools and share rewards to maintain stable incomes. Nowadays, the top six mining pools in Bitcoin control almost 75% of the total computation power.[1] The trend of centralization in computation power is not only against the goal of Bitcoin (i.e., a decentralized electronic cash system) but also raises concerns for incentive issues. Particularly, many studies show that a set of colluding miners who deviate from the protocol may obtain a revenue larger than their fair share from honest mining [11, 15, 17, 25, 33]. Such behavior is called *selfish mining* and the corresponding miners are called the selfish miners.

---

[1] Mining pool statistics: https://btc.com/stats/pool.

Fig. 2 An illustration of the chain structure stored in a miner. The miner tries to produce a block on the longest chain it observes

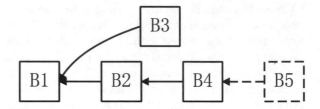

2.1 Overview of Bitcoin

Bitcoin relies on Nakamoto Consensus (NC) to make a group of distributed and mutually distrusting participants reach an agreement on a transparent and immutable ledger, also known as a blockchain. A blockchain is a list of blocks linked by hash values with each block containing a batch of ordered transactions. To make all participants agree on the same chain of blocks, NC leverages two components: the Proof-of-Work (PoW) mechanism and the longest chain rule (LCR). Each participant in NC (also referred to as a miner) collects valid and unconfirmed transactions from the network, then orders and packs these transactions into a block. A valid block needs to contain a proof of work, i.e., its owner needs to find a value of the nonce (i.e., a changeable data filed) such that the hash value of this block has required leading zeros [24]. The length of leading zeros is also known as the mining difficulty, which determines the probability to find a valid nonce of each computation try. Besides, the mining difficulty can be tuned by the system to make sure new blocks are mined every 10 min on average.

Once a new block is produced, it will be immediately broadcasted to the entire network. Ideally, the block should be accepted by all participants before the next block is produced. In reality, two new blocks might be mined around the same time, leading to a fork in which two "child" blocks share a common "parent" block, see Fig. 2 for an illustration. To resolve such a fork, an honest miner always accepts the longest chain as the valid one and mines after the last block of it. Particularly, if multiple chains have the same length, miners choose to mine after the one that it has first received. The forks do not last forever because the longest branch will win the competition and be accepted by all miners. The common prefix of all the longest chains of miners is called the *main chain*. In Bitcoin, a block miner will receive a block reward (if its block is eventually included in the system main chain) as well as transaction fees as another type of reward. These rewards encourage miners to devote their computational resources to the system.

2.2 Selfish Mining in Bitcoin

The selfish mining idea was first proposed in the Bitcoin forum [15]. Later, Eyal and Sirer formally describe and analyze the selfish mining attack [11]. In such an

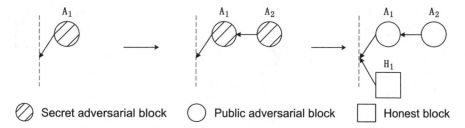

Fig. 3 The first example of the selfish mining attack in Bitcoin

attack, selfish miners first keep their newly discovered blocks private, creating a hidden branch. When some honest miners mine a new block, the selfish miners will immediately publish some private blocks, trying to make its branch the longest one. In this way, some honest miners' blocks will be abandoned (for not included in the longest chain), and the selfish miners could obtain a higher fraction of the block rewards than that from honest mining. Let us use two examples to better illustrate how the attack works. Figure 3 shows a case in which selfish miners first successfully mine two consecutive blocks A_1 and A_2 on the longest chain, and then it publishes these two blocks when receiving a block H_1 from an honest miner. In this way, the selfish miners will not only get two-block rewards (for blocks A_1 and A_2) but also make the honest block H_1 abandoned by all the miners. If the selfish miners can always replay this case, they certainly can obtain a higher fraction of block rewards than that from honest mining.

However, the selfish miners don't always take such a lead of two blocks and make their blocks accepted by all miners like the above case. Figure 4 shows a case in which the selfish miners successfully mine a block A_1 and then publish this block to match an honest block H_1. These two blocks form two forking branches of the same length, and miners cannot decide which one is the longest. To resolve the forking competition, miners shall continue mining and wait until a longer one wins. As said previously, under such a case, honest miners will mine after the first branch that it has received, whereas the selfish miners will mine after their branch. Therefore, the next block can be further divided into three subcases:

(a) The selfish miners produce the next blocks A_2 after their block and publish it. By LCR, all selfish miners' blocks are accepted by honest miners, whereas honest block H_1 is abandoned.
(b) Some honest miners produce the next block H_2 after the previous honest block H_1. If the selfish miners accept these two honest blocks H_1 and H_2 and mine after them, its block A_1 will be abandoned.
(c) Some honest miners produce the next block H_2 after the selfish miners' block A_1. Both blocks A_1 and H_2 will be accepted, whereas the block H_1 will be abandoned by LCR.

From this example, we can find that selfish miners' first block A_1 has the risk to be abandoned (shown in Fig. 4c). In other words, the selfish miners may suffer the loss of

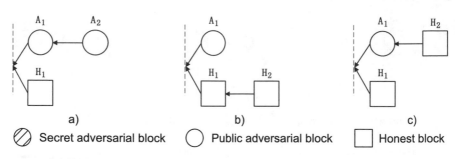

Fig. 4 The second example of the selfish mining attack in Bitcoin

block rewards in some cases, and so the attack is not always profitable. Moreover, we can see that the success of making the selfish mining attack profitable is determined by two parameters: the fraction of computation power owned by the selfish miners (denoted by α) and the fraction of honest computation power working on the selfish miners' blocks during a tie (denoted by γ). When α increases, the possibility for the selfish miners to take a lead of two blocks and to win the forking race during a tie also increases (in which the selfish miners have no loss). On the other hand, when γ is larger, the possibility that the selfish miners' blocks are abandoned is smaller. Particularly, when $\gamma = 1$ (all honest miners first receive the selfish miners' block A_1), the selfish miners' block A_1 is always accepted by all miners.

The above examples just illustrate several cases in the selfish mining attack. We can further design a complete selfish mining strategy covering all cases. In the following section, we will go through the selfish mining strategy in [11] and present the associated theoretical results. Also, note that the above strategy is not optimal under different α and γ. For example, in Fig. 4b, when α is small, the optimal strategy for the selfish miners is to accept these two honest blocks. But, when α is large, the optimal strategy for the selfish miners may be mining on its block A_1, trying to catch up (also called stubborn mining [25]). We refer our readers who are interested in the optimal selfish mining strategy to [17, 25, 33] for more details. Note that since the block generation rate is kept at a block every 10 min due to the mining difficulty adjustment, the number of mined blocks (and block rewards) is fixed during a period with or without attacks. Therefore, the higher fraction of rewards, the higher the mining revenue.

2.3   Theoretical Results on Selfish Mining in Bitcoin

Let us begin with the system model of selfish mining in Bitcoin. The system contains two types of miners: the honest, who follow the protocol, and the selfish, who may deviate from the protocol to maximize their profit. Let α denote the fraction of computation power controlled by selfish miners and β denote the fraction of total computation power controlled by honest miners. Without loss of generality, all selfish miners are assumed to be controlled by a single selfish miner. The mining process is modeled as a Poisson process with rate f. Accordingly, the selfish miner generates blocks at rate αf, and the honest miners generate blocks at rate βf. Blocks produced by honest miners are referred to as *honest* blocks, and blocks produced by the selfish miner are referred to as *adversarial* blocks.

The time an honest block takes to arrive at all miners is assumed to be negligible in the following analysis. This is because the block interval (i.e., 10 min on average in Bitcoin) is much larger than the propagation delay (usually tens of seconds). As introduced previously, γ is used to denote the ratio of honest miners that are mining on blocks produced by the selfish miner (rather than by the honest miners) whenever they observe a fork of two branches of equal length. Here, γ captures the selfish miner's communication capability and is assumed to be in the range [0, 1]. What is more, since transaction fees in a block are usually much smaller than the block reward, they are not considered in the following analysis.

Next, we present the selfish mining strategy in [11], as shown in Algorithm 1. We use $L_s(t)$ (resp., $L_h(t)$) to denote the length of the private branch (resp., public branches) seen by the selfish miner (resp., honest miners) at time t. In the beginning, we assume that the selfish miner and honest miners have the consensus of the same chain (lines 1). When the selfish miner mines a new block (see lines 2 to 8), it will

---

Algorithm 1 The selfish mining strategy in bitcoin

on Consensus
1: $(L_s, L_h) \leftarrow (0, 0)$

on The selfish miner mines a new block
2: $L_s \leftarrow L_s + 1$
3: **if** $(L_s, L_h) = (2, 1)$ **then**
4:    publish its private branch
5:    $(L_s, L_h) \leftarrow (0, 0)$ (since all the miners achieve a consensus)
6: **else**
7:    keep mining on its private branch
8: **end if**

on Some honest miners mine a new block
9: $L_h \leftarrow L_h + 1$
10: **if** $L_s < L_h$ **then**
11:    keep mining on this new block
12:    $(L_s, L_h) \leftarrow (0, 0)$
13: **else if** $L_s = L_h$ **then**
14:    publish the block of the private branch
15:    $(L_s, L_h) \leftarrow (1, 1)$
16: **else if** $L_s = L_h + 1$ **then**
17:    publish its private branch
18:    $(L_s, L_h) \leftarrow (0, 0)$ (since all the miners achieve a consensus)
19: **else**
20:    publish first unpublished block in its private branch
21:    set $(L_s, L_h) = (L_s - L_h, 0)$
22: **end if**

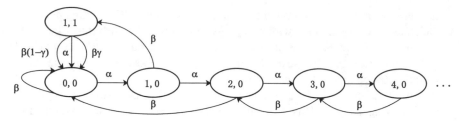

Fig. 5 The Markov process of the selfish mining in Bitcoin

keep this block private and continue mining on its private branch until its advantage is very limited (i.e., $(L_s, L_h) = (2, 1)$) which will be discussed later.

When some honest miners mine a new block, the length of a public branch will be increased by 1. We have the following cases. Case (1) If the new public branch is longer than the private branch, the selfish miner will adopt the public branch and mine on it. (That is why the selfish miner will set $(L_s, L_h) = (0, 0)$.) Case (2) If the new public branch has the same length as the private branch, the selfish miner will publish its private block immediately hoping that as many honest miners will choose its private branch as possible (since honest miners will see two branches of the same length when the private branch is published). Case (3) If the new public branch is shorter than the private branch by just one, the selfish miner will publish its private branch so that all the honest miners will adopt the private branch. Case (4) If the new public branch is shorter than the private branch by at least two, the selfish miner will publish the first unpublished block since the selfish miner still has a clear advantage. In addition, as the selfish miner will eventually win the competition (all its blocks are in the longest chain), (L_s, L_h) is set to $(L_s - L_h, 0)$.

With the above selfish mining strategy, we can use $(L_s(t), L_h(t))$ to capture the system state at time t. It is easy to verify that $(L_s(t), L_h(t))$ evolves as a Markov process under our selfish mining strategy, as illustrated in Fig. 5.[2] Moreover, the process $(L_s(t), L_h(t))$ is positive recurrent and so it has a unique stationary distribution. Let $\{\pi_{i,j}\}$ be the steady-state distribution of the Markov process $(L_s(t), L_h(t))$. By solving the above Markov process, we can compute the stationary distribution as shown in the following lemma.

Lemma 1 ([11]) *Given the selfish mining strategy in Algorithm 1, the stationary distribution of the states are*

---

[2] In Eyal and Sirer [11], the system states of the selfish mining process is denoted by using only $L_s(t)$. Here, we use $(L_s(t), L_h(t))$ as system states to keep consistent with that in Ethereum.

$$\pi_{0,0} = \frac{1 - 2\alpha}{2\alpha^3 - 4\alpha^2 + 1},$$

$$\pi_{1,0} = \frac{\alpha - 2\alpha^2}{2\alpha^3 - 4\alpha^2 + 1},$$

$$\pi_{1,1} = \frac{(1 - \alpha)(\alpha - 2\alpha^2)}{1 - 4\alpha^2 + 2\alpha^3}, \tag{1}$$

$$\pi_{i,0} = \left(\frac{\alpha}{1 - \alpha}\right)^{k-1} \frac{\alpha - 2\alpha^2}{1 - 4\alpha^2 + 2\alpha^3}, \; for\, i \geq 1,$$

With the stationary distribution of the state space, we next can analyze revenues obtained by the selfish miner and by the honest miners, respectively. To achieve this, we first need to determine the rewards for published blocks in each state transition. For example, given the state transition from $(2, 0) \rightarrow (0, 0)$, there are one honest block and two adversarial blocks that are published. By LCR, the selfish miner will obtain two-block rewards, whereas honest miners obtain nothing (see the case in Fig 3). For brevity, we don't provide the reward analysis and refer interested readers to [11] for more details. Finally, we can derive the relative revenue for the selfish miner in the following theorem.

Theorem 1 ([11]) *Given the selfish mining strategy in Algorithm 1, the relative revenue for the selfish miner is* $\frac{\alpha(1-\alpha)^2(4\alpha+\gamma(1-2\alpha))-\alpha^3}{1-\alpha(1+(2-\alpha)\alpha)}$.

When the selfish miner's revenue given in Theorem 1 is larger than α, the selfish miner will earn more from the selfish mining strategy than that from honest mining. Here, we usually assume $0 \leq \alpha < 1/2$. Since when $\alpha \geq 1/2$, the selfish miners may launch the famous double-spending attack, by which it can gain more [24]. We can derive the range of α for making selfish mining profitable in the following corollary.

Corollary 1 ([11]) *For a given* γ, *the selfish miner with* α *fraction of computation power can obtain a revenue larger than its honest mining in the following range:*

$$\frac{1 - \gamma}{3 - 2\gamma} < \alpha < \frac{1}{2}. \tag{2}$$

To better understand Theorem 1 and Corollary 1, we first plot the selfish miner's relative revenue for different γ with α ranging from 0 to 0.5 in Fig. 6a. As stated previously, the selfish miner has the risk of losing block rewards when it has only one private block and publishes this block to match an honest block. In this case, if the selfish miner always propagates its blocks more quickly than honest miners (i.e., $\gamma = 1$), all miners will mine on the selfish miner's block. This means that a selfish miner takes no risk when launching the selfish mining strategy and so the selfish miner with any fraction of computation power can benefit by launching the selfish mining strategy. The minimum fraction of computation power (referred to as

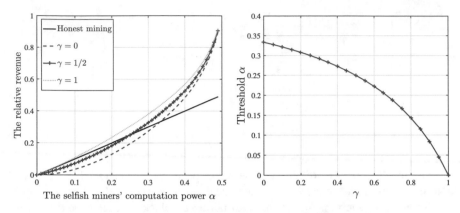

Fig. 6 The results of the selfish mining attack in Bitcoin. **a** The selfish miner's relative revenue with different α. **b** The threshold of the selfish minimg attack with different γ

threshold) making the selfish mining profitable is zero. contrary, when $\gamma = 0$ (honest miners always publish and propagate their block first), the threshold is at $1/3$. Hence, a selfish miner with more than one-third fraction of computation power can gain more revenue from the selfish mining attack. The threshold of selfish mining strategy is shown in Fig. 6b.

3 Incentive Design and Analysis in Ethereum

Ethereum is the second largest cryptocurrency by market capitalization and today's biggest decentralized platform that runs smart contracts. Ethereum currently uses a variant of NC as its underlying consensus but has a different reward design with Bitcoin by providing two additional uncle and nephew rewards.[3] In this section, we focus on studying the selfish mining attack in Ethereum, especially on analyzing the impact of these two rewards.

3.1 Overview of Ethereum

Ethereum is a distributed blockchain-based platform that runs smart contracts. Roughly speaking, a smart contract is a set of functions defined in a Turing-complete environment. The users of Ethereum are called clients. A client can issue transactions to create new contracts, to send Ether (internal cryptocurrency of Ethereum) to contracts, to other clients, or to invoke some functions of a contract. Clients' transactions are then collected into blocks by miners. Miners in Ethereum adopt a variant of NC

---

[3] Ethereum plans to gradually replace PoW with Proof of Stake (PoS).

Table 1 Mining rewards in Ethereum and Bitcoin

	Ethereum	Bitcoin	Purpose
Block reward	✓	✓	Compensate for miners' mining cost
Uncle reward	✓	×	Reduce centralization trend of mining
Nephew reward	✓	×	Encourage miners to reference uncle blocks
Transaction fee (gas cost)	✓	✓	Transaction execution; Resist network attack

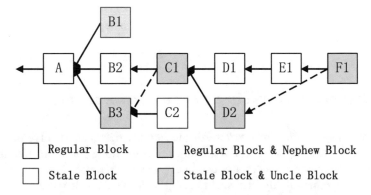

Fig. 7 Different block types in Ethereum. Here, regular blocks include $\{A, B2, C1, D1, E1, F1\}$ and stale blocks include $\{B1, B3, C2, D2\}$. Similarly, uncle blocks are $\{B1, B3, D2\}$ and nephew blocks are $\{C1, F1\}$. Uncle block $B3$ (uncle block $D2$, resp.) is referenced with distance one (two, resp.)

to reach an agreement on a growing sequence of blocks, i.e., blockchain.[4] Besides, Ethereum differs from Bitcoin in parameter settings; Ethereum sets its block size to about 20 KB and block interval to 10–20 s rather than 1 MB block size and 10 min block interval in Bitcoin for higher transaction throughput.

There are four types of rewards in Ethereum, namely, gas cost, block reward, uncle reward, and nephew reward [5, 9], as outlined in Table 1. On the one hand, Ethereum provides gas cost and block rewards, which are similar to rewards in Bitcoin. Specifically, the gas cost is used to reward miners to include and execute transactions in blocks, which is equivalent to transaction fees in Bitcoin. Also, like Bitcoin, miners in Ethereum can get block rewards, once their blocks are included in the blockchain accepted by all miners. On the other hand, Ethereum introduces two new rewards: uncle and nephew rewards.

---

[4] Although Ethereum claimed to apply the heaviest subtree rule [35], it seems to apply the longest chain rule instead to choose the main chain [17].

To explain these rewards, we shall introduce the concepts of *regular* and *stale* blocks. A block is called regular if it is included in the main chain, and is called stale block otherwise. In addition, an uncle block is a stale block that is a "direct child" of the main chain. In other words, the parent of an uncle block is always a regular block. An uncle block receives certain rewards if it is referenced by some future regular block, called a nephew block, through the use of reference links. See Fig. 7 for an illustration of uncle and nephew blocks. The values of uncle rewards depend on the "distance" between the uncle and nephew blocks. This distance is well defined because all the blocks form a tree. For instance, in Fig. 7, the distance between uncle block $B3$ (uncle block $D2$, resp.) and its nephew block is 1 (2, resp.). In Ethereum, if the distance is 1, the uncle reward is $\frac{7}{8}$ of the (static) block reward; if the distance is 2, the uncle reward is $\frac{6}{8}$ of the block reward; and so on. Once the distance is greater than 6, the uncle reward will be zero. By contrast, the nephew reward is always $\frac{1}{32}$ of the block reward.

3.2 Reward Design and Its Impact on Selfish Mining

In Ethereum, uncle and nephew rewards are initially designed to solve the mining centralization bias—miners form or join in some big mining pool. The mining pools with huge computation power are less likely to generate stale blocks and can be more profitable for mining. Thus, rewarding stale blocks can reduce the mining pools' advantage [8] and make them less attractive for small miners.

Unfortunately, uncle and nephew rewards also reduce the cost of launching selfish mining, and consequently lower the system security level. To see this, let us use one example to illustrate how the selfish miner can benefit from these rewards. Recall the case in which the selfish miner first produces one block, and then honest miners produce two subsequent blocks, as shown in Fig. 8a. In Bitcoin, the selfish miner will accept honest blocks H_1 and H_2 and lose the reward of block A_1 by the selfish mining strategy in Algorithm 1. By contrast, in Ethereum, this adversarial block A_1 can be referenced by the subsequent honest block A_2, by which the selfish miner can receive an uncle reward (7/8 of the block reward since the block distance is one), as shown in Fig. 8b. The additional uncle rewards reduce the loss of the selfish miner. Furthermore, the selfish miner can also obtain nephew rewards by referencing uncle blocks. In the next section, we will go through analytical results in [12], which systematically demonstrate the impact of these rewards on the selfish mining attack.

3.3 Theoretical Results on Selfish Mining in Ethereum

We follow the model introduced in Sect. 2.3. Let α denote the fraction of computation power controlled by selfish miners and β denote the fraction of total computation

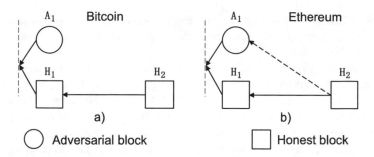

Fig. 8 A simple comparison of rewards obtained by the selfish miner in Bitcoin and Ethereum

power controlled by honest miners. Without loss of generality, all selfish miners are assumed to be controlled by a single selfish miner. The mining process is modeled as a Poisson process with rate f. Accordingly, the selfish miner generates blocks at rate αf, and the honest miners generate blocks at rate βf. We use K_s, K_u, and K_n to denote static, uncle, and nephew rewards, respectively. Without loss of generality,

---

Algorithm 2 A selfish mining strategy in ethereum

---
on Consensus
1: $(L_s, L_h) \leftarrow (0, 0)$

on The selfish miner mines a new block
2:   reference all (unreferenced) uncle blocks based on its private branch
3:   $L_s \leftarrow L_s + 1$
4:   **if** $(L_s, L_h) = (2, 1)$ **then**
5:     publish its private branch
6:     $(L_s, L_h) \leftarrow (0, 0)$ (since all the miners achieve a consensus)
7:   **else**
8:     keep mining on its private branch
9:   **end if**

on Some honest miners mine a new block
10:    The miner references all (unreferenced) uncle blocks based on its public branches
11:    $L_h \leftarrow L_h + 1$
12:    **if** $L_s < L_h$ **then**
13:      $(L_s, L_h) \leftarrow (0, 0)$
14:      keep mining on this new block
15:    **else if** $L_s = L_h$ **then**
16:      publish the last block of the private branch
17:    **else if** $L_s = L_h + 1$ **then**
18:      publish its private branch
19:      $(L_s, L_h) \leftarrow (0, 0)$ (since all the miners achieve a consensus)
20:    **else**
21:      publish first unpublished block in its private branch
22:      set $(L_s, L_h) = (L_s - L_h + 1, 1)$ if the new block is mined on a public branch that is a prefix of the private branch
23:    **end if**

---

we assume that $K_s = 1$ so that K_u (K_n, resp.) represents the ratio of uncle reward (nephew reward, resp.) to the static reward. In particular, $K_n < K_u < 1$ and K_u is a function of the distance.

We now describe the mining strategies for honest and selfish miners in Algorithm 2, which is based on the selfish mining strategy of Bitcoin in Algorithm 1. Similarly, we use $L_s(t)$ ($L_h(t)$) to denote the length of the private branch (resp., public branches) seen by the selfish miner (resp., honest miners) at time t. The main difference between Algorithms 1 and 2 is that when mining blocks, miners will include as many reference links as possible to (unreferenced) uncle blocks such that they can gain as many uncle and nephew rewards as possible (see Line 2 and 10). In addition, to track uncle and nephew rewards won by the selfish miner and honest miners, the lengths of private and public branches are kept (see Line 21 and 22). We refer our readers to [26] for more details.

With the algorithm, we can use $(L_s(t), L_h(t))$ to capture the system state at time t. The states $(L_s(t), L_h(t))$ evolve as a Markov process under the selfish mining strategy, as illustrated in Fig. 9. By comparing with Markov process of the selfish mining attack in Fig. 5, it is easy to see the increased complexity in system states because of the uncle and nephew rewards. Similarly, the process $(L_s(t), L_h(t))$ is positive recurrent and so it has a unique stationary distribution. Let $\{\pi_{i,j}\}$ be the steady-state distribution of the Markov process $(L_s(t), L_h(t))$. By solving the Markov process, we have the stationary distribution of states in the following lemma.

Lemma 2 ([12]) *Give the selfish mining strategy in Algorithm 2, the stationary distribution of the states is*

$$\pi_{0,0} = \frac{1 - 2\alpha}{2\alpha^3 - 4\alpha^2 + 1}, \quad \pi_{1,1} = \left(\alpha - \alpha^2\right)\pi_{0,0}, \quad \pi_{i,0} = \alpha^i \pi_{0,0}, \text{ for } i \geq 1$$

$$\pi_{i,j} = \alpha^i (1 - \alpha)^j (1 - \gamma)^j f(i, j, j)\pi_{0,0} + \alpha^{i-j} \gamma (1 - \gamma)^{j-1} \left(\frac{1}{(1 - \alpha)^{i-j-1}} - 1\right)\pi_{0,0} -$$

$$\gamma (1 - \gamma)^{j-1} \sum_{k=1}^{j} \alpha^{i-k} (1 - \alpha)^{j-k} f(i, j, j - k)\pi_{0,0}, \text{ for } i \geq j + 2, j \geq 1.$$

The function $f(x, y, z)$ is multiple summations and is defined as

$$f(x, y, z) = \begin{cases} \underbrace{\sum_{s_z=y+2}^{x} \sum_{s_{z-1}=y+1}^{s_z} \cdot\cdot \sum_{s_1=y-z+3}^{s_2} 1}_{z}, & z \geq 1, x \geq y + 2, \\ 0, & \text{otherwise.} \end{cases} \tag{3}$$

Next, we conduct the reward analysis for each state transition. Here, a probabilistic way is used to track various block rewards [26]. Specifically, in each state transition, there is a new block (mined by an honest miner or the selfish miner). It is impossible to decide the number of rewards associated with this new block when it is just created because the "destiny" of this new block depends on the evolution of the system. Hence, the expected rewards for the new block are computed based on possible

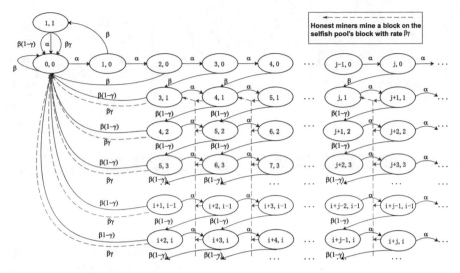

Fig. 9 The Markov process of the selfish mining in Ethereum

future events. In contrast, the selfish mining analysis in Bitcoin tracks published blocks associated with a state transition (whose destiny is already determined) rather than the new block and so it can compute the exact rewards [11]. This gives rise to the following two questions:

1. What is wrong with tracking published blocks?
2. How to compute the expected rewards for a new block at the time of its creation?

To answer the first question, one shall notice that tracking published blocks don't provide enough information to compute the uncle and nephew rewards. Recall from Sect. 3.1 that a published regular block can receive nephew rewards by referencing outstanding uncle blocks. The amount of nephew rewards depends on the number of outstanding uncle blocks. Therefore, it is necessary to keep track of all the outstanding uncle blocks in the system together with their depth information (which is needed to determine the number of uncle rewards). This greatly complicates the state space.

To answer the second question, one shall notice that it suffices to compute the expected rewards for a new block by using the following information: the probability that it becomes a regular block, the probability that it becomes an uncle block, the distance to its potential nephew block (if it indeed becomes an uncle block). Perhaps a bit surprisingly, all the information can be determined when this new block is generated for the selfish mining strategy in Algorithm 2.

To better understand this probabilistic method, we can see a simple example, as shown in Fig. 10. Assume that the selfish miner has already mined two blocks and kept them private at time t. Then, some honest miner generates a new block (case a in Fig. 10). According to Algorithm 2, the selfish miner publishes its private branch immediately. As such, this new block will become an uncle block with probability 1. Furthermore, we can see that this block will have a distance of 2 with its potential

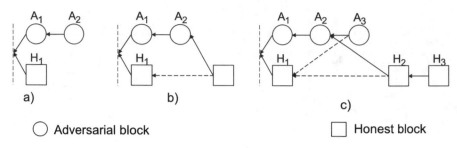

○ Adversarial block   □ Honest block

Fig. 10 A simple illustration of the probabilistic reward computing method

nephew block. Thus, this new block will receive an uncle reward of $K_u(2)$. Similarly, its potential nephew block will receive a nephew reward of $K_n(2)$. Moreover, this reward will belong to some honest miners with probability β (case b in Fig. 10) and $\alpha\beta^2(1 - \gamma)$ (case c in Fig. 10). Clearly, the selfish miner obtains this reward with probability $1 - \beta(1 + \alpha\beta(1 - \gamma))$. (See *Case 7* in Appendix B [12] for details.) Therefore, the expected rewards associated with this new block are $K_u(2) + K_n(2)$ in total among which $K_u(2) + K_n(2)\beta(1 + \alpha\beta(1 - \gamma))$ rewards will belong to honest miners (and the remaining will belong to the selfish miner). We refer interested readers to [12] for more details of the reward analysis.

With reward analysis, it is easy to compute the block rewards r_b^s (resp., r_b^s), uncle rewards r_u^s (resp., r_u^h), and nephew rewards r_n^s (resp., r_b^h) for the selfish miner (resp., honest miners) in the following theorem.

Theorem 2 ([12]) *Give the selfish mining strategy in Algorithm 2, the rewards for the selfish miner and honest miners are*

$$r_b^s = \frac{\alpha(1 - \alpha)^2(4\alpha + \gamma(1 - 2\alpha)) - \alpha^3}{2\alpha^3 - 4\alpha^2 + 1},$$

$$r_b^h = \frac{(1 - 2\alpha)(1 - \alpha)(\alpha(1 - \alpha)(2 - \gamma) + 1)}{2\alpha^3 - 4\alpha^2 + 1},$$

$$r_u^s = \frac{(1 - 2\alpha)(1 - \alpha)^2\alpha(1 - \gamma)}{2\alpha^3 - 4\alpha^2 + 1}K_u(1),$$

$$r_u^h = (\alpha\beta + \beta^2\gamma)K_u(1)\pi_{1,0} + \sum_{i=2}^{\infty}\beta K_u(i)\pi_{i,0} + \sum_{i=2}^{\infty}\sum_{j=1}^{\infty}\beta\gamma K_u(i)\pi_{i+j,j},$$

$$r_n^s = \alpha\beta K_s(1)\pi_{1,0} + \sum_{i=2}^{\infty}\sum_{j=1}^{\infty}\beta^{i-1}\gamma(\alpha - \alpha\beta^2(1 - \gamma))K_s(i)\pi_{i+j,j},$$

$$r_n^h = \alpha\beta^2(1 - \gamma)K_s(1)\pi_{0,0} + \beta^2\gamma K_s(1)\pi_{1,0} + + \sum_{i=2}^{\infty}\sum_{j=1}^{\infty}\beta^i\gamma(1 + \alpha\beta(1 - \gamma))K_s(i)\pi_{i+j,j}.$$

With this theorem, it is easy to derive the absolute revenue $U_s(\alpha, \gamma)$ of the selfish miner in Eq. (4). Note that the absolute revenue is equivalent to the relative revenue (i.e., the share $R_s(\alpha, \gamma)$) in Bitcoin, but it is different from the relative revenue in

Ethereum due to the presence of uncle and nephew rewards. This is because Bitcoin adjusts the mining difficulty level so that the regular blocks are generated at a stable rate, say one block per time unit. Thus, the long-term average total revenue is fixed to be one block reward per time unit with or without selfish mining. This makes the absolution revenue equivalent to the relative revenue. The situation is different in Ethereum. Even if the regular blocks are generated at a stable rate, the average total revenue still depends on the generation rate of uncle blocks, which is affected by the selfish mining attack. Indeed, Ethereum didn't take into account the generation rate of uncle blocks when adjusting the difficulty level until the Byzantium hard fork. Therefore, there are two scenarios: (1) the regular block generation rate is 1 block per time unit and (2) the regular and uncle block generation rate is 1 block per time unit.

In the above analysis, the regular block generation rate is $r_b^s + r_b^h$, which is smaller than 1 as explained before. Thus, the time can be re-scaled to make the regular block generation rate be 1 block per time unit. In this scenario, the long-term absolute revenue for the selfish pool is

$$U_s(\alpha, \gamma) = \frac{r_b^s + r_u^s + r_n^s}{r_b^s + r_b^h}, \tag{4}$$

and the long-term absolute revenue for honest miners is

$$U_h(\alpha, \gamma) = \frac{r_b^h + r_u^h + r_n^h}{r_b^s + r_b^h}. \tag{5}$$

Similarly, the time can be scaled to make the regular and uncle block generation rate to be 1 block per time unit and define long-term absolute revenues for the selfish pool and honest miners accordingly. Finally, the threshold of the computation power to make selfish mining profitable in Ethereum can be derived. Specifically, if the selfish miner follows the mining protocol, its long-term average absolute revenue will be α, since the network delay is negligible (and so no stale blocks will occur). If the miner applies the selfish mining strategy in Algorithm 2, its long-term absolute revenue is given by $U_s(\alpha, \gamma)$, which can be larger than α. So the thresholds $\min_\alpha \{U_s(\alpha, \gamma) > \alpha\} \min_\alpha$ and $\{R_s(\alpha, \gamma) > \alpha\}$ for both scenarios through numerical calculations can be derived, respectively.

To better the results, we plot the selfish miner's expected absolute revenue with α ranging from 0 to 0.5 in Fig. 11a. Here, due to the uniform tie-breaking policy, $\gamma = 1/2$. From it, we can see when the selfish miner controls more than 0.163 of the computation power, it can gain more from the selfish mining attack than that from honest mining. The threshold is lower than that (i.e., 0.25 when $\gamma = 1/2$) in Bitcoin. In other words, the additional uncle and nephew reward lowers the system security level. Figure 11b plots the threshold of the attack with different γ. Particularly, we compute the thresholds for the two scenarios: (1) the regular block generation rate is 1 block per time unit and (2) the regular and uncle block generation rate is 1 block per

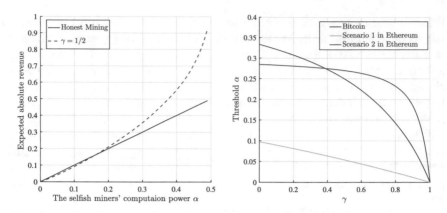

Fig. 11 The results of the selfish mining attack in Ethereum. **a** The selfish miner's relative revenue with different α. **b** The threshold of the selfish mining attack with different γ

time unit. From Fig. 11b, we can see that the higher γ is, the lower hash power needed for profitable selfish mining. Especially, when $\gamma = 1$, the selfish mining in Bitcoin and Ethereum can always be profitable regardless of their hash power. Besides that, the results show that the hash power thresholds of Ethereum in scenario 1 are always lower than in Bitcoin. By contrast, the hash power thresholds in scenario 2 are higher than Bitcoin when $\gamma \geq 0.39$. This is because the larger γ is, the more blocks mined by honest miners are uncle blocks. However, in scenario 2, the additional referenced uncle blocks will reduce the generation rate of the regular block, resulting in the decrease of selfish pools' block rewards. Thus, the selfish pool needs to have higher hash power to make selfish mining profitable. This suggests that Ethereum should consider the uncle blocks into the difficulty adjustment under the mining strategy given in Algorithm 2.

4 Incentive Design and Analysis in Bitcoin-NG

Bitcoin has suffered from low throughput (i.e., 7 transactions per second) and long latency (i.e., about one hour to confirm transactions) since inception. The poor performance significantly hinders blockchains' applications, and so a lot of scalable blockchain protocols are proposed [1, 10, 22, 29, 35, 42]. Among them, Bitcoin-NG (next generation) is among the first and the most prominent NC-based blockchains to approach the *near-optimal* throughput [10]. Bitcoin-NG creatively employs two types of blocks: (1) a *key block* that is very similar to a conventional block in Bitcoin except that it doesn't carry any transactions, and (2) a *microblock* that carries transactions. Every key block is generated through the leader election process (often known as the mining process) in NC. Each leader can issue multiple microblocks and receive the transaction fees until the next key block is generated. Unlike Bitcoin,

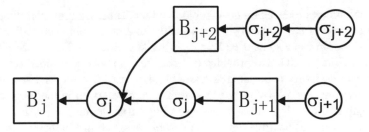

Fig. 12 An illustration of Bitcoin-NG. A square (respectively, circle) block denotes a key block (respectively, microblock). The microblocks are issued by the three key-block miners $\mathcal{B}_j, \mathcal{B}_{j+1}, \mathcal{B}_{j+2}$ with their signatures $\sigma_j, \sigma_{j+1}, \sigma_{j+2}$, respectively

Bitcoin-NG decouples leader election and transaction serialization. Intuitively, it is this decoupling that enables Bitcoin-NG to approach the near-optimal throughput, since the microblocks can be produced at a rate up to the network capacity. In this section, we focus on the incentive design of Bitcoin-NG, especially studying the impact of this decoupling idea on incentive-based attacks.

4.1 Overview of Bitcoin-NG

In Bitcoin, the mining of blocks has two functionalities: (1) electing leaders (i.e., the owners of valid blocks) by NC and (2) ordering and verifying transactions. By differentiating block functionalities, Bitcoin-NG decouples the leader election with the transaction serialization. Specifically, Bitcoin-NG uses key blocks mined through PoW to elect a leader at a stable rate (e.g., one key block per 100 s). Each leader can produce several microblocks containing unconfirmed transactions at another rate, often higher than the key-block rate (e.g., one microblock per 20 s). In a nutshell, a key block is very similar to a conventional block in Bitcoin except that it does not carry any transactions. On the other hand, microblocks contain transactions but do not contain any proof of work. Although the rate of microblocks is usually much larger than the key-block generation rate, it has to be bounded to prevent adversarial leaders from swamping the system with microblocks. This decoupling enables Bitcoin-NG to process many microblocks between two consecutive key blocks, which significantly increases its transaction throughput. Figure 12 illustrates these two types of blocks.

Bitcoin-NG adopts a similar fork choice rule as Bitcoin. In Bitcoin-NG, microblocks carry no weight, not even a secondary index for miners to choose which key block to mine. For instance, in Fig. 12, there are two forking branches with the same number of key blocks but different numbers of microblocks. Miners treat these two forking branches as equal, adopt a uniform tie-breaking rule to choose one branch, and then mine after the latest microblock in this branch [10]. To sum up, an honest miner still follows LCR to choose a "right" key block (i.e., the last key block in the longest chain consisting of key blocks only), and then mines on the latest microblock produced by the key-block miner. Thus, without microblocks, the mining process of key blocks is the same as the one in Bitcoin. The selfish mining attack in Bitcoin can also be used here to attack key blocks in Bitcoin-NG. Similar to Bitcoin, Bitcoin-NG also provides two rewards, namely, key-block reward and transaction fee. Every miner obtains a key-block reward if it mines a key block by successfully solving a PoW puzzle and its key block ends up in the longest chain. After mining a key block, miners can also obtain transaction fees by including as many transactions as possible (up to the microblock size limit) in their microblocks.

4.2 Microblocks and Its Incentive-Based Attacks

In Bitcoin-NG, miners should include as many transactions as possible in their microblocks and publish these microblocks to win transaction fees. This is called the transaction inclusion rule. In addition, miners should accept as many microblocks issued by the previous key-block miner as possible and mine on the latest received microblock, i.e., obeying the longest chain extension rule. It is easy to see when all miners obey these two rules, more produced microblocks could be included in the blockchain, and Bitcoin-NG could achieve better transaction throughput. By contrast, a selfish miner could break the transaction inclusion and the longest chain extension rules to maximize its profit from transaction fees as explained below:

- **Transaction inclusion attack.** When the selfish miner publishes one key block and generates multiple microblocks, it keeps the last several microblocks private. That is, the selfish miner continues to mine on top of its latest microblock chain, while honest miners can only mine on top of the last published microblock. Figure 13 shows the case in which the selfish miner withholds some of its microblocks mined after the key block B_j, and honest miners mine on the last public microblock of the selfish miner. This attack is incentivized if transaction fees in microblocks go primarily to the next key-block owner.
- **Longest chain extension attack.** When the selfish miner adopts an honest key block, it can reject some (or all) microblocks and mine directly on the last accepted microblock (or the last key block, respectively). In other words, the selfish miner rejects the transactions in these microblocks issued by the previous honest key-block miner. This attack is illustrated in Fig. 14. This attack is incentivized if transaction fees go primarily to the current key-block owner.

From the above cases, it is easy to see that the transaction fees in microblocks can neither go to the next key-block owner nor go primarily to the current key-block owner. Therefore, to resist these two attacks, Bitcoin-NG divides the transaction fees included in microblocks between two consecutive key-block miners into two parts. The first key-block miner gets the r fraction ($r \in [0, 1)$), while the second one

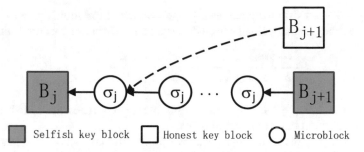

Selfish key block   Honest key block   Microblock

Fig. 13 An example of the transaction inclusion attack. The first two microblocks after the selfish B_j have been published and so they are public to honest miners. The other microblocks are kept private. A dashed square block denotes a future mined block

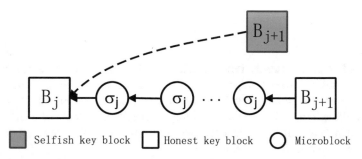

Selfish key block   Honest key block   Microblock

Fig. 14 An example of the longest chain extension attack. The selfish miner rejects all the microblocks and mines its key block on top of the honest B_j. A dashed square block denotes a future mined block

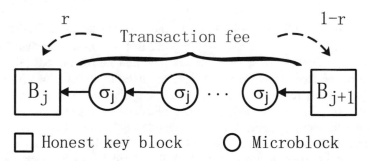

Honest key block   Microblock

Fig. 15 Bitcoin-NG fee distribution rule

obtains the remaining $1 - r$ fraction. Figure 15 illustrates this fee distribution rule. The following analysis shows how to decide the value of r to resist the microblock mining attacks.

Resisting Transaction Inclusion Attack. In this attack, the selfish miner can withhold a microblock to avoid sharing its transaction fees with the subsequent key-block miner. (We refer readers to the original paper [10] for more details.) Note that the probability for the selfish (respectively, honest) miner mines a block is α (respectively, $\beta = 1 - \alpha$). To guarantee the average revenue of the selfish miner launching the above attack is smaller than what it deserves, the distribution ratio r should satisfy

$$\overbrace{\alpha \times 100\%}^{\text{win 100\%}} + \overbrace{(1 - \alpha) \times \alpha \times (100\% - r)}^{\text{Lose 100\%, but mine after txn}} < r, \tag{6}$$

therefore $r > 1 - \frac{1-\alpha}{1+\alpha-\alpha^2}$. This ratio requirement encourages the selfish miner to place a transaction in a public microblock.

Later, Yin et al. [41] found that the above computation neglects a case: the incumbent leader can be re-elected as the next leader and gain an extra $\alpha(1 - r)$ fraction of the transaction fee. Thus, the distribution ratio r should satisfy

$$\overbrace{\alpha \times 100\%}^{\text{win 100\%}} + \overbrace{(1 - \alpha) \times \alpha \times (100\% - r)}^{\text{Lose 100\%, but mine after txn}} < r + \alpha(1 - r), \tag{7}$$

therefore $r > \frac{\alpha}{1-\alpha}$.

Resisting Longest Chain Extension Attack To increase revenue from some transactions, the selfish miner can ignore these transactions in an honest microblock and mine on a previous microblock. Later on, if the selfish miner mines a key block, it can place these transactions in its microblock. To resist this attack, the selfish miner's revenue in this case must be smaller than the revenue obtained by obeying the longest chain extension rule. Therefore, we have

$$\overbrace{\alpha \times r}^{\text{Mine next key block}} + \overbrace{\alpha^2 \times (100\% - r)}^{\text{Mine the third key Block}} < \overbrace{\alpha(100\% - r)}^{\text{Mine on microblock}}, \tag{8}$$

which leads to $r < \frac{1-\alpha}{2-\alpha}$. Taking the upper bound into consideration, the distribution ratio r satisfies $1 - \frac{1-\alpha}{1+\alpha-\alpha^2} < r < \frac{1-\alpha}{2-\alpha}$. In particular, when α is less than 25%, we obtain $37\% < r < 43\%$. Hence, $r = 40\%$ is chosen in the Bitcoin-NG [10].

Discussion. The above analysis provides theoretical results on the value of r, but it has several limitations. To better understand it, let us replay the longest chain extension attack, as shown in Fig. 16. Here, two simplifications are made to better illustrate the analysis limitation: 1) each leader is allowed to only create one microblock; 2) each microblock is allowed to only contain one transaction. Consider a scenario where an honest miner produces a key block B_j as well as a microblock containing a transaction tx. If the selfish miner obeys the longest chain extension rule and finds the next key block with probability α, it will get a $1 - r$ fraction of the transaction fee (which corresponds to the last item in Eq. (8)). However, the selfish miner can directly mine on the key block B_j, hoping to win a higher transaction fee of tx. If the selfish miner happens to create the next key block B_{j+1} with probability α, it can win r fraction of the transaction fee by including tx in its microblock (which corresponds to the first item in Eq. (8)). If the selfish miner is lucky to mine the next consecutive key block B_{j+2}, it will win the remaining $1 - r$ of the transaction fee. Combining all conditions leads to Eq. (8).

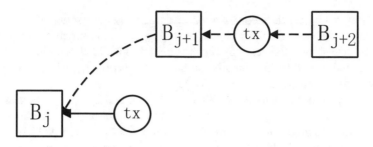

Fig. 16 A simple example to illustrate the limitation of the previous analysis. The selfish miner mines directly on block B_j and tries to include transaction tx in its future microblock

The analysis of the above example has two limitations. First, the simple analysis is quite reasonable if the selfish miner just hopes to get a higher fee from a targeted transaction tx. However, in reality, the selfish miner usually aims to increase its revenue from all transactions instead of just a targeted one. If the selfish miner applies the strategy to all the transactions rather than a single targeted one, it will quickly use up the space of its future microblocks. As a result, the selfish miner cannot include another transaction in its microblock, thereby losing the associated transaction fee. In other words, the above analysis ignores the impact of transaction size and microblock capacity, which magnifies the selfish miner's potential revenue from the attack. On the other hand, the above analysis works well for whale transactions with high fees, which are rare so that we don't need to worry about the space. But, most of the transactions in current blockchain systems have low fees.

Second, the selfish miner is assumed to always adopts honest miners' key blocks and immediately publishes its new key blocks (i.e., honest mining of key blocks). In other words, it does not consider the impact of key-block selfish mining. This assumption can only be justified when the selfish miner's computation power is less than the threshold of making key-block selfish mining profitable because the optimal mining strategy for key blocks is indeed honest mining [10, 11, 33]. However, once the selfish miner's computation power α is above the threshold, the selfish miner has the incentive to launch the key-block mining attack and so the impact of the key-block selfish mining cannot be ignored anymore. In the following, we will go through incentive analysis in [27], which has addressed the above two limitations.

4.3 Theoretical Results on Microblocks Mining

We use the mining models in Sect. 2.3. Let α denote the fraction of computation power controlled by selfish miners and β denote the fraction of total computation power controlled by honest miners. Without loss of generality, all selfish miners are assumed to be controlled by a single selfish miner. The key-block mining process is modeled as a Poisson process with rate f. Accordingly, the selfish miner generates key blocks at rate $f\alpha$, and the honest miners generate key blocks at rate $f\beta$. The miner of each key block becomes a leader and can issue a series of microblocks containing as many transactions as possible (up to the maximum microblock size) at a constant rate v until the next key block is mined. Specifically, a block (including key block and microblock) mined by an honest (respectively, the selfish) miner is referred to as *honest* (respectively, *selfish*) block.

Following the network model of Bitcoin [11, 12], we assume that honest miners are fully connected through the underlying network, and an honest miner spends negligible time to broadcast a key block or microblock in Bitcoin-NG.[5] In addition, we assume that the selfish miner can broadcast its private blocks immediately after it sees a new honest key block.

We assume two types of transactions according to their transaction fees: "whale" transactions with a high fee and regular transactions with a low fee. Also, we assume that the vast majority of transactions are regular ones. We assume that the transaction size is fixed, and so the maximum number of transactions included in a microblock is also fixed. We also assume that miners have enough pending transactions to be included in microblocks.[6] We call a microblock regular if it contains only regular transactions. In addition, we refer to the total transaction fees included in a regular microblock as the microblock fee and use R_t to denote it. In addition, we use R_b to denote the key-block reward. Let $k = R_b/R_t$ denote the ratio of the block reward to the microblock fee. This ratio k ranges from $(0, \infty)$. When k approaches 0 (respectively, ∞), it implies that the transaction fee (respectively, key-block reward) dominates the reward. The different values of k exhibit the various impact of rewards on the Bitcoin-NG system. As whale transactions are so rare that they use little microblock space. For this reason, we can ignore their space requirement (even under the network capacity constraints) and apply the analysis in Sect. 4.2.

Next, we present the incentive analysis of microblocks. In particular, the analysis does not consider the selfish mining of key blocks. That is, it assumes that the selfish miner always adopts honest miners' key blocks and immediately publishes its new key blocks. This assumption is justified shortly and will be relaxed by considering the joint mining of microblocks and key blocks in Sect. 4.4. In addition, as the propagation delay of key blocks is negligible, forked key blocks are also not considered in the following analysis.

We consider the revenue of transaction fees in terms of regular transactions for the selfish miner and honest miners during a time interval $[0, t]$. Without loss of generality, we assume that there exists a block B_0 that the selfish miner and honest miners both agree to mine on at the starting time. (For example, B_0 can be the genesis block.) Let $M(t)$ be the number of key blocks mined during the time interval $[0, t]$. Let X_i ($i \in [0, M(t)]$) denote an indicator random variable which equals one if the i-th key block is a selfish key block, as described below:

---

[5] This assumption is reasonable for key blocks because the inter-arrival time of two consecutive key blocks is often much larger than the block propagation delay. On the other hand, this assumption can be relaxed for microblocks, as we will show later.

[6] This assumption is reasonable in Bitcoin and Ethereum-like public blockchains. For instance, a mempool visualization website [21] shows that the number of pending transactions is around 136k in May 2021.

$$X_i = \begin{cases} 1, \text{selfish key block} \\ 0, \text{honest key block.} \end{cases}$$

Without loss of generality, we assume block B_0 is an honest key block. For other key blocks, the possibility that it is a selfish key block is equal to α.

After mining a key block, its owner can issue a series of microblocks at a constant rate v until the next key block is mined. Here, the rate v captures the network capacity constraints. Let Y_i denote the interval between the i-th key block and $(i + 1)$-th key block. Thus, the number of produced microblocks between i-th and $(i + 1)$-th key blocks is vY_i. In addition, each microblock contains a total fee of R_t because we only consider regular transactions here. We are now ready to compute the suitable value of r to resist the two microblock attacks for regular transactions.

Resisting Transaction Inclusion Attack. In this attack, the selfish miner hides some of its microblocks generated after a key block but keeps mining on top of the microblock chain. Hence, honest miners directly mine on top of the selfish miner's last published block. Let ρ denote the fraction of the unpublished microblocks among all the selfish microblocks between two consecutive key blocks. In particular, $\rho = 1$ means that the selfish miner hides all the microblocks it has generated between two consecutive key blocks. Thus, if any two consecutive key blocks satisfy $(X_i, X_{i+1}) = (1, 0)$, there are $(1 - \rho)vY_i$ microblocks between them from the view of an honest miner; otherwise, there are vY_i microblocks.

Let Z_i denote an indicator random variable equal to one if $\{X_i = 1, X_{i+1} = 0\}$, and equal to zero otherwise. Next, let $Z = \sum_{i=1}^{M(t)-1} Z_i$. Suppose $M(t) = m$. The following lemma will aid us to bound the value of Z with high probability.

Lemma 3 *For m consecutive key blocks, the number of block pairs $(X_i, X_{i+1}) = (1, 0)$ has the following Chernoff-type bound: For $0 < \delta < 1$,*

$$\Pr(|Z - \alpha\beta(m - 1)| > \delta\alpha\beta(m - 1)) < e^{-\Omega(\delta^2\alpha\beta m)}. \tag{9}$$

This lemma shows that as m increases, the number of key pairs $(X_i, X_{i+1}) = (1, 0)$ is between $(1 - \delta)\alpha\beta m$ and $(1 + \delta)\alpha\beta m$ with high probability.

Next, we compute the selfish miner's relative revenue for large m. On the one hand, the total amount of transaction fees for all the miners is given by $\sum_{i=1}^{m-1} (vY_i R_t - \rho v Z_i Y_i R_t)$. To see this, note that there are $\sum_{i=1}^{m-1} vY_i$ microblocks produced with associated transaction fees $\sum_{i=1}^{m-1} vY_i R_t$. Note also that once $Z_i = 1$, there are $\rho v Y_i$ microblocks not being included in the longest chain due to the transaction inclusion attack. Hence, the associated loss of transaction fees is $\sum_{i=1}^{m-1} \rho v Z_i Y_i R_t$. On the other hand, the total transaction fees for the selfish miner is given by $\sum_{i=1}^{m-1} (\alpha v Y_i R_t - r\rho v Z_i Y_i R_t)$. To see this, note that without any attack, the selfish miner can get α fraction of the total transaction fees given by $\sum_{i=1}^{m-1} \alpha v Y_i R_t$. Note also that with the transaction inclusion attack, the selfish miner will lose r

fraction of the total loss of transaction fees as the first leader. Combining the above analysis, we have the following lemma for large m.

Lemma 4 *The selfish miner's relative revenue u converges to $\frac{\alpha - r\alpha\beta\rho}{1 - \alpha\beta\rho}$ with high probability as $m \to \infty$.*

This lemma says that for large m, the selfish miner's relative revenue is $\frac{(\alpha - r\alpha\beta\rho)}{(1 - \alpha\beta\rho)}$. Recall that the key-block generation process is a Poisson process with rate f, and so $M(t)$ is a Poisson arrival process. Hence, when t tends to infinity, $M(t)/t \to f$ holds with high probability. Therefore, with high probability, the maximum relative revenue of the selfish miner during $[0, t]$ is

$$
\begin{aligned}
u &= \max_{0 \le \rho \le 1} \frac{\alpha - r\alpha\beta\rho}{1 - \alpha\beta\rho} \\
&= r + \max_{0 \le \rho \le 1} \frac{\alpha - r}{1 - \alpha\beta\rho}.
\end{aligned} \tag{10}
$$

If $r \le \alpha$, the optimal $\rho = 1$ and the corresponding

$$
u = r + \frac{\alpha - r}{1 - \alpha\beta}.
$$

In this case, u is always larger than α since $1 - \alpha\beta < 1$. This means that the selfish miner can always have a relative revenue greater than its fair share by utilizing this attack. On the other hand, if $r > \alpha$, the optimal $\rho = 0$ and $u = \alpha$. This means that the maximum relative revenue that the selfish miner can obtain is honest mining (i.e., $\rho = 0$). Therefore, we should set $r > \alpha$ to guarantee the adversary cannot gain more from the transaction inclusion attack.

Resisting Longest Chain Extension Attack. In this attack, the selfish miner can bypass some honest microblocks and mines directly on an old honest block. Similarly, let ρ denote the rejected microblock fraction. In particular, $\rho = 1$ means that the selfish miner rejects all honest microblocks and mines directly on the last honest key block. More precisely, if two consecutive key blocks are $(X_i, X_{i+1}) = (0, 1)$, there are $(1 - \rho)vY_i$ honest microblocks accepted by the longest chain. Let K_i denote an indicator random variable equal to one if $\{X_i = 0, X_{i+1} = 1\}$, and equal to zero otherwise. Let $K = \sum_{i=1}^{m-1} K_i$. The following lemma will aid us to bound the expectation of K for m blocks.

Lemma 5 *For the m block sequence, the number of block pair $(X_i, X_{i+1}) = (0, 1)$ has the following Chernoff-type bound: For $0 < \delta < 1$,*

$$
\Pr(|K - \alpha\beta(m - 1)| > \delta\alpha\beta(m - 1)) < e^{-\Omega(\delta^2 \alpha\beta m)}. \tag{11}
$$

Next, we compute the selfish miner's relative revenue for large m. On the one hand, the total amount of transaction fees for all the miners is given by $\sum_{i=1}^{m-1} (vY_i R_t - \rho v K_i Y_i R_t)$. To see this, recall that there are $\sum_{i=1}^{m-1} vY_i$ microblocks produced with associated transaction fees $\sum_{i=1}^{m-1} v Y_i R_t$. Once $K_i = 1$, there are $\rho v Y_i$ microblocks not being included in the longest chain due to the longest chain extension attack. Hence, the associated loss of transaction fees is $\sum_{i=1}^{m-1} \rho v K_i Y_i R_t$. On the other hand, the total transaction fees for the selfish miner is given by $\sum_{i=1}^{m-1} (\alpha v Y_i R_t - r \rho v Z_i Y_i R_t)$. To see this, recall that without any attack, the selfish miner can get α fraction of the total transaction fees given by $\sum_{i=1}^{m-1} \alpha v Y_i R_t$. With the longest chain extension attack, the selfish miner will lose $1 - r$ fraction of the total loss of transaction fees as the second leader. Combining the above analysis, we have the following lemma for larger m.

Lemma 6 *The selfish miner's relative revenue μ converges to $\frac{\alpha-(1-r)\alpha\beta\rho}{1-\alpha\beta\rho}$ with high probability as $m \to \infty$.*

This lemma says that for large m, the selfish miner's relative revenue is $\frac{\alpha-(1-r)\alpha\beta\rho}{1-\alpha\beta\rho}$. Similar with the previous analysis, we can show that as $t \to \infty$, with high probability, the maximum relative revenue of the selfish miner during $[0, t]$ is

$$
\begin{aligned}
u &= \max_{0 \le \rho \le 1} \frac{\alpha - (1 - r)\alpha\beta\rho}{1 - \alpha\beta\rho} \\
&= 1 - r + \max_{0 \le \rho \le 1} \frac{r - \beta}{1 - \alpha\beta\rho}.
\end{aligned}
\tag{12}
$$

If $r \ge \beta$, the optimal $\rho = 1$ and the corresponding

$$
u = 1 - r + \frac{r - \beta}{1 - \alpha\beta}.
$$

In this case, u is always larger than α since $1 - \alpha\beta < 1$. This means that the selfish miner can always have a relative revenue greater than its fair share by launching this attack. On the other hand, if $r < \beta$, the optimal $\rho = 0$ and $u = \alpha$. This means that the maximum relative revenue that the selfish miner can obtain is honest mining (i.e., $\rho = 0$). Therefore, we should set $r < \beta$ to guarantee the adversary cannot gain more from the longest chain extension attack. Combining the two incentive sub-mechanisms of transaction inclusion and longest chain extension, the value of r needs to satisfy that

$$
\alpha < r < \beta.
$$

Discussion. The analysis in [10, 41] can be used to bound the split ratio r for whale transactions, while the above analysis can provide a new bound for the ratio r for regular transactions under network capacity constraints. These bounds are depicted in Fig. 17. The figure shows that the new bound $\alpha < r < \beta$ (for regular transactions) contains the previous two bounds $1 - \frac{1-\alpha}{1+\alpha-\alpha^2} < r < \frac{1-\alpha}{2-\alpha}$ and $\frac{\alpha}{1-\alpha} < r < \frac{1-\alpha}{2-\alpha}$ (for whale transactions). This leads to several interesting implications.

First, introducing network capacity constraints doesn't make it harder to maintain the incentive compatibility of Bitcoin-NG. This is because the bounds for whale transactions are the same as the previous ones and the bound for regular transactions

Fig. 17 The comparison of
the transaction fee
distribution ratio

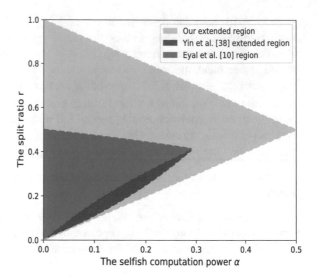

contains the previous ones. Second, when α is smaller than 29%, we can find a value of r that satisfies all the bounds. This means that the incentive compatibility of Bitcoin-NG can be maintained for all types of transactions even under network capacity constraints in this regime. Third, when α is larger than 29%, we cannot find a value of r that satisfies all the bounds, because the two bounds for whale transactions both become invalid. This means that the incentive compatibility of Bitcoin-NG can be maintained only for regular transactions but not for whale transactions in this regime. In other words, the presence of whale transactions might cause instability of the whole system in this regime. As such, some defense mechanisms should be designed accordingly.

4.4 Theoretical Results on Microblocks and Key-Block Mining in Bitcoin-NG

In this section, we go through the analysis that jointly considers microblock and key-block mining. Particularly, the Markov decision process (MDP) is applied to model various selfish mining strategies. To make our analysis tractable, two simplifications are made. First, the key-block interval is assumed to be $1/f$ and so the number of microblocks produced between two consecutive key blocks is v/f. (Note that the key-block interval is assumed to follow the exponential distribution with mean $1/f$.) Second, only binary choices: publishing or hiding all selfish microblocks in the transaction inclusion attack and accepting or rejecting all honest microblocks in the longest chain extension attack are considered. (This is consistent with the fact that $\rho = 0$ or 1 in Sect. 4.3.)

The MDP model can be presented by a 4-tuple $\mathcal{M} := (S, A, P, R)$, where S is the state space, A is the action space, P is the stochastic state transition matrix, and R is the reward matrix. Specifically, S contains all possible states in the selfish mining process; A includes the available actions (e.g., publishing or hiding blocks by the selfish miner) at each state; P contains the transition probabilities from the current state to the next state according to the taken action; and R records how much the selfish miner obtains when there are some state transitions. Table 2 illustrates the MDP of selfish mining in Bitcoin-NG. Note that blocks are assumed to be transmitted without delay (see Sect. 2.3), and so forks are not considered in the analysis. Below we will discuss each component of the 4-tuple:

Actions. The selfish miner has eight available actions.

- **Adopt and include.** The selfish miner accepts all honest key blocks and the corresponding honest microblocks. In other words, the selfish miner will mine its key block on the last honest block and abandon its private chain. This action is referred to as adopt.
- **Adopt and exclude.** The selfish miner accepts all honest key blocks and microblocks except for microblocks produced after the last honest key block. Specifically, the selfish miner directly mines on top of the last honest key block, which is referred to as adoptE.
- **Override and publish.** The selfish miner publishes all its key blocks and corresponding microblocks whenever its private chain is longer than the honest one. The chain length is counted by the key block. This action is denoted as override.
- **Override and hide.** The selfish miner publishes all its key blocks and the microblocks except for these mined after the last selfish key block whenever its private chain is longer than the honest one. This action is denoted as overrideH.
- **Match and publish.** When an honest miner finds one new key block, the selfish miner publishes its key block of the same height and the microblocks built after this key block. This action is available when the selfish miner has one block in advance and is referred to as match.

Table 2 State transition and reward matrices for the optimal selfish mining (Prob.: probability; Cond.: condition)

S

State × Action	State	Probability	Reward	Condition
$(l_a, l_h, \cdot, S_h)$, adopt			$(l_h, l_h, 0, 0)$	
$(l_a, l_h, \cdot, S_p)$, adopt	$(1, 0, \mathsf{noTie}, H_{\mathsf{in}})$	α	$(l_h, l_h - 1 + (1 - r), 0, r)$	–
$(l_a, l_h, \cdot, \{H_{\mathsf{in}}, H_{\mathsf{ex}}\})$, adopt	$(0, 1, \mathsf{noTie}, H_{\mathsf{in}})$	$1 - \alpha$	$(l_h, l_h - 1, 0, 0)$	

Table 2 (continued)

State × Action	State	Probability	Reward	Condition
$(l_a, l_h, \cdot, S_h)$, adoptE			$(l_h, l_h, 0, 0)$	
$(l_a, l_h, \cdot, S_p)$, adoptE	$(1, 0, \text{noTie}, H_{ex})$	α	$(l_h, l_h - 1 + (1 - r), 0, r)$	-
$(l_a, l_h, \cdot, \{H_{in}, H_{ex}\})$, adoptE	$(0, 1, \text{noTie}, H_{ex})$	$1 - \alpha$	$(l_h, l_h - 1, 0, 0)$	
$(l_a, l_h, \cdot, H_{ex})$, override	$(l_a - l_h, 0,$	α	$(0, 0, l_h + 1, l_h + 1)$	
$(l_a, l_h, \cdot, H_{in})$, override	$\text{noTie}, S_p)$ $(l_a - l_h - 1,$	$1 - \alpha$	$(0, r, l_h + 1, l_h + (1 - r))$	$l_a > l_h$
$(l_a, l_h, \cdot, \{S_p, S_h\})$, override	$1, \text{noTie}, S_p)$		$(0, 0, l_h + 1, l_h)$	
$(l_a, l_h, \cdot, H_{ex})$, overrideH	$(l_a - l_h, 0,$	α	$(0, 0, l_h + 1, l_h + 1)$	
$(l_a, l_h, \cdot, H_{in})$, overrideH	$\text{noTie}, S_h)$ $(l_a - l_h - 1, 1,$	$1 - \alpha$	$(0, r, l_h + 1, l_h + (1 - r))$	$l_a > l_h$
$(l_a, l_h, \cdot, \{S_p, S_h\})$, overrideH	$\text{noTie}, S_h)$		$(0, 0, l_h + 1, l_h)$	
$(l_a, l_h, \text{noTie}, \cdot)$, wait	$(l_a + 1, l_h, \text{noTie}, *)$	α	$(0, 0, 0, 0)$	-
	$(l_a, l_h + 1, \text{noTie}, *)$	$1 - \alpha$		
$(l_a, l_h, noTie, H_{in})$, match $(l_a, l_h, \text{tie}, H_{in})$, wait	$(l_a + 1, l_h, \text{tie}, H_{in})$	α	$(0, 0, 0, 0)$	
	$(l_a - l_h, 1, \text{noTie}, S_p)$	$\gamma(1 - \alpha)$	$(0, r, l_h, l_h - 1 + (1 - r))$	$l_a \geq l_h$
	$(l_a, l_h + 1, \text{noTie}, H_{in})$	$(1 - \gamma)(1 - \alpha)$	$(0, 0, 0, 0)$	
$(l_a, l_h, noTie, H_{ex})$, match $(l_a, l_h, \text{tie}, H_{ex})$, wait	$(l_a + 1, l_h, \text{tie}, H_{ex})$	α	$(0, 0, 0, 0)$	
	$(l_a - l_h, 1, \text{noTie}, S_p)$	$\gamma(1 - \alpha)$	$(0, 0, l_h, l_h - 1)$	$la \geq l_h$

Table 2 (continued)

State × Action	State	Probability	Reward	Condition
	$(l_a, l_h + 1,$ noTie, $H_{ex})$	$(1-\gamma)(1-\alpha)$	$(0,0,0,0)$	
$(l_a, l_h, \text{noTie},$ $\{S_p, S_h\})$, match $(l_a, l_h, \text{tie},$ $\{S_p, S_h\})$, wait	$(l_a+1, l_h, \text{tie}, *)$	α	$(0,0,0,0)$	$l_a \geq l_h$
	$(l_a - l_h, 1,$ noTie, $S_p)$	$\gamma(1-\alpha)$	$(0,0,l_h,l_h)$	
	$(l_a, l_h + 1,$ noTie, $*)$	$(1-\gamma)(1-\alpha)$	$(0,0,0,0)$	
$(l_a, l_h, \text{noTie}, H_{in}),$ matchH $(l_a, l_h, tie', H_{in}),$ wait	$(l_a + 1, l_h,$ $tie', H_{in})$	α	$(0,0,0,0)$	$l_a \geq l_h$
	$(l_a - l_h, 1,$ noTie, $S_h)$	$\gamma(1-\alpha)$	$(0, r, l_h, l_h - 1 + (1-r))$	
	$(l_a, l_h + 1,$ noTie, $H_{in})$	$(1-\gamma)(1-\alpha)$	$(0,0,0,0)$	
$(l_a, l_h, noTie, H_{ex}),$ matchH $(l_a, l_h, tie', H_{ex}),$ wait	$(l_a + 1, l_h,$ $tie', H_{ex})$	α	$(0,0,0,0)$	$l_a \geq l_h$
	$(l_a - l_h, 1,$ noTie, $S_h)$	$\gamma(1-\alpha)$	$(0,0,l_h,l_h - 1)$	
	$(l_a, l_h +$ $1, \text{noTie}, H_{ex})$	$(1-\gamma)(1-\alpha)$	$(0,0,0,0)$	
$(l_a, l_h, noTie,$ $\{S_p, S_h\})$, matchH $(l_a, l_h, tie',$ $\{S_p, S_h\})$, wait	$(l_a + 1, l_h,$ $tie', *)$	α	$(0,0,0,0)$	$l_a \geq l_h$
	$(l_a - l_h, 1,$ noTie, $S_h)$	$\gamma(1-\alpha)$	$(0,0,l_h,l_h)$	
	$(l_a, l_h +$ $1, \text{noTie}, *)$	$(1-\gamma)(1-\alpha)$	$(0,0,0,0)$	
$(l_a, l_h, tie', \cdot),$ revert	$(l_a, l_h, \text{tie}, *)$	1	$(0,0,0,0)$	—
$(l_a, l_h, \cdot, S_h),$ revert	$(l_a, l_h, *, S_p)$	1	$(0,0,0,0)$	$l_h = 0$
$(l_a, l_h, \cdot, H_{ex}),$ revert	$(l_a, l_h, *, H_{in})$	1	$(0,0,0,0)$	$l_a = 0$

$*$ denotes the state element remains the same in the state transition.

- **Match and hide.** When an honest miner generates a new key block, the selfish miner publishes its key block of the same height while hiding the microblocks built after this key block. This action is also available when the selfish miner has one block in advance. This action is denoted as matchH.
- **Wait.** In this action, the selfish miner does not publish any new key blocks and microblocks, while keeps mining on its private chain until a new key block and corresponding microblocks are found.
- **Revert.** The selfish miner reverts its previous actions. Specifically, the selfish miner can publish its hidden microblocks when there is no honest key block mined after its block; the selfish miner can include the honest microblocks (decided to exclude in the previous decision) or exclude the honest microblocks (decided to include in the previous decision) once there is no selfish key block mined on an honest block.

The adopt, override match, and wait actions include all possible actions on the selfish mining of key blocks, while hide, publish, and revert actions cover all possible actions on the transaction inclusion and longest chain extension attacks of microblocks. Note that in the match action, the selfish miner publishes its key block of the same height to match a key block produced by honest miners. Therefore, there are two forking branches of the same length. In Bitcoin-NG, honest miners adopt a uniform tie-breaking rule to choose which branch to mine on. In particular, the variable γ is introduced to denote the fraction of honest miners that mine on the selfish miner's branch.

State space. The state space S is also composed of 4-tuple $(l_a, l_h, \text{fork}, \text{lastMicroBlock})$.

- l_a accounts for the length of the chain mined by the selfish miner after the last common ancestor key block. More precisely, the last common ancestor key block is the last key block in the longest chain accepted by both the selfish miner and *all* honest miners and is updated once the selfish miner adopts the public chain or all honest miners adopt the selfish miner's chain. In addition, the chain length is counted by the selfish key blocks in this branch.
- l_h is the length of the public chain after the last common ancestor key block. This chain can be viewed by both the selfish miner and honest miners.
- **fork**. The field fork obtains three possible values, dubbed noTie, tie, and tie′. Specifically, tie means the selfish miner publishes l_h selfish key block and the corresponding microblocks; tie′ presents the selfish miner publishes l_h selfish key block and the corresponding microblocks except for these after the last selfish key block; noTie signifies that there are not two public branches with the equivalent length.
- **lastMicroBlock**. This field also includes four possible values, dubbed H_{in}, H_{ex}, S_p, and S_h. Specifically, H_{in} (respectively, H_{ex}) represents the common ancestor is an honest key block, and the corresponding microblocks are accepted (respectively, rejected) by the selfish miner. While S_p (or S_h) which stands for the common ancestor is a selfish key block, and the corresponding microblocks mined are published (or hidden, respectively) by the selfish miner.

State Transition and Reward. The rewards for the selfish miner and honest miners in the state transitions can be indicated by a 4-tuple (R_h, T_h, R_a, T_a). Specifically, R_h (respectively, R_a) is the key-block rewards for honest (respectively, the selfish) miners, while T_h (respectively, T_a) is the transaction fee for honest (respectively, the selfish) miners.

Recall that there are two types of transactions. Here, the analysis focuses on regular transactions and will discuss whale transactions later. Recall also that the microblock fee of a regular microblock is denoted by R_t. For convenience, instead of recording the number of rewards, each field only records the number of key-block rewards or transaction fees (the total transaction fee in v/f microblocks as one unit) won by miners. More importantly, the transaction fees included in the microblocks after the common ancestor key block are not assigned to miners until the next ancestor key block is decided. This is because these transaction fees are affected by some future actions of the selfish miner (see Sect 4.3).

In adopt or adoptE actions, the selfish miner accepts l_h honest key blocks and the microblocks mined before these key blocks. Honest miners obtain $l_h R_b$ key-block rewards and $(l_h - 1)v/f R_t$ transaction fees. In override or overrideH actions, the selfish miner publishes $l_h + 1$ selfish key blocks. Honest miners accept these key blocks and the microblocks produced before the key blocks. Thus, the selfish miner obtains $(l_h + 1)R_b$ key-block rewards and $l_h v/f R_t$ transaction fees. In the match actions, the next state depends on whether the next key block is created by the selfish miner (w.p. α), by some honest miners working on the honest branch (w.p. $(1 - \gamma)(1 - \alpha)$), or by the left honest miners mining on the selfish branch (w.p. $\gamma(1 - \alpha)$). In the latter case, the selfish miner effectively overrides the honest miners' branch. It can obtain $l_h R_b$ key-block reward and $(l_h - 1)v/f R_t$ transaction fees. Note that the value of γ is decided by the adopted fork solution (e.g., $\gamma = 0.5$ in the uniform tie-break policy).

Once the common ancestor key block is changed, the transaction fees in the microblocks produced after the previous ancestor key block will be assigned. There are two cases:

- The previous common ancestor key block is mined by an honest miner. This case can be further divided into two subcases: (1) the next key block is mined by honest miners, and honest miners get $v/f R_t$ transaction fees; (2) the next key block is mined by the selfish miner and **lastMicroBlock** $= H_{in}$, honest miners get $r v/f R_t$ transaction fees and the selfish miner gets $(1 - r)v/f R_t$ transaction fees.
- The previous common ancestor key block is mined by the selfish miner. This case can be further divided into two subcases: (1) the next key block is mined by the selfish miner, and the selfish miner gets $v/f R_t$ transaction fees; (2) the next key block is mined by some honest miners and **lastMicroBlock** $= S_p$, the selfish miner gets $r v/f R_t$ transaction fees and honest miners get $(1 - r)v/f R_t$ transaction fees.

Note that since whale transactions are rare and unpredictable, the microblock fee can be modeled as a random variable taking two values: R_t or R_t plus the fee of

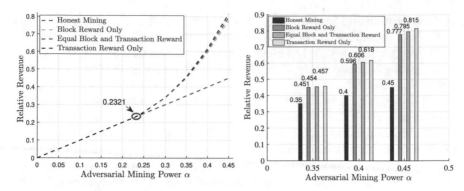

Fig. 18 The results of the selfish mining attack in Ethereum. **a** The selfish miner's relative revenue
with different α. **b** The threshold of the selfish mining attack with different β

a whale transaction. Let $\bar{R}_t$ be the expected microblock fee. Clearly, $\bar{R}_t > R_t$. The
long-term effect of whale transactions is to decrease the ratio k from R_b/R_t to $R_b/\bar{R}_t$.
Such an effect slightly increases the relative revenue of the selfish miner.

By using MDP toolbox [7] to solve the above MDP model, we can numerically
obtain the optimal policies under each scenario and obtain the selfish miner's relative
revenue. To better illustrate the results, we first plot the relative revenues of the
selfish miner when $r = 0.4$ (used in Bitcoin-NG [10]) with different α, as shown in
Fig. 18a. Here, three reward settings: $k \rightarrow 0$, $k = v/f$, and $k \rightarrow \infty$ are considered.
Specifically, in the first setting, the transaction fees dominate the miners' revenue;
in the second setting, the transaction fees included in v/f microblocks between
two consecutive key blocks have the same weight with one key-block reward; in
the third setting, the key-block rewards dominate miners' revenue. Note that the
key-block reward-dominated case has a similar reward distribution as Bitcoin, i.e.,
the microblock architecture does not impact the system. The figure shows that the
thresholds of making selfish mining profitable in these three settings are all 23.21%,
which is the same as the selfish mining threshold in Bitcoin. In other words, by
adopting the suitable r (i.e., $\alpha < r < 1 - \alpha$), the microblock architecture in Bitcoin-
NG does not affect the system security compared with Bitcoin. In addition, the selfish
miner's revenues in the three settings are still the same even when $\alpha > 29\%$, which
verifies the analysis in Sect. 4.3 and supports that Bitcoin-NG is as resilient as Bitcoin
under suitable settings. When $\alpha > 35\%$, the differences between the selfish miner's
revenues in the three settings and the honest revenue are exhibited in Fig. 18b. It's
easy to see that the selfish miner can obtain the highest revenue in the transaction fee-
dominated case. This implies that the microblock architecture can slightly increase
the selfish miner's revenue.

5 Further Reading

In this section, we provide more works on the incentive design and analysis of blockchains for interested readers.

Bitcoin. Eyal and Sirer are among the first to formally analyze selfish mining in Bitcoin [11]. However, the proposed selfish mining strategy isn't optimal. Later, Sapirshtein et al. [33] and Nayak et al. [25] demonstrated that by adopting optimized strategies, the threshold of the computation power to make selfish mining profitable can be reduced to 23.2% (instead of 25% in [11]) when honest miners adopt the uniform tie-breaking defense. In [17], Gervais et al. further extended the analysis to several variants of Bitcoin including Dogecoin, Litecoin, and Ethereum. Different with these works on block rewards, the selfish mining strategy of transaction fees was studied in [6]. The results showed that even an attacker with small computation power and a poor network connection can still gain more profits from the attack. Based on this work, Tsabary and Eyal [37] additionally studied the Bitcoin gap game between block reward and transaction fee. In [19], the propagation delay is considered for the analysis of selfish mining.

In addition to the incentive analysis, there are several works on defending the selfish mining attack in Bitcoin. Heilman proposed a defense mechanism called Freshness Preferred [20], in which by using the latest unforgeable timestamp issued by a trusted party, the threshold can be increased to 32%. Bahack in [2] introduced a fork-punishment rule to make selfish mining unprofitable. Specifically, each miner in the system can include fork evidence in their block. Once confirmed, the miner can get half of the total rewards of the winning branch. Solat and Potop-Butucaru [34] proposed a solution called ZeroBlock, which can make selfish miners' block expire and be rejected by all the honest miners without using forgeable timestamps. In [43], a backward-compatible defense mechanism called weighted FRP was proposed, in which the weights of the forked chains instead of their lengths are considered. In [28], Pass and Shi proposed Fruitchains, which distributes rewards to all recent fruits that are parallel products of block mining. Similar with Fruitchains, Szalachowski et al. [36] proposed a new protocol, called StrongChain, which enables miners to publish weak solutions, i.e., solutions with higher mining difficulty targets. Miners can include weak solutions in their blocks and always select the chain with the largest weighted count of blocks and weak solutions to mine on. Bissias and Levine [3] proposed Bobtail, which enables miners to publish and collect all PoW solutions with a higher target until the mean of the k smallest hashes is below a certain target.

Ethereum. In [30], Ritz and Zugenmaier conducted extensive simulations to study selfish mining in Ethereum. Wang et al. [38] analyzed two kinds of stubborn mining in Ethereum. Yang et al. [40] analyzed the impact of imperfect network on selfish mining in Ethereum. Besides, Ethereum has updated its transaction fee mechanism in EIP1559 [4], which make it quite different with that in Bitcoin. Several studies on this new transaction mechanism were conducted [13, 23, 31, 32].

Other Blockchains. Yin et al. [41] have extended the incentive analysis of Bitcoin-NG by considering a situation that the original paper omits [10]. Later, Wang at al. [39] considered advanced selfish mining strategies, i.e., stubborn mining strategies, when an attacker may manipulate the microblock chains between two honest parties. Fooladgaret al. [14] modeled the participation costs and rewards received within a strategic interaction scenario in Algorand [18]. They showed that the reward sharing approach in Algorand is not a Nash equilibrium and proposed a novel reward mechanism to fix it.

6 Conclusion

In this chapter, we revisit the incentive design and analysis of three influential blockchain protocols: Bitcoin, Ethereum, and Bitcoin-NG. In particular, compared with Bitcoin, Ethereum introduces two new rewards (i.e., uncle and nephew block rewards), while Bitcoin-NG redesigns the transaction fee distribution rule to accommodate the new consensus architecture. Throughout the studies, we first find that incentive designs are closely related with the system security. Hence, incentive design should be carefully evaluated before adoption. Second, we find that new consensus protocols also require new incentive designs. Therefore, consensus algorithms and incentive mechanisms should be jointly considered and evaluated for blockchain protocols. Third, we find that existing incentive analysis may not be feasible for new designs, and so tailored analysis is required. However, the process of modeling and theoretical analysis make it difficult to evaluate the incentive design for each blockchain protocol. Thus, AI-driven methods such as deep reinforcement learning (DRL) may help to automatically analyze incentive designs and to conduct the analysis. We notice that there are several works on this direction and leave discussion of these automatic analysis as future work.

References

1. Bagaria, V., Kannan, S., Tse, D., Fanti, G., Viswanath, P.: Deconstructing the blockchain to approach physical limits (2018). arXiv preprint arXiv:1810.08092
2. Bahack, L.: Theoretical Bitcoin attacks with less than half of the computational power (draft) (2013). arXiv preprint arXiv:1312.7013
3. Bissias, G., Levine, B.N.: Bobtail: improved blockchain security with low-variance mining. In: The Network and Distributed System Security Symposium (NDSS), NDSS '20 (2020)
4. Buterin, V., Conner, E., Dudley, R., Slipper, M., Norden, I., Bakhta, A.: EIP-1559: fee market change for ETH 1.0 chain (2019). https://eips.ethereum.org/EIPS/eip-1559
5. Buterin, V., et al.: A next-generation smart contract and decentralized application platform. White paper (2014)
6. Carlsten, M., Kalodner, H., Matthew Weinberg, S., Narayanan, A.: On the instability of Bitcoin without the block reward. In: Proceedings of the 2016 ACM SIGSAC Conference on Computer and Communications Security, CCS '16, pp. 154–167. ACM (2016)
7. Chadès, I., Chapron, G., Cros, M.-J., Garcia, F., Sabbadin, R.: Mdptoolbox: a multi-platform toolbox to solve stochastic dynamic programming problems. Ecography **37**(9), 916–920 (2014)
8. Ethereum. Design rationale: uncle incentivizatinon. github (2018)

9. Ethereum. Mining rewards (2018)
10. Eyal, I., Gencer, A.E., Sirer, E.G., Van Renesse, R.: Bitcoin-NG: a scalable blockchain protocol. In: 13th USENIX Symposium on Networked Systems Design and Implementation (NSDI 16), pp. 45–59 (2016)
11. Eyal, I., Sirer, E.G.: Majority is not enough: bitcoin mining is vulnerable. Commun. ACM **61**(7), 95–102 (2018)
12. Feng, C., Niu, J.: Selfish mining in ethereum. In: 2019 IEEE 39th International Conference on Distributed Computing Systems (ICDCS), pp. 1306–1316 (2019)
13. Ferreira, M.V.X., Moroz, D.J., Parkes, D.C., Stern, M.: Dynamic posted-price mechanisms for the blockchain transaction-fee market (2012). arXiv preprint arXiv:2103.14144
14. Fooladgar, M., Manshaei, M.H., Jadliwala, M., Rahman, M.A.: On incentive compatible role-based reward distribution in algorand. In: 2020 50th Annual IEEE/IFIP International Conference on Dependable Systems and Networks (DSN), pp. 452–463. IEEE (2020)
15. Forum, Bitcoin: mining cartel attack, December 2010
16. Gervais, A., Karame, G.O., Capkun, V., Capkun, S.: Is Bitcoin a decentralized currency? IEEE Secur. Privacy **12**(3), 54–60 (2014)
17. Gervais, A., Karame, G.O., Wüst, K., Glykantzis, V., Ritzdorf, H., Capkun, S.: On the security and performance of proof of work blockchains. In: Proceedings of the 2016 ACM SIGSAC Conference on Computer and Communications Security, CCS '16, pp. 3–16. ACM (2016)
18. Gilad, Y., Hemo, R., Micali, S., Vlachos, G., Zeldovich, N.: Algorand: scaling Byzantine agreements for cryptocurrencies. In: Proceedings of the 26th Symposium on Operating Systems Principles, SOSP '17, pp. 51–68. ACM (2017)
19. Göbel, J., Keeler, H.P., Krzesinski, A.E., Taylor, P.G.: Bitcoin blockchain dynamics: the selfish-mine strategy in the presence of propagation delay. Perform. Eval. **104**, 23–41 (2016)
20. Heilman, E.: One weird trick to stop selfish miners: fresh bitcoins, a solution for the honest miner. In: International Conference on Financial Cryptography and Data Security, pp. 161–162. Springer (2014)
21. Hoenicke, J.: Unconfirmed transaction count (mempool) (2020)
22. Kogias, E.K., Jovanovic, P., Gailly, N., Khoffi, I., Gasser, L., Bryan, F.: Enhancing bitcoin security and performance with strong consistency via collective signing. In: 25th USENIX Security Symposium (USENIX Security 16), pp. 279–296 (2016)
23. Leonardos, S., Monnot, B., Reijsbergen, D., Skoulakis, S., Georgios, P.: Dynamical analysis of the EIP-1559 ethereum fee market (2021). arXiv preprint arXiv:2102.10567
24. Nakamoto, S.: Bitcoin: a peer-to-peer electronic cash system. Working paper (2008)
25. Nayak, K., Kumar, S., Miller, A., Shi, E.: Stubborn mining: generalizing selfish mining and combining with an eclipse attack. In: 2016 IEEE European Symposium on Security and Privacy (EuroS P), pp. 305–320. IEEE (2016)
26. Niu, J., Feng, C., Dau, H., Huang, Y.-C., Jingge, Z.: Analysis of Nakamoto consensus, revisited (2019). arXiv preprint arXiv:1910.08510
27. Niu, J., Wang, Z., Gai, F., Feng, C.: Incentive analysis of Bitcoin-NG, revisited. In: Performance Evaluation: An International Journal, vol. 144, pp. 102–144. Elsevier (2020)
28. Pass, R., Shi, E.: Fruitchains: a fair blockchain. In: Proceedings of the ACM Symposium on Principles of Distributed Computing, PODC '17, pp. 315–324. ACM (2017)
29. Pass, R., Shi, E.: Hybrid consensus: efficient consensus in the permissionless model. In: 31st International Symposium on Distributed Computing (DISC 2017), vol. 91, pp. 39:1–39:16 (2017)
30. Ritz, F., Zugenmaier, A.: The impact of uncle rewards on selfish mining in Ethereum. In: 2018 IEEE European Symposium on Security and Privacy Workshops (EuroS PW), pp. 50–57. IEEE (2018)
31. Roughgarden, T.: Transaction fee mechanism design for the ethereum blockchain: An economic analysis of EIP-1559 (2020). arXiv preprint arXiv:2012.00854
32. Roughgarden, T.: Transaction fee mechanism design (2021). arXiv preprint arXiv:2106.01340
33. Sapirshtein, A., Sompolinsky, Y., Zohar, A.: Optimal selfish mining strategies in Bitcoin. In: International Conference on Financial Cryptography and Data Security, pp. 515–532. Springer (2016)

34. Solat, S., Potop-Butucaru, M.: Zeroblock: preventing selfish mining in Bitcoin. Technical report, Sorbonne Universites (2016)
35. Sompolinsky, Y., Zohar, A.: Secure high-rate transaction processing in Bitcoin. In: Financial Cryptography and Data Security, pp. 507–527. Springer (2015)
36. Szalachowski, P., Reijsbergen, D., Homoliak, I., Sun, S.: Strongchain: transparent and collaborative proof-of-work consensus. In: Proceedings of the 28th USENIX Conference on Security Symposium, SEC'19, pp. 819–836 (2019)
37. Tsabary, I., Eyal, I.: The gap game. In: Proceedings of the 2018 ACM SIGSAC Conference on Computer and Communications Security, CCS' 18, pp. 713–728 (2018)
38. Wang, Z., Liu, J., Qianhong, W., Zhang, Y., Hui, Yu., Zhou, Z.: An analytic evaluation for the impact of uncle blocks by selfish and stubborn mining in an imperfect ethereum network. Comput. Secur. **87**, 101581 (2019)
39. Wang, Z., Liu, J., Zhang, Z., Zhang, Y., Yin, J., Yu, H., Liu, W.: A combined micro-block chain truncation attack on Bitcoin-NG. In: Information Security and Privacy, pp. 322–339. Springer (2019)
40. Yang, R., Chang, X., Mišić, J., B Mišić, V.: Assessing blockchain selfish mining in an imperfect network: honest and selfish miner views. Comput. Secur. **97**, 101956 (2020)
41. Yin, J., Wang, C., Zhang, Z., Liu, J.: Revisiting the incentive mechanism of Bitcoin-ng. In: Information Security and Privacy, pp. 706–719. Springer (2018)
42. Yu, H., Nikolic, I., Hou, R., Saxena, P.: OHIE: Blockchain scaling made simple. In: Proceedings of the 41th IEEE Symposium on Security and Privacy, S&P. IEEE (2020)
43. Zhang, R., Preneel, B.: Publish or perish: a backward-compatible defense against selfish mining in Bitcoin. In: Cryptographers' Track at the RSA Conference, pp. 277–292. Springer (2017)

Cross-Blockchain Transactions: Systems, Protocols, and Topological Theory

Dongfang Zhao

Abstract In this chapter, we turn our focus to those applications touching multiple blockchains. Since a blockchain deals with its data in the form of transactions, the real technical question we want to answer is *how to handle cross-blockchain transactions (CBTs)?* We will first present the state-of-the-art systems, i.e., exchanging cryptocurrencies between Ethereum and Bitcoin, and discuss the challenges of extending existing approaches to a more general context, such as application-specific blockchains (instead of cryptocurrencies) and an arbitrary number of blockchains (instead of two). We then review two recent schools of thought about CBT protocols and discuss their properties in detail. Finally, we sketch an ongoing research effort on building a theoretical foundation for CBTs using topological machinery.

1 Introduction

In 2008, for the first time, cryptocurrency Bitcoin [6] introduced the concept of blockchain into practical applications. After more than a decade of development, blockchain is becoming a popular data management paradigm thanks to its immutability, decentralization, and autonomy. Various domains, such as digital health care [46], supply chains [33], big data analysis [2], and scientific computing [3], are actively launching blockchain-based projects. As a result, many believe that application-specific blockchains will emerge and, unsurprisingly, are concerned with the exchange among these heterogeneous blockchain systems.

In industry, a state-of-the-art production system for exchanging between cryptocurrencies is Cosmos [9], allowing for direct exchange between BTC (Bitcoin [6]) and ETH (Ethereum [11]). Although Cosmos is built upon an open cross-blockchain protocol named *sidechain* [34], no practical systems exist yet for exchanging assets between arbitrary blockchains other than cryptocurrency. Even for Cosmos and sidechain, criticisms have been widely received regarding the long latency: a cross-blockchain transaction usually takes hours, if not days [34], to complete. A number

D. Zhao (✉)
University of Nevada, Reno, Nevada 89557, United States
e-mail: dzhao@unr.edu

© The Author(s), under exclusive license to Springer Nature Switzerland AG 2022
D. A. Tran et al. (eds.), *Handbook on Blockchain*, Springer Optimization
and Its Applications 194, https://doi.org/10.1007/978-3-031-07535-3_5

of leading service providers such as IBM, Oracle, Azure Blockchain Services, and SAP have made a firm commitment to solving many of the technical challenges that currently plague the interoperability of blockchains. For example, the World Health Organization, in conjunction with the help of the aforementioned companies, was able to deploy a platform called MiPasa [27], which has been built atop the Hyperledger Fabric framework [23], to enable the "early detection of COVID-19 carriers and infection hotspots" [7].

In academia, researchers have been focusing on the cross-blockchain transaction (CBT) and its variants; similar terminology was used in the literature, to name a few: cross-chain swaps, cross-chain deals, etc. These studies are not limited to two-party transactions and aim to support transactions among an arbitrary number of distinct blockchains. At the writing of this chapter, two schools of thoughts prevail:

1. The first school was pioneered by Herlihy et al., who advocated to solve the cross-chain problem through a timestamp-based approach [16, 18]. The key idea is to introduce a timeout mechanism, known as *time lock* for the asset to be on hold until the recipient can provide proof that it qualifies to receive the asset within a predefined period of time. The approach was then criticized on *atomicity* and *scalability*: the timeout approach might render some of the parties "worse off"— an honest party who sends out its asset and cannot receive compensation due to the network delay (i.e., timeout); moreover, the timelock requires a sequence of linked smart contracts, leading to a time complexity proportional to the number of parties involved in the transaction—not scalable.
2. To address the atomicity and scalability challenges of the above approach, the second school of cross-blockchain studies, represented by Zakhary et al. [40], proposed approaches inspired by the conventional distributed commit protocol, namely, two-phase commit (2PC). Nevertheless, it is still unclear how to overcome the blocking scenario exhibited by 2PC, not to mention the possible forking exhibited by every single participating blockchain. Zhao et al. proposed a machine-learning-based mechanism [39] to prevent the possible blocking caused by 2PC and taxonomy of protocols to handle the possible forks [45].

Both of the aforementioned approaches on (arbitrary) cross-blockchain communications stay at the conceptual level without real implementations except for [39], which was implemented and evaluated on a blockchain emulator called *Blocklite* [38].

Arguably the most challenging obstacle for implementing a practical system supporting arbitrary CBTs stems from the possible forks from the participating blockchains: the complexity, delicacy, and fallibility of existing algorithms and protocols mentioned above, which are all based on the theory of replicated state machines, are necessary almost always due to the forked branches. We cannot render blockchains to eliminate forking, which is just part of its life, but we might find an alternative theory to, somehow, simplify the modeling and analyzing CBTs. To this end, Zhao et al. [42–44] proposed a series of techniques and tools based on mathematical topology, which were inspired by the seminal works [19–21] on a topological view of conventional distributed systems.

The remainder of this chapter is organized as follows. We will firstly present two state-of-the-art systems toward the so-called *Internet of blockchains* in Sect. 2. Specifically, we will discuss two representative systems, Cosmos and Polkadot, from various design perspectives. Notably, both Cosmos and Polkadot adopt a central master blockchain to manage the participating blockchains. We will then discuss three types of protocols for processing arbitrary cross-blockchain transactions without a centralized component in Sect. 3. These protocols have not been implemented in production systems except for some emulation results. We will then present a new methodology for modeling and analyzing CBTs through topology in Sect. 4. The methodology is rigorously built upon a series of axioms and results that have been extensively studied in point-set topology and algebraic topology. Finally, Sect. 5 provides a brief history of the development of this young research field—cross-blockchain transactions—in a chronicle order with pointers to important literature.

2 Internet-of-Blockchains Systems

This section will discuss the state-of-the-art production systems to achieve inter-operability among heterogeneous blockchains, usually coined as "Internet of Blockchains". As of the writing of this chapter, two leading production systems prevail, Cosmos [9] and Polkadot [30]. As we will discuss in this section, although both systems exhibit many differences such as programming interfaces and business models, they do share similar design spirits. For this reason, we will use Cosmos as the canonical example and occasionally compare it to Polkadot.

2.1 Background

The motivation of achieving interoperability among distinct blockchains is evident. For cryptocurrencies, attempts of exchanging between two types of currencies or even among more than two types of currencies are well justified. The implication of cross-blockchain interoperability is actually beyond cryptocurrencies. For instance, in a typical real estate transaction, there are at least three parties involved: the home owner's bank, the buyer's bank (either mortgage or cash or both), and the government that documents the title transfer. There are likely more parties such as the agent's company, the insurance company, etc., but the idea will not change: how to complete the transaction if each of these parties manages the data on its own blockchain? This is clearly a practical problem.

One of the earliest efforts for achieving interoperability was called *sidechain*, which was published in a white paper [34] not too long after the seminal paper of Bitcoin [6]. Primarily driven by cryptocurrency applications, sidechain was designed for the exchange between Bitcoin and another cryptocurrency, such as Ethereum [11]. Technically, data exchange between two parties has been extensively studied:

transaction processing has been mature enough to handle such data exchange and is offered in all commercial database systems (e.g., Oracle, Microsoft SQL Server); what makes data exchange among blockchains a new challenge lies in the possible *forks* within each participating blockchain. As a result, sidechain cannot simply commit the transaction, say between Bitcoin and Ethereum, once the blocks of both blockchains are verified and appended; at the very least, sidechain must wait for both blockchains to confirm that the blocks touched by the two-party transaction will not be rolled back due to forking. Sidechain is reported to take between a few hours and a couple of days to eventually complete a two-party transaction.

Nowadays, two production systems are clear leaders in the race toward the Internet of blockchains, or at least toward the "exchange between arbitrary cryptocurrencies" as both systems have tight relationships to specific cryptocurrencies. The first system, Polkadot [30], was co-founded by a co-founder of Ethereum [11]; the second system, Cosmos [9], was co-founded by co-founders of Tendermint [37]. Admittedly, cryptocurrency remains the driving force of blockchain applications, but our view of blockchains is more general: we envision that blockchains will become a competitive alternative to general data management systems. Therefore, the discussion in the subsequent subsections will treat both Polkadot and Cosmos as general blockchains from a system's point of view, as opposed to (over)emphasizing their cryptocurrency features.

2.2   Architecture

At the very core of Cosmos lies a backbone component called a *hub*. The hub, as the name suggests, is a central manager for the entities of the Cosmos *network*. The hub is implemented as an independent blockchain, or master blockchain (MB), and works as a broker between two or more participating blockchains (PBs), also called *zones*, through an inter-blockchain communication (IBC) protocol. In addition to the role as a broker, the MB itself is also involved in storing and managing the blocks of PBs, such as the intermediate balance during the transaction.

Figure 1 illustrates the hub-zone design with an oversimplified scenario where the hub is connected with four distinct zones, two of which are the popular cryptocurrencies: Bitcoin and Ethereum. It should be noted that a blockchain in Cosmos could be both a PB and MB, depending on from which angle we view it. That is, an MB for a specific set of PBs could be another MB's PB. This rule not only enables a hierarchy of management and blockchains in the entire network, but also allows for potential extensions (scale-out).

Polkadot shares a very similar architecture as Cosmos. The central (master) blockchain in Polkadot is called *relay chain*, and each participating blockchain is called *parallel chain*, or *parachain*. There is as well a hierarchical point of view in Polkadot: A parachain can be a relay chain of those participating blockchains at a lower level.

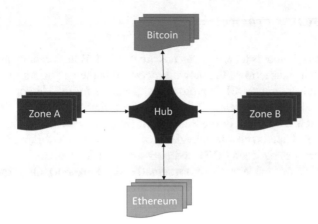

Fig. 1 The *hub-zone* architecture of a Cosmos network. This example network comprises one hub connected with Bitcoin, Ethereum, and two other arbitrary participating blockchains named Zone A and Zone B. This network has overall five distinct blockchains, including the hub itself. All communications must go through the hub, making the hub a broker, or master blockchain (MB), for the participating blockchains (PBs)

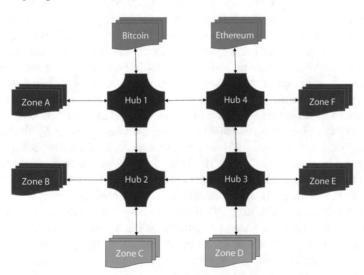

Fig. 2 The *nested hubs* of a Cosmos network. The hubs constitute a hierarchy of management and enable the scaling-out of participating blockchains. Logically, a hub at a lower level in the hierarchy is deemed as a zone of the hub at the level right above it

To make matters more concrete, Fig. 2 illustrates a series of *nested hubs* in Cosmos. In this case, Hub 1 can be assigned the role of the *global* MB, where Hub 2 and Hub 4 are considered as Zone 2 and Zone 4, respectively, for Hub 1. Similarly, Hub 3 serves as a zone of both Hub 2 and Hub 4. Therefore, here we have three layers of hubs, somewhat an analogue of routers in a networking system.

2.3   Consensus Protocol

The consensus protocols taken by Cosmos are twofold. When creating and finalizing a new block of transactions, the network needs to launch a variant of the practical Byzantine fault-tolerant (PBFT) protocol [8]. In order for the network to accept the new block, validators are selected based on the Delegated Proof-of-Stake (DPoS) protocol, which is a variant of the Proof-of-Stake (PoS) protocol [31].

While other chapters provide more detail of consensus protocols, we will articulate the key differences between PBFT and PoS such that readers can better understand the design decisions made by Cosmos (actually, also Polkadot). One crucial critique of the original consensus protocol Proof of Work (PoW) used by Bitcoin is energy consumption: the way (i.e., compete in a race of solving a hash-value puzzle) in which the node is selected for appending the new block is overly computation intensive from an energy's point of view. To this end, PoS represents a simpler way to select a node: the chance or probability of a node being elected is somewhat related to the node's "reputation", which is usually implemented as the balance in the cryptocurrency applications (hence the name "stake"). The reputation of a node is adjusted according to its performance: misconduct will deduct its reputation, and conversely, credits will be applied for honest work.[1] Although PoS can efficiently select a node to perform certain operations, it has a week fault tolerance: if the selected node is compromised, then there is not much we can do about it. To fix this, we need more nodes to be involved in the voting of an agreed value, hopefully, more than enough nodes to outnumber those compromised nodes. In the community of distributed computing and distributed systems, it is a well-known result that as long as the number of compromised nodes is below one-third (1/3) of the total number of nodes, the PBFT protocol [8] can guarantee the validity of the value agreed upon by the nodes. In fact, the Byzantine problem [25] was raised two decades ago before a practical protocol was proposed in [8]. The key idea of PBFT is to broadcast every node's local information (both its original local view and the messages it receives from other nodes) to all the peers in the network in multiple phrases. The very assumption that there are at least two-thirds (2/3) of honest nodes implies that arbitrary failures, including a coalition among up to one-third malicious nodes, will nonetheless be outnumbered by the honest nodes. The main critique of PBFT lies in its efficiency: PBFT incurs a quadratic number of messages (with respect to the number of nodes) when trying to reach a consensus on a value.

Cosmos takes a variant of PoS to select a *validator*, a node that validates the correctness (e.g., the hash value of the previous block) of the new block and initiates the voting in the network. The chance that a node is selected as the validator is proportional to its stake, according to the documentation of Cosmos. The validator then calls for voting using PBFT. After multiple rounds of broadcasting, the validator approves and appends the block if a consensus is finally reached.

---

[1] As a side note, the above procedure is also called a "leader election" algorithm in the literature of distributed systems.

Polkadot's consensus protocol works in a similar way except for implementation details. For instance, the chance of a node being picked as a validator is not as straightforward as a proportion to its stake; rather, Polkadot adopts a more delicate formula to compute the probability. Nonetheless, the two-phase mechanisms of a hybrid of PoS and PBFT variants remain unchanged.

2.4 Communication Model

In the literature of distributed computing [5], the communication model taken by a distributed systems has the following three main characteristics: (i) What forms of communication does the system take? Common choices include message passing, shared memory, and other variants. (ii) Is the communication synchronous, asynchronous, or some sort of hybrid? and (iii) What types of failures do we expect in the system? Would the failed node simply crash (and restart)? Is that possible that some nodes have been compromised and yet would work as they were "honest" nodes?

We summarize Cosmos's communication model from the aforementioned three perspectives in the following:

Message Passing Cosmos takes a message-passing model for cross-blockchain communication. The choice of such a model is well justified: the underlying infrastructure is assumed *shared-nothing*, as opposed to those high-performance computing systems where storage or even memory is shared among the nodes.

Semi-synchrony Like many other distributed systems, Cosmos assumes an unreliable interconnection, and the communication is "partially synchronous", a property usually called *semi-synchronous* in the literature [17]. In semi-synchronous communications, a message is expected to be delivered within some threshold, being a "reasonable" period of time or a "reasonable" number of rounds.

Arbitrary Failure As singular blockchains, Cosmos must ensure the fault tolerance of *arbitrary failures* from the nodes. An arbitrary failure is in contrast to a *crash failure*; in the former case, the nodes that have been compromised may, in the worst case, collaborate to sabotage or impede the progress of the applications.

Cosmos names its own communication protocol *inter-blockchain communication* (IPC). From a network's point of view, IPC is a transport-layer protocol (cf. TCP, UDP), meaning that features like security and application-related misbehavior do *not* interest IPC. In contrast, IPC assumes the participating blockchains (PBs) should provide sufficient application-layer features to complete the entire service or ecosystem. The remainder of this subsection will discuss the key features and design decisions of IPC; interested readers can find more details in IPC's white paper [12].

From a distributed system's point of view, IPC is implemented as a *wrapper* on the endpoints of communications. While IPC claims to be designed for "arbitrary

Fig. 3 An inter-chain message is being sent from `Blockchain A` to `Blockchain B`. The Inter-blockchain Communication (IPC) protocol wraps the message up on both ends of the communication. IPC wrappers delegate the wrapped message to the `relayer` for asynchronous message passing. IPC requires that at least one relayer process is available for inter-blockchain communication

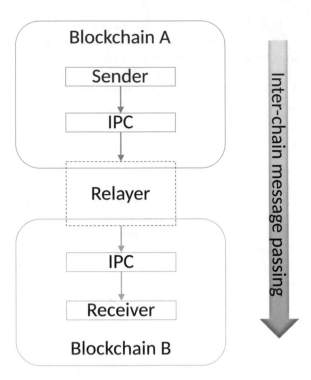

blockchains", it, conceptually, remains to connect only two endpoints of a communication channel. In addition, IPC's role of a wrapper is complemented with a *relayer*, which is a specific process that monitors and manages the messages being circulated between two blockchains. The relayer enables the asynchrony between the endpoints of the communication, as the former could temporarily store the messages even if the node at one endpoint fails. Indeed, Cosmos requires that at least one relayer process is live when IPC protocol is initiated. To make the matter more concrete, we illustrate the design decisions in an oversimplified example shown in Fig. 3.

Cosmos's IPC remains a communication protocol and has little to do with *transactions*: By no means an IPC wrapper bothers to attempt to ensure the *atomicity* of a data movement, not to mention the application layer's transactions. We want to remind the readers to differentiate between a communication protocol and a commit/transaction protocol: While both are called protocols referring to those algorithms designed for (the coordination among) multiple processes, the former is concerned with the data exchange, and the latter is focused on consistency before and after the communication. Usually, a transaction protocol can be designed with communication protocols as building blocks, such as the famous two-phase commit (2PC) protocol [1], but not conversely. For readers who are interested in transaction processing, we recommend the following books, one for practical system building [14] and the other for theoretical foundations [26].

Polkadot adopts a similar design for inter-chain communications. Polkadot names its protocol Cross-Chain Message Passing (XCMP) and uses the central blockchain, Relay Chain, as the relayer. XCMP, just as IBC, is a communication protocol for arbitrary messages and has nothing to do with transactions.

2.5 Programming Interface

Lastly, we want to discuss some of Cosmos's implementation detail, especially on the programming interface. Cosmos was implemented with the so-called Cosmos SDK primarily with Go and had a tight relationship with Tendermint. A library called *Starport* [35] is available for developers to build an application-specific blockchain from scratch, which can be integrated with Cosmos's API. An application developed through Starport and Cosmos SDK usually provides two types of interfaces: the command-line interface (CLI) and the REST interface.

Because Cosmos adopts Tendermint's consensus protocol and its codebase (the so-called "engine"), blockchain applications willing to join Cosmos must be compliant with the interface specified by Tendermint, which is called the Application Blockchain Interface (ABCI). The interaction between the Tendermint engine and the blockchain application is purely a pair of `request-response` messages. Cosmos adopts the `protobuf` library [32] to serialize (and deserialize/reconstruct) the (encoded) messages such that the application is not limited to use only Go for implementation. At the time of writing, the following programming languages are supported: C++ (native), Java, Python, Objective-C, C#, JavaScript, Ruby, Go, PHP, and Dart.

Polkadot is originally implemented with Rust (more than 98% lines of code). Other client bindings are available, such as Kagome [24] (C++) and Gossamer [13] (Go).

2.6 Limitation

Neither Cosmos nor Polkadot has reached the point where an Internet of blockchains really emerges as an analogue of the Internet of Things, at least not in their current forms. The intrinsic, centralized management adopted by both Cosmos and Polkadot is debatable: While it can be argued that the master blockchain of both Cosmos and Polkadot is a decentralized system, the nodes on the master blockchain are clearly superior to those nodes on participating blockchains—such a hierarchy is against the very philosophy of blockchains: Decentralization is not only about distribution but also about populism.

Another notable and more technical limitation of today's inter-chain ecosystems lies in the lack of support of cross-blockchain transactions. As we have discussed in Sect. 2.4, the communication protocols in state-of-the-art systems are (i) for only arbitrary data exchange (mostly cryptocurrency) and (ii) for only two-party communications. As a result, while an arbitrary number of participating blockchains can be simultaneously connected to the master chain, it is unclear how or whether more than three participants can be involved in the same transaction. On the other hand, the feature of supporting multi-party transactions is an indispensable building block for envisioned Internet of blockchains.

3 Protocols of Chain-to-Chain Federation

In sheer contrast to the paradigm of the Internet of blockchains through a master component as discussed in Sect. 2, this section presents recent advances for designing a federation of blockchains *without* a third-party's involvement. That is, the blockchains will be able to exchange data and complete transactions directly, just as the peer-to-peer (P2P) file sharing. We call this new paradigm of organizing a cluster of distinct blockchains (and their transactions) as chain-to-chain (C2C) federation.

3.1 C2C Blockchain Transactions Through Time Locks

In 2018, Herlihy [16] proposed the first protocol for processing a transaction among an arbitrary of blockchains without a third party's involvement. The key idea of [16] is to break[2] a transaction into a series of two-party sub-transactions, called *swaps*. Technically, the *atomicity* of a multi-party transaction is not guaranteed through such swaps, but practically, no participating blockchain is worse-off if following the specified swapping protocol.

While the swapping-based algorithm enables transactions among multiple transactions without an outsider's interference, it does rely on a timing mechanism to synchronize those two-party swaps. As a result, a global clock needs to be available in some form. The global clock is not necessarily a physical one and can be implemented as a logical one, such as vector clocks [36]. By and large, a transaction is broken down into a round-trip of message passing from an initiator to the last participant through every single intermediate participant involved in the transaction, and conversely back to the initiator. For each hop between a pair of adjacent participants, a *timeout* starts for sending out a message, and another *timeout* kicks in for receiving a reply from the recipient of the message (in the first timeout). If any of these two timeouts fails, the whole transaction is canceled *without* rolling back the possible partial changes. Of course, if a participating blockchain completes its job within the timeout period, it can be shown that the blockchain never experiences an asset penalty (i.e., "worse-off"). The timeout period can be implemented as part of the *smart contract*, which is widely supported in modern blockchain systems.

---

[2] In theory, a transaction cannot always be broken into a series of two-party sub-transactions unless we introduce *pseudo-transactions*, which are trivial from a practical point of view.

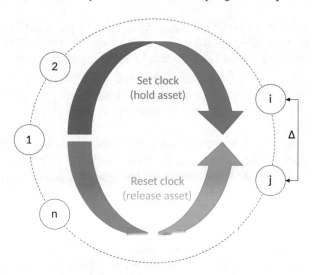

Fig. 4 An n-party transaction being split into $2n$ two-party swaps. Each swap between two adjacent blockchains takes a period of Δ. Counter-clockwise: The i-th blockchain holds its asset with a contract expiring in $2(n + 1 - i)\Delta$. Clockwise: If the j-th node has received a reply within $2(n + 1 - j)\Delta$ since its clock starts but fails to reply a message to i in Δ, the asset of the j-th node is liquidated without compensation from the i-th node

To make matters more concrete, we illustrate the procedure of completing a n-party transaction in Fig. 4. We can hypothetically think of a ring-like topology to represent the n blockchains involved in the transaction. For simplicity, we assume the transaction can be split into a series of two-party swaps between the i-th and the j-th nodes, where $1 \leq i < n$ and $j \equiv (i + 1) \bmod n$. Assuming the transaction is initiated by the first blockchain indexed by ①, the first round of message passing goes through every pair (i, j) of nodes all the way until $(n - 1, n)$ and $(n, 1)$. Once the first round is done, it is again ① initiating a reverse traversal: $(1, n) \rightarrow (n, n - 1) \rightarrow \cdots \rightarrow (j, i) \rightarrow \cdots \rightarrow (2, 1)$. Each blockchain in the first round, i.e., counter-clockwise direction, sets a timeout, also called a *time lock*, on its (smart) contract with the next node in the ring. For simplicity, we assume the time for each blockchain to complete its own local processing within a period of Δ. It follows that the i-th blockchain expects to receive a response from the j-th ($j = i + 1$) node within a time span of $2(n + 1 - i)\Delta$. In the second round, i.e., clockwise direction, if the j-th blockchain timely replies to the i-th blockchain, the latter will release its asset to the former. At this point, the j-th blockchain should have released its asset to the $[j + 1]_n$-th node, where $[j + 1]_n \equiv j + 1 \bmod n$, so the j-th blockchain's (local) transaction is complete. On the other hand, if ⓙ fails to respond to ⓘ within Δ, then ⓘ would not release its asset to ⓙ, who has released its asset to the node indexed by $[j + 1]_n$ and therefore ends up with a worse-off status.

Readers are reminded that each swap (between two adjacent blockchains) is *atomic* in its own right. However, the example we just show evidently illustrates that the atomicity of a swap-based transaction among n blockchains is violated. Specifically,

(j) possibly loses its asset if not meeting the specified time lock requirement: This should have been prohibited if the (global) transaction is atomic. That is, either (i) (j) does not release its asset to blockchain $[j + 1]_n$, or (ii) (j) releases its asset to blockchain $[j + 1]_n$ and receives the asset transferred from (i); there should be no other possibility. In subsequent works (e.g., [18]), such multi-party transactions are more deliberately handled, and yet their *atomicity* remains an open problem in this direction of research.

Since the publication [16] of swap-based approach to processing cross-blockchain transactions, two critiques have caught the most attention. The first one is concerned with performance, and the second one with functionality, or atomicity, to be more specific.

Performance Firstly, the time complexity of atomic swaps is linear to the number of participating blockchains, $O(n)$, where n denotes the number of distinct participating blockchains. Clearly, the latency caused by a constant factor of the total number of blockchains might not be acceptable for real-world applications and their scales. The root cause evidently lies at the core of the design: adopting a timeline (i.e., series of time locks) and allowing only one blockchain to make progress toward the n-party transaction inevitably incurs at least a linear time complexity. We will see how this can be overcome with the tools and techniques (e.g., 2PC) from the literature of distributed systems, which is elaborated in Sect. 3.2.

Atomicity There are two levels of atomicity concerning us: the atomicity of the multi-party transactions and that of two-party swaps. For the former one, conventional 2PC protocols can be incorporated as we will see in Sect. 3.2; the latter one needs a lot more work. The root cause of the latter problem lies in the fact that atomic swaps do not take into account the possible forks within each participating blockchain. Admittedly, the swap is considered atomic from a procedural point of view: the blocks touched by the transaction have no way to recall the transaction or break the transaction in the middle. However, it is possible, although unfortunate, for a specific blockchain to recall a fork that happens to hold the blocks touched by the n-party transaction. It is simply too much to ask the (atomic) swaps to catch or predict whether the fork will be recalled or not—theoretically, every fork can be recalled. We will discuss two groups of protocols that treat forks as first-class citizens in Sect. 3.3.

3.2 CBT Protocols Through Two-Phase Commits

In 2020, Zakhary et al. [40] proposed to achieve the atomicity and higher performance of cross-blockchain transactions (CBTs) using two-phase commit (2PC) protocols. The idea is to treat each participating blockchain of a CBT as a *node* in the context

of distributed systems. Unsurprisingly, adopting 2PC protocols over a set of heterogeneous blockchains entails a series of new challenges we need to overcome for
reasons that will become clearer in later subsections.
Two-phase Commit

Two-phase commit protocols are known to be *blocking*, meaning that in some scenarios, the protocol cannot make progress and is stuck, which has been theoretically
proven [36]. However, it is well accepted that the chance of blocking is relatively
low and, in fact, it can be ignored in practical systems if a strong fault-tolerance
mechanism is implemented.[3] The question becomes

> Is an Internet-of-blockchains system "practical" enough where 2PC is appro-
> priate with a negligible chance of blocking?

The answer is somewhat cliche: it depends. The federation of heterogeneous
blockchains is practical provided that the example designs shown in Figs. 2 and 4
are implemented in the system, which is convincingly the case. On the other hand,
the federation of distributed systems, each of which is treated as a conventional
"node" can be plausibly challenged: existing results and theorems from the literature
of distributed computing almost always assume the underlying node is an entity
on its own; extending the entity to a distributed system is possible only if we can
demonstrate that the distributed system (i.e., a participating blockchain) can be treated
as a conventional node from *every single* perspective. It turns out that they differ in
many ways, unfortunately:

Isolation In 2PC and conventional distributed computing, a node is an independent entity. However, it is possible for a workstation to participate
in more than one blockchains.

Determinacy A conventional node in a distributed system behaves like an automaton: Once the node enters a certain state, there is no way for the node
to "regret" the move.[4] As we discussed before, it is not uncommon
for a blockchain to recall a fork of blocks whose transactions (and
their states) are invalidated.

Fault Tolerance Let f denote the failure rate of a single node, then the chance for a
single node to follow the protocol is $1 - f$. For a n-node blockchain
network, however, assuming the failure rates are the same for all
nodes in a PBFT-like consensus protocol, the overall chance for a
blockchain to follow the protocol is

---

[3] There is a stronger protocol, called three-phase commit (3PC), which guarantees the *liveness* (i.e.,
no blocking) of the transaction. 3PC incurs much more message passing than 2PC and imposes
performance overhead overwhelmingly. As a result, 3PC is not often widely used in production
systems.

[4] Instead, the node at state s_i can choose to return to the previous state s_{i-1}, but technically that
state is a new one, s_{i+1}, which happens to be equal to s_{i-1}: $s_{i+1} = s_{i-1}$.

$$P = \sum_{k=0}^{\lfloor n/3 \rfloor} \binom{n}{k} \cdot f^k \cdot (1 - f)^{n-k},$$

which is clearly larger than f for large n's and practical f's. For example, when $n = 10$ and $f = 0.2$, $P \approx 0.9993 \gg 1 - f = (1 - 0.2) = 0.8$.

Communication   A single node is suitable to be applied with a semi-synchronous communication model—we can expect the node is responding within a certain period of time or number of rounds. For a blockchain to reach a consensus and respond to the requester, however, we may want to drop the somewhat strong assumption on the upper bound of delay (latency).

Leader Election

Distributed commit protocols like 2PC are not completely decentralized: for each of the two phases, a *coordinator* initiates the communication. In a more general sense, such a coordinator is called a *leader* in the literature of distributed systems. The question then becomes which participating blockchain should be elected as the leader in the 2PC protocol. Technically, any participant should qualify for the leader; but from an administrative perspective, there are many non-technical factors influencing the decision. The solution proposed in [40] advocates to specify a node or a blockchain to *witness* the 2PC protocol among blockchains. That is, it is the *witness blockchain*'s responsibility for managing the execution of 2PC.

The introduction of the witness blockchain, admittedly, resembles the concepts of hub and relay chain from Cosmos and Polkadot, respectively. There are a few key differences between the 2PC leader blockchain and the central master blockchain, though. First, the leader in 2PC is not involved in data storage and data exchange (e.g., token transfer). The fact that a leader being implemented as a blockchain is due to security reasons. Control messages pertaining to the 2PC protocol are expected to be circulated among the leading blockchain and participating blockchains. Second, the witness blockchain is part of the flat organization of blockchains. Although the witness blockchain is called a "leader" or "coordinator", it does not entail a hierarchy of blockchains as Cosmos or Polkadot. Note that in the original 2PC protocol, all nodes are equal, and the leader can be elected using arbitrary election algorithms. Third, the witness blockchain could have physical overlap with participating blockchains. As the paper [40] mentions, it is possible for the participating blockchains' nodes to contribute to the witness blockchain, which is not allowed in Cosmos and Polkadot.

Performance

The 2PC-based protocol is clearly more time-efficient than the swap-based protocol. In 2PC, the messages are *broadcasted* from the coordinator to all nodes, namely, the *participants*. Therefore, in a *civil* case—meaning that no blocking happens, there are two round trips of message passing or four hops. The time complexity is simply constant $O(1)$, by comparison to $O(n)$ as discussed in Sect. 3.1.

In terms of messages, 2PC does incur more overhead than swap. The total number of messages incurred by the 2PC protocol, assuming the civil case, is about $4n$, where n denotes the number of blockchains. By comparison, the swap-based protocol, assuming a sub-transaction or swap happens between a specific pair of blockchains, incurs $2n$ messages. Therefore, we can think of 2PC- and swap-based protocols as being a tradeoff between time efficiency and message overhead.

Atomicity

The 2PC-based protocol offers stronger atomicity than the swap-based protocol. It is impossible for a blockchain following 2PC to commit a local sub-transaction by itself, and as a result, a participating blockchain cannot be "worse-off" as in the swap-based protocol. At the inter-blockchain level, transactions are completed with the atomic property.

The atomicity at the level of individual blockchains, however, it is not guaranteed by 2PC. This is related to the fundamental question of whether we should treat a participating blockchain as a conventional node in distributed systems. Specifically, if a participating blockchain cancels a particular fork, then all the blocks and transactions should be rolled back, a scenario that cannot be handled by 2PC. We will discuss possible solutions to overcome the atomicity challenges in Sect. 3.3.

3.3 Atomicity of Forked Blockchains: A Taxonomy of Protocols

This section gives an overview of possible protocols under various communication models to achieve the atomicity of cross-blockchain transactions (CBTs). More technical details can be found in [45].

Assumption and Notation

We assume the crashed node will eventually be recovered and can be replaced by a functional node in a reasonable period, denoted by variable F following a normal distribution: $F \sim N(f, \sigma_F^2)$. Moreover, during a single transaction, the failures will not happen indefinitely but for finite times denoted by λ—a discrete variable indicating the number of failures during the transaction.

We assume the network transfer can be delayed but not indefinitely: the communication is asynchronous and persistent. That is, the messages can be *eventually* delivered in a reasonable time. The network latency L is assumed to follow a normal distribution with mean τ and standard derivation σ_L, respectively. That is, $L \sim N(\tau, \sigma_L^2)$.

We assume that a blockchain can *finalize* the main branch in *finite* time, after which the transactions cannot be rolled back. In Bitcoin, for example, the *pending* time is about 1 hour—six blocks of transactions. We denote the average pending time for C_i is δ_i, which also includes the *waiting* time for a transaction to be picked up by the system.

We assume there is an effective programmable way for different blockchains to communicate. This is mostly true for new blockchain implementations with *smart contracts*. For those old systems, e.g., Bitcoin, which do not support smart contracts, we assume a proxy is available on such systems for cross-blockchain communications.

Notations We denote the set of blockchains as $C = \{C_i\}$, where each C_i, $i \in \mathbb{Z}_+$, represents a specific blockchain in the consortium of blockchains. We use C_{-i} to denote the complement set $C \setminus \{C_i\}$, following the naming convention in game theory. The cardinality, or order, of the set, i.e., $|C|$, indicates the total number of blockchains involved in the transaction. Each blockchain C_i comprises a series of linked blocks, denoted as B_i^j, where the superscript j indicates the *index* of the block on blockchain C_i. Each block is filled with a series of transactions, denoted by T_k, where k implies a universally unique identifier (UUID) of each transaction since the inception of the blockchain consortium. It should be clear that, however, although k is unique globally, it will appear at least once on each C_i and possibly more than once if C_i has branches during the processing of T_k. For each C_i, there is a corresponding set $N_i \subseteq N$ denoting the set of *nodes* having joined the network of blockchain C_i. It is possible that a node joining multiple blockchains, e.g., $n \in N_i$, $n \in N_j$, and $i \neq j$.

Synchronous Cross-blockchain Transaction Protocols

The first category of protocols is called Synchronous Cross-Blockchain Transactions Protocol (SBP). SBP is designed to strictly enforce the ACID properties of cross-blockchain transactions. The targeted workloads include those that need to follow strong consistency models such as financial transactions. As a tradeoff, the performance, especially the latency, is not at the high end of the spectrum of candidate protocols.

SBP respects each individual blockchain's own branches and delays the global commit until no single blockchain can unilaterally rollback the transaction. As the conventional wisdom in distributed commit protocols, a specific blockchain initiates the multi-party transaction. In the literature, this initiator is usually called a *coordinator*, although we want to point out that this coordinator can be any participant C_i in the pool C. Many *leader election* algorithms can be applied to select the coordinator with the *proxies* on C's. The specific node $n \in N_i$ serving as the *endpoint* for the inter-blockchain communication can also be arbitrarily selected as long as the following conditions are met: (i) other nodes $N_i \setminus \{n\}$ are aware of the role of n and (ii) all the intra-blockchain transaction updates have been applied to n.

Suppose C_i initiates a transaction T_k among all elements in C, and $|C| \geq 3$. We will begin by describing the protocol in the civil case.

Phase I First, C_i broadcasts a PRE- COMMIT message to (the proxies of) C. It should be clear that C_i in this case serves as both the *coordinator* and a (local) *participant*. C_i then waits for a READY reply from each blockchain in C. A blockchain $C_j \in C$ (again, $j = i$ is allowed, implying a local message) replies a READY message to C_i after (i) all prerequisites are satisfied, e.g., the balance is higher than the funds to be deducted in a cryptocurrency application and (ii) more importantly, the entity is *locked*. The second action is crucial to avoid double-spending issues.

Phase II Second, C_i broadcasts a COMMIT message to C and waits for a DONE reply from each element in C. A blockchain $C_j \in C$ carries out its local operation, and wait for δ_j before returning a DONE message to C_i. The participants then should unlock the entities. Once C_i receives $|C|$ *done* replies, T_k is marked completed.

Therefore, the civil case of SBP runs much like a 2PC protocol except for the introduction of pending time δ_j. The period enforced by δ_j can only preclude the possible branches in blockchains and yet cannot avoid the possible blocking in the uncivil case where nodes do fail (up to crash failures) followed by possible blocking. One way to fix that is to introduce an additional phase, essentially extending the protocol into three phases, which has been extensively studied in the literature and is not a practical approach due to unacceptable performance. What is proposed in [39] to overcome the blocking issue is more lightweight: taking a passive heartbeat approach to effectively detect node failures. It should be noted that this approach becomes effective only because each node in a CBT is essentially a set of nodes, i.e., N_i for C_i, such that if the original proxy node $n \in N_i$ fails, we can quickly reselect $n' \in N_i$ to continue the SBP protocol.

Formally, suppose $n \in N_i$ is the endpoint of C_i, the proxies on other nodes $N_i \setminus \{n\}$ run a *heartbeat* probe to n, whose interval is denoted as σ_i. Let $\overline{\sigma} = \sup\{\sigma_i, 1 \leq i \leq |C|\}$, it is not hard to see that SBP can be blocked by up to $\overline{\sigma}$. In practice, we can set $\overline{\sigma} \ll \underline{\delta}$, where $\underline{\delta} = \inf\{\delta_j, 1 \leq j \leq |C|\}$, such that the heartbeat overhead is negligible.

Correctness We will go over the four required properties of a transaction.

Atomicity SBP takes a conservative approach to commit the requested transaction. At any point during the two-phase protocol, any states other than the expected ones mentioned in the protocol narrative results in a global ABORT. A more subtle yet rare case is that no qualified node can be found after the *heartbeat* protocol detects a crash failure, in which case the entire SBP also aborts the transaction.

Consistency The changes incurred by the transaction would be invisible to users until C_i marks the completion of the transaction. Thus, SBP implements a strong consistency model, and there are no dirty-write or repeated-read

Table 1 Symbols used in the analysis of synchronous cross-blockchain transaction protocols

Symbol	Meaning
C	Set of all blockchains
N	Set of nodes, i.e., miners
λ	Number of failures during the transaction
F	Recovery period, $F \sim N(f, \sigma_F^2)$
L	Network latency, $L \sim N(\tau, \sigma_L^2)$

issues during the course of distributed transaction processing. Indeed, this strong consistency is attributed to the locking approach with the price of suboptimal performance in transaction latency. We will speak more about performance in the complexity discussion shortly.

Isolation This can be trivially verified by the fact that locking and unlocking are implemented correctly, as discussed in the protocol.

Durability Updates are persisted on all the nodes in each involved blockchain.

Analysis The variables and notations used in the following analysis are summarized in Table 1.

Proposition 1 *The total number of messages passed in SBP is* $O(|N|)$.

Proof Obviously, the maximal number of messages are sent when the nodes are failed repeatedly for *finite* times, and the transaction eventually completes. It is crucial to note that the failure can happen for a limited time because otherwise, our assumption would not hold.

In phase I, the total number of messages between elements in C is

$$
\overbrace{2 \cdot \lambda \cdot (|C| - 1)}^{\text{inter-blockchain}} + \overbrace{\sum_{C_i \in C} (|N_i| - 1)}^{\text{intra-blockchain}}
$$

$$
\begin{aligned}
&\leq\ 2 \cdot \lambda \cdot (|C| - 1) + |C| \cdot |N| - |C| \\
&=\ |C| \cdot |N| + (2\lambda - 1)|C| - 2\lambda \\
&\leq\ |C| \cdot |N| + 2\lambda|C| \\
&=\ |C| \cdot (|N| + 2\lambda).
\end{aligned}
$$

The messages in phase II can be similarly calculated. The total number of messages is thus less than $2|C| \cdot (|N| + 2\lambda)$. The claim then follows if $|C|$ and λ are significantly smaller than N, which is part of our assumption. $\qquad\square$

Proposition 2 *The expectation of the longest period for a single transaction, i.e., the latency, is bounded by* $4\tau + \lambda(f + \overline{\delta})$, *where* $\overline{\delta} = \sup\{\delta_i, 1 \leq i \leq |C|\}$.

Proof The latency of phase I is calculated as

$$\Delta_1 = 2 \cdot L + \lambda_1 \cdot F,$$

and the latency of phase II is bounded by

$$\Delta_2 = 2 \cdot L + \underbrace{\lambda_2 \cdot F + \sum_{n \in N_i} \delta_i}_{\lambda_2},$$

where n indicates a failed node, λ_1 indicates the number of failures in Phase I, λ_2 indicates the number of failures in Phase II, and $\lambda = \lambda_1 + \lambda_2$. Therefore, the expectation of the overall latency $E(\Delta)$ has an upper bound:

$$E(\Delta) = E(\Delta_1 + \Delta_2)$$

$$= E\left(2L + \lambda_1 F + 2L + \lambda_2 F + \sum_{n \in N_i} \delta_i\right)$$

$$= E\left(4L + \lambda F + \sum_{n \in N_i} \delta_i\right)$$

$$= 4\tau + \lambda f + E\left(\sum_{n \in N_i} \delta_i\right)$$

$$\leq 4\tau + \lambda f + \lambda_2 \overline{\delta}$$

$$\leq 4\tau + \lambda(f + \overline{\delta}),$$

as claimed. □

In practice, τ can be easily measured in terms of milliseconds; f usually takes a few seconds, e.g., to reboot the failed node; δ is also well understood: in Bitcoin, for instance, it roughly takes an hour (i.e., six blocks, each of which takes about 10 min) to finalize a transaction. However, it is not trivial to estimate λ other than keeping an empirical log over the failure rate. We want to point out that a Poisson distribution can become a handy tool for quickly estimating the transaction delay. That is, the probability of k failures can be estimated by $\dfrac{\lambda^k e^{-\lambda}}{k!}$, where e is Euler's number.

RedoLog-based Protocols

While SBP discussed in the previous section achieves strong consistency, the price is a somewhat long delay. Therefore, SBP is ideal for those time-insensitive applications that are required to guarantee ACID properties. This subsection investigates the other end of the spectrum: What if the workload is highly time-sensitive and can tolerate temporary inconsistencies. For example, applications like emails can accept an *eventual consistency* semantically. To this end, a new type of protocols based on *redo logs*, namely, RBP [45], is recently proposed.

RBP makes a key change to the way how participants reply to the DONE messages back to the coordinator C_i. Instead of waiting for a period of δ_j, C_j replies C_i right after the local updates are completed. Indeed, the question then becomes what if C_j decides to cut off the branch comprising the completed transaction T_k between C_i and C_j's later on? To this end, blockchain C_i maintains a *sliding window* that records recent transactions completed in the past δ_i period. The rationale is that if any of these *pending* transactions are on the path of a shorter branch of C_i, C_i can take according actions such as (i) returning the transactions back to the request pool or (ii) immediately rescheduling the transactions. RBP takes the former approach: transactions on the shorter branches are recycled back into the pool of requests. Note that we cannot construct *complement* transactions to *undo* the changes because those transactions are invisible to the main branch of C_i: at any point, a transaction cannot be enclosed in more than one branch—once a transaction is packed within a block and appended to a branch, the request is considered completed, and the transaction will not be worked on by other miners. It should be clear that a branch is a sequence of blocks replicated at some nodes, but a node can only follow one such sequence. In doing so, double spending would be prevented in RBP.

Evidently, RBP still meets the *atomicity* requirement: there is no "partial" transaction committed. It is also trivial to check that both *isolation* and *durability* hold in RBP. For *consistency*, RBP implements an eventual consistency semantics: the transactions on shorter branches will eventually be reprocessed. It should be clear that those transactions that have been appended to a shorter branch were indeed verified by a quorum of miners. In blockchains, *verification* and *consensus* are two different procedures: there could be multiple branches, each of which has collected quorum—many (e.g., 51%) miners have verified the validity of the new block; when a specific miner itself starts its own mining, it will only pick the longest branch, and every other miner would do so, thus forming a consensus.

We conclude this subsection with a more detailed quantitative analysis in the following. We are particularly interested in the improved latency paid by the weak consistency semantics. Let the transactions in the sliding window of C_i be $\mathcal{T}_i$. Consequently, the throughput of blockchain C_i can be calculated by $\dfrac{|\mathcal{T}_i|}{\delta_i}$. Because of the possible *cascading effect* implied by the C_i's nondeterministic branching behavior, we cannot derive an upper bound over the latency of a transaction $T_k \in \mathcal{T}_i$. However, if no branching happens during T_k, the latency can be as low as $4\tau + \lambda f$. Recall that both τ and f are orders of magnitude smaller than δ, and λ represents a few failed nodes in a time unit; therefore, RBP is expected to deliver a significantly smaller latency than SBP. Again, this gain is traded by the (strong) consistency.

3.4 Limitation

One denominator shared by the protocols presented in this section is *time*. Each protocol, more or less, takes timing into the design space, which is understandable as timing indeed plays a critical role in a distributed system in terms of many aspects such as synchronization and fault tolerance. We usually call such algorithms and protocols with timing constraints *temporal analysis* or *procedural analysis*. As previous subsections demonstrate, temporal protocols might seem somewhat lengthy, often error prone, and almost always sensitive to the parameterization of end-users. Can we possibly design algorithms and protocols for cross-blockchain transactions using more static machinery without too much parameterization without compromising correctness and efficiency? We provide one such theory based on mathematical topology in Sect. 4.

4 A Topological Theory of Cross-Blockchain Transactions

We present a new approach to modeling and analyzing cross-blockchain transactions (CBTs) through the rigorous mathematical theory on topology. Our topological treatment of CBTs is motivated by the fact that the root cause of limitations of existing protocols lies in the sophisticated and nondeterministic interaction among processes in a distributed environment. By comparison, the topological method, which we will discuss in this section, offers a static "snapshot" and elegantly characterizes the processes (tasks) in CBTs. The very fact that we can *topologize* CBTs implies that we would be able to apply a vast number of tools in the literature of topology for designing more efficient and more reliable CBT protocols. More technical detail of the methods discussed in this section can be found in [39, 41–44].

4.1 Topological Preliminaries

We will briefly review the basic concepts and terminology in point-set topology (also called *general topology*) and algebraic topology. There are many excellent books on point-set topology, e.g., [4, 22, 28]; for algebraic topology, we recommended the following texts [10, 15, 29].

Point-set Topology

Point-set topology is naturally extended from set theory, as the name suggests. Mathematically speaking, a *topology* of a set S is a collection of subsets of S, denoted $\mathcal{T}$.

One example topology of S is then the power set of S, $\mathcal{P}(S)$, which consists of all the possible $2^{|S|}$ subsets of S. This is also called the *discrete topology* of S, which is the "largest" topology in the sense that it comprises the maximal number of subsets of S. The tuple $(S, \mathcal{T})$ is called the *topological space* of S. If the context is clear, we often refer to S to indicate space $(S, \mathcal{T})$. Each of the subsets U from $\mathcal{T}$ is called an *open set*, and the complement set $S \setminus U$ is a *closed set* by definition. A function g from space X to Y is called *continuous* if: $\forall v$ is an open set in Y, its *preimage* $g^{-1}(v)$ is an open set in X. The composition of two continuous functions is also continuous. If both g and g^{-1} are continuous and bijective (one-on-one mapping), we call g a *homeomorphism*. Because a homeomorphism is defined purely on open and closed sets, two topological spaces are considered equivalent if such homeomorphism exists. Usually, we migrate a complex problem in one topological space to another such that the problem can be solved more efficiently or more intuitively.

Point-set topology is not a handy tool for computational applications because many of its definitions and results (theorems) are qualitative; the whole theory is built upon a series of axioms. On the other hand, point-set topology is widely used in mathematical analysis (e.g., real analysis, functional analysis).

Simplex and Simplicial Complex

In addition to point-set topology, there are another branch of *algebraic-topological* methods usually categorized into *homotopy* groups and *homology* groups, which study the smaller pieces of the targeting objects and try to map the geometrical objects into algebraic objects, such as *groups*. The building blocks we are interested in are called (geometric) *simplices* (the plural form of simplex). In this chapter, by *simplex* we mean an *abstract simplex*[5], defined as a set S of vertices. Any subset of a simplex σ is also a simplex and is called a *face* of σ.

Geometrically speaking, a simplex σ consists of all the possible points, edges, triangles, tetrahedrons, and higher dimensional objects that can be composed of the vertices in the set of points, S. From a combinatorial perspective, a collection of σ's can be thought of as an object representing more sophisticated relationships among the vertices in S, which is called an *abstract simplicial complex*, denoted K. The definition of a simplicial complex can be easily violated in practice: It is "simplicial" in the sense that any $\sigma_1 \cap \sigma_2 \in \sigma_1, \sigma_2$, meaning that the simplices (including the empty set $\emptyset$) shared by σ_1 and σ_2 must also be valid simplices of both σ_1 and σ_2. Figure 5 shows that a triangulation[6] of Fig. "8" has the shared line segment $\{b, f\}$ that is not an element of the simplices of the upper triangle:

$$\{\emptyset, \{a\}, \{b\}, \{c\}, \{a, b\}, \{b, c\}, \{a, c\}, \{a, b, c\}\}.$$

---

[5] Historically, algebraic topology was called *combinatorial topology* when the focus was on the "counting" abstraction of the objects.

[6] There is a rigorous definition of triangulation with certain requirements. Here we use triangulation to refer to one possible abstraction with vertices and edges, i.e., a series of 1-simplices.

Fig. 5 Two overlapped
simplices, $\sigma_1 = \{a, b, c\}$ and
$\sigma_2 = \{d, e, f\}$, of Fig. "8".
The union of σ_1 and σ_2 is *not*
a simplicial complex: the
overlap of the two simplices
is a line segment that is *not*
an element of either simplex

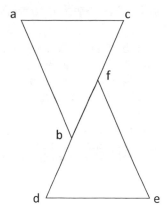

In the naming convention of topology, we usually call a simplex of $(n + 1)$ points
n-dimensional because the dimension reflects our geometrical intuition of the coalition among the given points. For instance, a two-dimensional plane can be uniquely
constructed by three points. The very fact that three points decide a two-dimensional
plane is critical; we can generalize this idea into other dimensions (we assume all the
points are pairwise linearly independent): two points can decide a one-dimensional
line, four points can decide a three-dimensional tetrahedron, and, of course, five points
can decide a four-dimensional object that cannot be visualized by human beings. That
is why the dimension, in topology, is defined as the number of underlying points -1:

$$dim\ \sigma = |\sigma| - 1,$$

where σ is an object called *simplex* comprised of n-dimensional points with additional
requirements (an arbitrary object will concern us later).

Formally, a simplex is a collection of points and all of their high-dimensional connections. For example, three points form a two-dimensional simplex that comprises
(i) three (zero-dimensional) points, (ii) three (one-dimensional) edges, and (iii) one
(two-dimensional) solid triangle. Similarly, if we have four points, then the induced
three-dimensional simplex will have four points, six edges, four solid triangles, and
one fulfilled tetrahedron. In contrast, a hollow tetrahedron is *not* a simplex but a union
of four two-dimensional simplices. In fact, the collection of arbitrary simplices is a
(abstract) simplicial complex.

Many properties can be carried from simplices to simplicial complexes. Let σ be
a simplex, and C be a simplicial complex. We define the dimension of the simplicial
complex as the dimension of the highest dimensional simplex in it:

$$dim\ \ C = dim\ \sigma, \sigma \in C \text{ and } \forall \tau \in C, dim\ \sigma \geq dim\ \tau.$$

Note that a simplex can also be thought of as a simplicial complex, the latter is
a collection of only one simplex. If no ambiguity arises, we also call a simplex a
(simplicial) complex.

4.2 Assumptions and Notations

Each blockchain is represented by a cluster of nodes. For this reason, we will use blockchain and cluster interchangeably. We will also use vertex and node interchangeably to refer to the computation entity in both the system and its low-dimensional topology (i.e., a two-dimensional graph). The set of n distinct blockchains is denoted C, where each blockchain is indexed as C_i, $0 \le i < n$. We assume each cluster can spawn an arbitrary number of forks. Although two forks are most commonly seen in cryptocurrency, the number of forks is not bounded from a theoretical perspective. Formally, a *blockchain fork* is defined as follows.

Definition 1 (*Blockchain Fork*) Physically, a blockchain fork is a continuous sublist of a node's local blockchain, whose values are different than those of other nodes in the same cluster. From the cluster's perspective, a blockchain fork is a *simple path* (i.e., no vertex is revisited) that has at least one non-spawning vertex untouched by other simple paths.

Let F denote the set of all forks in the cluster and $f_i \in F$ denote the ith fork of the cluster. We use $F_{-i} = F \setminus \{f_i\}$ to denote the complementary set of fork f_i in F, i.e., the forks followed by other nodes that do not follow f_i. From a practical perspective, each fork (in a specific round) has one of the following three states. f_i is called *eliminated* if any of other forks $f_j \in F_{-i}$ "surpasses" f_i according to the protocol[7], where $j \ne i$. f_i is *confirmed* if all other forks F_{-i} are eliminated. Any other forks are called *undecided*. Therefore, we can construct a *fork graph* among the elements of F with three types of vertices and their paths. Note that this graph is static; we speak of nothing (yet) about the timestamp or step number. Now we are ready to formalize the transactions among (forks of) blockchains.

Definition 2 (*Transaction Proxy*) A transaction proxy on a blockchain is the physical node where the transaction is initiated.

Accordingly, if there is no fork (i.e., a single branch) in the blockchain, it does not matter which node is chosen as the proxy for the transaction; however, multiple forks would imply the distinction between proxies and potentially lead to different outcomes of the transaction. A proxy is called *live* if its fork is undecided or confirmed. A transaction can thus be characterized by the states of the proxies in the fork graphs.

Although a conventional transaction in relational databases is time-oblivious since users are only interested in the final outcome of the transaction (*commit* or *abort*), a blockchain indeed grows over time. To this end, we introduce the extended concept of the fork graph, the so-called *growing fork graph* in the following.

Definition 3 (*Growing fork Graph*) A growing fork graph, F_i^ω, is a sequence of fork graphs associated with the cluster C_i. Let F_i^t denote the fork graph of cluster C_i at time t, then

---

[7] For example, in Bitcoin, a branch surpasses others when it first appends six blocks.

$$F_i^\omega = \left(F_i^0, F_i^1, \dots\right).$$

With the above terms defined, we are ready to construct the topological spaces associated with the transaction proxies, the static fork, and the real-time blockchain forks.

4.3 Topological Space of No-Fork Blockchains

We start with the simplest case where no forks are involved in the transaction. We call the blockchain, or cluster, to *commit* if it is ready to commit its local changes of the transaction. Similarly, a blockchain is said to *abort* if it rollbacks its pending local changes.

Definition 4 (*Cluster Distance*) Let C denote the set of clusters each of which represents a blockchain, $C = \{C_0, \dots, C_{n-1}\}$. Define the *ternary distance* between any pair of elements in C, C_i, and C_j as a function $d_t : C \times C \rightarrow \{1, \frac{1}{2}, 2\}$:

$$d_t(C_i, C_j) = \begin{cases} 2, & \text{both } C_i \text{ and } C_j \text{ commit;} \\ \frac{1}{2}, & \text{both } C_i \text{ and } C_j \text{ abort;} \\ 1, & \text{otherwise.} \end{cases}$$

Intuitively, the distances are selected as the 1st, the negative-1st, and the zeroth powers of base two. It should be noted that we assume $i \neq j$ in the definition; this assumption also holds in the remainder of this paper. The distance is trivially defined as 0 when $i = j$, thus satisfying the requirement of a *metric distance*.

We illustrate the three scenarios in Fig. 6. The two proxies in the green transaction can both commit since there is no fork at the moment. The two proxies in the orange transaction can safely abort because both forks from C_0 and C_1 are to be eliminated. We cannot decide the result of the yellow transaction because there are forks involved. A transaction might involve more than two clusters, say $n > 2, n \in \mathbb{Z}$. The states of a n-party transaction, i.e., an n-cluster transaction, can be naturally extended from the case of two parties.

In order for the above-mentioned transactions to form a nice topological space, we need to show that the distance is well defined. The following lemma proves that the distance is indeed a metric.

Lemma 1 d_t *is a metric on* C.

Proof Obviously, $d_t(C_i, C_j) = d_t(C_j, C_i) \geq 0$ by definition. It remains to show $d_t(C_i, C_j) \leq d_t(C_i, C_k) + d_t(C_k, C_j)$, for arbitrary $k \in \mathbb{Z}, 0 \leq k < n$.

If $d_t(C_i, C_j) = \frac{1}{2}$, it means that both C_i and C_j abort. If C_k commits, then

$$d_t(C_i, C_k) + d_t(C_k, C_j) = 1 + 1 > \frac{1}{2} = d_t(C_i, C_j).$$

Fig. 6 An oversimplified
example of three possible
results (distances) for a
two-blockchain transaction.
If neither of the two
blockchains has forks, we
can safely commit the CBT.
If both blockchains carry out
the transaction on recalled
forks, the transaction has to
be canceled (abort).
Otherwise, the transaction
has a pending state

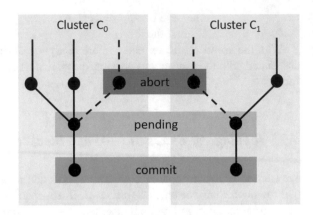

If C_k aborts, then

$$d_t(C_i, C_k) + d_t(C_k, C_j) = \frac{1}{2} + \frac{1}{2} > \frac{1}{2} = d_t(C_i, C_j).$$

If $d_t(C_i, C_j) = 1$, then we know one blockchain commits the transaction and the other aborts. Without loss of generality, we assume C_i commits and C_j aborts in the following. If C_k commits, then

$$d_t(C_i, C_k) + d_t(C_k, C_j) = 2 + 1 > 1 = d_t(C_i, C_j).$$

If C_k aborts, then

$$d_t(C_i, C_k) + d_t(C_k, C_j) = 1 + \frac{1}{2} > 1 = d_t(C_i, C_j).$$

Therefore, $d_t(C_i, C_j) \leq d_t(C_i, C_k) + d_t(C_k, C_j)$.

If $d_t(C_i, C_j) = 2$, meaning that both C_i and C_j commit, it is also easy to verify the triangle inequality. If C_k commits, then

$$d_t(C_i, C_k) + d_t(C_k, C_j) = 2 + 2 > 2 = d_t(C_i, C_j).$$

If C_k aborts, then

$$d_t(C_i, C_k) + d_t(C_k, C_j) = 1 + 1 = 2 = d_t(C_i, C_j).$$

Therefore, again, $d_t(C_i, C_j) \leq d_t(C_i, C_k) + d_t(C_k, C_j)$.   □

Now we have a well-defined metric, it is then natural to construct the topological space that is induced by this metric, as we have shown in Proposition 3.

Proposition 3 *Topological space $\mathcal{T}_{d_t}^C$ on C is induced by d_t.*

Proof Let $\epsilon = \frac{1}{4}$, then the open ball $B_{d_t}(u, \epsilon)$ around the origin u is an open set in the topology. Because all the possible distances between any pairs of clusters are at least $\frac{1}{2}$, this ϵ-ball $B_{d_t}(u, \epsilon)$ splits the cluster C into a collection of single-element sets, each of which is a set of a single cluster C_i, $0 \leq i < n$. That is,

$$\mathcal{B} = \{\{C_0\}, \ldots, \{C_{n-1}\}\}.$$

Evidently, $\mathcal{B}$ is basis of $\mathcal{T}_{d_t}^C$: every element in C exactly belongs to an element in $\mathcal{B}$, and an open set in the topology is simply an arbitrary union of $b \in \mathcal{B}$. □

Having constructed the topology of transactions without explicit forks, we have laid out the foundation of the topological treatment of cross-blockchain transactions and will start adding more real-world ingredients, namely, the forks. We will first study the easier ones where the forks are static, and then extend the idea into time-series forms of forks.

4.4   Topological Space of Static-Fork Complexes

Definition 5 (*Fork Distance*) When two transaction proxies are both live on two fork graphs F_i and F_j, respectively, the fork distance d_f is defined as

$$d_f(F_i, F_j) = \frac{1}{|F_i|} + \frac{1}{|F_j|}.$$

If any of the two proxies is not live, we define

$$d_f = \delta_f = \frac{1}{\sup\{|F_i| : F_i \in F\}}.$$

As an example, we calculate some fork distances of transactions on Fig. 6 as follows (F_0 and F_1 representing the fork graphs of C_0 and C_1, respectively):

$$d_f(F_0^{commit}, F_1^{commit}) = \frac{1}{|F_0^{commit}|} + \frac{1}{|F_1^{commit}|} = \frac{1}{2} + \frac{1}{2} = 1,$$

and

$$d_f(F_0^{abort}, F_1^{abort}) = \frac{1}{\sup\{|F_0^{abort}|, |F_1^{abort}|\}} = \frac{1}{\sup\{3, 2\}} = \frac{1}{3}.$$

Now, we will show that d_f is a metric on the set of fork graphs.

Lemma 2 *Fork distance d_f is a metric on F.*

Proof If any proxy is not live, say that on F_i, then we have

$$d_f(F_i, F_j) \triangleq \delta_f \geq 0,$$

$$d_f(F_i, F_j) = d_f(F_j, F_i) = \delta_f, \text{and}$$

$$d_f(F_i, F_j) = \delta_f \leq \delta_f + d_f(F_k, F_j) = d_f(F_i, F_k) + d_f(F_k, F_j).$$

If both proxies are live, then by definition

$$d_f > 0 \geq 0,$$

the distance between two forks is

$$d_f(F_i, F_j) = \frac{1}{|F_i|} + \frac{1}{|F_j|} = \frac{1}{|F_j|} + \frac{1}{|F_i|} = d_f(F_j, F_i),$$

and finally for triangle inequality:

$$\begin{aligned}
d_f(F_i, F_j) &< \frac{1}{|F_i|} + \frac{1}{|F_j|} + \frac{2}{|F_k|} \\
&= \left(\frac{1}{|F_i|} + \frac{1}{|F_k|} \right) + \left(\frac{1}{|F_k|} + \frac{1}{|F_j|} \right) \\
&= d_f(F_i, F_k) + d_f(F_k, F_j).
\end{aligned}$$

$\square$

We are ready to construct the topology of static-fork graphs, as demonstrated in the following proposition.

Proposition 4 (Fork Space) *The topological space $\mathcal{T}_{d_f}^F$ over the set of fork graphs F is induced by d_f.*

Proof Let $\epsilon = \dfrac{1}{1 + \sup\{|F_i| : F_i \in F\}}$. Let $B_{d_f}(F_i, \epsilon)$ be an open ball around the center of F_i with a radius of ϵ, namely, an ϵ-*ball*. We will show that these ϵ-balls form a *discrete topology*: every open set induced by an open ϵ-ball is a singleton subset of exactly one fork graph in F. That is, we need to show that $\epsilon < d_f$ for any pair of fork graphs. Note that by definition, $\epsilon < \delta_f$. Therefore, it suffices to show that $\delta_f < d_f$. Indeed, that is how we construct δ_f, because

$$\delta_f < 2\delta_f = \frac{2}{\sup\{|F_i| : F_i \in F\}}$$

$$= \frac{1}{\sup\{|F_i| : F_i \in F\}} + \frac{1}{\sup\{|F_j| : F_j \in F\}}$$

$$\leq \frac{1}{|F_i|} + \frac{1}{|F_j|} = d_f(F_i, F_j).$$

Therefore, ϵ is small enough to split the space into a series of elements whose distance exceeds the boundaries of those ϵ-balls. That is, we now have a collection $\mathcal{B}$ of singleton open sets, each of which comprises exactly one fork graph:

$$\mathcal{B} = \{\{F_0\}, \ldots, \{F_{n-1}\}\}.$$

$\mathcal{B}$ is a basis because any element in F belongs to a subset of $\mathcal{B}$ and the intersection between any two subsets is empty. $\square$

Now we are ready to study the time-varied topology of growing blockchain forks. Essentially, we extended the static-fork graph space with an additional time dimension that grows indefinitely.

4.5 Topological Space of Growing Fork Blockchains

We extend the fork graph with a timestamp to model the real-time fork topology. A fork graph at time t is denoted F^t; the set of infinite fork graphs is denoted F^ω, by the naming convention of point-set topology. That is, $F^\omega = (F^0, F^1, \ldots,)$. F_i^ω then represents the ever-growing fork topology of cluster C_i.

As before, we first define the metric over the growing forks.

Definition 6 (*Growing fork distance d_g*)

$$d_g(F_i^\omega, F_j^\omega) = \frac{1}{\inf\{|F_i^m|, |F_j^m|\}},$$

where $m = \inf\{t : |F_i^t| \neq |F_j^t|\}$, $t \in \mathbb{Z}$, $t \geq 0$. Essentially, d_g tracks the smallest possible number of forks between two growing fork graphs once any of them starts forking. Note that all blockchains initially have a single fork from the genesis block.

To make matters more concrete, we illustrate the metric in Fig. 7, where there are three blockchains whose fork graphs are denoted by F_0, F_1, and F_2, respectively, in the first three steps ($t = 0, 1, 2$). Here are some example calculations:

- $d_g(F_0^\omega, F_1^\omega) = \dfrac{1}{\inf\{|F_0^2|, |F_1^2|\}} = \dfrac{1}{\inf\{3, 2\}} = \dfrac{1}{2};$

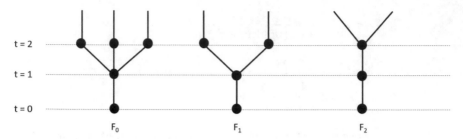

Fig. 7 Example of three blockchains' growing fork topology

- $d_g(F_1^\omega, F_2^\omega) = \dfrac{1}{\inf\{|F_1^2|, |F_2^2|\}} = \dfrac{1}{\inf\{2, 1\}} = \dfrac{1}{1} = 1;$
- $d_g(F_0^\omega, F_2^\omega) = \dfrac{1}{\inf\{|F_0^2|, |F_2^2|\}} = \dfrac{1}{\inf\{3, 1\}} = \dfrac{1}{1} = 1.$

As before, we will prove that d_g is indeed a metric:

$$\forall x, y, z \in F^\omega, \text{ then } d_g(x) \geq 0,$$
$$d_g(x, y) = d_g(y, x),$$
$$d_g(x, y) \leq d_g(x, z) + d_g(z, y).$$

Lemma 3 d_g *is a metric on* F^ω.

Proof By definition, we have

$$d_g(F_i^\omega, F_j^\omega) = d_g(F_j^\omega, F_i^\omega) = \frac{1}{\inf\{|F_i^m|, |F_j^m|\}} \geq 0.$$

It remains to show $d_g(F_i^\omega, F_j^\omega) \leq d_g(F_i^\omega, F_k^\omega) + d_g(F_k^\omega, F_j^\omega)$. Define m_i to be the smallest index such that $|F_i^{m_i}| > 1$.

If $m_i = m_j$, it implies $m = m_i = m_j$.

- If $m_k < m$, then we know that F_k^ω starts to fork earlier than F_i^ω and F_j^ω. Therefore,

$$d_g(F_i^\omega, F_k^\omega) = d_g(F_k^\omega, F_j^\omega) = \frac{1}{|F_i^{m_i}|} = 1.$$

Note that since both F_i^ω and F_j^ω start forking at the same time, we have

$$d_g(F_i^\omega, F_j^\omega) \leq \frac{1}{2} < 2 = d_g(F_i^\omega, F_k^\omega) + d_g(F_k^\omega, F_j^\omega).$$

- If $m_k > m$, this means F_k^ω starts to fork after F_i^ω and F_j^ω. Then the two distances $d_g(F_i^\omega, F_k^\omega)$ and $d_g(F_k^\omega, F_j^\omega)$ depend on $F_k^{m_k}$. We can similarly calculate the following:

$$d_g(F_i^\omega, F_k^\omega) = d_g(F_k^\omega, F_j^\omega) = \frac{1}{|F_i^{m_i}|} = 1,$$

and draw the same conclusion as above.

- If $m_k = m$, then all three growing forks start forking at the same time. Then we have

$$d_g(F_i^m, F_k^m) = \frac{1}{\inf\{|F_i^m|, |F_k^m|\}} \geq \frac{1}{|F_i^m|}.$$

Similarly, we also have

$$d_g(F_k^m, F_j^m) = \frac{1}{\inf\{|F_k^m|, |F_j^m|\}} \geq \frac{1}{|F_j^m|}.$$

Therefore, we have

$$d_g(F_i^m, F_k^m) + d_g(F_k^m, F_j^m) \geq \frac{1}{|F_i^m|} + \frac{1}{|F_j^m|}$$

$$> \frac{1}{\inf\{|F_i^m|, |F_k^m|\}} = d_g(F_i^m, F_j^m).$$

The triangular inequality is thus satisfied.

If $m_i \neq m_j$, without loss of generality we assume $m_i < m_j$ in the following:

- If $m_k < m_i$, $m_i < m_k < m_j$, or $m_k > m_j$, obviously we have

$$d_g(F_i^\omega, F_k^\omega) = d_g(F_k^\omega, F_j^\omega) = 1.$$

Then, indeed:

$$d_g(F_i^\omega, F_j^\omega) \leq 1 < 2 = d_g(F_i^\omega, F_k^\omega) + d_g(F_k^\omega, F_j^\omega).$$

- If $m_k = m_i < m_j$, then we know

$$d_g(F_k^\omega, F_j^\omega) = 1.$$

So, we have

$$d_g(F_i^\omega, F_j^\omega) \leq 1 < d_g(F_i^\omega, F_k^\omega) + 1 = d_g(F_i^\omega, F_k^\omega) + d_g(F_k^\omega, F_j^\omega).$$

- If $m_k = m_j > m_i$, then we know

$$d_g(F_i^\omega, F_k^\omega) = 1.$$

So, we have

$$d_g(F_i^\omega, F_j^\omega) \le 1 < 1 + d_g(F_k^\omega, F_j^\omega) = d_g(F_i^\omega, F_k^\omega) + d_g(F_k^\omega, F_j^\omega).$$

Therefore, again, the triangular inequality is satisfied.

□

Lastly, we show that d_g defined as such induces the topology over F^ω.

Proposition 5 *Topological space $\mathcal{T}_{d_g}^{F^\omega}$ on F^ω is induced by d_g.*

Proof Let $\epsilon = \dfrac{1}{1 + \sup\{|F_i^{m_i}| : F_i^\omega \in F^\omega\}}$, $i \in \mathbb{Z}$, $0 \le i < n$, where m_i is the smallest time index such that $|F_i^{m_i}| > 1$ in the infinite sequence F_i^ω. Then an open ball $B_{d_g}(u, \epsilon)$ is fine enough to isolate each element in F^ϵ because:

$$\epsilon < \frac{1}{\sup\{|F_i^{m_i}| : F_i^\omega \in F^\omega\}} \le \frac{1}{\inf\{|F_i^{m_i}| : F_i^\omega \in F^\omega\}}$$
$$\le \frac{1}{\inf\{|F_i^{m_i}|, |F_j^{m_j}|\}} = d_g(F_i^\omega, F_j^\omega),$$

for any $0 \le i, j < n$. Therefore, we have found a basis $\mathcal{B}_g$ of space $\mathcal{T}_{d_g}^{F^\omega}$:

$$\mathcal{B}_g = \{\{F_0^\omega\}, \ldots, \{F_{n-1}^\omega\}\},$$

and the topology can be constructed with this basis.

□

We have constructed three topological spaces for blockchain transactions with the increasing order of sophistication: no-fork transactions, static-fork graphs, and a time series of growing forks. The key insight is that a topological treatment is mathematically rigorous, and, therefore, we can apply existing well-known techniques and tools from topology to derive interesting results. We demonstrate one such result in Sect. 4.6.

4.6 Analyzing Blockchains Through Algebraic Topology

We use dim and skel^k as function operators of a simplex's dimension and k-skeleton, respectively. We use $|\sigma|$ to denote the geometric realization, i.e., the polygon, of (abstract) simplex σ. The N-time Barycentric and chromatic subdivisions are denoted Bary^N and ch^N, respectively. We assume an asynchronous, message-passing communication model among blockchains. Furthermore, we assume the number of faulty nodes t is less than 50% out of the total $(n + 1)$ nodes: $t < \frac{n+1}{2}$.

Task

The task of a cross-blockchain transaction (CBT) is represented by a triple $(\mathcal{I}, \mathcal{O}, \Delta)$, where $\mathcal{I}$ is the input simplical complex, $\mathcal{O}$ is the output simplical complex, and Δ is the carrier map $\Delta : \mathcal{I} \to 2^{\mathcal{O}}$. The triple is well defined because Sect. 4.5 has shown that a topological space can be constructed from any CBT and, therefore, can be abstracted into simplices and simplicial complexes.

Each vertex, i.e., 0-simplex, in $\mathcal{I}$ is a tuple in the form of (v_i^j, val_{in}), where v_i^j denotes block-j at blockchain-i and $val_{in} \in \{0, 1, \bot\}$ denotes the set of possible input values. The meaning in the input set is as follows:

- 0: local transaction not committed;
- 1: local transaction committed; and
- $\bot$: the branch where this block resides is suspended

There is an edge, i.e., 1-simplex, between every pair of vertices in $\mathcal{I}$ except that both vertices are the same block. In general, an l-simplex in $\mathcal{I}$ comprises a set of distinct $l + 1$ blocks as vertices and the higher dimensional k-skeletons, $1 \le k \le l$. Overall, for a $(n + 1)$-blockchain transaction, the input complex $\mathcal{I}$ comprises $3(n + 1)$ vertices and simplices of dimension up to n, i.e., $\dim(\mathcal{I}) = n$.

Each vertex in $\mathcal{O}$ is a tuple (v_i^j, val_{out}), where v_i^j is, again, a specific block and $val_{out} \in \{1, 0\}$ with the same semantics defined for val_{in}. Indeed, all of local transactions in T should only end up with either *committed* (1) or *aborted* (0), respecting the atomicity requirement. The 1-simplices of $\mathcal{O}$ are all the edges connecting vertices whose val_{out}'s are equal, either 0 or 1, among all blocks. Therefore, by definition, the output simplicial complex is disconnected and has two path-connected components: the global transaction is either (i) successfully committed or (ii) aborted without partial changes.

We now construct the carrier map Δ, which maps each simplex from $\mathcal{I}$ to a subcomplex of $\mathcal{O}$. Without loss of generality, pick any l-simplex $\sigma \in \mathcal{I}, 0 \le l \le n$, and Δ specifies the following:

- If all the l val_{in}'s in σ are 1, then $\texttt{skel}^0 \Delta(\sigma) = \{(v, 1) : v \in \texttt{skel}^0\sigma\}$.
- If any of the l val_{in}'s in σ is $\bot$, $\texttt{skel}^0 \Delta(\sigma) = \{(v, 0) : v \in \texttt{skel}^0\sigma\}$.
- For other cases, $\texttt{skel}^0 \Delta(\sigma) = \{(v, 0), (v, 1) : v \in \texttt{skel}^0\sigma\}$.
- Any k-face $\tau \in \sigma, 0 \le k \le l$, is similarly mapped.

Note that, by definition, Δ is rigid: In any of the above three cases, for any l-simplex $\sigma \in \mathcal{I}$, we have $\dim(\Delta(\sigma)) = l$. Evidently, Δ is monotonic: Adding new simplices into σ can only enlarge the mapped subcomplex in $\mathcal{O}$. Δ is also name-preserving as constructed.

Solvability

We start with a simpler version of the CBT task where the identity of the block is considered unimportant. In the literature of distributed computing, such tasks are called *colorless*. The protocol for solving a task is also defined as a triple $(\mathcal{I}, \mathcal{P}, \Xi)$, where $\mathcal{I}$ indicates the input complex, $\mathcal{P}$ indicates the *protocol complex*, and Ξ is the carrier map from simplices of $\mathcal{I}$ to subcomplexes of $\mathcal{P}$: $\Xi : \mathcal{I} \to 2^{\mathcal{P}}$. Informally, $\mathcal{P}$ is the union of all the resulting simplices of executions allowed (i.e., carried) by the protocol.

Definition 7 (*Colorless CBT*) A colorless version of CBT, $(\mathcal{I}, O', \Delta)$, is defined similarly as the general, "colored" CBT, $(\mathcal{I}, O, \Delta)$, without the block identities on vertices in O'. Also, no identity match is required for the carrier map $\Delta : \mathcal{I} \to 2^{O'}$.

Lemma 4 *For colorless CBT $(\mathcal{I}, O', \Delta)$, there exists no continuous map $|\mathtt{skel}^t \mathcal{I}| \to |O'|$ carried by Δ, where $0 < t < \frac{n+1}{2}$.*

Proof (Sketch) The condition $0 < t < \frac{n+1}{2}$ holds by assumption. The input simplicial complex $\mathcal{I}$ is pure of dimension n by construction, meaning that $\mathtt{skel}^t \mathcal{I}$ is $(t-1)$-connected. Because $t > 0$, $\mathtt{skel}^t \mathcal{I}$ is at least 0-connected (i.e., path-connected). As a result, the geometric realization $|\mathtt{skel}^t \mathcal{I}|$ must be connected. However, we know that O' has two disjoint connected components; so $|O'|$ is not connected. Therefore, a continuous map carried by Δ does not exist. $\qquad \square$

Lemma 5 *Colorless CBT $(\mathcal{I}, O', \Delta)$ does not have a t-resilient message-passing protocol.*

Proof (Sketch) For contradiction, suppose protocol $(\mathcal{I}, \mathcal{P}, \Xi)$ solves task $(\mathcal{I}, O', \Delta)$. That is to say, there exists a simplicial map $\delta : \mathcal{P} \to O'$. Then we know that, after N times of Barycentric subdivisions, the carrier map can be written in this form $\Xi(\sigma) = \mathtt{Bary}^N \mathtt{skel}^t \sigma$, for $\sigma \in \mathcal{I}$. That is, there exists a carrier map $\Phi : \mathtt{Bary}^N \mathtt{skel}^t \mathcal{I} \to 2^{O'}$. Taking the geometric realizations, we thus have a continuous map $|\Phi| : |\mathtt{Bary}^N \mathtt{skel}^t \mathcal{I}| \to |O'|$. Note that a subdivision does not change the geometric realization: $|\mathtt{Bary}^N \mathtt{skel}^t \mathcal{I}| = |\mathtt{skel}^t \mathcal{I}|$. Thus, we have a continuous map $|\mathtt{skel}^t \mathcal{I}| \to |O'|$ carried by Δ, a contradiction to Lemma 4. $\qquad \square$

Lemma 6 *A model for colorless CBT $(\mathcal{I}, O', \Delta)$ reduces to a model for general CBT $(\mathcal{I}, O, \Delta)$.*

Proof (Sketch) Suppose a protocol P solves $(\mathcal{I}, O, \Delta)$, we simulate P with P' for $(\mathcal{I}, O', \Delta)$ as follows. For any l-simplex in O, we drop the prefix of the l vertices with map $\varphi : \mathbb{Z} \times V \to V$ such that $(k, val_{out}) \mapsto (val_{out}) \in O'$, $0 \le k \le l$. The carrier map in the colorless counterpart is $\Xi = \Delta \circ \varphi$, such that for $\sigma \in \mathcal{I}$, $\Xi(\sigma) = \Delta(\varphi(\sigma)) \subseteq \Delta(\sigma)$, i.e., Ξ is carried by Δ. $\qquad \square$

Proposition 6 *For $t < \frac{n+1}{2}$, a general (colorful) CBT task $(\mathcal{I}, O, \Delta)$ does not have a t-resilient message-passing protocol.*

Proof The claim follows from Lemmas 5 and 6. $\qquad \square$

5 Bibliographic Notes

One of the earliest protocols, sidechain, for cross-blockchain transactions (CBTs), was published in 2014 [34]. While sidechain originally was not designed for a CBT in terms of a transaction's ACID properties, sidechain had heavily influenced the design and implementation of cross-chain communication in state-of-the-art systems, represented by Cosmos [9] and Polkadot [30], both of which adopt a similar approach of managing the CBTs through a central master blockchain.

While there are abundant works claiming to (partially) overcome the limitation rooted in the central role of a master blockchain, two directions seem most promising and, more importantly, are built upon solid theoretical foundations rather than ad hoc protocols. The first direction is led by Herlihy et al. [16, 18], where a CBT is treated as a series of two-party swaps. The atomicity is only provided at the swap level between two parties but not at the global transaction's level. The second direction is inspired by an old sub-branch of distributed computing called *distributed commit protocols*, such as two-phase commit (2PC) [1]. The 2PC-based protocol was first proposed in [40] and followed by [39, 45]. Both schools of thought treat CBTs through procedural analysis: each blockchain node is modeled as a replicated state machine.

Inspired by the topological approach to modeling and analyzing distributed systems [19–21], Zhao et al. [41–44] proposed a series of topological approaches to processing CBTs. Rather than modeling blockchains and their nodes with replicated state machines, the topological approach represents a more time-static and more geometrically understandable methodology to understand the intrinsic interaction among heterogeneous blockchains, thus opening the door to new protocols and systems of future cross-blockchain transactions.

References

1. 2PC Protocol: https://en.wikipedia.org/wiki/Two-phase_commit_protocol (Accessed 2021)
2. Al-Mamun, A., Li, T., Sadoghi, M., Zhao, D.: In-memory blockchain: toward efficient and trustworthy data provenance for hpc systems. In: IEEE International Conference on Big Data (BigData) (2018)
3. Al-Mamun, A., Yan, F., Zhao, D.: SciChain: blockchain-enabled lightweight and efficient data provenance for reproducible scientific computing. In: IEEE 37th International Conference on Data Engineering (ICDE) (2021)
4. Armstrong, M.: Basic Topology. McGraw-Hill Book Company Ltd (1983)
5. Attiya, H., Welch, J.: Distributed Computing: Fundamentals. Simulations and Advanced Topics. Wiley Inc, Hoboken (2004)
6. Bitcoin: https://bitcoin.org/bitcoin.pdf (Accessed 2021)
7. Blockchain Interoperability: https://cointelegraph.com/news/blockchain-interoperability-the-holy-grail-for-cross-chain-deployment (Accessed 2021)
8. Castro, M., Liskov, B.: Practical byzantine fault tolerance and proactive recovery. ACM Trans. Comput. Syst. **20**(4), 398–461 (2002)
9. Cosmos: https://cosmos.network (Accessed 2021)

10. Dieck, T.: Algebraic Topology. European Mathematical Society Publishing House (2008)
11. Ethereum: https://www.ethereum.org/ (Accessed 2021)
12. Goes, C.: The interblockchain communication protocol: an overview (2020). https://arxiv.org/abs/2006.15918
13. Gossamer: https://github.com/ChainSafe/gossamer (Accessed 2021)
14. Gray, J., Reuter, A.: Transaction Processing: Concepts and Techniques, 1st edn. Morgan Kaufmann Publishers Inc., San Francisco (1992)
15. Henle, M.: A Combinatorial Introduction to Topology. W. H, Freeman and Company (1979)
16. Herlihy, M.: Atomic cross-chain swaps. In: Proceedings of the 2018 ACM Symposium on Principles of Distributed Computing (PODC) (2018)
17. Herlihy, M., Kozlov, D.N., Rajsbaum, S.: Distributed Computing Through Combinatorial Topology. Morgan Kaufmann (2013)
18. Herlihy, M., Liskov, B., Shrira, L.: Crosschain deals and adversarial commerce. In: Proceedings of the 45rd International Conference on Very Large Data Bases (VLDB) (2019)
19. Herlihy, M., Rajsbaum, S.: The topology of shared-memory adversaries. In: Proceedings of the 29th Annual ACM Symposium on Principles of Distributed Computing (PODC), pp. 105–113. ACM (2010)
20. Herlihy, M., Shavit, N.: The asynchronous computability theorem for t-resilient tasks. In: Proceedings of the Twenty-Fifth Annual ACM Symposium on Theory of Computing (STOC), p. 111-120 (1993)
21. Herlihy, M., Shavit, N.: A simple constructive computability theorem for wait-free computation. In: Proceedings of the Twenty-Sixth Annual ACM Symposium on Theory of Computing (STOC), pp. 243–252 (1994)
22. Hocking, J., Young, G.: Topology. Addison-Wesley Publishing Company, Inc. (1961)
23. Hyperledger: https://www.hyperledger.org/ (Accessed 2021)
24. Kagome: https://github.com/soramitsu/kagome (Accessed 2021)
25. Lamport, L., Shostak, R., Pease, M.: The byzantine generals problem. ACM Trans. Program. Lang. Syst. **4**(3), 382–401 (1982). https://doi.org/10.1145/357172.357176
26. Lynch, N.A., Merritt, M., Yager, R.R.: Atomic transactions. In: Concurrent and Distributed Systems. Morgan Kaufmann Publishers Inc., San Francisco (1993)
27. MiPasa Blockchain: https://mipasa.org/ (Accessed 2021)
28. Munkres, J.: Topology. Pearson Education Limited (2003)
29. Munkres, J.R.: Elements of Algebraic Topology. Addison-Wesley (1984)
30. Polkadot: https://polkadot.network/ (Accessed 2021)
31. Proof of Stake: https://en.wikipedia.org/wiki/Proof_of_stake (Accessed 2021)
32. Protocol Buffers: https://github.com/protocolbuffers/protobuf (Accessed 2021)
33. Shen, H., Badsha, S., Zhao, D.: Consortium blockchain for the assurance of supply chain security. In: 27th Annual Network and Distributed System Security Symposium (NDSS) (2020). https://www.ndss-symposium.org/wp-content/uploads/2020/02/NDSS2020posters_paper_32-1.pdf
34. Sidechains: https://blockstream.com/sidechains.pdf (Accessed 2021)
35. Starport: https://github.com/tendermint/starport (Accessed 2021)
36. Steen, M.V., Tanenbaum, A.S.: Distributed Systems: Principles and Paradigms, 3rd edn. CreateSpace Independent Publishing Platform (2017)
37. Tendermint: https://tendermint.com (Accessed 2021)
38. Wang, X., Al-Mamun, A., Yan, F., Zhao, D.: Toward accurate and efficient emulation of public blockchains in the cloud. In: Proceedings of the 12th International Conference on Cloud Computing (CLOUD), Lecture Notes in Computer Science, vol. 11513, pp. 67–82. Springer (2019)
39. Wang, X., Tawose, O.T., Yan, F., Zhao, D.: Distributed nonblocking commit protocols for many-party cross-blockchain transactions (2020). CoRR arXiv:abs/2001.01174. http://arxiv.org/abs/2001.01174
40. Zakhary, V., Agrawal, D., El Abbadi, A.: Atomic commitment across blockchains. Proc. VLDB Endow. **13**(9), 1319-1331 (2020). https://doi.org/10.14778/3397230.3397231

41. Zhao, D.: Algebraic structure of blockchains: a group-theoretical primer (2020). CoRR arXiv:abs/2002.05973. https://arxiv.org/abs/2002.05973
42. Zhao, D.: An algebraic-topological approach to processing cross-blockchain transactions (2020). CoRR arXiv:abs/2008.08208
43. Zhao, D.: Completeness of cross-blockchain transactions: a combinatorial-algebraic-topological approach (2020). CoRR arXiv:abs/2004.08473
44. Zhao, D.: Topological properties of multi-party blockchain transactions (2020). CoRR arXiv:abs/2004.01045
45. Zhao, D., Li, T.: Distributed cross-blockchain transactions (2020). CoRR arXiv:abs/2002.11771 (2020)
46. Zhuang, Y., Sheets, L., Shae, Z., Chen, Y., Tsai, J., Shyu, C.: Applying blockchain technology to enhance clinical trial recruitment. AMIA Ann. Symp. Proc. 1276–1285 (2020). https://www.ncbi.nlm.nih.gov/pmc/articles/PMC7153067/

Scalability

Scaling Blockchains and the Case for Ethereum

Aditya Asgaonkar

Abstract This chapter provides a high-level introduction to scaling solutions for blockchains, with a special focus on Ethereum 2.0. Current blockchain capacity is a hurdle for the widespread adoption of Web3 and cryptocurrencies. First, we discuss the considerations and pitfalls of blockchain scaling strategies. We then explore the design landscape—layer-1 and layer-2 solutions—and discuss concepts in each category, namely sharding, rollups, and sidechains.

1 Introduction to the Scaling Problem

The two most popular blockchains—Bitcoin and Ethereum—are currently able to support between 5 and 15 transactions per second (TPS). With increasing mainstream Web3 adoption through Decentralized Finance (DeFi), Non-Fungible Tokens (NFTs), Decentralized Autonomous Organizations (DAOs), etc., blockchain scalability is becoming an increasingly important problem to solve. Ethereum has seen an explosive increase in usage since 2020, and at times network congestion has led to exorbitantly high transaction fees. Scaling blockchain capacity is a prerequisite for the widespread and commonplace adoption of Web3 systems.

This chapter will be focusing on scaling solutions for public, permissionless, and general-purpose blockchain systems. Let us define these terms:

Public: The blockchain can be used by the general public.

Permissionless: The requirements to participate in the decision-making process are defined by the protocol and are accessible to the general public. There is no gatekeeper entity that chooses the participants—the protocol admits all actors wishing to participate that satisfy the requirement.

General Purpose: The blockchain utilizes a general-purpose transaction system that supports smart contracts, e.g., the Turing-complete Ethereum Virtual Machine [1].

A. Asgaonkar (✉)
Ethereum Foundation, Bern, Switzerland
e-mail: aditya@ethereum.org

© The Author(s), under exclusive license to Springer Nature Switzerland AG 2022   197
D. A. Tran et al. (eds.), *Handbook on Blockchain*, Springer Optimization
and Its Applications 194, https://doi.org/10.1007/978-3-031-07535-3_6

1.1   Considerations

The main considerations while evaluating blockchain scaling solutions are:

1. Throughput and Latency;
2. Decentralization; and
3. Security.

Throughput and Latency

Throughput is the number of transactions per unit time that the blockchain system forms consensus over. Latency is the time required for a transaction to become a part of the chain under consensus. A higher throughput and a lower latency are desirable for any transaction system.

Decentralization

A key aspect of blockchain systems is decentralization. Qualitatively, the factors contributing to decentralization of a blockchain are:

- **Participation**: the participants of the decision-making process are not concentrated in a single group.
- **Verification**: a larger number of users are able to verify the output of the blockchain.

Nodes that verify all blocks and transactions are called *full nodes*. If the throughput of the blockchain is increased and full nodes have to verify a larger number of transactions at the same time, the minimum hardware requirements for operating a full node will be increased. This reduces the number of people that are able to run full nodes and verify the chain output, hence reducing the decentralization of the chain.

Security

The security of a blockchain system is quantified by the fraction of nodes that must misbehave to cause a safety or liveness failure of the decision-making process. Security is affected by the model of consensus employed, which involves factors such as:

- Honesty Assumptions: The assumptions that are made about the behavior of participants, e.g., unconditionally honest (actors that follow the protocol unconditionally), economically rational (actors that are willing to deviate from the protocol if profitable), etc.

- Quality of Safety: Consensus protocols differ in their guarantees about safety. Traditional BFT protocols provide a safety threshold—a minimum number of protocol-violating nodes required to cause a safety violation. In the context of blockchain consensus, it may be beneficial to consider the stricter notion of *accountable safety* threshold [11]—a minimum number of protocol-violating nodes that can be held accountable in a provable manner in case of a safety violation.

1.2   Naive Scaling Solutions

Bigger/Faster Blocks

A common naive solution to scaling a blockchain protocol is to increase the size and/or frequency of blocks. This has an effect on increasing the throughput and/or decreasing latency of the system. For example, if a proof-of-work blockchain increases the size of its blocks[1] by a factor of k, then the throughput of the system increases by k. However, this is not without tradeoffs. The minimum hardware requirements for verifying blocks will increase—the slowest processor that can verify the blockchain will need to be k times as fast as the earlier one. This reduces decentralization by limiting the number of nodes that can verify the chain and participate in the network. If the block size becomes huge, then only very large servers will be able to verify the chain.

1.3   Types of Scaling Solutions

Scaling solutions can be broadly classified into two categories:

- Layer-1: Scaling is achieved by employing fundamentally different architectures for the blockchain protocol. This includes changes to the consensus mechanism, network architecture, distribution of verification duties to subsets of the network, etc.
- Layer-2: Scaling is achieved by designing a transaction system such that a large number of transactions in the new system are executed with a small number of transactions on the base blockchain's transaction system. Such systems are designed for users to interact with the existing underlying blockchain, and rely on the security of the underlying blockchain protocol.

---

[1] In Ethereum, block size is defined by the *gas limit*—a limit on the total computing operations carried out by transactions included in a block.

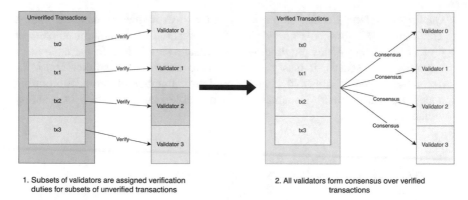

1. Subsets of validators are assigned verification   2. All validators form consensus over verified
   duties for subsets of unverified transactions   transactions

Fig. 1 Sharding designs improve throughput by distributing transaction verification tasks among subsets of validators and retain security by forming consensus over all transactions with the entire validator set

2 Layer-1 Scaling Solutions

2.1 Sharding

A popular layer-1 scaling solution is *sharding*, wherein the verification duties for the blockchain's transaction system are distributed among multiple smaller subsets of the participants, but consensus is formed by the entire set. Most sharded blockchains employ proof-of-stake[2] consensus in their design. Participants are called *validators* (Fig. 1).

2.2 Ethereum 2.0

Beacon Chain and Shard Chains

- **Beacon Chain**: The beacon chain is responsible for forming consensus over shard chain blocks and bookkeeping related to the consensus process (Fig. 2).
- **Shard Chains**: Shard chains are where the users' Ethereum transactions are executed. Each shard chain has an independent state and is responsible for validating and executing transactions concerning that piece of state.

---

[2] Proof-of-stake is a blockchain design that relies on intrinsic resources (such as its own cryptocurrency) to act as a mechanism to choose the consensus participant set—e.g., Ethereum 2.0 requires validators to deposit a certain minimum amount of Ether in order to be included as a consensus participant. In contrast, proof of work relies on some extrinsic resource to choose the participant set—e.g., Bitcoin limits its participant set to people with computing resources, and the probability of contributing to the chain is dependent on the speed and capacity of the computing resource.

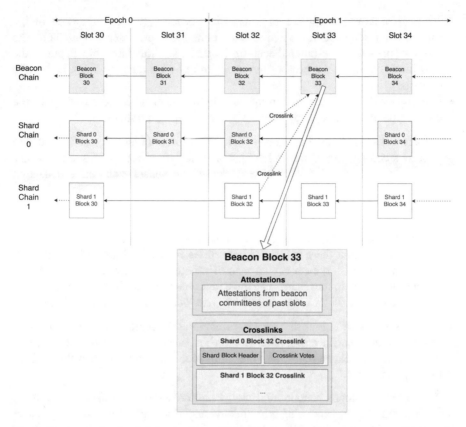

Fig. 2 Beacon chain, shard chains, and crosslinks

 Note

The design of shard chains is a work-in-progress effort. The most up-to-date design can be found in the Ethereum 2.0 specifications repository on GitHub: https://github. com/ethereum/consensus-specs/tree/dev/specs/sharding.

- **Slots and Epochs**: Time is divided into epochs, which are further divided into 32 slots. The current parameters are configured to 12 s per slot.
- **Validators**: Participants of Eth2's consensus process are called *validators*. Validators have two main duties:

  – Verifying and finalizing[3] beacon blocks
  – Verifying shard blocks

---

[3] A block is finalized when the validators decide using the consensus process that the block is a part of the canonical chain, and this block cannot be reverted in the future. Finalization is the consensus process to arrive at this decision.

- **Proposers**: At every slot in every chain (i.e., beacon chain and shard chains), a randomly chosen validator is assigned to be the block proposer in that chain. The proposer packages attestations[4] seen from other validators into a block, builds the new block on top of the head of the chain and gossips the new block in the p2p network.
- **Committees**: Committees are groups of validators that are assigned a duty. Based on the type of duty assigned, there are two types of committees:

  - **Beacon committees**: For every epoch, the entire validator set is divided into beacon committees such that there is one beacon committee assigned for every slot. At its assigned slot, each validator in the beacon committee makes an attestation for a beacon block. A validator's attestation for a particular beacon block indicates that the validator has verified the block, and constitutes a vote for the beacon block to be considered in the consensus process.
  - **Shard committees**: Similar to beacon committees, the entire validator set is divided such that there is a shard committee for some shard(s) in every slot. At its assigned slot, each validator in the shard committee makes a crosslink vote for a block in that shard. The crosslink vote indicates that the validator has verified the shard block, and should be *crosslinked* into the beacon chain.

- **Crosslinks**: A shard block that has crosslink votes from more than two-thirds of a shard committee can be crosslinked into the beacon chain—the shard block header is included in a beacon block, and the shard block is finalized when that beacon block gets finalized.

Beacon chain blocks contain attestations and crosslink information (i.e., shard block headers and corresponding crosslinks votes). Shard chain blocks contain user transactions, similar to Eth1 blocks today.

Consensus Process

Casper the Friendly Finality Gadget (FFG) [10] is a major component[5] of the Eth2.0 consensus process, and provides the following guarantees:

- **Accountable Safety**: If two conflicting blocks are finalized, then at least one-third of validators have broken the Casper FFG rules, and these validators can be identified.
- **Plausible Liveness**: In any state of the protocol, a deadlock is impossible and the validators can make new votes that progress the protocol (i.e., finalize a new block) without violating any Casper FFG rules.

---

[4] An attestation is a vote for a block from a validator and is used in the consensus process to finalize the block.

[5] Casper FFG in itself is not a full consensus protocol. It provides the rules for identifying when a state of consensus has been reached (the *finalization rule*) but does not describe how to achieve such a state. In this context, it is called a finality gadget.

Casper FFG

A brief description of the Casper FFG mechanism:

- **Justified Block**: A block is justified if it is the genesis block, or more than two-third of validators have made votes (A, B), where A is some ancestor of B and A is a justified block.

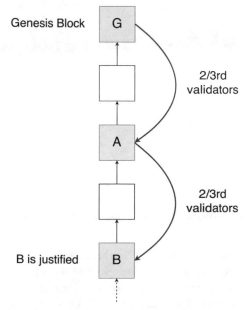

- **Finalized Block**: A block is finalized if it is the genesis block, or B is justified and more than two-third of validators have made votes (B, C), where C is the direct child of B (i.e., $height(C) = height(B) + 1$)

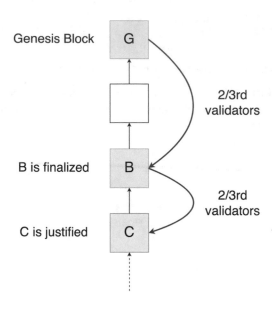

Only blocks at epoch boundaries are considered for Casper FFG in the beacon chain, for two reasons:

- Consensus processing is done at the interval of an epoch rather than every slot
- It allows for the entire validator set to communicate their consensus votes over the length of an epoch, which reduces p2p network congestion as compared to all validators communicating their votes in the same slot.

Fork Choice Rule

The fork choice rule describes how a canonical chain is chosen from a set of blocks which contain multiple different chains. Output from the Casper FFG mechanism is not enough to choose a canonical chain from a block tree—justification or finalization requires votes from the entire validator set, and only a sample of the validator set is heard from at each slot. Therefore, there is a need for an algorithm to choose a canonical chain from the unfinalized section of the block tree. The beacon chain uses the Hybrid Latest Message-Driven (LMD) GHOST fork choice rule, as described in Algorithm 3 (Fig. 3).

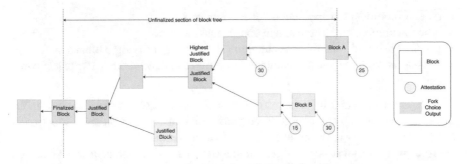

Fig. 3 Hybrid LMD GHOST chooses the chain defined by Block A in the above scenario

Fork Choice Algorithm

Algorithm 1 LMD GHOST Score

1: **procedure** LMD_GHOST_SCORE(b)
2:    $M \leftarrow$ list of latest attestations from all validators
3:    $score \leftarrow$ number of attestations in M voting for b
4:    **for all** child c of b **do**
5:        $score \leftarrow score +$ LMD_GHOST_SCORE(c)
6:    **end for**
7:    **return** $score$
8: **end procedure**

Algorithm 2 LMD GHOST Fork Choice

1: **procedure** LMD_GHOST(b)
2:    **while** b has children **do**
3:        $b \leftarrow \arg\max_{c\ \text{child of}\ b}$ LMD_GHOST_SCORE(c)
4:    **end while**
5:    **return** b
6: **end procedure**

Algorithm 3 Hybrid LMD GHOST Fork Choice

1: **procedure** HYBRID_LMD_GHOST(C)
2:    $b \leftarrow$ highest justified block in chain C
3:    **return** LMD_GHOST(b)
4: **end procedure**

Scalability Analysis

First, let us define some notation. Let

- c be the computations per second of a single node,
- n be the number of participants in consensus,
- *verification* (c) denote the computation required for verifying c blocks, and
- *consensus* (n, c) denote the computation required at each node for n participants and c blocks.

The verification task involves verifying digital signatures and executing transactions, i.e., looking up and operating on pieces of state. The *verification* function is assumed to be linear in the number of blocks to be verified.[6]

To come to consensus on a larger number of blocks with each consensus instance taking in a fixed amount of data, the number of times that the consensus process is run needs to be increased proportionally to the increase in blocks. So, the *consensus* function is assumed to be linear in the number of blocks.

In a single proof-of-stake chain with n participants, if the chain has a throughput of c blocks per second, then each node is performing $verification(c) + consensus(n, c)$ computations per second. If the rate of processing at each node becomes p times, each node can process $p \cdot (verification(c) + consensus(n, c)) = verification(p \cdot c) + consensus(n, p \cdot c)$ computations per second, i.e., the chain throughput becomes $p * c$ blocks per second.

In the Eth2 sharded proof-of-stake system with n participants and s shards, each having a throughput of c blocks per second:

- the throughput of the system is $s \cdot c$ blocks per second,
- the Beacon Chain has a throughput of s blocks per second, and
- each node performs $verification(c) + verification(s) + consensus(n, s)$ computations per second.

If the rate of processing at each node becomes p times, then each node will be able to process $p \cdot (verification(c) + verification(s) + consensus(n, s)) = verification(p * c) + verification(p \cdot s) + consensus(n, p \cdot s)$ computations per second. Thus, the system is able to support $p \cdot s$ number of shards each with a throughput of $p \cdot c$ blocks per second, leading to a total throughput of $(p \cdot c) \cdot (p \cdot s) = p^2 \cdot c \cdot s$ blocks per second. The throughput of the system increases quadratically proportional to the increase in rate of processing of each node.

Security Analysis

There are two relevant security analyses to be made:

- **Consensus Safety**: Safety against the creation of two conflicting finalized chains.
- **Shard Committee Safety**: Safety against an attacker-controlled committee submitting a malicious crosslink.

---

[6] Each block is assumed to be uniform in the number of operations that its transactions perform.

Consensus safety guarantees are inherited from the Casper FFG, as discussed in Sect. 2.2.

Shard committee safety prevents an attacker from submitting a crosslink for a maliciously created block using a shard committee that it controls. For example, an attacker may create a block that generates ETH from thin air, and this would go unnoticed because only the shard committee actually executes the block to verify its validity. Under a static adversary model,[7] security against these types of attacks is provided by the random sampling of committees. Before proceeding to the security analysis, these are some additional details about the shard committee sampling process:

- At present, there are 64 shards planned.
- A shard committee must be at least 128 validators. If there aren't enough validators to allow for a committee for each shard in every slot, then only some shards will have a committee in a slot.

Now, we can estimate the probability that an attacker controlling some fraction of validators is able to create a crosslink (i.e., control more than two-third of a shard committee). This probability can be derived from the CDF of the hypergeometric distribution, because:

- validators are sampled from the validator set without replacement,
- the attacker controls some fraction of the validator set, and
- the committee is broken if more than two-third of the sample is attacker-controlled

Let's define some notation:

For a random variable X that follows the hypergeometric distribution, let $Pr[X \leq k] = C(N, K, n, k)$ be the hypergeometric cumulative distribution function, where:

- N is the population size,
- K is the number of success states in the population,
- n is the number of draws in each trial, and
- k is the number of observed successes.

In our case, we observe the following:

- X is the number of attacker's validators in a sampled committee,
- N is the validator set size,
- K is the total number of validators under the attacker's control,
- $n = 128$ is the size of a committee,
- $k = \frac{2}{3} \times n$ is the minimum required number of attacker's validator in the sampled committee,

---

[7] A static adversary chooses which validators it controls before the protocol begins. A static adversary cannot, for example, choose to corrupt a validator after looking at the result of the random committee sampling.

- $Pr[X \leq k - 1]$ is the probability that the committee is not broken, i.e., the attacker controls less than two-third of the committee,
- $Pr[X \geq k] = 1 - Pr[X \leq k - 1]$ is the probability that the committee is broken, i.e., the attacker controls more than two-third of the committee.

So, the probability that a committee is broken is given by

$$Pr[X \geq k] = 1 - C(N, K, 128, \frac{2}{3} \cdot 128).$$

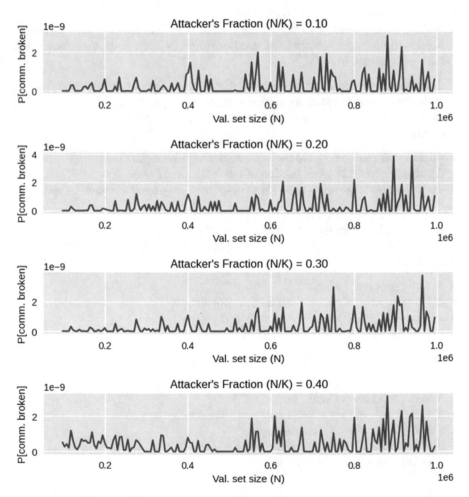

Fig. 4 Probability of a broken shard committee for various fractions of the validator set under the attacker's control

Figure 4 shows the graph of this function for varying values of N and K. The probabilities are lower than $4 \cdot 10^{-9}$, so an attacker is able to execute such an attack every $\frac{10^9}{4}$ slots, which is $\frac{10^9}{4}$ *slots* $\cdot$ 12 *s/slot* $\cdot \frac{1}{365 \cdot 24 \cdot 3600}$ *years/s* $= 95$ *years*.

3 Layer-2 Scaling Solutions

Aggregation-based scaling by relying on the security of L1 for consensus. Layer-2 scaling solutions allow users of a blockchain to execute their transactions on a faster system that operates in parallel to the base blockchain. The transaction system of the layer-2 solution and the base blockchain are able to interact through a smart contract on the base blockchain. Users deposit their assets from the base blockchain into the smart contract associated with the layer-2 system. The side chain maintains a separate state that tracks the ownership of these deposited assets. Users are now

At regular intervals, the state of the layer-2 system is committed to the smart contract on the base blockchain, which is called a checkpoint. It's useful for this state commitment to be in the form of an accumulator[8] with which facts about pieces of the state can be proved, e.g., a Merkle tree root. The smart contract also stores the rules for the layer-2 system's decision-making process. Using these rules, the smart contract ensures that all saved checkpoints are an outcome of the layer-2 system's decision-making process (Fig. 5).

The usual workflow for using a layer-2 scaling solution is as follows:

1. **Deposit**: User deposits their assets from the base blockchain into the smart contract associated with the layer-2 system. When a new asset is deposited in the smart contract, the asset is minted in the layer-2 state and assigned to the corresponding user.
2. **Transact**: Users on the layer-2 system are able to transact with each other using the deposited assets. This only changes the state of the layer-2 system.
3. **Withdraw**: Users wishing to withdraw an asset from the layer-2 system to the base blockchain have to:
   a. burn their asset from the layer-2 state,
   b. commit a checkpoint that contains this updated state to the base blockchain, and
   c. create a transaction on the base blockchain that proves the burning of the asset from the layer-2 state and requests the smart contract to transfer the asset to their address on the base block.

---

[8] An accumulator is a function that provides information about the membership of an item in a set. In our context, the set is the entire state, and users can check whether a particular piece of the state corresponding to a specific smart contract is indeed included in the current state of the layer-2 system.

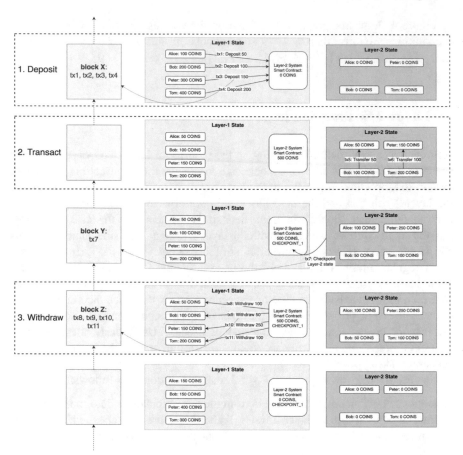

Fig. 5 Users of layer-2 scaling solutions follow the deposit, transact, and withdraw workflow. The scalability comes from aggregating multiple transactions on the layer-2 system (such as `tx5` and `tx6`) into a single checkpoint update transaction on the layer-1 system

3.1 Side Chains

Side chains are layer-2 scaling solutions for which the corresponding smart contract contains the rules of finalization for the chain, but does not contain the rules of its transaction system. The security of a side chain relies on an honest majority assumption about the participant set of its decision-making process.

For example, consider a multi-signature-based system, where any checkpoint update that appears in the smart contract along with a valid multi-signature is considered finalized on the side chain. If the participants of the multi-signature collude to sign on an invalid state transition (such as transferring funds without a valid transaction from the sender) and this update is made in the smart contract, the users have no way of protecting themselves.

A popular side chain on Ethereum is Polygon [7].

3.2   Rollups

Rollups are layer-2 scaling solutions for which the corresponding smart contract contains the rules of its transaction system. There is usually a single designated (but replaceable) actor called the *operator* who submits the checkpoint updates to the smart contract. The operator is expected to verify the validity of the state transition made by the checkpoint update.

There are two types of rollups based on how the state transition rules are verified:

- Optimistic Rollup.
- Zero-Knowledge (ZK) Rollup.

Optimistic Rollup

The rules of finalization of an optimistic rollup include a *challenge period*, which begins after the checkpoint update has been made on the smart contract. During this challenging period, any user can ask for proof of validity of the state transition (such as a valid, signed transaction) in that checkpoint update. When the proof is provided, the smart contract is able to check the validity of the state transition using the rules of the transaction system. If no proof can be produced, the checkpoint update is rejected. At the end of the challenge period, if the checkpoint update has not been rejected, it is deemed finalized.

Two popular optimistic rollups are live at the time of writing: Arbitrum [5] and Optimism [6].

An important distinction between Arbitrum and Optimism is in the way they check the validity of a challenged state transition:

- Optimism uses a single-round challenge [2], where the challenger specifies a pre-viously included transaction to be executed using the Optimism smart contract. The rollup uses EVM transactions, so the rollup's smart contract needs to know to execute EVM code, with appropriate changes to provide the Ethereum chain's context to opcodes that require it. The rollup's smart contract also needs to be able to correctly process the challenge, by executing the challenged transaction on the proper pre-state.[9] All of this is done using an implementation named the Optimistic Virtual Machine.

---

[9] The transaction defines a state transition, and the rollup's smart contract needs to apply this state transition on the pre-state of the transaction, which is the state of the rollup right before the challenged transaction was included in the rollup.

- Arbitrum uses a multi-round challenge [3], where the challenger and operator (who is defending the transaction they included) communicate back and forth to identify a specific opcode that was executed incorrectly inside a previously executed transaction. The Arbitrum smart contract provides the Arbitrum Virtual Machine, which is able to execute EVM code and support the multi-step challenge process.

Zero-Knowledge (ZK) Rollup

The state transition rules for a ZK Rollup are encoded into a proof system, such that proof of a correctly executed state transition can be made and is computationally cheap to verify.[10] The rules of verification of these proofs are then put into the smart contract for the rollup. Whenever a checkpoint is made on the smart contract, an associated proof is required for the state transition from the last checkpoint. Thus, all checkpoints are automatically verified to be the result of correctly executed state transition, without the need for challenge periods that appear to be optimistic rollups.

A number of ZK Rollups exist at the time of writing: Loopring [4], StarkNet [8], and zkSync [9].

4 Conclusion

Blockchain technology is a vast domain of computer science that is yet to be fully explored. Given the current levels of interest and resources being deployed into blockchain research, this space will undoubtedly see rapid development in the coming years. With the increasing adoption of Web3 in mainstream industries, blockchain scalability has become a crucial research area with immediate consequences. Over the course of this chapter, we've explored the considerations in scalability solutions, the general categories of scaling solutions, and then a further deeper exploration of the leading solutions. While this chapter aims to serve as a gentle introduction to the topic of blockchain scaling, the context provided through this chapter will also enable readers to understand and analyze other scaling solutions and related technologies.

References

1. Ethereum virtual machine (evm) | ethereum.org. https://ethereum.org/en/developers/docs/evm/. Accessed 02 Jan 2022

---

[10] Zero-Knowledge Succinct Non-Interactive Argument of Knowledge (ZK-SNARKS) are a popular choice for the proof system in ZK-Rollups. Although the zero-knowledge properties are not essential to this type of rollup, the name has stuck.

2. How does optimism's rollup really work? | paradigm research. https://research.paradigm.xyz/optimism. Accessed 31 Dec 2021
3. Inside arbitrum - offchain labs dev center. https://developer.offchainlabs.com/docs/inside_arbitrum. Accessed 31 Dec 2021
4. Loopring - zkrollup exchange and payment protocol. https://loopring.org. Accessed 22 Sept 2021
5. Offchain labs. https://offchainlabs.com/. Accessed 22 Sept 2021
6. Optimism. https://optimism.io/. Accessed 22 Sep 2021
7. Polygon | ethereum's internet of blockchains. https://polygon.technology/. Accessed 22 Sep 2021
8. Starknet - starkware industries ltd. https://starkware.co/product/starknet/. Accessed 22 Sept 2021
9. zksync - rely on math, not validators. https://zksync.io/. Accessed 22 Sept 2021
10. Buterin, V., Griffith, V.: Casper the friendly finality gadget (2017). CoRR, arXiv:abs/1710.09437
11. Neu, J., Tas, E.N., Tse, D.: The availability-accountability dilemma and its resolution via accountability gadgets (2021). CoRR, arXiv:abs/2105.06075

Building Protocols for Scalable Decentralized Applications

Kai Mast

Abstract Blockchain protocols are a promising technology in the abstract, but, in reality, fall short of the promise of supporting arbitrary decentralized applications. For example, Bitcoin supports < 10 transactions per second and Ethereum's gas limit prevents computationally expensive applications to execute on its chain. This chapter provides an overview of mechanisms that have been proposed to overcome these limitations. In particular, we describe novel consensus protocols, sharding mechanisms, state and payment channels, subchains, and federated protocols. Additionally, we give insight into the tradeoffs and benefits of the different approaches.

1 Introduction

Blockchains [45, 64], or more broadly decentralized ledgers, enable applications to execute across a trustless peer-to-peer infrastructure. We consider a system decentralized if individual nodes cannot influence its execution as long as they do not control a threshold of the network. This means that decentralized architectures protect against malicious adversaries in addition to simple crash failures. As a result, decentralized ledgers allow for online services to operate without reliance on a trusted party. Figure 1 outlines this stark contrast to previous architectures, where each user's data is in full control of a single organization.

While blockchain protocols are a promising technology in the abstract, they fall short in critical ways. For example, the Ethereum blockchain has roughly the processing power of a portable calculator or about 35k floating-point operations per

---

K. Mast (✉)

Department of Computer Sciences, University of Wisconsin-Madison, Madison, WI, USA

e-mail: kaimast@cs.wisc.edu

© The Author(s), under exclusive license to Springer Nature Switzerland AG 2022   215

D. A. Tran et al. (eds.), *Handbook on Blockchain*, Springer Optimization
and Its Applications 194, https://doi.org/10.1007/978-3-031-07535-3_7

second.[1] The culprit for these limitations is that decentralization requires massive replication of computation and data. This massive replication results in high computation, communication, and storage overheads, which in turn, hurts throughput and latency.

In this chapter, we first give an overview of the decentralized ledger model and the protocols that implement it. Then, we discuss different avenues for improving the performance of such protocols to support real-world workloads. Throughout the chapter, we will also give insight into the limitations and open problems of existing mechanisms.

We describe four different avenues for scaling blockchains. First, we discuss how the consensus protocol itself can be made faster. Second, we discuss sharding, which allows running multiple consensus protocols in parallel. Third, we provide an overview of layer-2 solutions, such as payment channels. Finally, we discuss federated blockchains, which can be viewed as a hybrid of layer-2 and sharding solutions. Our discussion is mainly focused on safety, i.e., that the consistency and integrity guarantees of the blockchain system are not broken, and availability, i.e., that the current and past states of a blockchain can always be retrieved and inspected.

2 Decentralized Ledger Abstraction

Each decentralized architecture, in essence, provides the abstraction of an append-only ledger with semantics that goes beyond the mere storage of data and execution of programs. These semantics are key to building applications with high integrity in a decentralized setting and it is important to understand them before modifying existing or creating new protocols for decentralized ledgers. For the rest of this chapter, we will refer to this abstraction as decentralized ledgers and the underlying protocols as decentralized ledger technologies (DLTs).

We extend the formalism of Adya [1], which defines a database $\mathcal{D}$ consisting of a history $\mathcal{H}_\mathcal{D}$ of transactions and a set of objects $\mathcal{O}_\mathcal{D}$, each associated with a totally ordered set of *object versions*. Each transaction is a set of operations applied to a particular object, such as a read, write, or append. Each object's version history initially only consists of the $\perp$ value, indicating that it has not been created yet.

Transactions affecting the same object(s) and their operations can be ordered with respect to each other. We say a transaction T precedes another transaction T' if it appears earlier in the database's history, denoted as $T \leftarrow T'$. This relationship is transitive, i.e., if $T_1 \leftarrow T_2$ and $T_2 \leftarrow T_3$ hold, then $T_1 \leftarrow T_3$ holds as well. Similarly, we say an operation op precedes another operation op' if the object versions it accesses precedes that of op'. Transactions affecting two disjoint sets of objects may

---

[1] With the current gas limit, Ethereum can do about two million floating-point multiplications per block, which are published about once a minute [21].

not be able to be ordered with respect to each other, denoted as $T \leftrightarrow T'$. Similarly, operations affecting two distinct objects cannot be ordered with respect to each other, denoted as $op \leftrightarrow op'$.

2.1 Consistency

Like many conventional database management systems, decentralized ledgers allow enforcing application-specific constraints on the data and provide strict serializability for all operations. Serializability ensures that all transactions execute atomically, i.e., in a serial or equivalent to serial order. In other words, if two transactions T_1 and T_2 are applied to two distinct objects, they must be applied in the same order to both objects. Strict serializability extends this notion of a real-time order: if T_1 started before T_2, its operations should also be applied before those of T_2 (see Eq. 1). This, in turn, not only ensures the integrity of a system's state but also makes it much easier for developers to build applications, because they do not have to reason about concurrency.

$$\forall T_1, T_2 \in \mathcal{H}, \forall op_1 \in T_1, op_2 \in T_2. \; T_1 \leftarrow T_2 \Rightarrow op_1 \leftarrow op_2 \vee op_1 \leftrightarrow op_2. \quad (1)$$

2.2 Immutability

Decentralized ledgers are eidetic: they maintain a record of all transactions ever processed by the system. From this record, any past state of the system can be regenerated and inspected. As a result, the ledger can serve as a notary or an impartial witness, by providing a reliable record of past information.

Formally, successfully applying a transaction T to a database $\mathcal{D}$ yields a new database $\mathcal{D}'$ with T appended to its history. Similarly, the version history of each object modified by T will be appended with its new version.

We then define immutability as a constraint on the allowed state transitions from $\mathcal{D}$ to $\mathcal{D}'$. Thus, if a database state $\mathcal{D}$ predates another state $\mathcal{D}'$, i.e., $\mathcal{D} \leftarrow \mathcal{D}'$, all of its transactions and object versions are contained in $\mathcal{D}'$. This means the successor state can only add new transactions and object versions and not remove or reorder them, denoted in Eq. (2).

$$\forall \mathcal{D}, \mathcal{D}'. \; \mathcal{D} \leftarrow \mathcal{D}' \Rightarrow \mathcal{H}_{\mathcal{D}'} \subseteq \mathcal{H}_{\mathcal{D}} \quad (2)$$

Immutable systems thus are, unlike the term *immutability* suggests, able to change their state, but will only allow state transitions that extend the state without removing existing information. Further, they might enforce other data policies to ensure the integrity of a particular application. For example, a cryptocurrency usually wants to ensure that no transaction is spent more than once.

2.3 Auditability

Auditability enables participants to join the network at any point in time and verify all states relevant to them up to the current point without having to trust a particular remote party. Formally, we say there exists a publicly available function $verify(\mathcal{D}, \mathcal{D}')$ that certifies a transition from $\mathcal{D}$ to $\mathcal{D}'$ is valid. Auditors can then recreate and verify the entire system execution by verifying all database state transitions, starting with the initial state consisting of an empty transaction history.

3 Decentralized Ledger Technologies

At the core of DLTs lies consensus protocols, used for *state machine replication (SMR)* to maintain a unified database. The definition of a state machine comprises a set of potential states the machine can be in and a set of admissible state transitions that allow moving from one state to another. SMR decides which state transitions to perform and replicates this decision across all participants of the protocol. As a result, all non-faulty participants maintain the same state at any point in time.

Most consensus protocols, while varying greatly in their implementation, are leader-based. This class of protocols first appoints a particular node to be a leader (sometimes called a primary), which then proposes state transitions to the system. These state transitions are then subject to approval by the rest of the network. The existence of a singular leader ensures that transactions are proposed in an order that ensures serializability. Finally, leader-based protocols can react to failures or bad performance at any point in time by appointing a different entity to be a leader.

3.1 Assumptions and Attack Model

Distributed ledgers are designed to be resilient against Byzantine failures, a model that encompasses both benign failures and those caused by malicious intent. A Byzantine actor may want to change the network's behavior to their advantage or break the network entirely. To achieve this, attackers may issue invalid or conflicting messages, and delay or hide communication. Correct nodes, on the other hand, follow the protocol as prescribed.

To ensure that correct nodes faithfully follow the protocol, distributed ledger protocols typically assume that the majority of network participants behave rationally and provide incentives for these rational actors to advance the protocol. These incentives can take the form of direct payments, where parties that process transactions receive compensation in the form of block rewards or transaction fees. Further, incentives can be based on collateral, where parties that misbehave are penalized financially.

DLTs may rely on different network assumptions. The three most common are synchronous, partially synchronous, and asynchronous [18]. In the synchronous setting, messages will be delivered within a known and fixed time-bound. In the partially synchronous setting, messages will be delivered in an unknown, but finite, time-bound. Finally, in the asynchronous model messages may take an unbounded amount of time to be delivered.

3.2   Data and Transaction Models

DLTs require a different data model than conventional databases because they execute in a trustless environment. Here, each user is associated with a set of cryptographic keys and must sign off transactions spending their cryptocurrencies with those keys. Nodes participating in a decentralized protocol need to prove that a transaction has been signed off by a particular set of users for it to be deemed valid. Additionally, transactions issued by a client might be conflicting. For example, Alice might request to spend $5 each on Bob and Claire, but only have $6 in her account.

Data models in decentralized ledgers are focused around the notion of payments and cryptocurrencies, as this was their initial application and cryptocurrencies are still the basis for incentive mechanisms in almost every DLT. We discuss the two most common ones: UTXO and Accounts.

The UTXO Model

Bitcoin represents a user's account balance as a set of *Unspent Transaction Outputs (UTXOs)*. Transactions in the UTXO model work similarly to a voucher system in which some input vouchers are exchanged for new vouchers of the same or lesser value. Figure 1 outlines how transactions consume UTXOs (the unspent outputs of a previous transaction) and produce new UTXOs. Note, that in a real system some transaction's input would go towards a transaction fee.

Bitcoin, like many other DLTs, relies on Merkle hash trees [40] to provide authentication of the blockchains state. Merkle trees can be generated for any arbitrary set of objects. To do this, these objects are first arranged in some pre-defined order. The tree is then constructed by recursively combining k hashes, where k is some branching factor of the tree, and generating a new hash value from the resulting value. A Merkle proof then allows verifying an object's state against the root of the tree without having access to the entire tree. The proof is the particular branch from the object to the tree root. An example for a Merkle proof and its associated tree is given in Fig. 2. The verifier here just recomputes and checks the correctness of every hash in the branch to ensure the proof's integrity. These proofs are virtually impossible to forge as it is very hard to find collisions, i.e., to input values that map to the same output value, for cryptographic hash functions [55].

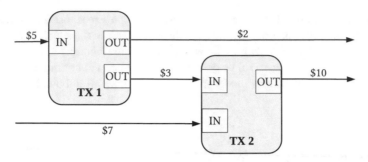

Fig. 1 Sketch of the UTXO model. Each transaction consumes one or multiple unspent transaction outputs and generates at least one new transaction output. Outputs are owned by a particular public key

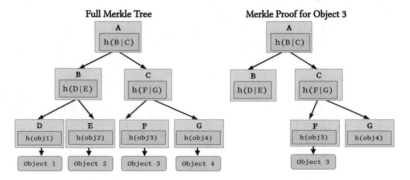

Fig. 2 Sketch of a Merkle tree and an associated proof. To prove the authenticity of Object 3 against the root A, we only need to provide a branch leading from A to the object

The key advantage of the UTXO model is that one can succinctly prove the existence of an unspent transaction output. Each block in Bitcoin contains the Merkle root of the current UTXO set, which allows verifying the current state without each block having to contain the entire set. Protocol participants just locally compute the current state by executing all previous blocks, generate the hash tree, and then verify the root against the public chain. Additionally, third parties that do not maintain the entire state of the blockchain, so-called "light clients", can verify the existence of a particular UTXO by verifying a Merkle proof.

The UTXO model significantly reduces the complexity of the data model that transactions execute on, but limits storing custom data. Participants in this protocol merely have to maintain the UTXO set to track the state of the blockchain and processing a transaction only involves adding and removing UTXOs to the set. As a result, platforms that are focused mostly on monetary transactions, such as Bitcoin or ZCash, often still rely on the UTXO model due to its simplicity.

The Accounts Model

Ethereum, in contrast to Bitcoin, relies on a data model focused on the notion of accounts. Intuitively, an account has a non-negative balance and can be owned by a particular user. Additionally, accounts can hold other data as well, which enable more complex applications.

Decentralized ledgers implementing the accounts model must provide additional measures for preventing double-spending and other conflicting transactions. First, instead of merely verifying the existence of a particular UTXO, the transaction must be verified against the account's state and ensure applying it to the account will not violate consistency constraints. Second, as long as there are sufficient funds in the spending account, a malicious node might use the same request to issue multiple transactions. To address this, Ethereum transaction requests contain a unique number, or nonce, and the protocol only admits one transaction for each combination of account identifier and nonce.

One drawback of the accounts model is that the authentication of state is more complex. Here, DLT nodes usually maintain three Merkle trees per block instead of just one as in the UTXO model. The first hash tree provides information about the resulting state of the system, the second hash tree represents the set of all transactions contained in this block, and the third hash tree represents the set of all changes.

In Ethereum and most other Account-based systems, state is represented in the form of Patricia Merkle trees [43]. Patricia trees have two key advantages over conventional Merkle trees: there exists a maximum bound on their height and updates are relatively inexpensive. This is achieved by storing data in a compact trie structure. Unlike when updating a conventional Merkle tree, where a new entry might reorder the set and rebuild the entire tree, inserting new values in Patricia trees only needs to update the affected subtree. An upper bound of the tree height is ensured by using hash values as object keys, which are guaranteed to be a certain length.

3.3   Smart Contracts

The client-server model, where applications perform computation locally and then write the resulting state to the database, does not apply to the decentralized setting. In the client-server model, Byzantine actors could attempt storing invalid application results in the globally replicated ledger and, thus, violate consistency. To prevent such attacks, DLTs provide the means to execute arbitrary applications directly on the ledger itself, similar to stored procedures in conventional database systems.

Ethereum introduced the notion of *Smart Contracts*, stateful programs that are stored and executed entirely on the decentralized ledger. While the previous system already provided some notion of programmability, such as Bitcoin Script, Ethereum smart contracts were the first to provide full Turing completeness and, as a result, the possibility to support arbitrary programs. Smart contracts are usually written in a high-level language, such as Solidity, and then compiled to byte code, such as EVM byte code or WebAssembly, before being stored and executed on the ledger.

Smart contracts reside on a particular address on the blockchain, analogous to how each user's account is assigned an address. Contracts may hold currency and contain a key-value store to store arbitrary data. Users call functions of a smart contract by issuing a transaction that contains a function call and some amount of currency to pay for the computation.

Ethereum replaced fixed transaction fees with a notion of "gas cost", which covers the cost of processing and executing the transaction. Each transaction request contains a gas limit, representing the maximum number of computational steps the issuer is willing to pay for. Like with transaction fees, the cost of a single unit of gas is determined by the market. If the transaction runs out of gas during execution, it aborts. Any unused gas is refunded to the party that issued the transaction.

Contracts can modify their local state directly while executing or invoking functions of other contracts. The latter allows reusing existing code and interaction between different applications. For example, one can implement a custom token on top of Ethereum that can be used as a form of payment by other contracts.

3.4 Committee-Based Consensus

Classical consensus protocols achieve state machine replication among a fixed set, or committee, of nodes. They were first introduced by Leslie Lamport [35], among others [47, 56]. These protocols now form the foundation for most fault-tolerant applications. For example, a web service might be implemented across three data centers. If one of the data centers fails, the consensus protocol ensures that operation can continue by shifting computation to the other two data centers.

While consensus protocols were initially intended to tolerate only benign failures, the introduction of Byzantine fault-tolerant consensus protocols allowed for more complex use cases. For example, a node in the committee might not simply become unavailable but encounter a software bug that makes it behave in ways not originally intended by the software developers. While such failures might be much more unlikely than a crash, it is still important to be resilient against them for safety-critical applications.

Recently, committee-based Byzantine fault-tolerant consensus protocols, such as *Practical Byzantine Fault Tolerance (PBFT)* [8], have received new attention in the context of decentralized ledgers. Because these protocols cannot only protect against software bugs or hardware failures but also against a malicious human adversary controlling a subset of the committee, they are suitable for implementing applications where mutually distrusting parties are trying to agree on a consistent state.

While committee-based, or *permissioned*, protocols allow for greater tolerance against Byzantine failures, they are not sufficient to provide full decentralization. In particular, such protocols often only work well with a small number of participants. As a result, small committees of nodes can quickly devolve into an oligopoly. Here, while not a single entity controls the system, a small number of participants can collude to take over control of the system. Similarly, committee members can collude to artificially increase transaction fees or impose censorship.

3.5 Sybil Detection

In the *permissionless*, or public, setting, such as that of Bitcoin or Ethereum, protocols must be resilient to Sybil attacks, where a single entity is creating multiple identities to gain more control over the system. These attacks are feasible because without a trusted third party there is no straightforward way to authenticate user identities.

Consensus protocols rely on either computational barriers or stakes to prevent such Sybil attacks. Stake-based systems manage membership information as part of their protocol. In committee-based consensus protocols stake usually is binary, which means only members of the committee are allowed to vote and each committee member has the same voting power. Here, each change to the committee must be approved by all participants. Recent protocols have introduced the notion of variable stake, often bound to how much cryptocurrency a certain party holds. In this setting, cryptocurrency can be passed on to other participants to dynamically reallocate voting power.

Systems that rely on computational barriers for Sybil detection do not manage any form of global membership information. Instead, participants have to perform a certain task to become, or have the chance to become, a leader. Most commonly, this task involves solving a cryptographic puzzle, where an input to a hash function has to be found such that the functions' output is below a specified threshold.

These particular cryptographic puzzles are better known as *Proof of Work (PoW)* [17]. The underlying intuition is simple: every attempt to solve the puzzle requires a constant amount of computation and the chance to solve the puzzle is independent of earlier attempts. PoW, thus, provides a very reliable means of Sybil detection, albeit being a very wasteful mechanism.

3.6 Nakamoto Consensus

The Bitcoin paper introduced the Nakamoto consensus, a consensus protocol that builds on top of *Proof of Work (PoW)*. There are two core differences between protocols described so far and the Nakamoto consensus. First, the use of PoW allows it to be a public protocol that allows participants to join and leave at any point in time. To take part in the protocol one does not have to register with some global mechanism, but merely starts attempting to solve the crypto-puzzle. Second, the Nakamoto consensus operates non-deterministically, where the current state is known to be agreed upon by the global network with some high, but not absolute, probability.

Systems based on Nakamoto consensus rely on Gossip protocols [15] to broadcast messages, such as transactions or blocks, because they execute across a peer-to-peer network with no pre-defined topology or membership. Instead of being connected to the full network, participants of a Gossip protocol only talk to a few peers. When receiving a new message, they forward it to all their peers. To make gossip efficient, participants usually keep track of which messages they already sent to or received from a particular peer. As a result, messages eventually spread to the entire network, without the network being fully connected.

Nakamoto consensus performs leader election using PoW through a process called mining. Once a party has solved the cryptographic puzzle, they forward their solution in form of a block to the network to become a leader. Instead of proposing transactions after becoming leader, they directly include a set of serialized transactions in the published block. Once participants start mining, their chance of becoming miner is directly proportional to the processing power available to them, because each attempt to solve the crypto-puzzle is independent of past attempts.

Nakamoto consensus achieves consensus by picking the longest chain of proposed chains. There can always be multiple competing blocks or chains because mining is a random process. Honest participants pick the longest chain of valid blocks they received and, as result, will all eventually converge on the same prefix of the blockchain. However, this means that one has to wait a significant amount of time for the block to be "buried" deep enough in the chain for it to be considered finalized and immutable. For example, in Bitcoin, one usually waits for depth for 10 blocks (about 60 min).

Figure 3 outlines how, at a particular point in time, there might be multiple competing chains. Here, while the prefix of the chain is considered stable and abandoned forks have been removed, at the head of the chain, multiple forks are competing for the longest chain. At any point in time, the protocol might switch to a different branch, potentially reverting multiple blocks. These switches are sometimes also called *reorganizations*.

Protocols based on Nakamoto consensus with open membership, usually assume a strong bound on the network latency. This ensures that a block will be visible to all network participants after some fixed time. More concretely, systems like Bitcoin assume that this bound is about 5 min. If this assumption was not made, there could potentially be an undetectable longer chain due to a network partition.

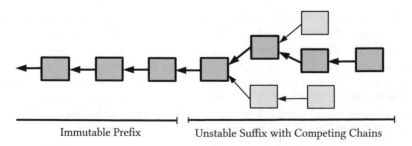

Immutable Prefix   Unstable Suffix with Competing Chains

Fig. 3 Sketched structure of a Bitcoin-like blockchain. The currently winning chain is highlighted

3.7 Bottlenecks

We now discuss the major bottlenecks of blockchains: *execution*, *verification*, and *communication*. Essentially, electing protocol leaders and ordering transactions in a globally replicated manner require massive replication of both data and computation. Thus, what the network can process as a whole is limited by the fact that every participant needs to process, forward, and execute all transactions.

Execution

Transactions in decentralized ledger systems differ significantly from those in conventional database systems. Every participant of the protocol maintains its local state in an authenticated data structure to be able to verify and process future blocks. In particular, DLT nodes usually calculate and store some form of hash tree of the state, and every block contains the root hash of the current state. These hash trees can be used both to verify blocks and to provide succinct proofs of some substate of the system. Executing transactions in such an authenticated manner requires more computation and storage. This is one of the reasons why systems such as Ethereum employ a limit on how many computational steps a block can contain ("gas limit"). Previous work has demonstrated that an improved storage engine can mitigate this bottleneck to some amount [50].

Verification

Blockchains rely on digital signatures to ensure the correctness and authenticity of messages. Intuitively, checking every transaction request and block generates a high computational workload as digital signatures are rather complex to verify. Increasing the frequency of transactions included by the system, thus, significantly increases the burden for every node in the network to participate in the protocol.

Privacy-preserving decentralized ledger may also rely on more advanced cryptography, such as zero-knowledge proofs, which increases the verification burden significantly.

Communication

Finally, for every node to be able to process every block and transaction, all transactions and blocks must be propagated to the entire network. Intuitively, this creates a high network communication overhead. Decentralized ledgers usually execute across a geo-distributed peer-to-peer network. Here, a larger state that needs to be synchronized will further increase the considerably high propagation latencies.

Even worse, scalability mechanisms may harm decentralization, a key promise of decentralized ledgers. For example, a naive attempt for increasing the throughput of a ledger is a higher block frequency or block size. Either, will cause a higher propagation delay of messages and, in turn, increase the likelihood of forks. Additionally, bigger block sizes raise the CPU and storage requirements for nodes participating in the network. This problem is especially salient for new nodes joining the network the need to verify *all* blocks in the chain before processing new transactions. As a result, only participants with strong hardware that is well connected may participate in the protocol, causing a more centralized network layout.

4 Improved and Novel Consensus Mechanisms

4.1 Improved Committee-Based Consensus Protocols

Figure 4 sketches the message exchange in the Paxos protocol [36], excluding its leader election, one of the most prominent mechanisms for SMR. Once a leader is elected (not shown in the figure), clients can submit transaction requests to it. The leader then proposes the transaction to its followers (`accept?`), which then each forwards their response (`accept!`) to the network. If a majority of the nodes accept the transaction, it is considered accepted by the system as a whole and the result is forwarded to the client. So-called Multi-Paxos pipelines this mechanism, by only electing a leader every so often and having that leader propose many transactions in sequence.

Practical Byzantine Fault Tolerance (PBFT) [8] was one of the first Byzantine fault-tolerant consensus protocols and is still widely used today. Figure 5 outlines how the protocol accepts a transaction. PBFT adds another round of messages to the

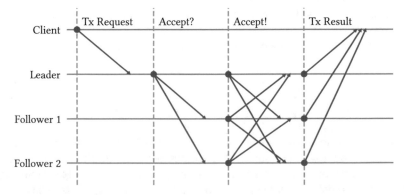

Fig. 4 The voting phase of the Paxos protocol: The leader proposes new transactions to the system, which then need to be approved by the majority of the network

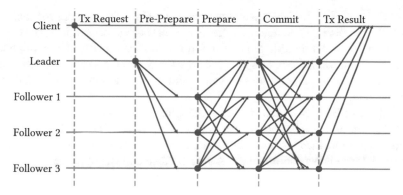

Fig. 5 Protocol diagram of PBFT: compared to Paxos an additional phase is added to account for Byzantine behavior. Note that also additional replicas are needed to tolerate Byzantine failures

protocol that confirms the receipt of the `prepare` message—which is equivalent to the `accept!` message in Paxos—by a majority of the network. This is necessary because nodes might send conflicting messages to different participants in the network. For example, follower 1 might send a `prepare` message for one transaction to the leader while also sending a prepare for a conflicting transaction to follower 2.

Similar to Multi-Paxos, PBFT can let a single leader propose many transactions to speed up the protocol. Leaders are usually only switches during failure. In the context of PBFT, this mechanism is usually called a *view change*.

Several minor improvements to PBFT have been proposed over the last few years. For example, Zyzzyva [32] avoids the third round of messages in the absence of failures using speculative execution. Aardvark [9] adds additional robustness by making clients digitally sign their requests and frequently rotating leadership. Tendermint [6] reduces communication complexity using Gossip protocols. We describe more complex modifications and entirely new permissioned protocols below.

Another key factor in making PBFT (and similar protocols) scale is *batching*. Similar to blocks in permissionless systems, a large set of transactions is bundled together. This allows to reduce the amount of communication required per transaction, but, in turn, increases the latency of the protocol.

Stellar

The Stellar Consensus Protocol (SCP) [37] is a variation Byzantine Agreement where each participant may have different levels of trust in other participants. Here, each participant picks a weight for peers they trust. The weight indicates their level of trust. Classical BFT consensus can be seen as a special case where each participant picks the same weight for all peers.

For SCP to reach consensus, a quorum must contain a *quorum slice* for each of its non-faulty members. Each node can define one or more quorum slices, of which at least one must be met for a valid quorum. A quorum slice consists of a set of nodes S and a certain threshold, for example, $\frac{3}{4}$, of how many members of S's members have to agree. Finally, for a quorum to be valid, each of its quorum slices must overlap with one another.

Similar to PBFT, SCP executes in three phases: NOMINATE, PREPARE, and COMMIT. The NOMINATE phase acts as a filter by identifying valid candidates for a consensus value. Each node can nominate multiple values, but must not nominate new values once it has confirmed the NOMINATE statements of a peer.

The NOMINATE phase is then followed by one or multiple rounds of ballots. Multiple rounds of ballots may be needed as it is not possible to determine whether a ballot got "stuck" due to a failure or if there is just a large network delay. Here, for the n-th ballot, the PREPARE(n,x)-message ensure that no value other than x is chosen for any ballot $\leq n$. A COMMIT(n,x)-message then states that the value x was chose at ballot n.

HotStuff

HotStuff [66] is a novel consensus protocol that improves upon PBFT by reducing message complexity by employing a star topology. HotStuff is of particular interest, as it is intended to be used by Facebook's Diem (formerly "Libra") cryptocurrency.

Figure 6 outlines the message exchange in HotStuff. HotStuff allows for a start communication pattern with linear complexity, not quadratic like PBFT, by passing all messages through the leader. To account for malicious leaders, another round of messages is needed compared to PBFT. In the so-called Decide-phase, the leader notifies all participants that a particular commit message has been received and accepted by a quorum.

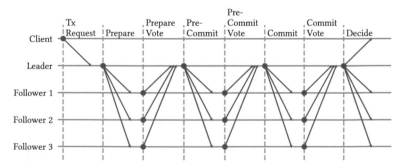

Fig. 6 Protocol diagram of HotStuff: the leader includes Quorum Certificates (signatures from all members of the quorum) from the previous phase to reduce message complexity

HotStuff reduces communication complexity through a primitive called *Quorum Certificates (QCs)*. Instead of having every participant exchange messages with each other, the committee communicates in a start topology with the leader at its center. The leader then includes QCs in its messages, which are certificates that prove the receipt and acceptance of a proposal by a quorum. This is achieved by having OCs contain a *threshold signature*, which, in essence, is a signature signed by multiple private keys that can be verified against a single public key. A key advantage of threshold signatures is that they do not grow in size with the size of the quorum.

To achieve even higher performance, the protocol can be pipelined, where multiple phases of the protocol (for different batches of transactions) execute in parallel. This is possible by including leader election in every prepare phase so that view changes do not interrupt pipelining.

Byzantine Ordered Consensus

For the committee-based setting, Zhang et al. [69] propose a mechanism that establishes transaction order outside of leader election. Their protocol, pompē, prevents the so-called "front-running", which allows malicious leaders to reorder transactions to their advantage. However, their results also indicate that pre-ordering transactions can enhance the performance of a permissioned system significantly. Here, instead of letting the leader decide the order of transactions directly, transactions are ordered before they are serialized.

To order transactions, a node n proposes a command c and asks a quorum of the participants to assign a timestamp for that command. Nodes first order all pending commands (or transactions) locally and then reply with a timestamp that honers this ordering. Node n then picks the median timestamp from the replies. Because there are at most f malicious nodes and a quorum consists of $2f + 1$ nodes, this median value is guaranteed to be within the valid range of time stamps.

HoneyBadger BFT

HoneyBadger [42] is a leaderless and asynchronous consensus protocol. At a high level, this protocol operates in three steps.

First, *each* node proposes a set of transactions, which are then broadcast to all other nodes. To do this, they maintain a local pool of transaction requests from client and sample some subset of this pool. The exact mechanism of how this subset is chosen is important for performance and safety but goes beyond the scope of this book. A *Reliable Broadcast Protocol (RBC)* ensures that each non-faulty node's proposal is propagated through the network.

Second, nodes send and acknowledgement for each valid set of transactions being proposed. They do this using threshold encryption, the same primitive as is used in the Stellar protocol. Encryption serves two purposes: it allows participants to sign on to a set without increasing the size of the message significantly and it prevents adversary to censor transactions they do not like.

Third, nodes agree on which sets to accept. This last step is important, because faulty nodes might not propose a transaction set and, as a result, could cause the protocol to wait forever or become inconsistent. To address this, nodes vote to accept a set of transactions once it has been signed by enough participants until at least $N - f$, where N is the number of nodes and f is the maximum allowed number of failures. Afterward, they will vote to reject all other transaction sets.

To retrieve the final batch of transaction to be accepted, nodes form a "common subset" of all accepted transaction sets. Nodes add all transactions for each accepted transaction set to the batch. Then, they sort the transactions lexicographically, e.g., by their transaction identifier, and remove duplicates.

This protocol is teared towards the UTXO model where the probability of two transactions conflicting is low. Thus, one potential drawback of this design can be in a setting where there might be multiple conflicting transactions.

4.2 Minor Changes to Nakamoto Consensus

Concurrency Inside Blocks

Bitcoin and Ethereum enforce a serial order of operations in a single block. This limits the execution of a block to a single logical core. TTOR (Topological Transaction Ordering Rule), which was used for some time in Bitcoin Cash, loosens this requirement and only enforces a partial order among transactions, and, thus, enables validating transactions of a single block concurrently.

However, in our experiments, we observed that the bulk of work performed by blockchain nodes is involved in validating and generating digital signatures, which can already be performed concurrently as transaction requests are usually received and generated out of band. Bitcoin Cash eventually switched from TTOR to another ordering mechanism as its benefits to block propagation were limited [53].

Block Size and Frequency

As mentioned in Sect. 3.7, one attempt to scale blockchains is to increase block size allowing for more transactions to be processed with the same number of blocks. The propagation delay depends on its size and performance of the underlying peer-to-peer network. In particular, a larger block takes longer to propagate between two peers as outlined in the equation below.

$$Latency_{block} = Latency_{network} + \frac{BlockSize}{Throughput_{network}}.$$

Block sizes that are too large might take longer to propagate than it takes to mine the next block [14, 24]. As a result, larger block sizes result in a higher likelihood of blockchain forks, which hurt performance. Once a fork is resolved, only one of the branches is considered part of the chain, and the rest is discarded.

Another intuitive attempt of increasing the throughput of a blockchain system is to increase the frequency of blocks. Similar to block sizes, a higher block frequency results in an increased chance of forks as blocks are created faster than they are being propagated through the network. This in turn also leads to more centralization of mining, as large-scale mining pools have a higher chance of receiving and processing blocks in time.

Ethereum relies on a mechanism called Greedy Heaviest Observed Subtree (GHOST) [58] to disincentivize centralization. Here, if a miner is aware of a fork will reference not only a block's direct predecessor but also the heads of competing chains (known as "uncle blocks"). As a result, miners receive a partial reward if they ended up mining on a fork.

Increasing Efficiency of Block Propagation

The nature of peer-to-peer protocols results in significant communication overheads when propagating data. Gossip communication inherently requires additional communication, because data does not flow in a straight path but spreads in multiple directions through a peer-to-peer network. As a result, peers might receive the same messages from multiple parties. This problem is exacerbated in Bitcoin as transactions are propagated through the network twice: as a transaction request and as part of a block.

One line of work is to improve the efficiency of Gossip protocols. Compact blocks [10] do not contain a full list of transactions as their payload but merely shortened transaction identifiers. Upon receiving a compact block, peers only request the transaction they have not seen yet.

Bloom filters can be used to efficiently keep track of which data a peer has already received [49]. Essentially, bloom filters are a lightweight data structure (usually only a few bytes) that provides a heuristic about whether a set contains some data item or not. When forwarding compact blocks, peers rely on Bloom filters to estimate which transactions the remote party already holds and forward only the ones that it probably does not have yet. This, in turn, avoids an additional round-trip time, where the remote party has to request transactions.

Another line of work is to augment peer-to-peer networks with a fast relay network. Relay networks are usually not a good fit for the decentralized ledger setting, as they have sparse topologies, and, thus, contain multiple points of failure. However, they can be used in addition to a fault-tolerant peer-to-peer network, to allow for faster propagation of blocks in the common case [11, 29].

4.3 Decoupling Mining from Transaction Serialization

Consensus protocols generally perform two distinct tasks. LEADER picks the next participant to be the leader of the protocol, i.e., the entity that propose the next block(s), and ORDER decides on the order of transaction inside those block(s).

In most permissionless protocols, these tasks are bundled together, which harms performance. In particular, blocks in Bitcoin or Ethereum can only hold a certain amount of transactions and are published at a low frequency. Additionally, as outlined in Sect. 4.2, increasing block size or frequency does not always result in higher throughput of the blockchain.

Bitcoin-NG

Bitcoin-NG [26] breaks down the process of mining in traditional Nakamoto consensus into its constituent processes to increase throughput. The Bitcoin-NG LEADER process proceeds as follows: Miners solve a PoW puzzle and broadcast a special block called a *keyblock* with the solution to the rest of the network, signaling their status as the protocol leader. At that point, the winning miner performs an ORDER process by grouping transactions into *microblocks* and broadcasting them into the network. This separation of key and microblocks is outlined in Fig. 7. The entity that mined the most recent keyblock creates and broadcasts microblocks so long as they are the leader. Solely the network speed and how quickly the leading miner can sequence them limit the flow of transactions.

While Bitcoin-NG improves throughput over the conventional Bitcoin protocol, it is still limited to the bandwidth of a single entity executing the ORDER process. Also, a single high-throughput chain harms decentralization as every participant of the protocol needs to possess the processing power and network bandwidth to process the chain in its entirety. Further, these long-lived leaders can be subject to Denial-of-Service (DoS) attacks.

ByzCoin

ByzCoin [30] follows the same observation as Bitcoin-NG, but, instead, establishes a committee of leaders. ByzCoin relies on some underlying *identity blockchain*. The committee is then chosen by picking the entities that mined the last n blocks on the

Fig. 7 Structure of the Bitcoin-NG blockchain: Keyblocks (square) hold leadership information, while microblocks (circle) hold serialized transactions

identity chain, where n is the size of the committee. The entity that mined the most recent block, is the designated leader. Each set of transactions proposed by the leader needs to be approved by a majority of the committee.

A key advantage of ByzCoin over Bitcoin-NG is that it has almost instant finality. Bitcoin-NG, on the other hand, has a similar latency as the unmodified Bitcoin protocol. The instant finality of ByzCoin is possible because, for a sufficiently large n, the majority of the committee is guaranteed to consist of honest miners.

4.4 Novel Proof-of-Stake Protocols

Proof of Stake (PoS) is a mechanism intended to be an energy-efficient replacement for PoW. At a high level, voting power here is not dependent on a party's processing capabilities but on the amount of funds they hold in the cryptocurrency, which, in turn, allows avoiding unnecessary computation. The key challenge in PoS is the "nothing at stake" problem: if block generation does not require mining, an adversary can easily generate many, potentially conflicting, blocks.

Ouroboros

Ouroboros [28] is a provably correct PoS protocol that powers the Cardano blockchain.[2] The protocol's execution is divided into constant-size epochs, each consisting of some number of time slots. At the beginning of an epoch, a seed is generated from the values of the previous epoch, to generate a pseudo-random assignment of participants to slots, where the likelihood of being assigned to a slot is directly proportional to the amount of currency a participant is holding.

In every slot, the selected participant is allowed to propose a block containing a set of transactions. Although an adversary here can still generate conflicting blocks, assuming the majority of the participants are honest, a sequence of correct blocks will eventually constitute the longest chain. However, in this scheme, it is still possible to anticipate who the next leader will be and launch a DoS attack on them. Like Bitcoin, Ouroboros requires blocks to propagate within a bounded amount of time and loosely synchronized clocks.

Algorand

Algorand [23] addresses some challenges in PoS using a verifiable random function (VRF). Here, each participant locally runs the VRF which takes some global data and their private key as an input. Depending on the input, the function may return

---

[2] Note that this section describes the initial version of Ouroboros outlined in the CRYPTO 2017 paper.

a certificate that the particular user is allowed to propose a block. Like in Bitcoin, multiple users may be allowed to propose blocks and Algorand provides a mechanism to sort certificates of concurrent blocks. Protocol participants then discard all blocks except the one with the highest priority to prevent forks.

Algorand prevents DoS attacks by making this random function unpredictable and switching participants after every round of voting. This unpredictability is achieved by taking the user's private key as an input. Later users can prove they executed the VRF correctly using their public key. Additionally, the protocol assumes the absence of network partitions to prevent malicious users from successfully proposing conflicting blocks.

Avalanche

Avalanche [54] is a probabilistic leaderless consensus mechanism with low communication overhead. Here, nodes periodically query a constant-size random set of peers about which transaction they accepted. Depending on their peers' responses, they adjust their confidence in the transaction being accepted by the network as a whole. Acceptance of a particular transaction will then eventually propagate through the network. Avalanche works well with the UTXO model as transactions do not need to be totally ordered and conflicting transactions are rare. The protocol needs to be combined with another mechanism, such as PoS, to ensure Sybil resistance.

```
def on_query(v, new_col):
    if col == None:
        col = new_col

    respond(v, col)

def slush_loop(u, col0 in [R, B, None]):
    # Initialize with red blue or nothing
    col = col0

    for _ in range(m):
        # if None, skip until on_query sets the color
        if col == None: continue

        K = sample_nodes(k)
        P = [query(v, col) for v in K]

        for ncol in [R, B]:
            if P.count(ncol) >= alpha*k:
                col = ncol

    accept(col)
```

Listing 1 Slush: a simplified version of the avalanche protocol

Listing 1 displays a simplified version of the Avalanche protocol without fault tolerance, dubbed "slush". Here, the network aims to decide on a single color `col`. To do this, each node queries, a random sample of `k` other nodes `m` times. In the outlined code, `on_query` is called whenever a node is queried by some other node and `slush_loop` is executed repeatedly by each node. After each set of queries, nodes participating in the slush protocol either decide to stick with the color it has currently accepted, or, if more than $\alpha * k$ (where $\alpha > 0.5$) other nodes have accepted a different color, switch over to that color.

A key advantage of this protocol is that it requires almost no state to be maintained at each node (only the currently accepted state) and that it involves communication with a small subset of, instead of a majority of, the network. More concretely, communication complexity per node is constant independent of the size of the network, because the sample size `k` does not grow with the size of the network.

Gasper and Ethereum 2.0

Casper is a "finality gadget": it allows ensuring that a block in Nakamoto consensus is finalized and cannot be undone due to a reorganization. Here, stakers endorse blocks they consider part of the longest chain using their stake. Because the total amount of stake is known, at some point, if a block is sufficiently endorsed, it can safely be considered final.

Ethereum 2.0 relies on *Gasper*, a protocol that combines mechanisms from Casper with GHOST. Gasper performs leader election similar to Ouroboros: time is segmented into slots, each having a leader that is defined by some randomized mechanism. For every k slots, the protocol uses the Casper mechanism to achieve finality. Like in Ouroboros and Bitcoin, this requires loosely synchronized blocks. Unlike those mechanisms, the protocol will not be unsafe in a partially synchronous setting, but may not make progress. Competing blocks can still exist, due to network delays or malicious leaders, but GHOST's notion of referencing "uncle blocks" allows quick convergence to a singular chain in this case.

Gasper relies employs a chain selection rule based on the amount of stake attached to a particular chain. Here, the "heaviest chain" is the chain with the most stake attached to it. This ensures that the protocol will converge on the chain that is considered finalized by the majority of the network, not adversarial chain that is potentially longer.

4.5 *Summary*

The Bitcoin protocol and derivatives are not sufficient to support any demanding workload and waste massive amounts of energy. We do need new protocols, or radically improve existing ones to overcome these limitations.

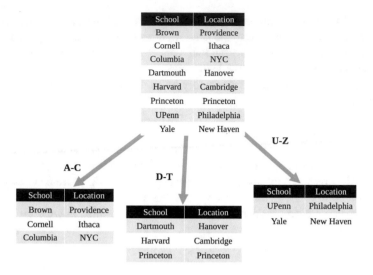

Fig. 8 Sketch of how a table of all Ivy League schools could be sharded alphabetically

Multiple potential contenders exist to replace outdated ledger technologies, each with different properties and tradeoffs. Further evaluation and benchmarking is necessary to determine the "winner" among these protocols.

5   Sharding Blockchains

At a high level, sharding breaks the keyspace of a database into multiple "shards".[3] Sharding is usually done using a hash function or breaking the keyspace into evenly sized pieces. Figure 8 sketches how a table can be broken apart by assigning different ranges of starting letters to different shards. Hash functions are usually the preferred mechanism of sharding as they assign objects to shards pseudo-randomly and thus spread the workload of a system more evenly across shards. Updates and queries for a particular shard can then be processed without involving other shards.

Sharding for blockchains is usually implemented in the following way [62]. Some mechanism keeps track of a set of identities, e.g., by examining the last k miners of a PoW chain [30]. The protocol then assigns shard some subset of these nodes. Each shard then locally runs a consensus mechanism, such as PBFT [8], and a distributed transaction protocol, such as a two-phase commit, handles cross-shard transactions. Finally, some scheme is in place to periodically "merge" the state of all shards.

---

[3] Note that some protocols shard transactions, not state. These mechanisms are significantly different and not covered by this section.

5.1 Challenges in Sharding Blockchains

Blockchain enthusiasts have long hoped that sharding will solve the scalability problem. In essence, sharding allows every participant to only process a subset of all transactions of the network. Ideally, this allows to linearly scale the throughput of the system *without* increasing the burden on any particular participant. However, so far, no sharding protocol has been deployed in a real-world setting. The reason for this is complex but, at a high level, sharding decentralized ledgers faces four major challenges: *reduced safety*, *reduced availability*, *loss of network decentralization*, *reduced consistency*, and *lack of economic incentives*.

Maintaining Safety

The essence of decentralized ledgers is that they protect some application, e.g., a cryptocurrency, against a strong Byzantine adversary. A basic requirement for protection against such an adversary is to have a Sybil detection mechanism, which is usually based on how much stake an entity has or how much computational work is done.

For example, a core assumption in Bitcoin is, that $<50\%$ (or 25% in some cases) of the entire network is controlled by adversaries. A shard intuitively has much less total stake (or computational work) than the system as a whole. Thus, some mechanisms must be in place to ensure that a single shard is safe as the network as a whole.

Ensuring Availability

When splitting upstate among subsets of the network, less participants hold a copy of a particular transaction. For example, if a transaction executes locally on a particular shard, there is no need for other shards to know about that transaction, let alone storing it on their machines. As a result, shard state could get lost during failure. This problem is exacerbated by the fact that an adversary might be incentivized to hide (parts of) a shard's state. For example, they might have misbehaved and want to hide the evidence, or they might want to revert the shard to a state that is more advantageous to them.

Ensuring Consistency

In systems such as Ethereum, a serial order of transactions is enforced to ensure consistent updates to contract states. Unfortunately, enforcing a total order for all transactions is very difficult if shards execute mostly independently.

To ensure consistency, a sharding protocol needs some mechanism to consistently apply transactions to multiple shards. Protocols for distributed transactions, which ensure consistent and atomic updates across shards, have been explored extensively in the systems community. However, adopting such protocols to a permissionless setting with Byzantine failures is a challenge.

Maintaining Decentralization

So far, we have discussed decentralization as an abstract system property. More concretely, ensuring decentralization means keeping the burden of joining the network and participating in the consensus protocol low. Ideally, anybody with a computing device should be able to join the system.

Even current non-sharded systems that promise decentralization are not very decentralized in practice. For example, Bitcoin and Ethereum are controlled by only a handful of entities [22]. The underlying reason for this is that decentralization often conflicts with the goal of scalability. If a network supports many participants of varying locations and processing power, data takes longer to be propagated across the network. As a result, control of many decentralized ledger protocols tends to centralize around nodes with access to large amounts of processing power and fast network connections.

Providing Sound Incentive Mechanisms

Miners (or stakers) participate in a consensus protocol because they receive some monetary reward or want to secure the value of their assets stored on the ledger. Bitcoin and Ethereum have fairly straightforward incentive mechanisms, where the miner of a new block gets some new currency and transaction fees as a reward.

Incentive mechanisms tend to get more complicated when introducing sharding, as there may exist distinct shard chains and transactions can execute across multiple shards. For example, OmniLedger [31], while providing a safe sharding protocol, does not provide sound incentives for the large set of validators required to power the protocol.

5.2 Foundations

Several sharding solutions have been proposed for permissionless and permissioned blockchain systems that are built on previous work on sharding databases and distributed transactions. In fact, the concepts of sharding and distributed transactions have been widely studied concerning conventional database systems. While the failure assumptions are vastly different in conventional systems, the underlying motivation of reducing coordination and increasing parallelism to achieve higher throughput is the same.

Sharding Databases

Sharding was first popularized by systems like Chord [59] and Mercury [5]. In such systems, usually, a hash function is applied to an object key to map it to a specific shard. Some systems also have the notion of virtual shards. Here, a large number of virtual shards is mapped to a considerably smaller number of nodes. Virtual shards can then be remapped to different nodes if the workload changes.

Later work introduced systems, such as Chubby [7], which provide serializable transactions on top of sharded systems. More recent work aims to improve performance by reducing coordination even further [13, 44] or relying on loosely synchronized clocks [12].

Distributed Transaction Protocols

The most prominent mechanism to apply transactions in a consistent and atomic fashion to multiple shards is a two-phase commit [4]. Here, in the first phase, a transaction first locks all relevant data objects, ensuring that no concurrent updates are made. In the second phase, the transaction applies all changes and releases the locks.

Two-phase commit can be separated into two variants. First, a conventional (or pessimistic) two-phase commit acquires locks gradually while executing a transaction. If a lock is already held by another transaction, some mechanism such as wound-wait must be in place to avoid deadlocks. Optimistic concurrency control [33], on the other hand, first executes transactions without holding locks, then submits the transactions as a set of operations to the involved servers in the first phase of the protocol. The main advantage of OCC is that it keeps the time a lock is held short allowing for higher concurrency. Pessimistic concurrency control usually works better in update heavy workloads and in settings where latencies are high.

5.3   Public Blockchain Sharding Protocols

To our knowledge, virtually all blockchain sharding protocols apply to public (or permissionless) blockchains. While private blockchains can leverage sharding as well to increase performance, the problem of low performance is less severe there as they operate committee-based protocols with a small number of participants. For example, HotStuff can process thousands of transactions per second, while Ethereum can only process tens.

Monoxide

Monoxide [63] breaks up the workload across independent consensus zones, each having its own set of miners. Monoxide does not support generalized transactions, but only money transfers between exactly two accounts. For a cross-zone transaction, the transactions are first processed in the source account's zone and then forwarded to the target account's zone together with a Merkle proof of the transaction's inclusion. At some point, the transaction will be included in the source and the target zone, however, the protocol does not provide an upper time-bound for this.

Furthermore, the transaction processing scheme proposed in monoxide is susceptible to recursive invalidation of dependent transactions in the case of zone-forks. Another challenge with Monoxide's design is that its independent zones naturally partition the mining power of the blockchain system, which dilutes the overall security of the system. The authors address this by assuming the majority of miners will work in all zones at the same time, which requires miners to possess large amounts of processing power for verification to maintain the same security guarantees as Bitcoin. This encourages mining centralization for high throughput, giving up the key property of blockchains.

Elastico, RapidChain, and OmniLedger

Elastico [38] and OmniLedger [31] in a similar class of scalability solutions that propose dividing the nodes in a system into small committees, each of which performs a Byzantine consensus protocol for intra-shard consensus. In these protocols, there exists a single identity blockchain, similar to that in ByzCoin, as well as, a distinct blockchain for each shard. The Elastico protocol, the first of such solutions, proceeds in the following fashion: protection against Sibyls is achieved using an identity chain based on PoW. It then pseudo-randomly assigns nodes to committees that perform PBFT in rounds until all the nodes in the system agree on a final changeset to be committed. The protocol then re-assigns committees and restarts the process for the next set of transactions.

OmniLedger makes further improvements on top of Elastico, such as using RandHound [60] to better seed for randomness in shard assignments and helps ameliorate some security compromises introduced by Elastico's small committee sizes. However, OmniLedger still adds several layers of complexity to public blockchains. This complexity is especially salient when examining the need for OmniLedger to have day-long epochs because of the amount of overhead required for bootstrapping at the beginning of an epoch, which makes it susceptible to quick-responding attackers.

Additionally, OmniLedger allows for atomic cross-shard transactions using Atomix, a variant of a two-phase commit. Here, clients have to first lock funds of the affected shards in phase one. They collect proofs of inclusion of the lock message in the shard, or, respectively, proofs that the transaction could not be included in the shard. In phase two, if all shards lock the transaction successfully, they issue a unlock message that commits the transaction. Otherwise, they issue an unlock message that will abort the transaction. Elastico, on the other hand, has no notion of atomic cross-shard transaction.

Note that, in the Atomix protocol, if a client fails, the transaction is "stuck". The authors argue that the client has an incentive to finalize their transaction as, in the UTXO model, their funds are locked while the transaction is in progress. This is similar to how Avalanche incentivizes clients to not issue conflicting transactions, as it would lock up their own funds. However, this makes it difficult to implement a similar mechanism for smart contracts, where the incentive structure is not as clear.

RapidChain [67], among other changes, replaces the Atomix protocol with one that does not rely on the behavior of particular clients. Instead, the transaction is assigned to a particular shard by hashing its identifier. The output of the transaction, i.e., the generated UTXOs, are then also stored on that particular shard. The shard then contacts all shards that hold inputs for the particular transaction. To make this scheme efficient, shards are not connected to every other shard, but instead, route messages through a shard network.

Zilliqa

Zilliqa [70] shards transactions, but not state. This protocol relies on a similar mechanism as OmniLedger for assigning nodes to shards but uses a different cross-shard commit protocol. Instead of splitting the state of the system across shards, they only split the transaction workload and replicate state among all nodes.

Each shard then processes a subset of all transactions for a specific epoch and merges their resulting state with other shards at certain checkpoints. At a high level, the protocol allows a particular shard to lock parts of the state to prevent concurrent modification of the same data entries. Zilliqa employs a data flow-based programming model to implement this scheme efficiently.

Ethereum 2.0

Ethereum 2.0 [61] introduces a sharding scheme among other major changes to the protocol. This mechanism borrows ideas from both off-chain mechanisms and randomness-based protocols like OmniLedger. Here, nodes participate in the protocol by putting down a deposit. A verified random function then assigns each node to a particular shard.

Additionally to the random assignment, the protocol ensures safety and availability by punishing nodes that sign an invalid block or respond too slowly. At the time of writing this chapter, the Ethereum developers have not yet decided on a protocol for cross-shard transactions and it is unclear whether the protocol will support serializable cross-shard transactions.

5.4 Summary

Sharding is almost certainly necessary to make decentralized ledgers scale. However, it is a problem that is still in the process of being solved without losing any of the core guarantees that blockchains provide.

6 Layer-2 Solutions

Instead of scaling the blockchain protocol itself, the so-called "Layer 2" protocols can be layered on top of existing systems to improve performance. These protocols are usually orthogonal to previously mentioned approaches, such as sharding, as they build on top of an existing DLT.

Payment channels lock funds on the global ledger and facilitate fast transactions between parties through an off-chain protocol. Only the amount locked on the base chain is allowed to be exchanged in these systems, and a tally of balances is kept for when it is time to settle. On settling, the amount apportioned to the settler as denoted by her balance in the subchain is unlocked on the main chain and returned to the settler. State channels extend this scheme from cryptocurrency funds to the arbitrary state.

6.1 Building Blocks

Layer-2 solutions rely on a common set of cryptographic primitives to implement their functionality securely. We outline the most important ones here.

Merkle Proofs

Merkle trees allow creating a succinct tree of cryptographic hashes that represent a system's state. Such trees are constructed by hashing all objects of the state and then recursively merging hashes by applying the hash function to them again. Usually, only the root of such trees are stored on the blockchain and the rest of the tree can either be constructed on the fly by clients or is provided in the form of Merkle proofs.

Merkle proofs then allow showing the validity of a system's (sub-)state by providing the branch of the tree from the affect objects to the root. This proof can then be verified against the root hash on the blockchain. Because these proofs rely on cryptographic hashes, it is virtually impossible to forge a Merkle proof against the same root for a different state.

Cryptographic Commitments and Fraud Proofs

Analogous to a promise in real life, participants can provide a *commitment* in the form of a statement that is signed with their private key. For example, one can generate a hash $H = h(S)$ of the current state S of a system and sign it with a cryptographic key. This enables any holder of the commitment to prove later that the state of the system was indeed S.

If a party detects misbehavior, they can then raise a *fraud proof* that shows two conflicting cryptographic commitments by a particular party. Layer-2 protocol often rely on fraud proofs to punish misbehaving party, as well as, to recover from failure.

Time Locks

Time locks allow for a certain transaction or statement to become invalid if not included on the blockchain in a specified timebound. In Bitcoin, this mechanism is implemented using the `CheckSequenceVerify` and `CheckLockTimeVerify` protocol extensions. Here, time can be expressed either as real-world time or as the length of the blockchain. Again, in systems like Ethereum, similar functionality can be implemented using a smart contract.

This mechanism enables layer-2 protocols to recover in case of participants becoming unresponsive. For example, in a payment or state channel funds could be lost if a party refuses to cooperate. Some protocols, thus, release funds after a certain amount of time if no progress is made.

Hash-Locked Transactions

In Bitcoin, *Hash-Locked Transactions* allow locking funds, which can then be retrieved using a custom key. More concretely, such hash locks define under which conditions a particular transaction output can be spent. These locks are implemented using Bitcoin Script and they can be implemented similarly in other programmable blockchains, such as Ethereum.

In the context of layer-2 protocols, this primitive is especially useful, as it allows to lock funds for a channel or subchain and later release it only if a certain condition is met. For example, it allows extending time-locks with a fraud-proof mechanism from Sect. 6.1, which we will outline later.

6.2 Payment Channels

At a high level, payment channels lock funds on some existing systems and facilitate fast transactions between parties through an off-chain protocol. Only the amount locked on the base chain is allowed to be exchanged in these systems, and a tally of balances is kept for when it is time to settle. This flow is outline in Fig. 9. At the time of writing, the most prominent payment channel protocols are Lightning [52] for Bitcoin and Plasma [51] for Ethereum, respectively. Payment channels do not rely on additional consensus mechanisms but, instead, on cryptographic commitments and time locks.

We now outline the Lightning protocol at a high level.

Creating and Updating Channels

First, a `funding`-transaction is created, which records the initial funds deposited into the payment channel. This initial funding transaction creates a single transaction output that can only be unlocked using a transaction signed by both parties.

Then, whenever the balance of the payment channel is updated, a new `commitment`-transaction is created that records the new state. `commitments` are not immediately stored on the blockchain, but saved by the participating parties for later use. Each new commitment contains an `revocation` for the previous commitment.

The `funding`-transaction is signed and stored on the main chain once the first `commitment` has been created. The latter ensures that channels can always be terminated, as termination requires a `commitment`-transaction to exist.

Terminating Channels (One-Sided)

In Lightning, either party can close the channel at any time by storing the most recent `commitment`-transaction on the main chain. While the other party has immediate access to the released funds, the closing party needs to wait for a certain amount of time before their funds can be used.

This waiting time is implemented using hashed time-locks. There are two potential outcomes of this hashed time-lock. First, if the closing party has attempted to close the channel with an outdated `commitment`, the other party can reveal the `revocation` for the outdated `commitment`, which serves as a fraud proof. If

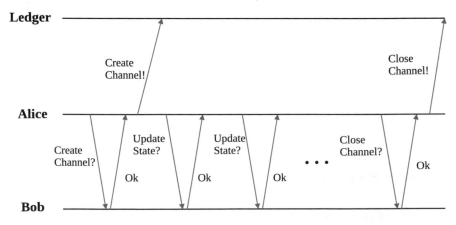

Fig. 9 Flow of a payment channel: only the opening and closure of the channel are recorded on the global chain

they do so, they will claim all funds stored in the channel and, thus, punish the misbehaving party.

Terminating Channels (Cooperatively)
If both parties are available and well behaving, they can cooperatively close the channel. Because the original funding transaction is a single UTXO that can be unlocked by a transaction signed by both parties, this is much easier to do.

A cooperative channel termination is then just a regular transaction that consumes the channel's UTXO and distributes the funds among the participants. No contest time or hash lock is required.

Increasing or Decreasing a Channel's Funds
In Lightning, a channel must be closed and recreated in order to change the number of funds it has access too. Some other solutions, such as that provided by Miller at al. [41] allow restocking funds.

Payment Networks

In most implementations, payment channels allow an arbitrary number of payments, with only two transactions stored on the blockchain. However, one-to-one channels, like the ones described in the previous section, require a new channel to be established whenever one wants to transact with a new participant. This makes their use somewhat limited as establishing a new payment channel is costly. Payment networks address this limitation.

At a high level, payment networks allow transacting with another party through many intermediates. For example, if Alice wants to send money to Bob, but they do not have a channel established between each other, Alice can rely on a third party that has a channel established with her and Bob. Payment networks then provide a protocol to send money through that third party, or a series of third parties in the general case. This protocol has to allow these third parties to be *untrusted* to maintain decentralized properties of a blockchain system.

In Lightning, this mechanism is ensured as follows. Consider the topology from Fig. 10. Here, Alice wants to pay Bob but does not have a direct channel established with him. To do this, Alice first notifies Bob about her intention to pay him and Bob responds with a value H, where H is the cryptographic hash of some other value R. Bob keeps R secret. Alice then promises, through a cryptographic commitment, that she will pay Carol once she reveals R to her. Carol similarly tells Dave she will pay him once he reveals R to her. Dave tells the same to Bob, except Bob knows the

Fig. 10 Example topology for a payment network: Alice wants to pay Bob through Carol and Dave

value of R. Bob can then reveal R to Dave to initiate the payment process. The other participants do the same to pass the money through the chain.

This mechanism is safe as R will not be revealed before a chain of commitments has been established. If Bob reveals R ahead of time, the money would be sent through the network partially and not reach him. As a result, Bob is financially incentivized to wait until Dave has created a cryptographic commitment to him.

Payment networks need to provide a routing mechanism, that allows discovering the current topology and establishing a path between two parties. This problem is exacerbated by the fact that not all paths are valid for a particular payment, as a particular path may contain nodes that do not hold sufficient funds to process the payment. One promising approach to this problem is that of Sivaraman et al. [57], which, among other mechanisms, improves the flow of payments by breaking them into smaller "packets".

A challenge with payment networks is to prevent them from becoming too centralized. For example, the network could devolve into a topology consisting of a few large nodes that route most payments. Such large nodes would introduce single points of failure that could harm the reliability of the network.

6.3   State Channels

Payment channels can generalize to *state channels* [41], that support arbitrary smart contracts. For example, one can implement a chess game using a state channel, where all moves are processed by the channel and only the final result is stored on the blockchain itself. In most implementations, similarly as before, participants sign off every state change using cryptographic commitment.

At the time of writing, Dziembowski et al. [19] provide the only sound mechanism for state channel networks, which relies on the notion of virtual channels. While, in regular state channels, one relies on the blockchain to resolve conflicts, in virtual channels a third entity serves in this role. The key challenge here is that, unlike the blockchain, this third party is untrusted. As a result, virtual channels can still fall back to the underlying blockchain if the third party is faulty.

Figure 11 outlines how virtual channels can be constructed on top of other virtual channels, as well, which allows building a more complex state channel networks. Regular state channels, such as that between Alice and Carol or Dave and Bob, are constructed using a *State Channel Contract (SCC)* on the blockchain itself. Here, for example, if Alice becomes unresponsive, Carol will rely on the SCC to "forcefully" close the channel. Virtual channels, such as y_1, are then constructed using what the authors call a *Virtual State Channel Contract (VSCC)*. Similar to an SCC, Carol only becomes involved, during the creation and closing of y_1. In the case that Alice becomes unresponsive, Dave can "forcefully" close the channel using Carol. If Carol is unresponsive as well, he can leverage the SCC between him and Carol to recursively close both channels.

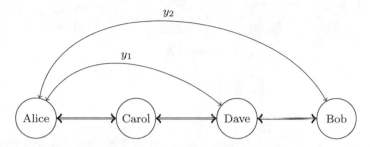

Fig. 11 Sketch of how virtual channels are constructed recursively. Alice first constructs a virtual channel y_1 to Dave, via Carol. Then she uses y_1 to construct a second virtual channel y_2 to Bob via Dave

Note, that this concept of virtual channels is tailored towards applications that have exactly two participants. To our knowledge, no sound state channel (network) construction exists yet that allows for a larger number of users to interact.

6.4 Watchtowers

A major drawback of layer-2 solutions is that they rely on a constant audit of the blockchain to prevent malicious behavior. In many implementations, if a party misbehaves, other participants must raise a fraud proof within a certain time-bound. However, not all participants might be online and actively participating in the blockchain consensus at all times.

Watchtowers [2, 3, 16, 39] enable outsourcing this task to a third party. At a high level, these mechanisms provide a reward to third parties, the watchtowers, if they detect misbehavior. More advanced solutions link this reward to the payment channel itself, to prevent malicious parties to bribe the watchtower(s).

6.5 Subchains

Systems such as BlockchainDB [20], Plasma [51] or Arbitrum [27] maintain authenticated data structures outside the blockchain and solely rely on it in case of failures. Such an authenticated data structure can be a Merkle tree, where the root is stored on the parent blockchain, as outlined in Fig. 12. This design aims to combine the advantages of a centralized system with that of a decentralized one. In the common case, the state of the system can be updated at a small number of sites without involving the blockchain. If the system fails, users issue fraud proofs to the main chain.

A core challenge with many subchain solutions is ensuring availability of blocks. As usually, only the Merkle root of the subchains state is stored on the blockchain

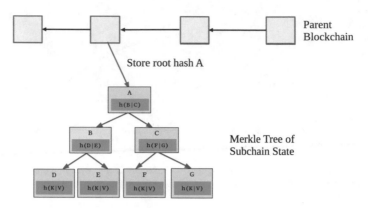

Fig. 12 Simplified concept of a subchain. The side chain's state is recorded by a Merkle tree and the tree's root stored on the parent blockchain

itself, the provider(s) of the subchain can hide the state to prevent audits. To address this, Arbitrum assumes that at least one participant in a database replica set behaves honestly and always remains available. Similarly, BlockchainDB assumes clients trust the particular database instance they are connected to.

6.6 Optimistic Rollups

Optimistic rollups are a special kind of subchain, where transactions are recorded and ordered on the main chain, but executed on a subchain. The key advantage here is that no tradeoff in terms of availability is made: if needed the subchains state can be recovered by re-executing all transactions. Examples of such systems are Optimistic Ethereum [48] and Arbitrum One [46].[4]

Rollups are mainly useful for computationally expensive transactions, such as complex smart contracts. While there exist mechanisms to batch transactions together, they usually still require all transaction data to be on-chain. This means that for simple payment transactions, the performance gain is negligible.

6.7 Summary

Layer-2 solutions are a great mechanism to *augment* other scaling solutions. For example, payment and state channels allow for instant confirmation through the exchange of cryptographic commitments. Subchains enable bundling many transactions into one on-chain transaction for efficiency. Additionally, subchains can enable

---

[4] Arbitrum One is not to be confused with the version of Arbitrum described in the previous section.

even high performance by making availability tradeoffs. The latter might be acceptable for applications of low financial value. Finally, optimistic rollups can overcome the computational limitations of current blockchains.

7 Federated Chains

Federated chains attempt to scale blockchains by allowing multiple separate chains to work together through a global relay chain or cross-chain swaps. While sharding usually has a fixed mechanism that dictates how and when shards are created and who they are assigned to, blockchain federation allows creating new chains organically.

While very similar to sharding, at first sight, this mechanism is much more similar to sidechains. In most designs, there exist a global chain that takes a similar role as the blockchain in side-chain protocol, in that it processes cross-chain communication and handles failures. However, one common point with sharding is that there usually exists a mechanism to update the state on multiple chains atomically, similar to cross-shard transactions (Fig. 13).

7.1 Cross-Chain Swaps

Cross-chain swaps [25, 68] allow exchanging cryptocurrencies between two separate blockchains without the involvement of a third party. Here, funds are locked on both chains for a certain amount of time. If a chain is provided with proof that the funds on the other chain are locked as well, it considers the transaction as successful,

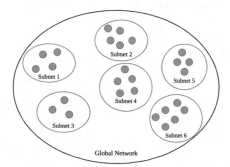
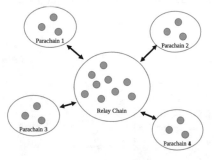

(a) Avalanche Subnetworks: All nodes participate in the global network, while some may also participate in other networks.

(b) Polkadot: A set of validators maintain a global relay chain. Each parachain has its own "collators" that bundle and forward parachain state to the validators.

Fig. 13 Comparision of Polkadot and Avalanche subnetworks

otherwise, it aborts and releases the funds on timeout. This, again, assumes a strong bound on network latency.

A key advantage of cross-chain swaps is that it allows to federate existing blockchains with minimal modifications. As a result, the primitive can be leveraged to build decentralized cryptocurrency exchanges. For example, one can trade Ethereum for Avalanche tokens using cross-chain swaps.

7.2   Polkadot

Polkadot [65] is a self-described "scalable heterogeneous multi-chain". It provides a single *relay chain* that handles cross-chain transactions and multiple *parachains* that rely on the relay chain for security. Polkadot additionally introduces the notion of *bridges*, special subchains that connect to other blockchain systems, such as Ethereum.

The Polkadot architecture relies on four different roles for nodes: nominators, validators, collators, and fishermen. We outline these roles in Fig. 14. Nominators are entities holding tokens on the relay chain, who appoint validators to process the relay chain. The validators then form the committee for the underlying consensus algorithm of the relay chain. Collators serve a similar role as validators, but for a specific parachain. Finally, fishermen check validators for correctness.

Each parachain is then assigned a random subset of all validators. These validators do not have to process the entire state of that parachain. In fact, they might not be able to as they are frequently reassigned to a different parachain. Instead, they rely on parachain collators to propose block candidates to them.

Collators do not need to run a consensus protocol for a parachain. Instead, collators can compete for the validators' "trust", e.g., through a history of good behavior or by providing the blocks containing the most transaction fee revenue. In addition to the

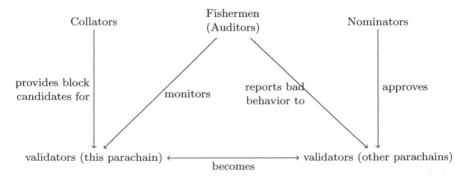

Fig. 14 Roles in the Polkadot network. Collators propose blocks for a specific parachain, which is then approved by a subset of the relay chain's validator. This approval is then inspected by the fishermen who report misbehavior to the validator set as a whole

block itself, collators provide a zero-knowledge proof that the contents of the block are correct and do not violate the parachain state. Additionally, they can provide funds that can be withheld if the block turned out to be faulty. The validators then include the headers of the accepted parachain block in the relay chain.

Because only a subset of validators processes a particular relay chain, they need to be checked by fishermen for correctness. Fishermen are somewhat similar to watchtowers in layer-2 protocols, in that, they look for misbehavior in a particular parachain or bridge, and generate a fraud proof if needed. As a result, parachains operate mostly independently from the relay chain, except when recovering from failure.

Validators additionally participate in the relay chain consensus. Here, they process and approve relay chain block, which contains parachain block headers and cross-chain transaction information, as a whole. Periodically, these validators are (re-)appointed by the nominators.

To ensure consistency of cross-chain communication, the relay chain processes all messages between parachain. Each parachain block contains an egress set of messages sent by the particular chain, and an ingress set of those messages received and process by the chain. This ensures that cross-chain transactions execute on all involved shards, and are correctly ordered. However, depending on how many cross-chain transactions there are, this can constitute a bottleneck of the system.

7.3 *Avalanche Subnetworks and Cosmos Zones*

Avalanche also provides the notion of federated chains through its subnetworks concept. Similar to Polkadot, there exists a global chain handling cross-chain and global transactions. Additionally, any set of entities can create a new subchain that operates independently from the global chain.

Cosmos [34] is a federated blockchain system leveraging the Tendermint consensus protocol. Here, a global "hub" processes cross-chain transactions, while there can be many "zones" that operate independently from each other. Each zone and the hub run their own instance of the Tendermint protocol and can have a different set of validators. Similar to Avalanche and Polkadot, there then exists a mechanism for cross-shard messages and coin swaps.

Both, Avalanche and Cosmos, to our knowledge, have no mechanism to recover from subnetwork (or zone) failure. This means that these systems' subnetworks (or zones) have weaker availability and safety guarantees than its global chain. On the other hand, this design can potentially allow for higher performance, as the global chain is less involved in the particular shard execution.

7.4   Summary

The key advantage of the federation is that it allows connecting blockchain systems that rely on different consensus protocols, currencies, and even different virtual machines for smart contract execution.

On the other hand, federation usually trades for safety by splitting stake or mining power into multiple independent systems. Thus, it might be safe to federate a handful of large blockchains, but not hundreds or thousands.

8   Conclusion

This chapter gave an overview of different scaling approaches to distributed ledger protocols, from the network level to off-chain solutions. Each of these mechanisms has unique advantages, disadvantages, and challenges. Note that, aside from scalability, availability, and safety, decentralized ledger technologies face many challenges that were not discussed in this chapter at all, such as ensuring user privacy or providing mechanisms for governance.

As none of the described mechanisms is a solution to all problems, we believe only a combination of multiple mechanisms can address the scalability limitations of current decentralized applications. The ledger's underlying network needs to be fast to allow for low latency transmission of blocks and transactions. The consensus mechanism must have high throughput to enable managing the global state and processing fraud proofs. Sharding allows processing even more global state for applications that cannot be executed well on layer-2. Finally, layer-2 protocols are necessary to achieve low-cost low-latency transactions with high throughput for end-users, and blockchain federation to allow for interoperability between different architectures and virtual machines.

References

1. Adya, A.: Weak consistency: a generalized theory and optimistic implementations for distributed transactions. Ph.D. thesis, Massachusetts Institute of Technology (1999)
2. Avarikioti, Z., Litos, O.S.T., Wattenhofer, R.: Cerberus channels: incentivizing watchtowers for Bitcoin. In: Financial Cryptography and Data Security, pp. 346–366. Kota Kinabalu, Sabah, Malaysia, February 2020
3. Avarikiotia, Z., Kokoris-Kogias, E., Wattenhofer, R., Zindros, D.: Brick: asynchronous incentive-compatible payment channels. In: International Conference on Financial Cryptography and Data Security (2021)
4. Bernstein, P.A., Goodman, N.: Concurrency control in distributed database systems. ACM Comput. Surv. (CSUR) 13(2), 185–221 (1981)
5. Bharambe, A.R., Agrawal, M., Seshan, S.: Mercury: supporting scalable multi-attribute range queries. In: SIGCOMM Conference, pp. 353–366, Portland, Oregon, August 2004

6. Buchman, E., Kwon, J., Milosevic, Z.: The latest gossip on BFT consensus (2018). arXiv preprint arXiv:1807.04938
7. Burrows, M.: The Chubby lock service for loosely-coupled distributed systems. In: Symposium on Operating System Design and Implementation, pp. 335–350, Seattle, Washington, November 2006
8. Castro, M., Liskov, B.: Practical byzantine fault tolerance. In: Symposium on Operating System Design and Implementation, pp. 173–186, New Orleans, Louisiana, February 1999
9. Clement, A., Wong, E.L., Alvisi, L., Dahlin, M., Marchetti, M.: Making Byzantine fault tolerant systems tolerate byzantine faults. In: Symposium on Networked System Design and Implementation, pp. 153–168, Boston, Massachusetts, April 2009
10. Corallo, M.: Compact block relay. https://github.com/TheBlueMatt/bips/blob/master/bip-0152.mediawiki. Accessed June 2020
11. Corallo, M.: The fast Internet Bitcoin Relay Engine (FIBRE). http://www.bitcoinfibre.org/. Accessed June 2020
12. Corbett, J.C., Dean, J., Epstein, M., Fikes, A., Frost, C., Furman, J.J., Ghemawat, S., Gubarev, A., Heiser, C., Hochschild, P., Hsieh, W.C., Kanthak, S., Kogan, E., Li, H., Lloyd, A., Melnik, S., Mwaura, D., Nagle, D., Quinlan, S., Rao, R., Rolig, L., Saito, Y., Szymaniak, M., Taylor, C., Wang, R., Woodford, D.: Spanner: Google's globally distributed database. ACM Trans. Comput. Syst. **31**(3), 8-1 (2013)
13. Cowling, J., Liskov, B.: Granola: low-overhead distributed transaction coordination. In: USENIX Annual Technical Conference (2012)
14. Croman, K., Decker, C., Eyal, I., Gencer, A.E., Juels, A., Kosba, A.E., Miller A., Saxena, P., Shi, E., Sirer, E.G., Song, D., Wattenhofer, R.: On scaling decentralized blockchains—(A Position Paper). In: Financial Cryptography and Data Security, pp. 106–125, Christ Church, Barbados, February 2016
15. Demers, A.J., Greene, D.H., Hauser, C., Irish, W., Larson, J., Shenker, S., Sturgis, H.E., Swinehart, D.C., Terry, D.B.: Epidemic algorithms for replicated database maintenance. In: ACM Symposium on Principles of Distributed Computing, pp. 1–12, Vancouver, Canada, August 1987
16. Drya, T. (2016) Unlinkable outsourced channel monitoring. In: Scaling Bitcoin Milan (2016)
17. Dwork, C., Naor, M.: Pricing via processing or combatting junk mail. In: Annual International Cryptology Conference, pp. 139–147, Santa Barbara, California, August 1992
18. Dwork, C., Lynch, N.A., Stockmeyer, L.J.: Consensus in the presence of partial synchrony. J. ACM **35**(2), 288–323 (1988)
19. Dziembowski, S., Faust, S., Hostáková, K.: General state channel networks. In: Computer and Communications Security, pp. 949–966, Toronto, Canada, October 2018
20. El-Hindi, M., Binnig, C., Arasu, A., Kossmann, D., Ramamurthy, R.: BlockchainDB—a shared database on blockchains. Proc. VLDB Endowm. **12**(11), 1597–1609 (2019)
21. Etherscan: Ethereum average gas limit chart. https://etherscan.io/chart/gaslimit. Accessed June 2020
22. Gencer, A.E., Basu, S., Eyal, I., van Renesse, R., Sirer, E.G.: Decentralization in Bitcoin and Ethereum networks. In: Financial Cryptography and Data Security, pp. 439–457, Porta Blancu, Curaçao, February 2018
23. Gilad, Y., Hemo, R., Micali, S., Vlachos, G., Zeldovich, N.: Algorand: scaling byzantine agreements for cryptocurrencies. In: Symposium on Operating Systems Principles, pp. 51–68, Shanghai, China, October 2017
24. Göbel, J., Krzesinski, A.E.: Increased block size and Bitcoin blockchain dynamics. In: 27th International Telecommunication Networks and Applications Conference, ITNAC 2017, Melbourne, Australia, November 22–24, 2017, pp. 1–6 (2017)
25. Herlihy, M.: Atomic cross-chain swaps. In: ACM Symposium on Principles of Distributed Computing, pp. 245–254, London, United Kingdom, July 2018
26. Ittay E., Gencer, A.E., Sirer, E.G., van Renesse, R.: Bitcoin-NG: a scalable blockchain protocol. In: Symposium on Networked System Design and Implementation, pp. 45–59, Santa Clara, California, March 2016

27. Kalodner, H.A., Goldfeder, S., Chen, X., Matthew Weinberg, S., Felten. Arbitrum, E.W.: Scalable, private smart contracts. In: USENIX Security Symposium, pp. 1353–1370, Baltimore, Maryland, August 2018
28. Kiayias, A., Russell, A., David, B., Oliynykov, R.: Ouroboros: a provably secure proof-of-stake blockchain protocol. In: Annual International Cryptology Conference, pp. 357–388, Santa Barbara, California, August 2017
29. Klarman, U., Basu, S., Kuzmanovic, A., Sirer, E.G.: bloXroute: A scalable trustless blockchain distribution network. Bloxroute White Paper (2018)
30. Kogias, E.K., Jovanovic, P., Gailly, N., Khoffi, I., Gasser, L., Ford, B.: Enhancing bitcoin security and performance with strong consistency via collective signing. In: USENIX Security Symposium, pp. 279–296, Austin, Texas, August 2016
31. Kogias, E.K., Jovanovic, P., Gasser, L., Gailly, N., Syta, E., Ford, B.: OmniLedger: a secure, scale-out, decentralized ledger via sharding. In: IEEE Symposium on Security and Privacy, pp. 583–598, San Francisco, California, May 2018
32. Kotla, R., Alvisi, L., Dahlin, M., Clement, A., Wong, E.L.: Zyzzyva: speculative byzantine fault tolerance. In: Symposium on Operating Systems Principles, pp. 45–58, Stevenson, Washington, October 2007
33. Kung, H.T., Robinson, J.T.: On optimistic methods for concurrency control. In: International Conference on Very Large Data Bases, p. 351, Rio de Janeiro, Brazil, October 1979
34. Kwon, J., Buchman, E.: Cosmos whitepaper. https://cosmos.network/resources/whitepaper. Accessed March 2021
35. Lamport, L.: Using time instead of timeout for fault-tolerant distributed systems. ACM Trans. Program. Lang. Syst. **6**(2), 254–280 (1984)
36. Lamport, L.: The part-time parliament. ACM Trans. Comput. Syst. **16**(2), 133–169 (1998)
37. Lokhava, M., Losa, G., Mazières, D., Hoare, G., Barry, N., Gafni, E., Jove, J., Malinowsky, R., McCaleb, J.: Fast and secure global payments with Stellar. In: Symposium on Operating Systems Principles, pp. 80–96, Huntsville, Ontario, Canada, October 2019
38. Luu, L., Narayanan, V., Zheng, C., Baweja, K., Gilbert, S., Saxena, P.: A secure sharding protocol for open blockchains. In: Computer and Communications Security, pp. 17–30, Vienna, Austria, October 2016
39. McCorry, P., Bakshi, S., Bentov, I., Meiklejohn, S., Miller, A.: Pisa: arbitration outsourcing for state channels. In: Advances in Financial Technologies, pp. 16–30, Zürich, Switzerland, October 2019
40. Merkle, R.C.: A digital signature based on a conventional encryption function. In: Annual International Cryptology Conference, pp. 369–378, Santa Barbara, California, August 1987
41. Miller, A., Bentov, I., Kumaresan, R., McCorry, P.: Sprites: payment channels that go faster than lightning (2017). arXiv:1702.05812
42. Miller, A., Xia, Y., Croman, K., Shi, E., Song, D.: The honey badger of BFT protocols. In: Computer and Communications Security, pp. 31–42, Vienna, Austria, October 2016
43. Morrison, D.R.: PATRICIA—practical algorithm to retrieve information coded in alphanumeric. J. ACM **15**(4), 514–534 (1968)
44. Mu, S., Cui, Y., Zhang, Y., Lloyd, W., Li, J.: Extracting more concurrency from distributed transactions. In: Symposium on Operating System Design and Implementation, pp. 479–494, Broomfield, Colorado, October 2014
45. Nakamoto, S.: Bitcoin: a peer-to-peer electronic cash system (2008)
46. Offchain Labs: Arbitrum rollup basics. https://developer.offchainlabs.com/docs/rollup_basics. Accessed January 2022
47. Oki, B.M., Liskov, B.: Viewstamped replication: a general primary copy. In: ACM Symposium on Principles of Distributed Computing, pp. 8–17, Toronto, Canada, August 1988
48. Optimism PBC: Optimistic ethereum documentation. https://community.optimism.io/. Accessed Jan 2022
49. Pinar Ozisik, A., Andresen, G., Levine, B.N., Tapp, D., Bissias, G., Katkuri, S.: Graphene: efficient interactive set reconciliation applied to blockchain propagation. In: SIGCOMM Conference, pp. 303–317, Beijing, China, August 2019

50. Ponnapalli, S., Shah, A., Tai, A., Banerjee, S., Chidambaram, V., Malkhi, D., Wei, M.: Scalable and efficient data authentication for decentralized systems (2019). arXiv preprint arXiv:1909.11590
51. Poon, J., Buterin, V.: Plasma: scalable autonomous smart contracts. White Paper (2017)
52. Poon, J., Dryja, T.: The bitcoin lightning network: scalable off-chain instant payments (2016). https://lightning.network/lightning-network-paper.pdf
53. Redman, J.: BCH upgrades: What's new and What's next (2018). https://news.bitcoin.com/bch-upgrades-whats-new-and-whats-next/. Accessed June 2020
54. Rocket, T., Yin, M., Sekniqi, K., van Renesse, R., Sirer, E.G.: Scalable and probabilistic lead erless BFT consensus through metastability (2019). arXiv preprint arXiv:1906.08936
55. Rogaway, P.: Formalizing human ignorance: collision-resistant hashing without the keys. IACR Cryptol. ePrint Arch. **2006**, 281 (2006)
56. Schneider, F.B.: Implementing fault-tolerant services using the state machine approach: a tutorial. ACM Comput. Surv. **22**(4), 299–319 (1990)
57. Sivaraman, V., Venkatakrishnan, S.B., Ruan, K., Negi, P., Yang, L., Mittal, R., Fanti, G.C., Alizadeh, M.: High throughput cryptocurrency routing in payment channel networks. In: Symposium on Networked System Design and Implementation, pp. 777–796, Santa Clara, California, February 2020
58. Sompolinsky, Y., Zohar, A.: Secure high-rate transaction processing in bitcoin. In: Financial Cryptography and Data Security, pp. 507–527, San Juan, Puerto Rico, January 2015
59. Stoica, I., Morris,R.T., Karger, D.R., Frans Kaashoek,M., Balakrishnan, H.: Chord: a scalable peer-to-peer lookup service for internet applications. In:SIGCOMM Conference, pp. 149–160, San Diego, California, August 2001
60. Syta, E., Jovanovic, P., Kokoris-Kogias, E., Gailly, N., Gasser, L., Khoffi, I., Fischer, M.J., Ford, B.: Scalable bias-resistant distributed randomness. In:IEEE Symposium on Security and Privacy, pp. 444–460, San Jose, California, May 2017
61. The Ethereum Foundation. Ethereum 2.0 Specifications. https://github.com/ethereum/eth2.0-specs. Accessed August 2020
62. Wang, G., Shi, Z.J., Nixon, M., Han, S.: SoK: sharding on blockchain. In: Advances in Financial Technologies, pp. 41–61, Zürich, Switzerland, October 2019
63. Wang, J., Wang, H.: Monoxide: scale out blockchains with asynchronous consensus zones. In: Symposium on Networked System Design and Implementation, pp. 95–112, Boston, Massachusetts, February 2019
64. Wood, G.: Ethereum: a secure decentralised generalised transaction ledger. Ethereum project yellow paper **151**, 1–32 (2014)
65. Wood, G.: Polkadot: vision for a heterogeneous multi-chain framework. White Paper (2016)
66. Yin, M., Malkhi, D., Reiter, M.K., Golan-Gueta, G., Abraham, I.: HotStuff: BFT consensus with linearity and responsiveness. In: ACM Symposium on Principles of Distributed Computing, pp. 347–356, Toronto, Canada, July 2019
67. Zamani, M., Movahedi, M., Raykova, M.:Rapidchain: scaling blockchain via full sharding. In: Computer and Communications Security, pp. 931–948, Toronto, Canada, October 2018
68. Zamyatin, A., Harz, D., Lind, J., Panayiotou, P., Gervais, A., Knottenbelt, W.J.: XCLAIM: trustless, interoperable, cryptocurrency-backed assets. In: IEEE Symposium on Security and Privacy, pp. 193–210, San Francisco, California, May 2019
69. Zhang, Y., Setty, S.T.V., Chen, Q., Zhou, L., Alvisi, L.: Byzantine ordered consensus without byzantine oligarchy. In: Symposium on Operating System Design and Implementation, pp. 633–649 (2020)
70. Zilliqa Team: The Zilliq technical whitepaper. Technical Report (2017)

Information-Theoretic Approaches to Blockchain Scalability

Ravi Kiran Raman and Lav R. Varshney

Abstract Blockchain systems fundamentally provide an environment of distributed trust in networks by creating individual copies of cryptographically secure ledgers of all transactions on the network at each node in the network. This redundant storage when combined with democratized transaction validation and the security from recording the ledgers as hash chains enable a self-sustainable system of distributed trust. However, the principal source of security and fairness of blockchain systems is from every participating node maintaining a local record of all transactions in the network. This in turn implies a significant amount of storage cost that scales prohibitively with larger block sizes, higher transaction volume, greater size of the network, and time in use. In this chapter, we will take a few blockchain applications as examples and highlight the storage and communication demands for maintaining a full node in the network. We then study some approaches with roots in coding theory that aim to reduce this cost and enable network scaling. Finally, we study some practical use cases in establishing distributed trust in computational systems using coding-theoretic methods.

1 Introduction

Transactions of any nature in business need a reliable notion of trust between the participating parties, and an acceptable common proof of the transaction that can be validated at a later time, possibly at a moment of conflict between the participants. Enforcement and validation of such contracts have conventionally required a trusted third party whose decisions on the enforcement of terms are agreed upon by the

R. K. Raman (✉)
Analog Devices Inc., 125 Summer Street, Boston, MA 02140, USA
e-mail: ravi.raman@analog.com

L. R. Varshney
University of Illinois Urbana-Champaign, 314 Coordinated Science Laboratory, 1308 W. Main St., Urbana, IL 61801, USA
e-mail: varshney@illinois.edu

parties. The invention of blockchains, starting with its application in Bitcoin [44], has ushered in an era of distributed trusted transactional networks.

Fundamentally, blockchains rely on a combination of distributed transaction validation among peers of the network, reproducible verification of transaction validity among individual peers, secure transactions records in the form of hash chain-based ledgers available at every peer in the network, and byzantine fault-tolerant consensus mechanism to ensure consistency of transaction history across peers to establish distributed consensus across the peer to peer network. To elaborate, transactions on a blockchain network are validated and enacted by some peer in the network and communicated to all other peers in the system. Each peer maintains a ledger comprising the historical record of all transactions in the network in a consistent and cryptographically secure form. When a validated transaction (or set of transactions) is reported to the peers of the network, they correspondingly update their ledger. Finally, as the ledgers are cryptographically secured using a hash chain, they may be referred to at a later point and verified for consistency.

The invention of blockchains has led to the emergence of a new environment of business transactions and self-regulated cryptocurrencies [9, 45]. Due to such favorable properties, blockchains are being adopted extensively outside cryptocurrencies in a variety of novel application domains [71, 73] such as medicine [2], supply chain management and global trade [12], and government services [70]. Blockchains are expected to revolutionize the way financial/business transactions are done [27], for instance, in the form of smart contracts [32].

As an exemplary domain, consider that a variety of potential applications of blockchains have been identified in the healthcare sector, such as shared digital health records, health insurance, shared biomedical research records, and digitizing the supply chain of drugs from production to procurement [52]. Currently, the healthcare system faces significant challenges with access to concurrent and complete healthcare records of patients, linking and accessing patient insurance policies, and reducing prescription forgeries. A blockchain-based solution can help solve some of these challenges by providing a systematic, consistent, secure, and trusted record of transactions without the need for a trusted third party [64]. To elaborate with an example, consider the field of maintaining digital healthcare records. A blockchain shared among doctors, healthcare workers, pharmacies, and patients, with written access provided to the healthcare professionals, and encrypted using private keys for each patient enables a seamless shared record of each patient, as an immutable, secure record. A patient visit to a clinic or the results of a health checkup can just be recorded as a block on the blockchain and made available to doctors to enable easy, immutable access at a later point. Similarly, it also seamlessly records prescriptions made and accessed, and can be integrated with insurance policies to reduce friction in correspondences.

Likewise, blockchains also enable seamless access to research outcomes and publications that can be validated in a distributed fashion across the network [53]. With regard to computational platforms, blockchains and secure data sharing provide access to trusted distributed computing networks over which complex computations such as simulations and machine learning could be performed in a transparent manner [54]. Focusing on the healthcare sector can provide access to research data on vaccines and drug development that can accelerate the discovery process. The fact that IP rights can be embedded onto the blockchain framework, as highlighted through the emergence of non-fungible tokens (NFTs) [28], further highlights the practical feasibility of the application [25].

Blockchains have also promised disruptive transformations to supply chain management [12] by enabling a transparent transactional network over which various parts of a product's supply chain from sourcing to distribution may be traced. Several research efforts are underway to identify sources of friction in the global supply chain and mechanisms through which blockchain systems may be incorporated to enable seamless transaction records [37, 61]

Blockchains can also empower developing economies by providing access to trusted and reliable resource ownership tracking platforms. One such instance is in the maintenance of land titling records in developing countries [72]. Recording land ownership rights in blockchain networks helps protect the underprivileged from their land being poached, particularly in places like India where land ownership documents are often lost or untraceable [34].

Thus blockchain as a technology promises disruptive applications in several fields well beyond the creation of cryptocurrencies. Several such applications have in fact been realized in platforms such as Ethereum and Hyperledger. In this chapter, we study the cost involved in implementing such solutions over blockchain systems. In particular, we first highlight the storage and communication costs associated with blockchain solutions and then highlight solutions inspired by information and coding theory to reduce the cost associated with the implementations.

First, in Sect. 2, we formally introduce blockchains by defining the system model mathematically. In Sect. 3, we highlight the storage costs and the difficulty in developing scalable applications of blockchain systems. Then in Sect. 4, we introduce an information-theoretic off-chain solution, called the dynamic distributed storage, to reduce the cold storage cost of blockchain ledgers while enhancing the security of the stored data from targeted corruption. This solution is application-agnostic, and though modeled after the Hyperledger fabric, extends directly to other blockchain platforms as well. In Sect. 5, we describe how a blockchain-based solution may be developed by molding the application to cater to the storage costs of the blockchain ledger. To be precise, we consider the problem of trusted sharing of computational results among a peer-to-peer network and highlight a protocol that enables compressed, trusted sharing of computational results among peers. Finally, in Sect. 6, we describe some potential areas of future work directed toward developing scalable blockchain solutions.

2   Blockchain System Primer

The term blockchains have been used to broadly describe technology that resembles that of the Bitcoin network. Broadly speaking, blockchain networks involve the participation of several peers interconnected through a peer-to-peer network, individual copies of ledgers of transactions (hash chains) maintained in a cryptographically-secure manner, and a protocol that governs the distributed validation of transactions between peers in the network. For ease of description, we now abstract blockchain systems with a mathematical model defining the peer network and the hash chain primarily based on the Hyperledger Fabric [1].

2.1   The Blockchain Network

Blockchains generally comprise a connected peer-to-peer network of nodes that are functionally characterized as:

1. **Clients**: nodes that invoke or are involved in a transaction, have the blocks validated by endorsers and communicate them to the orderers.
2. **Peers**: nodes that commit transactions and maintain a current version of the ledger. Peers may also adopt endorser roles.
3. **Endorsers**: for a given data block, a peer acts as an endorser if it validates the contents of the transaction prior to its inclusion in the blockchain ledger.
4. **Orderer**: nodes that communicate the transactions to the peers in chronological order to ensure consistency of the hash chain.

Note that the classification is only based on function, and individual nodes in the network can serve multiple roles.

The distributed ledger of the blockchain maintains a current copy of the sequence of transactions. A transaction is initiated by the participating clients and is verified by endorsers. Subsequently, the verified transaction is communicated to the orderer, who then broadcasts them to peers to store in the ledger. The nodes in the blockchain are as depicted in Fig. 1. Here, nodes C_i are clients, P_i are peers, and O is the set of orderers in the system, categorized by function.

The transaction and the nature of the data associated with it are application-specific such as proof of fund transfer across clients in Bitcoin-like cryptocurrency networks, smart contracts in business applications, patient diagnoses/records in medical record storage, and raw data in cloud storage. We use the term *transaction* broadly to represent all such data. A transaction is initiated by participating clients, verified by endorsers (select peers), and broadcast to peers through orderers.

2.2   Ledger Construction

The blockchain ledger stores the sequence of transactions securely in the form of a (cryptographic) *hash chain*. Hash chains are constructed using hash functions that are defined as follows.

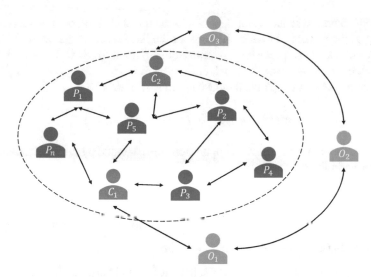

Fig. 1 Architecture of a blockchain network: Here, the network is categorized by functional role into clients C_i, peers P_i, and orderers O. As mentioned earlier, the clients initialize transactions. Upon validation by the endorsers of this transaction, the transactions are communicated to peers by orderers. The peers maintain an ordered copy of the ledger of transactions

Definition Let M be a set of messages of arbitrary lengths, $\mathcal{H}$ the set of (fixed-length) *hash values*. A *cryptographic hash function* family is a function $h : \mathcal{I} \times M \to \mathcal{H}$, where $\mathcal{I}$ is the set of parameters that dictate the deterministic map that is employed. $\square$

Good hash functions satisfy salient properties [60] such as

1. **Computational ease**: Hash values are easy to compute.
2. **Pre-image resistance**: Given $H \in \mathcal{H}$, it is computationally infeasible to find $M \in M$ such that $h(M) = H$. To be precise, given a randomized and computationally limited adversary who samples the message $M' = A(H, I)$, we consider the pre-image resistance in terms of the hitting probability

$$P_{pre-image} = \mathbb{P}\left[h(I, A(H, I)) = H\right]. \tag{1}$$

3. **Collision resistance**: It is computationally infeasible to find $M_1, M_2 \in M$ such that $h(M_1) = h(M_2)$. Again, to be precise, given a randomized and computationally limited adversary who sample messages $(M, M') = A(I)$, we consider collision resistance in terms of the hitting probability

$$P_{collision} = \mathbb{P}\left[\{M \neq M'\} \cap \{h(I, M) = h(I, M')\}\right]. \tag{2}$$

A hash chain is a sequence of data blocks such that each block includes a header pointing to the previous block in the form of the hash value of the previous (header included) block. To be precise, if the transaction data at time t is B_t, and its header is H_{t-1}, then the header for the block at time $t + 1$ is $H_t = h(I_t, (H_{t-1}, B_t))$. Here, $h(\cdot)$ is the hash function. Thus, the hash chain is stored as

$$(H_0, \mathbf{B}_1) - (H_1, \mathbf{B}_2) - \cdots - (H_{t-1}, \mathbf{B}_t).$$

Here, we presume the following operations are executed by the nodes, depending on their role in the transaction:

- WRITE(B): initiate a block B of data that includes transaction data which are verified and appended to the blockchain ledger.
- READ(t): call with index t to recover block B_t from the blockchain ledger.

The endorsers perform the following operations:

- VERIFY(B): check the details of the transaction and verify authenticity.
- MINE(B, t): recover hash value H_{t-1} and use B to compute hash H_t and report to orderer to include block in blockchain ledger.

The operations of the orderer are:

- VALIDATE(B, H, t): validate block and hash value reported by endorser.
- APPEND(B, H, t): encode data and hash, and communicate to peers, to append block at index t in the ledger.

Blockchain systems use a variety of endorsement mechanisms invoking different numbers of validations from network peers to add blocks to the chain. Transactions on cryptographic systems such as Bitcoin are validated by miners who verify the contents of the transactions and approve them using a method called Proof-of-Work (PoW). This entails composing a block of valid transactions to be added to the blockchain such that the hash value of the block when added to the current ledger falls below a pre-set target value. The first valid block mined is added to the longest chain in the ledger. Fundamentally, Bitcoin-like platforms leverage independent peers of the network to validate and add transactions onto the ledger in an honest manner. The trusted growth of the chain arises from a combination of data replication across peers and the fact that the miner spends computational resources and competes with other peers to add a valid block to the ledger.

Alternative mechanisms such as Proof-of-Stake (PoS) have also been proposed. Here transaction validation is assigned to peers at random in a volume proportional to their personal stake in the system. This mechanism has been proposed to be used on Ethereum. The fundamental idea exploited here is that people with a larger stake in the cryptocurrency also have an interest in promoting value through transactions on the framework, and allow only valid transactions.

Fundamentally, every blockchain-based system relies on the participation of independent peers of the network to validate, add, and maintain the ledgers in a consistent manner. Typically some resources are spent in the process of proving such validity.

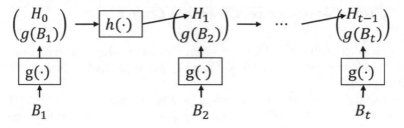

Fig. 2 Hash chain structure for the ledger. The chain is constructed by hashing a hash value of the data for easier recovery and consistency verification

In the case of PoW, this is the computational resources spent in a competition against peers. In PoS, this is the personal stake in the system. A more rudimentary and simple alternative would be to imagine a simple majority vote on the validity of the transactions that are to be added to the ledger. That is, peers of the network vote on the validity of transactions before they are added, or even after the transaction is being referred or challenged. Such a mechanism can also be used in principle in the construction and maintenance of blockchain ledgers. In fact, private blockchains that are used to record business transactions in corporations among a select subset of participants could very well benefit from such a validation mechanism. To simplify and present a single mechanism for this chapter, let us presume that some form of a voting mechanism is used in validating transactions on the blockchain. Much of the descriptions and claims extend *mutatis mutandis*.

Bitcoin-type blockchains use more complicated data structures such as the Merkle tree to group transactions into blocks, fasten the validation process, and reduce the total number of blocks in the ledger. In such systems, the hash chain is constructed using the Merkle root as the data in the block [13]. We consider a simple form of this as shown in Fig. 2. Let $\mathbf{B}_t$ be the data block corresponding to the t-th transaction. Let g, h be two hash functions. Let $W_t = (H_{t-1}, g(\mathbf{B}_t))$ be the concatenation of the previous hash and a hash of the current data. Then, $H_t = h(I_t, W_t)$ is the hash value stored with the $(t + 1)$th block, where the index I_t is sampled uniformly.

Using such a hashed form to construct the chain simplifies consistency verification and reduces recovery costs while retaining all the salient features of the hash chain that directly hashed blocks would have. In more general forms, the data block can be replaced by a Merkle tree structure with the results extending directly.

Thus, a typical blockchain ledger consists of a hash chain that grows in length over time with the addition of validated transactions. We shall now take a closer look at the costs incurred in maintaining a copy of the ledger.

2.3 Costs of Maintaining Blockchain Ledger

Let us now formally introduce the various costs involved in maintaining the blockchain ledger, and expound on the security that is guaranteed through the hash chain structure.

Let us take a closer look at the storage cost per transaction block. For all t, let $\mathbf{B}_t \sim \mathrm{Unif}(\mathbb{F}_q)$ and $g(B_t)$, $H_t \in \mathbb{F}_p$, where $q, p \in \mathbb{N}$ and $\mathbb{F}_q, \mathbb{F}_p$ are finite fields of orders q and p, respectively. We assume uniform sampling of blocks without loss of generality so as to consider the worst case, i.e., maximum entropy, storage scenario. Thus, the cost of storage per peer per transaction in the conventional implementation is

$$\tilde{R}_s = \log_2 q + 2\log_2 p \text{ bits.} \tag{3}$$

That is, each block caters peer stores $\tilde{R}_s$ bits per block added to the ledger.

Transactions stored in the ledger may at a later point be recovered, in order to validate claims or verify details, by nodes that have read access to the data. Different implementations of the blockchain invoke different recovery mechanisms depending on the application. One such method is authentication where select peers return the data stored in the ledger and the other peers validate (sign) the content. Depending on the application, one can envision varying the number of authorizations necessary to validate the content. As described earlier, for ease of description, we restrict this study to the majority rule, i.e., in order to recover the tth transaction, each peer returns its copy of the transaction and the majority rule is applied to recover the block.

The user may also verify that the data block has not been corrupted by checking the correctness of the hash values on the blockchain. That is, if the data block is uncorrupted, the hash of the data and the hash value at the time of storage should match the following hash value stored in the chain. Thus, a user wishing to refer to prior transactions can always perform a local check to validate the data. Such recovery requires at least $\tilde{R}_s$ bits to be communicated to be able to validate the transaction.

This is a fundamental reason for the data security on the ledger. A malicious adversary can not corrupt the data at random as the corruption may be identified by checking for hash consistency on the hash chain, nor is it easy for a malicious adversary to infer the nature of corruption that maintains hash consistency, owing to pre-image resistance of the hash function. Further, from the definition of the hash function, the pre-image and collision resistance characteristics of the hash family in use are [68]

$$P_{pre-image} \approx \frac{1}{p}, \quad P_{collision} \approx \frac{1}{p}.$$

Thus, it is computationally infeasible for an individual adversary to corrupt recorded transactions on the ledger. Corruption of transactions thus requires collusion among a large group of peers.

3 The Storage Problem with Blockchain Systems

The size of the Bitcoin blockchain ledger as of the time of writing this chapter is roughly 350 GB, growing at an average linear rate of 70 GB per year, as shown in Fig. 3 [7, 8]. This face value does not seem significant. However, the average transaction rate on the Bitcoin network has roughly remained stable at around 3.5 transactions per second. In comparison entities such as Visa process roughly 2000 transactions per second. In blockchain systems, the average transaction rate may be increased by increasing the block sizes and/or decreasing the hash difficulty. Either approach increases the rate of increase of the size of the blockchain network considerably, with a simple extrapolation implying an average addition of 93 GB of data to the blockchain on a daily basis. The average user hoping to retain a full node on the blockchain network can not practically sustain such storage and communication costs. This, in turn, presents the possibility of a large portion of the network being composed of light nodes that contain only summary information, thereby resulting in an undesirable concentration of control in the blockchain network.

Let us take a look at the Ethereum blockchain for comparison. The Ethereum blockchain is roughly 240 GB in size and is growing at an average linear rate of 125 GB per year, as shown in Fig. 3 [8]. Further, the linear rate of increase continues to grow over time, as reflected in the number of transactions as shown in Fig. 4. More critically, the size of an archival node of Ethereum is already 7.5 TB, and has grown by 5 TB in the last 2 years. Whereas it isn't always essential for every node to be an archival node in a typical use of a cryptocurrency, business entities might require archival nodes to maintain historical records of smart contract states across

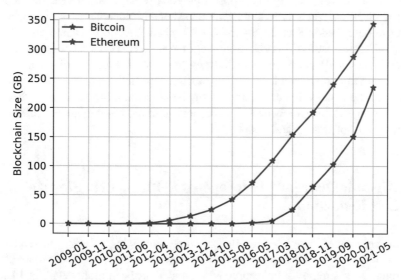

Fig. 3 Growth of Blockchain sizes: size of the Bitcoin and Ethereum blockchains over time. Both blockchains are several hundred GB in size and are growing at an increasing pace with increased adoption of the cryptocurrencies

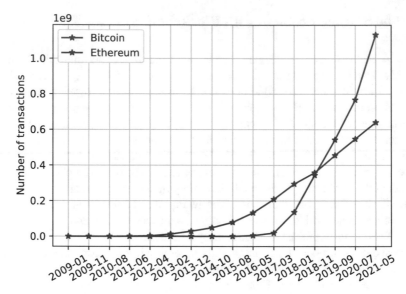

Fig. 4 Growth of number of transactions: Number of transactions on the Bitcoin and Ethereum blockchains over time. The number of transactions on the Ethereum blockchain is increasing at a higher rate

all nodes of the network locally. The scale of growth of the ledger, reflected even more significantly on the sizes of the archival node sizes, emphasizes the growing concern with regard to the size of the blockchain ledgers.

Whereas the prohibitive storage cost of cryptocurrencies such as Bitcoin and Ethereum arise from having to support a potentially large volume of transactions, other applications such as recording land ownership documents on a blockchain ledger, as described in Sect. 2, may involve a limited number of voluminous transactions. This again results in high communication costs on the network, and voluminous ledgers to be stored by each node. Such storage costs might limit individual users from maintaining full nodes over long periods of time, especially in developing countries, leading to the undesirable concentration of the blockchain. In fact, game-theoretic analysis suggests that simply providing higher transaction fees might not sufficiently offset the cost of storage and mining [38].

4 Dynamic Distributed Storage—On Blockchain Storage Cost Reduction

The concerns of storage have resulted in some interesting solutions to reduce the portion of the blockchain that is stored at individual nodes. In this section, we'll first highlight some common simple approaches, and describe in more detail the ideas of sharding and coded compression for blockchain ledgers.

The most common approach used for the cryptocurrency blockchains is to allow users the choice of a *light node*. Light nodes are essentially copies of a full node, without actually downloading the entire ledger locally. To be precise, light nodes have the ability to verify the successful inclusion of transactions and contents of blocks by pinging another full node on the network but don't locally store the sequence of transactions. This provides a simple way to transact over the network without updating a local copy of the ledger. On the downside, the nodes rely on the information passed down to them by the full nodes to derive trust in the transactions, i.e., they are no longer trustless participants of the blockchain. Additionally, the absence of the transaction archive greatly limits their ability to contribute to the validation/mining and development of the blockchain network.

Another common approach used to reduce storage costs of the blockchain ledgers is to *prune* old transactions on the networks. The fundamental idea behind transaction pruning is to remove data related to old validated transactions that are perhaps not likely to be referenced in the future. Whereas this in principle seems doable, this can be complicated by the application use-cases. For instance, each bitcoin is by design traced back to its source, therein necessitating the retention of all exchanges of each coin. Another instance in which old transactions might need to be referenced much later in time is in supply chain blockchains. Thus transaction pruning requires a careful consideration of the blockchain platform, the nature of the transaction, the role of the transaction in the use-case, and a trustless method to prune the data across the blockchain network [41, 42, 47], i.e., the data that is pruned should in no way provide additional incentives from retaining them. Whereas pruning is to be customized to the application and platform, it certainly can prove highly effective in reducing the storage costs of blockchain.

Cold storage records of transactions stored on the blockchains such as land ownership records or supply chain logs are often used for reference at a later point and in cases of conflicts between clients. It is thus important to record the contents of the ledger in a secure manner, such that malicious adversaries may not be able to corrupt the contents of the ledger. In this subsection, we consider the problem of storing the blockchain ledgers in a distributed manner, that is secure from adversaries who may be interested in modifying the contents of transactions on the ledger by colluding with other peers. This problem is more telling especially in blockchains with a smaller number of participating agents such as private blockchains between a few businesses. We will delve deeper into the problem through the lens of dynamic distributed storage codes for blockchains [56].

4.1 Problem Description

Let us consider a blockchain system that records transactions as described in the model in Sect. 2.2. Consider the presence of *active adversaries* in the network who alter contents of a transaction B_t to a desired value B_t'. Let us explicitly define the semantic rules of a valid corruption for such an adversary. If an adversary corrupts a peer, then he can

1. learn the contents stored in the peer;
2. alter block contents if he is in the access list of the corresponding block; and
3. alter hash values as long as chain integrity is preserved, i.e., an attacker cannot invalidate the transaction of another node in the process.

We assume the active adversary is aware of the contents of the hash chain and the block it wishes to corrupt. This adversary is stronger than a typical adversary who is unaware of the contents of the transaction and only wishes to corrupt data. Further, this adversary aims to not only corrupt the block but also to precisely alter it to some preferred content B'_t that would be validated by the network. In a land ownership blockchain, this might involve changing land ownership to the adversary itself.

Moreover, the active adversary adapts to the deployed storage algorithm and can further learn from the contents shared by corrupted peers. We elaborate on the integrity of our coding scheme against such active adversaries. We also briefly discuss data confidentiality as guaranteed by our system against local information leaks.

A simple distribution of the ledger among the peers of the network so as to reduce the storage cost per peer makes the data vulnerable to such adversaries. Thus, the goal is to develop a coding mechanism that constructs secure shares of the data that may be distributed among peers and yet safe from adversaries.

For this section, assume that at any point of time t, there exists a partition $\mathcal{P}_t$ of the set of peers $[n]$ into sets of size m each. For ease of description, let us presume that n is divisible by m. Let each set of the partition be referred to as a *zone*. Without loss of generality, the zones are referred to by indices $1, \ldots, \frac{n}{m}$. At each time t, for each peer $i \in [n]$, let $p_t^{(i)} \in [\frac{n}{m}]$ be the index that represents the zone that includes peer i.

4.2   Coding Data Block

In our coding scheme, a single copy of each data block is stored in a distributed fashion across each zone. Consider the data block $\mathbf{B}_t$ corresponding to time t. We use a technique inspired by [33]. First a private key K is generated at each zone and the data block is encrypted using the key. The private key is then stored by the peers in the zone using Shamir's secret key sharing scheme. Finally, the encrypted data block is distributed among peers in the zone using a distributed storage scheme. The process involved in storage and recovery of a block, given a zone division is shown in Fig. 5.

More generally, we can allow the zone sizes at time t to be chosen by the client. For ease, however, we describe the coding scheme for constant zone sizes m. To customize the storage, we just need to replace the zone sizes by m_t and use a corresponding key space $\mathcal{K}_{m_t}$. The coding scheme is given by Algorithm 1. In this discussion, we will assume that the distributed storage scheme just distributes the components of

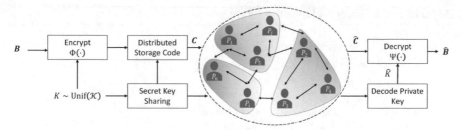

Fig. 5 Encryption and decryption process for a given zone allocation. The shaded regions represent individual zones in the peer network. The data is distributed among peers in each zone and the data from all peers in a zone are required to recover the transaction data

---

Algorithm 1 Coding scheme for data block

---

for $z = 1$ to $\frac{n}{m}$ **do**

    Generate private key $K_t^{(z)} \sim \text{Unif}(\mathcal{K})$

    Encrypt block with key $K_t^{(z)}$ as $\mathbf{C}_t^{(z)} = \Phi(\mathbf{B}_t; K_t^{(z)})$

    Distribute $\mathbf{C}_t$ and store among peers in $\{i : p_t^{(i)} = z\}$

    Use Shamir's (m, m) secret sharing on $K_t^{(z)}$ and distribute shares $(K_1^{(z)}, \ldots, K_{m_t}^{(z)})$ among peers in the zone

end for

---

the code vector $\mathbf{C}_t$ among the peers in the zone. The theory extends naturally to other distributed storage schemes.

To preserve the integrity of the data, we use secure storage for the hash values as well. In particular, at time t, each zone $Z \in \mathcal{P}_t$ stores a secret share of the hash value H_{t-1} generated using Shamir's (m, m) secret sharing scheme.

The storage per transaction per peer is thus given by

$$R_s = \frac{1}{m} \log_2 |C| + 2 \log_2 |\mathcal{K}| + 2 \log_2 p \text{ bits,} \qquad (4)$$

where $|C| \geq q$ depending on the encryption scheme. In particular, when the code space of encryption matches the message space, i.e., $|C| = q$, the gain in storage cost per transaction per peer is given by

$$\text{Storage Gain} = \tilde{R}_s - R_s = \frac{m-1}{m} \log_2 q - \log_2(p|\mathcal{K}|^2) \text{ bits.} \qquad (5)$$

Thus, when the size of the private key space is much smaller than the size of the blocks, we have a storage reduction.

Algorithm 2 Recovery scheme for data block

$\mathcal{N} \leftarrow [n]$

Compute $K_t^{(z)}$, for all z, by polynomial interpolation

Decode blocks $B_t^{(z)} \leftarrow \Psi \left(\mathbf{C}_t^{(z)}; K_t^{(z)} \right)$, for all $z \in [\frac{n}{m}]$

if $|\{B_t^{(z)} : z \in [\frac{n}{m}]\}| > 1$ **then**

    for $\tau = t$ to min $\{t + d_t, T\}$ **do**

        Compute $H_\tau^{(z)}$, for all z, by polynomial interpolation

        Determine $W_\tau^{(i)} = \left(g(B_\tau^i), H_{\tau-1}^i \right)$, for all $i \in [n]$

        $\mathcal{I} \leftarrow \left\{ i : h(W_\tau^{(i)}) \neq H_\tau^{(z)}, z = p_{\tau+1}^{(i)} \right\}$

        $\mathcal{N} \leftarrow \mathcal{N} \backslash \mathcal{I}$

        if $|\{B_t^{(p_t^{(i)})} : i \in \mathcal{N}\}| = 1$ **then**

            break

        end if

    end for

end if

return Majority in $\{\{B_t^{(p_t^{(i)})} : i \in \mathcal{N}\}\}$

4.3 Recovery Scheme

We now describe the algorithm to retrieve a data block B_t in a blockchain with a total of T blocks. However, instead of exploring the entire length until we identify a unique consistent version, we provide the client the freedom to choose the depth d_t of blocks that follow in the hash chain and return the majority consistent version. The algorithm to recover block B_t is in Algorithm 2.

According to Algorithm 2, each peer first communicates the codeword corresponding to the data block and the secret share of the encryption key. This corresponds to $\frac{1}{m_t} \log_2 q + m_t(2 \log_2 m_t + 1)$ bits. Additionally, each peer also communicates the secret shares of hash values corresponding to the next d blocks, each of which contributes $2 \log_2 p$ bits, and the corresponding data blocks for consistency check. Thus, the total worst-case cost of recovering the t-th data block is

$$R_r^{(t)} = C_r \left(\frac{1}{m} \log_2 q + \log_2 |\mathcal{K}| + 4d_t \log_2 p \right), \tag{6}$$

where C_r is the cost per bit of communication.

The recovery algorithm exploits information-theoretic security in the form of the coding scheme and also invokes the hash-based computational integrity check established in the chain. First, the data blocks are recovered from the distributed, encrypted storage from each zone. In case of a data mismatch, the system inspects the chain for consistency in the hash chain. The system scans the chain for hash values and eliminates peers that have inconsistent hash values. A hash value is said to be inconsistent if the hash value corresponding to the data stored by a node in the previous instance does not match the current hash value. Through the inconsistency

---

Algorithm 3 Encryption scheme

$T \leftarrow \text{Unif}(\mathcal{T})$, $K \leftarrow \text{Key}(T)$; $\mathbf{b} \leftarrow \text{Binom}(n, 1/2)$
Assign peers to vertices, i.e., peer i is assigned to node θ_i
For all $i \neq v_0$, $\tilde{C}_i \leftarrow B_i \oplus B_{\mu_i}$; flip bits if $b_i = 1$.
$\tilde{C}_{v_0} \leftarrow \left(\oplus_{j \neq v_0} \tilde{C}_j \right) \oplus B_{v_0}$
if $b_{v_0} = 1$ **then**
    Flip the bits of $\tilde{C}_{v_0}$
end if
Store $C_i \leftarrow \tilde{C}_{\theta_i}$ at each node i in the zone
Store (K, θ) using Shamir's secret sharing at the peers
Store the peer assignment θ_i locally at each peer i

---

check, the system eliminates some, if not all corrupted peers. Finally, the majority of consistent data is returned.

In the implementation, we presume that all computation necessary for the recovery algorithm is done privately by a black box. In particular, we presume that the peers and clients are not made aware of the code stored at other peers or values stored in other blocks. Specifics of practical implementation of such a black box scheme are beyond the scope of this work.

4.4 Feasible Encryption Scheme

The security of the coding scheme from corruption by active adversaries depends on the encryption scheme used. We now describe an encryption scheme that is order optimal in the size of the private key space up to log factors. Let $\mathcal{T}$ be the set of all rooted, connected trees defined on m nodes. Then, by Cayley's formula [16], $|\mathcal{T}| = m^{(m-1)}$. Let us define the key space by the entropy-coded form of uniform draws of a tree from $\mathcal{T}$. Hence, the encryption scheme presumes that given the private key K, we are aware of all edges in the tree. Let $V = [m]$ be the nodes of the tree and v_0 be the root. Let the parent of a node i in the tree be μ_i.

Consider the encryption function, Φ, given in Algorithm 3. Here, $\text{Key}(T)$ is the sampling function that generates a key K from the set of all keys corresponding to the chosen tree architecture T. Without loss of generality, we assume the keys are sampled uniformly at random. The encryption algorithm proceeds by first selecting a rooted, connected tree uniformly at random on m nodes. Then, each peer is assigned to a particular node of the tree. For each node other than the root, the codeword is created as the modulo 2 sum of the corresponding data block and that corresponding to the parent. Finally, the root is encrypted as the modulo 2 sum of all codewords at other nodes and the corresponding data block. The bits stored at the root node are flipped with a probability of half. The encryption scheme for a sample data block is shown in Fig. 6.

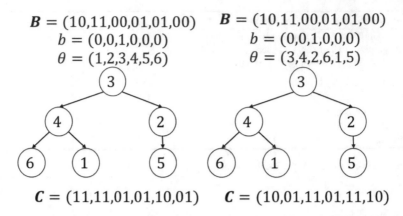

$$\mathbf{B} = (10,11,00,01,01,00) \qquad \mathbf{B} = (10,11,00,01,01,00)$$
$$b = (0,0,1,0,0,0) \qquad b = (0,0,1,0,0,0)$$
$$\theta = (1,2,3,4,5,6) \qquad \theta = (3,4,2,6,1,5)$$

$$\mathbf{C} = (11,11,01,01,10,01) \qquad \mathbf{C} = (10,01,11,01,11,10)$$

Fig. 6 Encryption examples for a zone with six peers. The data block, parameters, tree structure, and corresponding codes are shown. The two cases consider the same rooted tree with varying peer assignments. The corresponding change in the code is shown

---

Algorithm 4 Decryption scheme

Use polynomial interpolation to recover $(K, \mathbf{b}, \theta)$
Define $\tilde{\theta}_i \leftarrow j$ if $\theta_j = i$
Flip the bits of $C_{\tilde{\theta}_{v_0}}$, if $b_{v_0} = 1$
$B_{v_0} \leftarrow C_{\tilde{\theta}_{v_0}} \oplus_{j \neq \tilde{\theta}_{v_0}} C_j$
For all $i \in [n] \backslash \{v_0\}$, flip bits of $\mathbf{C}_i$ if $b_i = 1$
Iteratively compute $B_i \leftarrow C_{\tilde{\theta}_i} \oplus B_{\mu_i}$ for all $i \neq v_0$
return B

The decryption of the stored code is as given in Algorithm 4. That is, we first determine the private key, i.e., the rooted tree structure, the bit, and peer assignments. Then we decrypt the root node by using the codewords at other peers. Then we sequentially recover the other blocks by using the plain text message at the parent node.

Let us now describe a way to enhance the security offered by the coding scheme by incorporating a method to distribute successive data blocks wider across the blockchain network.

4.5 Dynamic Zone Allocation

Earlier we presumed the existence of a zone allocation strategy over time. Here, we make it explicit. The distributed secure encoding process ensures that corrupting a transaction or a hash requires an adversary to corrupt all peers in the zone. This can be exploited to ensure that with each transaction following the corrupted transaction, the client would need to corrupt an increasing set of peers to maintain a valid hash chain.

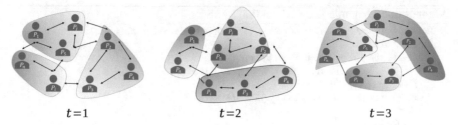

$$t=1 \qquad\qquad t=2 \qquad\qquad t=3$$

Fig. 7 Dynamic zone allocation: Iterate zones among peers so that an increasing number of peers need to be corrupted to maintain a valid hash chain

In particular, let us assume a blockchain in the state

$$(H_0, \mathbf{B}_1) - (H_1, \mathbf{B}_2) - \cdots - (H_{t-1}, \mathbf{B}_t).$$

Let us assume without loss of generality that an adversary wishes to corrupt the transaction entry $\mathbf{B}_1$ to $\mathbf{B}'_1$. The validated, consistent version of such a corrupted chain is

$$(H_0, \mathbf{B}'_1) - (H'_1, \mathbf{B}_2) - \cdots - (H'_{t-1}, \mathbf{B}_t).$$

If the zone segmentation used for encoding is static, the adversary can easily maintain such a corrupted chain at half the peers to validate its claim. If each peer is paired with varying sets of peers across blocks, then, for sufficiently large t, each corrupted peer eventually is grouped with an uncorrupted peer.

Let us assume this occurs at slot $\tau + 1$. Then, in order to successfully corrupt the hash H_τ to H'_τ, the adversary would need to corrupt the uncorrupted peers in this zone. If not, the hash values reveal an inconsistency at the corrupted peers.

Thus, it is evident that if the zones are sufficiently well distributed, corrupting a single transaction would eventually require corruption of the entire network and not just a majority. A sample allocation scheme is shown in Fig. 7.

However, the total number of feasible zone allocations is

$$\text{No. of zone allocations} = \frac{n!}{(m!)^{\frac{n}{m}}} \approx \frac{\sqrt{2\pi n}}{\left(\sqrt{2\pi m}\right)^{\frac{n}{m}}} \left(\frac{n}{m}\right)^n,$$

which increases exponentially with the number of peers and is monotonically decreasing in the zone size m. This indicates that naive deterministic cycling through this set of all possible zone allocations is practically infeasible.

To ensure that every corrupted peer is eventually grouped with an uncorrupted, one we need to ensure that every peer is eventually grouped with every other. Further, the allocation needs to ensure uniform security for every transaction.

A deterministic and fair zone allocation algorithm addressing these requirements may be designed by studying a K-way handshake combinatorial problem. Let $r = \frac{n}{m}$.

Algorithm 5 Dynamic Zone Allocation Strategy

Let $v_2 \ldots, v_{2r}$ be the vertices of a $2r - 1$ regular polygon, and v_1 its center
for $i = 2$ to $2r$ **do**
   Let L be the line passing through v_1 and v_i
   $M \leftarrow \{(v_j, v_k) : \text{line through } v_j, v_k \text{ perp. to } L\}$
   $M \leftarrow M \cup \{(v_1, v_i)\}$
   Construct zones as $\{v_j \cup v_k : (v_j, v_k) \in M\}$
end for
restart for loop

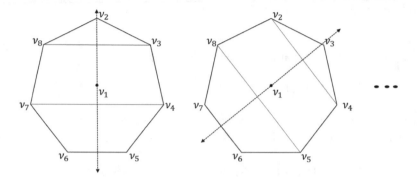

Fig. 8 Dynamic zone allocation when $n = 4m$. Algorithm cycles through matchings of K_n by viewing them as regular polygon orientations

Partition the peers into $2r$ sets, each containing $m/2$ peers. Let these sets be given by $v_1, \ldots, v_{2r}$. Then, we can use matchings of K_{2r} to perform the zone allocation.

Algorithm **??** provides a constructive method to create zones such that every peer is grouped with every other over time. The functioning of the algorithm is as in Fig. 8.

This scheme is order optimal in the number of slots required to achieve coverage across the entire network and is also fair in its implementation to all transactions over time.

4.6 Security Enhancement

From Algorithm 2, we know that inconsistent peers are removed from consideration for data recovery. We know that an adversary who wishes to corrupt a block corrupts at least $n/2$ nodes originally.

Lemma 1 *Consider an adversary who successfully corrupts W_t to W_t'. Further, let us assume the adversary requires successful corruption with a probability of at least $1 - \epsilon$, where $1 - \epsilon > \frac{1}{p}$. Then under the cyclic zone allocation scheme, the adversary needs to corrupt at least m new nodes with a probability of at least $1 - \frac{1}{p}$, in order to successfully alter H_t.*

Suppose the adversary corrupts a peer independently with probability $P_{tc} \in (0, 1)$. This probability represents the ability of the adversary to corrupt other peers in the network.

Theorem 1 *Let the consistency check depth be d. Then, the successful targeted corruption probability is*

$$\mathbb{P}\left[Targeted\ Corruption\right] \leq \binom{r}{\frac{r}{2}} \rho^{\left(\frac{r}{2}\right)} \left[\frac{\left[1 - \left(\rho(1 - \frac{1}{p})\right)^d\right]}{p\left[1 - (1 - \frac{1}{p})\rho\right]} + (1 - \frac{1}{p})^d \rho^d\right],$$

(7)

where $r = n/m$ and $\rho = P_{tc}^{2m}$.

The proof of the theorem is given in [56]. Naturally, this implies that in the worst case, with $\frac{n}{2m}$ transactions, the data becomes completely secure in the network. That is, only a corruption of *all peers* (not just a majority) leads to a consistent corruption of the transaction.

We now characterize the scaling of the probability of targeted corruption by its upper bound in the presence of an active adversary. We presume the probability of successful corruption of a node by the peer is P_{tc}. Naturally, the probability of targeted corruption decays exponentially with n, and thus, a large network is practically incorruptible.

From Fig. 9, we note that the probability of successful corruption decays with the number of peers per zone. Naturally, when each zone is comprised of more peers, the

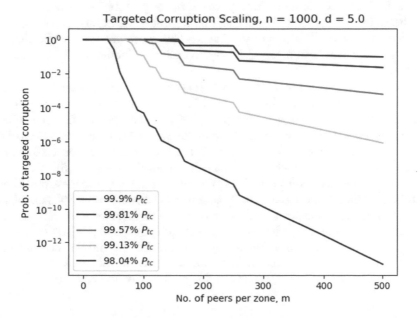

Fig. 9 Targeted corruption scaling with number of peers per zone

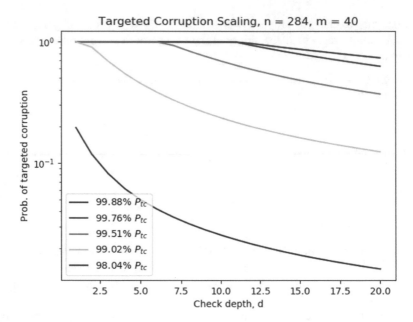

Fig. 10 Targeted corruption scaling with recovery depth

harder it is for the active adversary to corrupt the data. This is because corruption of a zone fails even if a single peer in the network refuses to corrupt the data as dictated by the adversary.

Finally, in Fig. 10, we study the targeted corruption probability as a function of the number of blocks, d, used to check data validity. For better exposition, we consider a network with $n = 284$ and $m = 40$. The probability decays with d, indicating that the more blocks we check for consistency, the harder it is for the adversary to corrupt the data block.

Thus, as can be noted here, it is beneficial to choose larger numbers of peers in each zone, and check deeper into the chain for consistency at recovery, as it makes the data more robust to targeted corruptions by active adversaries. On the contrary, larger zone sizes result in higher data loss as there are fewer individual shares of the transactions on the blockchain that are available, and hence are easier to be affected by DoS adversaries. The problem of data loss is studied in detail in [56].

Thus, selecting the coding parameters for the algorithm requires these tradeoffs to be balanced based on the application. A natural application inspired by the distributed storage code over blockchains is secure data storage for archived retrieval. One such storage scheme with a data insurance against corruption is elaborated in detail in [57]. Thus, the dynamic distributed storage highlights not only a mechanism to enable scalable archival storage enabling full node peers on the blockchain but also inspires the development of new and novel applications on the blockchain.

5 Application-Based Scalable Blockchain Methods

In the previous section, we studied the storage problem on the blockchain for archival storage of transactions by identifying mechanisms to distribute the storage cost among peers of the network. Alternatively, one can envision developing applications that have built-in mechanisms to reduce the content that is stored on the blockchain. In this section, we study one such application of blockchain that demands high storage and communication costs and discuss the off-chain implementation protocol that reduces the effective cost of storage on the blockchain.

Machine learning, data science, and large-scale computation have created an era of computation-driven inference, applications, and policymaking [51, 63]. Decisions with far-reaching consequences are increasingly based on data-driven computational frameworks. Often, multiple people and organizations are tasked with collaboratively making decisions by interactively sharing data and results of computations. However, when such organizations are independent and do not trust each other, they might suspect the validity of computations reported by others and may not collaborate with them. These computations are also expensive and time-consuming, and thus infeasible for recomputation by the doubting peer as a general course of action. In such systems, creating an environment of trust, accountability, and transparency in the local computations of individual agents promotes collaborative operations among entities and ultimately better decisions. Blockchain functions as the ideal tool for establishing trust in such complex, long-running computations of interest. In this section, we will study the protocol to establish trust in computations by efficiently recording validated computations on a blockchain ledger [10, 53].

5.1 Problem Statement

Let us highlight the problem of establishing trust in distributed computational systems through a couple of examples. Consider training a deep neural network with a given architecture using Stochastic Gradient Descent (SGD). Here, the model and computations are deterministic given the data used for gradient computation. Data scientists are primarily interested in using the trained model represented by the weights of the trained network. But, if they lack trust in the training agent, they have no simpler way to verify the network than to retrain it. This is often impractical since the (re)training process consumes extensive amounts of time and tends to require the use of specialized hardware like GPUs or TPUs that often only the largest technology companies have access to [11]. It is thus important to establish trust in the computations involved in the training phase.

Another important example of a multi-agent sociotechnical decision-making system is composed of health ministries, non-governmental organizations, and other agencies along with epidemiological simulation models for malaria eradication policymaking [5, 67]. OpenMalaria (OM) is an open-source simulation environment, collaboratively developed to study malaria epidemiology and the effectiveness of control mechanisms [67]. It is used extensively to design policies to tackle the disease. Here, individual agencies propose hypotheses regarding the disease and/or intervention policies, and study them by simulating them under specific environments [49]. Considering the potential impact of such work in designing disease control policies, it is important to establish accountability and transparency in the process so as to facilitate the trusted adoption of results. Calls have been made for accountability and transparency in multi-agent computational systems, especially in high-impact fields such as health [46]. A framework for decision provenance helps track the source of results, transparent computational trajectories, and a unified, trusted platform for information sharing. However, there exists significant disparity and inconsistency in current information-sharing mechanisms that not only hinder access but also lead to questionable informational integrity [48]. Here, trust and transparency are critical, but absent in current practice.

Establishing trust in computations translates to guaranteeing the correctness of *individual steps* of the simulation and the integrity of the overall computational process leading to the reported results. Importantly, when computational models and parameters along with intermediate results of individual steps are shared, these steps can be validated by other agents who can recompute them (or approximately validate using heuristic approximations), thereby validating the entire computation in a distributed manner.

Reproducing results from research papers in AI has also been found to be challenging as a significant fraction of hyperparameters and model considerations are not documented [24]. In another paper focused on the reproduction of results in deep learning [26], the authors explore the possible reasons and cite variability in evaluation metrics and reporting among different algorithms and implementations.

Accountability and transparency are increasingly sought after in large-scale computational platforms, with a particular focus on establishing tractable, consistent computational pipelines [74]. The problem of establishing provenance in decision-making systems has been considered [66] through the use of an audit mechanism. Distributed learning in a federated setting with security and privacy limitations has also been considered recently [75].

Recent efforts toward federated learning of machine learning models have further emphasized the need for trusted adoption of training updates provided by independent peers in the network. Adversaries intent on gaming the system or biasing the outcomes of the models, such as recommendation systems, for instance, can provide biased updates that corrupt the model learnt over the network. The possibilities of adversarial corruption, methods to mitigate them, and emergent biased models are being studied extensively [6, 19]. This environment further emphasizes the need for a reliable training platform for the peers to collaboratively train and share models [29, 50].

Such requirements highlight the need for a reliable, scalable system that can establish an environment of distributed trust and transparency in multi-agent systems that share data and models arising from large-scale computations. The notion of trust has been considered from a variety of standpoints [18] and has contextually varied definitions [40]. A qualitative definition of trust in multi-agent computational systems can be adapted from [14, 59] as:

> *Trust* is the belief an agent has that the other party will execute an agreed upon sequence of actions and reports an accurate representation of computed result (being honest and reliable).

The problem of trust in multi-agent computational systems was considered at the beginning of the twentieth century from the viewpoint of reducing errors in complex calculations performed by human workers [23]. Large-scale computational problems were solved using redundant evaluation of smaller sub-tasks assigned to human workers and verified using computational checkpoints. We can draw significant insight into reliable distributed computing from these practices.

5.2 Computation Model

Let us now mathematically formalize the computation model and trust requirements under consideration. We limit our study to iterative computational algorithms in this chapter. System design for enumerative computations can be found in [55].

Consider a computational process that updates a system state, $X_t \in \mathbb{R}^d$, over iterations $t \in \{1, 2, \ldots\}$, depending on the current state and an external source of randomness $\theta_t \in \mathbb{R}^{d'}$, according to an atomic operation $f : \mathbb{R}^d \times \mathbb{R}^{d'} \to \mathbb{R}^d$ as

$$X_{t+1} = f(X_t, \theta_t). \tag{8}$$

For simplicity, let us assume that θ_t is shared by all agents. This can easily be generalized as elaborated in [53]. We also assume that the function $f(\cdot)$ is L-Lipschitz continuous in the system state and the randomness, without loss of generality, under the Euclidean norm. That is,

$$\|f(X_1, \theta) - f(X_2, \theta)\| \leq L \|X_1 - X_2\|, \text{ for all } \theta \in \mathbb{R}^{d'}, \tag{9}$$

$$\|f(X, \theta_1) - f(X, \theta_2)\| \leq L \|\theta_1 - \theta_2\|, \text{ for all } X \in \mathbb{R}^d. \tag{10}$$

Essentially, minor modifications to the inputs of the atomic operation result in correspondingly bounded variation in the outputs. This is expected, for instance, in simulations of physical or biological processes, as seen in epidemiological and meteorological simulations, as most physical systems governing behavior in nature are smooth.

For such a computational process, we decompose trust into the following two components:

- **Validation**: The individual atomic computations of the simulation are guaranteed and accepted to be correct.
- **Verification**: The integrity of the overall simulation process can be checked by other agents in the system.

The two elements respectively ensure local consistency of computation and post hoc corroboration of audits by independent peers of the network. Their mathematical characterization is provided below following a formal characterization of the network.

Consider a multi-agent system where one agent, referred to as the *computing client*, runs the computational algorithm. The other agents in the system, called *peers*, are aware of the atomic operation $f(\cdot)$ and share the same external randomness and hence can recompute the iterations. Validation of intermediate states is performed by independent peers referred to as *endorsers* through an iterative, possibly approximate, recomputation of the reported states from the most recent validated state using the atomic operation $f(\cdot)$. The process of validation is referred to as an *endorsement* of the state.

A reported state, $\tilde{X}_t$ is *valid* if it lies within a margin, Δ_{val}, of the state $\hat{X}_t$ as recomputed by the endorser from the last valid report $\tilde{X}_{t-1}$, i.e.,

$$\left\| \tilde{X}_t - \hat{X}_t \right\| = \left\| \tilde{X}_t - f(\tilde{X}_{t-1}) \right\| \leq \Delta_{\text{val}}. \tag{11}$$

Verification of the process involves checking integrity of the reported results from the frequent audits of validated states recorded on the blockchain. Thus, if the audits record the states $\left\{ \tilde{Y}_1, \tilde{Y}_2, \ldots \right\}$, then verification corresponds to guaranteeing that a recomputation of the state, $\hat{Y}_t$, is within a margin, Δ_{ver}, of the recorded version, i.e.,

$$\left\| \hat{Y}_t - \tilde{Y}_t \right\| \leq \Delta_{\text{ver}}. \tag{12}$$

We now describe a protocol, called Multiagent Blockchain Framework (MBF), to establish trust for a synthetic example before elaborating on the design of the system.

5.3 Validation Protocol

Let us now describe the computation validation protocol by elaborating the functions of the different peers of the network. For ease of our discussion, let us consider a deterministic iterative algorithm for computation, $X_{t+1} = f(X_t)$.

The peer-to-peer network and their interaction are as shown in Fig. 11. The validation protocol at a high level is as follows:

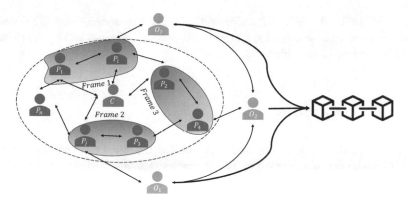

Fig. 11 Functional categorization of peer-to-peer network: clients run the iterative algorithm; multiple independent frames are validated in parallel by non-overlapping subsets of endorsers; orderers check consistency and append valid frames to the blockchain

1. Client runs the computations to iteratively compute states $\{X_1, X_2, \ldots\}$.
2. The client group states sequentially into frames and compresses and communicates the frames to sets of endorsers.
3. The endorsers decompress frames, validate states by recomputing them iteratively, and report endorsements to orderers.
4. The orderers subsample and add valid frames to the blockchain if all prior frames have been validated and added.
5. The peers update their copy of the blockchain ledger.

Such a validation protocol in conjunction with the secure storage of validated audits effectively makes up the MBF. MBF in turn enables the following benefits:

- **Accountability**: MBF guarantees provenance through the immutable record of computations. Thus, we can not only detect the source of potential conflicts but also trace ownership.
- **Transparency**: MBF establishes trust among agents through a transparent record of the validated trajectories of computation.
- **Adaptivity**: The frame design, endorsement, and validation methods adapt according to the state evolution. Further, the validity margins can be altered across time by dynamically varying the quantizers. In convergent simulations/algorithms, the system can thus use monotonically decreasing margins to obtain stricter guarantees at convergence.
- **Computation universality**: The design is agnostic to computational process specifics and can be implemented for a diverse set of applications.
- **Scientific reproducibility**: By storing intermediate states MBF guarantees reliable data and model sharing, and collaborative research, facilitating scientific reproducibility in large-scale computations.

Note that the policy requires a large volume of communication of intermediate states between the peers, a recomputation cost for state endorsements, and a storage cost for recording the validated audits on the blockchain at every peer. These overheads are reduced by employing effective compression schemes and by distributing the validation pipeline as elaborated below.

5.4   Client Operations

Let us now construct the compression scheme that the client can employ to reduce the communication cost. Owing to the Lipschitz continuity,

$$\|X_{t+1} - X_t\| \leq L \|X_t - X_{t-1}\|.$$

Thus, state updates (differences) across iterates are bounded to within a factor of the deviation in the previous iteration. This property can be leveraged to compress state updates using delta encoding [22], where states are represented in the form of differences from the previous state. Then, it suffices to store the state at certain checkpoints of the computational process with the iterates between checkpoints represented by the updates.

We describe the construction inductively, starting with the initial state X_0, the first checkpoint. Let us assume that the state reported at time t is $\tilde{X}_t$ and the true state is X_t. Then, if $X_{t+1} = f(X_t)$, define the update as

$$\Delta X_{t+1} = X_{t+1} - \tilde{X}_t.$$

The cost of communication (for validation) and storage (for verification) of these updates is reduced by performing lossy compression (vector quantization [20]). Let the maximum quantization error magnitude of our quantizer, $Q(\cdot)$, be ϵ, i.e., if the client reports $\tilde{\Delta} X_t = Q(\Delta X_t)$, then,

$$\left\|\tilde{\Delta} X_t - \Delta X_t\right\| \leq \epsilon. \tag{13}$$

Additionally, the checkpoints can also be compressed using a Lempel-Ziv-like dictionary-based lossy compressor. Here, a dictionary of unique checkpoints is maintained on the blockchain. For each new checkpoint, we search for an entry in the dictionary that is within a margin ϵ from the state and report its index if one exists. If not, this state is added to the dictionary and this index and value are reported. We denote this quantizer by $\tilde{Q}(\cdot)$. Other universal vector quantizers can also be used instead.

Checkpoints are created when there is either a significant change in the state, or if the length of the frame exceeds a preset maximum $\bar{M}$ (to limit the validation overhead). Let Δ_{quant} be the maximum magnitude of an acceptable state update within a frame, i.e., if $\|\Delta X_t\| > \Delta_{\text{quant}}$, the client creates a checkpoint at $t + 1$ and reports

Frame ID	Timestamp	Client ID	Compute ID
Endorsement List		Computation Metadata	
Frame Checkpoint: $\tilde{Q}(X_t)$			
$\tilde{\Delta}X_{t+1}$	$\tilde{\Delta}X_{t+2}$	$\tilde{\Delta}X_{t+3}$	$\tilde{\Delta}X_{t+4}$	$\tilde{\Delta}X_{t+5}$	$\tilde{\Delta}X_{t+6}$

Fig. 12 Structure of frames: each frame includes a header followed by compressed updates of successive iterates

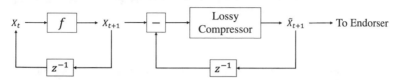

Fig. 13 Operations performed by the client within a frame

$\tilde{X}_{t+1} = \tilde{Q}(X_{t+1})$. Alternatively, if the current frame includes M states, then $t + 1$ is assigned as a checkpoint for the next frame. Thus, X_{t+1} is reported as

$$\tilde{X}_{t+1} = \begin{cases} \tilde{Q}(X_{t+1}), & \text{if } t+1 \text{ is a checkpoint} \\ \tilde{X}_t + \tilde{\Delta}X_{t+1}, & \text{o/w} \end{cases}. \tag{14}$$

The resulting sequence of frames is as shown in Fig. 12. The sequence of tasks performed by the client is shown in Fig. 13.

The choice of design parameters, ϵ, Δ_{quant}, are to be made such that the reports are accurate enough for validation. First, the choice of ϵ requires that a correct computation by a client is not invalidated by an honest endorser owing to the inaccuracy of the report from the compression.

Lemma 2 *If the state X_{t+1}, correctly computed by the client, is reported after compression as $\bar{X}_{t+1}$, then*

$$\left\| \hat{X}_{t+1} - \bar{X}_{t+1} \right\| \leq (L+1)\epsilon.$$

Thus, choosing $\epsilon \leq \frac{\Delta_{val}}{L+1}$ implies that an honest endorser does not invalidate a correctly computed state.

We can further derive the necessary and sufficient conditions in terms of the deviation of the recomputed state from the true state as shown below.

Theorem 2 *Let $f(\cdot)$ be L-Lipschitz continuous, and $\epsilon \leq \frac{\Delta_{val}}{L+1}$.*

1. *A report $\tilde{X}_t$ is invalidated by an honest endorser only if $\left\| \tilde{X}_t - X_t \right\| \geq \epsilon$.*
2. *If $\left\| \tilde{X}_t - X_t \right\| \geq \Delta_{val} + L\epsilon$, then $\tilde{X}_t$ is invalidated.*

The necessary and sufficient conditions for invalidation in Theorem 2 highlight the fact that computational errors of magnitude less than ϵ are missed, and any error of magnitude at least $\Delta_{\text{val}} + L\epsilon$ is detected. When the approximation error is made arbitrarily small, all errors beyond the tolerance are detected. A variety of vector quantizers, satisfying Theorem 2 can be used for lossy delta encoding—we employ lattice vector quantizers (LVQ) [62]. From hereon, presume that the quantizer $Q(\cdot)$ is an LVQ defined using some chosen lattice in $\mathbb{R}^d$.

Under this compression framework, we can derive the cost of communication and storage as shown below.

Theorem 3 *Let $\mathcal{B}(\Delta_{quant}) = \left\{x \in \mathbb{R}^d : \|x\| \leq \Delta_{quant}\right\}$ be the set of possible state updates and let $\Delta X_t \sim Unif(\mathcal{B}(\Delta_{quant}))$. Then the communication and storage cost per state update within the frame is $O\left(d \log\left(\frac{\Delta_{quant}}{\epsilon}\right)\right)$ bits.*

This follows directly from the covering number of $\mathcal{B}(\Delta_{\text{quant}})$ using $\mathcal{B}(\epsilon)$ balls; a similar cost is incurred for other standard lattices.

Theorem 4 *For any frame n, with checkpoint at T_n, maximum number of states in the frame, M_n, is bounded as*

$$M_n \leq \min\left\{\frac{\log\left(\Delta_{quant} - \epsilon\right) - \log \delta_n}{\log L}, \bar{M}\right\}, \tag{15}$$

where $\delta_n = \left\|X_{T_n+1} - X_{T_n}\right\|$, is the first update in the frame.

This provides a simple sufficient condition on the size of each frame in terms of the magnitude of the first update in the frame. Naturally, a small first iterate implies the possibility of accommodating more iterates in the frame. This lower bound can be used in identifying the typical frame size and the corresponding costs of communication and computation, prior to scheme design.

Whereas longer frames result in fewer blocks to validate and store, therein resulting in simpler verification, it also implies that a validation failure at a state also invalidates subsequent states in that frame, increasing validation time and recomputation cost. Shorter frames on the other hand result in excessive communication and storage costs.

5.5 Endorser and Orderer Operations

We now define the role of an endorser in validating a frame. A summary of the operations is depicted in Fig. 14. For preliminary analysis, we assume that endorsers are honest and are homogeneous in terms of communication latency and computational capacity. A more refined allocation policy can be designed easily to account for the case of variabilities in communication and computational costs.

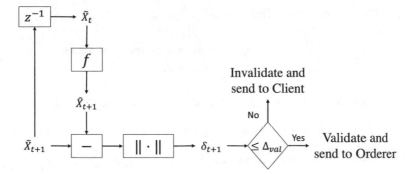

Fig. 14 Operations performed by the endorser for a single frame

Each endorser involved in validating a frame, sequentially checks the state updates by recomputing from the last valid state, i.e., to validate the report $\tilde{X}_{t+1}$, the endorser computes $\hat{X}_{t+1} = f(\tilde{X}_t)$ and checks for the validity criterion (11). The frame is validated if and only if all updates are valid in the frame. The endorsements are then reported to the orderer.

Individual state validations can be performed in parallel and finally verified for sequential consistency. Such parallelism can be performed either at the individual endorser level or in the form of a distribution of the sub-frames across endorsers through coded computing. This results in a reduction of the time required for validating a frame.

Upon receiving frame endorsements, the orderer checks for consensus among the endorsers and for consistency of the checkpoints. It then adds the frame to the ledger once prior frames have been added, and broadcasts it to other peers as in Fig. 15. Since the state updates are stored on the immutable blockchain ledger, they provide an avenue for independent verification of the computations at a later stage through a simple check for consistency of the hash chain.

As described in (12), the verification requirements are not as strict as the validation requirements. Thus, it suffices to subsample the updates in a frame and store only one for every K iterates, such that the block records the sum of the K intermediate updates. Thus, a block stored on the blockchain comprises validated audits that are either checkpoints or the cumulative updates corresponding to K successive iterates.

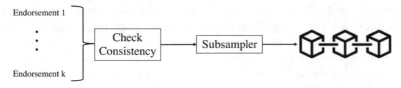

Fig. 15 Operations performed by the orderer for frames. The orderer sequentially adds valid frames to the blockchain after checking for consistency

The audits $\tilde{Y}_\tau$ are then defined by

$$\tilde{Y}_{\tau+1} = \begin{cases} \tilde{Y}_\tau + \sum \tilde{\Delta} X_{t+1}, & \text{if no checkpoint in } K \text{ iterates} \\ \tilde{X}_{t'}, & \text{otherwise} \end{cases} , \qquad (16)$$

where the sum is over the intermediate iterates and t' is the next checkpoint.

Theorem 5 *For subsampled storage at a frequency* $1/K$ *a Lipschitz constant* L *of* $f(\cdot)$ *and quantization error* ϵ,

$$\left\| \hat{Y}_{\tau+1} - \tilde{Y}_{\tau+1} \right\| \leq \left(L^K + 1 \right) K\epsilon, \qquad (17)$$

where $\hat{Y}_{\tau+1} = f^K(\tilde{Y}_\tau)$.

A viable subsampling frequency can be determined by finding a K such that $(L^K + 1)K \leq \frac{\Delta_{\text{ver}}}{\epsilon}$. This further reduces the storage cost on the blockchain at the expense of the rigorousness of recorded audits. If the agents are interested in increasing the accuracy of the records over time, then the quantizers and samplers can be dynamically adjusted accordingly.

5.6 Parameter Agnostic Design

As discussed in Sect. 5.3, we used vector quantizers based on the Lipschitz constant L. In practice, such parameters of the computation are unknown *apriori*. Underestimating L can result in using a larger quantization error, that could cause errors in validation even when the client computes correctly. In such cases, it is essential to be able to identify the cause of the error. One option is to estimate L from computed samples. This translates to estimating the maximum gradient magnitude for the atomic operation which might be expensive in sample and computational complexity, depending on the application. Thus, we propose an alternative compression scheme.

   We draw insight from video compression strategies and propose the use of *successive refinement coding* of the state updates [17]. A compressed stream is generated for each state update such that the accuracy can be improved by sending additional bits from the bit stream. Successive refinement allows clients to update reports such that the accuracy can be iteratively improved, in the event of invalidation. That is, if a frame is invalidated, the client has two options—checking the computations and/or refining the reported state through successive refinement. Depending on the computation-communication tradeoff, the client appropriately chooses the more economical alternative. Through successive refinement, the client provides more accurate descriptions of the state vector, and thus reduces the possibility of validation errors caused by report inaccuracy.

A simple way to implement successive refinement is by extending the lattice vector quantizers [39, 43]. This scheme also reduces the size of the codebook, if the refinement lattices are assumed to be of the same geometry, because the client only needs to communicate the scaling corresponding to the refinement. This allows for improved adaptability in the refinement updates. More efficient quantizers can also be defined if additional information regarding the application and state updates are available.

5.7 Computations with External Randomness

As described in Sect. 5.2, such computational algorithms in practice typically evolve iteratively as a function of the current state x_t, and an external randomness θ_t. When this randomness is not shared across agents and is inaccessible to the client, reproduction of the reported results by an individual endorser becomes infeasible. This could also emerge in cases where the client is unwilling to share private data associated used by the algorithm [75].

Whereas the exact values of these random variables are unavailable for reproduction, the source of such randomness is often common, i.e., $\theta_t \stackrel{i.i.d.}{\sim} P_\theta$, and P_θ is known (or accessible). In this context, we redefine validation as guaranteeing (11) with probability at least $1 - \rho$, i.e., the probability that the estimate recomputed by a set of M endorsers, $\hat{X}_t$, deviates from the report $\bar{X}_t$ by Δ_{val} satisfies

$$\mathbb{P}\left[\left\|\tilde{X}_t - \hat{X}_t\right\| \geq \Delta_{\text{val}}\right] \leq \rho. \tag{18}$$

This requirement removes outliers in the computation process and only allows trajectories close to the expected behavior.

One possible way to endorse such randomized computations is to use the average behavior across independent endorsers,

$$\hat{X}_{t+1} = \frac{1}{m} \sum_{i=1}^{m} f(\tilde{X}_t, \theta_i),$$

where $\theta_i \stackrel{i.i.d.}{\sim} P_\theta$. By choosing a sufficiently large number of endorsers, depending on ρ, we can assure (18). The role of the endorsers is appropriately modified and the system calls for higher coordination among the endorsers. We derive a sufficient condition on the number of endorsers using multivariate concentration inequalities.

Theorem 6 *Let $\epsilon < \frac{\Delta_{val}}{L+1}$. For a state at time t, if the average of m endorsers is used for validation,*

$$\mathbb{P}\left[\left\|\tilde{X}_t - \hat{X}_t\right\| \geq \Delta_{val}\right] \leq \frac{2d\tilde{\lambda}^2}{(\Delta_{val} - (L+1)\epsilon)^2}\left(1 + \frac{1}{m\tilde{\lambda}}\right)^2, \tag{19}$$

where $\tilde{\lambda}$ is the maximum eigenvalue of the covariance matrix of the quantized state vector.

Corollary 1 *To guarantee validation with probability at least $1 - \rho$, for a margin of deviation of Δ_{val}, where $\rho \leq \frac{2d\tilde{\lambda}^2}{(\Delta_{val}-(L+1)\epsilon)^2}$, it suffices to use*

$$m = \left\lceil \left[\left(\sqrt{\frac{\rho}{2d}}(\Delta_{val} - (L+1)\epsilon) - \tilde{\lambda} \right) \right]^{-1} \right\rceil \tag{20}$$

endorsers.

This sufficient condition follows directly from Theorem 6. Thus, in the case of randomized experiments, we can choose the set of endorsers according to the validation criterion and the second-order statistics of the atomic computation.

5.8 Iterative Experiments with MNIST Training

In this experiment, we run some simple synthetic experiments using the MNIST database [35]. These synthetic experiments highlight the efficacy of the MBF protocol in the domain of NN training that is widely familiar to the research community.

Let us consider a simple 3-layer neural network (NN), trained on the MNIST database, with 25 neurons in the hidden layer. Consider a client training the NN using mini-batch stochastic gradient descent (SGD) with limited resources. Specifically, the client is limited to using only small batch sizes of 10 samples per iteration and performs 1000 iterations for the training. The average precision of such a neural network trained with gradient descent is 97.4%. We now wish to establish trust in the training. Whereas this configuration is far from state of the art, it does help understand the trust environment better.

Since the training uses stochastic gradients, exact recomputation of the iterates is infeasible. Hence, we compare deviations from the average computed over $m = 5$ endorsers per state for validation. We evaluate the computation and communication cost of validation as a function of the tolerance chosen for validation. Since the neural network eventually converges to a local minimum, we use a reduce the tolerance over iterations as $\Delta_{val}(t) = \frac{\Delta_{max}}{\log(t+1)}$. This ensures that we impose more stringent requirements for the trained networks.

We consider three main cases of the simulation:

1. **Base case**: Compression error is less than validation tolerance, i.e., $\epsilon \leq \Delta_{max}$, and maximum frame size is 10% of the total number of iterations.
2. **Coarse compression**: Large compression error, i.e., $\epsilon \geq \Delta_{max}$, and same frame size as base case.
3. **Large frames**: Same base compression error and maximum frame size is 20% of the number of iterations.

MBF helps eliminate spurious gradients in this setup. These three cases in particular help understand the communication-computation tradeoff well. In the case of coarse compression, in addition to spurious gradients, inaccurately reported gradients are also invalidated by the endorsers. Thus, coarse compression increases the recomputation cost significantly in comparison to the base case. In the case of using large frames with small compression error, even if the reported gradients are accurate, one spurious gradient that is invalidated in a frame requires all subsequent iterations to be recomputed as well. This in turn increases the computation cost in comparison to the base case.

On the other hand, in terms of the communication cost, coarse compression naturally consumes a smaller bandwidth in comparison to the other two schemes. In comparison, the base case and large frame cases cost more. In particular, the base case requires more frames to be communicated as well which in turn implies a higher communication cost in comparison to the large frame case.

Figure 16 shows the average number of gradient recomputations per iteration for the three cases. As expected, this decays sharply as we increase the tolerance. Note that at either extreme, the three cases converge in the number of recomputations. This is owing to the fact that at one end all gradients are accepted, whereas, at the stricter end, most gradients are rejected with high probability, irrespective of the compression parameters. In the moderate tolerance range, we observe the tradeoffs as explained above. The corresponding communication cost tradeoff is shown in Fig. 17.

Figure 18 shows the precision of the neural network trained under the validation requirement as compared to the networks trained with standard mini-batch SGD of batch sizes 10, 30, and 50. We note that the network trained with trust outperforms the case of vanilla SGD with the same batch size as it eliminates spurious gradients at validation. Decreasing the tolerance results in improved precision of the model.

Fig. 16 Depicts the average number of recomputations of gradients per iteration for varying validation requirements

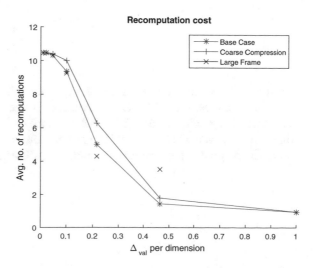

Fig. 17 Depicts the average number of bits per dimension communicated by clients to endorsers for varying validation tolerance

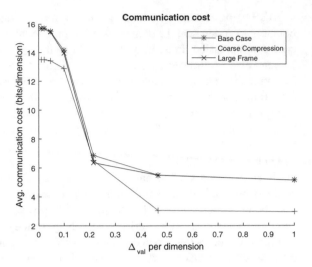

Fig. 18 Precision of the trained neural network satisfying the local validation criterion. Eliminating spurious gradients through validation enhances training process

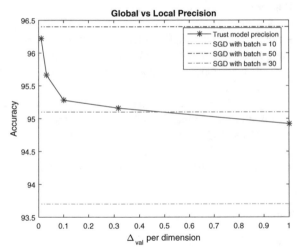

In particular, it is worth noting that the strictest validation criterion results in performance that is almost as good as training with a batch size of 50. This is understandable as the validated gradients are the ones that are close to those averaged across 50 data points. In fact, even when the trust requirements are fairly relaxed, just eliminating outliers in the gradients enhances the training significantly.

5.9 Extensions of Distributed Trust Protocol

In this section, we detailed the development of the MBF protocol to establish distributed trust among peers for iterative computational systems. Alternatively, one could consider enumerative computational systems such as hyperparameter tuning of neural networks and doing "what-if analyses" with various choices of interventions that affect disease spread. The MBF protocol as stated here leverages the sequential nature of the computations to leverage successive refinement encoding and thus compressed storage. Thus, it isn't immediately clear if and how one could imbibe distributed trust among enumerative computational systems.

Compression of data is made effective by grouping similar computational instances together. This may be done by using a distance-based clustering algorithm and compression mechanism that is detailed in [53]. Fundamentally, the protocol is modified to first identify closeness relationships among the computational instances in the form of a tree and group them by distance into frames that are encoded effectively using a Lempel-Ziv-style compression algorithm.

Another common attribute of the computations that is leveraged in this section in order to define effective compression schemes is the fact that the atomic computation functions are smooth and the state and output variables are in the Euclidean space. Often computational systems operate on discrete, Boolean, or mixed domains in which the definitions of closeness and continuity do not directly extend. One could leverage ideas from functional compression [15] to develop equivalent compression methods, depending on the function to be computed.

However, this isn't always feasible as functional compression methods are often strictly limited to certain sub-classes of functions. To elaborate this idea, let us consider an example. Consider the task of computing the hash value corresponding to a set of digital files. The files are in the binary vector space, and the hash functions, by definition, are not smooth in their output. Thus, any algorithm that is able to compress the computations of this nature implicitly also implies the ability to predict the input-output pairs of the hash function. This stands in contradiction to the definition of the cryptographic hash function. Thus, it is evident that not all computational systems can be subject to such compression protocols to enable efficient storage on the blockchain.

6 Future Work

Blockchains promise disruptive revolutions in a variety of sectors. Enabling scalable solutions on the blockchain platform, however, does require a careful consideration of the costs of storage, communication, consensus, and security. And in this respect, several open problems still need to be solved in these avenues.

In Sect. 4, we introduced the Dynamic Distributed Storage scheme that reduces the cost of cold storage of blockchain ledgers without compensating on the security of the transactions. The tradeoff between the storage cost and block security is emphasized by (4) and (7). Thus the design of applications that leverage the DDS scheme, in turn, require optimized code constructions that balance the tradeoff between storage cost and the resultant security. As elaborated in [57, 58], designing codes that optimize this tradeoff enable the creation of secure distributed storage systems for data sharing and data insurance mechanisms. Designing codes optimized for the applications is a core part of future work in the development of scalable blockchain systems.

In the future, a closer coding-theoretic analysis can help develop efficient and robust codes such as [21, 31, 65]. The codes can also be reinforced with additional capabilities such as repairing faulty nodes, given that individual peers in the network can undergo failure or data loss and might thus need repairing the code. The DDS scheme enables fault detection using consistency checks on the blockchains. However, fault tolerance is currently only enabled through redundancy across the zones of the network. However, we could integrate fault tolerance into the code construction to allow transaction recovery in the presence of errors in the stored codewords. Alternatively, the codes can also be secured from the possibility of data loss by incorporating erasure correction in the coding scheme.

Another area rife with challenging problems is on-chain solutions to enable the scalability of blockchain systems. On-chain solutions to reduce storage costs and amplify throughput have also been explored through the idea of sharding the peer-to-peer network [36]. Enhancements to the proof of work (PoW) mechanisms governing the transaction validation in blockchain systems also improve network throughput and enable faster transaction validation and system updates [3].

In Sect. 5, we introduced the application of secure data sharing among nodes of a distributed computational network. In this chapter, we restricted the application to atomic operations that are Lipschitz-continuous. One immediate extension would be to derive generalized adaptations of the protocol for sharing computational outcomes for generic functional forms. This can be enabled perhaps using cryptographic mechanisms like probabilistically checkable proofs [4, 30, 69].

7 Conclusion

In this chapter, we studied some of the costs associated with blockchain systems. Focusing on the storage challenge involved with scalable blockchain solutions, we highlighted the increasing costs of maintaining a full ledger of the blockchain. We then highlighted two examples of solutions targeted at reducing the storage cost. First, we considered an off-chain solution in the form of the Dynamic Distributed Storage scheme for enabling cold storage of blockchain ledgers. We studied the tradeoff between storage cost and the security offered by the storage scheme in the presence of active and denial of service adversaries. We then considered the example of an application that is optimized through a protocol to enable its scalable implementation on the blockchain system. To this end, we studied the problem of establishing distributed trust in a peer-to-peer network of computational nodes and evaluated the costs associated with the compression schema that enables a scalable implementation of the solution.

This chapter is meant to provide an introduction to the complexity of the scalability problem associated with practical blockchain systems. Through the two examples, one from the blockchain protocol and one in the form of optimization for an application, we aim to highlight the fact that this problem can be addressed on multiple axes, and the solutions may be as varied as the applications themselves.

Acknowledgements The authors would like to thank Kush R. Varshney, Roman Vaculin, Michael Hind, Nelson Bore, and Sekou Remy from IBM Research for their contributions towards the work on trusted distributed computing in the Multiagent Blockchain Framework.

References

1. Androulaki, E., Barger, A., Bortnikov, V., Cachin, C., Christidis, K., Caro, A.D., Enyeart, D., Ferris, C., Laventman, G., Manevich, Y., Muralidharan, S., Murthy, C., Nguyen, B., Sethi, M., Singh, G., Smith, K., Sorniotti, A., Stathakopoulou, C., Vukolić, M., Cocco, S.W., Yellick, J.: Hyperledger fabric: a distributed operating system for permissioned blockchains. In: Proceedings of 13th EuroSys Conference, EuroSys '18, pp. 30:1–30:15 (2018)
2. Azaria, A., Ekblaw, A., Vieira, T., Lippman, A.: Medrec: using blockchain for medical data access and permission management. In: 2nd International Conference Open Big Data (OBD 2016), pp. 25–30 (2016)
3. Bagaria, V., Kannan, S., Tse, D., Fanti, G., Viswanath, P.: Prism: deconstructing the blockchain to approach physical limits. In: Proceedings of 2019 ACM SIGSAC Conference Computer and Communication Security, pp. 585–602. Association for Computing Machinery (2019)
4. Bellare, M., Goldwasser, S., Lund, C., Russell, A.: Efficient probabilistically checkable proofs and applications to approximations. In: Proceedings of 25th Annual ACM Symposium Theory Computing (STOC'93), STOC '93, pp. 294–304. Association for Computing Machinery, New York, NY, USA (1993)
5. Bent, O., Remy, S.L., Roberts, S., Walcott-Bryant, A.: Novel exploration techniques (NETs) for malaria policy interventions. In: Proceedings of 32nd AAAI Conference Artificial Intelligence (2018)
6. Bhagoji, A.N., Chakraborty, S., Mittal, P., Calo, S.: Analyzing federated learning through an adversarial lens. In: Chaudhuri, K., Salakhutdinov , R.(eds.) Proceedings of 36st International Conference Machine Learning (ICML 2019), Proceedings of Machine Learning Research, vol. 97, pp. 634–643. PMLR (2019)
7. Blockchain info. https://blockchain.info/home
8. Blockchair. https://blockchair.com
9. Bonneau, J., Miller, A., Clark, J., Narayanan, A., Kroll, J.A., Felten, E.W.: SoK: research perspectives, challenges for bitcoin and cryptocurrencies. In: Proceedings of 2015 IEEE Symposium Security Privacy, pp. 104–121 (2015)
10. Bore, N.K., Raman, R.K., Markus, I.M., Remy, S.L., Bent, O., Hind, M., Pissadaki, E.K., Srivastava, B., Vaculin, R., Varshney, K.R., Weldemariam, K.: Promoting distributed trust in machine learning and computational simulation. In: Proceedings IEEE International Conference Blockchain Cryptocurrency. Seoul, Korea (2019)
11. Brundage, M., Avin, S., Wang, J., Belfield, H., Krueger, G., Hadfield, G., Khlaaf, H., Yang, J., Toner, H., Fong, R., Maharaj, T., Koh, P.W., Hooker, S., Leung, J., Trask, A., Bluemke, E., Lebensold, J., O'Keefe, C., Koren, M., Ryffel, T., Rubinovitz, J., Besiroglu, T., Carugati, F., Clark, J., Eckersley, P., de Haas, S., Johnson, M., Laurie, B., Ingerman, A., Krawczuk, I.,

Askell, A., Cammarota, R., Lohn, A., Krueger, D., Stix, C., Henderson, P., Graham, L., Prunkl, C., Martin, B., Seger, E., Zilberman, N., hÉigeartaigh, S.Ó., Kroeger, F., Sastry, G., Kagan, R., Weller, A., Tse, B., Barnes, E., Dafoe, A., Scharre, P., Herbert-Voss, A., Rasser, M., Sodhani, S., Flynn, C., Gilbert, T.K., Dyer, L., Khan, S., Bengio, Y., Anderljung, M.: Toward trustworthy AI development: mechanisms for supporting verifiable claims (2020). arXiv:2004.07213 [cs.CY]

12. Casey, M.J., Wong, P.: Global supply chains are about to get better, thanks to blockchain. Harvard Bus. Rev. (2017). https://hbr.org/2017/03/global-supply-chains-are-about-to-get-better-thanks-to-blockchain

13. Croman, K., Decker, C., Eyal, I., Gencer, A.E., Juels, A., Kosba, A., Miller, A., Saxena, P., Shi, E., Sirer, E.G., Song, D., Wattenhofer, R.: On scaling decentralized blockchains. In: Clark, J., Meiklejohn, S., Ryan, P.Y.A., Wallach, D., Brenner, M., Rohloff, K. (eds.) Financial Cryptography and Data Security. Lecture Notes in Computer Science, vol. 9604, pp. 106–125. Springer, Berlin (2016)

14. Dasgupta, P.: Trust as a commodity. In: Trust: Making and Breaking Cooperative Relations, vol. 4, pp. 49–72 (2000)

15. Doshi, V., Shah, D., Médard, M., Jaggi, S.: Distributed functional compression through graph coloring. In: Proceedings IEEE Data Compression Conference (DCC 2007), pp. 93–102 (2007). https://doi.org/10.1109/DCC.2007.34

16. Durrett, R.: Random Graph Dynamics. Cambridge Series in Statistical and Probabilistic Mathematics. Cambridge University Press (2006). https://doi.org/10.1017/CBO9780511546594

17. Equitz, W.H.R., Cover, T.M.: Successive refinement of information. IEEE Trans. Inf. Theory 37(2), 269–275 (1991)

18. Falcone, R., Singh, M., Tan, Y.H.: Trust in Cyber-Societies: Integrating the Human and Artificial Perspectives, vol. 2246. Springer Science & Business Media (2001)

19. Fung, C., Yoon, C.J., Beschastnikh, I.: Mitigating sybils in federated learning poisoning. arXiv:1808.04866 [cs.LG]

20. Gersho, A., Gray, R.M.: Vector Quantization and Signal Compression, vol. 159. Springer Science & Business Media (2012)

21. Gopalan, P., Huang, C., Simitci, H., Yekhanin, S.: On the locality of codeword symbols. IEEE Trans. Inf. Theory 58(11), 6925–6934 (2012)

22. Granger, C.W., Joyeux, R.: An introduction to long-memory time series models and fractional differencing. J. Time Ser. Anal. 1(1), 15–29 (1980)

23. Grier, D.A.: Error identification and correction in human computation: lessons from the WPA. In: Human Computation (2011)

24. Gundersen, O.E., Kjensmo, S.: State of the art: reproducibility in artificial intelligence. In: Proceedings of 32nd AAAI Conference Artificial Intelligent New Orleans, USA (2018)

25. Gurkaynak, G., Yilmaz, I., Yesilaltay, B., Bengi, B.: Intellectual property law and practice in the blockchain realm. Comput. Law Secur. Rev. 34(4), 847–862 (2018)

26. Henderson, P., Islam, R., Bachman, P., Pineau, J., Precup, D., Meger, D.: Deep reinforcement learning that matters (2017). arXiv:1709.06560v2 [cs.LG]

27. Iansiti, M., Lakhani, K.R.: The truth about blockchain. Harvard Bus. Rev. 95(1), 118–127 (2017)

28. Jones, N.: How scientists are embracing NFTs (2021). https://www.nature.com/articles/d41586-021-01642-3

29. Kang, J., Xiong, Z., Niyato, D., Zou, Y., Zhang, Y., Guizani, M.: Reliable federated learning for mobile networks. IEEE Trans. Wirel. Commun. 27(2), 72–80 (2020)

30. Kilian, J., Petrank, E., Tardos, G.: Probabilistically checkable proofs with zero knowledge. In: Proceedings of 29th Annual ACM Symposium Theory Computing (STOC'97), STOC '97, pp. 496–505. Association for Computing Machinery (1997)

31. Kim, Y., Raman, R.K., Kim, Y., Varshney, L.R., Shanbhag, N.R.: Efficient local secret sharing for distributed blockchain systems. IEEE Commun. Lett. 23(2), 282–285 (2019)

32. Kosba, A., Miller, A., Shi, E., Wen, Z., Papamanthou, C.: Hawk: the blockchain model of cryptography and privacy-preserving smart contracts. In: Proceedings of 2016 IEEE Symposium Security Privacy, pp. 839–858 (2016)

33. Krawczyk, H.: Secret sharing made short. In: Stinson, D.R. (ed.) Advances in Cryptology— CRYPTO '93. Lecture Notes in Computer Science, vol. 773, pp. 136–146. Springer, Berlin (1994)

34. Kshetri, N., Voas, J.: Blockchain in developing countries. IT Professional **20**(2), 11–14 (2018)

35. Lecun, Y., Bottou, L., Bengio, Y., Haffner, P.: Gradient-based learning applied to document recognition. Proc. IEEE **86**(11), 2278–2324 (1998)

36. Li, S., Yu, M., Yang, C.S., Avestimehr, A.S., Kannan, S., Viswanath, P.: Polyshard: coded sharding achieves linearly scaling efficiency and security simultaneously. IEEE Trans. Inf. Forensics Secur. **16**, 249–261 (2021)

37. Litkc, A., Anagnostopoulos, D., Varvarigou, T.: Blockchains for supply chain management: architectural elements and challenges towards a global scale deployment. Logistics **3** (2019)

38. Liu, Y., Fang, Z., Cheung, M.H., Cai, W., Huang, J.: Economics of blockchain storage. In: Proceedings of IEEE International Conference Communication (ICC 2020) (2020)

39. Liu, Y., Pearlman, W.A.: Multistage lattice vector quantization for hyperspectral image compression. In: Conference Record 41st Asilomar Conference Signals, Systems Computing, pp. 930–934 (2007)

40. Marsh, S.P.: Formalising trust as a computational concept. Ph.D. thesis, University of Stirling (1994)

41. Matzutt, R., Kalde, B., Pennekamp, J., Drichel, A., Henze, M., Wehrle, K.: Coinprune: Shrinking bitcoin's blockchain retrospectively. IEEE Trans. Netw. Serv. Manag. (2021)

42. Matzutt, R., Kalde, B., Pennekamp, J., Drichel, A., Henze, M., Wehrle, K.: How to securely prune bitcoin's blockchain. In: 2020 IFIP Networking Conference, pp. 298–306 (2020)

43. Mukherjee, D., Mitra, S.K.: Successive refinement lattice vector quantization. IEEE Trans. Signal Process. **11**(12), 1337–1348 (2002)

44. Nakamoto, S.: Bitcoin: a peer-to-peer electronic cash system (2008). http://bitcoin.org/bitcoin.pdf

45. Narayanan, A., Bonneau, J., Felten, E., Miller, A., Goldfeder, S.: Bitcoin and Cryptocurrency Technologies: A Comprehensive Introduction. Princeton University Press, Princeton (2016)

46. Nelson, J.: The operation of non-governmental organizations (NGOs) in a world of corporate and other codes of conduct. Corporate Social Responsibility Initiative (2007)

47. Palm, E., Schelén, O., Bodin, U.: Selective blockchain transaction pruning and state derivability. In: 2018 Crypto Valley Conference on Blockchain Technology (CVCBT), pp. 31–40 (2018)

48. Panhuis, W.G.V., Paul, P., Emerson, C., Grefenstette, J., Wilder, R., Herbst, A.J., Heymann, D., Burke, D.S.: A systematic review of barriers to data sharing in public health. BMC Pub. Health **14**(1), 1144 (2014)

49. Piette, J.D., Krein, S.L., Striplin, D., Marinec, N., Kerns, R.D., Farris, K.B., Singh, S., An, L., Heapy, A.A.: Patient-centered pain care using artificial intelligence and mobile health tools: protocol for a randomized study funded by the US Department of Veterans Affairs Health Services Research and Development Program. JMIR Res. Protocols **5**(2) (2016)

50. Pokhrel, S.R., Choi, J.: Federated learning with blockchain for autonomous vehicles: analysis and design challenges. IEEE Trans. Commun. (2020)

51. Power, D.J.: Data science: supporting decision-making. J. Decis. Sys. **25**(4), 345–356 (2016)

52. Radanović, I., Likić, R.: Opportunities for use of blockchain technology in medicine. Appl. Health Econ., Health Policy **16**(5), 583–590 (2018)

53. Raman, R.K., Vaculin, R., Hind, M., Remy, S.L., Pissadaki, E.K., Bore, N.K., Daneshvar, R., Srivastava, B., Varshney, K.R.: A scalabale blockchain approach for trusted computation and verifiable simulation in multi-party collaboration. In: Proceedings of IEEE International Conference Blockchain Cryptocurrency. Seoul, Korea (2019)

54. Raman, R.K., Vaculin, R., Hind, M., Remy, S.L., Pissadaki, E.K., Bore, N.K., Daneshvar, R., Srivastava, B., Varshney, K.R.: Trusted multi-party computation and verifiable simulations: a scalable blockchain approach (2018). arXiv:1809.08438 [CS.DC]

55. Raman, R.K., Varshney, K.R., Vaculin, R., Bore, N.K., Remy, S.L., Pissadaki, E.K., Hind, M.: Constructing and compressing frames in blockchain-based verifiable multi-party computation. In: Proceedings of IEEE International Conference on Acoustics, Speech, and Signal Processing (2019)

56. Raman, R.K., Varshney, L.R.: Coding for scalable blockchains via dynamic distributed storage. IEEE/ACM Trans. Netw. 1–14 (2021)
57. Raman, R.K., Varshney, L.R.: Distributed storage meets secret sharing on the blockchain. In: Proceedings of 2018 Information Theory and Applications Workshop (2018)
58. Raman, R.K., Varshney, L.R.: Dynamic distributed storage for blockchains. In: Proceedings of 2018 International Symposium on Information Theory (2018)
59. Ramchurn, S.D., Huynh, D., Jennings, N.R.: Trust in multi-agent systems. Knowl. Eng. Rev. **19**(1), 1–25 (2004)
60. Rogaway, P., Shrimpton, T.: Cryptographic hash-function basics: definitions, implications, and separations for preimage resistance, second-preimage resistance, and collision resistance. In: Roy, B., Meier, W. (eds.) Fast Software Encryption, pp. 371–388. Springer, Berlin Heidelberg, Berlin, Heidelberg (2004)
61. Saberi, S., Kouhizadeh, M., Sarkis, J., Shen, L.: Blockchain technology and its relationships to sustainable supply chain management. Int. J. Prod. Res. **57**(7), 2117–2135 (2019)
62. Servetto, S.D., Vaishampayan, V.A., Sloane, N.J.A.: Multiple description lattice vector quantization. In: Proceedings of IEEE Data Compression Conference (DCC 1999), pp. 13–22 (1999)
63. Shah, D.: Data science and statistics: opportunities and challenges. Technol. Rev. (2016)
64. Shuaib, K., Saleous, H., Shuaib, K., Zaki, N.: Blockchains for secure digitized medicine. J. Personalized Med. **9**(3) (2019)
65. Silberstein, N., Rawat, A.S., Koyluoglu, O.O., Vishwanath, S.: Optimal locally repairable codes via rank-metric codes. In: Proceedings of 2013 IEEE International Symposium on Information Theory, pp. 1819–1823 (2013)
66. Singh, J., Cobbe, J., Norval, C.: Decision provenance: capturing data flow for accountable systems (2018). arXiv:1804.05741 [cs.CY]
67. Smith, T., Maire, N., Ross, A., Penny, M., Chitnis, N., Schapira, A., Studer, A., Genton, B., Lengeler, C., Tediosi, F., Savigny, D.D., Tanner, M.: Towards a comprehensive simulation model of malaria epidemiology and control. Parasitology **135**(13), 1507–1516 (2008)
68. Stinson, D.R.: Some observations on the theory of cryptographic hash functions. Des. Codes Cryptogr. **38**(2), 259–277 (2006)
69. Sudan, M.: Probabilistically checkable proofs. Commun. ACM **52**(3), 76–84 (2009)
70. Swan, M.: Blockchain: Blueprint for a New Economy. O'Reilly Media Inc., Sebastopol, CA (2015)
71. Tapscott, D., Tapscott, A.: Blockchain Revolution: How the Technology behind Bitcoin is Changing Money, Business, and the World. Penguin, New York (2016)
72. Thakur, V., Doja, M., Dwivedi, Y.K., Ahmad, T., Khadanga, G.: Land records on blockchain for implementation of land titling in India. Int. J. Inform. Manag. **52**, 101940 (2020)
73. Underwood, S.: Blockchain beyond bitcoin. Commun. ACM **59**(11), 15–17 (2016)
74. Veale, M., Binns, R., Edwards, L.: Algorithms that remember: model inversion attacks and data protection law. Phil. Trans. Royal Soc. A: Math., Phys., Engg. Sci. **376**(2133) (2018)
75. Verma, D., Calo, S., Cirincione, G.: Distributed AI and security issues in federated environments. In: Proceedings of the Workshop Program of the 19th International Conference on Distributed Computing and Networking (2018)

Trust and Security

On Trust, Blockchain, and Reputation Systems

Bruno Rodrigues, Muriel Franco, Christian Killer, Eder J. Scheid, and Burkhard Stiller

Abstract Trust management in distributed systems has always been a topic of active interest in the research community to understand how to foster and manage aspects. In this sense, Distributed Ledger Technologies (DLT) and, among them, Blockchains (BC), emerge as an alternative for shifting trust assumptions between users to the protocol that regulates the interaction, fostering trust in distributed systems. Especially reputation management systems have enabled several applications to be revisited as an application running based on an underlying distributed system. Thus, a clear understanding of major properties, threats and vulnerabilities, and challenges of reputation systems based on different types of DLT and BC (i.e., permissioned and permissionless) are key to determine their usefulness and optimization potentials. In this sense, a use case of a BC-based reputation system within the context of cooperative network defenses illustrates such benefits and drawbacks of exploiting DLTs for reputation systems.

1 Introduction

In the context of distributed systems, research on how to manage and establish trust relationships is fundamental. An initial approach minimizes the requirement for trust among users based on a communication protocol being open and verifiable

---

B. Rodrigues (✉) · M. Franco · C. Killer · E. J. Scheid · B. Stiller
Communication Systems Group CSG, Department of Informatics IfI, University of Zürich UZH
Binzmühlestrasse 14, CH—8050, Zürich, Switzerland
e-mail: rodrigues@ifi.uzh.ch

M. Franco
e-mail: franco@ifi.uzh.ch

C. Killer
e-mail: killer@ifi.uzh.ch

E. J. Scheid
e-mail: scheid@ifi.uzh.ch

B. Stiller
e-mail: stiller@ifi.uzh.ch

© The Author(s), under exclusive license to Springer Nature Switzerland AG 2022
D. A. Tran et al. (eds.), *Handbook on Blockchain*, Springer Optimization
and Its Applications 194, https://doi.org/10.1007/978-3-031-07535-3_9

by members (e.g., via a Blockchain, BC). However, it is not always possible to eliminate the need for trust between communication partners. At this point, reputation systems have a significant impact on establishing a quantifiable measure of trust through reputation evaluations. For example, approaches on how to establish trust were widely studied in the early days of e-commerce [65, 66, 82] to understand how to encourage people to buy products from a previously unknown merchant in a new, unfamiliar environment. Before e-commerce, relationships between merchants and customers were direct, person-to-person, and trust had been (or not) established without intermediaries. Since e-commerce leveraged the Internet as an intermediary, a peculiar environment with unknown characteristics compared to direct relations, the study on the development of a reliable, i.e., a trustful, environment was of paramount importance for today's e-commerce to be considered a regular Internet activity.

In the case of e-commerce, [15] noted that one of the most relevant factors for establishing trust was the "general credibility of the seller" (i.e., the seller's reputation), which represents a strong influence on the likelihood of a buyer to do online shopping. However, the likelihood was also related to other factors as well, e.g., the level of education of buyers, in which individuals with a higher educational level were more likely to worry about e-commerce technology and risks of the platform, e.g., or confidentiality of personal information, such as credit card number or home addresses. Otherwise, trust was more likely to rely on the reputation built offline, thus, relying on the general credibility of the seller. Interestingly, the consistency of actions between individuals of the same ethic background was highlighted as important for building trust.

Tracking reputation plays a vital role in online communities to examine how trust develops in long-term relationships, avoiding abuses, and offering indications of content quality in varying contexts. The basic principle of the reputation mechanisms is to allow consumers to evaluate services (i.e., producers) and leave a feedback rating after the completion of an interaction. In this regard, a reputation system is used to aggregate ratings provided by consumers and derive a reputation score for the producer, which can assist other consumers in deciding whether to interact with the specific service in the future.

In this sense, applications based on Distributed Ledgers (DL), and especially Blockchains (BC), whose characteristics involve refraining from trusted entities managing users' interaction, need to consider the design and security aspects of the reputation management system. BCs impacted the way trust is established, especially moving from a centralized to a decentralized model, where trust is shifted to the protocol, where its transparency is based on a fully transparent action history [63].

Using BCs and their major ingredient, cryptography, enables the creation of time-stamped [33] certifications of actions, which are now "under the control" of all members' participating. Recipients of information persisted in a BC can share a digital "proof" with a counterpart, while being trustworthy, since the proof of such an information, e.g., a degree [32], was in fact issued to the person presenting it. Initially, digital credentials had been maintained by an "Open Badges" system that became inclusive and recognizes a wider variety of accomplishments [62].

BCs extended this approach to share and verify credentials in a decentralized approach, which was enabled mainly due to Bitcoin's consensus mechanism, namely Proof-of-Work (PoW). PoW ensures decentralization while relying on strong cryptography to ensure that (a) no entity should gather more than 50% of the network's computational power and (b) participants are incentivized to maintain the network [63] of members. While reputation mechanisms are not needed in the main application domain of cryptocurrencies, for which the Bitcoin BC was developed, the creation of several applications in different areas (e.g., Networking [25], Insurance [23], and Education [74]) has expanded such need of tracking reputation and a wider adoption toward a closer and more detailed application relation.

In its purest form, a BC acts like a decentralized, public digital ledger that transparently and permanently persists blocks of information in transactions stored across a network of computers, which communicate via a peer-to-peer (P2P) protocol based on a consensus algorithm, without modifying any of the antecedent blocks [60, 77]. However, technical details, such as permissions to write and read, as well as the participation in the block-validation process, can be differently designed and implemented within a BC [73]. Thus, different types of BCs exist, such as Distributed Ledger Technologies (DLT), which show flexible characteristics suitable for individual needs of transparency and confidentiality applied to each use case. For example, the sharing of sensitive data between stakeholders (e.g., patient data in the healthcare industry) could be based on a BC that should not be publicly accessible, since sensitive data could leak and impose privacy issues, whereas tracking vaccine temperatures in a cold-chain scenario could be based on a public BC to ensure publicly auditability [78].

This chapter analyzes the interplay of trust, BC, and reputation systems by clarifying how the different types of DLTs (including BC) impact trust and the challenges that arise in the context of DLT-based reputation systems. Therefore, it is imperative to distinguish and classify the different types of DLTs to understand their trust assumptions and thus understand how the characteristics of each type of DLT impact the properties of reputation management systems. For example, the verifiability of the scoring engine may be facilitated by the fundamental transparency feature of DLTs. However, aspects related to account identification (and therefore vulnerability to specific threats listed in Sect. 2) may present a drawback.

This chapter is organized as follows. While the background on trust, BC, and reputation is presented in Sects. 2, 3 overviews existing tools and methods to enable the tracking of reputation in BC/DLT-based systems. Section 4 discusses a relevant use case of a BC-based reputation system within the context of cooperative network defense. Lastly, Sect. 5 draws conclusions and provides considerations outlining a future perspective over DLT-based reputation systems.

2 Definitions and Fundamentals

It is essential to base a justified discussion on measurable aspects of a BC-based reputation systems. Thus, the overview of concepts related to trust (cf. Sect. 2.1) is followed by BCs and their interplay with trust (cf. Sect. 2.2) leading to properties of reputation systems (cf. Sect. 2.3), further clarifying how the different types of DLTs (including BCs) do and can impact the establishment of trust. There, especially major challenges of DLT-based reputation systems are outlined.

2.1 Definitions of Trust

Various definitions of trust exist in literature. Trust is either defined as a (i) perception, (ii) a belief, or (iii) the probability that a third party exploits a vulnerability. Based on these different definitions and nuances, which often show a philosophical nature, it is possible to conceive trust models to for different types of BCs and DLT. Therefore, the focus here is on discussing and understanding the fundamental meaning of trust, which includes the semantic appreciation of key terms to understand their meaning for different types of actors involved (e.g., someone or something). Table 1 lists the most relevant and different definitions of trust, which are analyzed in sequence according to their keywords. These terms are presented firstly according to their meaning in the dictionary [51], and secondly, their similarities and differences are clarified by drawing parallels with BCs and DLTs:

- **Belief and Expectation**: While "belief" is an acceptance that something exists or is true, especially one without proof, "expectation" is a strong belief that something will happen or be the case.
- **Reliance and Assurance**: Whereas "reliance" is related to a dependence on someone or something, "assurance" refers to a confidence in someone or something.
- **Subjective Probability and Accepted Vulnerability**: Whilest a "subjective probability" is related to someone's opinion about the probability of an event, "accepted vulnerability" is associated to a perception of risk that is accepted by someone.

In these three main subjective variations on the definition of trust, there is a common characteristic of the verification that a certain fact, opinion, or belief has a certain likelihood to happen within a confidence interval. Therefore, trust requires verifiability, non-repudiation, and integrity of certain actions that are found as fundamental features in blockchains and distributed ledger-based systems. While verifiability, corresponding to transparency and integrity, is provided as a main characteristic of BL and DC, non-repudiation may vary depending on the permissioned or permissionless model, in which accounts can be created unrestricted by their users. In this sense, repudiation may occur in permissionless blockchains due to the absence of a strong correlation between real identity and account, making it impossible to be accountable for actions taken.

Table 1 Definitions of "Trust" and their respective keywords

Author	Trust definition	Keyword
[20]	"A confidence in the integrity of an entity for **reliance** on that entity to fulfill specific responsibilities"	Reliance
[27]	"A particular level of the subjective **probability** with which an agent will perform a particular action"	Probability
[66]	"**Belief** that the other party will behave in a socially responsible manner, and, by so doing, will fulfill the trusting party's expectations without taking advantage of its vulnerabilities"	Belief
[50]	"An individual's **belief** in, and willingness to act on the basis of, the words, actions, and decisions of another"	Belief
[31]	"An assumed **reliance** on some person or thing. A confident dependence on the character, ability, strength, or truth of someone or something"	Reliance
[42]	"The property of a business relationship, such that **reliance** can be placed on the business partners and the business transactions developed with them"	Reliance
[51]	"Firm **belief** in the reliability, truth, or ability of someone or something"	Belief
[57]	"**Expectation** that partners in interaction will carry out their fiduciary obligations and responsibilities, that is, their duties in certain situations to place others' interests before their own"	Expectation
[4]	"Accepted **vulnerability** to another's possible but not expected lack of good will"	Vulnerability
[59]	"The willingness of a party to be vulnerable to the actions of another party based on the **expectation** that the other will perform a particular action important to the trustor, irrespective of the ability to monitor or control that other party"	Expectation

Belief and Expectation

Pavlou and Fygenson [66] defines trust in an e-commerce context as a "**belief** *that the other party will behave in a socially responsible manner, and, by so doing, will fulfill the trusting party's expectations without taking advantage of its vulnerabilities.*" Similarly, trust is defined by [50] as *"an individual's **belief** in, and willingness to act based on, the words, actions, and decisions of another.".* Besides, [50] identifies two bases for trust or distrust, whereas (1) is more common in professional relationships being related to assessments of costs and rewards for violating or sustaining trust, and (2) is based on the mutual identification of parties, where mutual understanding and affinity plays an important role. Drawing a parallel with BCs and DLTs, it is essential to note that while (1) is related to trust in cryptographic mechanisms that compose a BC, among other technical elements, (2) is related to the direct trust in which individuals know previous actions of other individuals, therefore, being able to trust a reputable third party—a more frequent characteristic in DLTs. Furthermore, the Oxford Dictionary [51] proposes a definition in the same sense, trust as a *"firm* **belief** *in the reliability, truth, or ability of someone or something."*

The keyword "**belief**" expresses an acceptance of someone or something as true (or the truth) in a third party, without any guarantee, that such a third party will behave as agreed. Reference [57] offers a similar definition, wherein trust is defined as *"a* **expectation** *that partners in interaction will carry out their fiduciary obligations and responsibilities, that is, their duties in certain situations to place others' interests before their own."* In this sense, the similarity between these two keywords is based on the definition of "expectation", which is based on the Oxford Dictionary [51] as a strong belief about the way something should happen or how someone should behave. Also, expectancy can be seen as a rigid clinging to a belief, i.e., a strong belief which, however, in a detailed semantic analysis [9], there is a difference in terms of proportions of magnitude, but not in a sense. For instance, the degree to which one believes a BC to be a trustworthy platform is not the same as to which one expects it to be. Thus, belief and expectancy are two vectors pointing to the same direction in different semantic magnitudes.

Reliance and Assurance

Other authors follow a slightly different definition, based on the keyword "reliance". For instance, [42] defines trust as *"the property of a business relationship, such that* **reliance** *can be placed on the business partners and the business transactions developed with them."* Similarly, [31] states that trust is an *"assumed* **reliance** *on some person or thing. A confident dependence on the character, ability, strength, or truth of someone or something."* While belief requires a perception without proof that something or someone fulfills a deal, reliance provides a notion of dependence or confidence based on prior knowledge that something or someone fulfills a deal.

Furthermore, [29] states *"to build a relationship of trust, there must be an* **assurance** *that the other player will act in a predictable manner,"* and uses the term "assurance" to define a condition in which trust can be established. In the sense used by the authors, reliance and assurance are used in terms of the level of confidence in which an agent can be trusted. However, [4, 39] agree that there must be a distinction between trust and reliance. For instance, [39] states that relying on something to happen is different from relying on someone to make something happen, i.e., one can rely on someone to do something, but trusting someone to do something is different. In order to illustrate this sentence, it is possible to rely upon the Bitcoin protocol [63] to create a new block every 14 s, but it is not possible to trust someone to generate a block in the same period. While the former involves an indirect trust model utilizing a deterministic protocol, i.e., something transparent and verifiable to everyone, the latter involves a direct trust in someone.

Subjective Probability and Accepted Vulnerability

Definitions exist as well, based on an acceptance or a perception of risk. Thus, it is possible to observe the probability of a vulnerability to be exploited or the probability of an agent to fulfill an agreement. Reference [27] defines trust as *"a particular level of the subjective* **probability** *with an agent will perform a particular action"*, i.e., a type of probability derived from someone's judgment or experience that someone or something (an agent) will act [41]. Vulnerability, however, is employed by [4] to define trust in terms of *"accepted* **vulnerability** *to another's possible, but not expected lack of goodwill."*, i.e., the probability with an agent will not perform a particular action by exploiting a weakness or lack of a countermeasure, as defined by NIST (National Institute of Standards and Technology) [85]. The difference between subjective probability and the (acceptance of) vulnerability is that these definitions assess a probability that an event will happen in opposite directions. While the former observes the likelihood of success, the latter analyses analyzes the likelihood of failure. These definitions, based on a probability, apply to one of the categories proposed by [50], the (1) the calculus-based trust, which relies on the assessments of costs and rewards violating (acceptance of risk) or sustaining trust (probability of success). However, the common aspect between these definitions is that they require prior knowledge of something or someone that is the object of trust.

Perception of Trust and its Interplay with Blockchain and Reputation

Between unknown parties, such as in an e-commerce case, trust is not straightforward to be established, since parties involved do not know past histories or the prospect of future interactions, and they are not subject to a network of informed individuals that can punish bad or reward good behavior [69]. This is especially the case for open, on-line collaborations, where parties involved are "strange to each other" or are potential adversaries such that their relation faces several uncertainties. While "stranger" is typically defined as a new peer or a user with no information available concerning their reputation, an adversary may not necessarily be new, but has the intention to disrupt another user or the application (e.g., an e-commerce's Web site).

While there are different (and valid) interpretations of defining trust, there is also a valid viewpoint [30, 83] concerning the need to focus on trust relationships as much as possible, therefore, minimizing uncertainties. This view opposes the common sense about the need for trust, considering that vulnerabilities mostly stem from trust relationships. It is exactly in this sense that BCs and DLs deliver a measurable contribution, since they eliminate or minimize the need for trust in Trusted Third Parties (TTP). In the case of DLs, the need for trust in a single trusted party (in the case of centralized databases) is transparently distributed among (pre-)selected partners. Similarly, BCs also show a logic (protocol) that is transparent to their members, but there exists no need to trust even a single selected partner at all, since participation in the consensus is free (i.e., permissionless).

It is also relevant to analyze the role that reputation and associated mechanisms for reputation management show with respect to trust composition. Definitions discussed as above converge in the sense that it takes repeated interactions over a period of time to create a positive or negative perception of trust. In this context, reputation management systems provide a mechanism by which perceptions of these interactions can be systematically evaluated. In addition, BCs play a fundamental role in providing transparency through the open and unrestricted recording of perceptions.

2.2   Blockchains as a Trust Enabler Platform

Trust is a concept involving human relationships, and it builds the foundation for decision-making in different contexts or communities [13]. Trust factors exert a fundamental relevance in the influence of decision-making processes, forming the understanding, establishment, and management of trust in these contexts. However, trust is a highly subjective concept, often relying on an individual's behavior within particular contexts, which are governed by a set of rules [27]. For individuals, the consistent ethical behavior among peers enables the creation and establishment of trust, whereas the subset of ethics may break out of the given set of rules, but still remain within the context itself.

With the creation of Bitcoin [63] BCs changed the perception and establishment of trust. Instead of relying on a direct trust relationship, in which a need to know the identity and reputation of entities exists, BCs follow a permissionless, or trustless,[1] and fully decentralized trust model, where it is not necessary to maintain knowledge of entities, but only their actions, which are immutably and transparently available on the platform [8]. Therefore, the paradigm shift from placing trust directly in an entity to placing trust in the platform is measurable, whose capabilities enable transparency and public verifiability. Therefore, individuals can verify, whether actions of another individual are in accordance with their morals even without revealing their identity.

For example, belief and expectation are similar in its meaning but different in magnitude, considering that they require an individual's perception of something or someone. However, it is observed in [39] that believing may not be enough to create trust in the context of human relationships (and not someone-something). Therefore, a public and permissionless BC whose consensus mechanism is the same for all its participants, in a transparent and verifiable manner, should not fit into the same trust model as a permissioned DLT. Whereas the latter requires that certain participating members or those responsible for the creation and propagation of the blocks be trusted, whilst the former trust must relay on the algorithm or Smart Contract (SC) defining the rules of their interaction (cf. Fig. 1). However, it needs to be stated that different types of BC (e.g., permissioned, permissonless) exist [73], in which each

---

[1] Trust "remains" to exist from BC participants in all underlying cryptographic means and mechanisms, since otherwise, manipulations would become possible and not detectable. Thus, "trustless" refers typically to participants themselves, who do not need to trust anyone else (cf. below).

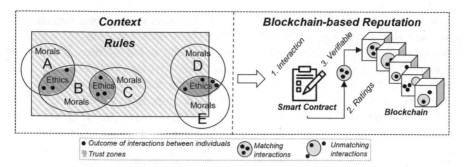

Fig. 1 Blockchains favor transparency over individuals' relations

type follows a very specific trust model, making it of utmost importance to evaluate these different flavors concerning the trust modeling.

Although a BC does not provide an utterly trustless solution, the inherent disintermediation contributes through its transparency to an increase of trust among those participants, typically labeled stakeholders, involved [74]. Information once persisted in a BC is immutable and transparent to participants, e.g., those of the cooperative defense [75]. For example, while transparency has a positive impact on trust, information leaks about attacks signaled could result in potential damages toof the public image of a domain. In addition, its participants are identified by an address provided by the use of public- key cryptography, ensuring the pseudonymity of participants. Therefore, the paradigm shift, in which trust as previously directly reliant on an entity is now transferred to the platform itself, whose capabilities allow for a visualization of actions of an account related to an unknown entity. The use of hashes and a Merkle Tree allows for the integrity of transactions (i.e., user actions) and blocks (i.e., data structure grouping transactions) to reach a verified integrity. Since its proposal and as of today, neither a bug nor vulnerabilities were disclosed in the Bitcoin code itself, the BC precursor. Therefore, it is observed that the platform's consistency over time is a decisive factor in the trustworthiness of its operation.

Such a verification of data integrity is a crucial step in reputation systems, because the system is based on a consistent view of the same data stored in different peers. This is possible through cryptography tools, such as public-key infrastructure, digital signatures, and hashes, that enable the creation of a unique identifier in time of the information present in a block, which includes the unique identifier of the previous block. Once data (i.e., transactions) are submitted to a BC, they are digitally signed by the issuer and, eventually, stored in a block. This block, containing other transactions, carries a hash that is calculated based on the hash of other information stored in the block. Thus, changing a single piece of information within the block would result in a different block hash, indicating that the block has been modified and, therefore, resulting in a new branch of the BC, i.e., a fork.

Trust is a fundamental aspect of any cooperative environment and, at the same time, difficult to obtain, since it may rely on many non-technical aspects [37]. Also, the process of building trust between entities has no relation to a specific technology. Several non-technical and specific aspects of each organization are required. *BC has a role of a "trust-enabler" in this context, providing transparency between cooperative organizations and, thus, possibly increasing trust- levels based in on their interactions* [37]. However, it is not possible to quantify the role of BCs as a trust enabler, being not possible to determine a "probability" in which the use of BCs is a determining factor in ensuring trust between organizations. The role of BCs in building trust has been studied by [35], in which the authors present solutions on how these conflicting notions may be solved. They explore the potential of BCs for dissolving the trust issue. According to [68, 86], the main characteristics of trust are defined as:

- **Dynamicity**: as it applies only in a given time period, and it may change as time goes by. For example, a history of security data sharing between two or more companies does not guarantee that these companies will always share data at any time. Trust can only be built during a time- frame.
- **Context Dependency**: the degree of trust on in different contexts is significantly different. For example, organization A may share threat indicators but may not disclose actual malware intelligence due to several factors (e.g., legal issues). Thus, trust may exist between organizations A and B only for sharing "threat indicators" context.
- **Non-transitivity**: if organization A trusts organization B, and organization B trusts organization C, then organization A may not trust organization C. However, A may trust any organization that organization B trusts in, given a certain context.
- **Asymmetry**: trust is a non-mutual reciprocal in nature. That means, if entity A trusts organization B, then the statement *entity B trusts entity A* is not always true.

Among various (non-technical) facets of trust, trust plays a crucial role in a cooperative platform. This has been demonstrated in e-commerce studies [47, 56], where online shoppers must necessarily rely on the functioning of the online store to make the purchase (i.e., , use of a credit card in a potentially unknown online store). These studies suggest to measure trust as the belief that a platform is honest, reliable, and competent.

Mapping these dimensions to BCs, a permissioned deployment model with a consensus necessarily open to the participation of members within the joint application context, e.g., a cooperative defense, meets these requirements. The capability to create an immutable, mutually agreed upon, and publicly (within that context) available record of transactions is seen as an enabler of trust in the platform [35]. In addition, the definition of rules between participants through SCs does allow for parties being involved to verify the execution of such SC code that defines the cooperation. It is important to note, however, that the algorithmic trust is not only limited to the correct functioning of the algorithm, but also includes a variety of sociotechnical factors, such as its formal and legal correctness, that goes beyond any technical solution.

2.3 Taxonomy of Reputation

Human interactions determine a complex process involving numerous aspects of human relationships, which altogether impacts "trust" between parties in a given context or community [13]. In turn, trust in its different interpretations exerts a fundamental relevance in the decision-making processes, and therefore the establishment and management of trust. In this sense, the management of reputation in a decentralized system adds a layer of complexity in to its design by taking into account not only different human behavior, but also technical flaws and vulnerabilities that can impact the trust perception of its users. Therefore, different aspects and characteristics that make up a reputation system as well as possible threats (i.e., flaws and vulnerabilities) were explored in the state -of- the -art [3, 58, 69]. Thus, major reputation properties and threats need a brief discussion.

Reputation Properties

Reputation systems have been deployed in different business areas and application contexts. A multitude of use cases for reputation and incentive mechanisms exists, ranging from crowdsourcing and sensing [17, 90], MANETs (Mobile and Ad- hoc Networks) [18, 46], Border Gateway Protocol (BGP) routing [22], e-commerce [76], file-sharing [87], mobile data plan sharing [55], prediction markets [67], various P2P and BC networks, to cooperative DDoS (Distributed Denial-of-Service) defense [71, 72, 75]. General properties of reputation systems are identified as follows [22]:

- **Foundation of Trust**: Repeated interactions and a clear interaction history build a reliable foundation of trust.
- **Self-policing nature**: The system should be self-policing, with the social norms defined by the users and not by a central authority [45].
- **Carrots and Sticks**: With incentives and penalties, users are more likely to behave according to the social norm.
- **Robustness**: The system should be robust against gaming attempts in reputation systems [19]. It needs not be foolproof, but reasonably secure against collusion and Sybil attacks, bad-mouthing (bad ratings), ballot stuffingballot-stuffing and identity whitewashing through re-entry. Current literature examines rating fraud, as a type of information fraud. The goal of an attacker domain is to either increase the reputation of itself (ballot stuffingballot-stuffing) or to decrease the reputation of others (bad-mouthing). The attacker can achieve its goals using a variety of attack models. A constant attacker behaves consistently evil, whereas a disturbing agent camouflages its actions skillful.
- **Accurate and Verifiable Scoring Engine**: The scoring engine should provide accurate and verifiable metrics.
- **Anonymity and Privacy**: The personal user feedback should be collected anonymously or bound to a pseudonym, to ensure honest feedback and guarantee some degree of privacy [17]. In a decentralized, anonymous marketplace (DAM) [84] the

reputation system requires additional properties to ensure a fair and complete listing of items, payments, and reviews. Most importantly, such a marketplace ensures the unlinkability of reviews and payments (with associated customer information), without compromising the legitimacy of the review [84]. The fully anonymous reputation system developed for the DAM "Beaver" [84] makes use of advanced cryptographic techniques, such as ring-signatures and zero-knowledge proofs, to uphold stringent security and anonymity requirements. These techniques allow raters to stay anonymous by veiling the source of the reputation claim, but without compromising the validity of the rating.

Within the different business areas and application contexts, these properties define essential aspects of a reputation management system, but not necessarily of the underlying architecture (e.g., centralized, decentralized, or distributed). On one hand, while specific properties may disregard the underlying architecture (carrots and sticks, self-policing nature), others are impacted by the underlying architecture. On the other hand, aspects related to the foundation of trust, verifiable scoring engine, anonymity, and privacy are influenced by the type of the underlying system the reputation management system is implemented on. For example, the foundation of trust and the ease of checking the scoring engine can be given by the main characteristic of BC, which is transparency.

Ratings submitted by users are stored in a transparent and immutable way, just as the logic of the reputation management system can be seen in the SC. In this sense, the use of BCs as the underlying system for reputation management positively impacts these properties. However, such transparency features may harm anonymity and privacy, depending on how the system manages identities in the SC. In general, BCs utilize pseudo-anonymous identities in which users are identified by accounts, which may be explicitly authorized in the SC reputation management logic. In this way, even if it is impossible to connect a hash-identified account with a real user identity, it is possible to control access and interactions of this account in the reputation system. Thus, it is possible to prevent abuse (specified in Sect. 2.3), but it is not possible to prevent users with a negative reputation from creating new accounts in the case of a public BC. However, in case of DLs, where participants show known identities, the so-called whitewashing behavior can be prevented.

Node and Network Properties

A reputation management system must take into account the characteristics of the network in which it will operate, and also those requirements desired by its users. These aspects are defined as follows [58]:

- **Node Churn** defines the rate in which nodes (peers) enters and leaves the network. Greater node churn implies greater complexity in the P2P network for content redistribution and message routing.
- **Reliability** concerns the node's availability and the data replication model used in the P2P system. It is also affected by the node churn's rate.

- **Scalability** defines the ability of the reputation management system to maintain, without deficit in service quality, the properties as the number of users increases.
- **Privacy** relates to the protection of content from unauthorized access. A common solution is to encrypt data before storing or transmitting.
- **Node Integrity** describes the ability or measures of a node to ensure that a reputation assessment is not inadvertently altered.
- **Anonymity** concerns the guarantees of user's privacy by ensuring a certain level of anonymity. It may vary from pseudonyms bounded to identities or peers, or based in real identities.

These aspects show different characteristics, considering the type of infrastructure that the reputation management system is built upon. A major aspect is the premise of the existence of a TTP, which allows for using a centralized reputation system (e.g., in case of e-commerce that centrally manages the ranking of stores and users). Furthermore, the possibility to rely on a consortium of entities managing a distributed reputation system has to be mentioned. From an organizational point of view, trust is still delegated to an entity (e.g., constituted by the consortium in a DLT), in which technical aspects related to node churn, reliability, and CIA (Confidentiality, Integrity, and Availability) must be observed to guarantee homogeneity among nodes managed by different entities of the consortium. In this sense, the consortium must present uniformity concerning guarantees (i.e., minimum technical requirements) of the operation of nodes that compose the reputation system. The observance of these aspects gains more relevance in case of a P2P network (e.g., BC), where the reputation system is implemented on nodes that freely participate in the application. In this sense, the higher the node churn, the greater the indication that the network has a high node retention and likely a larger reliability.

While reputation systems built on permissionless BCs rely on the number of nodes that compose the chain and its respective node churn, aspects related to a BCs availability are considered excellent due to the full replication of all data between nodes. In contrast, it is possible to consider a permissioned BC in which certain super-nodes have control over data replication to create partitioning. Such scenario shows advantages in terms of network performance (e.g., less communication overhead and data traversing the network), but disadvantages in terms of data availability once data is partitioned among these super-nodes. For instance, aspects related to the scalability of BC-based reputation management systems were discussed in [5]. The authors also point out the differentiation between DLT types in that permissioned DLTs achieve high throughput and scalability compared to permissionless (i.e., BC).

The transparency of transactions submitted (e.g., a reputation assessment) is a generally positive aspect for reputation systems. Once ratings are submitted for a particular system or service, the entire rating record is available to users. However, in case of a public BC, there is no identity control (i.e., users can create new identities) and it is also not possible to trace an account to a real identity. There is a need, in this sense, for BC-based applications to implement identity control at the application layer to avoid such threats as discussed in the following.

Reputation Threats

Threats derive from different natures, such as from flaws and vulnerabilities in the infrastructure (e.g., BC or DLT) or the reputation management system software itself. Nevertheless, different types of users are considered: adversary, identified, and unknown. This extended definition based on [58] considers known users as well as malicious and unknown ones. Known users are previously identified (even if by aliases) within the management system and have a reputation history (negative or positive) to allow those users to transmit feedback. An opposite and less effective approach is to allow unknown or unidentified users to allow reputation ratings, which have not interacted (yet) with other peers, and therefore no trust information is available. For example, a reputation system based on an SC allows unknown accounts (i.e., not previously registered) to submit reputation ratings. Finally, the users considered adversaries can be either known or unknown, but have in common their malicious actions to exploit vulnerabilities in the reputation system or manipulate reputations to benefit themselves or others. Possible actions of malicious users are listed in [58]:

- **Traitors or Malicious Feedback** are peers that behave properly for a period of time to build up a strongly positive reputation, then begin defecting.
- **Front-peers** refer to malicious peers colluding in order to increase the reputation of itself (ballot stuffingballot-stuffing) or to decrease the reputation of others (bad-mouthing) [69].
- **Collusion** refers to two or more malicious nodes conspiring to give each other high local reputation values and (or) give other nodes low local reputation values in order to gain high global reputation [81].
- **Whitewashers** exists, when a peer with poor reputation/trust changes its identity to start afresh and escapes from the consequences of its bad actions accumulated under its previous identity.
- **Ballot-stuffing** refers to peers, which may provide feedback for interactions that never took place. Such feedback can be either originated from a newly created account or an existing account that had no interaction with the other party.
- **Free-riding** denotes a characteristic of users, who use a certain service without contributing. In case of reputation systems, without evaluating services or products of others, such as the case of P2P sharing networks, it means that users may download media without uploading files.
- **Denial-of-Service** in a BC-based reputation system mostly involves application layer attacks that prevent the application to work. In this sense, vulnerabilities can occur both in the front-end logic provided to users and in the SC implementing main functions of the reputation system.

The category of DoS attacks has a stronger connection to the underlying infrastructure on which the reputation system is deployed, with the possibility of exploiting vulnerabilities rendering the application unavailable for a period of time [24]. In case of a BC-based reputation system, this is related to possible vulnerabilities in the SC code. The remaining threats are related to malicious user behavior, which requires relatively more complex modeling to deploy detection and mitigation actions. For instance, [81] proposes collusion detection methods to thwart collusion behavior-based clustering algorithms (e.g., K-means) to detect orchestrated reputation ratings. In this direction of research, there is work such as [26, 70], which proposes different models for the detection of reputation system manipulations (e.g., malicious feedbacks, front-peers, or white washers) requiring an extended analysis of user evaluations in order to establish a pattern of behavior that can be compared with what is considered "honest" behavior. In this sense, the use of BCs and DLs favor the mapping of user behavior (accounts), because evaluations are transparent to users. On the contrary, in public BCs, where the application developed in the SC does not control users by sending evaluations, it is difficult to control or avoid manipulations based on actions of white-washers and ballot stuffingballot-stuffing.

2.4 Discussion of Trust and Reputation in Blockchains and Distributed Ledgers

Whilest trust is an abstract concept varying according to the individual perception of a collaboration of individuals, reputation brings quantitative objectivity to this notion, utilizing different mechanisms and metrics to evaluate the perception of trust. BCs developed in this sense fundamental importance as a platform to provide essential properties (cf. Sect. 2.3) that allow these individuals to transparently verify the logic involving metrics and mechanisms being implemented (e.g., based on SCs). However, it is essential to remember again that there are not only different types of reputation systems, but also different types of BCs, i.e., DLTs, which represent the platforms on which these reputation systems can be implemented as an application. Thus, by varying the platform's underlying characteristics, also different characteristics of reputation systems are observed as summarized within Table 2.

Table 2 clarifies the key differences between the use of BCs and DLs as a platform for implementing reputation systems. Although most characteristics and threats are influenced at the application level, i.e., depending on how the logic of the reputation system is implemented in the SC, it is possible to observe fundamental differences between BCs and DLs that mainly impact the threats. For example, the permissionless character of BCs makes it easy for members to create new identities, a determining factor in not encouraging good behavior. Thus, the logic implemented in the SC must be restrictive to control, which accounts can interact in the SC to avoid fraud, such as whitewashing. Nonetheless, an issue that affects both BCs and DLs is the lack of confidentiality as a side effect of their high transparency. Thus, information disclosed in the reputation system is available, being essential to note that members participate in the BC/DL consensus on an equal basis, preventing members from e.g., having the ability to censor certain transactions. For example, while transparency favors a trust-free platform, it is needed to strike a balance with the confidentiality requirements of each member in order to exchange information securely.

Table 2 Comparison of reputation aspects for Blockchains (BC) and Distributed Ledgers (DL)

Property/threat	Blockchain	Distributed ledger
Reputation system properties		
Foundation of Trust	Provide a transparent trail of records for any peer	Provide a transparent trail of records for selected peers within a context
Self-policing nature	Depends on the implementation of the reputation system	Depends on the implementation of the reputation system
Carrots and sticks	Typically provided on a smart contract feature. Contracts are typically more limited on BC than DL	Typically implemented in a smart contract feature
Robustness	High data availability, but poor control of identities	Low, once it is distributed to selected peers
Accurate and verifiable engine score	Fully transparent	Fully transparent
Anonymity and privacy	Peers are antonym but scoring is not confidential	Peers are identified and scoring is not confidential
Reputation threats		
Traitors or malicious feedback	Not easily identified	Easily to identify and punish such behavior
Front-peers	Likely due to its permissionless nature where any peer can interact	Less likely due to its permissioned nature where only selected peers interact
Collusion	Higher chance of collusion once identities are not controlled	Lower chance of collusion once identities are controlled
Whitewashers	Peers can create new accounts (identities) freely	Peers cannot create new accounts without permission
Ballot-stuffing	Peers may manipulate ballots off-chain	Peers may manipulate ballots off-chain
Free-riding	Peers are not incentivized	Peers are not incentivized
Denial-of-Service	Resistant to DoS due to the higher number of peers	Prone to DDoS due to the lower number of peers

3 Tools and Methods for Blockchain Reputation Tracking

Reputation and incentive-based mechanisms have been explored widely for decentralized systems. However, BCs introduce a novel way to decentralize trust creating a certification infrastructure that records traces of transactions, allowing users to look up and record histories of transaction outcomes. By relying on the public ledger to form reputation scores and reward contributors, reputation and rewards of users are tightly coupled to their actual network activity. BCs provide the underlying structure toward a financial reward scheme through cryptocurrency. This currency represents an incentive to behave reputable, if actions are rewarded accordingly. Financial rewards and reputation can be treated independently or in a combined manner.

This section highlights those tools and methods that stood out in the BC use of DLs, since reputation systems overviews are available as surveys [36, 38, 44] for a more comprehensive view on this specific state- of- the- art. These systems have already been proven useful for e-commerce Web-sites to incentivize peers to contribute with relevant information and establish fairness among peers. Similarly, situations where incentive mechanisms are needed to stimulate the good behavior, where incentive mechanisms are needed to stimulate the good behavior, appears, for example, in socio-economic aspects in network management involving end-users, internet service providers, and telecommunication operators [40]. Moreover, similar social dilemmas exist in other research areas, e.g., crowdsourcing [90]. Even in large P2P networks, peers maintain lasting business relations and transact repeatedly [87]. This increases the potential benefit of such type of systems in P2P-related domains. In Mobile Ad- hoc Networks (MANET), researchers also identified the same need, to provide incentives and credit-based mechanisms for cooperation among peers [18, 46]. Thus, incentives or reward mechanisms are required.

3.1   Reputation Tokens

Electronic tokens determine a common approach to represent reputation. Reference [54] proposes an identity and reputation management system on Ethereum. A reputation coin termed *RpCoin* can be earned by completing tasks, which is like reputation itself, a non-transferable coin. The authors define two different types of tasks:

1. The reputation task contains a reputation claim that is beneficial to the publisher. Each task participant is allowed to cast a binary vote and votes corresponding to the final result of the task will increase the reputation of the voter. The reputation of the task publisher is only increased, if the result of the voting is positive.
2. The incentive task is used to discipline peers in a peer-to-peer environment such as a decentralized market [34]. Without these tasks, the system is vulnerable to ballot stuffingballot-stuffing and speculation of voting peers about final results. Alternatively, in a centralized system, there is the need to trust a central authority which may determine and enforce the rules of the market. Therefore, these incentive tasks contain a negative claim about another peer in the system. Voting rules are similar to the reputation task, but voters are rewarded for detecting fraudulent peers.

Work related to reputation and reward isare omnipresent in academia, citations, and degrees. This can be interpreted as indicators for its relevance in a myriad of decentralized applications. Reference [80] analyzed the potential of BCs for educational record management. In the educational economy, credits and degrees are stored in the BC [32]. This economy is fueled by Kudos, an educational reputation currency. In comparison to *RpCoins*, Kudos are envisioned to be traded according to rules enforced by SCs. Projects to implement such systems have already been undertaken in the Ethereum community. Furthermore, from an academic point of view, the Work.nation project is designing a similar infrastructure for a decentralized portfolio management and skill attestation for working professionals [88], being built on Ethereum. Contributions of work during projects are verified by team members, and the transparent system allows for a rapid team building based on attested skills.

As of another example targeting the decentralized prediction market, Augur is used to weight the reports about real worlreal-world events [67]. Much of the theory behind Augur was developed in the work on Truthcoin [64]. A prediction market in Truthcoin is called "oracle corporation" and consists of a customer and an employee layer with different currencies. Reputation is represented by VoteCoins (VTC) and is the currency transacted on at the employee level. Another coin CashCoin on the customer level is solely used to buy/sell shares of an expected outcome. The purpose of prediction markets is to forecast outcomes of real worlreal-world events based on share prices. The market price of these shares reflect the expected probability of the outcome. VTCs are tradable between the employees of the oracle corporation, but the total supply of reputation tokens in the system is fixed. VTCs are gained when a report about a real-world event is consistent with the consensus, and lost otherwise. Only reputation owners can create reports about such events. VTCs are also withdrawn, if the user does no longer participate in event-reporting. In such a case, the voter can maximize its utility by selling the VTCs, because they are liability as much as asset.

3.2 Event Reputation Factors

In the trust module proposed by [61] BC payloads trigger particular events. The module is applied to establish trust in decentralized sensor networks. Trust is computed by estimating the nodes' reputation over time (cf. Eq. 1):

$$\forall event(N, Blk(t)) : C_{n,t} = \sum C_e vt, \tag{1}$$

where t_{first} represents the first block in the BC after the node is authenticated. C_t defines the global coefficient for events involving the node at time t (i.e., the sum of C_{event} at time t). Every event is associated with a positive or a negative reputation factor, depending on the nature of the event from the originating transaction. The cumulative reputation score can be calculated by stepping through past events of blocks and process the associated reputation factors (cf. Eq. 2) (Fig. 2).

$$Reputation(N, t_{now}) = C_{auth} + \sum_{t=t_{now}}^{t=t_{first}} C_n \times \epsilon^{\frac{-t_{now}-t}{256}}. \tag{2}$$

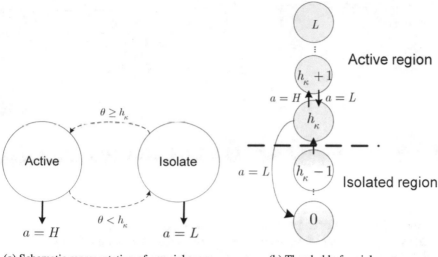

(a) Schematic representation of a social norm   (b) Threshold of social norm

Fig. 2 **a** State-machine **b** Threshold of social normal state changes, based on [90]

Reference [61] uses an exponentially decaying function to present recent events strongly (reputation decay). With this approach, the required minimum reputation of an agent to execute a specific action can be calculated based on reputation factors. These minimum requirements can be designed trust defaulting for small sized networks and scale well with the network size. A practical benefit of events for reputation aggregation is that common BCs and SC languages have built in support for event management (e.g., events in Ethereum or Hyperledger Fabric chaincode event listeners).

3.3 Reputation Thresholds

In mobile crowd-sensing, workers' smartphone sensors are used to aggregate knowledge. The reputation can indicate reported measurement data quality and help to validate data [17]. Reference [90] proposes a threshold-based incentive protocol, which is designed to be robust against "false-reporting" and "free-riding". The incentive protocol works like a state machine (cf. Fig. 3).

Assuming that a newly activated user starts with a reputation θ at the social norm threshold h_k, at the end of each time period the requester evaluates the action (a) of that user. If the service requester is satisfied with the action, the evaluation will result in $a = H$ and the user is given one additional unit of reputation ($\theta = h_k + 1$). If the task is not solved at the end of the time period, the service requester rates $a = L$ and one unit of reputation is deducted from the user. An ex-ante payment ca

Fig. 3 MAD transaction, based on [49]

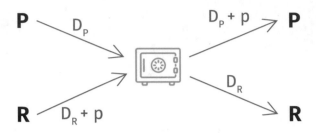

be assumed and, thus, a truth-telling requester. As long as the user's reputation (θ) is in the active region above the social norm threshold ($\theta >= h_k$), the user is allowed to work and actions will be evaluated by the requester. In this active region, the worker needs to be in compliance with the social norm. Still, whenever $\theta = h_k$ and $a = L$, the user will be forbidden to solve tasks and its reputation will be reset. An isolated user receives one unit of reputation each time period and automatically builds up reputation over h_k time periods. The user is reactivated, and its actions are evaluated. The user's reputation is again adjusted by the requester from this point onward.

3.4 Multi Signature Transaction (Multi-Sig)

The social dilemma still exists without the assumption of an ex-ante payment. On one hand, the requester of the service may not confirm the job and refuses to pay the requested (free-riding). On the other hand, the requested would rather pretend to show effort in order to minimize costs (false-reporting). Contractual agreements to ensure the order of payments and voting rights can be seen as a solution to mitigate this dilemma. However, malicious requesters might still rate untruly.

An SC protected by a multi-sig schema presents another solution to mitigate the same problem. Considering that a service provider (P) and requester (R) both pay a deposit (D_R and D_P). Additionally, the requester has to pay the service price to the service provider. Then funds D_R and D_P are protected with this multi-sig schema, which means that they can only be spent, if all parties agree. Reference [49] calls this concept "Mutual Assured Destruction (MAD)" transaction and uses it to build a decentralized file storage with financial incentives. As both parties store a safety deposit in the contract, they both have the incentive to resolve the transaction and retrieve the locked funds. As depicted in Fig. 3, the price (p) stays locked inside the contract until user (U) and provider (P) agree to payout. Again, under the assumption of rational agents who want to retrieve the deposit, this contractual agreement can reduce counter-party risks. After the successful termination of the contract, the service provider receives the price (p) for the service and both parties retrieve their deposits.

3.5 *Optimistic Fair Exchange*

The order of service delivery, reward payment, and reputation change needs to be clearly defined. Otherwise, misbehaving peers could delay delivery or reward payments. Reference [53] proposes a BC-based payment-for-receipt protocol to design a fair exchange between producer and consumer, which requires timeliness and fairness. Timeliness means that either party should be able to abort the protocol at any time (strong-timeliness). If that happens, a fair protocol requires the exchange of service and reputation/reward to be completed, or both parties return to the same state before the service request (strong-fairness). Deadlines and predefined timeouts can serve as means to achieve timeliness. However, configuring a short timeout (e.g., a few milliseconds) can render the exchange impossible, thus, ineffective. Also, the BC needs time to include transactions in the next block, which can delay the protocol and introduce time lags. Furthermore, the strong-timeliness is not guaranteed, because the parties may wait until the deadlines to proceed in the exchange.

The weaknesses of timeouts and deadlines in fair exchanges can be mitigated with an Optimistic Fair Exchange (OFE) protocol [53]. This protocol usually depends on a TTP to mediate between the two parties. This TTP knows the key to items i_A and i_B of both parties, Alice (A) and Bob (B), participating in the fair exchange. Such a basic scenario is presented in [53] and illustrated in Fig. 4.

In this OFE, first, Alice sends Bob a verifiable encryption of her item (c_A) and her expectations about Bob's item (e_A). Bob verifies c_A in the second step. In the third step, both of them are free to abort the protocol. If Bob sends c_B to Alice in the fourth step, she can verify the item i_B in the fifth step. Afterwards, both parties are free to abort or resolve the protocol. It is essential to notice that the TTP knows

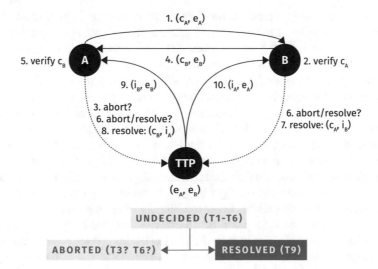

Fig. 4 Optimistic fair exchange with a TTP, based on [53]

about the items and has to maintain the state. Once one party decides to resolve or abort, the other party cannot choose otherwise. In the scenario depicted in Fig. 4, Bob can resolve in step seven and sends his item i_B to the TPP. The TPP decrypts c_A and safely stores both items as long as not both parties are resolved. When Alice resolves and sends i_A in the eighth step, the TPP can verify and compare both items with these expectations and hand out both items to the respective new owners.

The BC allows for an OFE without a TTP. The exchange can be automated with an SC similar to the MAD transaction (cf. Fig. 3) with the difference that the provider (P) also needs to include the message expected by the user (U). This message can be a sign of a past transaction, namely a receipt. As the signed receipt, which confirms service delivery, needs to be fully confirmed on the BC first, the provider may experience a slight delay before receiving the automated payout from the SC.

3.6 Anonymous Feedback

Some applications require the reputation system to be privacy-preserving [76]. In e-commerce or anonymous marketplaces [84], it might pose an issue that the producer knows which consumer gave a particular lousy rating. Also, anonymous feedback mechanisms allow for honest feedback and remove the bias toward positive ratings [44]. An anonymous reputation system is not a contradiction, since reputation can also be bound to temporary pseudonyms [17]. Reference [76] designed a BC-based trustless reputation system that preserves the privacy and anonymity of the party giving the rating. This is achieved with blind signatures on key pairs created for each transaction. In the end, the service provider does not know from whom he received the rating.

First, the consumer (c) prepares a public/private key pair for the transaction. Before the transaction happens, the consumer requires a blind signature on the public key for that transaction from the service client (s). The service client cannot relate the public key to either transaction or consumer, since only the client knows the secret and random parts of the blinded key. This random number allows the client to verify and unblind the token in combination with the public key of the service client. After a successful service delivery, the consumer is advised to wait until other consumers require service. Otherwise, single feedback could still be linked to a single consumer. In the end, the consumer publishes on the BC a rating (r) about the service client (s). This rating is valid, since the blinded token was issued and made public by the service client beforehand, and there is only one rating per token. The token from the producer gives the consumer the right to give feedback to a transaction. However, neither producer nor other clients know the identity of the consumer.

Limited token issuance for s can be an effective method to fight against ballot stuffingballot-stuffing [76], since according to [11], "this creates a trade-off between rating and profit for the seller" s. For every token, s can either choose to use it for a real transaction or pay the opportunity cost and use it to inflate its reputation [76]. The opportunity cost equals the unrealized profit from a legit sale of goods or services. Even though a limited supply of tokens can incentivize s to use these tokens for genuine purposes, s can never be discouraged completely from "buying" its reputation [76].

3.7 Insurance Models

Trust can be established based on reputation information [44]. A trust model helps the user to take decisions in concrete situations [52]. A reputation model however is holistic, since it includes the view of peers in the system about a given user [44].

Some trust models work in a similar fashion to insurance models. To better understand these models, the example of a supply chain with upstream suppliers and downstream customers is helpful. Peers trust each other up to some amount of money. In TrustDavis for example, peers can obtain weighted references from their neighbors [16]. If a user gives a reference, she would become liable to pay the reference price to any customer, if the supplier she gave the reference to cannot deliver the product or service. These references can be modeled as edges in a trust graph. In a trust graph, each edge represents the maximum amount of money, that the originator is liable for or trusts the target of that reference with. A similar approach has been presented recently by [52]. Trust and reputation are not directly linked to rewards, but can be expressed in monetary terms using such schemes.

Reference [52] derive indirect trust by relying on the peers that are trusted by your trusted peers (i.e., transitive trust, whereas if a trusts b and b trusts c, then a trusts c). If an intermediary in the trust chain defects, she is free to either take a loss or steal the amount from a friend in the trust chain. Since this game can be played transitively, the financial loss can be carried over to other trusted peers. The maximum trust (and equally potential maximal loss) of two peers is limited by the maximum flow in the trust graph between them.

3.8 Reputation Engines

Besides reputation aggregation protocols, another important aspect of a reputation system is the reputation computation engine, where scores and metrics are calculated based on the inputs [44]. Most reputation systems in the P2P space are single-dimensional, meaning that only one factor (e.g., number of contributions) serves as input for the system [19]. For the output metric, a broad range of different models exist. Surveys and classification about the different types of reputation engines are presented in [44] and [79]. The overview here is mainly based on these two approaches, where the emphasis on engines that are considered suitable for the design of reputation systems in a cooperative DDoS defense, thus, especially serving the use case as of Sect. 4.

- **Probabilistic Engines**: The reputation score is mostly accumulated linearly, exponentially or calculated with help of probabilistic density functions (e.g., the Beta reputation system) [3, 7].
- **Fuzzy Engines**: A peer's behavior can be analyzed through fuzzy queries on data stores holding this reputation, storing multidimensional reputation data [44].

In probabilistic engines, reputation is expressed as a probability. The expected value of the beta distribution forecasts the probability of a positive, future event x, based on the past binary events x and $\bar{x}$. Binary events represent positive and negative historical reputation ratings. The expected value of this distribution can be interpreted as a reputation score [3]. One particular example of a Beta reputation system applied to BC is found in the Topl protocol [14]. Topl is a proposed BC protocol to create profit sharing agreements with producers in emerging and frontier markets. The protocols' reputation engine "Divine" builds upon a Beta reputation engine that facilitates due diligence and reduces counter-party risk [48].

The peers' behavior can be analyzed through fuzzy queries on data stores holding this reputation data. A reputation score is stored in tuple-form. $rep = (a, b, i, d, v)$ is the rating from a about b, in interaction or transaction i, on dimension or skill d, with v being the actual rating value (e.g., in range $[-1, 1]$). With a fuzzy query $q = (a, \_, \_, quality, \_)$ an agent can retrieve and aggregate ratings concerning the $quality$ dimension from peer a, to any other peer ("_" is the wildcard) in any interaction and with any score value. Insights on the different interaction dimensions like quality, price and service time can be gained by executing and aggregating the results of such queries on the data store.

Beta Distribution—Binary Events

The Beta distribution is defined by [3] to provide a model of continuous random variables, whose range is between 0 and 1, e.g., a binomial distribution (a series of successes and failures), such as positive or negative scoring of reputations. In these cases, the most suitable approach to observe the prior binary rankings over a variable is to use the Beta distribution. The Probability Density Function (PDF) of a beta distribution is given by:

$$f(x) = \frac{(x - \alpha)^{p-1}(\beta - x)^{q-1}}{B(p, q)(b - \alpha)^{p+q-1}}, \tag{3}$$

where p and q are shape parameters (i.e., numerical parameter of a parametric probability distribution), α and β are defined as lower and upper bounds determined as in the range $\alpha < y < \beta$; $p, q > 0$. Hence, the distribution and $B(\alpha, \beta)$ is the Beta function defined as follows:

$$B(a, b) = \frac{\Gamma(a)\Gamma(b)}{\Gamma(a + b)}, \tag{4}$$

where $\Gamma(a)$ is the Gamma function defined as:

$$\Gamma(\alpha) = \int_{t=0}^{\infty} t^{\alpha-1} e^{-t} dt. \tag{5}$$

Finally, the expectation of the Beta distribution can be determined as $\frac{\alpha}{\alpha+\beta}$, where parameters α and β are derived from the method of moments, that is, by setting the above-mentioned mean and variance equal to the sample mean $\hat{x}$ and sample variance s^2 and solving them for α and β. The respective solutions in this case are:

$$\hat{a} = \hat{x} \left[\frac{\hat{x}(1-\hat{x})}{s^2-1} \right], \hat{b} = (1-\hat{x}) \left[\frac{\hat{x}(1-\hat{x})}{s^2-1} \right]. \tag{6}$$

As an example, a process with two possible outcomes that has produced an outcome 5 seven times and an outcome 5 only once, will have a Beta function expressed as plotted in Fig. 5 [3]. This curve expresses the uncertain probability that the process will produce outcome 5 during future observations. The expectation is given by $E(p) = 0.8$, which is interpreted as the relative frequency of outcome x in the future, where the most likely value is 0.8.

Beta reputation is defined considering a parameter p in range of [0, 1], where 0.5 is the initial, neutral score of every new customer. In the acceptance mode, it would be possible to assume a threshold, where no customer is willing to work with a counterparty that has a Beta reputation lower than 0.3. The Beta reputation for a customer c was calculated as the expected value $E(p) = \frac{\alpha}{\alpha+\beta}$, with $\alpha = positive(c) + 1$ and $\beta = negative(c) + 1$ being the accumulated positive and negative reputation values at one point in time [3]. This value reflects the probability of a future positive interaction with customer c, based on its past positive and negative ratings. Likewise to

Fig. 5 Beta function of event x after 7 observations of x and 1 observation of $\hat{x}$, based on [3]

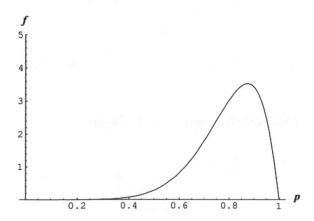

the raw metrics (i.e., positive, negative, or neutral reputation), the Beta reputation is averaged here by time and customer strategy.

Dirichlet Distribution—Multiple Events

While Beta distributions are ideal to map binary events, more complex reputation systems that require different degrees of positive or negative rankings are not handled (e.g., bad, average, excellent). According to [43], even if it would still be possible to express a wide range of ratings by splitting a binary range into partially positive or negative ratings, the mathematical approach would remain not straightforward. Henceforth, [43] proposes the Dirichlet reputation system to enable a multivariate probability distribution. The Dirichlet model requires $k \geq 2$ variables $x_1, \ldots, x_k$, such that each variable $x_i \in (0, 1)$ and $\sum_N i01 X_i = 1$, i.e., a parameterized vector of positive-valued parameters $\alpha = (\alpha_1, \ldots, \alpha_k)$. Henceforth, the Dirichlet distribution is a generalization of the Beta distribution that considers multiple dimensions instead of binary ratings. The PDF is defined as follows [43]:

$$f(\vec{p}|\vec{\alpha}) = \frac{\Gamma(\sum_{i=1}^{k} \alpha(\theta_i))}{\prod_{i=1}^{k} \Gamma\alpha(\theta_i))} \prod_{i=1}^{k} p(\theta_i)^{\alpha(\theta_i)-1}, \tag{7}$$

where $p(\theta_1), \ldots, p(\theta_k) \geq 0$, $\sum_{i=1}^{k} \alpha(\theta_i) = 1$, and $\alpha(\theta_1), \ldots, \alpha(\theta_k) \geq 0$. In turn, the expectation of probability of any of the k random variables can be defined as:

$$E(p(\theta_i)|\vec{\alpha} = \frac{\alpha(\theta_i)}{\sum_{i=1}^{k} \alpha(\theta_0))}. \tag{8}$$

The Dirichlet distribution provides a flexible basis to build reputation systems extending the Binomial distribution defined in the Beta distribution. Thus, reputation scores can be represented as point estimates or multinomial probabilities that represent a more fine-grained measure of grades in between binary positive or negative scores. Although the mathematical formulation seems more complex than for the Beta distribution, its computation is straightforward, once it computes probabilities instead of a distribution being simply fetched in a search query.

3.9 Reaction and Service Differentiation

Reputation can be used to trigger a certain reaction in the system or to differentiate services. These mechanisms should be applied with caution, since rewarding or penalizing users in the system based on their reputation value could lead to negative feedback loops [21]. E.g., if job applicants are selected according to their credit rating, this can lead to a vicious cycle, where applicants never find a job. But if job applicants are selected based on work performance scores, such a reliable reputation signal is based on an objective assessment of the applicants' work history. INDAPSON [55] and PaySense [17] are incentive systems that define objective reputation metrics, which allow to couple reputation and rewards. INDAPSON prices depend on reputation, whereas PaySense rewards are interpreted as reputation.

INDAPSON is an incentive data plan sharing system based on a self-organizing network [55]. The incentive mechanism in this system is the Reputation Adaptive Pricing (RAP), defined in four stages. In the first stage, it is possible to retrieve the price P_1 considering a flat unit price P_o, where F represents the target data $P_1 = F.P_o$. In the second stage, a metric is defined to present the reputation of the price unit in INDAPSON $M = \frac{D}{R}$, where D, R denotes the user's total amount of data downloaded, such that the server can calculate the total adapted price P_a to pay in the task $P_a = (1 + \gamma)^M.P_1$. γ represents the punishment factor as a positive constant defined on the reputation management server. If a sufficient amount of credits is reached in stage three, a client decides whether the price offered is acceptable and will proceed with the download in stage four, or stops the download. In stage four, the revised pricing is calculated at the management server for each $Client_i$ ($i = 1, ..., N$):

$$P_i = \frac{P_o}{1 + A.E_i}, \; E_i \in [0, 1],$$

where E represents the percentage of $Client_i$ at the start of the download and A defines a positive constant. The credit given to the client is defined as:

$$P_{C_i} = \frac{p_i f_i}{\sum_{j=1}^{N} p_j f_f} . P_A$$

This pricing was developed to enable fair sharing of mobile data plans. The reputation of a user is the ratio of downloaded and relayed (or shared) data. The price depends on the remaining energy level and on past reputation of contributing mobile devices, which share their data plan with other users in the network. This credit system makes it possible to soften the rigid time and quantity constraints via mobile data plans.

In contrast, PaySense is an integrated reward and reputation system for mobile crowd sensing [17]. Integrated means that reputation and rewards are coupled. If a sensor earns a high reward, it also gains reputation and vice versa. In this case, the money flow (reward) in the transaction history determines the reputation of a sensor. This principle is also applied by the feedback system on top of Bitcoin [12], where the reputation increase is equal to the amount paid.

3.10 Graph and Flow Engines

A graph structure is a suitable format to capture BC transactions and associated reputation scores. Reputation metrics can be calculated by following a path or flow

Table 3 Comparison of flow engines

	Network flow [7, 10]	EigenTrust [45]	Proof-of-importance [1]
Goal	Find selfish peers or compute maximum reputation between two peers	Detect malicious peers	BC consensus
Input	Bitcoin transaction graph or reputation graph with reputation constraints	Local trust values	NEM transaction graph
Procedure	Net flow convergence rate or max flow	Aggregation of local trust values	NCDawareRank

in such graphs. This is why flow and graph based reputation engines are particularly interesting for incentive systems on BCs. Current reputation management systems use centrality measures to indicate the importance of nodes. Table 3 summarizes goal, input, and procedure of such flow algorithms.

In EigenTrust each user holds a local trust vector, which is the net total of satisfactory and unsatisfactory transactions [45]. By weighting local trust values with other values of trusted peers, the EigenTrust algorithm can assign each member in the system a trust value that converges to a global value after iterating through the peers (iterative trust). The global trust vector contains reputation scores of peers in the system and is the same for every peer in the system. The convergence property of the algorithm enables each peer to have the same, global view of the network, independent of their local trust vector used as an input for the algorithm.

The New Economy Movement (NEM) is a novel approach, DL that leverages a modified version of EigenTrust to identify malicious nodes and reduce unsuccessful transactions [1]. Additionally, the algorithm adjusts for node credibility. This "balance factor of trust", proposed by [89], is calculated based on the number of similar feedback ratings from two nodes. This approach protects the reputation system against attacks stemming from colluded local trust values, which are still subjective user data. The Proof-of-Importance (PoI) consensus algorithm in NEM determines, who forges the next block (also known as harvesting or mining). The importance of a peer is calculated with the NCDawareRank algorithm, which works similar to PageRank [2]. This algorithm requires the transaction graph as input. PoI is a metric that indicates how much a peer makes use of the network and with whom it interacts. Nodes that are active in the network are assigned a higher importance and have a higher probability of harvesting new blocks.

Reference [10] shows that an analysis of network flow in a transaction graph can produce reliable measures for reputation. In comparison to EigenTrust, the net flow algorithm does not depend on subjective user input (ratings), but is applied on the transaction graph produced by the Bitcoin transaction network. The goal of the algorithm is to find selfish peers which transact only among themselves.

The idea of the net flow algorithm reads as follows. The Directed Acyclic Graph (DAG) of the Bitcoin transaction network has a net in- and out flow (i.e., net flow) of zero, because no Bitcoin can enter or leave the closed system. Consequently, there exists a positive or negative net flow in every subgraph of the DAG. The net flow convergence rate indicates, how fast the net flow converges to zero, when an incident edge is added to the edge set of the subgraph of interest. The assumption is that a fast convergence toward zero is an indication of selfish peers that only transact among themselves. While iteratively increasing the edge set of the subgraph, the money flows of colluding subgraphs will converge faster toward zero than it is the case for honest peers and subgraphs. This holds, because honest peers generally transact more with other peers, which results in a higher net flow. One benefit of using net flow to compute reputation scores is the lower computational complexity in comparison to other graph algorithms like Dijkstra's shortest path between two nodes [10].

The max flow algorithm calculates the maximum flow between two nodes in a graph network. When weighted edges between nodes are interpreted as reputation constraints, this algorithm can return the maximum reputation between two distant nodes (or peers) in the graph [7]. The interpretation of such a metric would be comparable to the interpretation of the maximum trust in an insurance model. Other engines average or blur reputation scores over time. Amazon builds an average over star ratings from consumers for a specific product [44]. Engines with multidimensional scores could also average across different dimensions. In blurred engines, recent ratings have relatively more weight. If only the latest reputation value is valid, the engine is called an "only-last" system. The reputation engine in [61] serves as an example for such a blurred system, because reputation decay is modeled with an exponentially decreasing function. In adaptive engines, the newly assigned score depends on the current reputation of the contributor [79]. E.g.,, the Spora system defines a damping function, which slows the reputation increase for reputable peers [28]. Thus, the most reputable peers converge toward an upper reputation limit.

4   Case Study of Blockchain-Based Reputation in a Cooperative Defense

A cooperative defense defines an alternative to cope with large-scale DDoS attacks, where the mitigation takes place at the egress point of the attack. Advantages over traditional/on-premise defenses have been widely recognized in the literature [75]. It allows, for example, to combine detection/mitigation capabilities of different domains to reduce the detection/mitigation overhead at a single point and to block malicious traffic near its source.

The use of BC as a platform allows not only for the full replication of attack information, but also for the creation of a market of DDoS mitigation services as a fundamental pillar to foster cooperation between service providers. A reputation scheme within such cooperative defense allows contributors and consumers of the

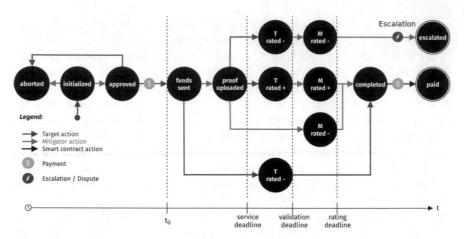

Fig. 6 BC-based cooperative defense workflow and reputation system

network to rate entities that request protection in a cooperative defense. Since reputation is earned in interactions between peers, it can be attached to transactions preventing arbitrarily manipulations or gaming attempts. The cooperative defense BloSS [75] is based on Ethereum and implemented as a dApp (Decentralized Application) providing REST interfaces for a network management system to interact with the cooperative system by requesting or offering mitigation services. An overview of BloSS protocol including its reputation mechanism is provided in Fig. 6.

Once a mitigation service is accepted, a deadline to upload an evidence of completion is started (t0). Data exchange is done off-chain exchanging the encrypted data (e.g., blacklisted addresses) via the Inter Planetary File System (IPFS) [6] ensuring confidentiality as well as integrity of the attack information based on a per-message signature bundled with the attack information. The Mitigator can act rationally and upload an evidence or miss the upload by expiring the validation deadline. A Target can rate the service of the Mitigator and based on this rating funds initially locked in the SC are released to the Mitigator. In case of no feedback (i.e., Target is selfish) a rational Mitigator is allowed to rate negatively.

Varying time-windows were chosen for deadlines, such that they might lead to situations, where customers could fail to meet a deadline. E.g., a mitigator uploading a proof with the minimum service deadline of three blocks might not be executed fast enough, because the block time of the BC increases faster due to other transactions being processed. The service, validation, and final rating/escalation deadlines are sampled randomly for each task. Service deadlines could be chosen in a range of [3, 13] blocks and validation deadlines in the range of [17, 27] blocks. Final rating deadlines, for instance, could be sampled in the range of [32, 42] blocks. Before the new mitigation contract is created, all peers are funded with 10 Ethers each, while the contract price was fixed to 1 Ether for every mitigation task.

4.1 Analysis of Reputation Properties

The properties of this reputation system were analyzed, considering both the platform on which the application was developed and details of the protocol implemented in the SC. BloSS is based on a Proof-of-Authority-based implementation of an Ethereum permission, in which all peers participating in the cooperative defense are authorities for one round within the protocol.

Transaction censorship can occur by any member, but can be verified by signing the transaction block. Furthermore, participation in submitting transactions and reading contents of BloSS is allowed and confidential to the participating members. Table 4 summarizes the analysis, which is followed by a discussion on each property.

- **Foundation of Trust** is provided by design in the BloSS protocol, considering repeated interactions with verifiable results lay the necessary foundation for establishing trust.
- **Self-policing Nature**: BC and SCs are reactive, i.e., once an action is performed by one customer M or T, the other customer must perform a subsequent action within the established period. Thus, the platform is not auto-policing, requiring peers to check the outcome of actions in the protocol periodically.
- **Carrots and Sticks** are provided in the protocol for both T and M after the mitigator performs the service requested by the target. Thus, depending on the evaluation of T and M, the incentive for the service is released by the protocol and

Table 4 Assessment of BloSS' reputation properties

Property	Short Description	Achieved
Foundation of Trust	Records are transparent to all members of BloSS, which are able to verify the outcome of each step in the protocol	●
Self-policing nature	Smart contracts are reactive by nature meaning that policing should be done on a peer-side and not on-chain	✗
Carrots and sticks	BloSS protocol provides the possibility to rate both the target T and mitigator M, which satisfies the incentives and punishments	●
Robustness	The protocol is partially vulnerable to actions held off-chain such as peer colluding to simulate an interaction to boost their own reputation. However, new identities cannot be created and malicious behavior can be identified due to the inherent transparency	◖
Accurate and verifiable engine score	All outcomes of all steps of the protocol are visible not only to the interacting peers but all members of BloSS	●
Anonymity and privacy	Partially satisfied once the chain is permissioned to a set of participants but transparent between those participants	✗

● = property; ◖ = property partially provided; ✗=property not provided

Table 5 Assessment of reputation frauds

Target	Threats	Short description	Achieved
System	Free-riding	Incentives are required to request mitigation services	●
	False-reporting	M is not incentivized to provide false-reports on T but the protocol allows such behavior, which can be tracked on future interactions	◑
Rating	Sybil and Collusion	Whitewashing (re-entry) of identities is not prevent in a permissionless deployment	◑
	Ballot Stuffing	Malicious M's and T's can collude to elevate their reputation	✗
	Bad-mouthing	Unfair ratings are not incentivized by design once a service is paid upfront by T and M only rates, when the service is completed	●

● = property ; ◑ = property partially provided; ✗=property not provided

M can have either a positive evaluation or an incentive for the mitigation service to be performed.

- **Robustness** depends on a combination of factors listed in the threats' analysis subsection. In this sense, BloSS partially meets these characteristics, since peers can collude off-chain to either boost their ratings (and fraud the reputation protocol) or harm a particular peer's reputation.
- The **Accurate and Verifiable Scoring Engine** provides by design this characteristic, since all peers' steps can be observed on-chain by all members of the cooperative defense, allowing no discrepancies in the calculation of scores.
- **Anonymity and Privacy** is a property not guaranteed by the platform design, since the permissioned setting only restricts the visibility of actions to a selected (trusted) set of peers. However, all peers have known identity and their actions are verifiable within the platform.

4.2 Analysis of Reputation Threats

By analyzing the design of the cooperative protocol, it is possible to make an analysis of different types of fraud possible. An evaluation with customers M and T with different profiles (e.g., honest, malicious, lazy, and others) was performed in previous work [75]. Table 5 summarizes this analysis and discussed as follows.

- **Free-riding** is an activity prevented by design in BloSS by requiring T's to deposit the incentive required by M's into the SC. Since the SC is designed as a state machine, it is not possible to circumvent this step, making the mitigation service start before funds are locked into the SC.
- **False-reporting** fraud can happen, when a malicious M assigns a false rate to an honest T at the end of the interaction. Although the protocol allows for this,

no rational incentive exists, since actions are recorded on the BC. Thus, future interactions of a malicious M can be tracked by all T's.

- **Sybil- and Collusion Attacks:** BloSS excludes the possibility where a customer can boost its reputation by creating mitigation SCs with itself. A possible deployment on a public ledger would enable actors, i.e., a M or T, to maintain multiple account pseudonyms on the BC and transacting between them to inflate reputation.
- **Ballot StuffingBallot-stuffing:** BloSS is not immune against ballot stuffingballot-stuffing. Besides transactions recorded on the BC, customers can agree on discounts and benefits over alternative communication channels. For instance, two malicious T and M would be able to rate each other positively independent of the mitigation outcome in rounds where both can perform the role of T and M.
- **Bad-mouthing:** A BC-based reputation system design impedes bad-mouthing, in which a T or M can only provide feedback for transactions completed. This elevates costs of bad-mouthing a competitor, since a transaction has to be committed for each fraudulent reputation statement.

Reputation and reward schemes integrated into the system prevent free-riding (attack targets) and false- reporting (mitigators). These mechanisms incentivize the rational behavior of operators in the long run. Selfish members are identified by looking at their past interactions on with the BC. Furthermore, the payment of rewards provides a highly suitable countermeasure to dis-incentivize selfish customers. Mitigators are incentivized to execute the final service rating step, since otherwise, they would deprive themselves of payments. Further, it prevents Sybil and collusion attacks by mapping customer accounts to real-world identities, preventing customers from creating several identities to manipulate reputation scores. However, ballot-stuffing and bad-mouthing are not prevented yet, but are clearly discouraged due to the cost to deploy a mitigation contract only to manipulate reputation scores.

5   Chapter Considerations

While theory about reputation and the respective reputation management systems are consolidated in the state- of- the- art, DLTs, among them BC, are relatively new and showroom for innovation. In this sense, aspects that are relatively easy to implement in a centralized reputation system (e.g., identity and time management or increased storage and performance) are more complex due to limitations encountered by SCs to develop complex logic. At this point, it is necessary to note limitations in terms of performance, such as the number of transactions per second and on-chain storage capacity (and possibly tools for off-chain storage), and the paradigm shift related to the development of SCs. The following considerations are observed explicitly based on the respective overview provided in this chapter:

- **There is no technical approach to guarantee the existence of trust only by its use**, which also applies to BC- and DL-based approaches. Promoting trust relies on the outcome of actions between individuals, which need to occur according to the expectations of those being involved and are external to any technical approach.

- **BCs and DLTs play a major role in increasing levels of an existing trust.** Based on transparency and non-repudiation characteristics, BCs and DLTs allow to quickly verify the outcome of these actions, creating a favorable environment for trust establishment. Henceforth, they play a major role in boosting trust levels in settings where an initial level of trust exists.

- **BCs and DLTs become especially useful in case of the need to dis-intermediate the control** of the reputation system, shifting from a paradigm based on a trusted third-party model to a "trust in the protocol" model that mediates the interaction between two or more users.

- **BCs and DLTs provide a convenient platform for distributing incentives** via cryptocurrencies in a native way, mapped onto an on-chain interaction protocol. In this sense, (properly acknowledged in the protocol) rewards for services can be authorized and initiated in a decentralized and immediate way.

- **BCs are not entirely vulnerable to existing threats of reputation systems**. Malicious actions, such as false- reporting of ratings and collusion, can still happen off-chain by two or more colluding peers in order to fraud the reputation system. While this type of action has the potential to be recognized relatively easier compared to traditional reputation systems, such external actions based on non-native assets still generate the need for a behavioral assessment of peers.

- **SC-oriented development poses new challenges to developers**, especially concerning code maintenance and challenges related to the paradigm shift from centralized to decentralized control and maximizing efficiency aspects considering the cost of on-chain operations. Furthermore, the need for determinism limits the scope of arithmetic operations, requiring external oracles that do require trust on the oracle level themselves (external to the BC/DLT in use).

Thus, the advantages and drawbacks of using DLTs as an underlying reputation system need to be evaluated on a case-by-case basis, considering all synergies concerning the ease of verification of scoring engines and drawbacks related to maintaining privacy and performance. In conclusion, the world of DLTs, including BCs, indicates that fully decentralized reputation systems can be reliably built in a distributed manner. However, performance efficiency, security, and legality in many different facets have to be addressed explicitly in the near future to reach a viable long-term balance.

Acknowledgements This paper was supported partially by *(a)* the University of Zürich UZH, Switzerland and *(b)* the European Union's Horizon 2020 Research and Innovation Program under Grant Agreement No. 830927, the CONCORDIA project.

References

1. NEM: Technical Reference. Technical report (2015). https://bit.ly/3DwAIkY
2. Amintoosi, H., Kanhere, S.S.: A reputation framework for social participatory sensing systems. Mob. Netw. Appl. **19**(1), 88–100 (2014) 00053. https://doi.org/10.1007/s11036-013-0455-x. https://link.springer.com/article/10.1007/s11036-013-0455-x
3. Josang, A., Ismail, R.: The beta reputation system. In: The 15th Bled Electronic Commerce Conference, vol. 5, pp. 2502–2511. Bled, Slovenia (2002). 01484
4. Baier, A.: Trust and antitrust. Ethics **96**(2), 231–260 (1986). https://doi.org/10.1086/292745
5. Battah, A., Iraqi, Y., Damiani, E.: Blockchain-based reputation systems: implementation challenges and mitigation. Electronics **10**(3), 289 (2021)
6. Benet, J.: IPFS-content addressed, versioned, P2P file system (2014). arXiv preprint arXiv:1407.3561
7. Bocek, T., Shann, M., Hausheer, D., Stiller, B.: Game theoretical analysis of incentives for large-scale, fully decentralized collaboration networks. In: 2008 IEEE International Symposium on Parallel and Distributed Processing, pp. 1–8. Washington (2008). https://doi.org/10.1109/IPDPS.2008.4536195. 00020
8. Bocek, T., Stiller, B.: Smart contracts–blockchains in the wings. In: Digital Marketplaces Unleashed, pp. 169–184. Springer (2018)
9. Broome, J.: Desire, belief and expectation. Mind **100**(2), 265–267 (1991)
10. Buechler, M., Eerabathini, M., Hockenbrocht, C., Wan, D.: Decentralized reputation system for transaction networks. Technical report, Department of CIS - Senior Design, University of Pennsylvania, Philadelphia (2015). https://bit.ly/3wlp9IM
11. Cai, Y., Zhu, D.: Fraud detections for online businesses: a perspective from blockchain technology. Financ. Innovat. **2**(1), 20 (2016). https://doi.org/10.1186/s40854-016-0039-4. https://bit.ly/3q3MqN5
12. Carboni, D.: Feedback based reputation on top of the bitcoin blockchain (2015). arxiv:abs/1502.01504. 00008
13. Cho, J.H., Chan, K., Adali, S.: A survey on trust modeling. ACM Comput. Surv. (CSUR) **48**, 28–40 (2015)
14. Georgen, C.: Topl, empowering growth by enabling investment (2017). https://github.com/Topl/whitepaper/blob/master/Whitepaper.pdf
15. Corbitt, B.J., Thanasankit, T., Yi, H.: Trust and E-commerce: a study of consumer perceptions. Electron. Commerce Res. Appl. **1**(3), 203–215 (2003)
16. DeFigueiredo, D.B., Barr, E.T.: Trustdavis: a non-exploitable online reputation system. In: Seventh IEEE International Conference on E-Commerce Technology, 2005. CEC 2005, pp. 274–283. IEEE, Munich, Germany (2005). 00057
17. Delgado-Segura, S., Tanas, C., Herrera-Joancomartí, J.: Reputation and reward: two sides of the same bitcoin. Sensors **16**(6), 776 (2016). https://doi.org/10.3390/s16060776. http://www.mdpi.com/1424-8220/16/6/776
18. Denko, M.K.: Detection and prevention of denial-of-service (DoS) attacks in mobile ad hoc networks using reputation-based incentive scheme. J. Syst. Cybern. Inf. **3**(4), 1–9 (2005)
19. Dennis, R., Owen, G.: Rep on the block: a next generation reputation system based on the blockchain. In: 2015 10th International Conference for Internet Technology and Secured Transactions (ICITST), pp. 131–138. London, United Kingdom (2015). https://doi.org/10.1109/ICITST.2015.7412073. 00006
20. ETSI: Network Functions Virtualisation (NFV), NFV Security, Security and Trust Guidance (2014)
21. Farmer, F.R., Glass, B.: Building Web Reputation Systems, 1st edn. O'Reilly, Sebastopol (2010)
22. Felten, H.Y.J.R.E.W.: A distributed reputation approach to cooperative internet routing protection. In: 1st IEEE ICNP Workshop on Secure Network Protocols, 2005. (NPSec), pp. 73–78 (2005). https://doi.org/10.1109/NPSEC.2005.1532057. 00045

23. Franco, M., Berni, N., Scheid, E., Killer, C., Rodrigues, B., Stiller, B.: SaCI: a blockchain-based cyber insurance approach for the deployment andmanagement of a contract coverage. In: Economics of Grids, Clouds, Systems, and Services. Springer International, Virtually (2021)
24. Franco, M., Sula, E., Rodrigues, B., Scheid, E., Stiller, B.: ProtectDDoS: a platform for trustworthy offering and recommendation of protections. In: Economics of Grids, Clouds, Systems, and Services. Springer International, Izola, Slovenia (2020)
25. Franco, M.F., Scheid, E., Granville, L., Stiller, B.: BRAIN: blockchain-based reverse auction for infrastructure supply in virtual network functions-as-a-service. In: IFIP Networking (Networking 2019), pp. 1–9. IEEE, Warsaw (2019)
26. Friedman, E., Resnick, P., Sami, R.: Manipulation-resistant reputation systems. Algorithm. Game Theory **677** (2007)
27. Gambetta, D., et al.: Can we trust trust? Trust: Making Break. Cooper. Relat. **13**, 213–237 (2000)
28. Zacharia, G.: Collaborative reputation mechanisms for online communities. Ph.D. thesis, Massachusetts Institute of Technology (1999). https://dspace.mit.edu/bitstream/handle/1721.1/9379/44870919-MIT.pdf?sequence=2
29. Goldberg, I., Hill, A., Shostack, A.: Trust, ethics, and privacy. BUL Rev. **81**, 407 (2001)
30. Gollmann, D.: Why trust is bad for security. Electron. Notes Theor. Comput. Sci. **157**(3), 3–9 (2006). https://doi.org/10.1016/j.entcs.2005.09.044. https://www.sciencedirect.com/science/article/pii/S1571066106002891. Proceedings of the First International Workshop on Security and Trust Management (STM 2005)
31. Grandison, T., Sloman, M.: A survey of trust in internet applications. IEEE Commun. Surv. Tutor. **3**(4), 2–16 (2000)
32. Gresch, J., Rodrigues, B., Scheid, E., Kanhere, S.S., Stiller, B.: The proposal of a blockchain-based architecture for transparent certificate handling. In: International Conference on Business Information Systems, pp. 185–196. Springer (2018)
33. Haber, S., Stornetta, W.S.: How to time-stamp a digital document. In: Conference on the Theory and Application of Cryptography, pp. 437–455. Springer (1990)
34. Haussheer, D., Stiller, B.: Decentralized auction-based pricing with PeerMart. In: 2005 9th IFIP/IEEE International Symposium on Integrated Network Management, 2005. IM 2005, pp. 381–394. IEEE (2005)
35. Hawlitschek, F., Notheisen, B., Teubner, T.: The limits of trust-free systems: a literature review on blockchain technology and trust in the sharing economy. Electron. Commer. Res. Appl. **29**, 50–63 (2018)
36. Hendrikx, F., Bubendorfer, K., Chard, R.: Reputation systems: a survey and taxonomy. J. Parallel Distrib. Comput. **75**, 184–197 (2015)
37. Henshel, D., Cains, M., Hoffman, B., Kelley, T.: Trust as a human factor in holistic cyber security risk assessment. Proc. Manufact. **3**, 1117–1124 (2015)
38. Hoffman, K., Zage, D., Nita-Rotaru, C.: A survey of attack and defense techniques for reputation systems. ACM Comput. Surv. (CSUR) **42**(1), 1–31 (2009)
39. Holton, R.: Deciding to trust, coming to believe. Australas. J. Philos. **72**(1), 63–76 (1994)
40. Hoßfeld, T., Haussheer, D., Hecht, F.V., Lehrieder, F., Oechsner, S., Papafili, I., Racz, P., Soursos, S., Staehle, D., Stamoulis, G.D., et al.: An economic traffic management approach to enable the TripleWin for users, ISPs, and overlay providers. In: Future Internet Assembly, pp. 24–34 (2009)
41. Jeffrey, R.: Subjective Probability the Real Thing. Princeton (2002)
42. Jones, S., Wilikens, M., Morris, P., Masera, M.: Trust requirements in e-business. Commun. ACM **43**(12), 81–87 (2000). https://doi.org/10.1145/355112.355128. http://doi.acm.org/10.1145/355112.355128
43. Josang, A., Haller, J.: Dirichlet reputation systems. In: The Second International Conference on Availability, Reliability and Security (ARES'07), pp. 112–119. IEEE (2007)
44. Jøsang, A., Ismail, R., Boyd, C.: A survey of trust and reputation systems for online service provision. Decis. Support Syst. **43**(2), 618–644 (2007)

45. Kamvar, S.D., Schlosser, M.T., Garcia-Molina, H.: The eigentrust algorithm for reputation management in p2p networks. In: Proceedings of the 12th International Conference on World Wide Web, pp. 640–651. ACM (2003). 04580
46. Khan, R., Vatsa, A.: Detection and control of DDOS attacks over reputation and score based MANET. Manet (2017). 00024
47. Kim, J., Yoon, Y., Zo, H.: Why people participate in the sharing economy: a social exchange perspective. In: The 19th Pacific Asia Conference on Information Systems (PACIS 2015), p. 76. Singapore (2015)
48. Kindy, M.: Divine: a blockchain reputation system for determining good market actors (2017). https://bit.ly/3hBE4IA
49. Kopp, H., Mödinger, D., Hauck, F., Kargl, F., Bösch, C.: Design of a privacy-preserving decentralized file storage with financial incentives. In: 2017 IEEE European Symposium on Security and Privacy Workshops (EuroS PW), pp. 14–22 (2017). https://doi.org/10.1109/EuroSPW.2017.45
50. Lewicki, R.J., Wiethoff, C.: Trust, trust development, and trust repair. Handbook Conflict Resolut. Theory Pract. 1(1), 86–107 (2000)
51. Lexico: Lexico Dictionary (2019). https://www.lexico.com. Dictionary powered by Oxford Press and Lexico.com
52. Litos, O.S.T., Zindros, D.: Trust is risk: a decentralized financial trust platform. Technical report, 156, National Technical University of Athens (2017). http://eprint.iacr.org/2017/156
53. Liu, J., Li, W., Karame, G.O., Asokan, N.: Towards fairness of cryptocurrency payments (2016). arXiv:1609.07256 [cs]. 00002
54. Liu, Y., Zhao, Z., Guo, G., Wang, X., Tan, Z., Wang, S.: An identity management system based on blockchain (2017). https://www.ucalgary.ca/pst2017/files/pst2017/paper-8.pdf
55. Yu, T., Zhou, Z., Zhang, D., Wang, X., Liu, Y., Lu, S.: INDAPSON: An incentive data plan sharing system based on self-organizing network. In: IEEE INFOCOM 2014 - IEEE Conference on Computer Communications, pp. 1545–1553 (2014). https://doi.org/10.1109/INFOCOM.2014.6848090. 00032
56. Lu, Y., Zhao, L., Wang, B.: From virtual community members to C2C E-commerce buyers: trust in virtual communities and its effect on consumers' purchase intention. Electron. Commer. Res. Appl. 9(4), 346–360 (2010)
57. Marsh, S.P.: Formalising trust as a computational concept (1994)
58. Marti, S., Garcia-Molina, H.: Taxonomy of trust: categorizing P2P reputation systems. Comput. Netw. 50(4), 472–484 (2006)
59. Mayer, R.C., Davis, J.H., Schoorman, F.D.: An integrative model of organizational trust. Acad. Manag. Rev. 20, 709–734 (1995)
60. Mazieres, D., Shasha, D.: Building secure file systems out of byzantine storage. In: Proceedings of the Twenty-First Annual Symposium on Principles of Distributed Computing, pp. 108–117 (2002)
61. Moinet, A., Darties, B., Baril, J.L.: Blockchain based trust and authentication for decentralized sensor networks (2017). arXiv:1706.01730 [cs]
62. Mozilla: Open Badges (2018). https://openbadges.org/
63. Nakamoto, S.: Bitcoin: a peer-to-peer electronic cash system (2008). https://bitcoin.org/bitcoin.pdf
64. Sztorc, P.: Truthcoin whitepaper (2015). http://www.truthcoin.info/papers/truthcoin-whitepaper.pdf
65. Pavlou, P.A.: Consumer acceptance of electronic commerce: integrating trust and risk with the technology acceptance model. Int. J. Electron. Comm. 7, 101–134 (2003). https://doi.org/10.1080/10864415.2003.11044275
66. Pavlou, P.A., Fygenson, M.: Understanding and predicting electronic commerce adoption: an extension of the theory of planned behavior. MIS Quart. 115–143 (2006)
67. Peterson, J., Krug, J.: Augur: a decentralized, open-source platform for prediction markets (2015). arXiv preprint arXiv:1501.01042. https://arxiv.org/pdf/1501.01042.pdf

68. Pranata, I., Skinner, G., Athauda, R.: A holistic review on rrust and reputation management systems for digital environments. Int. J. Comput. Inf. Technol. **1**, 44–53 (2012)
69. Resnick, P., Kuwabara, K., Zeckhauser, R., Friedman, E.: Reputation systems. Commun. ACM **43**(12), 45–48 (2000)
70. Resnick, P., Sami, R.: The influence limiter: provably manipulation-resistant recommender systems. In: Proceedings of the 2007 ACM conference on Recommender systems, pp. 25–32 (2007)
71. Rodrigues, B., Bocek, T., Lareida, A., Hausheer, D., Rafati, S., Stiller, B.: A blockchain-based architecture for collaborative DDoS mitigation with smart contracts. In: IFIP International Conference on Autonomous Infrastructure. Management, and Security (AIMS 2017), Lecture Notes in Computer Science, vol. 10356, pp. 16–29. Springer, Zürich (2017)
72. Rodrigues, B., Bocek, T., Stiller, B.: Multi-domain DDoS mitigation based on blockchains. In: IFIP International Conference on Autonomous Infrastructure. Management and Security, pp. 185–190. Springer, Zürich (2017)
73. Rodrigues, B., Bocek, T., Stiller, B.: The Use of Blockchains: Application-Driven Analysis of Applicability, Advances in Computers, vol. 111, pp. 163–198. Springer, Waltham (2018). https://www.sciencedirect.com/science/article/pii/S006524581830024X
74. Rodrigues, B., Franco, M., Scheid, E., Stiller, B., Kanhere, S.: A technology-driven overview on blockchain-based academic certificate handling. IGI Glob. 1–290 (2020). https://doi.org/10.4018/978-1-5225-9478-9. https://www.igi-global.com/book/blockchain-technology-applications-education/221313
75. Rodrigues, B., Scheid, E.J., Killer, C., Franco, M., Stiller, B.: Blockchain signaling system (BloSS): cooperative signaling of distributed denial-of-service attacks. J. Netw. Syst. Manag. **28**(3), 1–27 (2020). https://doi.org/10.1007/s10922-020-09559-4
76. Schaub, A., Bazin, R., Hasan, O., Brunie, L.: A trustless privacy-preserving reputation system. In: ICT Systems Security and Privacy Protection. IFIP Advances in Information and Communication Technology, pp. 398–411. Springer, Cham (2016)
77. Scheid, E.J., Rodrigues, B., Killer, C., Franco, M., Rafati, S., Stiller, B.: Blockchains and distributed ledgers uncovered: clarifications, achievements, and open issues. In: Advancing Research in Information and Communication Technology. IFIP AICT Festschrifts, pp. 1–29. Springer, Cham (2021)
78. Scheid, E.J., Rodrigues, B., Stiller, B.: Toward a policy-based blockchain agnostic framework. In: IFIP/IEEE Symposium on Integrated Network and Service Management (IM 2019), pp. 609–613. Washington, D.C., USA (2019)
79. Schlosser, A., Voss, M., Brückner, L.: Comparing and evaluating metrics for reputation systems by simulation. In: The IEEE Workshop on Reputation in Agent Societies (2004). 00031
80. Sharples, M., Domingue, J.: The blockchain and kudos: a distributed system for educational record, reputation and reward. In: Adaptive and Adaptable Learning. Lecture Notes in Computer Science (LNCS), pp. 490–496. Springer, Cham (2016)
81. Shen, H., Lin, Y., Sapra, K., Li, Z.: Enhancing collusion resilience in reputation systems. IEEE Trans. Parallel Distrib. Syst. **27**(8), 2274–2287 (2015)
82. Siau, K., Shen, Z.: Building customer trust in mobile commerce. Commun. ACM **46**, 91–94 (2003)
83. Singer, A., Bishop, M.: Trust-based security; or, trust considered harmful. In: New Security Paradigms Workshop 2020, NSPW '20, pp. 76–89. Association for Computing Machinery, New York (2020). https://doi.org/10.1145/3442167.3442179
84. Soska, K., Kwon, A., Christin, N., Devadas, S.: Beaver: a decentralized anonymous marketplace with secure reputation. IACR Cryptol. ePrint Arch. **2016**, 464 (2016)
85. Stoneburner, G., Goguen, A., Feringa, A.: Risk management guide for information technology systems. NIST Spec. Publ. **800**(30), 800–30 (2002)
86. Um, T.W., Lee, G.M., Choi, J.K.: Strengthening trust in the future social-cyber-physical infrastructure: an ITU-T perspective. IEEE Commun. Mag. **54**, 36–42 (2016)
87. Wang, Y., Vassileva, J.: Trust and reputation model in peer-to-peer networks. In: The Third International Conference on Peer-to-Peer Computing (P2P2003), pp. 150–157. Linkoeping, Sweden (2003). https://doi.org/10.1109/PTP.2003.1231515

88. WorkNation: Work.nation: Decentralized skill attestations using uPort, Ethereum and IPFS (2017). https://github.com/worknation/work.nation
89. Xiong, L., Liu, L.: Building trust in decentralized peer-to-peer electronic communities. In: International Conference on Electronic Commerce Research (ICECR-5) (2003)
90. Zhang, Y., van der Schaar, M.: Reputation-based incentive protocols in crowdsourcing applications (2011). arXiv:1108.2096 [physics]

Blockchain for Trust and Reputation Management in Cyber-Physical Systems

Guntur Dharma Putra, Volkan Dedeoglu, Salil S Kanhere, and Raja Jurdak

Abstract The salient features of blockchain, such as decentralization and transparency, have allowed the development of Decentralised Trust and Reputation Management Systems (DTRMS), which mainly aim to quantitatively evaluate the trustworthiness of network participants and help to protect the network from adversaries. In the literature, proposals of DTRMS have been applied to various Cyber-physical Systems (CPS) applications, including supply chains, smart cities, and distributed energy trading. In this chapter, we outline the building blocks of a generic DTRMS and discuss how it can benefit from blockchain. To highlight the significance of DTRMS, we present the state-of-the-art DTRMS in various fields of CPS applications. In addition, we also outline challenges and future directions in developing DTRMS for CPS.

1 Introduction

Trust is a subjective and intangible belief about the behavior of a particular entity or individual, which is built up from consecutive interactions [1]. Trust is context-related and thus cannot be generalized, as it is linked to a specific behavior or trait. According to Gambetta, trust is defined as a subjective probability that an individual

G. D. Putra (✉) · S. S. Kanhere
School of Computer Science and Engineering, UNSW Sydney, 2052 Kensington, NSW, Australia
e-mail: gdputra@unsw.edu.au

S. S. Kanhere
e-mail: salil.kanhere@unsw.edu.au

Cyber Security Cooperative Research Centre (CSCRC), Joondalup, Australia

V. Dedeoglu
Commonwealth Scientific and Industrial Research Organisation's Data 61, Pullenvale, Australia
e-mail: volkan.dedeoglu@data61.csiro.au

R. Jurdak
School of Computer Science, Queensland University of Technology (QUT), Brisbane, Australia
e-mail: r.jurdak@qut.edu.au

© The Author(s), under exclusive license to Springer Nature Switzerland AG 2022   339
D. A. Tran et al. (eds.), *Handbook on Blockchain*, Springer Optimization
and Its Applications 194, https://doi.org/10.1007/978-3-031-07535-3_10

expects from another individual on performing an expected action [2]. Occasionally, trust and reputation are referred interchangeably in the literature. However, there is a subtle difference between these two terms. Trust refers to a subjective belief towards the behavior of an entity that builds up as more interactions happen, while reputation can be seen as the aggregated opinion or trust degree of an entity from other entities that have prior interaction with the entity.

Trust and Reputation Management Systems (TRMS) aim to assess the accountability or trustworthiness of each participant in distributed systems by means of a quantitative approach. In TRMS, trustworthiness is derived from direct experience or recommendations from other peers and is represented as numerical scores using which the trustworthiness level can be conveniently measured. In general, the trust and reputation score can be used as a safeguard to manage the associated risk in communicating with other peers in a distributed system, which might be very dynamic and hostile.

There have been many applications of TRMS in Cyber-Physical Systems (CPS) and Internet of Things (IoT). For example, TRMS is deployed in the context of social IoT, which is used to assess the trustworthiness of each participating node in the network [3]. In e-commerce sites, TRMS is implemented to help customers determine the credibility of the sellers in the marketplace [4].

However, several challenges exist in building a TRMS. For instance, traditional TRMS architectures rely on a centralized actor to manage the collection of feedback and calculation of trust scores, which raises the risk of data loss and manipulation by the centralized party. When the centralized actor is compromised, an adversary may maliciously alter the trust computation thus undermining the use of these metrics. In addition, authentication and identification of users in TRMS may expose the actual identities of the users, which should be concealed and protected.

Blockchain, the underpinning technology behind Bitcoin, has seen a lot of interest, due to its inherent characteristics, such as traceability, tamper-resilience, trustless environment, programmability, immutability, and transparency. These characteristics show promise in addressing the aforementioned issues of TRMS. For instance, blockchain may replace the trusted centralized actor that assesses the trustworthiness of participants in traditional TRMS. The adoption of blockchain in TRMS is referred to in the literature as Decentralized TRMS (DTRMS) [5]. While blockchain enables trustless interaction between participants, a reputation system is still required to provide some degree of trust quantification for off-chain operations.

In this chapter, we present and discuss how blockchain technology can be incorporated into TRMS for enhancing its effectiveness for CPS applications. We outline the building blocks of a generic TRMS that delineate how trust is empirically built up by collecting and aggregating evidence of direct and indirect interactions to obtain a quantifiable trust measure. Then, we describe the blockchain properties that can help address challenges in building TRMS for CPS applications. The chapter also highlights the latest developments of DTRMS for CPS by presenting some recent implementations of DTRMS across various CPS application domains. We also outline the challenges and future directions for DTRMS that still need to be addressed.

The rest of the chapter is organized as follows. Section 2 presents the concept of DTRMS. Section 3 outlines several implementations of DTRMS for CPS. We discuss the open challenges for future research directions in Sect. 4 and give a conclusion of our chapter in Sect. 5.

2 Blockchain-Based Trust and Reputation Management Systems

In this section, we first discuss the necessity of a TRMS for CPS along with the general properties of CPS applications, including trust derivation, types of trust, evidence aggregation approach, and trust dimensions. We also illustrate the salient properties of blockchain that hold the potential to enhance TRMS.

2.1 Trust and Reputation Management Systems for CPS

In general, CPS applications involve a group of agents collecting data from physical environments and performing specific tasks based on the collected data, which includes interactions with other agents in the network. While we can assume that the majority of agents are honest, some agents may behave opportunistically to maximize their gains through dishonest behavior. In addition, the collected data may also be noisy, faulty, or maliciously tampered with. Ideally, an agent should not blindly trust other agents due to these risks that may degrade the quality of service of their interaction. TRMS are designed to quantitatively assess the trustworthiness of a particular agent or data in a system through numerical and tangible values. In CPS, TRMS acts as an intermediary between service providers and requesters by providing protocols that guarantee trustworthiness in each interaction by means of authentication, resource management, and access control. We discuss the general properties of a generic TRMS for CPS in this subsection.

Trust Derivation and Application

Similar to real-life social interactions, computational trust is built gradually from successive interactions between entities that correspond to positive or negative experiences affecting the overall belief of the trustworthiness level. In a generic TRMS, the interactions are assessed empirically, which includes four steps for collecting and applying trust computation, depicted in Fig. 1 [6].

Information Gathering The first step in TRMS is defining the input parameters and attributes for quantifying or computing the trustworthiness level, which in general is highly application-specific. Some of the examples include adherence to

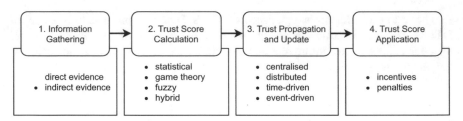

Fig. 1 The steps for trust derivation in a TRMS

communication protocol, quality of service, and degree of satisfaction towards a service. The TRMS should gather all of these input values either by (1) direct observations or interactions or (2) recommendations from other entities if prior interactions are unavailable.

Trust Score Calculation The next step includes the actual calculation of the trustworthiness level as quantifiable values or scores according to the preferred trust or reputation model. The TRMS may use various computation models that suit the application requirements, for instance, statistical, game theory, fuzzy computation, or hybrid. Note that, the input attributes also determine the appropriate computation model, e.g., sum and mean models are suitable for continuous input values, while the Bayesian model is more suited for discrete binary values [6]. In addition, trust score calculation should also take into account the types of trust (see Sect. 2.1).

Trust Propagation and Update Typically, trust propagation can be performed in a centralized, distributed, or semi-distributed fashion, depending on the underlying architecture of the system. The trust computation should be initiated based on temporal dynamics depending on the specific application, which includes time-driven and event-driven approaches. In the time-driven approach, the trust score is updated on a regular basis, while the event-driven approach only requires updating the trust values upon new interactions and events.

Trust Score Application The specific manner in which the trust score is used depends on the requirements and operation of the application. Generally, the trust score is employed to give certain quantified and fair measures for providing incentives or enforcing penalties, which may include certain privileges and monetary incentives or some restrictions and punishments. Section 3 discusses specific examples of how the trust scores are manifested in various CPS applications.

Types of Trust

As discussed earlier, CPS applications rely on the data collected, processed, and transferred in the system, and the interactions among entities. Thus, we can categorize the computation of trust as follows:

Behavior-based Trust Computation In behavior-based trust computation, the trustworthiness level of an entity is derived from how the entity behaves in the system as perceived by a subject during its interaction with the entity. A subject identifies a positive behavior if the observed entity conforms to the prescribed protocols and expectations, while negative behavior corresponds to a deviation from protocols and expected behavior.

Data-based Trust Computation In data-based trust computation, the trust values are calculated based on the quality of data provided by an entity. For instance, in service-oriented CPS applications, trust can be derived from the quality of the data acquired from the data provider. Here, trust grows with the authenticity of the data, i.e., deliberate manipulation, noise, or anomaly in the data would degrade the trust. In this type of trust, data validation plays an integral role, and one approach to validate the data quality may include using correlated observations obtained from other entities in proximity.

Hybrid Approach Relying on a single type of trust may not be sufficient for deriving trust in certain CPS applications, for instance, mobile crowdsourcing, wherein trustworthy agents are seen as those who provide reliable data and conform to the predetermined governance. In such scenarios, trust can be computed considering both the data-based and behavior-based characteristics. In the hybrid approach, weightings are used to give favorable emphasis on either data or behavior characteristics.

Evidence Aggregation Approach

TRMS may adopt one of the evidence aggregation approaches to accumulate trust evidence and calculate the final trust and reputation score. While there is an exhaustive list of aggregation approaches [7], the following approaches are among the most widely adopted:

Sum and Mean The most intuitive and popular aggregation approach in TRMS is the summation or average of the aggregated trust evidence [8]. Due to its simple operation, this method can also be validated manually to provide an objective confirmation. Some weighting parameters may also be incorporated to give more weight to recent or more important evidence. One of the challenges with this approach is the determination of appropriate weights, which would have an impact on the performance of the TRMS.

Flow Network This approach is proposed in Advogato [9], wherein each participant is seen as a node in the network, while the interactions between participants are modeled as network flows. Consequently, the trust is derived from the number of flows a participant obtained from others. This approach is relatively robust to trust-related attacks, as the total active flows in the network are assumed to be constant and strictly regulated by the TRMS.

Markov Chain As implemented in EigenTrust [10], the Markov chain approach works based on probability modeling of a user's feedback reaching a particular participant. The feedback from one user to another is modeled as a probability function of a transition from source to target user, using which the reputation score is derived.

Bayesian In this approach, the trust and reputation scores are computed using statistics. The trustworthiness score is described as a beta distribution of two parameters where α and β denote positive and negative recommendations, respectively. To calculate and update the score, an update to the provided beta distribution is performed, through which unfair ratings can also be removed [11].

Trust Dimensions

In general, trust is strongly attached to a particular context and generally cannot be transferred to another context without rigorous adjustment and re-calculation. With this regard, context-awareness is an important factor to consider in designing a TRMS. A TRMS can work on a single context or multiple-context awareness in deriving trust from collected evidence depending on its initial design [5].

Single-dimension A lightweight TRMS with a simplified trust and reputation model might only incorporate a single-dimension trust evaluation for the sake of limited resources in CPS applications. While single-dimension trust model may not be comprehensive, it may be preferred depending on the application design, e.g., when a majority of resource-constrained devices are in use.

Multi-dimension On the other hand, multidimensional trust and reputation model represent trust and reputation scores in multiple parameters or a single value derived from multiple parameters with appropriate weightings. In practice, this type of TRMS may require heavier computation and may not be suited for constrained devices.

2.2 Adopting Blockchain for TRMS

Since its initial inception in 2009 as a pioneer in decentralized cryptocurrency, blockchain has also been applied in many non-monetary applications, one of which includes TRMS. While blockchain may introduce some overheads, blockchain has promising potential to be implemented in TRMS. Here, we describe the inherent properties of blockchain that would bring enhancements and benefits to TRMS.

Decentralization

Conventional TRMS relies on a third-party aggregator to collect trust evidence and calculate trust scores. Trusting a third-party aggregator actually introduces significant risks, for instance, when the aggregator is compromised of any underlying processes of trust computation could be maliciously altered and the sensitive data

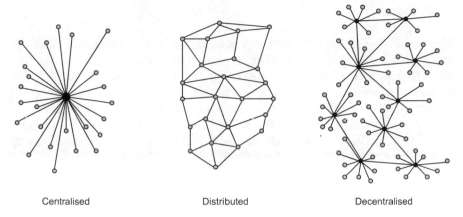

Centralised   Distributed   Decentralised

Fig. 2 Different types of architecture: centralized, distributed, and decentralized

could be in danger. On the other hand, blockchain removes any Trusted Third Party (TTP) and comes with a decentralized architecture, as seen in Fig. 2, which eliminates associated risks of employing third-party aggregators in TRMS. Blockchain employs certain consensus algorithms, such as Proof-of-Work and Proof-of-Stake, to enforce collaborative execution and validation of transactions, using which fraudulent manipulations can be avoided. The ledgers, in which the transactions or data are stored, are replicated among all participants in the network that enhances availability. In addition, blockchain can also be incorporated in a distributed TRMS to enhance the mechanism, for instance, by utilizing smart contracts, which we describe in the following subsections.

Smart Contract

Smart contract is a form of execution code agreed by a set of users, which allows deterministic and trusted execution of business logic with reliable guarantees that the process would be accomplished and validated collaboratively in the network. Smart contracts can be embedded into a blockchain-based TRMS to perform collection and calculation of trust scores which can offload the trust computation from the CPS devices. For instance, a node may submit feedback to the smart contract about the experience interacting with a service provider, which later will be used to calculate the service provider's reputation score. Another node in the network can also query the smart contract to obtain the reputation scores of particular service providers. That is, a smart contract acts as a reliable intermediary for computing and querying trust-related information.

Pseudonyms

There is an inherent risk of leaking sensitive information in conventional TRMS, as the authentication mechanism may link the identification details to real-life identities. Blockchain introduces an elliptic curve public key cryptography mechanism which utilizes pseudonyms, i.e., public key, for identification purposes, resulting in higher privacy preservation, as real identities are not used. The use of pseudonyms is, to some extent, beneficial for protecting users' privacy, which is a desirable property in designing a TRMS. In a blockchain-based TRMS, each node is identifiable by its public key which hides any personal data, such as device ownership details. We discuss more challenges and opportunities in privacy preservation for DTRMS in Sect. 4.2.

Immutable Storage

Traditional TRMS stores trust evidence and interaction history on each device's internal memory, which may overwhelm the devices, especially if there is a large amount of information in a network with thousands of nodes. As discussed earlier, a traditional TRMS can also rely on a TTP to keep track of the trust-related information, but with the fundamental risk of data loss and manipulation linked to the centralized approach. Blockchain data structure, as depicted in Fig. 3, enforces immutability as it is difficult if not almost impossible to tamper with the stored data on the blockchain. To tamper with the data, an attacker should break the hash cryptography and may need to traverse all the way back to the genesis block. With proper removal of any Personally Identifiable Information (PII), blockchain is a perfect and safe solution to store interaction evidence that would later be used to calculate the trust score.

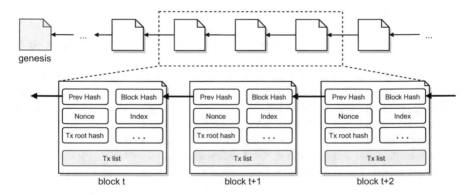

Fig. 3 An overview of blockchain's immutable storage

Transparency

Recall that in conventional centralized TRMS, any underlying mechanism in trust and reputation calculation is performed by a centralized aggregator, which conceals the actual process from other participants in the network. On the other hand, blockchain offers transparent mechanisms for collaborative trusted execution of business logic via smart contracts and transparent immutable storage via a transparent shared ledger. With precautions in handling and storing sensitive information in the ledger, this type of transparent mechanism is preferred as it enables a traceable source of evidence where any participant can ascertain the integrity of a trust calculation by examining the ledger.

3 Use Cases

In this section, we outline various implementations of DTRMS for CPS across different application domains to demonstrate how blockchain enhances and brings benefits to these applications. We also present a summary of the use cases in Table 1.

Table 1 Summary of DTRMS use cases

Application fields	Goals	Blockchain features
Generic CPS	end-to-end IoT trust establishment; securing CPS/IoT	IS, AC, PS, SC
Supply chain managements	managing traders and sellers; maintaining product quality	SC, IS, PS
Crowdsourcing	selecting reliable workers	AC, IS, PS
Robotic and autonomous systems	selecting reliable service providers and detecting Byzantine nodes	SC, IS, TR
Vehicular ad hoc network	validating exchanged messages to avoid malicious messages	PS, AC, IS
IoT data marketplace	curating traded data and providing fair payments	SC, PS, IS
Distributed energy trading	managing prosumers' reliability	SC, AC, IS

IS = Immutable Storage, AC = Adaptive Consensus, SC = Smart Contracts, PS = Pseudonyms

3.1 Generic CPS Trust Architecture

Blockchain ensures trusted, persistent and immutable storage for keeping observational data which achieves tamper-proof storage and prevents unwanted or malicious modification. However, the fundamental problem about establishing trust in the data itself cannot be solved only by using blockchain. In [12], the authors propose an end-to-end trust architecture for IoT. The authors also proposed a dynamic block validation mechanism, wherein trust management reduces the computation load on the nodes by reducing the number of transactions that need to be validated from trusted nodes.

In [13], the authors proposed a blockchain-based trust evaluation system for Pervasive Social Networking (PSN), which helps to protect a node from unfamiliar or unknown acquaintances in a trustless decentralized environment. In the proposed solution, the trustworthiness of a participant is derived from its social behavior, represented as trust evidence stored on the blockchain. The authors also proposed a trust-based consensus algorithm that aims to reduce resource consumption and accelerate block generation in the consensus process. A new block is confirmed if it is approved by an adequate number of miners above a certain threshold.

In [14], a layered architecture, called BC-Trust, is proposed to provide a scalable DTRMS solution for devices with high mobility in fog computing. In BC-Trust, the trustworthiness of a particular service provider is derived from a user's direct experience and recommendations from known peers in the network. Fog nodes with sufficient computing resources maintain the blockchain and perform all required trust computation, offloading the trust computation to high resource devices, which help to reduce the load in the constrained devices. BC-Trust is also designed to be robust against known trust-related attacks, such as ballot-stuffing and bad-mouthing (see Sect. 4.4).

A trust architecture can also be utilized to protect important CPS resources from illegitimate access by unauthorized service consumers. In [15], the authors proposed a decentralized Attribute-based Access Control (ABAC), in which each service consumer is associated with a certain trust score based on its behavior in the network. The trust score of the service consumer is then included in the required attributes to access resources. The authors introduced three smart contracts, namely attribute provider, trust and reputation system, and policy smart contract to administer and operate the trust-based access control. In [16], the authors extended their solution to include privacy preservation and more efficient trust computation, where the computation is offloaded to the blockchain entirely.

3.2 Supply Chain Management

In general, Supply Chain Management (SCM) systems demand for traceability, which is fulfilled by the blockchain. In addition, blockchain allows multiple writers to the system which is suitable in most instantiations of supply chain, whereas multiple stakeholders are involved. DTRMS is used in supply chains to provide additional trustworthiness and benefits, as the quantified trust score could be associated with certain product qualities or producers.

Bai et al. [17] proposed a trust management scheme in an e-agriculture supply chain scenario where a network of smart greenhouses act as miners and form the blockchain network. Each greenhouse manages a set of sensors that monitor the condition of the farm. The farmers can use the network to query the sensors to get some agriculture-related information, such as the probability of whether the farm needs to be fertilized or watered. The authors proposed a DTRMS in which a game-theoretic approach is used to determine the trustworthiness of each sensor's reading by associating a trust score to each sensor. When a farmer queries for sensor data, the greenhouse will also look for other readings from related sensors and perform Bayesian inference to report an estimate along with a trust score. A low trust score indicates that the readings are incorrect and the sensor may need to be replaced.

Although blockchain can solve the immutability and traceability issues in supply chain applications, the issue of the integrity of inserted data remains unsolved. In [18], the authors proposed TrustChain, a three-layered trust management framework tailored for supply chains. The architecture consists of three layers, namely data layer (for data input), blockchain layer (where all process happens), and application layer (for transacting with blockchain). In general, the reputation system assesses the quality of the commodities based on multiple observations within the supply chain. The solution adopts smart contracts to automate reputation calculation and specifically deploys two contracts: (1) a quality contract to assess the quality of each supply chain commodity based on sensor readings (e.g. temperature to keep the food quality) and (2) a rating contract to compute the reputation of the traders. Each trader has an inherent trust score, which is derived from all ratings with customizable weightings. The method also utilizes time-varying and amnesic trust calculation where more emphasis is given to recent observations.

In [19], the authors proposed Reputation-based Trustworthy Blockchain Supply Chain Management (RTB-SCM) to address trustworthiness issues in supply chains. The solution is based on a consortium blockchain, named Reputation Assessment Blockchain (RAB), that stores trade records and commodity information. RAB introduces a token-based reputation system, which is based on a crypto-asset governed by the trusted regulator that runs the consortium blockchain. In addition, the design utilizes a smart contract-based reputation rating model that utilizes tokens for rewards and punishments. The authors proposed two algorithms; (1) Quality Status Generation (QSG) for quantifying $QInfo$, the quality of trade with regard to a commodity type derived automatically by the sensors; and (2) Token-based Reputation Reward/Punishment (TR2P) for determining the appropriate reward or punishment based on $QInfo$ from QSG.

3.3   Crowdsourcing

In [20], the authors proposed a hybrid blockchain architecture to enhance data validation in crowdsourcing, wherein a private consortium blockchain is used as the
backbone of the network, while the public blockchain acts as a method of transaction validation for the novel consensus protocol. In this work, trust management is
incorporated into a consensus algorithm called Proof-of-Trust (PoT). The intuition
is to select reliable validators to validate collected data based on the trustworthiness
score of the participants in the crowdsourcing service. Combined with RAFT leader
election and Shamir's secret sharing algorithm, the framework calculates the trust
score based on three independent parameters: (1) the number of transactions the user
has on the platform, (2) the total time the user has been involved in the validation
process, and (3) the number of complaints the user receives. The PoT protocol splits
the consensus process into four phases, each of which is conducted by different roles,
which ensures performance and consistency of the consensus process while greatly
improving scalability.

In [21], Feng and Yan proposed MCS-Chain, which is a fully decentralized trust
management for Mobile Crowdsourcing (MCS) purposes without relying on trusted
actors. The architecture consists of end-users, workers, and miners. The blockchain
acts as the MCS platform, where all the procedures are recorded and trust scores
are evaluated. The miners in this case are the cell towers which are responsible for
managing the blockchain. A trust evaluation scheme is designed to help the end-users
choose appropriate workers based on reliability. End-users post some particular tasks
to the blockchain by invoking blockchain transactions, which then broadcast the task
to the workers for bidding. In this step, the trust score helps the end-users to pick
the preferred workers. The trust score of the workers is then updated based on the
feedback from the end-users about certain tasks. To avoid unfair rating, the mechanism applies deviation between personal and average feedback and also considers
the previous trust score of the submitter. The authors implemented their solution on
Android and Windows to evaluate the performance and highlight the efficiency of
the proposed system.

3.4   Robotic and Autonomous Systems

DTRMS has also been implemented in the area of robotics and autonomous systems,
where blockchain overcomes several reliability issues in information sharing and aids
the selection of service providers in a transparent way.

In [22], Alowayed et al. proposed a custom DTRMS which enables Autonomous
Systems to evaluate network providers based on their ability to provide reliable
interconnection service as per the pre-approved Service-Level Agreement (SLA). In
this framework, the network performance measurements are stored as transactions
on a permissioned blockchain. A smart contract then quantifies the trustworthiness

of each network provider, called the SLA score, and analyzes if the provider has submitted a misleading performance report. They propose to use the SLA scores to select network providers to ensure that clients are provided the required quality of service. The framework requires each SLA score to be written on the blockchain with a transparent and publicly-agreed SLA score calculation method between network participants, while privacy preservation is achieved by adopting an order-preserving encryption mechanism [23].

Strobel and Dorigo proposed a blockchain-based knowledge-sharing architecture for swarm robotics [24]. In this framework, a DTRMS is employed to identify Byzantine or malicious robots which may hinder the overall performance of the system due to misleading data measurements. A permissioned blockchain network is utilized, wherein each robot serves as an Ethereum node, through which the robot could exchange knowledge with other robots within 50 cm of proximity via blockchain transactions. The reputation for each robot is calculated based on the absolute difference between reported observation and the average of other observations from other robots in the proximity.

3.5 Vehicular Ad Hoc Networks

Several works have been proposed in the field of Vehicular Ad Hoc Network (VANET) to incorporate DTRMS for enhancing the security or avoiding malicious events in the network. The typical architecture of VANET includes mobile and fixed nodes, e.g., the smart vehicles and Road Side Units (RSU), respectively.

The authors of [25] proposed a privacy-preserving announcement protocol for the Internet of Vehicles (IoV) called PBTM. The authors also designed a blockchain-based trust management system to ascertain the authenticity and synchronize the timing of the exchanged messages. RSUs play an important role in maintaining the blockchain and calculating the trustworthiness score of each vehicle using the weighted sums method according to the validity of the transmitted message. Privacy preservation is achieved by adopting an identity-based group signature, which realizes the anonymity of each vehicle.

Wang et al. proposed BSIS: Blockchain-based Secure Incentive Scheme, a reputation-based consensus protocol to obtain efficient consensus within a Vehicular Energy Network (VEN) [26]. In the proposed consensus algorithm, the validators are selected based on their trust score and each validator would receive incentives upon successful execution of the consensus mechanism. Consequently, the higher the trust score, the more chance a node has to be selected as a validator and subsequently obtain the reward. The authors classify two types of trustworthiness scores, namely local trust, and reputation value. The local trust score is derived from ratings obtained from each interaction with other energy nodes, while the final reputation value is calculated by aggregating all local trust scores that a node has obtained.

In [27], the authors proposed a blockchain-based trust management for VANET, wherein RSUs are the only approved actors to compute the trust values for each vehicle. In this framework, each vehicle may send messages to other vehicles from which the trust score of the vehicle is calculated. Each message receiver generates ratings that represent the credibility of the corresponding message. Due to high mobility and limited storage capacity, each vehicle is expected to periodically submit the ratings to nearby RSUs, in which ratings are aggregated and grouped to calculate the trust score of each vehicle using Bayesian inference. Once the score is stored on the blockchain, any vehicle may query the data if necessary. The authors argue that the proposed DTRMS would give guidance to the vehicles about the quality of the received messages and also provide the underlying evidence for reward and punishment mechanisms.

Lu et al. proposed a trust management system for VANETs, called BARS: a Blockchain-based Anonymous Reputation System [28], with an emphasis on anonymity by avoiding linkability between real and pseudo-identities. In this framework, the trustworthiness of each vehicle is determined by the authenticity of the broadcast message and the reported opinion from other vehicles. The authors introduced the concept of the Law Enforcement Authority (LEA) which is responsible for managing the framework and resolving disputes.

3.6 IoT Data Marketplace

The proliferation of CPS/IoT deployments has generated an enormous amount of data, using which the owner can obtain financial benefits by selling useful data to specific consumers. In an IoT data marketplace, both sellers and consumers can communicate and share data. However, the customers do not trust the sellers as the data quality cannot be guaranteed. Blockchain in this case can enhance the IoT data marketplace by utilizing DTRMS to evaluate the trustworthiness of each participant and providing a decentralized payment mechanism using built-in cryptocurrency.

Camilo et al. proposed a blockchain-based data trading platform, in which trust and reputation play an important role in helping customers determine the quality of the sellers [29]. The authors define a distinction between trust and reputation, where trust corresponds to a buyer's view of a seller based on his trading experience, while reputation is an aggregated view of a particular seller from multiple individual trust scores across different buyers. In this platform, the data owners or sellers may advertise the metadata via a smart contract on the blockchain, which can be explored by data buyers to select the preferred sellers. The data buyers are required to submit a feedback that rates the seller and the data quality through a feedback transaction to the blockchain, using which the smart contract evaluates the trust and reputation score of the sellers accordingly.

In [30], the authors proposed an automatic review system to assess the quality of the data, which is used for monetization of IoT data. The system adopts a publish-subscribe mechanism, where an MQTT broker plays an important role along with the permissioned blockchain. A rating is associated to each data on each topic in the platform. Any data buyer may request the smart contract to browse available data based on subscribed topics, with reviews associated with each data sale. Upon completing the data access, each buyer is required to submit reviews about the accessed data for an incentive. The system utilizes a smart contract to achieve automated payment and incentive mechanisms that eliminate the need for a TTP, while also reducing the associated risks of trusting an external party.

The authors in [31] proposed a reputation system for online marketplaces, which is based on hashcash PoW algorithm, originally designed to reduce spams in email [32]. The reputation system is designed to assess and incentivize watchtowers to behave rationally. A watchtower is an independent entity that preserves the client's interest for specific purposes, e.g., lightning payment network [33], by continuously monitoring the blockchain network on behalf of the clients that may frequently be offline. The watchtowers offer a monitoring service in the open market, where clients would tend to pick the watchtowers with the highest reputation score. During the negotiation phase with a client, a watchtower publishes a smart contract as a persistent proof that bonds the terms and conditions of payments and services with the client. The reputation of the watchtowers is derived from successful transactions, while a proof-of-breach is generated from the contract if a watchtower does not fulfill its obligation.

In [34], the authors proposed a payment mechanism for IoT marketplace, called Secure Pub-Sub (SPS), where blockchain is utilized to provide fair payments and reliability. In SPS, blockchain performs as a payment gateway between a publisher and subscriber that do not necessarily have to trust each other, wherein a subscriber can deposit some funds prior to subscribing to a particular publisher to access the data. In addition, a reputation system is employed so that each subscriber can assess the publisher after accessing its service, where a smart contract transparently maintains the process. Here, the reputation system can help the subscribers to pick appropriate publishers based on certain reputation scores that are higher than a threshold for determining reliable publishers.

3.7   Distributed Energy Trading

Distributed or peer-to-peer energy trading is a marketplace where each prosumer, a user who consumes and produces energy, can transact energy to the end-user directly without the need of a central entity. It has been demonstrated by recent works that TRMS can help improve the efficiency and enhance the fairness of energy trading.

In [35], the authors proposed a secure blockchain-based energy trading platform with a built-in reputation system to enhance reliability and encourage honest behavior among blockchain nodes. The authors also incorporate reputation scores into the Proof-of-Work consensus algorithm, called PoWR, to reduce the block creation time and overall latency. Each participant utilizes the blockchain network as a communication channel, through which each participant exchanges information about direct and indirect trust experiences. In this reputation framework, a higher reputation score corresponds to a higher probability of participating in the PoWR, hence more chance of getting incentives.

Khorasany et al. proposed a peer-to-peer framework for energy trading in [36], where blockchain and smart contracts are employed to build a decentralized trading mechanism. The authors also associate a reputation factor to each energy agent that represents its reliability in fulfilling the obligations. In addition, the authors also introduced an algorithm, called Anonymous Proof of Location (A-PoL), to anonymously prove a user's location. In performing the energy trading, each user may select a partner based on their preference on both reputation factor and agent's location, which is handled by an automated algorithm. A Dispute Resolution smart contract is in charge of calculating each agent's reputation factor which is based on prior commitments in delivering energy to the trading counterpart.

A framework called Reputation for Blockchain-based energy Trading (RBT) is proposed in [37]. RBT utilizes blockchain as a traceable and immutable storage for reputation scores and smart contract for automated reputation calculation. Here, the reputation is derived from the behavior of each node according to its role in the P2P process via three parameters, namely role, rule, and reputation. A matchmaking strategy based on the k-double auction algorithm is used to connect both buyers and sellers and to decide trading prices that are more beneficial to both parties. The matchmaking strategy also includes a fairness indicator, which is a ratio between reputation score and the average income and cost for sellers and buyers, respectively.

4   Challenges and Future Directions

In this section, we discuss several issues and challenges that need to be addressed for future research on DTRMS, which include scalability, privacy, excessive resource consumption, security, and interoperability.

4.1   Scalability

Blockchain requires a transparent shared ledger which is replicated between blockchain participants to enforce redundancy and maintain consistency. While the replication removes a single point of failure and increases availability, the shared ledger may grow significantly due to the append-only nature of blockchain storage mechanism. In the long run, the explosion of storage requirements may hinder the performance of the network as it demands high memory requirements, which causes high synchronization times for new node initialization. For instance, as of June 2021,

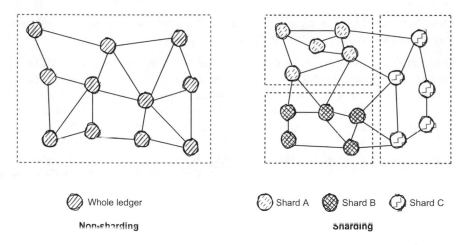

Whole ledger   Shard A   Shard B   Shard C

Non-sharding   Sharding

Fig. 4 The intuition of sharding for improving scalability

Ethereum blockchain size has reached approximately 820GB for a default full node[1] and 7.5 TB for a full archival node[2] and is expected to grow by approximately 75GB per year.

In addition, blockchain is known to have a limitation in block generation time, which contributes to scalability issues. While the Visa payment network could achieve up to 47,000 transactions per second (tps), Bitcoin is only able to cater approximately 7 tps with a limited block size of 1 megabyte [33]. The fixed rate of block generation time introduces a bottleneck and could be exacerbated if there are more transactions to be processed.

DTRMS is thus faced with potential scalability issues inherent to the underlying blockchain. The growing number of participants in the network would also deteriorate the scalability of DTRMS, as accommodating a large number of nodes results in large storage requirements. It is known that addressing scalability for blockchain is still an open research problem and the community is still actively proposing new solutions. Some methods to overcome scalability issues in a DTRMS may include:

Sharding In this mechanism, a full copy of the shared ledger that represents the current state of the blockchain is separated into several chunks which are distributed to different nodes. That is, each node would retain different parts of the whole ledger which significantly reduces the storage requirement to store the entire blockchain state (see Fig. 4).

Scalable Reputation Scheme The reputation model can be adjusted to circumvent the scalability issue by storing a collective or aggregated trust information that resembles a group of devices instead of storing trust information for all devices, which may be redundant [6]. These storage mechanisms may help reducing the amount of data stored in the ledger, which helps to alleviate the scalability issue. For example, the

---

[1] https://etherscan.io/chartsync/chaindefault.

[2] https://etherscan.io/chartsync/chainarchive.

reputation model may average the trust score of a group of devices from the same owner prior to storing the record in the ledger.

Off-chain computations As demonstrated in the Lightning network of Bitcoin [33], off-chain computations would result in a significant reduction of processing latency, as the transactions are processed off-chain without going through excessive consensus processes. In this scenario, two parties retain a signed contract that resembles the current token balance of each party and continue the transactions off-chain by keeping a signed log of balance transfers. An overlay network can also be constructed from the multiple-signed contract of different owners, i.e., the so-called Lightning network. The contract is ceased when a party submits a final transaction to the main ledger that transfers the tokens according to what is recorded in the signed logs.

4.2   Privacy

One of the main motivations for incorporating blockchain for TRMS is the inherent use of pseudonyms instead of real identities for authentication purposes. Using public keys as pseudonyms for authentication would conceal the actual identities of the users and make it difficult to link the public keys to the real user identities, a preferred requirement for privacy preservation. However, research has shown that it is possible to track user behavior from pseudonym-based transaction logs and link them back to the real identities [38]. Ideally, a DTRMS should achieve privacy preservation by addressing the following concerns [8]:

User anonymity The goal of user anonymity is to conceal the actual identity and prevent linkage attacks. For example, a user can be represented by more than one pseudonym, e.g., replaceable public keys, that would allow the users to continue transacting in the system without allowing a malicious entity to link the multiple pseudonyms to their actual identities.

Feedback confidentiality In practice, total user anonymity is difficult to achieve and some information regarding the identities or interactions of users might be inferred from the data revealed by the applications. However, it is a critical requirement to ensure the confidentiality of feedback information submitted by the users in order to encourage users to submit truthful feedback.

The research community in the field of security and privacy still actively proposes new methods for privacy preservation, some of which can be implemented for DTRMS:

Zero-Knowledge Proofs Using Zero-Knowledge Proofs (ZKP), one can verify the validity of a statement by getting a plausible proof without exposing any additional information other than the validity of a statement, hence preserving the privacy [39]. ZKP can be implemented to provide a means to validate a particular pseudonym without any risk of revealing sensitive information that might be linked back to the real identity of a user.

Homomorphic Encryption Typically, a generic encryption algorithm provides a means to conceal sensitive information to preserve privacy. However, the encrypted data, i.e., ciphertext, is practically unusable and decryption is required for the ciphertext to be usable, which may expose sensitive information to public. Homomorphic encryption, on the other hand, allows computations on encrypted data without the need for decryption, which may help to keep the sensitive information hidden. For instance, in DTRMS homomorphic encryption permits several operations on encrypted feedback to achieve feedback confidentiality.

Secure Multi-Party Computation Hiding sensitive information can be achieved by utilizing Secure Multi-Party Computation (SMPC), using which several input values can be aggregated into an output value without revealing the individual input values. SMPC can be implemented in a DTRMS to calculate trust or reputation scores by aggregating several feedback values while keeping the feedback values private.

4.3   Resource Consumption

The traditional centralized TRMS approach utilizes a single TTP that acts as a single authoritative entity for managing resources and making decisions. While there is an inherent risk in trusting a TTP, this centralized architecture results in a relatively low resource consumption. On the other hand, the absence of TTPs in DTRMS requires a distributed consensus algorithm that typically consumes high resources while also sacrificing latency and throughput. For instance, Bitcoin is notorious for its low block generation rate of 10 minutes and high carbon footprint from the coin mining process [40].

In general, CPS applications may consist of thousands of interconnected constrained devices that demand high throughput and low latency. One possible solution to overcome the high resource consumption of blockchain is to devise a tailored consensus algorithm for CPS that suits the constraints of typical CPS devices [41]. In [42], the authors proposed a blockchain platform tailored for CPS/IoT architecture, called Lightweight Scalable Blockchain (LSB), where the computation and storage capacity are typically constrained. LSB employs a lightweight consensus algorithm and distributed throughput management specifically designed for CPS/IoT.

As discussed in previous sections, a reputation system can be implemented to create an efficient consensus algorithm with less overheads. In the reputation-based consensus algorithm, a Proof-of-Work consensus algorithm that demands solving a mathematical puzzle for verifying a block is replaced with a Proof-of-Stake-like consensus, where the block validator is selected based on its reputation in the network [43]. While the reputation-based approach is still prone to trust or reputation attacks (see Sect. 4.4), it can reduce the computational load for achieving a distributed consensus to determine the next block to be mined.

4.4   Security

As both traditional and decentralized TRMS rely on multiple untrusted parties to gain collective knowledge for building reputation scores, any entity in the system can launch attacks that may impact the normal operation of the system. There are known attacks in both traditional and decentralized TRMS, which are commonly referred to as trust-related attacks. A DTRMS should be resilient to these attacks, as noted in the following:

Sybil Attack In this type of attack, an adversary creates several forged identities that can be utilized to gain disproportionate influence against a benign user with a single truthful identity. The adversary may illegitimately utilize the forged identities to launch attacks such as ballot-stuffing or bad-mouthing for its own benefits. Sybil attacks can be avoided by increasing the cost of creating new identities [44]. Also, when privacy is not a key requirement, linking to real identity is proven to be effective in preventing Sybil attack [45].

Ballot-stuffing An adversary can illegitimately increase its reputation score by launching a Sybil attack to submit multiple fake feedback or by colluding with other adversaries. This attack is called ballot-stuffing and is also often referred to as a self-promotion attack. The risk of a ballot-stuffing attack can be partly reduced by using coins to submit feedback [5], which would impose a significant cost to the adversaries for submitting multiple feedback by themselves, while the advantage of the attack may not be worthwhile.

Bad-mouthing In contrast to ballot-stuffing, bad-mouthing is an attack that aims to ruin another honest user's reputation score by providing negative feedback regardless of the behavior of the target user. Bad-mouthing attacks could lead to severe damaging effects, especially for sensitive applications, such as monetary systems [8]. While mitigating this effect is a non-trivial task, one possible protection could be to compare feedback from unknown entities to those from highly trusted nodes [46].

Whitewashing When the adversaries have low or negative reputation scores, they can rejoin the system with a new identity resulting in a fresh reputation score. This type of attack is attractive to the adversaries especially when the cost of re-entering the system is very minimal. As a mitigation scheme, the system may require users to link the identity or pseudonym with a real-world identity, e.g., a website, that would incur a significant cost for modification [47].

On-off Attack The adversary may act opportunistically by providing alternating feedback, i.e., positive and negative, to maintain the reputation score at a safe level to avoid detection. For instance, the adversary may constantly provide positive services to get selected as a service provider, but once selected, the adversary launches the attack by providing poor services to the selected target nodes [48]. Appropriate weighting according to temporal dynamics in the reputation formula would help to reduce the risk of this attack [8]. In this approach, higher weights are applied to recent or more important interaction evidence, resulting in a significant decline in the trust score in case of an attack.

4.5 Interoperability

In practice, it is common for CPS applications to have a wide range of technological implementations with different types of hardware and protocols for communications. Consequently, these implementations tend to work in isolation with very minimal cross-platform collaboration. Interoperability is an important factor to consider in designing CPS solutions for achieving efficient collaboration across platforms and applications. For instance, in blockchain-based smart city architecture, interoperability would allow the transfer of digital assets and values across different blockchain platforms [49]. There are several initiatives in blockchain interoperability, which include Cosmos[3] and Interledger.[4]

In addition, carefully modeled interoperability schemes would also allow the transfer of trust and reputation scores across different platforms. In any TRMS application, the lack of prior interactions or recommendations is a challenge in quantifying reputation, which could be mitigated by transferring reputation values from a separate platform that has already gathered some evidence about the trustworthiness of a particular entity. Note that context-awareness should also be taken into account in this mechanism, as trust is highly associated with the underlying context and may not be transferable from one context to the other.

In the context of DTRMS for CPS, interoperability would also mean an ability to appropriately derive trust and reputation scores across heterogeneous constrained devices. As typical CPS applications include a variety of hardware types with different capabilities in power, computation, and storage, the trust and reputation model should take into account these conditions of heterogeneity [6]. For instance, when the trust score is derived from the computational power of a device, the model should apply an appropriate weighting for high-performance and constrained devices to overcome the heterogeneity in computational power.

5 Conclusion

The goal of a Trust and Reputation Management Systems (TRMS) is to assess and quantify the trustworthiness of each participant in the system for safeguarding users from the risk of interacting with untrustworthy entities. Adoption of blockchain to TRMS can benefit and enhance TRMS. In this chapter, we presented how the salient features of blockchain can enhance TRMS. We specifically focused on four features of blockchain, namely decentralization, smart contracts, pseudonyms, immutable storage, and transparent mechanism. We described several implementations of DTRMS across different fields of applications. We also discussed several open challenges and future directions for research in the area of DTRMS.

---

[3] https://cosmos.network/.

[4] https://interledger.org/.

Acknowledgements The work has been supported by the Cyber Security Research Centre Limited whose activities are partially funded by the Australian Government's Cooperative Research Centres Programme.

References

1. Batwa, A., Norrman, A.: Blockchain technology and trust in supply chain management: a literature review and research agenda. Oper. Supply Chain Manag. Int. J. **14**(2), 203–220 (2021). https://doi.org/10.31387/oscm0450297
2. Gambetta, D., et al.: Can we trust trust. In: Trust: Making and Breaking Cooperative Relations, vol. 13, pp. 213–237 (2000)
3. Chen, I.R., Bao, F., Guo, J.: Trust-based service management for social internet of things systems. IEEE Trans. Depend. Secure Comput. **13**(6), 684–696 (2016). ISSN: 19410018. https://doi.org/10.1109/TDSC.2015.2420552
4. Avyukt, A., Ramachandran, G.S., Krishnamachari, B.: A decentralized review system for data marketplaces. In: 2021 IEEE International Conference on Blockchain and Cryptocurrency (ICBC), pp. 1–9 (2021)
5. Bellini, E., Iraqi, Y., Damiani, E.: Blockchain-based distributed trust and reputation management systems: a survey. In: IEEE Access, vol. 8, pp. 21127–21151 (2020). ISSN: 21693536. https://doi.org/10.1109/ACCESS.2020.2969820
6. Sharma, A., et al.: Towards trustworthy internet of things: a survey on trust management applications and schemes. In: Computer Communications, vol. 160, pp. 475–493 (2020). ISSN: 1873703X. https://doi.org/10.1016/j.comcom.2020.06.030
7. Jøsang, A., Ismail, R., Boyd, C.: A survey of trust and reputation systems for online service provision. Decis. Support Syst. **43**(2), 618–644 (2007)
8. Hasan, O., Brunie, L., Bertino, E.: Privacy preserving reputation systems based on blockchain and other cryptographic building blocks: a survey. In: University of Lyon Research Report, pp. 1–65 (2020)
9. Levien, R., Aiken, A.: Attack-resistant trust metrics for public key certification. In: Usenix Security Symposium, pp. 229–242 (1998)
10. Kamvar, S.D., Schlosser, M.T., Garcia-Molina, H.: The eigentrust algorithm for reputation management in p2p networks. In: Proceedings of the 12th International Conference on World Wide Web 2003, pp. 640–651 (2003)
11. Whitby, A., Jøsang, A., Indulska, J.: Filtering out unfair ratings in bayesian reputation systems. In: Proceedings of the 7th International Workshop on Trust in Agent Societies, vol. 6, pp. 106–117. Citeseer (2004)
12. Dedeoglu, V., et al.: A trust architecture for blockchain in IoT. In: Proceedings of the 16th EAI International Conference on Mobile and Ubiquitous Systems: Computing, Networking and Services. MobiQuitous'19, pp. 190–199. Association for Computing Machinery, Houston (2019). ISBN: 9781450372831. https://doi.org/10.1145/3360774.3360822
13. Yan, Z., et al.: Social-chain: decentralized trust evaluation based on blockchain in pervasive social networking. ACM Trans. Internet Technol. **21**(1) (2021). ISSN: 15576051. https://doi.org/10.1145/3419102
14. Kouicem, D.E., et al.: A decentralized blockchain-based trust management protocol for the internet of things. IEEE Trans. Depend. Secure Comput. 1 (2020). https://doi.org/10.1109/TDSC.2020.3003232
15. Putra, G.D., et al.: Trust management in decentralized IoT access control system. In: 2020 IEEE International Conference on Blockchain and Cryptocurrency (ICBC), pp. 1–9 (2020). https://doi.org/10.1109/ICBC48266.2020.9169481

16. Putra, G.D., et al.: Trust-based blockchain authorization for IoT. IEEE Trans. Netw. Serv. Manag. 1 (2021). https://doi.org/10.1109/TNSM.2021.3077276
17. Bai, Y., et al.: Blockchain-based trust management for agricultural green supply: a game theoretic approach. J. Cleaner Product. 127407. ISSN: 09596526. https://doi.org/10.1016/j.jclepro.2021.127407
18. Malik, S., et al.: TrustChain: trust management in blockchain and IoT supported supply chains. In: 2019 IEEE International Conference on Blockchain (Blockchain), pp. 184–193 (2019). https://doi.org/10.1109/Blockchain.2019.00032
19. Li, H., et al.: Reputation-Based Trustworthy Supply Chain Management Using Smart Contract. LNCS, vol. 12454, pp. 35–49. Springer International Publishing (2020). ISBN: 9783030602475. https://doi.org/10.1007/978-3-030-60248-2_3
20. Zou, J., et al.: A proof-of-trust consensus protocol for enhancing accountability in crowdsourcing services. IEEE Trans. Serv. Comput. **12**(3), 429–445 (2019). https://doi.org/10.1109/TSC.2018.2823705
21. Feng, W., Yan, Z.: MCS-chain: decentralized and trustworthy mobile crowdsourcing based on blockchain. In: Future Generation Computer Systems, vol. 95, pp. 649–666 (2019). ISSN: 0167739X. https://doi.org/10.1016/j.future.2019.01.036
22. Alowayed, Y., et al.: Picking a partner: a fair blockchain based scoring protocol for autonomous systems. In: Proceedings of the Applied Net24 working Research Workshop. ANRW'18, pp. 33–39. Association for Computing Machinery, Montreal (2018). ISBN: 9781450355858. https://doi.org/10.1145/3232755.3232785
23. Boldyreva, A., et al.: Order-preserving symmetric encryption. In: Annual International Conference on the Theory and Applications of Cryptographic Techniques, pp. 224–241. Springer (2009)
24. Strobel, V., Dorigo, M.: Blockchain technology for robot swarms: a shared knowledge and reputation management system for collective estimation. In: Swarm Intelligence-Proceedings of ANTS 2018-Eleventh International Conference, pp. 425–426. Springer (2018)
25. Zhao, Y., et al.: PBTM: a privacy-preserving announcement protocol with blockchain-based trust management for IoV. In: IEEE Syst. J. 1–10 (2021). ISSN: 1932-8184. https://doi.org/10.1109/JSYST.2021.3078797. https://ieeexplore.ieee.org/document/9442949/
26. Wang, Y., Su, Z., Zhang, N.: BSIS: blockchain-based secure incentive scheme for energy delivery in vehicular energy network. In: IEEE Trans. Ind. Inf. **15**(6), 3620–3631 (2019). ISSN: 15513203. https://doi.org/10.1109/TII.2019.2908497
27. Yang, Z., et al.: Blockchain-based decentralized trust management in vehicular networks. IEEE Int. Things J. **6**(2), 1495–1505 (2019). ISSN: 23274662. https://doi.org/10.1109/JIOT.2018.2836144
28. Lu, Z., et al.: BARS: a blockchain-based anonymous reputation system for trust management in VANETs. In: Proceedings - 17th IEEE International Conference on Trust, Security and Privacy in Computing and Communications and 12th IEEE International Conference on Big Data Science and Engineering, Trustcom/BigDataSE 2018, pp. 98–103 (2018). ISSN: 2324-9013. https://doi.org/10.1109/TrustCom/BigDataSE.2018.00025. arXiv:1807.06159
29. Camilo, G.F., et al.: A secure personal-data trading system based on blockchain, trust, and reputation. In: 2020 IEEE International Conference on Blockchain (Blockchain), pp. 379–384 (2020). https://doi.org/10.1109/Blockchain50366.2020.00055
30. Javaid, A., et al. Reputation system for IoT data monetization using blockchain. In: Barolli, L., Hellinckx, P., Enokido, T., (eds.), Advances on Broad-Bandwireless Computing, Communication and Applications, pp. 173–184. Springer International Publishing, Cham (2020). ISBN: 978-3-030-33506-9
31. Rahimpour, S., Khabbazian, M.: Hashcashed reputation with application in designing watch towers. In: 2021 IEEE International Conference on Blockchain and Cryptocurrency (ICBC). 2021, pp. 1–9 (2021)
32. Back, A.: Hashcash-a denial of service counter-measure (2002)
33. Poon, J., Dryja, T.: The bitcoin lightning network: scalable off-chain instant payments (2016)

34. Zhao, Y., et al.: Secure pub-sub: blockchain-based fair payment with reputation for reliable cyber physical systems. In: IEEE Access, vol. 6, pp. 12295–12303 (2018). ISSN: 21693536. https://doi.org/10.1109/ACCESS.2018.2799205
35. Yahaya, A.S., et al.: Blockchain-based energy trading and load balancing using contract theory and reputation in a smart community. In: IEEE Access, vol. 8, pp. 222168–222186 (2020). ISSN: 21693536. https://doi.org/10.1109/ACCESS.2020.3041931
36. Khorasany, M., et al.: Lightweight blockchain framework for locationaware peer-to-peer energy trading. Int. J. Electrical Power Energy Syst. **127**, 106610 (2021). ISSN: 01420615. https://doi.org/10.1016/j.ijepes.2020.106610. arXiv:2005.14520
37. Wang, T., et al.: RBT: a distributed reputation system for blockchainbased peer-to-peer energy trading with fairness consideration. In: Applied Energy, vol. 295 (2021). ISSN: 03062619. https://doi.org/10.1016/j.apenergy.2021.117056
38. Dorri, A., et al.: On the activity privacy of blockchain for IoT. In: 2019 IEEE 44th Conference on Local Computer Networks (LCN), 2019, pp. 258–261. https://doi.org/10.1109/LCN44214.2019.8990819
39. Goldwasser, S., Micali, S., Rackoff, C.: The knowledge complexity of interactive proof systems. SIAM J. Comput. **18**(1), 186–208 (1989)
40. Stoll, C., Klaaßen, L., Gallersdrfer, U.: The carbon footprint of bitcoin. Joule **3**(7), 1647–1661 (2019)
41. Dorri, A., Kanhere, S.S., Jurdak, R.: Blockchain in internet of things: challenges and solutions. (2016). arXiv: 1608.05187 [cs.CR]
42. Dorri, A., et al.: LSB: a lightweight scalable blockchain for IoT security and anonymity. J. Parallel Distrib. Comput. **134**, 180–197 (2019)
43. Yu, J., et al.: RepuCoin: your reputation is your power. IEEE Trans. Comput. **68**(8), 1225–1237 (2019). https://doi.org/10.1109/TC.2019.2900648
44. Douceur, J.R.: The sybil attack. In: International Workshop on Peer-Topeer Systems, pp. 251–260. Springer (2002)
45. Haifeng, Y.: Sybillimit: a near-optimal social network defense against Sybil attacks'. In: IEEE Symposium on Security and Privacy (sp 2008), vol. 2008, pp. 3–17. IEEE (2008)
46. Jøsang, A., Golbeck, J.: Challenges for robust trust and reputation systems. In: Proceedings of the 5th International Workshop on Security and Trust Management (SMT 2009), Saint Malo, France, vols. 5, 9. Citeseer (2009)
47. Hoffman, K., Zage, D., Nita-Rotaru, C.: A survey of attack and defense techniques for reputation systems. ACM Comput. Surv. (CSUR) **42**(1), 1–31 (2009)
48. Chahal, R.K., Kumar, N., Batra, S.: Trust management in social internet of things: a taxonomy, open issues, and challenges. In: Computer Communications 150, November 2019, pp. 13–46 (2020). ISSN: 1873703X. https://doi.org/10.1016/j.comcom.2019.10.034
49. Dedeoglu, V., et al.: Blockchain technologies for IoT. In: Kim, S., Deka, G.C., (eds.), Advanced Applications of Blockchain Technology, pp. 55–89. Springer Singapore, Singapore (2020). https://doi.org/10.1007/978-981-13-8775-3_3

Advances in Blockchain Security

Truc Nguyen, Tre' R. Jeter, and My T. Thai

Abstract Blockchain, the technology that underpins the great success of Bitcoin and various other cryptocurrencies, has incredibly emerged as a trending research topic in both academic institutes and industry associations in recent years. With great potential and benefits, the blockchain technology can stimulate a new decentralized platform for various applications such that the possibility of censorship, monopoly, and single point of failures can be eliminated. However, the blockchain is still in its early stages and not yet ready to realize that vision, since there are many security vulnerabilities that can be exploited to obstruct blockchain systems. In this chapter, we present fundamental challenges and recent advancements in the blockchain technology, especially in terms of security. In particular, we investigate the security threats of blockchain, effectively capturing the recent attacks, and review some security enhancement solutions for blockchain.

1 Introduction

Since the original paper in 2009 [1], Bitcoin has gained much attention from both academic institutes and industry associations. With a market capitalization of more than one hundred million dollars [2], Bitcoin is undoubtedly one of the most successful cryptocurrencies, averaging thousands of transactions per day. Blockchain is the technology that underpins the success of Bitcoin. At its core, blockchain is essentially a distributed ledger of transactions maintained by a set of nodes that do not trust one another. By using a consensus mechanism, nodes in a blockchain network agree on an ordered set of linked data blocks that each contains multiple valid and digitally signed transactions. The main selling point of blockchain is a decentralized nature

T. Nguyen (✉) · T. R. Jeter · M. T. Thai
University of Florida, Gainesville, USA
e-mail: truc.nguyen@ufl.edu

T. R. Jeter
e-mail: t.jeter@ufl.edu

M. T. Thai
e-mail: mythai@cise.ulf.edu

© The Author(s), under exclusive license to Springer Nature Switzerland AG 2022
D. A. Tran et al. (eds.), *Handbook on Blockchain*, Springer Optimization
and Its Applications 194, https://doi.org/10.1007/978-3-031-07535-3_11

in which applications can operate efficiently without the need of a central authority. From the perspective of database systems, blockchain can also be viewed as a distributed database for transaction management. While traditional databases assume a trusted environment, nodes in a blockchain network can behave in arbitrary manner.

One of the core components of blockchain systems is a consensus mechanism that is used to achieve verifiable decentralized consensus in the presence of malicious nodes. This is also referred to as making blockchain Byzantine Fault Tolerant, or BFT. A consensus mechanism can take the form of a probabilistic (e.g., Proof of Work/Proof of Stake) or deterministic (e.g., Practical BFT [3]) algorithm. Finality achieved via probabilistic consensus algorithms is temporary, nonetheless, as more blocks are added to the chain over time, the probability of overturning the previous blocks become smaller, approaching zero. By design, blockchain can tolerate Byzantine failure, thus it offers stronger security than conventional database systems.

Bitcoin, in its original design, is a blockchain that stores coins and is limited to facilitating financial transactions that move coins from one address to another. Since then, blockchain has evolved beyond cryptocurrencies to support any arbitrary, programmable transaction logic [4]. For example, Ethereum is a blockchain that enables any decentralized applications in the form of smart contracts. In the context of blockchain, *smart contracts* are defined as self-executing and self-enforcing programs that are stored on chain. They are intended to facilitate and verify the execution of terms and conditions of a contract within the blockchain system. By employing this technology, applications that previously require a trusted intermediary can now operate in a decentralized manner while achieving the same functionality and certainty. For that reason, blockchain and smart contracts together have inspired many decentralized applications and stimulated scientific research in diverse domains [5–9].

Unfortunately, due to its popularity and the value of cryptocurrencies, efforts have been made to exploit the weaknesses and vulnerabilities of blockchain. As a result, it is known to be susceptible to various security issues [2, 10] and was attacked multiple times in the last 10 years. For this reason, the blockchain technology is still in its early stage and not yet ready to realize its full potential.

In this chapter, we conduct a comprehensive survey on recent advances in blockchain that aims to make the technology more practical and deployable. In specific, we present some security threats of blockchain, especially the vulnerabilities in the blockchain network and smart contracts, which effectively capture past attacks to the blockchain. Then, we review the security enhancement solutions for blockchain and how they would affect the scalability and decentralization. Finally, we survey some other significant advances in blockchain including blockchain anonymity, consensus protocols, and the use of secure hardwares in blockchain.

Organization. The rest of the chapter is structured as follows. Section 2 establishes some background knowledge on cryptology and blockchain technology. Section 3 describes some recent security threats of blockchain, especially on the blockchain network and smart contracts and also shows some security enhancement solutions. In Sect. 4, we present other notable advances in blockchain in terms of privacy and consensus protocols. Finally, Sect. 5 concludes the chapter.

2 Background

This section covers necessary background knowledge for discussing security issues in blockchain. Specifically, we present some cryptographic primitives, including public-key cryptography and cryptographic hash functions, and a general explanation of blockchain technology and transitions from known cryptographic practices to more advanced and practical security methods used in blockchain.

2.1 Cryptographic Primitives

Cryptographic Hash Functions. A cryptographic hash function is generated by a mathematical function that compresses information in a string of letters and numbers of a fixed size. Cryptographic hashes are one-way functions. A one-way function is a function that can be computed easily and quickly for any input, but is very difficult to revert back to the original input [11]. A cryptographic hash function follows this same method. Any input can be "hashed", but the computational complexity to revert it back to the original input proves to be exceedingly difficult.

Cryptographic hash functions should be (1) pre-image resistant, (2) collision resistant, and (3) second pre-image resistant. Pre-image resistance directly corresponds to the one-way functionality of hashing functions. For any hash value, it should be very difficult and nearly impossible to read the corresponding message related to that hash. Denoting $H : \{0, 1\}^* \rightarrow \{0, 1\}^l$ as a public cryptographic hash function that maps a bitstring of arbitrary length to a bitstring of fixed-length l, the pre-image resistance property states that, given a hash value h, it is infeasible to find any message m such that $h = H(m)$. Formally speaking, for any probabilistic polynomial algorithm A_1, we have

$$\Pr[m \leftarrow A_1(1^\lambda, h)|h = H(m)] < negl(\lambda) \tag{1}$$

where $negl(\cdot)$ denote some negligible function, and λ is a security parameter.

A hash function that attains second pre-image resistant must be so complex that it is computationally difficult to find a second input message that will result in an identical hash output. Specifically, given a message m_1, second pre-image resistance makes it computationally infeasible to find a message $m_2 \neq m_1$ such that $H(m_1) = H(m_2)$. This property can be defined formally as follows

$$\Pr[m_2 \leftarrow A_2(1^\lambda, m_1)|m_1 \neq m_2 \wedge H(m_1) = H(m_2)] < negl(\lambda) \tag{2}$$

for any probabilistic polynomial algorithm A_2.

Collisions in hash functions refer to the chance that two inputs' hash values are equivalent to one another. A cryptographic hash function should be collision resistant in that it is difficult to find two distinct messages $m_1 \neq m_2$ such that $H(m_1) = H(m_2)$. In other words, it ensures that, for any probabilistic polynomial algorithm A_3, the following holds:

$$\Pr[(m_1, m_2) \leftarrow A_3(1^\lambda)|m_1 \neq m_2 \wedge H(m_1) = H(m_2)] < negl(\lambda) \qquad (3)$$

Moreover, it can be shown that collision resistance implies second pre-image resistance. Suppose there exists a polynomial algorithm A_2 that can violate equation (2), an adversary can devise a polynomial algorithm A_3 as follows: pick a random message m_1 and obtain $m_2 \leftarrow A_2(1^\lambda, m_1)$ in polynomial time. This results in $m_2 \neq m_1$ and $H(m_1) = H(m_2)$, thus violating equation (3).

The "birthday paradox" places an upper bound on the computational difficulty of a collision-finding algorithm: if the output length of a hash function is l bits, then an attacker who computes hashes of $2^{l/2}$ random inputs can find a collision with probability greater than 0.5. A hash function is considered flawed if a collision can be found by a method easier than this brute-force attack.

Public Key Cryptography. Public key cryptography was the solution to two problems: key distribution and signatures [12]. Sometimes called asymmetric cryptography, this cryptographic system involves a pair of keys: a public key and a private key. The public key can be distributed to all users sending encrypted messages to one another and the private key is always kept private. When a message is sent over an insecure network, it is encrypted using the public key. The recipient will then use their private key to decrypt the message.

Public key cryptography is also used to authenticate users. A message can be combined with a user's private key to generate a digital signature on top of the message. Another user with the related public key can also combine the same message with a known signature. If the generated signature matches the message that was sent from the first user, then that user is said to be authenticated and trusted.[1] The message itself is also verified.

Asymmetric cryptographic algorithms are slower than symmetric cryptographic algorithms, but still useful. Some algorithms are built for key distribution and privacy such as the Diffie-Hellman key exchange. There are algorithms such as the Digital Signature Algorithm that only create digital signatures. However, when algorithms like the two above-mentioned are combined, the Rivest-Shamir-Adleman (RSA) algorithm is the result. This algorithm allows for users to openly share encrypted files, data, or other information through the internet or email for example. The public key encrypts the message and only the recipient's private key can decrypt the message [13]. RSA is widely used today for secure data transmission and an easy way to implement multi-factor authentication into secure systems.

---

[1] https://www.ibm.com/docs/en/ztpf/1.1.0.14?topic=concepts-digital-signatures.

Taking these algorithms further in application, implementing a Public Key Infrastructure (PKI) would further verify and authenticate users. A PKI is a system of a third-party user called a Certificate Authority (CA) that certifies the ownership of a set of keys. This system is good for avoiding attacks because it is a set of protocols that manage overall public-key encryption and the creation, distribution, use, storing, and disabling of digital certificates.

2.2 Blockchain Primer

Peer-To-Peer Network. A peer-to-peer network is a decentralized network that is maintained by a distributed group of users that can act as servers and clients. Unlike the traditional Client-Server network, a peer-to-peer network does not have a central server. By definition, a peer-to-peer network consists of each node/peer within that network providing and making their personal resources accessible to others on the network without the need for an intermediary entity like a central server [14]. The peer-to-peer network plays a huge role in blockchain technology because it allows for transactions of any kind without the need for a middle-man or central server. In this distributed network, any user can verify or validate transactions and be a part of the process of creating new blocks by simply setting up a node on the blockchain.

The decentralized nature of blockchain technology makes it easily accessible and available. The peer-to-peer make-up also ensures resiliency. If one peer goes down, the other peers within the network are not affected and can maintain work. Blockchain technology has consensus constraints as does a stand-alone peer-to-peer network which mitigates a blockchain from many malicious activities. This network is also nearly impossible to execute a Denial-of-Service attack because there is no central server to attack.

Blockchain Architecture. A blockchain is essentially a database of records (transactions) shared across a peer-to-peer network [15]. Because of its immutable nature, changing a block on the blockchain is very difficult to do without being noticed. A record on a blockchain can be any piece of information such as bank transactions, health information, purchases, voting results, cryptocurrency, etc. A block on the blockchain is made up of a group of records. The actual blockchain itself is all of the blocks linked to one another.

Each record lists the details of each transaction and appends a digital signature from all who were involved with the transaction. The record is then verified by the network by every participant (node) to check the validity of the transaction. This process is called consensus. A decentralized network of nodes must come to a consensus before a block is added to the blockchain by any of the consensus algorithms (i.e., PoW, PoS, PoA, PoAh, etc).[2] Once a record is verified by the network, it is added to a block.

---

[2] https://www.investopedia.com/terms/c/consensus-mechanism-cryptocurrency.asp.

Every block in a blockchain has a cryptographic hash as its unique identifier. Each block contains the hash of the previous block and its own unique hash. Because these hash values match with each subsequent block, it is very difficult (nearly impossible) to change information on the blockchain without being noticed. Altering a block will change the hash of that block and break the chain. If the hashes of the previous block and current block do not match, then the blockchain has been altered. To restore the altered block, one would have to recalculate the original hash and each hash of the following blocks.

Double-Spending and Reaching Consensus. The Bitcoin and blockchain technology in general were motivated by the double-spending attack. In digital currency systems, double spending is the act of successfully spending some coins more than once.

In conventional systems, all the transactions are validated and recorded in a centralized manner. However, in blockchain, every node processes transactions and keeps a copy of the ledger. Due to the fact that multiple copies of blockchain are stored at different nodes in the network, it is challenging to maintain a consistent global view of the blockchain in the whole system. In particular, a node can simultaneously issue two different transactions on the same set of coins as input, to two different receivers. If both the receivers successfully validate the transaction independently based on their local view of the blockchain, then their copies of the blockchain become different and the blockchain ends up with **forks**. Specifically, the nodes will have different views of the global state and the network will no longer be consistent unless we resolve this fork. This type of malicious behavior makes decentralized currency particularly vulnerable to double spending.

Therefore, a distributed consensus mechanism is needed in a blockchain network to tackle this issue [6]. Intuitively, all nodes in the blockchain could vote on the order of transactions for each block, and the result is decided by the majority. Unfortunately, in an open network where anyone can participate, this mechanism would not be secure due to the **Sybil attack**: a single entity can generate various identities, vote several times, and thereby becoming the majority of the network. In other words, any adversary can easily take over the blockchain.

Bitcoin tackles this issue by proposing the proof-of-work mechanism [1], where each node has to solve a computationally expensive puzzle to vote for a block. This is also referred to as *mining*. As a result, generating several sybil identities on the blockchain is futile, as the computing resources of any single node are limited. In the event of a fork, the proof-of-work mechanism ensures that the nodes choose the fork that contains the most amount of work, that is, choosing the longest fork. Hence, this enables the network of untrusted nodes to reach consensus on the proper order of transactions.

Bitcoin Mining. Mining is where cryptographic hashes come into play with Bitcoin and blockchain technology. Miners maintain the consistency and immutability of the blockchain by placing each new transaction into a new block to be placed on the blockchain. Miners send this new block to the entire network for validation and consensus before the block is added to the blockchain. Every block has a tag in the form of a SHA-256 cryptographic hash [16] of the previous block and its own cryptographic hash.

Blocks are only accepted if they contain a valid proof-of-work (PoW). The PoW is generated by double hashing a block header, which contains a nonce, using the SHA-256 function. To make a valid PoW of a block, the miners need to find a nonce so that the resulting hash output is smaller than the difficulty target determined by the network. The higher the target value, the easier it is to find the correct PoW combination.[3] A nonce is a number that is only used once and is usually a sequence of natural numbers. This process gives the blockchain its immutable nature. Depending on the difficulty target, it usually takes an immense amount of computing power to generate a valid PoW for a new block. Furthermore, because the block will have the cryptographic hash of the previous block and its own attached, the block is very difficult to alter without invalidating the chain. Altering the transaction information of the block will change the hash of the block which will render the block compromised.

Ethereum and Smart Contracts. Ethereum is another decentralized network similar to Bitcoin that was created in 2013 by Vitalik Buterin. Its cryptocurrency is called Ether and is only second in value to Bitcoin. Ethereum allows for the creation of non-fungible tokens (NFTs). These tokens can represent anything that has an actual value as a unique item like art, photos, or any digital files. The distributed ledger within blockchain is then used to verify ownership of these items creating a safe and efficient marketplace. The main difference between Ethereum and Bitcoin is Ethereum's use of smart contracts.

Smart contracts are programs that are stored and executed by all nodes in the Ethereum blockchain using the Ethereum Virtual Machine (EVM). The EVM is basically a stack-based virtual machine that supports a Turing-complete programming language. Smart contracts can be deployed and triggered by the blockchain transactions. Each operation on the EVM costs some amount of gas that determines the fee needed to execute the smart contract. The transaction is assigned with a bounded amount of gas, and when that amount is exceeded, the entire execution is terminated and the operations are reversed. In contrast to conventional programs, smart contracts are immutable by design. Therefore, programming mistakes or vulnerabilities on the smart contract cannot be reversed or fixed.

In Ethereum, a smart contract can utilize three memory regions to perform data operations during execution: stack, memory, and storage. A (data) stack is a virtual stack that can be used to store data. Note that EVM also has a call stack, which is different from the data stack. The memory is a byte-addressable region allocated at run-time. Storage is implemented using a key-value store. The stack and memory are both volatile, meaning that the data stored are cleared after each execution. However, the storage is persistent, which can be used to store data across transactions.

---

[3] https://medium.com/geekculture/the-implication-of-bitcoins-proof-of-work-algorithm-40921bb13530.

3   Blockchain Security: Attacks and Counter-measures

This section presents a comprehensive survey on some security issues of blockchain, especially focusing on attacks and threats on the blockchain network and smart contract.

3.1   Blockchain Network

Attacks on Peer-to-Peer Network. By design, the blockchain peer-to-peer network is open, decentralized, and independent of a public-key infrastructure. Hence, it does not employ cryptographic authentication between nodes, and they are identified purely by IP addresses. In the Bitcoin network, each node is implemented to use a randomized protocol to select eight peers to create outgoing connections. Nodes with public IPs accept up to 117 incoming connections from any IP addresses. Nodes exchange their local views of the state of the blockchain with their connected peers.

However, this open nature of blockchain makes it feasible for adversaries to join and attack the peer-to-peer network. Heilman et al. [17] investigate an eclipse attack on the bitcoin network where the attacker takes control over all of the victim's incoming and outgoing connections, thereby isolating the victim from the rest of its peers in the network. After that, the attacker is free to manipulate the victims' view of the blockchain, force the victim to waste computing power on obsolete views of the blockchain, or exploit the victims' mining power for its own malicious purposes. The authors present an off-path attack in which the attacker only controls endhosts but not key network infrastructure between the victim and the rest of the Bitcoin network. The attack mainly forms incoming connections to the victim from a set of controlled endhosts, sends fake network information, and waits until the victim restarts. With high probability, the victim then creates all eight of its outgoing connections to attacker-controlled endhosts. Additionally, the attacker also monopolizes the victims' 117 incoming connections.

Some security implications of this attack include: (1) An attacker can hoard blocks discovered by eclipsed miners, and release blocks to both the eclipsed and non-eclipsed miners once a competing block has been found, thereby making the eclipsed miners waste computing resources on orphan blocks; (2) selfish mining [18]; (3) eclipsing miners eliminates their mining power from the rest of the network, making it easier to for the attacker to becomes the majority in the network; (4) double spending [19].

On the other hand, Apostolaki et al. [20] exploit the Bitcoin hosting centralization issue to conduct a routing attack. Although one can run a Bitcoin node, the nodes that form the Bitcoin network today are far from being distributed uniformly around the globe. Specifically, their experimental results illustrate that few Internet Service providers (ISPs) host most of the Bitcoin nodes. Specifically, 13 ISPs, which is about 0.026% of all ISPs, host roughly 30% of the entire Bitcoin network. Furthermore, a majority of the network traffic between Bitcoin nodes propagate over only a few ISPs. Indeed, their experiment shows that 60% of all possible Bitcoin connections cross three ISPs. Simply speaking, three ISPs observe, drop, or modify 60% of all Bitcoin traffic.

The authors [20] present how an adversary can utilize the network infrastructure to perform (1) an eclipse attack and (2) a delay attack. The eclipse attack works by intercepting network traffic between blockchain nodes. To do so, an attacker can leverage the fact that Border Gateway Protocol (BGP), an Internet routing protocol, does not verify the source of routing announcements. This attack involves getting a router to spread false announcements that it has a shorter path to certain IP prefixes, thereby maliciously rerouting Internet traffic. From that, the attacker can proceed with hijacking all the IP prefixes associated with the nodes in one component, effectively intercepting all the network traffic exchanged between that component and the rest of the network. This is commonly referred to as a BGP hijacking attack. Once the path is hijacked, the attacker can drop all these connections to disconnect that component from the rest of the network, thus eclipsing the miners. The network centralization of Bitcoin nodes as above-mentioned further exacerbates the issue as few IP prefixes need to be hijacked, making eclipse attacks particularly feasible. In fact, their study shows that 39 prefixes, accounting for 0.007% of all Internet prefixes, host 50% of Bitcoin mining power. This implies that, by hijacking only those 39 prefixes, an attacker is able to isolate roughly 50% of the mining power. BGP hijacking attacks that involve orders of magnitude more IP prefixes are routinely seen in the Internet today.

Besides eclipse attack, a delay attack can be conducted based on the fact that Bitcoin nodes are implemented to send a request for blocks to only one peer to prevent the network from being overwhelmed with the transmissions of blocks. If the peer is not responsive for 20 minutes, an alternative peer will be selected to send the request to. This implementation, together with the fact that Bitcoin messages are exchanged in plaintexts, allows for an effective attack where attackers try to prolong block transmissions by delaying or dropping those requests for blocks. Specifically, the attacker can simply modify to the content of the Bitcoin messages that they intercept. Since the Bitcoin protocol does not offer protection for those messages, both the receiver and the sender become oblivious of the fact that the message has been tampered with, thus enabling a very stealthy attack. The implication is that the attacker can then conduct other attacks like double spending or try to waste computing power of honest miners. What makes such delay attacks feasible and practical is the centralization of Bitcoin nodes in a small number of networks and prefixes, as well as the centralization of mining power in some certain mining pools. The authors discover that three ISPs control a majority of all Bitcoin traffic. This implies that these ISPs can stealthily interfere with Bitcoin traffic. In contrast to eclipse attacks, delay attacks could not disrupt the whole blockchain system, but rather reduce the performance of the network. Thus, even if many nodes are slowed down under attack, the Bitcoin system would still be able to function, but at a lower performance and less secure.

Saad et al. [21] propose some potential attacks based on spatial and temporal characteristics of the Bitcoin network. They investigate three different levels of attacks, emphasizing the network centralization. At the network level, due to the increasing centralization of the Bitcoin network, the authors are able to empirically demonstrate that an attacker can easily partition the network spatially through BGP hijacking by controlling only a few ASes, thus causing a hard fork. At the AS level, they discover that in certain cases, by hijacking roughly 20 prefixes, the adversary can gain control over more than 80% of the Bitcoin nodes that are placed inside the same AS. At the organization level, they show that multiple ISPs control more than one AS, which results in even more centralization, and facilitating new attack avenues. Furthermore, they leverage the non-uniform consensus among connected nodes to propose temporal attacks. They observe that there is a significant delay in consensus and block propagation because of the latency and adversarial peer behavior. Their study suggests that even after a few minutes from the publication of a block, about 62.7% of nodes in the network are not up-to-date and still remain behind the latest block by one or two blocks. As a result, it is suggested that such a behavior can be leveraged to optimize an attack where false blocks are fed to nodes, thereby temporally partitioning the network.

Since those above-mentioned attacks are based on BGP hijacking, however, due to the openness of BGP operations, such a hijacking attempt can be observed globally, thereby enabling instant attack detection and attacker identification. Specifically, the real identity of the attacker (i.e., the malicious AS) is instantly revealed to the public. As such, this can be a deal-breaker for large ASes since attempting the attack can potentially damage their reputation. Tran et al. [22] present a more stealthy Bitcoin attack, which is referred to as EREBUS, that enables a network attacker to control the peer connections of a victim Bitcoin node without manipulating the network routing protocol, thereby eliminating control-plane evidence of attacks. This is possible because the attack strategy only exploits data-plane attack messages, so it remains invisible to any control-plane monitoring systems. Furthermore, even if data-plane traces of the attack are detected, the attack still offers plausible deniability. The authors demonstrate that Tier-1 or large Tier-2 ISPs can conduct this attack to target a majority of thousands of Bitcoin nodes in the system that accept incoming connections from other nodes. Consequently, attackers who control large ISPs (such as nation-state adversary), are capable of launching the EREBUS attack stealthily.

At a high level, EREBUS works as follows. Without interfering with the underlying routing protocols, the adversary AS alters the existing outgoing peering connections of a victim node to the new connections with the Bitcoin nodes whose victim-to-node inter-domain paths include the adversary AS. Eventually, the malicious AS will be placed on the paths of all the peer-to-peer connections of the victim node. The attack is feasible not because of the implementation of Bitcoin nodes but the inherent topological advantage of being a network adversary. In specific, as a man-in-the-middle adversary, the EREBUS malicious AS can exploit an enormous amount of network addresses reliably over a long period of time.

Counter-measures. Two counter-measures are typically recommended for this type of network attack: (1) disable incoming connections and (2) only make outgoing connections to well-connected or known/whitelisted miners. However, there are several problems with scaling this to the full Bitcoin network. First, if incoming connections are disabled on all current nodes, how do new nodes join the network? Second, how does one decide which peers to connect to? Who determines the whitelist of miners? In [17], the authors propose a set of counter-measures that partially preserve openness by allowing unsolicited incoming connections, while raising the threshold for eclipse attacks. The counter-measures ensure that, with high probability, if a victim stores enough legitimate miners that accept incoming connections, then the victim cannot be eclipsed regardless of how many IP addresses the attacker controls.

In [23], the authors propose the SABRE network to secure Bitcoin against the BGP hijacking attacks. SABRE is a Bitcoin relay network that relays blocks worldwide through a set of connections that are resilient to routing attacks. SABRE is designed to be secure and scalable and is able to run alongside the existing peer-to-peer network and can be deployed easily. SABRE is specifically designed to protect both relay-to-relay and relay-to-client connections. At a high level, to secure relay-to-relay connections, SABRE places nodes in ISPs that connect directly to one another, creating a fully connected graph of direct links and also in /24 prefixes. To secure relay-to-client connections, relay nodes are placed in a way that most nodes have for each potential attacker at least one route to SABRE that is more preferable than any route that this attacker can advertise, thereby tackling the BGP hijacking attacks. The main technical insight is that SABRE leverages fundamental properties of BGP policies to host relay nodes in networks that are essentially protected against routing attacks, and on network routes that are preferable by the majority of Bitcoin nodes. These properties are generic and can be used to protect other blockchain networks. However, this approach introduces a trusted entity to the system to control the network connections between nodes, which violates the trust model of blockchain.

The authors in [22] propose a set of counter-measures to defend against stealth BGP hijacking attacks. First, some third-party proxies can be used to verify the reachability of IP addresses. However, this approach has limited scalability because creating multiple proxies at different locations for thousands of potentially vulnerable nodes in the Bitcoin network would be difficult in practice. Furthermore, because of the limited scalability, any proxy-based approaches could eventually result in few centralized proxies. Another solution is increasing the amount of outgoing connections that a Bitcoin node is able to make. According to the authors, the increase can in fact potentially sabotage the network if it is not deployed properly. This is because it may instantly boost the amount of network connections and the volume of network traffic in the system. This sudden increase can potentially exacerbate the delay of transactions and blocks in the system. Therefore, the practicality of these counter-measures remains questionable.

3.2 Smart Contracts

Ensuring the correctness of smart contracts is a critical and urgent security concern. Nowadays, billions of dollars are handled by smart contracts, and only in the past couple of years, millions of these have been lost by adversaries who exploited subtle flaws in the logic of the contracts [24, 25]. In fact, Ethereum already encountered a lot of disastrous attacks on vulnerable smart contracts. The most notable ones are the DAO hack in 2016[4] and the Parity Wallet hack in 2017,[5] together resulting in a loss of over 300 million US dollars. The problem is exacerbated as the smart contracts become immutable once placed on the blockchain, hence bugs and flaws found after deployment cannot be fixed.

Real-World Vulnerabilities of Smart Contracts. Below is a list of vulnerabilities in Ethereum smart contracts according to [7].

- *Airdrop hunting.* Airdrop is a method to reward new users a small amount of tokens as a way of promoting attention and appealing to more users. Airdrop hunting is an attack strategy that leverages the weaknesses of airdrop and bypasses the identity verification of new users to keep generating new sybil users to obtain a large amount of free tokens.
- *Call injection.* Call injection is a method that allows any contract to call any function in a vulnerable contract. It is often used to modify ownership and trigger money transfers.
- *Reentrancy.* A reentrancy attack happens a function is created that makes an external call to another untrusted contract before it updates its own state. A reentrancy attack may lead to a repeated transfer of money from the victim to the adversary, thereby exhausting the balance of the victim contract.
- *Honeypot.* A honeypot is a bait that lures a victim into losing tokens.
- *Call-after-destruct.* Call-after-destruct is the act of calling a function in a destructed contract with tokens, resulting in the loss of these tokens.

Common Attacks on Smart Contracts. In [7], the authors conduct an analysis of real-world attacks based on the log of transactions generated by "uninstrumented" Ethereum Virtual Machine (EVM). In specific, they capture two essential behaviors of a malicious transaction: (1) it attempts to exploit a vulnerable contract and (2) it often results in ether or token transfers. The results unveil a large volume of attacks that is greater than what have been discovered in the literature. In particular, airdrop hunting and zero-day variants of known vulnerabilities are often the targets of those attacks.

One of the most common attacks is luring victims into traps. This type is also commonly referred to as *honeypot*, as it often involves setting up a bait to attract victims. Honeypots are smart contracts that seem to have some apparent flaws and

---

[4] https://www.coindesk.com/understanding-dao-hack-journalists.

[5] https://medium.com/@Pr0Ger/another-parity-wallet-hack-explained-847ca46a2e1c.

bugs in their design and implementation. For instance, several Ethereum smart contracts enable any malicious user to retrieve ether (Ethereum's cryptocurrency) from the contract's balance, given that the user previously transfers a certain amount of ether to the contracts in the first place. However, once the user tries to take advantage of this obvious vulnerability, a second trapdoor (unknown to the user) opens and prevents the draining of ether from succeeding. The key observation here is that the user only pays attention to the obvious flaw and does not think of the possibility that some other vulnerabilities might be concealed within the smart contract. In the same manner as other types of fraud, honeypots exploit the fact that human beings are usually greedy and easily manipulated.

In [25], the authors investigate the incidents of such honeypot smart contracts in Ethereum and introduce HONEYBADGER, a toolbox that uses a combination of symbolic execution and precise heuristics to automatically detect various types of honeypots. By using HONEYBADGER, users have the capability of providing interesting insights on some properties of honeypots that are being hidden in smart contracts on the Ethereum blockchain.

Another attack on Ethereum smart contracts is to exploit several flaws in the *metering* mechanism of Ethereum to conduct a DoS attack. This metering mechanism is used to assign a gas cost to smart-contract execution in order to incentivize miners to operate the blockchain system and protect it against DoS attacks. In the past, several problems in the implementation of Ethereum metering mechanism allowed several DoS attacks.

In [26], the authors unveil a number of issues in the Ethereum metering model, especially some substantial discrepancies in the pricing of the Ethereum instructions. Additionally, they found that the correlation between the gas cost and the utilized computing resources, such as CPU and memory consumption, is very small. To conduct this study, they use a large amount of Ethereum smart contracts to determine some critical edge cases that point out several problems in EVM metering. First, there are several EVM instructions that cost significantly less gas than their actual resource consumption. Second, there are cases where the cache substantially impacts the execution time.

From these findings, the authors present a new DoS attack called Resource Exhaustion Attack targeting Ethereum smart contracts, which uses these flaws to generate low-performance contracts in terms of throughput. The challenging part is how to produce well-formed EVM contracts that minimize the throughput. The proposed attack combines empirical data and a genetic algorithm so as to create low-throughput contracts on Ethereum. As a result, the authors are able to generate contracts that are about a hundred times slower in average than typical contracts. They also show that most current Ethereum client implementations are vulnerable to this attack and those clients would not be able to stay in sync with the rest of the network when under attack. The authors have disclosed this vulnerability to the Ethereum Foundation and were awarded 5,000 USD [26].

Formal Verification of Smart Contracts. Researchers believe that smart contracts, similarly to any safety-critical system, must be formally verified before deployment

[8, 24, 27]. ZEUS [8] is a practical framework for automatic formal verification of smart contracts using abstract interpretation and symbolic model checking. At a high-level view, ZEUS works as follows. With smart contracts that are programmed in high-level languages, ZEUS leverages user assistance to formulate the criteria relating to correctness and fairness. These contracts and the policy specification are then translated into a low-level intermediate representation (IR) that encodes the execution semantics to properly inspect the behavior of the contract. After that, static analysis is performed based on the IR to identify the points at which the verification predicates (as defined in the policy) must be asserted. Finally, the modified IR is fed to a verification engine that ensures the safety of the smart contract.

In [24], the authors list out two crucial challenges of building an automated verifier for smart contracts. First, via function calls, smart contracts that we want to verify usually communicate with some external contracts. Consequently, as we do not know the code of the external contracts, these external contracts may eventually trigger the original contract in some arbitrary ways. It is very challenging to devise automated verification when there are potentially a large number of arbitrary callbacks from unknown external contracts. Second, the number of transactions that smart contracts process is unbounded. Considering processing a single transaction as an iteration in a loop, the functions in smart contracts are indeed implicitly executed in an infinite loop. Thus, even though smart contracts often do not have loops, the verifier still needs to soundly handle loops.

To address those challenges, the authors in [24] propose VERX, an automated verifier of functional requirements for Ethereum smart contracts. VERX is mainly motivated by the practical challenges that emerge when assessing real-world smart contracts. One of the main insights is that most practical contracts use a defensive strategy against external callbacks by making sure that these do not create any new behaviors. Specifically, any behavior with external callbacks is considered as another behavior without external callbacks; these are referred to as external callback free (EECF) contracts. VERX focuses on verifying EECF contracts as they offer two essential benefits. First, formalization of requirements is simplified, as auditors can write the specification without explicitly considering all possible external callbacks. Second, exploring all possible external callbacks is not necessary, thereby enabling precise and scalable analysis.

Ensuring Privacy of Smart Contracts. When implementing applications in smart contracts, one of the major concerns is data privacy. Since smart-contract transactions are processed by the blockchain's nodes, transaction data have to be made available to all nodes. Hence, it is not trivial to preserve data privacy on smart contracts without violating the security model of blockchain. This is a major problem for applications that deal with sensitive data such as voting or healthcare applications.

Most approaches to enforcing privacy use cryptographic protocols to both secure secret data and validate the integrity of computations on blockchains like Ethereum without altering their trust model. In particular, Non-Interactive Zero-Knowledge (NIZK) proofs allow a prover to prove statements involving private data without revealing any information other than the correctness of the statements. NIZK basically satisfies four properties: (1) completeness (if the statement is correct, the probability that an honest verifier accepting the proof from an honest prover is 1); (2) soundness (if the statement is incorrect, with a probability less than some small soundness error, an honest verifier can accept the proof from a dishonest prover showing that the statement is correct); (3) zero-knowledge (during the execution of the ZKP protocol, the verifier cannot learn anything other than the fact that the statement is correct); and (4) non-interactive. Practical NIZK proof constructions have been proposed and made available in Ethereum.

The paper [9] presents Hawk, a decentralized smart-contract system that does not store blockchain transactions in the clear on the blockchain, thereby preserving data privacy for the transactions, effectively concealing them from the public view. The main advantage of Hawk is that a smart-contract developer can program a private smart contract in a simple manner without having to develop any cryptographic schemes. Then, the Hawk compiler will generate an efficient cryptographic protocol in which contractual parties interact with the blockchain, using cryptographic primitives such as NIZK proofs.

Another approach to a decentralized smart-contract system is the *zkay* language proposed in [27]. The authors introduce privacy types that define owners of private values. Zkay contracts are statically type checked to ensure they are realizable using NIZK proofs and to prevent unexpected information leakage. To enforce zkay contracts, the compiler automatically converts them into contracts that have the same functionalities, retain the same privacy properties, and are executable on Ethereum.

3.3 Other Security Issues

Denial-of-Service. By design, blockchain platforms are appealing victims for Denial-of-Service (DoS) attacks: the rivalry among cryptocurrencies is very intense, and there are potential gains from short selling [28]. However, in practice, DoS attacks receive less attention comparing to other types of attack. This is due to the fact that traditional, network-based DoS attacks cannot scale to large decentralized systems, and that known DoS attacks on the mining process [29] are enormously costly. Specifically, mining-based DoS attacks require that the attacker's computing resources need to be greater than those of other miners combined, which is not practical.

Mirkin et al. [28] propose a Blockchain Denial of Service (BDoS) sabotage attack that is based on incentives: the underlying mechanism of the blockchain platform is targeted and the attacker tries to violate its incentive compatibility. In specific, the adversary uses its computing resources in order to convince honest miners to stop mining. In other words, the attacker can cause a blockchain system to stop its normal operation with only a fraction of other miners' resources. The key main insight in conducting this attack is that an attacker can manipulate the miners into thinking that the system is in a state that diminishes their revenue. The attack leverages the fact that the adversary can generate a block and broadcast only the block header as a proof to show that they mined it. The purpose is to show that they have an advantage over other miners, but do not have to reveal the block's content. The profit of a honest miner may decrease if they are oblivious of the block header, and thus they would be willing to receive the block headers. Therefore, miners are motivated to accept block headers. Simply ignoring the block header is not an effective defense strategy, since a miner is encouraged to receive block headers to maximize their payoff, such a defense strategy will not be employed by the miners.

In detail, the attack works in the following manner. The adversary generates a block B and broadcasts only the header of B. A miner may disregard the header of B and create a block following its previous block in the current chain, resulting in an additional branch of blockchain. Next, the adversary publishes the contents of B, resulting in two forks. Depending on the parameters and the state of the system, the miner's block may or may not be added to the main chain. The main idea is that when the expected profitability of the honest miners decreases, suppose that it is lower than some threshold, it is better for them to stop the mining process. If the decrease in profitability is substantial enough so that all miners decide to pause the mining process, the adversary can also stop mining. As a result, the blockchain mining comes to a complete halt, and new transactions will not be processed.

Mining Pool. Mining pools are formed by miners with the purpose of increasing the computing resource which may shorten the mining time of a block. Thus, it boosts the probability of obtaining the mining reward. Motivated by this benefit, a large number of mining pools have been formed in recent years, and many different mining strategies have been devised. In general, mining pools are managed by pool managers that forward unsolved work units to its members. The members are essentially miners of the Bitcoin network who decide to join a pool. Once a member mines a new block, the miner submits the block and the full proofs-of-work (FPoWs) to the manager. The manager sends the block to the Bitcoin network so as to obtain the mining reward. The reward is then distributed by the manager to participating miners based on how much they contribute to solving the mining puzzle. In specific, participants are rewarded based on the partial proofs-of-work (PPoWs) submitted to the manager. There are some open pools that allow participation from any miners, and private pools that only allow some authorized miners [2].

Due to the financial benefits of mining pool, the attack vector that targets the vulnerabilities in mining pool has been explored. Etay et al. [18] propose a selfish mining strategy to abuse Bitcoin's forks mechanism to obtain an unfair reward. Recall that only one branch of a fork can be accepted and others will be invalidated. In selfish mining, an attacker as a pool does not broadcast a block immediately, but instead builds a private chain internally. When the length of the public chain approaches its private chain, the attacker broadcasts the private chain, forcing other miners to accept this longer chain. Since the mining pool has large computing power, the attacker can earn a greater reward by invalidating blocks of honest miners, this also makes honest miners waste their computing resources.

Block Withholding (BWH) Attack. Different from the selfish mining, this attack is considered as an internal attack inside a mining pool. In this BWH attack, a malicious miner shares with the pool manager only PPoWs and keeps all the computed FPoWs to herself [30]. The pool manager is unaware of the blocks that were withheld and thinks that the attacker is still trying to use her computing resources to mine the block like other miners. The pool, being oblivious of this malicious behavior of the attacker, distributes its mining reward to her. Therefore, the malicious miner earns rewards without contributing anything useful to the pool. This is at the expense of the honest miners of the pool. On June 13, 2014, it was reported that a large-scale Block Withholding Attack attack was launched against Eligius, a popular mining pool, resulting in a loss of 5 million US dollar at the expense of honest miners.[6]

The authors in [30] propose a "sponsored block withholding attack". It can be observed that by conducting a BWH attack on a victim pool, the attacker indirectly increases the probability of wining the mining process for another pool. Thus, she can collude with some other pools to use a portion of her computing resources for attacking one pool and diminish the victim pool's chance of winning. In that scenario, she can be rewarded by the malicious pool for targeting the victim pool. The amount of reward can be determined according to the increase of profit to the malicious pool resulted from attacking the victim pool.

Kwon et al. [31] describe another attack called a fork after withholding (FAW) attack, which combines a BWH attack with intentional forks. In the same manner as the BWH attack, the FAW attack is always profitable regardless of an attacker's computing resources. In addition, the FAW attack provides much more rewards compared to the BWH attack. Particularly, the BWH attacker's reward is only the lower bound of the FAW attacker's. The authors propose two scenarios for this attack: single-pool and multi-pool.

In a single-pool FAW attack, in the same manner as a BWH attacker, an FAW attacker participates in the target pool and conducts an FAW attack against it. FPoWs are submitted to the pool manager by the attacker only when there is another miner who is not in the same pool submits a block. If the pool manager accepts the submitted FPoW and broadcasts the block, then a fork will be created. Because of the forks, all Bitcoin network participants will agree on only one branch. If the attacker's block is selected, the target pool will receive the mining reward, and thus, the pool will also reward her as well. In any case, the attacker is entitled to the extra rewards. The lower bound of the extra reward is the same for a BWH attacker.

On the other hand, to increase the reward, the attacker can conduct a multi-pool attack by simultaneously attacking n pools. The analysis shows that, as in the single-pool case, the FAW attack is always profitable, and the reward for an FAW attacker is greater than that for a BWH attacker. If the attacker executes the FAW attack against four currently popular pools, she will earn roughly 56% more reward than a BWH attacker does.

---

[6] https://bitcointalk.org/?topic=441465.msg7282674.

4   Other Significant Advances in Blockchain

Besides research efforts in preventing certain types of attacks on the blockchain network and smart contracts, this section shows some other notable advances in blockchain. Particularly, we focus on the privacy of blockchain transactions, consensus protocols, and the use of secure hardware in blockchain.

4.1   Anonymous Transactions

Most of anonymity vulnerabilities in blockchain arise because of the fact that Bitcoin, and many other blockchain platforms, associate each user with a pseudonym, and these pseudonyms are linked to financial transactions issued to the public blockchain. If an attacker can identify the user behind a pseudonym, the attacker may learn the user's transaction history. In practice, there are several ways to associate a user with her Bitcoin pseudonym. The most common method is to analyze transaction patterns in the public blockchain, and link those patterns using external information [32, 33].

Fanti et al. [34] investigate a lower-layer vulnerability: the networking stack. Whenever a user issues a transaction sending coins to another user, she first creates a transaction that contains the sender's pseudonym, receiver's pseudonym, and the transaction amount. This transaction is then broadcasted over the peer-to-peer network, which allows other users to validate her transaction and include it in the global chain. The authors demonstrate that, by using simple estimators to infer the source IP of each transaction broadcast, an eavesdropper adversary can link IP addresses to Bitcoin pseudonyms with an accuracy of up to 30%.

To address the anonymity issue in blockchain, Ben-Sasson et al. [35] propose Zerocash, a decentralized anonymous payments scheme for Bitcoin, that leverages recent advances in zero-knowledge Succinct Non-interactive ARguments of Knowledge (zk-SNARKs) [36]. The proposed payment scheme enables users to directly pay each other in a private manner: the transaction does not reveal the payment's origin, destination, and transferred amount. Zerocash extends and upgrades the Bitcoin protocol and software with anonymous transactions supporting privacy-preserving payments. As a result, despite using some of the same technology and software as Bitcoin, Zerocash becomes a new system that is distinct from Bitcoin. This new protocol introduces two types of coins: zerocoins (anonymous coins), and basecoins (non-anonymous coins). Comparing to Bitcoin's transactions, payment transactions created by the Zerocash protocol conceal any information that can be used to infer payment's origin, destination, or amount. Furthermore, the validity of the transaction can be verified on constant time via the use of a zk-SNARK. Users can convert from basecoins to zerocoins, send zerocoins to other users, and split or merge zerocoins they own in any way that preserves the total value, just as it is with Bitcoin.

However, it is worth noting that anonymous transactions take away the traceability of blockchain transactions. Basically, without knowing a transaction's origin and destination, it is impossible to trace back the transaction history. Some applications like supply-chain require a high degree of traceability, which means the transactions cannot be anonymous. The anonymity is also criticized for limiting accountability, regulation, and oversight. However, by using zk-SNARK, Zerocash is not limited to enforcing only the basic monetary invariants of a currency system. A wide range of policies can be supported by the underlying zk-SNARK cryptographic proof protocol. For instance, a user can prove in zero-knowledge that he paid his due taxes on all transactions without revealing those transactions, their amounts, or even the amount of taxes paid. In principle, if the policy can be specified by NP statements, it can be implemented using zk-SNARKs, and included in Zerocash.[7]

4.2   Consensus Protocols

Gilad et al. [37] present Algorand, a new consensus protocol that is designed to confirm transactions as fast as one minute. The core of Algorand uses a Byzantine agreement protocol, called BA, that scales to a large number of users, thereby allowing nodes in Algorand to agree on a new block in a short amount of time and without the possibility of forks. Algorand decides to employ BA due to the fact that it uses of verifiable random functions (VRFs) to randomly select users in a private, verifiable, and non-interactive way. Algorand mainly tackles three challenges: (1) it must avoid Sybil attacks, (2) it should scale to millions of users, and (3) it must be resilient to DoS attacks, and robust to users dropping out.

Algorand addresses these challenges in the following manner. First, Algorand assigns a weight to each user to prevent Sybil attacks. BA is designed to ensure consensus as long as a weighted fraction of the users are honest. Second, BA improves scalability by choosing a small committee that is formed by randomly selecting from the total set of users, to run each step in the protocol. All other users observe the protocol messages that allow them to learn the block that was agreed upon. Third, to hinder an adversary from manipulating committee selection, they are selected in a private, verifiable, and non-interactive way by the BA. In specific, each user in the system can independently and reliably determine whether they are chosen as a committee member, by computing a VRF that takes as input their private key and some information from the blockchain. Finally, to hinder an adversary from targeting a committee member after that member sends a message, BA requires committee members to speak only once. Therefore, once a committee member sends his message, hence revealing his identity to the adversary, the BA discards any further messages coming from that committee member.

In [38], the authors present Bitcoin-NG, a scalable blockchain protocol, that uses the same trust model as Bitcoin. Bitcoin-NG's latency and throughput are limited only by the propagation delay of the network and the processing capacity of the individual Bitcoin nodes, respectively. The key idea in designing Bitcoin-NG is decoupling

---

[7] http://zerocash-project.org/q_and_a.

Bitcoin's blockchain operation into two planes: leader election and transaction serialization. In particular, time is divided into epochs, where each epoch has a single leader. In the same manner as Bitcoin, a leader is elected randomly and infrequently. Once a leader is chosen, the leader is able to to serialize transactions at his or her discretion until the election of a new leader, which marks the end of the former's epoch. While this approach is substantially different from that of Bitcoin, the authors claim that Bitcoin-NG still maintains Bitcoin's security properties. In fact, leader election is already taking place in Bitcoin, though it is implicit. However, in Bitcoin, the task of the leader is serializing history, thereby freezing the system during the time between leader elections. On the contrary, leader election in Bitcoin-NG is forward-looking and ensures that the system is still able to process incoming transactions continuously.

Miller et al. [39] present an alternative to the Practical BFT [3] protocol, called HoneyBadgerBFT, the first practical asynchronous BFT protocol, which ensures liveness without making any timing assumptions. The authors make major efficiency improvements on the best state-of-the-art asynchronous atomic broadcast protocol that requires each node to transmit $O(N^2)$ bits for each committed transaction, thereby significantly limiting its throughput for all but the smallest networks. The cause of this efficiency is twofold. First, there is redundant work among the parties. However, naively eliminating the redundancy negatively impacts the fairness property, and paves the way for targeted censorship attacks. A solution is invented to overcome this problem by using a threshold public-key encryption scheme to tolerate these attacks. The second cause of the efficiency is the use of a suboptimal instantiation of the Asynchronous Common Subset (ACS) subcomponent. The authors show how to efficiently instantiate ACS by combining existing but overlooked techniques: (1) employ erasure codes for an efficient and reliable broadcast and (2) reduce ACS to reliable broadcast in the context of multi-party computation.

4.3 Trusted Execution Environments (TEE) in Blockchain

TEE in a computer system is realized as a module that performs some verifiable executions in such a way that no other applications, even the OS, can interfere [40]. Simply speaking, a TEE module is a trusted component within an untrusted system. Memory regions in TEE are transparently encrypted and integrity-protected with keys that are only available to the processor. TEE's memory is also isolated by the CPU hardware from the rest of the host's system, including high-privilege system software. Thus, this isolation protects the integrity and confidentiality of the enclave's execution from any malicious software running on the same system and ensures that the operating system, hypervisor, and other users cannot access the TEE's memory. Among available implementations of TEE, Intel SGX [41] supports generating remote attestations that are used to prove the correct execution of programs running inside TEE.

The authors in [42] offer a key observation that TEEs and blockchains have complementary properties. On the one hand, a blockchain can guarantee strong availability and persistence of its state, whereas a TEE cannot guarantee availability, since the host can arbitrarily terminate TEEs. Additionally, it cannot reliably access the network or persistent storage. On the other hand, a blockchain requires a huge amount of computing power, and exposes its entire state for public verification, while computation in TEE only incurs negligible overhead compared with native computation. TEE also offers verifiable computation with confidential state via remote attestation (e.g., SGX). Thus it is intuitive to build hybrid protocols that combine the advantages of both, in a way that we can exploit the immutability of blockchain to overcome the shortcomings of TEEs, and offload on-chain computation to TEE. However, note that using TEE also introduces a trusted entity to the blockchain system, which alters the trust model.

Cheng et al. [42] propose Ekiden, a system for highly performant and privacy-preserving smart contracts. The key idea behind the design of Ekiden is a secure and principled combination of blockchains and trusted hardware. Ekiden combines any desired underlying blockchain system with TEE-based execution. The design uses an architecture in which computation and consensus are separated. There are two main entities in the Ekiden architecture: compute nodes and consensus nodes. Compute nodes in Ekiden are tasked with performing smart-contract computation over private data off-chain in TEEs, then attesting the integrity of their execution on chain. In addition, the consensus nodes in Ekiden maintain the underlying blockchain, which do not need to use trusted hardware. Ekiden can be applied on top of any consensus mechanisms, in fact, it only requires a blockchain that can validate remote attestations from compute nodes. Therefore, the main advantage of Ekiden is that it can scale consensus and compute nodes independently according to performance and security needs.

In [43], the authors use TEE to improve the privacy of Bitcoin lightweight clients, in terms of concealing clients' addresses and transactions, without compromising the performance of the assisting full nodes. Specifically, they propose BITE, a solution in which an SGX enclave is run within an untrusted full node. The SGX enclave is tasked with validating transactions sent by clients. Since SGX provides code integrity and data confidentiality for enclaves, such a solution can preserve privacy and integrity of client requests. However, the authors also show that, although SGX can prevent a malicious software from directly accessing the enclave's memory, certain secret-dependent access patterns to external storage can still reveal the client's address. An example of such external storage is the transaction database. SGX is also susceptible to side-channel attacks, in which secret-dependent enclave data access patterns or control flow can be inferred by malicious software running in the same host. In specific, the adversary can monitor shared resources, such as caches, to gain insight into the execution of an SGX enclave. Taking into consideration such limitations of SGX, the authors devise a solution based on primitives such as oblivious transfer mechanisms, that enables client requests to be processed privately, even in the presence of the enclave's privacy leakage, without compromising the system's overall performance.

Lind et al. [44] leverage TEE to address the availability problem of state channels. The main insight is that, rather than having the parties to rely on the blockchain system to detect dishonest behaviors during off-chain transactions, they propose a design for a payment network in which parties use TEEs as a trusted entity to ensure correct protocol execution. In particular, they propose Teechain, a new payment network that supports highly secure and instant payments on existing blockchains. The main advantage of Teechain is that it only requires asynchronous blockchain access, that is, it makes no assumption on the timing of reading and writing transactions on the blockchain. Teechain maintains fund deposits for off-chain payment channels by using secure and trusted treasuries, which are protected by implementing them inside TEEs. By trusting the TEEs, treasuries can adopt a new efficient off-chain payment protocol that simplifies both payment and finalizing payment. To make Teechain robust against TEE failures or compromises, the state of each treasury is replicated among a small committee. In each committee of treasuries, a treasury must obtain approvals from a subset of other committee treasuries to be able to issue an off-chain transaction or finalize a payment channel. Hence, the efficiency of payment channels as a whole is improved by the TEEs, but the security guarantees of Teechain do not depend on each individual TEE.

5 Conclusions

In this chapter, we have surveyed existing literature on recent advances in the security of blockchain. In particular, we have shown several recent attacks, especially on network and smart contacts, and reviewed some security enhancement solutions for blockchain. It is suggested that the blockchain technology is still susceptible to various attacks that could obstruct an entire system and potentially cost hundreds of millions of dollars. Therefore, despite the great potential of blockchain, it is still in its early stage and a lot of research effort is needed to realize the vision of a decentralized platform for various applications.

References

1. Nakamoto, S.: Bitcoin: A peer-to-peer electronic cash system (2009). http://www.bitcoin.org/bitcoin.pdf
2. Conti, M., Kumar, E.S., Lal, C., Ruj, S.: A survey on security and privacy issues of bitcoin. IEEE Commun. Surv. Tutor. **20**(4), 3416–3452 (2018)
3. Castro, M., Liskov, B., et al.: Practical byzantine fault tolerance. OSDI **99**(1999), 173–186 (1999)
4. Androulaki, E., Barger, A., Bortnikov, V., Cachin, C., Christidis, K., De Caro, A., Enyeart, D., Ferris, C., Laventman, G., Manevich, Y., et al.: Hyperledger fabric: a distributed operating system for permissioned blockchains. In: Proceedings of the Thirteenth EuroSys Conference, pp. 1–15 (2018)

5. Nguyen, T.D., Thai, M.T.: A blockchain-based iterative double auction protocol using multi-party state channels. ACM Trans. Internet Technol. (TOIT) **21**(2), 1–22 (2021)
6. Christidis, K., Devetsikiotis, M.: Blockchains and smart contracts for the internet of things. IEEE Access **4**, 2292–2303 (2016)
7. Zhou, S., Möser, M., Yang, Z., Adida, B., Holz, T., Xiang, J., Goldfeder, S., Cao, Y., Plattner, M., Qin, X., et al.: An ever-evolving game: evaluation of real-world attacks and defenses in ethereum ecosystem. In: 29th {USENIX} Security Symposium ({USENIX} Security 20), pp. 2793–2810 (2020)
8. Kalra, S., Goel, S., Dhawan, M., Sharma, S.: Zeus: analyzing safety of smart contracts. In: NDSS (2018)
9. Kosba, A., Miller, A., Shi, E., Wen, Z., Papamanthou, C.: Hawk: the blockchain model of cryptography and privacy-preserving smart contracts. In: IEEE Symposium on Security and Privacy (SP), vol. 2016. IEEE, 839–858 (2016)
10. Zhang, M., Zhang, X., Zhang, Y., Lin, Z.: {TXSPECTOR}: uncovering attacks in ethereum from transactions. In: 29th {USENIX} Security Symposium ({USENIX} Security 20), pp. 2775–2792 (2020)
11. Merkle, R.: One way hash functions and des. In: Conference on the Theory and Application of Cryptology, pp. 428–446. Springer (1989)
12. Diffie, W.: The first ten years of public-key cryptography. In: Proceedings of the IEEE, pp. 560–577 (1988)
13. Garfinkel, S.: Public key cryptography. Computer **29**(6), 101–104 (1996)
14. Schollmeier, R.: A definition of peer-to-peer networking for the classification of peer-to-peer architectures and applications. In: Proceedings First International Conference on Peer-to-Peer Computing, pp. 101–102. IEEE (2001)
15. Badreddin, O., Gomez Rivera, A., Malik, A.: Blockchain fundamentals and development platforms. In: Proceedings of the 28th Annual International Conference on Computer Science and Software Engineering, pp. 377–379. ACM (2018)
16. Penard, W., van Werkhoven, T.: On the secure hash algorithm family. In: Cryptography in Context, pp. 1–18 (2008)
17. Heilman, E., Kendler, A., Zohar, A., Goldberg, S.: Eclipse attacks on bitcoin's peer-to-peer network. In: 24th {USENIX} Security Symposium ({USENIX} Security 15), pp. 129–144 (2015)
18. Eyal, I., Sirer, E.G.: Majority is not enough: Bitcoin mining is vulnerable. In: International Conference on Financial Cryptography and Data Security, pp. 436–454. Springer (2014)
19. Karame, G.O., Androulaki, E., Capkun, S.: Double-spending fast payments in bitcoin. In: Proceedings of the 2012 ACM Conference on Computer and Communications Security, pp. 906–917 (2012)
20. Apostolaki, M., Zohar, A., Vanbever, L.: Hijacking bitcoin: routing attacks on cryptocurrencies. In: IEEE Symposium on Security and Privacy (SP), vol. 2017, pp. 375–392. IEEE (2017)
21. Saad, M., Cook, V., Nguyen, L., Thai, M.T., Mohaisen, A.: Partitioning attacks on bitcoin: colliding space, time, and logic. In: 2019 IEEE 39th International Conference on Distributed Computing Systems (ICDCS), pp. 1175–1187. IEEE (2019)
22. Tran, M., Choi, I., Moon, G.J., Vu, A.V., Kang, M.S.: A stealthier partitioning attack against bitcoin peer-to-peer network. In: IEEE Symposium on Security and Privacy (S&P) (2020)
23. Apostolaki, M., Marti, G., Müller, J., Vanbever, L.: Sabre: protecting bitcoin against routing attacks. In: NDSS Symposium (2019)
24. Permenev, A.,Dimitrov, D., Tsankov, P., Drachsler-Cohen, D., Vechev, M.: Verx: safety verification of smart contracts. In: 2020 IEEE Symposium on Security and Privacy, SP, pp. 18–20 (2020)
25. Torres, C.F., Steichen, M., et al.: The art of the scam: Demystifying honeypots in ethereum smart contracts. In: 28th {USENIX} Security Symposium ({USENIX} Security 19), pp. 1591–1607 (2019)
26. Perez, D., Livshits, B.: Broken metre: attacking resource metering in evm. In: NDSS Symposium (2019)

27. Steffen, S., Bichsel, B., Gersbach, M., Melchior, N., Tsankov, P., Vechev, M.: zkay: specifying and enforcing data privacy in smart contracts. In: Proceedings of the 2019 ACM SIGSAC Conference on Computer and Communications Security, pp. 1759–1776 (2019)
28. Mirkin, M., Ji, Y., Pang, J., Klages-Mundt, A., Eyal, I., Jules, A.: Bdos: blockchain denial of service. In: Proceedings of the 2020 ACM SIGSAC Conference on Computer and Communications Security (2020)
29. Bonneau, J.: Hostile blockchain takeovers (short paper). In: International Conference on Financial Cryptography and Data Security, pp. 92–100. Springer (2018)
30. Bag, S., Ruj, S., Sakurai, K.: Bitcoin block withholding attack: analysis and mitigation. IEEE Trans. Inf. Forens. Secur. **12**(8), 1967–1978 (2016)
31. Kwon, Y., Kim, D., Son, Y., Vasserman, E., Kim, Y.: Be selfish and avoid dilemmas: fork after withholding (faw) attacks on bitcoin. In: Proceedings of the 2017 ACM SIGSAC Conference on Computer and Communications Security, pp. 195–209 (2017)
32. Ron, D., Shamir, A.: Quantitative analysis of the full bitcoin transaction graph. In: International Conference on Financial Cryptography and Data Security, pp. 6–24. Springer (2013)
33. Androulaki, E., Karame, G.O., Roeschlin, M., Scherer, T., Capkun, S.: Evaluating user privacy in bitcoin. In: International Conference on Financial Cryptography and Data Security, pp. 34–51. Springer (2013)
34. Fanti, G., Viswanath, P.: Deanonymization in the bitcoin p2p network. In: Advances in Neural Information Processing Systems, pp. 1364–1373 (2017)
35. Sasson, E.B., Chiesa, A., Garman, C., Green, M., Miers, I., Tromer, E., Virza, M.: Zerocash: decentralized anonymous payments from bitcoin. In: IEEE Symposium on Security and Privacy, vol. 2014, pp. 459–474. IEEE (2014)
36. Bitansky, N., Chiesa, A., Ishai, Y., Paneth, O., Ostrovsky, R.: Succinct non-interactive arguments via linear interactive proofs. In: Theory of Cryptography Conference, pp. 315–333. Springer (2013)
37. Gilad, Y., Hemo, R., Micali, S., Vlachos, G., Zeldovich, N.: Algorand: scaling byzantine agreements for cryptocurrencies. In: Proceedings of the 26th Symposium on Operating Systems Principles, pp. 51–68 (2017)
38. Eyal, I., Gencer, A.E., Sirer, E.G., Van Renesse, R.: Bitcoin-ng: a scalable blockchain protocol. In: 13th {USENIX} Symposium on Networked Systems Design and Implementation ({NSDI} 16), pp. 45–59 (2016)
39. Miller, A., Xia, Y., Croman, K., Shi, E., Song, D.: The honey badger of bft protocols. In: Proceedings of the 2016 ACM SIGSAC Conference on Computer and Communications Security, pp. 31–42 (2016)
40. Nguyen, T., Thai, M.T.: Denial-of-service vulnerability of hash-based transaction sharding: attack and countermeasure. In: (2020). arXiv preprint arXiv:2007.08600
41. Anati, I., Gueron, S., Johnson, S., Scarlata, V.: Innovative technology for CPU based attestation and sealing **13**, 7 (2013)
42. Cheng, R., Zhang, F., Kos, J., He, W., Hynes, N., Johnson, N., Juels, A., Miller, A., Song, D.: Ekiden: a platform for confidentiality-preserving, trustworthy, and performant smart contracts. In: IEEE European Symposium on Security and Privacy (EuroS&P), vol. 2019, pp. 185–200. IEEE (2019)
43. Matetic, S., Wüst, K., Schneider, M., Kostiainen, K., Karame, G., Capkun, S.: {BITE}: Bitcoin lightweight client privacy using trusted execution. In: 28th {USENIX} Security Symposium ({USENIX} Security 19), pp. 783–800 (2019)
44. Lind, J., Naor, O., Eyal, I., F. Kelbert, Sirer, E.G., Pietzuch, P.: Teechain: a secure payment network with asynchronous blockchain access. In: Proceedings of the 27th ACM Symposium on Operating Systems Principles, pp. 63–79 (2019)
45. Dinh, T.T.A., Liu, R., Zhang, M., Chen, G., Ooi, B.C., Wang, J.: Untangling blockchain: a data processing view of blockchain systems. IEEE Trans. Knowl. Data Eng. **30**(7), 1366–1385 (2018)
46. Rogaway, P., Shrimpton, T.: Cryptographic hash-function basics: definitions, implications, and separations for preimage resistance, second-preimage resistance, and collision resistance. In: International Workshop on Fast Software Encryption, pp. 371–388. Springer (2004)

47. Yaga, D., Mell, P., Roby, N., Scarfone, K.: Blockchain Technology Overview. Cornell University (2019). arXiv preprint arXiv:1906.11078
48. Hafid, A., Hafid, A.S., Samih, M.: Scaling blockchains: a comprehensive survey. In: IEEE Access, vol. 8, pp. 125 244–125 262 (2020)
49. Tschorsch, F., Scheuermann, B.: Bitcoin and beyond: a technical survey on decentralized digital currencies. IEEE Commun. Surv. Tutorials **18**(3), 2084–2123 (2016)
50. Kogias, E.K., Jovanovic, P., Gailly, N., Khoffi, I., Gasser, L., Ford, B.: Enhancing bitcoin security and performance with strong consistency via collective signing. In: 25th {usenix} Security Symposium ({usenix} Security 16), pp. 279–296 (2016)
51. Luu, L., Narayanan, V., Zheng, C., Baweja, K., Gilbert, S., Saxena, P.: A secure sharding protocol for open blockchains. In: Proceedings of the 2016 ACM SIGSAC Conference on Computer and Communications Security, pp. 17–30 (2016)
52. Kokoris-Kogias, E., Jovanovic, P., Gasser, L., Gailly, N., Syta, E., Ford, B.: Omniledger: A secure, scale-out, decentralized ledger via sharding. In: IEEE Symposium on Security and Privacy (SP), vol. 2018. IEEE, 583–598 (2018)
53. Zamani, M., Movahedi, M., Raykova, M.: Rapidchain: scaling blockchain via full sharding. In: Proceedings of the 2018 ACM SIGSAC Conference on Computer and Communications Security, pp. 931–948 (2018)
54. Dziembowski, S., Faust, S., Hostáková, K.: General state channel networks. In: Proceedings of the 2018 ACM SIGSAC Conference on Computer and Communications Security, pp. 949–966 (2018)
55. Miller, A., Bentov, I., Bakshi, S., Kumaresan, R., McCorry, P.: Sprites and state channels: payment networks that go faster than lightning. In: International Conference on Financial Cryptography and Data Security, pp. 508–526. Springer (2019)
56. Malavolta, G., Moreno-Sanchez, P., Kate, A., Maffei, M., Ravi, S.: Concurrency and privacy with payment-channel networks. In: Proceedings of the 2017 ACM SIGSAC Conference on Computer and Communications Security, pp. 455–471 (2017)
57. Dziembowski, S., Eckey, L., Faust, S., Malinowski, D.: Perun: virtual payment hubs over cryptocurrencies. In: IEEE Symposium on Security and Privacy (SP), pp. 106–123. IEEE (2019)
58. Malavolta, G., Moreno-Sanchez, P., Schneidewind, C., Kate, A., Maffei, M.: Anonymous multi-hop locks for blockchain scalability and interoperability. In: NDSS (2019)
59. Mavroudis, V., Wüst, K., Dhar, A., Kostiainen, K., Capkun, S.: Snappy: fast on-chain payments with practical collaterals. In: NDSS (2020)
60. Li, P., Miyazaki, T., Zhou, W.: Secure balance planning of off-blockchain payment channel networks. In: IEEE INFOCOM 2020-IEEE Conference on Computer Communications, pp. 1728–1737. IEEE (2020)
61. Khalil, R., Gervais, A.: Revive: rebalancing off-blockchain payment networks. In: Proceedings of the 2017 ACM SIGSAC Conference on Computer and Communications Security, pp. 439–453 (2017)
62. Yu, H., Nikolić, I., Hou, R., Saxena, P.: Ohie: blockchain scaling made simple. In: IEEE Symposium on Security and Privacy (SP), pp. 90–105. IEEE (2020)

Formal Verification of Blockchain Byzantine Fault Tolerance

Pierre Tholoniat and Vincent Gramoli

Abstract To implement a blockchain, the trend is now to integrate a non-trivial Byzantine fault-tolerant consensus algorithm instead of the seminal idea of waiting to receive blocks to decide upon the longest branch. After a dozen years of existence, blockchains trade now large amounts of valuable assets and a simple disagreement could lead to disastrous losses. Unfortunately, Byzantine consensus solutions used in blockchains are at best proved correct "by hand" as we are not aware of any of them having been automatically verified. We propose two contributions: (i) we illustrate the severity of the problem by listing six vulnerabilities of blockchain consensus including two new counter-examples; (ii) we then formally verify two Byzantine fault-tolerant components of Red Belly Blockchain (Crain et al. in Red belly: a secure, fair and scalable open blockchain, 2021, [32]) using the ByMC model checker. First, we specify its simple broadcast primitive in 116 lines of code that is verified in 40 s on a 2-core Intel machine. Then, we specify its blockchain consensus algorithm in 276 lines of code and assume a round-rigid adversary to verify in 17 minutes on a 64-core AMD machine using MPI. To conclude, we argue that it has now become both possible and crucial to formally verify the correctness of blockchain consensus protocols.

1 Introduction

As blockchain is a popular abstraction to handle valuable assets, it has become one of the cornerstones of promising solutions for building critical applications without requiring trust. Unfortunately, after a dozen years of research in the space, the blockchain still appears in its infancy, unable to offer the guarantees that are needed by the industry to automate critical applications in production. The crux of the problem

P. Tholoniat
Columbia University, New York, NY, USA
e-mail: pierre@cs.columbia.edu

V. Gramoli (✉)
University of Sydney, Sydney, Australia
e-mail: vincent.gramoli@sydney.edu.au

© The Author(s), under exclusive license to Springer Nature Switzerland AG 2022
D. A. Tran et al. (eds.), *Handbook on Blockchain*, Springer Optimization
and Its Applications 194, https://doi.org/10.1007/978-3-031-07535-3_12

is the difficulty of having remote computers agree on a unique block at a given index of the chain when some of them are malicious. The first blockchains [61] allow disagreements on the block at an index of the chain but try to recover from these disagreements before assets get stolen through double spending: with disagreement, an asset owner could be fooled when they observe that they received the asset. Instead the existence of a conflicting block within a different branch of the chain may indicate that the asset belongs to a different user who can re-spend it. This is probably why most blockchains now build upon some form of Byzantine fault-tolerant consensus solutions [17, 18, 31] that guarantee agreement despite malicious, also known as *Byzantine*, participants.

Solving the Byzantine consensus problem, defined four decades ago [65], is needed to guarantee that machines agree on a common block at each index of the chain. The consensus was recently shown to be necessary in the general scenario where conflicting transactions might be requested from distributed machines [41]. Various solutions to the consensus problem were proposed in the last four decades [8, 22, 30, 48, 49, 52, 69]. Most of these algorithms were proved correct "by hand", often listing a series of lemmas and theorems in prose leading the reader to the conclusion that the algorithm solves agreement, validity, and termination in all possible distributed executions. In the worst case, these algorithms are simply described with text on blog post [43, 52]. In the best case, a mathematical specification is offered, like in TLA+, but without machine-checked proofs [74]. Unfortunately, such a formal specification that is not machine-checked remains error prone [73].

Formal verification techniques are often limited while blockchain consensus protocols are complex and expected to run on hundreds or thousands of nodes. Theorem provers [3, 23, 53] check proofs but not algorithms. Proofs by refinement exist [50] but do not show liveness. Symbolic model checkers checked consensus algorithms but for up to 10 processes [75, 76]. Parameterized model checking [33] already proved Bosco [51], the Ben-Or consensus algorithm [12] and the condition-based consensus algorithm [9] for any number of processes. Unfortunately, Bosco [71] is a wrapper on top of another consensus that needs to be proven, Ben-Or's does not tolerate Byzantine failures and the condition-based consensus algorithm [59, 60] solves consensus only with specific sets of input values. As a result, none of these solutions fit blockchains. Only recently was a variant of the DBFT consensus algorithm proved live with any number of processes [11] using a decomposition.

In this paper, we first survey important problems that recently affected blockchain consensus. In particular, we propose two new counter-examples explaining why the Casper FFG algorithm, which should be integrated in phase 0 of Ethereum 2.0 and the HoneyBadgerBFT, which is being integrated into one of the most popular blockchain software, called `parity`, may not terminate. We also list four additional counter-examples from the literature to illustrate the amplitude of the problem for blockchains. While there exist alternative solutions to some of these problems that could be implemented it does not prevent other problems from existing. Moreover, proving "by hand" that the fixes solve the bugs may be found unconvincing, knowing that these bugs went unnoticed when the algorithms were proven correct, also "by hand", in the first place.

We then build upon modern tools and equipments at our disposal to formally verify components of the Red Belly Blockchain [32] consensus that do not assume synchrony under the assumption that $t < n/3$ processes are Byzantine (or *faulty*) among n processes. Red Belly Blockchain [32] is a fast blockchain that solves consensus deterministically and performs reasonably well on one thousand geodistributed replicas. Its scalability stems from the superblock optimization that combines multiple proposed blocks into one decision. Using Red Belly Blockchain as an example, we explain how the Byzantine model checker ByMC [47] can be used by distributed computing scientists to verify blockchain consensus components. The idea is to convert the distributed algorithm into a threshold automaton [51] that represents a state as a group of all the states in which a *correct* (or non-faulty) process resides until this process receives sufficiently many messages to transition. We offer the threshold automaton specification of a Byzantine fault-tolerant broadcast primitive that is key to few blockchains [28, 30, 56]. Finally, we also offer the threshold automaton specification of a slight variant of the Byzantine consensus algorithm [30] of Red Belly Blockchain that we prove safe and live under the round-rigidity assumption [13] that helps modeling a fair scheduler [15], hence allowing other distributed computing scientists to reproduce the verification with this publicly available model checker.

Various specification languages (e.g., [54, 79]) were proposed for distributed algorithms before threshold automata, but they did not allow the simplification needed to model check algorithms as complex as the Byzantine consensus algorithms needed in blockchain. As an example, in Input/Output Automata [54], the number of specified states accessible by an asynchronous algorithm before the threshold is reached could be proportional to the number of permutations of message receptions. Executing the automated verification of an invariant could require a computation proportional to the number of these permutations. More dramatically, the Byzantine fault model typically allows some processes to send arbitrarily formed and arbitrarily many messages— making the number of states to explore potentially infinite. As a result, this is only with the recent progress in parameterized model checking that we were able to verify our blockchain consensus components.

The remainder of the paper is organized as follows. Section 2 presents new and existing problems affecting known blockchain Byzantine consensus. In Sect. 3, we explain how we verified a Byzantine fault-tolerant broadcast abstraction common to multiple blockchains. In Sect. 4, we list the pseudocode, specification, and verification experiments of the Byzantine consensus used in Red Belly Blockchain. Section 5 presents the related work and Sect. 6 discusses our verifications and concludes the paper.

2 The Problem of Proving Blockchain Consensus Algorithms by Hand

In this section, we illustrate the risk of trying to prove blockchain consensus algorithms by hand by describing a list of safety and liveness limitations affecting the

Table 1 Some consensus algorithms that experienced liveness or safety limitations

Algorithms	Ref.	Limitation	Counter-example	Alternative	Blockchain
Randomized consensus	[57]	Liveness	[new]	[58]	HoneyBadger [56]
Casper	[18]	Liveness	[new]	[80]	Ethereum v2.0 [38]
Ripple consensus	[69]	Safety	[7]	[24]	xRapid [16]
Tendermint consensus	[17]	Safety	[6]	[5]	Tendermint [49]
Zyzzyva	[48]	Safety	[1]	[8]	SBFT [39]
IBFT	[52]	Liveness	[68]	[68]	Quorum [25]

Byzantine fault-tolerant algorithms implemented in actual blockchain systems. These limitations, depicted in Table 1, are not necessarily errors in the proofs but stem from the ambiguous descriptions in prose rather than formal statements and the lack of machine-checked proofs. As far as we know, until now no Byzantine fault-tolerant consensus algorithms used in a blockchain had been formally verified automatically.

2.1 The HoneyBadger and Its Randomized Binary Consensus

HoneyBadger [56] builds upon the combination of three algorithms from the literature to solve the Byzantine consensus with high probability in an asynchronous model. This protocol is being integrated in one of the most popular blockchain software, called Ethereum `parity`.[1] First, it uses a classic reduction from the problem of multi-value Byzantine consensus to the problem of binary Byzantine consensus working in the asynchronous model. Second, it reuses a randomized Byzantine binary consensus algorithm [57] that aims at terminating in expected constant time by using a common coin that returns the same unpredictable value at every process. Third, it uses a common coin implemented with a threshold signature scheme [19] that requires the participation of correct processes to return a value.

Randomized binary consensus. In each asynchronous round of this randomized consensus [57], the processes "binary value broadcast"—or "BV-broadcast" for short—their input binary value. The binary value broadcast (detailed later in Sect. 3.1) simply consists of broadcasting (including to oneself) a value, then rebroadcasting (or *echoing*) any value received from $t + 1$ distinct processes and finally bv-delivering any value received from $2t + 1$ distinct processes. These delivered values are then broadcast to the other processes and all correct processes record, into the set *values*, the values received from $n - t$ distinct processes that are among the ones previously delivered. For any correct process p, if *values* happen to contain only the value c returned by the common coin then p decides this value, if *values* contains only the other binary value $\neg c$, then p sets its estimate to this value and if *values* contains two values, then p sets its estimate to c. Then p moves to the next round until it decides.

---

[1] https://forum.poa.network/t/posdao-white-paper/2208.

Liveness issue. The problem is that in practice, as the communication is asynchronous, the common coin cannot return at the exact same time at all processes. In particular, if some correct processes are still at the beginning of their round r while the adversary observes the outcome of the common coin for round r then the adversary can prevent progress among the correct processes by controlling messages between correct processes and by sending specific values to them. Even if a correct process invokes the common coin before the Byzantine process, then the Byzantine can prevent correct processes from progressing.

Counter-example. To illustrate the issue, we consider a simple counter-example with $n = 4$ processes and $t = 1$ Byzantine process. Let p_1, p_2, and p_3 be correct processes with input values 0, 1, 1, respectively, and let p_4 be a Byzantine process. The goal is for process p_4 to force some correct processes to deliver $\{0, 1\}$ and another correct process to deliver $\{\neg c\}$ where c is the value returned by the common coin in the current round. As the Byzantine process has control over the network, it prevents p_2 from receiving anything before guaranteeing that p_1 and p_3 deliver $\{0, 1\}$. It is easy to see that p_4 can force p_1 and p_3 to bv-deliver 1 so let us see how p_4 forces p_1 and p_3 to deliver 0. Process p_4 sends 0 to p_3 so that p_3 receives value 0 from both p_1 and p_4, and thus echoes 0. Then p_4 sends 0 to p_1. Process p_1 then receives value 0 from p_3, p_4 and itself, hence p_1 echoes and delivers 0. Similarly, p_3 receives value 0 from p_1, p_4 and itself, hence p_3 delivers 0. To conclude p_1 and p_3 deliver $\{0, 1\}$. Processes p_1, p_3, and p_4 invoke the coin and there are two cases to consider depending on the value returned by the coin c.

- **Case $c = 0$:** Process p_2 receives now 1 from p_3, p_4 and itself, so it delivers 1.
- **Case $c = 1$:** This is the most interesting case, as p_4 should prevent some correct process, say p_2, from delivering 1 even though 1 is the most represented input value among correct processes. Process p_4 sends 0 to p_2 and p_3 so that both p_2 and p_3 receive value 0 from p_1 and p_4 and thus both echo 0. Due to p_3's echo, p_2 receives $2t + 1$ 0s and p_2 delivers 0.

At least two correct processes obtain $values = \{0, 1\}$ and another correct process can obtain $values = \{\neg c\}$. It follows that the correct processes with $values = \{0, 1\}$ adopt c as their new estimate while the correct process with $values = \{\neg c\}$ takes $\neg c$ as its new estimate and no progress can be made within this round. Finally, if the adversary (controlling p_4 in this example) keeps this strategy, then it will produce an infinite execution without termination.

Alternative and counter-measure. The problem would be fixed if we could ensure that the common coin always returns at the correct processes before returning at a Byzantine process; however, we cannot distinguish a correct process from a Byzantine process that acted correctly. We are thankful to the authors of the randomized algorithm for confirming our counter-example, they also wrote a remark in [58] indicating that both a fair scheduler and a perfect common coin were actually needed for the consensus of [57] to converge with high probability; however, no counter-example motivating the need for a fair scheduler was proposed. The intuition behind the fair scheduler is that it requires to have the same probability of receiving messages in any order [15] and thus limits the power of the adversary on the network. A new algorithm [58] does not suffer from the same problem and offers the same asymptotic complexity in message and time as [57] but requires more communication steps, it could be used as an alternative randomized consensus in HoneyBadger to cope with this issue. Cachin and Zanolini [21] detailed recently the aforementioned counter-example and proposed a fix to [57] that retains its simplicity. Finally, a similar bug report to the aforementioned counter-example was also reported by Ethan MacBrough[2] who proposes a patch but we are unaware of any proof.

2.2   The Ethereum Blockchain and Its Upcoming Casper Consensus

Casper [18, 80] is an alternative to the existing longest branch technique to agree on a common block within Ethereum. It is well known that Ethereum can experience disagreement when different processes receive distinct blocks for the same index. These disagreements are typically resolved by waiting until the longest branch is unanimously identified. Casper aims at solving this issue by offering consensus.

The Casper FFG consensus algorithm. The FFG variant of Casper is intended to be integrated to Ethereum v2.0 during phase 0 [38]. It is claimed to ensure finality [18], a property that may seem, at first glance, to result from the termination of consensus. The model of Casper assumes authentication, synchrony and that strictly less than $1/3$ stake is owned by Byzantine processes. Casper builds a "blockchain tree" consisting of a partially ordered set of blocks. The genesis block as well as blocks at indices multiple of 100 are called *checkpoints*. Validator processes vote for a link between checkpoints of a common branch and a checkpoint is *justified* if it is the initial, so-called *genesis*, block, or there is a link from a justified checkpoint pointing to it voted by a supermajority of $\lfloor \frac{2n}{3} \rfloor + 1$ validators.

Liveness issue. Note first that Casper executes speculatively and that there is not a single consensus instance per level of the Casper blockchain tree. Each time an agreement attempt at some level of the tree fails due to the lack of votes for the same checkpoint, the height of the tree grows. Unfortunately, it has been observed that nothing guarantees the termination of Casper FFG [28] and we present below an example of infinite execution.

Counter-example. To illustrate why the consensus does not terminate in this model, let h be the level of the highest block that is justified.

1. Validators try to agree on a block at level $h + k$ $(k > 0)$ by trying to gather $\lfloor \frac{2n}{3} \rfloor + 1$ votes for the same block at level $h + k$ (or more precisely the same link from level h to $h + k$). This may fail if, for example, $\frac{n}{3}$ validators vote for one of three distinct blocks at this level $h + k$.

---

[2] https://github.com/amiller/HoneyBadgerBFT/issues/59.

2. Upon failure to reach consensus at level $h + k$, the correct validators, who have voted for some link from height h to $h + k$ and are incentivized to abstain from voting on another link from h to $h + k$, can now try to agree on a block at level $h + k'$ ($k' > k$), but again no termination is guaranteed.

The same steps (1) and (2) may repeat infinitely often. Note that plausible liveness [18, Theorem 2] is still fulfilled in that the supermajority "can" always be produced as long as you have infinite memory, but no such supermajority link is ever produced in this infinite execution.

Alternative and counter-measure. Another version of Casper, called CBC, has also been proposed [80]. It is claimed to be "correct by construction", hence the name CBC. This could potentially be used as a replacement to FFG Casper for Ethereum v2.0 even in phase 0 for applications that require consensus, and thus termination.

2.3   Known Problems in Blockchain Byzantine Consensus Algorithms

To show that our two counter-examples presented above are not isolated cases in the context of blockchains, we also list below four counter-examples from the literature that were reported by colleagues and affect the Ripple consensus algorithm, Tendermint and Zyzzyva. This adds to the severity of the problem of proving algorithm by hand before using them in critical applications like blockchains.

The XRP ledger and the quorums of the Ripple consensus. The Ripple consensus [69] is a consensus algorithm originally intended to be used in the blockchain system developed by the company Ripple. The algorithm is presented at a high level as an algorithm that uses unique node lists as a set of *quorums* or mutually intersecting sets that each individual process must contact to guarantee that its request will be stored by the system or that it can retrieve consistent information about asset ownership. The original but deprecated white paper [69] assumed that quorums overlap by about 20%.

Later, some researchers published an article [7] indicating that the algorithm was inconsistent and listing the environmental conditions under which consensus would not be solved and its safety would be violated. They offered a fix in order to remedy this inconsistency through the use of different assumptions, requiring that quorums overlap by strictly more than 40%. Finally, the Ripple consensus algorithm has been replaced by the XRP ledger consensus protocol [24] called ABC-Censorship-Resilience under synchrony in part to fix this problem.

The Tendermint blockchain and its locking variant to PBFT. Tendermint [49] has similar phases as PBFT [22] and works with asynchronous rounds [35]. In each round, processes propose values in turn (phase 1), the proposed value is prevoted (phase 2), precommitted when prevoted by sufficiently many[3] processes (phase 3)

---

[3] "Sufficiently many" processes stand for at least $\lfloor \frac{2n}{3} \rfloor + 1$ among n processes.

and decided when precommitted by sufficiently many processes. To progress despite failures, processes stay in a phase only for up to a timeout period. A difference with PBFT is that a correct process produces a proof-of-lock of v at round r if it precommits v at round r. A correct process can only prevote v' if it did not precommit a conflicting value $v \neq v'$.

As we restate here, there exists a counter-example [5] that illustrates the safety issue with four processes p_1, p_2, p_3, and p_4 among which p_4 is Byzantine that propose in the round of their index number. In the first round, correct processes prevote v, p_1, and p_2 lock v in this round and precommit it, p_1 decides v while p_2 and p_3 do not decide, before p_1 becomes slow. In the second round, process p_4 informs p_3 that it prevotes v so that p_3 prevotes, precommits, and locks v in round 2. In the third round, p_3 proposes v locked in round 2, forcing p_2 to unlock v and in the fourth round, p_4 forces p_3 to unlock v in a similar way. Finally, p_1 does not propose anything and p_2 proposes another value $v' \neq v$ that gets decided by all. It follows that correct processes p_1 and p_2 decide differently, which violates agreement. Since this discovery, Tendermint kept evolving and the authors of the counter-example acknowledged that some of the issues they reported were fixed [6], the authors also informed us that they notified the developers but ignore whether this particular safety issue has been fixed.

Zyzzyva and the SBFT concurrent fast and regular paths. Zyzzyva [48] is a Byzantine consensus that requires view-change and combines a fast path where a client can learn the outcome of the consensus in three message delays and a regular path where the client needs to collect a commit-certificate with $2f + 1$ responses where f is the actual number of Byzantine faults. The same optimization is currently implemented in the SBFT permissioned blockchain [39] to speed up termination when all participants are correct and the communication is synchronous.

There exist counter-examples [1] that illustrate how the safety property of Zyzzyva can be violated. The idea of one counter-example consists of creating a commit-certificate for a value v, then experiencing a first view-change (due to delayed messages) and deciding another value v' for a given index before finally experiencing a second view-change that leads to undoing the former decision v' but instead deciding v at the same index. SBFT is likely to be immune to this issue as the counter-example was identified by some of the authors of SBFT. But a simple way to cope with this issue is to prevent the two paths from running concurrently as in the simpler variant of Zyzzyva called Azyzzva [8].

The Quorum blockchain and its IBFT consensus. IBFT [52] is a Byzantine fault-tolerant consensus algorithm at the heart of the Quorum blockchain designed by JP Morgan. It is similar to PBFT [22] except that it offers a simplified version of the PBFT view-change by getting rid of new-view messages. It aims at solving consensus under partial synchrony. The protocol assumes that no more than $t < n/3$ processes—usually referred by IBFT as "validators"—are Byzantine.

As reported in [68], IBFT does not terminate in a partially synchronous network even when failures are crashes. More precisely, IBFT cannot guarantee that if at least one honest validator is eventually able to produce a valid finalized block then the transaction it contains will eventually be added to the local transaction ledger of any other correct process. IBFT v2.x [68] fixes this problem but requires a transaction to be submitted to all correct validators for this transaction to be eventually included in the distributed permissioned transaction ledger. The proof was made by hand and we are not aware of any automated proof of this protocol as of today.

3   A Methodology for Verifying Blockchain Components

In this section, we explain how we verified the binary value broadcast blockchain component using the Byzantine model checker. Then we explain how this helped us verify the consistency of a slight variant of the binary consensus of DBFT used in Red Belly Blockchain under the round-rigid adversary assumption. Note that the DBFT binary consensus algorithm was since then proven safe and live without this assumption [11].

3.1   Preliminaries on ByMC and BV-Broadcast

Byzantine model checker. Fault-tolerant distributed algorithms, like the Byzantine fault-tolerant broadcast primitive presented below, are often based on parameters, like the number n of processes, the maximum number of Byzantine faults t, or the number of Byzantine faults f. Threshold-guarded algorithms [45, 46] use these parameters to define threshold-based guard conditions that enable transitions to different states. Once a correct process receives a number of messages that reaches the threshold, it progresses by taking some transition to a new state. To circumvent the undecidability of model checking on infinite systems, Konnov, Schmid, Veith, and Widder introduce two parametric interval abstractions [44] that model *(i)* each process with a finite-state machine independent of the parameters and *(ii)* the whole system with abstract counters that quantify the number of processes in each state in order to obtain a finite-state system. Finally, they group a potentially infinite number of runs into an execution schema in order to allow bounded model checking, based on an SMT solver, over all the possible execution schemas [46]. ByMC [47] verifies threshold automata with this model checking and has been used to prove various distributed algorithms, like atomic commit or reliable broadcast. Given a set of safety and liveness properties, it outputs traces showing that the properties are satisfied in all the reachable states of the threshold automaton. Until 2018, correctness properties were only verified on one round but more recently the threshold automata framework was extended to randomized algorithms, making possible to verify algorithms such as Ben-Or's randomized consensus under round-rigid adversaries [13].

Binary value broadcast. The binary value broadcast [57], also denoted BV-broadcast, is a Byzantine fault-tolerant communication abstraction used in blockchains [31, 56] that works in an asynchronous network with reliable channels where the maximum number of Byzantine failures is $t < n/3$. The BV-broadcast guarantees that no values broadcasted exclusively by Byzantine processes can be delivered by correct processes. This helps limiting the power of the adversary to make sure that a Byzantine consensus algorithm converges toward a value. In particular, by requiring that all correct processes BV-broadcast their proposals, one can guarantee that all correct processes will eventually observe their proposals, regardless of the values proposed by Byzantine processes. The binary value broadcast finds applications in blockchains: first, it is implemented in HoneyBadger [56] to detect that correct processes have proposed diverging values in order to toss a common coin that returns the same result across distributed correct processes, to make them converge to a common decision. Second, Red Belly Blockchain [31] and the accountable blockchain that is derived from it [26, 27] implement the BV-broadcast to detect whether the protocol can converge toward the parity of the round number by simply checking that it corresponds to one of the values that were "bv-delivered".

The BV-broadcast abstraction satisfies the four following properties:

1. BV-Obligation. If at least $(t + 1)$ correct processes BV-broadcast the same value v, v is eventually added to the set $conts_i$ of each correct process p_i.
2. BV-Justification. If p_i is correct and $v \in conts_i$, v has been BV-broadcast by some correct process. (Identification following from receiving more than t 0s or 1s.)
3. BV-Uniformity. If a value v is added to the set $conts_i$ of a correct process p_i, eventually $v \in conts_j$ at every correct process p_j.
4. BV-Termination. Eventually the set $conts_i$ of each correct process p_i is not empty.

3.2 Automated Verification of a Blockchain Byzantine Broadcast

In this section, we describe how we used threshold automaton to specify the binary value broadcast algorithm and ByMC in order to verify the protocol automatically. We recall the BV-broadcast algorithm as depicted in Algorithm 1. The algorithm consists of having at least $n - t$ correct processes broadcasting a binary value. Once a correct process receives a value from $t + 1$ distinct processes, it broadcasts it if it did not do it already. Once a correct process receives a value from $2t + 1$ distinct processes, it delivers it. Here the delivery is modeled by adding the value to the set

---

Algorithm 1 The binary value broadcast algorithm

1: bv-broadcast(MSG, *val*, *conts*, i): // *bv-broadcast filters out values proposed only by Byzantine*
2:   broadcast(BV, $\langle val, i \rangle$) // *broadcast binary value* val
3:   **repeat:** // *re-broadcast a received value only if it is sufficiently represented*
4:     **if** (BV, $\langle v, * \rangle$) received from $(t + 1)$ distinct processes but not yet broadcast **then**
5:       broadcast(BV, $\langle v, i \rangle$) // *echo* v
6:     **if** (BV, $\langle v, * \rangle$) received from $(2t + 1)$ distinct processes **then** // *from correct majority*
7:       $conts \leftarrow conts \cup \{v\}$ // *deliver* v

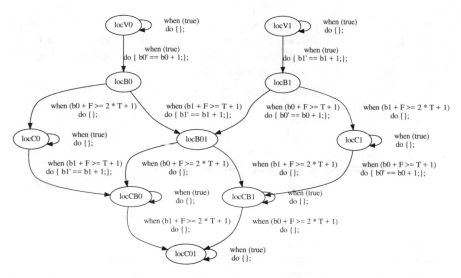

Fig. 1 The threshold automaton of the binary value broadcast algorithm

conts, which will simplify the description of our slight variant of the DBFT binary consensus algorithm in Sect. 4.

Specifying the distributed algorithm in a threshold automaton. Let us describe how we specify Algorithm 1 as a threshold automaton depicted in Fig. 1. Each state of the automaton or node in the corresponding graph represents a local state of a process. A process can move from one state to another thanks to an edge, called a *rule*. A rule has the form $\phi \mapsto u$, where ϕ is a guard and u an action on the shared variables. When the guard evaluates to true (e.g., more than $t + 1$ messages of a certain type have been sent), the action is executed (e.g., the shared variable s is incremented).

In Algorithm 1, we can see that only two types of messages are exchanged: process i can only send either $(\mathsf{BV}, \langle 0, i \rangle)$ or $(\mathsf{BV}, \langle 1, i \rangle)$. Each time a value is sent by a correct process, it is actually broadcasted to all processes. Thus, we only need two shared variables $b0$ and $b1$ corresponding to the value 0 and 1 in the automaton (cf. Fig. 1). Incrementing $b0$ is equivalent to broadcasting $(\mathsf{BV}, \langle 0, i \rangle)$. Initially, each correct process immediately broadcasts its value. This is why the guard for the first rule is true: a process in $locV0$ can immediately move to $locB0$ and send 0 during the transition.

We then enter the *repeat* loop of the pseudocode. The two *if* statements are easily understandable as threshold guards. If more than $t + 1$ messages with value 1 are *received*, then the process should broadcast 1 (i.e., increment $b1$) since it has not already been done. Interestingly, the corresponding guard is $b1 + f \geq t + 1$. Indeed, the shared variable $b1$ only counts the messages *sent* by correct processes. However, the f faulty processes might send messages with arbitrary values. We want to consider all the possible executions, so the earliest moment a correct process can move from $locB0$ to $locB01$ is when the f faulty processes and $t + 1 - f$ correct processes have sent 1. The other edge leaving $locB0$ corresponds to the second *if* statement, that is, satisfied when $2t + 1$ messages with value 0 have been received. In state $locC0$, the value 0 has been delivered. A process might stay in this state forever, so we add a self-loop with guard condition set to *true*.

After the state $locC0$, a process is still able to broadcast 1 and eventually deliver 1 after that. After the state $locB01$, a process is able to deliver 0 and then deliver 1, or deliver 1 first and then deliver 0, depending on the order in which the guards are satisfied. Apart from the self-loops, we remark that the automaton is a directed acyclic graph. On every path of the graph, we can verify that a shared variable is incremented only once. This is because in the pseudocode, a value can be broadcasted only if it has not been broadcasted before.

Finally, the states of the automaton correspond to the following (unique) situations for a correct process:

- **locV0**. Initial state with value 0, nothing has been broadcasted nor delivered.
- **locV1**. Initial state with value 1, nothing has been broadcasted nor delivered.
- **locB0**. Only 0 has been broadcasted, nothing has been delivered.
- **locB1**. Only 1 has been broadcasted, nothing has been delivered.
- **locB01**. Both 0 and 1 have been broadcasted, nothing has been delivered.
- **locC0**. Only 0 has been broadcasted, only 0 has been delivered.
- **locCB0**. Both 0 and 1 have been broadcast, only 0 has been delivered.
- **locC1**. Only 1 has been broadcasted, only 1 has been delivered.
- **locCB1**. Both 0 and 1 have been broadcast, only 1 has been delivered.
- **locC01**. Both 0 and 1 have been broadcast, both 0 and 1 have been delivered.

Once the pseudocode is converted into a threshold automaton depicted in Fig. 1, one can simply write the corresponding specification in the threshold automata language to obtain the specification listed below (Listing 1) for completeness.

Defining the correctness properties and fairness assumptions. The above automaton is only the first half of the verification work. The second half consists in specifying the correctness properties that we would like to verify on the algorithm. We use temporal logic on the algorithm variables (number of processes in each location, number of messages sent, and parameters) to formalize the properties. In the case of the BV-broadcast, the BV-Justification property of the BV-broadcast is "If p_i is correct and $v \in conts_i$, v has been BV-broadcast by some correct process". Given $\lozenge$, $\rightarrow$ and $||$ with the LTL semantics of "eventually", "implies", and "or", respectively, we translate this property in the following conjunction:

$$\begin{cases} justification0 : (\lozenge(locC0 \neq 0 \;||\; locC01 \neq 0)) \rightarrow \\ \qquad\qquad\qquad (locV0 \neq 0), \\ justification1 : (\lozenge(locC1 \neq 0 \;||\; locC01 \neq 0)) \rightarrow \\ \qquad\qquad\qquad (locV1 \neq 0). \end{cases}$$

Liveness properties are longer to specify, because we need to take into account some fairness constraints. Indeed, a threshold automaton describes processes evolving in an asynchronous setting without additional assumptions. An execution in which a process stays in a state forever is a valid execution, but it does not make any progress. If we want to verify some liveness properties, we have to add some assumptions in the specification. For instance, we require that processes eventually leave the states of the automaton as long as they have received enough messages to enable the condition guarding the outgoing rule. In other words, a liveness property will be specified as follows: *liveness_property* : *fairness_condition* → *property*.

Note that this assumption is natural and differs from the round-rigidity assumption that requires the adversary to eventually take any applicable transition of an infinite execution. Finally, we wrote a threshold automaton specification whose .ta file is presented in Listing 1 in only 116 lines.

Experimental results. On a simple laptop with an Intel Core i5-7200U CPU running at 2.50GHz, verifying all the correctness properties for BV-broadcast takes less than 40 s. For simple properties on well-specified algorithms, such as the ones of the benchmarks included with ByMC, the verification time can be less than one second. This result encouraged us to verify a complete Byzantine consensus algorithm in Sect. 4 that builds upon the binary value broadcast.

Debugging the manual conversion of the algorithm to the automaton. It is common that the specification does not hold at first try, because of some mistakes in the threshold automaton model or in the translation of the correctness property into a formal specification. In such cases, ByMC provides a detailed output and a counter-example showing where the property has been violated. We reproduced such a counter-example in Fig. 2 with an older preliminary version of our specification. This specification was wrong because a liveness property did not hold. ByMC gave parameters and provided an execution ending with a loop, such that the condition of the liveness was never met. This trace helped us understand the problem in our specification and allowed us to fix it to obtain the correct specification we

```
1   N:=34; T:=11; F:=1;
2   0 (F 0) x 0: b0:=0; b1:=0; K[pc:0]:=21; K[pc:1]:=12; K[*]:=0;
3   1 (F 1) x 1: b0:=1; K[pc:0]:=20; K[pc:2]:=1;
4
5                           (...)
6
7   24 (F 52) x 1:  b1:=21; K[pc:5]:=12; K[pc:7]:=21;
8   ****************
9   b0:=33; b1:=21; K[pc:0]:=0; K[pc:1]:=0; K[pc:2]:=0;
10  K[pc:3]:=0; K[pc:4]:=0; K[pc:5]:=12; K[pc:6]:=0; K[pc:7]:=21;
11  K[pc:8]:=0; K[pc:9]:=0;
12
13  ****** LOOP *******
14  N:=34; T:=11; F:=1;
15  25 (F 83) x 1:  <self-loop>
16  ****************
17  K[pc:2]:=0; K[pc:4]:=0; K[pc:5]:=12; K[pc:7]:=21; K[pc:8]:=0;
18  K[pc:9]:=0;
```

Fig. 2 Truncated counter-example produced by ByMC for a faulty specification of BV-broadcast

illustrated before in Fig. 1. Building upon this successful result, we specified a more complex Byzantine consensus algorithm that uses the same broadcast abstraction but we did not encounter any bug during this process and our first specification was proved correct by ByMC. The pseudocode, threshold automaton specification, and experimental results are presented in Sect. 4.

Listing 1 Threshold automaton specification for the binary value broadcast communication primitive

```
1   thresholdAutomaton Proc {
2     local pc; shared b0, b1;
3     parameters N, T, F;
4
5     assumptions (0) { N>3*T; T>=F; T>=1; }
6
7     locations (0) {
8       locV0:[0]; locV1:[1]; locB0:[2];
9       locB1:[3]; locB01:[4]; locC0:[5];
10      locC1:[6]; locCB0:[7];
11      locCB1:[8]; locC01:[9];
12    }
13
14    inits (0) {
15      (locV0+locV1)==N-F;
16      locB0==0; locB1==0; locB01==0;
17      locC0==0; locC1==0; locCB0==0;
18      locCB1==0; locC01==0; b0==0; b1==0;
19    }
20
21    rules (0) {
22    % for v in [0, 1]:
23      1: locV${v} -> locB${v}
24        when (true)
25        do { b${v}'==b${v}+1;
26          unchanged(b${1-v}); };
27
28      2: locB${v} -> locB01
29        when (b${1-v}+F>=T+1)
30        do { b${1-v}'==b${1-v}+1;
31          unchanged(b${v}); };
32
33      3: locB${v} -> locC${v}
34        when (b${v}+F>=2*T+1)
35        do { unchanged(b0, b1); };
36
37      2: locC${v} -> locCB${v}
38        when (b${1-v}+F>=T+1)
39        do { b${1-v}'==b${1-v}+1;
40          unchanged(b${v}); };
41
42      3: locB01 -> locCB${v}
43        when (b${v}+F>=2*T+1)
44        do { unchanged(b0, b1); };
45
46      3: locCB${v} -> locC01
47        when (b${1-v}+F>=2*T+1)
48        do { unchanged(b0, b1); };
49
50    /* self loops */
51      10: locV${v} -> locV${v}
52        when (true) do {unchanged(b0, b1);};
53
54      10: locC${v} -> locC${v}
55        when (true) do {unchanged(b0, b1);};
56
57      10: locCB${v} -> locCB${v}
58        when (true) do {unchanged(b0, b1);};
59    % endfor
60
61      10: locC01 -> locC01
62        when (true) do {unchanged(b0, b1);};
63    }
```

```
1
2    specifications (0) {
3
4    % for v in [0,1]:
5      obligation${v}:
6      <>[]((locV0==0) && (locV1==0) &&
7      (locB0==0 || b1<T+1) && (locB1==0 || b0<T+1) &&
8      (locB0==0 || b0<2*T+1) && (locB1==0 || b1<2*T+1) &&
9      (locB01==0 || b0<2*T+1) && (locB01==0 || b1<2*T+1) &&
10     (locC0==0 || b1<T+1) && (locC1==0 || b0<T+1) &&
11     (locCB0==0 || b1<2*T+1) && (locCB1==0 || b0<2*T+1))
12     ->
13       ((locV${v}>=T+1)
14     ->
15     <>(locV0==0 && locV1==0 &&
16         locB0==0 && locB1==0 &&
17         locB01==0 && locC${1-v}==0 &&
18         locCB${1-v}==0));
19
20     justification${v}: (<>(locC${v}!=0
21     || locCB${v}!=0 || locC01!=0))
22     -> (locV${v}!=0);
23
24     uniformity${v}:
25     <>[]((locV0==0) && (locV1==0) &&
26     (locB0==0 || b1<T+1) && (locB1==0 || b0<T+1) &&
27     (locB0==0 || b0<2*T+1) && (locB1==0 || b1<2*T+1) &&
28     (locB01==0 || b0<2*T+1) && (locB01==0 || b1<2*T+1) &&
29     (locC0==0 || b1<T+1) && (locC1==0 || b0<T+1) &&
30     (locCB0==0 || b1<2*T+1) && (locCB1==0 || b0<2*T+1))
31     ->
32     (<>(locC${v}!=0 || locCB${v}!=0 || locC01!=0)
33     ->
34     <>[](locC${1-v}==0 && locCB${1-v}==0));
35     % endfor
36
37     termination:
38     <>[]((locV0==0) && (locV1==0) &&
39         (locB0==0 || b1<T+1) &&
40         (locB1==0 || b0<T+1) &&
41         (locB0==0 || b0<2*T+1) &&
42         (locB1==0 || b1<2*T+1) &&
43         (locB01==0 || b0<2*T+1) &&
44         (locB01==0 || b1<2*T+1) &&
45         (locC0==0 || b1<T+1) &&
46         (locC1==0 || b0<T+1) &&
47         (locCB0==0 || b1<2*T+1) &&
48         (locCB1==0 || b0<2*T+1))
49     ->
50     <>(locV0 ==0 && locV1 ==0 &&
51         locB0 ==0 && locB01==0);
52    }
53  } /* Proc */
```

4   Verifying a Blockchain Byzantine Consensus Algorithm

The Democratic Byzantine Fault-Tolerant consensus algorithm [30] is a Byzantine consensus algorithm that does not require a leader. It was implemented in the recent Red Belly Blockchain [32] to offer high performance through multiple proposers and was used in Polygraph [26, 27] to detect malicious participants responsible of disagreements when $t \geq n/3$ and in the Long-Lasting Blockchain [67] to recover from forks by excluding misbehaving participants. As depicted in Algorithm 2, a slight variant of its binary consensus, made simpler than the original algorithm by omitting timeouts, proceeds in asynchronous rounds that correspond to the iterations of a loop where correct processes refine their estimate value.

Algorithm 2 A variant of the DBFT binary Byzantine consensus algorithm

Notation: "Received k messages" is a shortcut for "Received k messages from different processes in the same round r as the current round."

1: propose(v):
2:     $est \leftarrow v$ // *initial estimate is the proposed value*
3:     $r \leftarrow 0$ // *initialize the round number*
4:     **repeat:** // *repeat in asynchronous rounds*
5:         $r \leftarrow r + 1$; // *increment the round number*
6:         broadcast($tag = $ BV, $round = r$, $value = est$) // *initial broadcast*
7:         **while** true **do** // *start of binary value broadcast phase*
8:             **if** received $(t + 1)$ BV messages with value w and w not broadcast yet **then**
9:                 broadcast($tag = $ BV, $round = r$, $value = w$) // *rebroadcast legitimate estimates*
10:            **if** received $(2t + 1)$ BV messages with value w **then** // *recvd from correct majority*
11:                broadcast($tag = $ ECHO, $round = r$, $value = w$) // *broadcast ECHO message*
12:                **break** // *exit the while loop to proceed to next phase*
13:        **while** true **do** // *wait to have received enough messages*
14:            $echoes \leftarrow \{w \in \{0, 1\} : $ received $(2t + 1)$ BV messages with value $w\}$
15:            **if** received $(n - t)$ ECHO messages with value $w \in echoes$ **then**
16:                $est \leftarrow w$ // *refine estimate*
17:                **if** $w = r$ mod 2 and not decided yet **then** // *depending on the singleton value w...*
18:                    decide(w) // *...decide the parity of the round*
19:                **break** // *exit the while loop to proceed to next round*
20:            **if** received $(n - t)$ ECHO messages and $echoes = \{0, 1\}$ **then** // *all bv-delivered*
21:                $est \leftarrow r$ mod 2 // *set estimate to round parity*
22:                **break** // *exit the while loop to proceed to next round*
23:        **if** decided in round $r_i - 2$ **then** exit // *exit the consensus only after having helped others*

Initially, each correct process sets its estimate to its input value. Correct processes broadcast these estimates and rebroadcast only values received by $t + 1$ distinct processes because they are proposed by correct processes. Each value received from $2t + 1$ distinct processes (and from a majority of correct processes) is stored in the *echoes* set and is broadcasted as part of an ECHO message. The ECHO value received from $n - t$ distinct processes that also belongs to *echoes* becomes the new estimate (line 16) for the next round. If this value corresponds to the parity of the round, then the correct process decides this value. If *echoes* contain both values, then the estimate for the next round becomes the parity of the round. As opposed to the original and partially synchronous deterministic version [30], this variant uses one less broadcast phase and offers termination in an asynchronous network under round-rigidity that requires the adversary to eventually perform any applicable transition within an infinite execution. This assumption was previously used to show termination of another algorithm with high probability [13]. The specification of our consensus algorithm in threshold automata is depicted in Listing 2.

Listing 2 Variant of the DBFT binary Byzantine consensus

```
1  thresholdAutomaton Proc {
2
3     local pc;
4
5     /* Messages sent by correct proc. */
6     /* First round */
7     shared b0, b1;
8     shared e0, e1;
9     /* Second round */
10    shared b0x, b1x;
11    shared e0x, e1x;
12
13    parameters N, T, F;
14
15    assumptions (0) {
16       N > 3 * T;
17       T >= F;
18       T >= 1;
19    }
20
21    locations (0) {
22       locV0:    [0];
23       locV1:    [1];
24       locB0:    [2];
25       locB1:    [3];
26       locB01:   [4];
27       locC:     [5];
28       locE0:    [6];
29       locE1:    [7];
30       locD1:    [8];
31       locB0x:   [9];
32       locB1x:   [10];
33       locB01x:  [11];
34       locCx:    [12];
35       locE0x:   [13];
36       locE1x:   [14];
37       locD0:    [15];
38    }
39
40    inits (0) {
41       (locV0 + locV1) == N - F;
42
43       locB0 == 0;
44       locB1 == 0;
45       locB01 == 0;
46       locC == 0;
47       locE0 == 0;
48       locE1 == 0;
49       locD1 == 0;
50       locB0x == 0;
```

```
1
2
3    rules (0) {
4    % for v in [0, 1]:
5       1: locV${v} -> locB${v}
6          when (true)
7          do { b${v}' == b${v} + 1;
8             unchanged(b${1-v}, e0, e1);
9             unchanged(b0x, b1x, e0x, e1x);
10         };
11   % endfor
12
13   % for v in [0, 1]:
14      2: locB${v} -> locB01
15         when (b${1-v} + F >= T + 1)
16         do { b${1-v}' == b${1-v} + 1;
17            unchanged(b${v}, e0, e1);
18            unchanged(b0x, b1x, e0x, e1x);
19         };
20   % endfor
21
22   % for v in [0, 1]:
23      3: locB${v} -> locC
24         when (b${v} + F >= 2 * T + 1)
25         do { e${v}' == e${v} + 1;
26            unchanged(b0, b1, e${1-v});
27            unchanged(b0x, b1x, e0x, e1x);
28         };
29   % endfor
30
31   % for v in [0, 1]:
32      4: locB01 -> locC
33         when (b${v} + F >= 2 * T + 1)
34         do { e${v}' == e${v} + 1;
35            unchanged(b0, b1, e${1-v});
36            unchanged(b0x, b1x, e0x, e1x);
37         };
38   % endfor
39
40      5: locC -> locD1
41         when (e1 + F >= N - T
42   && b1 + F >= 2 * T + 1)
43         do {
44            unchanged(b0, b1, e0, e1);
45            unchanged(b0x, b1x, e0x, e1x);
46         };
47
48      6: locC -> locE0
49         when (e0 + F >= N - T
50   && b0 + F >= 2 * T + 1)
51         do {
52            unchanged(b0, b1, e0, e1);
53            unchanged(b0x, b1x, e0x, e1x);
54         };
55
```

```
51       locB1x == 0;
52       locB01x == 0;
53       locCx == 0;
54       locE0x == 0;
55       locE1x == 0;
56       locD0 == 0;
57
58       b0 == 0;
59       b1 == 0;
60       e0 == 0;
61       e1 == 0;
62       b0x == 0;
63       b1x == 0;
64       e0x == 0;
65       e1x == 0;
66     }
```

```
56     7: locC -> locE1
57         when (e0 + e1 + F >= N - T
58      && b0 + F >= 2 * T + 1
59      && b1 + F >= 2 * T + 1)
60         do {
61             unchanged(b0, b1, e0, e1);
62             unchanged(b0x, b1x, e0x, e1x);
63         };
64
65     % for v in [0, 1]:
66     8: locE${v} -> locB${v}x
67         when (true)
68         do { b${v}x' == b${v}x + 1;
69             unchanged(b0, b1, e0, e1);
70             unchanged(b${1-v}x, e0x, e1x);
71         };
72     % endfor
```

```
1
2    % for v in [0, 1]:
3    9: locB${v}x -> locB01x
4        when (b${1-v}x + F >= T + 1)
5        do { b${1-v}x' == b${1-v}x + 1;
6            unchanged(b0, b1, e0, e1);
7            unchanged(b${v}x, e0x, e1x);
8        };
9    % endfor
10
11   % for v in [0, 1]:
12   10: locB${v}x -> locCx
13       when (b${v}x + F >= 2 * T + 1)
14       do { e${v}x' == e${v}x + 1;
15           unchanged(b0, b1, e0, e1);
16           unchanged(b0x, b1x, e${1-v}x);
17       };
18   % endfor
19
20   % for v in [0, 1]:
21   11: locB01x -> locCx
22       when (b${v}x + F >= 2 * T + 1)
23       do { e${v}x' == e${v}x + 1;
24           unchanged(b0, b1, e0, e1);
25           unchanged(b0x, b1x, e${1-v}x);
26       };
27   % endfor
28
29   12: locCx -> locD0
30       when (e0x + F >= N - T
31    && b0x + F >= 2 * T + 1)
32       do {
33           unchanged(b0, b1, e0, e1);
34           unchanged(b0x, b1x, e0x, e1x);
35       };
36
37   13: locCx -> locE1x
38       when (e1x + F >= N - T
39    && b1x + F >= 2 * T + 1)
40       do {
41           unchanged(b0, b1, e0, e1);
42           unchanged(b0x, b1x, e0x, e1x);
43       };
44
45   14: locCx -> locE0x
46       when (e0x + e1x + F >= N - T
47    && b0x + F >= 2 * T + 1
48    && b1x + F >= 2 * T + 1)
49       do {
50           unchanged(b0, b1, e0, e1);
51           unchanged(b0x, b1x, e0x, e1x);
52       };
53
54
55
56   /* self loops */
57
58   % for v in [0, 1]:
59   10: locV${v} -> locV${v}
60       when (true)
```

```
1
2    % for v in [0, 1]:
3    10: locE${v}x -> locE${v}x
4        when (true)
5        do {
6            unchanged(b0, b1, e0, e1);
7            unchanged(b0x, b1x, e0x, e1x);
8        };
9    % endfor
10   }
11   specifications (0) {
12
13   % for v in [0, 1]:
14       validity${v}:
15       (locV${1-v} == 0) ->
16       [](locD${1-v} == 0 && locE${1-v}x == 0);
17   % endfor
18
19   % for v in [0, 1]:
20       agreement${v}:
21       []((locD${v} != 0) ->
22       [](locD${1-v} == 0 && locE${1-v}x == 0));
23   % endfor
24
25       round_termination:
26   <>[](
27           (locV0 == 0) &&
28           (locV1 == 0) &&
29           (locB0
30   == 0 || (b1 < T + 1 && b0 < 2 * T + 1)) &&
31           (locB1
31   == 0 || (b0 < T + 1 && b1 < 2 * T + 1)) &&
32           (locB01
32   == 0 || (b0 < 2 * T + 1 && b1 < 2 * T + 1)) &&
33           (locC == 0 ||
34           ((e1 < N - T || b1 < 2 * T + 1) &&
35           (e0 < N - T || b0 < 2 * T + 1) &&
36           (e0 + e1 < N - T ||
37               b0 < 2 * T + 1 ||
38               b1 < 2 * T + 1) )) &&
38           (locE0 == 0) &&
39           (locE1 == 0) &&
40           (locB0x
41   == 0 || (b1x < T + 1 && b0x < 2 * T + 1)) &&
41           (locB1x
42   == 0 || (b0x < T + 1 && b1x < 2 * T + 1)) &&
42           (locB01x == 0 ||
43               (b0x < 2 * T + 1 && b1x < 2 * T + 1)) &&
44           (locCx == 0 ||
45           ((e1x < N - T || b1x < 2 * T + 1) &&
46           (e0x < N - T || b0x < 2 * T + 1) &&
47           (e0x + e1x < N - T ||
48               b0x < 2 * T + 1 ||
49               b1x < 2*T+1)))
50       )
51   ->
52   <>(
53           locV0 == 0 &&
54           locV1 == 0 &&
55           locB0 == 0 &&
```

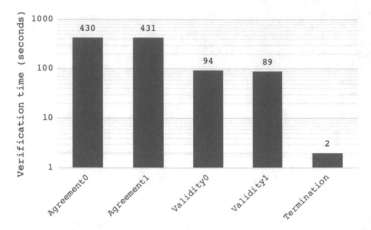

Fig. 3 Time to verify the Byzantine consensus of Algorithm 2

```
61        do {                                    56        locB1 == 0 &&
62            unchanged(b0, b1, e0, e1);          57        locB01 == 0 &&
63            unchanged(b0x, b1x, e0x, e1x);      58        locC == 0 &&
64        };                                      59        locE0 == 0 &&
65    % endfor                                    60        locE1 == 0 &&
66                                                61        locB0x == 0 &&
67    % for v in [0, 1]:                          62        locB1x == 0 &&
68      10: locD${v} -> locD${v}                  63        locB01x == 0 &&
69          when (true)                           64        locCx == 0
70          do {                                  65      );
71              unchanged(b0, b1, e0, e1);        66
72              unchanged(b0x, b1x, e0x, e1x);    67    }
73          };                                    68  } /* Proc */
74    % endfor
```

4.1   Experimental Results

The Byzantine consensus algorithm has far more states and variables than the BV-broadcast primitive and it is too complex to be verified on a personal computer. We ran the parallelized version of ByMC with MPI on a 4 AMD Opteron 6276 16-core CPU with 64 cores at 2300 MHz with 64 GB of memory. The verification times for the five properties are listed in Fig. 3 and sum up to 17 min and 26 s.

5   Related Work

The observation that some of the blockchain consensus proposals have issues is not new [20, 40]. It is now well known that the termination of existing blockchains like Ethereum requires an additional assumption like synchrony [40]. Our Ethereum counter-example differs as it considers the upcoming consensus algorithm of Ethereum v2.0. In [20], the conclusions are different from ours as they generalize on other Byzantine consensus proposals, like Tangaroa, not necessarily in use in blockchain systems. Our focus is on consensus used in blockchains that are critical due to trading valuable assets. Note that other consistency violations related to the consensus offered in Ethereum v1.x and v2.0 have been concurrently reported [36, 37, 62].

Threshold automata already proved helpful to automate the proof of existing consensus algorithms [47]. They have even been useful in illustrating why a specification of the King-Phase algorithm [10] was incorrect [72] (due to the strictness of a lower symbol), later fixed in [14]. We did not list this as one of the inconsistency problems that affects blockchains as we are not aware of any blockchain implementation that builds upon the King-Phase algorithm. In [51], the authors use threshold guarded automata to prove two broadcast primitives and the Bosco Byzantine consensus correct; however, Bosco offers a fast path but requires another consensus algorithm for its fallback path so its correctness depends on the assumption that it relies on a correct consensus algorithm.

In general, it is hard to formally prove algorithms that work in a partially synchronous model while there exist tools to reduce the state space of synchronous consensus to finite-state model checking [4]. Part of the reason is that common partially synchronous solutions attempt to give sufficient time to processes in different asynchronous rounds by incrementing a timeout until the timeout is sufficiently large to match the unknown message delay bound. PSync [34] and ConsL [55] are languages that help reasoning formally about partially synchronous algorithms. In particular, ConsL was shown effective at verifying consensus algorithms but only for the crash fault-tolerant model. Here we used the ByMC model checker [45] for asynchronous Byzantine fault-tolerant systems and require the round-rigidity assumption to show a variant of the binary consensus of DBFT [30].

Interactive theorem provers [66, 70, 77] were used to prove consensus algorithms. In particular, the Coq proof assistant helped prove distributed algorithms [2] like two-phase commit [70], Raft [78] and the Algorand consensus algorithm [3] while Dafny [42] proved MultiPaxos. Isabelle/HOL [64] was used to prove byzantine fault-tolerant algorithms [23] and was combined with Ivy to prove the Stellar consensus protocol [53]. Theorem provers check proofs, not the algorithms. Hence, one has to invest efforts into writing detailed mechanical proofs.

In [79], the authors present TLC, a model checker for debugging a finite-state model of a TLA+ specification. TLA+ is a specification language for concurrent and reactive systems that build upon the temporal logic TLA. One limitation is that the TLA+ specification might comprise an infinite set of states for which the model checker can only give a partial proof. In order to run the TLC model checker on a TLA+ specification, it is necessary to fix the parameters such as the number of processes n or the bounds on integer values. In practice, the complexity of model checking explodes rapidly and makes it difficult to check anything beyond toy examples with a handful of processes. TLC remains useful—in particular in industry—to prove that some specifications are wrong [63]. TLA+ also comes with a proof system called TLAPS. TLAPS supports manually written hierarchically structured proofs, which are then checked by backend engines such as Isabelle, Zenon, or SMT solvers [29]. TLAPS is still being actively developed but it is already possible—albeit technical and lengthy—to prove algorithms such as Paxos (Fig. 4).

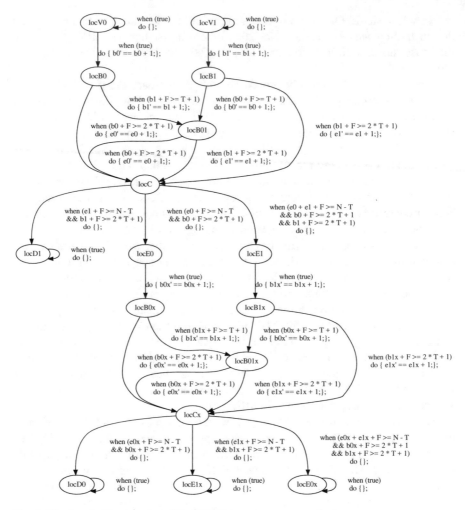

Fig. 4 The threshold automaton of the DBFT binary consensus variant

Recently, the binary consensus of DBFT [30] was formally proved safe and live using parameterized model checking [11] but without any round-rigid adversary assumption. To this end, the specification of the Byzantine consensus algorithm was split into multiple threshold automata.

6   Discussion and Conclusion

In this paper, we argued for the formal verification of blockchain Byzantine fault-tolerant algorithms as a way to reduce the numerous issues resulting from non-formal proofs for such critical applications as blockchains. In particular, we illustrated the problem with new counter-examples of algorithms at the core of widely deployed blockchain software.

We show that it is now feasible to verify blockchain Byzantine components on modern machines thanks to the recent advances in formal verification. We illustrate it with relatively simple specifications of a broadcast abstraction common to multiple blockchains as well as a variant of the Byzantine consensus algorithm of the Red Belly Blockchain.

To verify the Byzantine consensus, we assumed a round-rigid adversary that schedules transitions in a fair way. This is not new as in [13] the model checking of the randomized algorithm from Ben-Or required a round-rigid adversary. Interestingly, we do not need this assumption to verify the binary value broadcast abstraction that works in an asynchronous model. A concomitant result replaces the round-rigidity assumption by a deterministic fairness assumption to formally verify the liveness and safety properties of the consensus algorithm of DBFT [11].

As future work, we would like to prove other Byzantine fault-tolerant algorithmic components of blockchain systems.

Acknowledgements Parts of the content of this chapter have been presented in the non-archiving workshops FRIDA'19 and ConsensusDays'21. We wish to thank Igor Konnov and Josef Widder for helping us understand the syntax and semantics of the threshold automata specification language and for confirming that ByMC verified the agreement1 property of our initial specification. We thank Tyler Crain, Achour Mostéfaoui, and Michel Raynal for discussions of the HoneyBadger counter-example, and Yackolley Amoussou-Guenou, Maria Potop-Butucaru, and Sara Tucci for discussions on the Tendermint counter-example. This research is supported under Australian Research Council Discovery Projects funding scheme (project number 180104030) entitled "Taipan: A Blockchain with Democratic Consensus and Validated Contracts" and Australian Research Council Future Fellowship funding scheme (project number 180100496) entitled "The Red Belly Blockchain: A Scalable Blockchain for Internet of Things".

References

1. Abraham, I., Gueta, G.G., Malkhi, D., Alvisi, L., Kotla, R., Martin J.-P.: Revisiting fast practical byzantine fault tolerance. Technical report (Dec 2017). arXiv
2. Altisen, K., Corbineau, P., Devismes, S.: A framework for certified self-stabilization. In: FORTE, pp. 36–51 (2016)
3. Alturki, M.A., Chen, J., Luchangco, V., Moore, B.M., Palmskog, K., Peña, L., Rosu, G.: Towards a verified model of the algorand consensus protocol in coq. In: International Workshops on Formal Methods (FM), pp. 362–367 (2019)
4. Aminof, B., Rubin, S., Stoilkovska, I., Widder, J., Zuleger F.: Parameterized model checking of synchronous distributed algorithms by abstraction. In: Proceedings of the 19th International Conference on Verification, Model Checking, and Abstract Interpretation, VMCAI, pp. 1–24 (2018)
5. Amoussou-Guenou, Y., Pozzo, A.D., Potop-Butucaru, M., Piergiovanni, S.T.: Correctness and fairness of tendermint-core blockchains. Technical Report (2018). arXiv:1805.08429
6. Amoussou-Guenou, Y., Pozzo, A.D., Potop-Butucaru, M., Tucci-Piergiovanni, S.: Dissecting tendermint. In: Proceedings of the 7th Edition of The International Conference on Networked Systems (2019)

7. Armknecht, F., Karame, G.O., Mandal, A., Youssef, F., Zenner, E.: Ripple: overview and outlook. In: International Conference on Trust and Trustworthy Computing, pp. 163–180. Springer (2015)
8. Aublin, P.-L., Guerraoui, R., Knežević, N., Quéma, V., Vukolić M.: The next 700 BFT protocols. ACM Trans. Comput. Syst. **32**(4), 12:1–12:45 (2015). Jan
9. Balasubramanian, A.R., Esparza, J., Lazic, M.: Complexity of verification and synthesis of threshold automata. In: ATVA, pp. 144–160 (2020)
10. Berman P., Garay, J.A.: Asymptotically optimal distributed consensus (extended abstract). In: ICALP, pp. 80–94 (1989)
11. Bertrand, N., Gramoli, V., Konnov, I., Lazic, M., Tholoniat, P., Widder, J.: Compositional verification of byzantine consensus. Technical Report hal-03158911v1 (2021). HAL
12. Bertrand, N., Konnov, I., Lazic, M., Widder, J.: Verification of randomized consensus algorithms under round-rigid adversaries. In: CONCUR, pp. 33:1–33:15 (2019)
13. Bertrand, N., Konnov, I., Lazic, M., Widder, J.: Verification of randomized distributed algorithms under round-rigid adversaries. In: CONCUR (2019)
14. Biely, M., Schmid, U., Weiss, B.: Synchronous consensus under hybrid process and link failures. Theor. Comput. Sci. **412**(40), 5602–5630 (2011). Sept
15. Bracha, G., Toueg, S.: Asynchronous consensus and broadcast protocols. J. ACM **32**(4), 824–840 (1985). Oct
16. Brown, B.: xRapid: everything you need to know about ripple's crypto service (now live) (Jan 2019). https://blockexplorer.com/news/what-is-xrapid/
17. Buchman, E., Kwon, J., Milosevic, Z.: The latest gossip on BFT consensus. Technical report, Tendermint (2018)
18. Buterin, V., Griffith, V.: Casper the friendly finality gadget. Technical Report (Jan 2019). arXiv:1710.09437v4
19. Cachin, C., Kursawe, K., Shoup, V.: Random oracles in constantipole: practical asynchronous byzantine agreement using cryptography (extended abstract). In: PODC, pp. 123–132 (2000)
20. Cachin, C., Vukolić, M.: Blockchains consensus protocols in the wild (2017). arXiv:1707.01873
21. Cachin, C., Zanolini, L.: Asymmetric byzantine consensus. Technical Report (2020). arXiv:2005.08795
22. Castro, M., Liskov, B.: Practical byzantine fault tolerance and proactive recovery. ACM Trans. Comput. Syst. **20**(4), 398–461 (2002). Nov
23. Charron-Bost, B., Debrat, H., Merz, S.: Formal verification of consensus algorithms tolerating malicious faults. In: Stabilization, Safety, and Security of Distributed Systems-13th International Symposium, SSS 2011, Grenoble, France, October 10-12, 2011. Proceedings, pp. 120–134 (2011)
24. Chase, B., MacBrough, E.: Analysis of the xrp ledger consensus protocol. Technical Report (2018). arXiv:1802.07242v1. (Feb. 2018)
25. Chase, J.M.: Quorum whitepaper (Aug 2018). https://github.com/jpmorganchase/quorum/blob/master/docs/Quorum%20Whitepaper%20v0.2.pdf
26. Civit, P., Gilbert, S., Gramoli, V.: Brief announcement: Polygraph: accountable byzantine agreement. In: DISC (2020)
27. Civit, P., Gilbert, S., Gramoli, V.: Polygraph: accountable byzantine agreement. In: ICDCS (Jul. 2021)
28. Civit, P., Gramoli, V., Gilbert, S: Polygraph: accountable byzantine agreement. Technical Report 2019/587, ePrint (2019). https://eprint.iacr.org/2019/587.pdf
29. Cousineau, D., Doligez, D., Lamport, L., Merz, S., Ricketts, D., Vanzetto, H.: TLA + proofs. In: FM, pp. 147–154 (2012)
30. Crain, T., Gramoli, V., Larrea, M., Raynal, M.: DBFT: efficient leaderless Byzantine consensus and its applications to blockchains. In NCA, IEEE (2018)
31. Crain, T., Natoli, C., Gramoli, V.: Evaluating the Red Belly blockchain. Technical Report (2018). arXiv:1812.11747
32. Crain, T., Natoli, C., Gramoli, V.: Red belly: a secure, fair and scalable open blockchain. In: Proceedings of the 42nd IEEE Symposium on Security and Privacy (S&P'21), pp. 1501–1518 (May 2021)

33. Downey, R.G., Fellows, M.R.: Parameterized Complexity. Monographs in Computer Science. Springer (1999)
34. Dragoi, C., Henzinger, T.A., Zufferey, D.: PSync: a partially synchronous language for fault-tolerant distributed algorithms. In: POPL, pp. 400–415 (2016)
35. Dwork, C., Lynch, N., Stockmeyer, L.: Consensus in the presence of partial synchrony. J. ACM **35**(2), 288–323 (1988). Apr
36. Ekparinya, P., Gramoli, V., Jourjon, G.: The attack of the clones against proof-of authority. In: Community Ethereum Development Conference (EDCON'19) (2019). (Apr. 2019, Presentation)
37. Ekparinya, P., Gramoli, V., Jourjon G.: The Attack of the clones against proof-of-authority. In: Proceedings of the Network and Distributed Systems Security Symposium (NDSS'20). Internet Society (Feb. 2020)
38. Ethereum: Ethereum 2.0 (serenity) phases (2019). https://docs.ethhub.io/ethereum-roadmap/ethereum-2.0/eth-2.0-phases/. (23 Aug. 2019)
39. Golan-Gueta, G., Abraham, I., Grossman, S., Malkhi, D., Pinkas, B., Reiter, M.K., Seredinschi, D., Tamir, O., Tomescu, A.: SBFT: a scalable decentralized trust infrastructure for blockchains. Technical Report (2018). arXiv:1804.01626
40. Gramoli, V.: On the danger of private blockchains. In: Workshop on Distributed Cryptocurrencies and Consensus Ledgers (2016)
41. Guerraoui, R., Kuznetsov, P., Monti, M., Pavlovič, M., Seredinschi, D.-A.: The consensus number of a cryptocurrency. In: PODC, pp. 307–316 (2019)
42. Hawblitzel, C., Howell, J., Kapritsos, M., Lorch, J.R., Parno, B., Roberts, M.L., Setty, S.T.V., Zill, B.: Ironfleet: proving practical distributed systems correct. In: SOSP, pp. 1–17 (2015)
43. Igor Barinov, P.K.: Viktor Baranov. POA network white paper (Sept. 2018). https://github.com/poanetwork/wiki/wiki/POA-Network-Whitepaper
44. John, A., Konnov, I., Schmid, U., Veith, H., Widder, J.: Parameterized model checking of fault-tolerant distributed algorithms by abstraction. In: FMCAD, pp. 201–209 (2013)
45. Konnov, I., Lazić, M., Veith, H., Widder, J.: A short counter example property for safety and liveness verification of fault-tolerant distributed algorithms. In: POPL, pp. 719–734 (2017)
46. Konnov, I., Veith, H., Widder, J.: SMT and POR beat counter abstraction: parameterized model checking of threshold-based distributed algorithms. In: CAV, vol. 9206. LNCS, pp. 85–102 (2015)
47. Konnov, I., Widder, J.: ByMC: byzantine model checker. In: ISoLA, pp. 327–342 (2018)
48. Kotla, R., Alvisi, L., Dahlin, M., Clement, A., Wong, E.: Zyzzyva: speculative byzantine fault tolerance. ACM Trans. Comput. Syst. **27**(4), 7:1–7:39 (2010). Jan
49. Kwon, J.: Tendermint: consensus without mining-draft v.0.6 (2014)
50. Lamport, L.: Byzantizing paxos by refinement. In: DISC, pp. 211–224 (2011)
51. Lazic, M., Konnov, I., Widder, J., Bloem, R.: Synthesis of distributed algorithms with parameterized threshold guards. In: OPODIS, pp. 32:1–32:20 (2017)
52. Lin, Y.-T.: Istanbul byzantine fault tolerance-eip 650 (2019). https://github.com/ethereum/EIPs/issues/650. (21 Aug. 2019)
53. Losa, G., Dodds, M.: On the formal verification of the stellar consensus protocol. In: 2nd Workshop on Formal Methods for Blockchains, FMBC@CAV 2020, pp. 9:1–9:9 (2020)
54. Lynch, N.: Input/output automata: basic, timed, hybrid, probabilistic, dynamic,... In: Amadio R.L.D. (ed.) Proceedings of the Conference on Concurrency Theory (CONCUR), vol. 2761. Lecture Notes in Computer Science (2003)
55. Maric, O., Sprenger, C., Basin, D.A.: Cutoff bounds for consensus algorithms. In: Proceedings fo the Computer Aided Verification Conference, CAV, pp. 217–237 (2017)
56. Miller, A., Xia, Y., Croman, K., Shi, E., Song, D.: The honey badger of BFT protocols. In: CCS (2016)
57. Mostéfaoui, A., Moumen, H., Raynal, M.: Signature-free asynchronous Byzantine consensus with $T < N/3$ and $O(N^2)$ messages. In: PODC, pp. 2–9 (2014)
58. Mostéfaoui, A., Moumen, H., Raynal, M.: Signature-free asynchronous binary Byzantine consensus with $t < n/3$, $O(n^2)$ messages and $O(1)$ expected time. J. ACM (2015)

59. Mostéfaoui, A., Mourgaya, E., Parvédy, P.R., Raynal, M.: Evaluating the condition-based approach to solve consensus. In: DSN, pp. 541–550 (2003)
60. Mostéfaoui, A., Rajsbaum, S., Raynal, M.: Conditions on input vectors for consensus solvability in asynchronous distributed systems. J. ACM **50**(6), 922–954 (2003). Nov
61. Nakamoto, S.: Bitcoin: a peer-to-peer electronic cash system (2008)
62. Neu, J., Tas, E.N., Tse, D.: Ebb-and-flow protocols: a resolution of the availability-finality dilemma. In: Proceedings of the 42nd IEEE Symposium on Security and Privacy (S& P'21) (2021). May 2021
63. Newcombe, C.: Why amazon chose TLA+. In: ABZ, pp. 25–39 (2014)
64. Nipkow, T., Paulson, L.C., Wenzel, M.: Isabelle/HOL-A Proof Assistant for Higher-Order Logic, vol. 2283. Lecture Notes in Computer Science. Springer (2002)
65. Pease, M.C., Shostak, R.E., Lamport, L.: Reaching agreement in the presence of faults. J. ACM **27**(2), 228–234 (1980)
66. Rahli, V., Guaspari, D., Bickford, M., Constable, R.L.: Formal specification, verification, and implementation of fault-tolerant systems using EventML. *ECEASST*, 72, 2015
67. Ranchal-Pedrosa, A., Gramoli, V.: Blockchain is dead, long live blockchain! accountable state machine replication for longlasting blockchain. Technical Report (2020). arXiv:abs/2007.10541
68. Saltini, R.: Correctness analysis of IBFT. Technical Report (Jan. 2019). arXiv:1901.07160v1
69. Schwartz, D., Youngs, N., Britto, A.: The ripple protocol consensus algorithm, vol. 5. Ripple Labs Inc., White Paper (2014)
70. Sergey, I., Wilcox, J.R., Tatlock, Z.: Programming and proving with distributed protocols. In: PACMPL, 2(POPL), 28:1–28:30 (2018)
71. Song, Y.J., van Renesse, R.: Bosco: one-step byzantine asynchronous consensus. In: DISC, pp. 438–450 (2008)
72. Stoilkovska, I., Konnov, I., Widder, J., Zuleger, F.: Verifying safety of synchronous fault-tolerant algorithms by bounded model checking. In: TACAS, pp. 357–374 (2019)
73. Sutra,P.: On the correctness of egalitarian Paxos. Inf. Proc. Lett. **156** (2020)
74. Thomas, S., Schwartz, E.: A protocol for interledger payments (2015). https://interledger.org/interledger.pdf
75. Tsuchiya, T., Schiper, A.: Using bounded model checking to verify consensus algorithms. In: Taubenfeld, G. (ed.) Distributed Computing, pp. 466–480 (2008)
76. Tsuchiya, T., Schiper, A.: Verification of consensus algorithms using satisfiability solving. Distributed Comput. **23**(5–6), 341–358 (2011)
77. von Gleissenthall, K., Kici, R.G., Bakst, A., Stefan, D., Jhala, R.: Pretend synchrony: synchronous verification of asynchronous distributed programs. In: PACMPL, vol. 3(POPL), pp. 59:1–59:30 (2019)
78. Wilcox, J.R., Woos, D., Panchekha, P., Tatlock, Z., Wang, X., Ernst, M.D., Anderson, T.E.: Verdi: a framework for implementing and formally verifying distributed systems. In: PLDI, pp. 357–368 (2015)
79. Yu, Y., Manolios, P., Lamport, L.: Model checking TLA$^+$ specifications. In: CHARME, pp. 54–66 (1999)
80. Zamfir, V., Rush, N., Asgaonkar, A., Piliouras, G.: Introducing the "minimal" cbc casper family of consensus protocols (2018). https://github.com/cbc-casper/cbc-casper-paper/blob/master/cbc-casper-paper-draft.pdf. (21 Aug. 2019)

Decentralized Finance

Constant Function Market Makers: Multi-asset Trades via Convex Optimization

Guillermo Angeris, Akshay Agrawal, Alex Evans, Tarun Chitra, and Stephen Boyd

Abstract The rise of Ethereum and other blockchains that support smart contracts has led to the creation of decentralized exchanges (DEXs), such as Uniswap, Balancer, Curve, mStable, and SushiSwap, which enable agents to trade cryptocurrencies without trusting a centralized authority. While traditional exchanges use order books to match and execute trades, DEXs are typically organized as constant function market makers (CFMMs). CFMMs accept and reject proposed trades based on the evaluation of a function that depends on the proposed trade and the current reserves of the DEX. For trades that involve only two assets, CFMMs are easy to understand, via two functions that give the quantity of one asset that must be tendered to receive a given quantity of the other, and vice versa. When more than two assets are being exchanged, it is harder to understand the landscape of possible trades. We observe that various problems of choosing a multi-asset trade can be formulated as convex optimization problems and can therefore be reliably and efficiently solved.

1 Introduction

In the past few years, several new financial exchanges have been implemented on blockchains, which are distributed and permissionless ledgers replicated across networks of computers. These *decentralized exchanges* (DEXs) enable agents to

G. Angeris (✉) · A. Agrawal · S. Boyd
Stanford University, Stanford, USA
e-mail: angeris@stanford.edu

A. Agrawal
e-mail: akshayka@stanford.edu

S. Boyd
e-mail: boyd@stanford.edu

A. Evans
Bain Capital Crypto, Charlotte, USA

T. Chitra
Gauntlet Networks, New York, USA
e-mail: tarun@gauntlet.network

trade cryptocurrencies, i.e., digital currencies with account balances stored on a blockchain, without relying on a trusted third party to facilitate the exchange. DEXs have significant capital flowing through them; the four largest DEXs on the Ethereum blockchain (Curve Finance [Ego19], Uniswap [ZCP18, AZS+21], SushiSwap [Sus20], and Balancer [MM19]) have a collective trading volume of several billion dollars per day.

Unlike traditional exchanges, DEXs typically do not use order books. Instead, most DEXs (including Curve, Uniswap, SushiSwap, and Balancer) are organized as *constant function market makers* (CFMMs). A CFMM holds reserves of assets (cryptocurrencies), contributed by liquidity providers. Agents can offer or tender baskets of assets to the CFMM, in exchange for another basket of assets. If the trade is accepted, the tendered basket is added to the reserves, while the basket received by the agent is subtracted from the reserves. Each accepted trade incurs a small fee, which is distributed pro-rata among the liquidity providers.

CFMMs use a single rule that determines whether or not a proposed trade is accepted. The rule is based on evaluating a *trading function*, which depends on the proposed trade and the current reserves of the CFMM. A proposed trade is accepted if the value of the trading function at the post-trade reserves (with a small correction for the trading fee) equals the value at the current reserves, i.e., the function is held constant. This condition is what gives CFMMs their name. One simple example of a trading function is the product [Lu17, But17], implemented by Uniswap [ZCP18] and SushiSwap [Sus20]; this CFMM accepts a trade only if it leaves the product of the reserves unchanged. Several other functions can be used, such as the sum or the geometric mean (which is used by Balancer [MM19]).

For trades involving just two assets, CFMMs are very simple to understand, via a scalar function that relates how much of one asset is required to receive an amount of the other, and vice versa. Thus the choice of a two-asset trade involves only one scalar quantity: how much you propose to tender (or, equivalently, how much you propose to receive).

For general trades, in which many assets may be simultaneously exchanged, CFMMs are more difficult reason about. When multiple assets are tendered, there can be many baskets that can be tendered to receive a specific basket of assets, and vice versa, there are many choices of the received basket, given a fixed one that is tendered. Thus the choice of a multi-asset trade is more complex than just specifying an amount to tender or receive. In this case, the trader may wish to tender and receive baskets that are most aligned with their preferences or utility (e.g., one that maximizes their risk-adjusted return).

In all practical cases, including the ones mentioned above, the trading function is concave [AC20]. In this paper, we make use of this fact to formulate various multi-asset trading problems as convex optimization problems. Because convex optimization problems can be solved reliably and efficiently (in theory and in practice) [BV04], we can solve the formulated trading problems exactly. This gives a practical solution to the problem of choosing among many possible multi-asset trades: the trader articulates their objective and constraints, and a solution to this problem determines the baskets of assets to be tendered and received.

Outline. We start by surveying related work in Sect. 1.1. In Sect. 2, we give a complete description of CFMMs, describing how agents may trade with a CFMM, as well as add or remove liquidity. In Sect. 3, we study some basic properties of CFMMs, many of which rely on the concavity of the trading function. In Sect. 4 we examine trades involving just two assets, and show how to understand them via two functions that give the amount of asset received for a given quantity of the tendered asset. Finally, in Sect. 5, we formulate the general multi-asset trading problem as a convex optimization problem and give some specific examples.

1.1   Background and Related Work

Blockchain. CFMMs are typically implemented on a *blockchain*: a decentralized, permissionless, and public ledger. The blockchain stores accounts, represented by cryptographic public keys, and associated balances of one or more cryptocurrencies. A blockchain allows any two accounts to securely transact with each other without the need for a trusted third party or central institution, using public-key cryptography to verify their identities. Executing a *transaction*, which alters the state of the blockchain, costs the issuer a fee, typically paid out to the individuals providing computational power to the network. (This network fee depends on the amount of computation a transaction requires and is paid in addition to the CFMM trading fee mentioned above and described below.)

Blockchains are highly tamper resistant: they are replicated across a network of computers and kept in consensus via simple protocols that prevent invalid transactions such as double-spending of a coin. The consensus protocol operates on the level of *blocks* (bundles of transactions), which are verified by the network and chained together to form the ledger. Because the ledger is public, anyone in the world can view and verify all account balances and the entire record of transactions.

The idea of a blockchain originated with a pseudonymously authored whitepaper that proposed Bitcoin, widely considered to be the first cryptocurrency [Nak08].

Cryptocurrencies. A cryptocurrency is a digital currency implemented on a blockchain. Every blockchain has its own native cryptocurrency, which is used to pay the network transaction fees (and can also be used as a standalone currency).

A given blockchain may have several other cryptocurrencies implemented on it. These additional currencies are sometimes called *tokens*, to distinguish them from the base currency. There are thousands of tokens in circulation today, across various blockchains. Some, like the Uniswap token UNI, give holders rights over the governance of a protocol, while others, like USDC, are *stablecoins*, pegged to the market value of some external or real-world currency or commodity.

Smart contracts. Modern blockchains, such as Ethereum [But13, Woo14], Polkadot [Woo16], and Solana [Yak18], allow anyone to deploy arbitrary stateful programs called *smart contracts*. A contract's public functions can be invoked by anyone, via a transaction sent through the network and addressed to the contract. (The term 'smart contract' was coined in the 1990s, to refer to a set of promises between agents codified in a computer program [Sza95].) Because creators are free to compose deployed contracts or remix them in their own applications, software ecosystems on these blockchains have developed rapidly.

CFMMs are implemented using smart contracts, with functions for trading, adding liquidity, and removing liquidity. Their implementations are usually simple. For example, Uniswap v2 is implemented in just 200 lines of code. In addition to DEXs, many other financial applications have been deployed on blockchains, including lending protocols (e.g., [aav21, com21]) and various derivatives (e.g., [uma21, dyd21]). The collection of financial applications running on blockchains is known as decentralized finance, or DeFi for short.

Exchange-traded funds. CFMMs have some similarities to exchange-traded funds (ETFs). A CFMM's liquidity providers are analogous to an ETF's authorized participants; adding liquidity to a CFMM is analogous to the creation of an ETF share, and subsequently removing liquidity is analogous to redemption. But while the list of authorized participants for an ETF is typically very small, anyone in the world can provide liquidity to a CFMM or trade with it.

Comparison to order books. In an order book, trading a basket of multiple assets for another basket of multiple assets requires multiple separate trades. Each of these trades would entail the blockchain fee, increasing the total cost of trading to the trader. In addition, multiple trades cannot be done at the same time with an order book, exposing the trader to the risk that some of the trades go through while others do not, or that some of the trades will execute at unfavorable prices. In a CFMM, multiple asset baskets are exchanged in one trade, which either goes through as one group trade, or not at all, so the trader is not exposed to the risk of partial execution.

Another advantage of CFMMs over order book exchanges is their efficiency of storage, since they do not need to store and maintain a limit order book, and their computational efficiency, since they only need to evaluate the trading function. Because users must pay for computation costs for each transaction, and these costs can often be nonnegligible in some blockchains, exchanges implementing CFMMs can often be much cheaper for users to interact with than those implementing order books.

Previous work. Academic work on automated market makers began with the study of scoring rules within the statistics literature, e.g., [Win69]. Scoring rules furnish probabilities for baskets of events, which can be viewed as assets or tokens in a prediction market. The output probability from a scoring rule was first proposed as a pricing mechanism for a binary option (such as a prediction market) in [Han03]. Unlike CFMMs, these early automated market makers were shown to be computationally complicated for users to interact with. For example, Chen [CFL+08] demonstrated that computing optimal arbitrage portfolios in logarithmic scoring rules (the most popular class of scoring rules) is #P-hard.

The first CFMM on Ethereum (the most commonly used blockchain for smart contracts) was Uniswap [ZCP18, AZS+21]. The first formal analysis of Uniswap was first done in [AKC+20] and extended to general concave trading functions in [AC20]. Evans [Eva20] first proved that constant mean market makers could replicate a large set of portfolio value functions. The converse result was later proven, providing a mechanism for constructing a trading function that replicates a given portfolio value function [AEC21b]. Analyses of how fees [EAC21, TW20] and trading function curvature [AEC20, Aoy20, AI21] affect liquidity provider returns are also common in the literature. Finally, we note that there exist investigations of privacy in CFMMs [AEC21a], suitability of liquidity provider shares as a collateral asset [CAEK21], and the question of triangular arbitrage [WCDW21] in CFMMs.

1.2 Convex Analysis and Optimization

Convex analysis. A function $f : D \rightarrow \mathbf{R}$, with $D \subseteq \mathbf{R}^n$, is convex if D is a convex set and

$$f(\theta x + (1 - \theta)y) \leq \theta f(x) + (1 - \theta) f(y),$$

for $0 \leq \theta \leq 1$ and all $x, y \in D$. It is common to extend a convex function to an extended-valued function that maps $\mathbf{R}^n$ to $\mathbf{R} \cup \{\infty\}$, with $f(x) = +\infty$ for $x \notin D$. A function f is concave if $-f$ is convex [BV04, Chap. 3].

When f is differentiable, an equivalent characterization of convexity is

$$f(z) \geq f(x) + \nabla f(x)^T (z - x),$$

for all $z, x \in D$. A differentiable function f is concave if and only if for all $z, x \in D$ we have

$$f(z) \leq f(x) + \nabla f(x)^T (z - x). \tag{1}$$

The right-hand side of this inequality is the first-order Taylor approximation of the function f at x, so this inequality states that for a concave function, the Taylor approximation is a global upper bound on the function.

By adding (1) and the same inequality with x and z swapped, we obtain the inequality

$$(\nabla f(z) - \nabla f(x))^T (z - x) \leq 0, \tag{2}$$

valid for any concave f and $z, x \in D$. This inequality states that for a concave function f, $-\nabla f$ is a monotone operator [RB16].

Convex optimization. A *convex optimization problem* has the form

$$\begin{aligned}
&\text{minimize} \quad f_0(x) \\
&\text{subject to} \quad f_i(x) \leq 0, \quad i = 1, \ldots, m \\
&\qquad\qquad\quad g_i(x) = 0, \quad i = 1, \ldots, p,
\end{aligned}$$

where $x \in \mathbf{R}^n$ is the optimization variable, the objective function $f_0 : D \to \mathbf{R}$ and inequality constraint functions $f_i : D \to \mathbf{R}$ are convex, and the equality constraint functions $g_i : \mathbf{R}^n \to \mathbf{R}$ are affine, i.e., have the form $g_i(x) = a_i^T x + b_i$ for some $a_i \in \mathbf{R}^n$ and $b_i \in \mathbf{R}$. (We assume the domains of the objective and inequality functions are the same for simplicity.) The goal is to find a *solution* of the problem, which is a value of x that minimizes the objective function, among all x satisfying the constraints $f_i(x) \leq 0, i = 1, \ldots, m$, and $g_i(x) = 0, i = 1, \ldots, p$ [BV04, Chap. 4]. In the sequel, we will refer to the problem of maximizing a concave function, subject to convex inequality constraints and affine equality constraints, as a convex optimization problem, since this problem is equivalent to minimizing $-f_0$ subject to the constraints.

Convex optimization problems are notable because they have many applications, in a wide variety of fields, and because they can be solved reliably and efficiently [BV04]. The list of applications of convex optimization is large and still growing. It has applications in vehicle control [SB08, Bla16, LB14], finance [CT06, BBD+17], dynamic energy management [MBBW19], resource allocation [ABN+21], machine learning [FHT01, BPC+11], inverse design of physical systems [AVB21], circuit design [HBL01, BKPH05], and many other fields.

In practice, once a problem is formulated as a convex optimization problem, we can use off-the-shelf solvers (software implementations of numerical algorithms) to obtain solutions. Several solvers, such as OSQP [SBG+20], SCS [OCPB16], ECOS [DCB13], and COSMO [GCG19], are free and open source, while others, like MOSEK [ApS19], are commercial. These solvers can handle problems with thousands of variables in seconds or less, and millions of variables in minutes. Small to medium-size problems can be solved extremely quickly using embedded solvers [DCB13, SBG+20, WB10] or code generation tools [MB12, CPDB13, BSM+17]. For example, the aerospace and space transportation company SpaceX uses CVX-GEN [MB12] to solve convex optimization problems in real-time when landing the first stages of its rockets [Bla16].

Domain-specific languages for convex optimization. Convex optimization problems are often specified using domain-specific languages (DSLs) for convex optimization, such as CVXPY [DB16, AVDB18] or JuMP [DHL17], which compile high-level descriptions of problems into low-level standard forms required by solvers. The DSL then invokes a solver and retrieves a solution on the user's behalf. DSLs vastly reduce the engineering effort required to get started with convex optimization, and in many cases are fast enough to be used in production. Using such DSLs, the convex optimization problems that we describe later can all be implemented in just a few lines of code that very closely parallel the mathematical specification of the problems.

2   Constant Function Market Makers

In this section, we describe how CFMMs work. We consider a DEX with $n > 1$ assets, labeled $1, \ldots, n$, that implements a CFMM. Asset n is our numeraire, the asset we use to value and assign prices to the others.

2.1   CFMM State

Reserve or pool. The DEX has some *reserves* of available assets, given by the vector $R \in \mathbf{R}_+^n$, where R_i is the quantity of asset i in the reserves.

Liquidity provider share weights. The DEX maintains a table of all the *liquidity providers*, agents who have contributed assets to the reserves. The table includes weights representing the fraction of the reserves each liquidity provider has a claim to. We denote these weights as $v_1, \ldots, v_N$, where N is the number of liquidity providers. The weights are nonnegative and sum to one, i.e., $v \geq 0$, and $\sum_{i=1}^{N} v_i = 1$. The weights v_i and the number of liquidity providers N can change over time, with addition of new liquidity providers, or the deletion from the table of any liquidity provider whose weight is zero.

State of the CFMM. The reserves R and liquidity provider weights v constitute the state of the DEX. The DEX state changes over time due to any of the three possible *transactions*: a *trade* (or *exchange*), *adding liquidity*, or *removing liquidity*. These transactions are described in Sects. 2.2 and 2.6.

2.2   Proposed Trade

A *proposed trade* (or *proposed exchange*) is initiated by an agent or trader, who proposes to trade or exchange one basket of assets for another. A proposed trade specifies the *tender basket*, with quantities given by $\Delta \in \mathbf{R}_+^n$, which is the basket of assets the trader proposes to give (or tender) to the DEX, and the *received basket*, the basket of assets the trader proposes to receive from the DEX in return, with quantities given by $\Lambda \in \mathbf{R}_+^n$. Here Δ_i (Λ_i) denotes the amount of asset i that the trader proposes to tender to the DEX (receive from the DEX). In the sequel, we will refer to the vectors that give the quantities, i.e., Δ and Λ, as the tender and receive baskets, respectively.

The proposed trade can either be rejected by the DEX, in which case its state does not change, or accepted, in which case the basket Δ is transferred from the trader to the DEX, and the basket Λ is transferred from the DEX to the trader. The DEX reserves are updated as

$$R^+ = R + \Delta - \Lambda, \tag{3}$$

where R^+ denotes the new reserves. A proposed trade is accepted or rejected based on a simple condition described in Sect. 2.3, which always ensures that $R^+ \geq 0$.

Disjoint support of tender and receive baskets. Intuition suggests that a trade would not include an asset in both the proposed tender and receive baskets, i.e., we should not have Δ_i and Λ_i both positive. We will see later that while it is possible to include an asset in both baskets, it never makes sense to do so. This means that Δ and Λ can be assumed to have disjoint support, i.e., we have $\Delta_i \Lambda_i = 0$ for each i. This allows us to define two disjoint sets of assets associated with a proposed or accepted trade:

$$\mathcal{T} = \{i \mid \Delta_i > 0\}, \qquad \mathcal{R} = \{i \mid \Lambda_i > 0\}.$$

Thus $\mathcal{T}$ are the indices of assets the trader proposes to give to the DEX, in exchange for the assets with indices in $\mathcal{R}$. If $j \notin \mathcal{T} \cup \mathcal{R}$, it means that the proposed trade does not involve asset j, i.e., $\Delta_j = \Lambda_j = 0$.

Two-asset and multi-asset trades. A very common type of proposed trade involves only two assets, one that is tendered and one that is received, i.e., $|\mathcal{T}| = |\mathcal{R}| = 1$. Suppose $\mathcal{T} = \{i\}$ and $\mathcal{R} = \{j\}$, with $i \neq j$. Then we have $\Delta = \delta e_i$ and $\Lambda = \lambda e_j$, where e_i denotes the ith unit vector, and $\lambda \geq 0$ is the quantity of asset j the trader wishes to receive in exchange for the quantity $\delta \geq 0$ of asset i. (This is referred to as exchanging asset i for asset j.) When a trade involves more than two assets, it is called a *multi-asset* trade. We will study two-asset and multi-asset trades in Sect. 4 and Sect. 5, respectively.

2.3 Trading Function

Trade acceptance depends on both the proposed trade and the current reserves. A proposed trade (Δ, Λ) is accepted only if

$$\varphi(R + \gamma\Delta - \Lambda) = \varphi(R), \tag{4}$$

where $\varphi : \mathbf{R}^n_+ \to \mathbf{R}$ is the *trading function* associated with the CFMM, and the parameter $\gamma \in (0, 1]$ introduces a *trading fee* (when $\gamma < 1$). The "constant function" in the name CFMM refers to the acceptance condition (4).

We can interpret the trade acceptance condition as follows. If $\gamma = 1$, a proposed trade is accepted only if the quantity $\varphi(R)$ does not change, i.e., $\varphi(R^+) = \varphi(R)$. When $\gamma < 1$ (with typical values being very close to one), the proposed trade is accepted based on the devalued tendered basket $\gamma\Delta$. The reserves, however, are updated based on the full tendered basket Δ as in (3).

Properties. We will assume that the trading function φ is concave, increasing, and differentiable. Many existing CFMMs are associated with functions that satisfy the additional property of homogeneity, i.e., $\varphi(\alpha R) = \alpha\varphi(R)$ for $\alpha > 0$.

2.4 Trading Function Examples

We mention some trading functions that are used in existing CFMMs.

Linear and sum. The simplest trading function is linear,

$$\varphi(R) = p^T R = p_1 R_1 + \cdots + p_n R_n,$$

with $p > 0$, where p_i can be interpreted as the price of asset i. The trading condition (4) simplifies to

$$\gamma p^T \Delta = p^T \Lambda.$$

We interpret the right-hand side as the total value of received basket, at the prices given by p, and the left-hand side as the value of the tendered basket, discounted by the factor γ.

A CFMM with $p = \mathbf{1}$, i.e., all asset prices equal to one, is called a *constant sum market maker*. The CFMM mStable, which held assets that were each pegged to the same currency, was one of the earliest constant sum market makers.

Geometric mean. Another choice of trading function is the (weighted) geometric mean,

$$\varphi(R) = \prod_{i=1}^{n} R_i^{w_i},$$

where total $w > 0$ and $\mathbf{1}^T w = 1$. Like the linear and sum trading functions, the geometric mean is homogeneous.

CFMMs that use the geometric mean are called *constant mean market makers*. The CFMMs Balancer [MM19], Uniswap [ZCP18], and SushiSwap [Sus20] are examples of constant mean market makers. (Uniswap and SushiSwap use weights $w_i = 1/n$, and are sometimes called *constant product* market makers [AKC+20, AC20].)

Other examples. Another example combines the sum and geometric mean functions,

$$\varphi(R) = (1 - \alpha)\mathbf{1}^T R + \alpha \prod_{i=1}^{n} R_i^{w_i},$$

where $\alpha \in [0, 1]$ is a parameter, $w \geq 0$, and $\mathbf{1}^T w = 1$. This trading function yields a CFMM that interpolates between a constant sum market (when $\alpha = 0$) and a constant geometric mean market (when $\alpha = 1$). Because it is a convex combination of the sum and geometric mean functions, which are themselves homogeneous, the resulting function is also homogeneous.

The CFMM known as Curve [Ego19] uses the closely related trading function

$$\varphi(R) = \mathbf{1}^T R - \alpha \prod_{i=1}^{n} R_i^{-1},$$

where $\alpha > 0$. Unlike the previous examples, this trading function is not homogeneous.

2.5 Prices and Exchange Rates

In this section, we introduce the concept of asset (reported) prices, based on a first-order approximation of the trade acceptance condition (4). These prices inform how liquidity can be added and removed from the CFMM, as we will see in Sect. 2.6.

Unscaled prices. We denote the gradient of the trading function as $P = \nabla \varphi(R)$. We refer to P, which has positive entries since φ is increasing, as the vector of *unscaled prices*,

$$P_i = \nabla \varphi(R)_i = \frac{\partial \varphi}{\partial R_i}(R), \quad i = 1, \ldots, n. \tag{5}$$

To see why these numbers can be interpreted as prices, we approximate the exchange acceptance condition (4) using its first-order Taylor approximation to get

$$0 = \varphi(R + \gamma \Delta - \Lambda) - \varphi(R) \approx \nabla \varphi(R)^T (\gamma \Delta - \Lambda) = P^T (\gamma \Delta - \Lambda),$$

when $\gamma \Delta - \Lambda$ is small, relative to R. We can express this approximation as

$$\gamma \sum_{i \in \mathcal{T}} P_i \Delta_i \approx \sum_{i \in \mathcal{R}} P_i \Lambda_i. \tag{6}$$

The right-hand side is the value of the received basket using the unscaled prices P_i. The left-hand side is the value of the tendered basket using the unscaled prices P_i, discounted by the factor γ.

Prices. The condition (6) is homogeneous in the prices, i.e., it is the same condition if we scale all prices by any positive constant. The *reported prices* (or just *prices*) of the assets are the prices relative to the price of the numeraire, which is asset n. The prices are

$$p_i = \frac{P_i}{P_n}, \quad i = 1, \ldots, n.$$

(The price of the numeraire is always 1.) In general, the prices depend on the reserves R. (The one exception is with a linear trading function, in which the prices are constant.) In terms of prices, the condition (6) is

$$\gamma \sum_{i \in \mathcal{T}} p_i \Delta_i \approx \sum_{i \in \mathcal{R}} p_i \Lambda_i. \tag{7}$$

We observe for future use that the prices for two values of the reserves R and $\tilde{R}$ are the same if and only if

$$\nabla\varphi(\tilde{R}) = \alpha\nabla\varphi(R), \tag{8}$$

for some $\alpha > 0$.

Geometric mean trading function prices. For the special case $\varphi(R) = \prod_{i=1}^{n} R_i^{w_i}$, with $w_i > 0$ and $\sum_{i=1}^{n} w_i = 1$, the unscaled prices are

$$P = \nabla\varphi(R) = \varphi(R)(w_1 R_1^{-1}, w_2 R_2^{-1}, \ldots, w_n R_n^{-1}),$$

and the prices are

$$p_i = \frac{w_i R_n}{w_n R_i}, \quad i = 1, \ldots, n. \tag{9}$$

Exchange rates. In a two-asset trade with $\Delta = \delta e_i$ and $\Lambda = \lambda e_j$, i.e., we are exchanging asset i for asset j, the *exchange rate* is

$$E_{ij} = \gamma\frac{\nabla\varphi(R)_i}{\nabla\varphi(R)_j} = \gamma\frac{P_i}{P_j} = \gamma\frac{p_i}{p_j}.$$

This is approximately how much asset j you get for each unit of asset i, for a small trade. Note that $E_{ij}E_{ji} = \gamma^2 < 1$, when $\gamma < 1$, i.e., round-trip trades lose value.

These are first-order approximations. We remind the reader that the various conditions described above are based on a first-order Taylor approximation of the trade acceptance condition. A proposed trade that satisfies (7) is not (quite) valid; it is merely close to valid when the proposed trade baskets are small compared to the reserves. This is similar to the midpoint price (average of bid and ask prices) in an order book; you cannot trade in either direction exactly at this price.

Reserve value. The value of the reserves (using the prices p) is given by

$$V = p^T R = \frac{\nabla\varphi(R)^T R}{\nabla\varphi(R)_n}. \tag{10}$$

When φ is homogeneous we can use the identity $\nabla\varphi(R)^T R = \varphi(R)$ to express the reserves value as

$$V = p^T R = \frac{\varphi(R)}{\nabla\varphi(R)_n}. \tag{11}$$

2.6 Adding and Removing Liquidity

In this section, we describe how agents called *liquidity providers* can add or remove liquidity from the reserves. When an agent adds liquidity, she adds a basket $\Psi \in \mathbf{R}_+^n$ to the reserves, resulting in the updated reserves $R^+ = R + \Psi$. When an agent removes liquidity, she removes a basket $\Psi \in \mathbf{R}_+^n$ from the reserves, resulting in the updated reserves $R^+ = R - \Psi$. (We will see below that the condition for removing liquidity ensures that $R^+ \geq 0$.) Adding or removing liquidity also updates the liquidity provider share weights, as described below.

Liquidity change condition. Adding or removing liquidity must be done in a way that preserves the asset prices. Using (8), this means we must have

$$\nabla\varphi(R^+) = \alpha\nabla\varphi(R), \tag{12}$$

for some $\alpha > 0$. (We will see later that $\alpha > 1$ corresponds to removing liquidity, and $\alpha < 1$ corresponds to adding liquidity.) This liquidity change condition is analogous to the trade exchange condition (4). We refer to Ψ as a *valid liquidity change* if this condition holds.

The liquidity change condition (12) simplifies in some cases. For example, with a linear trading function the prices are constant, so any basket can be used to add liquidity, and any basket with $\Psi \leq R$ can be removed. (The constraint comes from the requirement $R^+ \geq 0$, the domain of φ.)

Liquidity change condition for homogeneous trading function. Another simplification occurs when the trading function is homogeneous. For this case, we have, for any $\alpha > 0$,

$$\nabla\varphi(\alpha R) = \nabla\varphi(R),$$

(by taking the gradient of $\varphi(\alpha R) = \alpha\varphi(R)$ with respect to R). This means that $\Psi = \nu R$, for $\nu > 0$, is a valid liquidity change (provided $\nu \leq 1$ for liquidity removal). In words: you can add or remove liquidity by adding or removing a basket proportional to the current reserves.

Liquidity provider share update. Let $V = p^T R$ denote the value of the reserves before the liquidity change, and $V^+ = (p^+)^T R^+ = p^T R^+$ the value after. The change in reserve value is $V^+ - V = p^T \Psi$ when adding liquidity, and $V^+ - V = -p^T \Psi$ when removing liquidity. Equivalently, $p^T \Psi$ is the value of the basket a liquidity provider gives, when adding liquidity, or receives when removing liquidity. The fractional change in reserve value is $(V^+ - V)/V^+$.

When liquidity provider j adds or removes liquidity, all the share weights are adjusted pro-rata based on the change of value of the reserves, which is the value of the basket she adds or removes. The weights are adjusted to

$$v_i^+ = \begin{cases} v_i V/V^+ + (V^+ - V)/V^+ & i = j \\ v_i V/V^+ & i \neq j. \end{cases} \tag{13}$$

Thus the weight of liquidity provider j is increased (decreased) by the fractional change in reserve value when she adds (removes) liquidity. These new weights are also nonnegative and sum to one.

When φ is homogeneous and we add liquidity with the basket $\Psi = \nu R$, with $\nu > 0$, we have $V_+ = (1 + \nu) p^T R$, so

$$V/V^+ = 1/(1 + \nu), \qquad (V^+ - V)/V^+ = \nu/(1 + \nu).$$

The weight updates for adding liquidity $\Psi = \nu R$ are then

$$v_i^+ = \begin{cases} (v_i + \nu)/(1 + \nu) & i = j \\ v_i/(1 + \nu) & i \neq j. \end{cases}$$

For removing liquidity with the basket $\Psi = \nu R$, we replace ν with $-\nu$ in the formulas above, along with the constraint $\nu \leq v_j$.

2.7 Agents Interacting with CFMMs

Agents seeking to trade or add or remove liquidity make proposals. These proposals are accepted or not, depending on the acceptance conditions given above. A proposal can be rejected if another agent's proposed action is accepted (processed) before their proposed action, thus changing R and invalidating the acceptance condition.

Slippage thresholds. One practical and common approach to mitigating this problem during trading is to allow agents to set a *slippage threshold* on the received basket. This slippage threshold, represented as some percentage $0 \leq \eta \leq 1$, is simply a parameter that specifies how much slippage the agent is willing to tolerate without their trade failing. In this case, the agent presents some trade (Δ, Λ) along with a threshold η, and the contract accepts the trade if there is some number α satisfying $\eta \leq \alpha$ such that the trade $(\Delta, \alpha\Lambda)$ can be accepted. In other words, the agent allows the contract to devalue the output basket by at most a factor of η. If no such value of α exists, the trade fails.

Maximal liquidity amounts. While setting slippage thresholds can help with reducing the risk of trades failing, another possible failure mode can occur during the addition of liquidity. A simple solution to this problem is that the liquidity provider specifies some basket Ψ to the CFMM contract, and the contract accepts the largest possible basket Ψ^- such that $\Psi^- \leq \Psi$, returning the remaining amount, $\Psi - \Psi^-$, to the liquidity provider. In other words, Ψ can be seen as the maximal amount of liquidity a user is willing to provide.

3 Properties

In this section, we present some basic properties of CFMMs.

3.1 Properties of Trades

Non-uniqueness. If we replace the trading function φ with $\tilde{\varphi} = h \circ \varphi$, where h is concave, increasing, and differentiable, we obtain another concave increasing differentiable function. The associated CFMM has the same trade acceptance condition, the same prices, the same liquidity change condition, and the same liquidity provider share updates as the original CFMM.

Maximum valid receive basket. Any valid trade satisfies $\varphi(R + \gamma\Delta - \Lambda) = \varphi(R)$, so in particular $R + \gamma\Delta - \Lambda \geq 0$. Since we assume Δ and Λ have non-overlapping support, it follows that

$$\Lambda \leq R.$$

A valid trade cannot ask to receive more than is in the reserves.

Non-overlapping support for valid tender and receive baskets. Here we show why a valid proposed trade with $\Delta_k > 0$ and $\Lambda_k > 0$ for some k does not make sense when $\gamma < 1$, justifying our assumption that this never happens. Let $(\tilde{\Delta}, \tilde{\Lambda})$ be a proposed trade that coincides with (Δ, Λ) except in the kth components, which we set to

$$\tilde{\Delta}_k = \Delta_k - \tau/\gamma, \qquad \tilde{\Lambda}_k = \Lambda_k - \tau,$$

where $\tau = \min\{\gamma\Delta_k, \Lambda_k\} > 0$. Evidently $\tilde{\Delta} \geq 0$, $\tilde{\Lambda} \geq 0$, and

$$R + \gamma\Delta - \Lambda = R + \gamma\tilde{\Delta} - \tilde{\Lambda},$$

so the proposed trade $(\tilde{\Delta}, \tilde{\Lambda})$ is also valid. If the trader proposes this trade instead of (Δ, Λ), the net change in her assets is

$$\tilde{\Lambda} - \tilde{\Delta} = \Lambda - \Delta + \left(\frac{1}{\gamma} - 1\right)\tau e_k.$$

The last vector on the right is zero in all entries except k, and positive in that entry. Thus the valid proposed trade $(\tilde{\Delta}, \tilde{\Lambda})$ has the same net effect as the trade (Δ, Λ), except that the trader ends up with a positive amount more of the kth asset. Assuming the kth asset has value, we would always prefer this.

Trades increase the function value. For an accepted nonzero trade, we have

$$\varphi(R^+) = \varphi(R + \Delta - \Lambda) > \varphi(R + \gamma\Delta - \Lambda) = \varphi(R),$$

since φ is increasing and $R + \Delta - \Lambda \geq R + \gamma\Delta - \Lambda$, with at least one entry being strictly greater, whenever $\gamma < 1$.

We can derive a stronger inequality using concavity of φ, which implies that

$$\varphi(R + \gamma\Delta - \Lambda) \leq \varphi(R + \Delta - \Lambda) + (\gamma - 1)\nabla\varphi(R + \Delta - \Lambda)^T \Delta.$$

This can be rearranged as

$$\varphi(R^+) \geq \varphi(R) + (1 - \gamma)(P^+)^T \Delta,$$

where $P^+ = \nabla\varphi(R^+)$ are the unscaled prices at the reserves R^+. This tells us the function value increases at least by $(1 - \gamma)$ times the value of tendered basket at the unscaled prices.

Trading cost is positive. Suppose (Δ, Λ) is a valid trade. The net change in the trader's holdings is $\Lambda - \Delta$. We can interpret $\delta = p^T(\Delta - \Lambda)$ as the decrease in value of the trader's holdings due to the proposed trade, evaluated at the current prices. We can interpret δ as a trading cost, evaluated at the pre-trade prices, and now show it is positive.

Since φ is concave, we have

$$\varphi(R + \gamma\Delta - \Lambda) \leq \varphi(R) + \nabla\varphi(R)^T(\gamma\Delta - \Lambda).$$

Using $\varphi(R + \gamma\Delta - \Lambda) = \varphi(R)$, this implies

$$0 \leq \nabla\varphi(R)^T(\gamma\Delta - \Lambda) = P^T(\gamma\Delta - \Lambda).$$

From this we obtain

$$P^T(\Delta - \Lambda) = P^T(\gamma\Delta - \Lambda) + (1 - \gamma)P^T\Delta \geq (1 - \gamma)P^T\Delta.$$

Dividing by P_n gives

$$\delta \geq (1 - \gamma)p^T\Delta.$$

Thus the trading cost is always at least a factor $(1 - \gamma)$ of $p^T\Delta$, the total value of the tendered basket.

The trading cost δ is also the *increase* in the total reserve value, at the current prices. So we can say that each trade increases the total reserve value, at the current prices, by at least $(1 - \gamma)$ times the value of the tendered basket.

3.2   Properties of Liquidity Changes

Liquidity change condition interpretation. One natural interpretation of the liquidity change condition (12) is in terms of a simple optimization problem. We seek a basket Ψ that maximizes the post-change trading function value subject to a given total value of the basket at the current prices,

$$
\begin{aligned}
\text{maximize} & \quad \varphi(R^+) \\
\text{subject to} & \quad p^T(R^+ - R) \le M.
\end{aligned}
\tag{14}
$$

Here the optimization variable is $R^+ \in \mathbf{R}_+^n$, and M is the desired value of the basket Ψ at the current prices, for adding liquidity, or its negative, for removing liquidity. The optimality conditions for this convex optimization problem are

$$
p^T(R^+ - R) \le M, \qquad \nabla\varphi(R^+) - \nu p = 0,
$$

where $\nu \ge 0$ is a Lagrange multiplier. Using $p = \nabla\varphi(R)/\nabla\varphi(R)_n$, the second condition is

$$
\nabla\varphi(R^+) = \frac{\nu}{\nabla\varphi(R)_n}\nabla\varphi(R),
$$

which is (12) with $\alpha = \nu/\nabla\varphi(R)_n$. We can easily recover the trading basket Ψ from R^+ since $\Psi = R^+ - R$.

Liquidity provision problem. When the trading function is homogeneous, it is easy to understand what baskets can be used to add or remove liquidity: they must be proportional to the current reserves. In other cases, it can be difficult to find an R^+ that satisfies (12). In the general case, however, the convex optimization problem (14) can be solved to find the basket Ψ that gives a valid liquidity change, with M denoting the total value of the added basket (when $M > 0$) or removed basket (when $M < 0$).

Liquidity change and the gradient scale factor α. Suppose that we add or remove liquidity. Since φ is concave (2) tells us that

$$
(\nabla\varphi(R^+) - \nabla\varphi(R))^T(R^+ - R) \le 0.
$$

Using $\nabla\varphi(R^+) = \alpha\nabla\varphi(R)$, this becomes

$$
(\alpha - 1)\nabla\varphi(R)^T(R^+ - R) \le 0.
$$

We have $\nabla\varphi(R) > 0$. If we add liquidity, we have $R^+ - R \ge 0$ and $R^+ - R \ne 0$, so $\nabla\varphi(R)^T(R^+ - R) > 0$. From the inequality above we conclude that $\alpha < 1$. If we remove liquidity, a similar arguments tells us that $\alpha > 1$.

4   Two-Asset Trades

Two-asset trades, sometimes called *swaps*, are some of the most common types of trades performed on DEXs. In this section, we show a number of interesting properties of trades in this common special case.

4.1   Exchange Functions

Suppose we exchange asset i for asset j, so $\Delta = \delta e_i$ and $\Lambda = \lambda e_j$, with $\delta \geq 0$, $\lambda \geq 0$. The trade acceptance condition (4) is

$$\varphi(R + \gamma \delta e_i - \lambda e_j) = \varphi(R). \tag{15}$$

The left-hand side is increasing in δ and decreasing in λ, so for each value of δ there is at most one valid value of λ, and for each value of λ, there is at most one valid value of δ. In other words, the relation (15) between λ and γ defines a one-to-one function. This means that two-asset trades are characterized by a single parameter, either δ (how much is tendered) or λ (how much is received).

Forward exchange function. Define $F : \mathbf{R}_+ \to \mathbf{R}$, where $F(\delta)$ is the unique λ that satisfies (15). The function F is called the *forward exchange function*, since $F(\delta)$ is how much of asset j you get if you exchange δ of asset i. The forward exchange function F is increasing since φ is componentwise increasing and nonnegative since $F(0) = 0$. We will now show that the function F is concave.

Concavity. Using the implicit function theorem on (15) with $\lambda = F(\delta)$, we obtain

$$F'(\delta) = \gamma \frac{\nabla \varphi(R')_i}{\nabla \varphi(R')_j}, \tag{16}$$

where we use $R' = R + \gamma \delta e_i - F(\delta)e_j$ to simplify notation. To show that F is concave, we will show that, for any nonnegative trade amounts $\delta, \delta' \geq 0$, the function F satisfies

$$F(\delta') \leq F'(\delta)(\delta' - \delta) + F(\delta), \tag{17}$$

which establishes that F is concave.

We write $R'' = R + \gamma \delta' e_i - F(\delta')e_j$, and note that $\varphi(R) = \varphi(R') = \varphi(R'')$ from the definition of F. Since φ is concave it satisfies

$$\varphi(R'') \leq \nabla \varphi(R')^T (R'' - R') + \varphi(R'),$$

so $\nabla \varphi(R')^T (R'' - R') \geq 0$. Using the definitions of R'' and R', we have

$$0 \leq \gamma(\delta' - \delta)\nabla \varphi(R')_i - (F(\delta') - F(\delta))\nabla \varphi(R')_j.$$

Dividing by $\nabla \varphi(R')_j$ and using (16), we obtain (17).

Reverse exchange function. Define $G : \mathbf{R}_+ \to \mathbf{R} \cup \{\infty\}$, where $G(\lambda)$ is the unique δ that satisfies (15), or $G(\lambda) = \infty$ is there is no such δ. The function G is called the *reverse exchange function*, since $G(\lambda)$ is how much of asset i you must exchange, to receive λ of asset j. In a similar way to the forward trade function, the reverse exchange function is nonnegative and increasing, but this function is convex rather than concave. (This follows from a nearly identical proof.)

Forward and reverse exchange functions are inverses. The forward and reverse exchange functions are inverses of each other, i.e., they satisfy

$$G(F(\delta)) = \delta, \qquad F(G(\lambda)) = \lambda,$$

when both functions are finite.

Analogous functions for a limit order book market. There are analogous functions in a market that uses a limit order book. They are piecewise linear, where the slopes are the different prices of each order, while the distance between the kink points is equal to the size of each order. The associated functions have the same properties, i.e., they are increasing, inverses of each other, F is concave, and G is convex.

Evaluating F and G. In some important special cases, we can express the functions F and G in a closed form. For example, when the trading function is the sum function, they are

$$F(\delta) = \min\{\gamma\delta, R_j\}, \qquad G(\lambda) = \begin{cases} \lambda/\gamma & \lambda/\gamma \le R_j \\ +\infty & \text{otherwise.} \end{cases}$$

When the trading function is the geometric mean, the functions are

$$F(\delta) = R_j \left(1 - \frac{R_i^{w_i/w_j}}{(R_i + \gamma\delta)^{w_i/w_j}} \right), \qquad G(\lambda) = \frac{R_i}{\gamma} \left(\frac{R_j^{w_j/w_i}}{(R_j - \lambda)^{w_j/w_i}} - 1 \right),$$

whenever $\lambda < R_j$, and $G(\lambda) = \infty$ otherwise.

On the other hand, when the forward and reverse trading functions F and G cannot be expressed analytically, we can use several methods to evaluate them numerically [PTFV92, Sect. 9]. To evaluate $F(\delta)$, we fix δ and solve for λ in (15). The left-hand side is a decreasing function of λ, so we can use simple bisection to solve this nonlinear equation. Newton's method can be used to achieve higher accuracy with fewer steps. Exploiting the concavity of φ, it can be shown an undamped Newton iteration always converges to the solution. With superscripts denoting iteration, this is

$$\lambda^{k+1} = \lambda^k + \frac{\varphi(R + \gamma\delta e_i - \lambda^k e_j) - \varphi(R)}{\nabla\varphi(R + \gamma\delta e_i - \lambda^k e_j)_j},$$

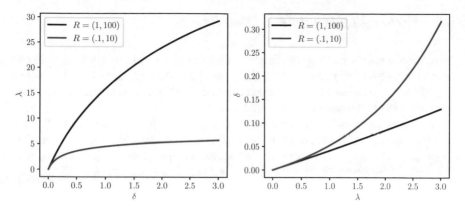

Fig. 1 *Left.* Forward exchange functions for two values of the reserves. *Right,* Reverse exchange functions for the same two values of the reserves

with starting point based on the exchange rate,

$$\lambda^0 = \delta E_{ij} = \delta \frac{\gamma p_i}{p_j}.$$

(It can be shown that the convergence is monotone decreasing.) We note that one of the largest CFMMs, Curve, uses a trading function that is not homogeneous and uses this method in production [Ego19].

Slope at zero. Using (16), we see that $F'(0^+) = E_{ij}$, i.e., the one-sided derivative at 0 is exactly the exchange rate for assets i and j. Since F is concave, we have

$$F(\delta) \le F'(0^+)\delta = E_{ij}\delta. \tag{18}$$

This tells us that the amount of asset j you will receive for trading δ of asset i is no more than the amount predicted by the exchange rate.

The one-sided derivative of the reverse exchange function G at 0 is $G'(0^+) = E_{ji}$. The analog of the inequality (18) is

$$G(\lambda) \ge G'(0^+)\lambda = \gamma^{-2} E_{ji}\lambda, \tag{19}$$

which states that the amount of asset i you need to tender to receive an amount of asset j is at least the amount predicted by the exchange rate.

Examples. Figure 1 shows the forward and reverse exchange functions for a constant geometric mean market with two assets and weights $w_1 = 0.2$ and $w_2 = 0.8$, and $\gamma = 0.997$. We show the functions for two values of the reserves: $R = (1, 100)$ and $R = (0.1, 10)$. The exchange rate is the same for both values of the reserves and equal to $E_{12} = \gamma w_1 R_2 / w_2 R_1 = 25$.

4.2   Exchanging Multiples of Two Baskets

Here we discuss a simple generalization of two-asset trade, in which we tender and receive a multiple of fixed baskets. Thus, we have $\Delta = \delta\tilde{\Delta}$ and $\Lambda = \lambda\tilde{\Lambda}$, where $\lambda \geq 0$ and $\delta \geq 0$ scale the fixed baskets $\tilde{\Delta}$ and $\tilde{\Lambda}$. When $\tilde{\Delta} = e_i$ and $\tilde{\Lambda} = e_j$, this reduces to the two-asset trade discussed above.

The same analysis holds in this case as in the simple two-asset trade. We can introduce the forward and reverse functions F and G, which are inverses of each other. They are increasing, F is concave, G is convex, and they satisfy $F(0) = G(0) = 0$. We have the inequality

$$F(\delta) \leq E\delta,$$

where E is the exchange rate for exchanging the basket $\tilde{\Delta}$ for the basket $\tilde{\Lambda}$, given by

$$E = \gamma \frac{\nabla\varphi(R)^T \tilde{\Delta}}{\nabla\varphi(R)^T \tilde{\Lambda}}.$$

There is also an inequality analogous to (19), using this definition of the exchange rate. We mention two specific important examples in what follows.

Liquidating assets. Let $\Delta \in \mathbf{R}_+^n$ denote a basket of assets we wish to liquidate, i.e., exchange for the numeraire. We can assume that $\Delta_n = 0$. We then find the $\alpha > 0$ for which $(\Delta, \alpha e_n)$ is a valid trade, i.e.,

$$\varphi(R + \gamma\Delta - \alpha e_n) = \varphi(R). \tag{20}$$

We can interpret α as the *liquidation value* of the basket Δ. We can also show that the liquidation value is at most as large as the discounted value of the basket; i.e., $\alpha \leq \gamma p^T \Delta$.

To see this, apply (1) to the left-hand side of (20), which gives, after canceling $\varphi(R)$ on both sides,

$$\nabla\varphi(R)^T (\gamma\Delta - \alpha e_n) \geq 0.$$

Rearranging, we find:

$$\alpha \leq \frac{\gamma\nabla\varphi(R)^T \Delta}{\nabla\varphi(R)_n} = \gamma p^T \Delta.$$

Purchasing a basket. Let $\Lambda \in \mathbf{R}_+^n$ denote a basket we wish to purchase using the numeraire. We find $\alpha > 0$ for which $(\alpha e_n, \Lambda)$ is a valid trade, i.e.,

$$\varphi(R + \gamma\alpha e_n - \Lambda) = \varphi(R).$$

We interpret α as the *purchase cost* of the basket Λ. It can be shown that $\alpha \geq (1/\gamma)\, p^T \Lambda$, i.e., the purchase cost is at least a factor $1/\gamma$ more than the value of the basket, at the current prices. This follows from a nearly identical argument to that of the liquidation value.

5 Multi-asset Trades

We have seen that two-asset trades are easy to understand; we choose the amount we wish to tender (or receive), and we can then find the amount we will receive (or tender). Multi-asset trades are more complex, because even for a fixed receive basket Λ, there are many tender baskets that are valid, and we face the question of which one should we use. The same is true when we fix the tendered basket Δ: there are many baskets Λ we could receive, and we need to choose one. More generally, we have the question of how to choose the proposed trade (Δ, Λ). In the two-asset case, the choice is parameterized by a scalar, either δ or λ. In the multi-asset case, there are more degrees of freedom.

Example. We consider an example with $n = 4$, geometric mean trading function with weights $w_i = 1/4$ and fee $\gamma = 0.997$, with reserves $R = (4, 5, 6, 7)$. We fix the received basket to be $\Lambda = (2, 4, 0, 0)$. There are many valid tendered baskets, which are shown in Fig. 2. The plot shows valid values of (Δ_3, Δ_4), since the first two components of Δ are zero.

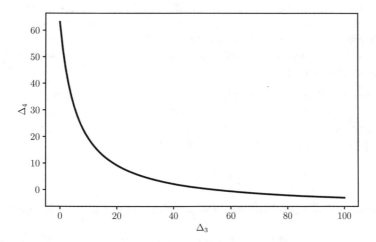

Fig. 2 Valid tendered baskets (Δ_3, Δ_4) for the received basket $\Lambda = (2, 4, 0, 0)$

5.1   The General Trade Choice Problem

We formulate the problem of choosing (Δ, Λ) as an optimization problem. The net change in holdings of the trader is $\Lambda - \Delta$. The trader judges a net change in holdings using a utility function $U : \mathbf{R}^n \to \mathbf{R} \cup \{-\infty\}$, where she prefers (Δ, Λ) to $(\tilde{\Delta}, \tilde{\Lambda})$ if $U(\Lambda - \Delta) > U(\tilde{\Lambda} - \tilde{\Delta})$. The value $-\infty$ is used to indicate that a change in holdings is unacceptable. We will assume that U is increasing and concave. (Increasing means that the trader would always prefer to have a larger net change than a smaller one, which comes from our assumption that all assets have value.)

To choose a valid trade that maximizes utility, we solve the problem

$$
\begin{aligned}
&\text{maximize} && U(\Lambda - \Delta) \\
&\text{subject to} && \varphi(R + \gamma\Delta - \Lambda) = \varphi(R), \quad \Delta \geq 0, \quad \Lambda \geq 0,
\end{aligned}
\tag{21}
$$

with variables Δ and Λ. Unfortunately, the constraint $\varphi(R + \gamma\Delta - \Lambda) = \varphi(R)$ is not convex (unless the trading function is linear), so this problem is not in general convex.

Instead we will solve its convex relaxation, where we change the equality constraint to an inequality to obtain the convex problem

$$
\begin{aligned}
&\text{maximize} && U(\Lambda - \Delta) \\
&\text{subject to} && \varphi(R + \gamma\Delta - \Lambda) \geq \varphi(R), \quad \Delta \geq 0, \quad \Lambda \geq 0,
\end{aligned}
\tag{22}
$$

which is readily solved. It is easy to show that any solution of (22) satisfies $\varphi(R + \gamma\Delta - \Lambda) = \varphi(R)$ and so is also a solution of the problem (21). (If a solution satisfies $\varphi(R + \gamma\Delta - \Lambda) > \varphi(R)$, we can decrease Δ or increase Λ a bit, so as to remain feasible and increase the objective, a contradiction.)

Thus we can (globally and efficiently) solve the non-convex problem (21) by solving the convex problem (22).

No-trade condition. Assuming $U(0) > -\infty$, the solution to the problem (22) can be $\Delta = \Lambda = 0$, which means that trading does not increase the trader's utility, i.e., the trader should not propose any trade. We can give simple conditions under which this happens for the case when U is differentiable. They are

$$
\gamma p \leq \alpha \nabla U(0) \leq p,
\tag{23}
$$

for some $\alpha > 0$. We can interpret the set of prices p for which this is true, i.e.,

$$
K = \{p \in \mathbf{R}^n_+ \mid \gamma p \leq \alpha \nabla U(0) \leq p \text{ for some } \alpha > 0\},
$$

as the *no-trade cone* for the utility function U. (It is easy to see that K is a convex polyhedral cone.)

We interpret $\nabla U(0)$ as the vector of marginal utilities to the trader, and p as the prices of the assets in the CFMM. For $\gamma = 1$, the condition says that we do not trade when the marginal utility is a positive multiple of the current asset prices; if this does not hold, then the solution of the trading problem (22) is nonzero, i.e., the trader should trade to increase her utility. When $\gamma < 1$, the trader will not trade when the prices are in K.

To derive condition (23), we first derive the optimality conditions for the problem (22). We introduce the Lagrangian

$$L(\Delta, \Lambda, \lambda, \omega, \kappa) = U(\Lambda - \Delta) + \lambda(\varphi(R + \gamma\Delta - \Lambda) - \varphi(R)) + \omega^T \Delta + \kappa^T \Lambda,$$

where $\lambda \in \mathbf{R}_+$, $\omega \in \mathbf{R}_+^n$, and $\kappa \in \mathbf{R}_+^n$ are dual variables or Lagrange multipliers for the constraints. The optimality conditions for (22) are feasibility, along with

$$\nabla_\Delta L = 0, \qquad \nabla_\Lambda L = 0.$$

The choice $\Delta = 0$, $\Lambda = 0$ is feasible and satisfies this condition if

$$\nabla_\Delta L(0, 0, \lambda, \omega, \kappa) = 0, \qquad \nabla_\Lambda L(0, 0, \lambda, \omega, \kappa) = 0.$$

These are

$$-\nabla U(0) + \lambda\gamma\nabla\varphi(R) + \omega = 0, \qquad \nabla U(0) - \lambda\nabla\varphi(R) + \kappa = 0,$$

which we can write as

$$\nabla U(0) \geq \lambda\gamma\nabla\varphi(R), \qquad \nabla U(0) \leq \lambda\nabla\varphi(R).$$

Dividing these by λP_n, we obtain (23), with $\alpha = 1/(\lambda P_n)$.

5.2 Special Cases

Linear utility. When $U(z) = \pi^T z$, with $\pi \geq 0$, we can interpret π as the trader's private prices of the assets, i.e., the prices she values the assets at. From (23), we see that the trader will not trade if her private asset prices satisfy

$$\gamma p \leq \alpha\pi \leq p \qquad\qquad (24)$$

for some $\alpha > 0$.

In the special case where π satisfies

$$(\pi_2, \ldots, \pi_n) = \lambda(p_2, \ldots, p_n),$$

for $\lambda \geq 0$, i.e., π is collinear with p except in the first entry, (24) is satisfied if and only if

$$\lambda \gamma p_1 \leq \pi_1 \leq \lambda \gamma^{-1} p_1.$$

If $\lambda = 1$, then this simplifies to the condition

$$\gamma p_1 \leq \pi_1 \leq \gamma^{-1} p_1.$$

(This will arise in an example we present below.)

Markowitz trading. Suppose the trader models the return $r \in \mathbf{R}^n$ on the assets over some period of time as a random vector with mean $\mathbf{E} r = \mu \in \mathbf{R}^n$ and covariance matrix $\mathbf{E}(r - \mu)(r - \mu)^T = \Sigma \in \mathbf{R}^{n \times n}$. If the trader holds a portfolio of assets $z \in \mathbf{R}_+^n$, the return is $r^T z$; the expected portfolio return is $\mu^T z$ and the variance of the portfolio return is $z^T \Sigma z$. In Markowitz trading, the trader maximizes the risk-adjusted return, defined as $\mu^T z - \kappa z^T \Sigma z$, where $\kappa > 0$ is the *risk-aversion parameter* [Mar52, BBD+17]. This leads to the Markowitz trading problem

$$
\begin{aligned}
\text{maximize} \quad & \mu^T z - \kappa z^T \Sigma z \\
\text{subject to} \quad & z = z^{\text{curr}} - \Delta + \Lambda \\
& \varphi(R + \gamma \Delta - \Lambda) \geq \varphi(R) \\
& \Delta \geq 0, \quad \Lambda \geq 0,
\end{aligned}
\tag{25}
$$

with variables z, Δ, Λ, where z^{curr} is the trader's current holdings of assets. This is the general problem (22) with concave utility function

$$U(Z) = \mu^T(z^{\text{curr}} + Z) - \kappa(z^{\text{curr}} + Z)^T \Sigma(z^{\text{curr}} + Z).$$

A well-known limitation of the Markowitz quadratic utility function U, i.e., the risk-adjusted return, is that it is not increasing for all Z, which implies that the trading function relaxation need not be tight. However, for any sensible choice of the parameters μ and Σ, it is increasing for the values of Z found by solving the Markowitz problem (25), and the relaxation is tight. As a practical matter, if a solution of (25) does not satisfy the trading constraint, then the parameters are inappropriate.

Expected utility trading. Here the trader models the returns $r \in \mathbf{R}^m$ on the assets over some time interval as random, with some known distribution. The trader seeks to maximize the expected utility of the portfolio return, using a concave increasing utility function $\psi : \mathbf{R} \to \mathbf{R}$ to introduce risk aversion. (Thus we use the term utility function to refer to both the trading utility function $U : \mathbf{R}_+^n \to \mathbf{R}$ and the portfolio return utility function $\psi : \mathbf{R} \to \mathbf{R}$, but the context should make it clear which is meant.) This leads to the problem

$$
\begin{aligned}
\text{maximize} \quad & \mathbf{E}\,\psi(r^T z) \\
\text{subject to} \quad & z = z^{\text{curr}} - \Delta + \Lambda \\
& \varphi(R + \gamma\Delta - \Lambda) \geq \varphi(R) \\
& \Delta \geq 0, \quad \Lambda \geq 0,
\end{aligned}
\tag{26}
$$

where the expectation is over r. This is the general problem (22), with utility

$$
U(Z) = \mathbf{E}\,\psi(r^T(z^{\text{curr}} + Z)),
$$

which is concave and increasing.

This problem can be solved using several methods. One simple approach is to replace the expectation with an empirical or sample average over some Monte Carlo samples of r, which leads to an approximate solution of (26). The problem can also be solved using standard methods for convex stochastic optimization, such as projected stochastic gradient methods.

5.3 Numerical Examples

In this section, we give two numerical examples.

Linear utility. Our first example involves a CFMM with six assets, geometric mean trading function with equal weights $w_i = 1/6$, and trading fee parameter $\gamma = 0.9$. (We intentionally use an unrealistically small value of γ so the no-trade condition is more evident.) We take reserves

$$
R = (1, 3, 2, 5, 7, 6).
$$

The corresponding prices are given by (9),

$$
p = (R_6/R_1, R_6/R_2, \ldots, 1) = (6, 2, 3, 6/5, 6/7, 1).
$$

We consider linear utility, with the trader's private prices given by

$$
\pi = (tp_1, p_2, \ldots, p_n),
$$

where t is a parameter that we vary over the interval $t \in [1/2, 2]$. For $t = 1$, we have $\pi = p$, i.e., the CFMM prices and the trader's private prices are the same (and not surprisingly, the trader does not trade). As we vary t, we vary the trader's private price for asset 1 by up to a factor of two from the CFMM price.

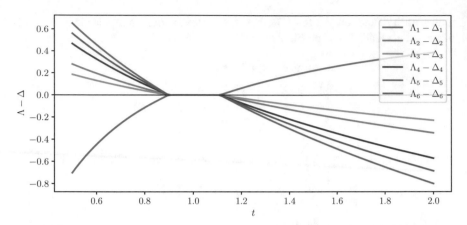

Fig. 3 Solutions $\Lambda - \Delta$ for the linear utility maximization problem, as the private price for asset 1 is varied by the factor t from the CFMM price. The blue curve shows asset 1

The family of optimal trades is shown in Fig. 3, as a function of the parameter t. We plot $\Lambda - \Delta$ versus t, which shows assets in the tender basket as negative and the received basket as positive. The blue curve shows asset 1, which we tender when t is small, and receive when t is large. The no-trade region is clearly seen as the interval $t \in [0.9, 1.1]$.

Markowitz trading. Our second example uses nearly the same CFMM and reserves as the previous example, but with a more realistic trading fee parameter $\gamma = 0.997$. (This is a common choice of trading fee for many CFMMs.) We solve the Markowitz trading problem (25), with current holdings

$$z^{\mathrm{curr}} = (2.5, 1, 0.5, 2.5, 3, 1),$$

mean return

$$\mu = (-0.01, 0.01, 0.03, 0.05, -0.02, 0.02),$$

and covariance $\Sigma = V^T V / 100$, where the entries of $V \in \mathbf{R}^{6 \times 6}$ are drawn from the standard normal distribution. We solve the optimal trading problem for values of the risk aversion parameter κ varying between 10^{-2} and 10^1. (For all of these values, the trading constraint is tight.) These optimal trades are shown in Fig. 4. It is interesting to note that depending on the risk aversion, we either tender or receive assets 2 and 3.

The CVXPY code for the Markowitz optimal trading problem is given below. In this snippet, we assume that mu, sigma, gamma, kappa, R, and z_curr have been previously defined. Note that the code closely follows the mathematical description of the problem given in (25).

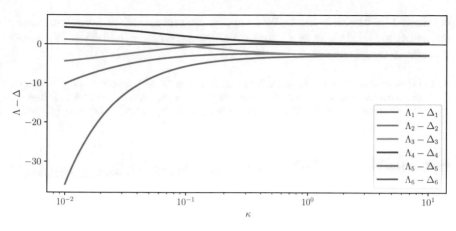

Fig. 4 Solutions $\Lambda - \Delta$ for instances of an example Markowitz trading problem as the risk-aversion parameter κ is varied

```
import cvxpy as cp

delta = cp.Variable(6)
lam = cp.Variable(6)

z = z_curr - delta + lam
R_new = R + gamma*delta - lam

objective = cp.Maximize(z.T @ mu - kappa*cp.quad_form(z, sigma))
constraints = [
    cp.geo_mean(R_new) >= cp.geo_mean(R),
    delta >= 0,
    lam >= 0
]

problem = cp.Problem(objective, constraints)
problem.solve()
```

Listing 1 Markowitz trading CVXPY code.

6 Conclusion

We have provided a general description of CFMMs, outlining how users can interact with a CFMM through trading or adding and removing liquidity. We observe that many of the properties of CFMMs follow from concavity of the trading function. In the simple case where two assets are traded or exchanged, it suffices to specify the amount we wish to receive (or tender), which determines the amount we tender (receive), by simply evaluating a convex (concave) function. Multi-asset trades are more complex, since the set of valid trades is multi-dimensional, i.e., multiple tender or received baskets are possible. We formulate the problem of choosing from among these possible valid trades as a convex optimization problem, which can be globally and efficiently solved.

Acknowledgements The authors would like to acknowledge Shane Barratt for useful discussions. Guillermo Angeris is supported by the National Science Foundation Graduate Research Fellowship under Grant No. DGE-1656518. Akshay Agrawal is supported by a Stanford Graduate Fellowship.

References

[aav21]   Aave. https://aave.com, 2021
[ABN+21]   Akshay Agrawal, Stephen Boyd, Deepak Narayanan, Fiodar Kazhamiaka, and Matei
           Zaharia. Allocation of fungible resources via a fast, scalable price discovery method.
           arXiv preprint arXiv:2104.00282, 2021
[AC20]   Guillermo Angeris and Tarun Chitra. Improved price oracles: Constant function mar-
           ket makers. In *Proceedings of the 2nd ACM Conference on Advances in Financial
           Technologies*, pages 80–91, New York NY USA, October 2020. ACM
[AEC20]   Guillermo Angeris, Alex Evans, and Tarun Chitra. When does the tail wag the dog?
           Curvature and market making. *arXiv preprint* arXiv:2012.08040, 2020
[AEC21a]   Guillermo Angeris, Alex Evans, and Tarun Chitra. A note on privacy in constant
           function market makers. *arXiv preprint* arXiv:2103.01193, 2021
[AEC21b]   Guillermo Angeris, Alex Evans, and Tarun Chitra. Replicating market makers. *arXiv
           preprint* arXiv:2103.14769, 2021
[AI21]   Jun Aoyagi and Yuki Ito. Liquidity implications of constant product market makers.
           Available at SSRN 3808755, 2021
[AKC+20]   Guillermo Angeris, Hsien-Tang Kao, Rei Chiang, Charlie Noyes, and Tarun Chitra.
           An analysis of Uniswap markets. *Cryptoeconomic Systems*, November 2020
[Aoy20]   Jun Aoyagi. Liquidity provision by automated market makers. *Available at SSRN
           3674178*, 2020
[ApS19]   MOSEK   ApS.   MOSEK   Optimizer   API   for   Python   9.1.5.
           https://docs.mosek.com/9.1/pythonapi/index.html, 2019
[AVB21]   Angeris, G., Vučković, J., Boyd, S.: Heuristic methods and performance bounds for
           photonic design. Optics Express **29**(2), 2827 (2021)
[AVDB18]   Agrawal, A., Verschueren, R., Diamond, S., Boyd, S.: A rewriting system for convex
           optimization problems. Journal of Control and Decision **5**(1), 42–60 (2018)
[AZS+21]   Hayden Adams, Noah Zinsmeister, Moody Salem, River Keefer, and Dan Robinson.
           Uniswap v3 core. Technical report, 2021
[BBD+17]   Boyd, S., Busseti, E., Diamond, S., Kahn, R., Koh, K., Nystrup, P., Speth, J.: Multi-
           period trading via convex optimization. Foundations and Trends in Optimization **3**(1),
           1–76 (2017)
[BKPH05]   Stephen Boyd, Seung-Jean Kim, Dinesh Patil, and Mark Horowitz. Digital circuit
           optimization via geometric programming. *Operations Research*, 53(6), 2005
[Bla16]   Lars Blackmore. Autonomous precision landing of space rockets. *The BRIDGE*, 26(4),
           2016
[BPC+11]   Stephen Boyd, Neal Parikh, Eric Chu, Borja Peleato, and Jonathan Eckstein. Dis-
           tributed optimization and statistical learning via the alternating direction method of
           multipliers. *Foundations and Trends® in Machine learning*, 3(1):1–122, 2011
[BSM+17]   Goran Banjac, Bartolomeo Stellato, Nicholas Moehle, Paul Goulart, Alberto Bempo-
           rad, and Stephen Boyd. Embedded code generation using the OSQP solver. In *IEEE
           Conference on Decision and Control*, 2017

[But13]   Vitalik Buterin. Ethereum: A next-generation smart contract and decentralized application platform, 2013
[But17]   Vitalik Buterin. On path independence (2017). https://vitalik.ca/general/2017/06/22/marketmakers.html
[BV04]   Boyd, S., Vandenberghe, L.: Convex Optimization. Cambridge University Press, Cambridge, UK; New York (2004)
[CAEK21]   Tarun Chitra, Guillermo Angeris, Alex Evans, and Hsien-Tang Kao. A note on borrowing constant function market maker shares. 2021
[CFL+08]   Yiling Chen, Lance Fortnow, Nicolas Lambert, David Pennock, and Jennifer Wortman. Complexity of combinatorial market makers. In *Proceedings of the 9th ACM Conference on Electronic Commerce*, pages 190–199, 2008
[com21]   Compound. https://compound.finance, 2021
[CPDB13]   Eric Chu, Neal Parikh, Alexander Domahidi, and Stephen Boyd. Code generation for embedded second-order cone programming. In *European Control Conference*, pages 1547–1552. IEEE, 2013
[CT06]   Gerard Cornuejols and Reha Tütüncü. *Optimization Methods in Finance*. Cambridge University Press, 2006
[DB16]   Diamond, S., Boyd, S.: CVXPY: A Python-embedded modeling language for convex optimization. Journal of Machine Learning Research **17**(83), 1–5 (2016)
[DCB13]   Alexander Domahidi, Eric Chu, and Stephen Boyd. ECOS: An SOCP solver for embedded systems. In *2013 European Control Conference (ECC)*, pages 3071–3076, Zurich, July 2013. IEEE
[DHL17]   Dunning, I., Huchette, J., Lubin, M.: JuMP: A modeling language for mathematical optimization. SIAM review **59**(2), 295–320 (2017)
[dyd21]   dydx. https://dydx.exchange, 2021
[EAC21]   Alex Evans, Guillermo Angeris, and Tarun Chitra. Optimal fees for geometric mean market makers. *arXiv preprint* arXiv:2104.00446, 2021
[Ego19]   Michael Egorov. StableSwap - efficient mechanism for Stablecoin liquidity. page 6, 2019
[Eva20]   Alex Evans. Liquidity provider returns in geometric mean markets. *arXiv preprint* arXiv:2006.08806, 2020
[FHT01]   Jerome Friedman, Trevor Hastie, and Robert Tibshirani. *The Elements of Statistical Learning*, volume 1. Springer Series in Statistics, 2001
[GCG19]   Michael Garstka, Mark Cannon, and Paul Goulart. COSMO: A conic operator splitting method for large convex problems. In *2019 18th European Control Conference (ECC)*, pages 1951–1956, Naples, Italy, June 2019. IEEE
[Han03]   Hanson, R.: Combinatorial information market design. Information Systems Frontiers **5**(1), 107–119 (2003)
[HBL01]   Hershenson, M., Boyd, S., Lee, T.: Optimal design of a CMOS op-amp via geometric programming. IEEE Transactions on Computer-aided design of integrated circuits and systems **20**(1), 1–21 (2001)
[LB14]   Lipp, T., Boyd, S.: Minimum-time speed optimisation over a fixed path. International Journal of Control **87**(6), 1297–1311 (2014)
[Lu17]   Alan Lu. Building a decentralized exchange in Ethereum. https://blog.gnosis.pm/building-a-decentralized-exchange-in-ethereum-eea4e7452d6e, 2017
[Mar52]   Markowitz, H.: Portfolio selection. The. Journal of Finance **7**(1), 77–91 (1952)
[MB12]   Mattingley, J., Boyd, S.: CVXGEN: A code generator for embedded convex optimization. Optimization and Engineering **13**(1), 1–27 (2012)
[MBBW19]   Nicholas Moehle, Enzo Busseti, Stephen Boyd, and Matt Wytock. Dynamic energy management. *arXiv preprint* arXiv:1903.06230, 2019
[MM19]   Fernando Martinelli and Nikolai Mushegian. Balancer: A non-custodial portfolio manager, liquidity provider, and price sensor. 2019
[Nak08]   Satoshi Nakamoto. Bitcoin: A peer-to-peer electronic cash system, 2008

[OCPB16]   O'Donoghue, B., Chu, E., Parikh, N., Boyd, S.: Conic optimization via operator splitting and homogeneous self-dual embedding. Journal of Optimization Theory and Applications **169**(3), 1042–1068 (2016)

[PTFV92]   Press, W., Teukolsky, S., Flannery, B.: and William Vetterling. The Art of Scientific Computing. Cambridge University Press, Numerical Recipes (1992)

[RB16]   Ernest Ryu and Stephen Boyd. A primer on monotone operator methods. *Applied Computational Math*, 2016

[SB08]   Gregory Stewart and Francesco Borrelli. A predictive control framework for industrial turbodiesel engine control. In *IEEE Conference on Decision and Control (CDC)*, pages 5704–5711, 2008

[SBG+20]   Bartolomeo Stellato, Goran Banjac, Paul Goulart, Alberto Bemporad, and Stephen Boyd. OSQP: An operator splitting solver for quadratic programs. *Mathematical Programming Computation*, February 2020

[Sus20]   Sushi. The SushiSwap project, 2020

[Sza95]   Nick Szabo. Smart contracts. *Extropy: Journal of Transhumanist Thought*, 16, 1995

[TW20]   Martin Tassy and David White. Growth rate of a liquidity provider's wealth in $xy = c$ automated market makers, 2020

[uma21]   UMA project. https://umaproject.org, 2021

[WB10]   Wang, Y., Boyd, S.: Fast evaluation of quadratic control-Lyapunov policy. IEEE Transactions on Control Systems Technology **19**(4), 939–946 (2010)

[WCDW21]   Ye Wang, Yan Chen, Shuiguang Deng, and Roger Wattenhofer. Cyclic arbitrage in decentralized exchange markets. *Available at SSRN 3834535*, 2021

[Win69]   Winkler, R.: Scoring rules and the evaluation of probability assessors. Journal of the American Statistical Association **64**(327), 1073–1078 (1969)

[Woo14]   Gavin Wood. Ethereum: A secure decentralised generalised transaction ledger, 2014

[Woo16]   Gavin Wood. Polkadot: Vision for a heterogeneous multi-chain framework, 2016

[Yak18]   Anatoly Yakovenko. Solana: A new architecture for a high performance blockchain, 2018

[ZCP18]   Yi Zhang, Xiaohong Chen, and Daejun Park. Formal specification of constant product $(xy = k)$ market maker model and implementation. 2018

Stablecoins: Reducing the Volatility of Cryptocurrencies

Ayten Kahya, Bhaskar Krishnamachari, and Seokgu Yun

Abstract In the wake of financial crises, stablecoins are gaining adoption among digital currencies. We discuss how stablecoins help reduce the volatility of cryptocurrencies by surveying different types of stablecoins and their stability mechanisms. We classify different approaches to stablecoins in three main categories (i) fiat or asset backed, (ii) crypto-collateralized, and (iii) algorithmic stablecoins, giving examples of concrete projects in each class. We assess the relative tradeoffs between the different approaches. We also discuss challenges associated with the future of stablecoins and their adoption, their adoption and point out future research directions.

1 Introduction

The introduction of Bitcoin in 2009 revolutionized the world of finance by offering the first truly decentralized peer-to-peer protocol for digital cash. However, even as Bitcoin has been growing in popularity, spawning many other cryptocurrencies in its wake, their use as a medium of exchange has been challenging because they show high volatility, fluctuating greatly in price on a monthly, weekly, daily, sometimes even hourly basis. To address these challenges, researchers and developers have started to focus on the design of "stablecoins."

A stablecoin is a digital token on a blockchain that is designed to minimize price volatility with respect to a stable fiat currency or asset. The majority of stablecoins are pegged to fiat currencies such as USD, followed by assets such as gold or a basket of assets. This allows stablecoins to be utilized as primarily a unit of exchange as well as a unit of account and a store of value (if the underlying asset maintains value in the

A. Kahya (✉) · B. Krishnamachari
Viterbi School of Engineering, University of Southern California, Los Angeles, CA, USA
e-mail: betul@dlt.mobi

B. Krishnamachari
e-mail: bkrishna@usc.edu

S. Yun
SovereignWallet Network Pte. Ltd, Singapore, Singapore

© The Author(s), under exclusive license to Springer Nature Switzerland AG 2022
D. A. Tran et al. (eds.), *Handbook on Blockchain*, Springer Optimization
and Its Applications 194, https://doi.org/10.1007/978-3-031-07535-3_14

long term) compared to highly speculative volatile cryptocurrencies. Stablecoins are currently used for payments, trading, lending, investing, remittances, and purchases. Volatility of cryptocurrencies worldwide in recent years enabled stablecoins to gain popularity across users and increase competition in financial markets. Indeed, in 2020, stablecoins have shown dramatic growth as various platforms experienced exponential growth in stablecoins use [1].

Stablecoins can be designed in various ways depending on the desired utility. They are commonly useful for retail payments, international money transfer, while some stablecoins are designed for settlements between banks or sustaining an ecosystem around an activity. Depending on the design, stablecoins can increase efficiency of payments [2]. Stablecoins can be classified in three main categories as (i) fiat or asset backed, (ii) crypto-collateralized, and (iii) algorithmically stabilized stablecoins. There are also hybrid approaches, which may involve more than one type of backing such ad crypto and fiat backing. The degree of automation and centralization varies across the stablecoin types and use cases. Meanwhile, stablecoins market share grew during the impact of COVID-19 to global markets and cryptocurrency market crash following Bitcoin's large drop on March 12 2020 [3]. Investors turned to stablecoins amidst the market turmoil as the combined transfer of all stablecoins tracked by Coin Metrics reached $444.21 M on March 13. In January 2021, the US Office of Comptroller of Currency allowed national banks and federal savings associations to use stablecoins for bank-permissible functions [4] (Fig. 1).

2   Types of Stablecoins

We classify different types of stablecoins based on how they try to stabilize the price. The first class we consider is fiat or asset-backed stablecoins, which directly connect the number of coins in circulation with either fiat currency or assets being held in reserve by some entity. We then discuss crypto-collateralized stablecoins, in which the asset being collateralized is itself a (potentially volatile) cryptocurrency. Finally, we discuss algorithmic stablecoins, which aim to utilize sophisticated smart contracts driven by external price feeds to automate the process of minting and withdrawing coins from circulation to stabilize the price.

3   Fiat or Asset Backed Stablecoins

Fiat or asset backed stablecoins are generally pegged to and backed one to one by an asset that is held in a reserve by a private bank. The most common type is USD pegged fiat backed stablecoins such as Tether, which has the highest market capitalization among all stablecoins [5]. Other approaches include backing by traditional assets such as gold (e.g., PAXG) or a basket of currencies such as one of the options considered in Diem (formerly called Libra, this project also includes individual currency-backed

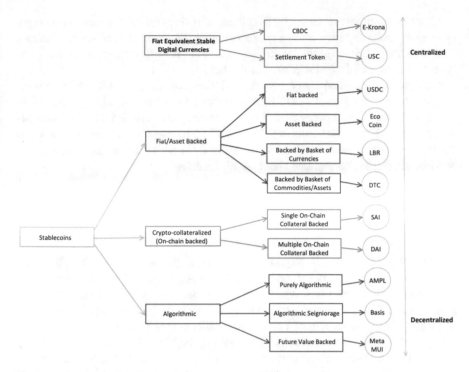

Fig. 1 Stablecoins categorization

stablecoins). Although top ranked fiat backed stablecoin projects are generally able to minimize volatility in regards to the peg, they can be impacted by fluctuations in the underlying asset's value. In the following, we describe five types of asset/fiat backed stablecoins through exemplary projects.

3.1 Single Fiat-backed—USDC

Launched in 2018, USD Coin (USDC) is one of top fiat backed stablecoins [5]. USDC is an ERC20 token (also a token on the Stellar blockchain network) pegged to the USD [6]. Unlike decentralized stablecoins such as Dai, USDC is centralized as the token is issued by Centre, a consortium that is founded by the companies Coinbase and Circle [7]. It features multi-issuer scheme where eligible financial institutions need to meet various requirements such as being Anti Money Laundering (AML) audited and compliance with FATF standards. Center aims to achieve a broad ecosystem where USDC can be integrated into other services and apps with this membership scheme and open-source framework. USDC can be used for trading, payments, cross-border transactions, lending and investment.

Centre guarantees that each USDC is backed by one USD held in reserve by regulated institutions and they are always redeemable [6]. A significant feature claimed by Center about USDC is that it is audited by a well-known independent organization (Grant Thornton LLP) and regulated in the US [7].

Coins are issued when a user requests USDC in exchange for USD. After the user transfers funds for tokens, CENTRE network verifies, mints and validates the USDC tokens to be transferred to the user. When a user requests a redemption, the network verifies and validates then removes the USDC by burning the tokens and returning the backing fiat to the user. Stability is ensured through this minting and burning process and the one-to-one USD backing of the coins.

3.2 Multi-Fiat-backed—Diem

Formerly known as Libra (LBR), Diem is undoubtedly one of the most well-known stablecoin projects. It is developed by the Diem Association based in Switzerland (previously called the Libra Association), co-founded by the social networking giant Facebook [8]. Libra was planned to be backed by and pegged to a basket of assets initially consisting of USD, GBP, EUR, and JPY with the Libra association governing the Libra network and managing the Libra Reserve for the backing assets [9]. According to its whitepaper, "Libra's mission is to enable a simple global currency and financial infrastructure that empowers billions of people" [8]. Due to the large outreach of Facebook and other well-established founding members including financial services and payments platforms companies, Libra was believed to emerge as a significant competitor in the global financial markets. Regulatory uncertainty of Libra's classification (as currency/derivative/security/commodity pool) in addition to other regulatory concerns such as fraud prevention received regulatory backlash and led to some of the big co-founders of the association such as Visa, Mastercard and eBay dropping out [10]. Consequently, the project has made various changes to the whitepaper and plans to launch at the end of 2020 [11]. The novel version of the whitepaper states that Libra's vision has been to complement the fiat currencies rather than competing with them [8]. Concerns rose about Libra having potential to interfere with monetary policy and sovereignty rose if it were to scale up significantly and large volume of payments were to be made in LBR. Some viewed it as having potential to reduce reliance on a single currency for international trade (un-dollarization) [9]. Hence, Libra moved forward with adding single-currency stablecoins from their proposed currency basket (e.g., LibraUSD, LibraEUR, LibraGBP or LibraSGD) to their platform, which will be fully backed by the Reserve [8]. While LBR will not be a separate asset from the single-currency stablecoins, it will rather be a digital composite of single-currency stablecoins defined with reference to fixed nominal weights. The team intends to work with regulators, financial institutions and central banks to increase the single currency stablecoins on their platform. Furthermore, Libra also aims to support the UN's Sustainable Development Goals.

3.3 Single Asset-backed—ECO

ECO coin is a unique ecological cryptocurrency that is asset backed by trees [12]. Based on a circular economy concept, it acts an alternative digital currency that encourages environmentally sustainable actions through financial incentives [13]. The project was developed by Next Nature Network in Amsterdam and launched in 2017 with an implementation at a popular music festival [14]. Eco Coin is currently operating a pilot in a community [13].

Users earn Eco Coins (ECOs) by sustainable actions at individual or organizational level; for instance, riding a bike to work or switching to green energy providers [12]. The value of these sustainable actions is determined according to their relative offsets of Carbon Dioxide emissions equivalent [13]. Additionally, ECOs can also be obtained via verifying sustainable actions as ECO inspectors, backing the token by contributing trees to the system, being a certified vendor or being a part of the technical development [12]. They can also be bought at the Initial Coin Offering. ECO inspectors, certified vendors and sensor-integrated systems verify sustainable actions to prevent malicious actors from gaming the system. The platform is governed via the Decentralized Autonomous Charity (DAC) where ECO holders and stakeholders participate (by running a node) in votes for the development and decisions regarding ECOs.

Eco Coin has a unique issuance process. While each ECO is backed by a tree, one ECO coin is earned in exchange for contributing 10 trees, while the other 9 can be redeemed through sustainable actions. The trees are kept in escrow through the ECO Coin Foundation (ownership still belongs to the original owner). Since the lifespan of a tree is finite, the lifespan of an ECO is finite based on average tree lifespan. Thus, to make the system practical, the coin deteriorates by a small percentage every year. If the average lifespan of the tree was 100 years, 1 ECO would represent a 1-year-old tree and 0.01 ECO represent a hundred-year-old tree. A tree owner can only cut down and plant another as a replacement when the average tree lifespan is over.

After they are earned, ECOs can be spent in exchange for goods or services. The project aims to expand the sustainable marketplace where ECOs can be earned and used [13]. As the platform grows, the ECOs will be exchangeable with Euros according to the developers.

3.4 Multi-Asset-backed—DTC

Digital Trade Coin (DTC) is an example of asset backed stablecoin, which employs a unique approach to system design [15, 16]. Currently, in development at MIT, the project aims to explore an efficient and reliable digital currency that is trade-oriented, scalable, fast, and environmentally friendly. DTC is pegged to real-world assets such as energy, crops, and minerals, which are supplied to the platform by a consortium of sponsors as reserve collaterals (backing) in exchange for DTC tokens. Sponsors may include alliances of small nations, commercial trades, business or farmers etc. In the DTC ecosystem, DTCs are traded amongst sponsors while non-sponsors (users) can obtain "e-Cash" from the consortium that is backed by DTCs in exchange for fiat money. E-Cash serves as a stable payment method for everyday transactions and can store value over time. It is important to note that financial transactions involving fiat currencies are carried out through a narrow bank. In addition to e-Cash, if a participant wants to obtain newly minted DTC; they transfer cash to the narrow bank, which transfers money to the sponsors who in turn release DTC to the participant. This way, the participant turns to a shareholder in the pool. To redeem fiat money, the participant can return the DTC to the administrator who sells assets to return the cash and burns the DTC.

The system is governed by the consortium and its delegated administration responsible with carrying out monetary policies of the consortium and controlling various system functions. Stability of the DTC is also maintained by the consortium. When the market price of DTC falls significantly below the market price of the relevant asset pool, economic agents will return DTC to the administrator. The administrator will sell the corresponding amount of assets to return the proceeds to these agents. Conversely, if the market price of DTC is significantly above the market price of the relevant asset pool, sponsors will contribute more assets to the pool. So, the administrator can issue more DTCs to sponsors that sell them to other participants for cash thereby pushing the DTC price down.

According to the developers of DTC, the complexity of system design and various system functions depend on specific applications of the concept. There are three layers of ledgers within the system architecture; recording of the assets is done through the Assets Ledger while the coin transactions are enabled through the Coins Ledger. Lastly, E-Cash transactions take place on the Transactions Ledger. Additionally, DTC ledger can be designed as semi-private to meet AML/Know Your Customer (KYC) standards. In order to establish a more efficient system and avoid energy-intensive mining methods, DTC network will utilize a set of trusted nodes as validators. DTC concept is currently being explored through two pilot projects related to international commerce and commodity markets.

3.5 Settlement Coin—FTC

Fnality, formerly known as Utility Settlement Coin (USC), utilizes a similar approach to fiat-backed stablecoins, albeit focused on the problem of bank settlements. The project is developed by a consortium of banks including some of the world's major banks and financial institutions [17]. Fnality aims to establish a decentralized Financial Market Infrastructure in each currency on its platform to deliver means of payment for wholesale banking markets via its tokenized settlement asset USC. USC will operate on a private ledger on an Enterprise Ethereum blockchain and based on the jurisdiction of the relevant central bank money, it will act as a digital representation of an entitlement, claim or interest [18]. USC will serve as a medium-of-exchange for the wholesale market and as a store of value meant solely to help settlement.

The primary distinction of USC from other coins is that the aforementioned digital representation is backed by corresponding assets at the respective central banks. The initial currencies on USC's platform will be CAD, EUR, GBP, JPY, and USD while more currencies might be added in the future [19]. Furthermore, USC plans to be fully backed with guarantee of exchangeability into fiat currency anytime. The key aspect of Fnality's innovation is the facilitated finality of settlements. Settlement is achieved in compliance with local settlement finality laws and regulations. Thus, the finality and irreversibility (by court) of the settlement are ensured locally, for each jurisdiction [18]. Developers believe Fnality will reduce liquidity needs and facilitate cash management by removing the need of "having many separate accounts at Corre-spondents and Custodians". This also reduces the settlement time and complexity by enabling international banks to easily transfer ownership of USC [20].

4   Crypto-Collateralized Stablecoins

Crypto-collateralized stablecoins (a.k.a. on-chain backed stablecoins) are backed by other cryptocurrencies on the blockchain. The core component is over-collateralizing the backing cryptocurrencies so that their volatilities have minimal impact on the stablecoin's price. However, they may be impacted by severe changes in collaterals' price. Various projects mitigate this problem by multiple on-chain asset backing to reduce the dependence on a single type of collateral. In this section, we describe how collateralized stablecoins work through analyzing MakerDAO.

4.1   MakerDAO

Launched in 2017 by MakerDAO, Dai is a crypto-collateralized token soft-pegged to the USD [5]. The Maker Protocol is amongst the largest de-fi applications on Ethereum as well as the leading crypto collateralized stablecoin by market capitalization. Dai has no fiat backing and there is no central authority in the Maker Protocol issuing the tokens [21]. It can be traded on various exchanges, used for payments and transactions, lent or held for savings via Dai Savings Rate (DSR). Although the Maker Foundation founded MakerDAO and bootstrapped the Maker Governance, they plan to dissolve once the DAO (Decentralized Autonomous Organization) is ready to fully govern the platform. The initial single collateral Dai on the platform (backed by ETH) was called "Sai" after transitioning to the new Maker Protocol with multiple collateral types. Sai officially shutdown in May 2020 [22].

Dai is generated when a user locks in excess collateral in a "Vault" [21]. During this process, the collateralization ratio needs to be set above the liquidation ratio at which collateral becomes too risky. Liquidation ratio is a key risk parameter that is determined by the governance according to the risk characteristics of collaterals that helps keep stability of the token. Maker Protocol currently accepts Ether (ETH), Basic Attention Token (BAT), USD Coin (USDC) and Wrapped Bitcoin (WBTC) tokens as collateral for Dai. MakerDAO community is also considering including tokenized trade invoices and music streaming loyalties as collaterals for Dai [23]. When the collateral debt is paid with the stability fee, the vault is then closed while the collateral is returned to the user and the corresponding Dai is burnt from supply. Stability fee acts as an interest rate and is one of the primary features of Dai's stability mechanism. Lower stability fee encourages users to open more Vaults and borrow Dai, thus increasing the Dai in circulation and lowering the price when Dai's market price is above the target price of 1 USD. Similarly, higher Stability Fee incentivizes users to close Vaults, thus removing Dai from circulation and increasing the price when Dai's market price is less than the target price. Each type of collateral has a specific stability fee determined by Maker Governance. DSR also helps maintain stability through active governance by MKR holders. When the market price of Dai is above the target price governance can vote to decrease DSR to reduce demand thereby reducing the price of Dai and vice versa.

Liquidation is a significant concern for Dai users and it also encourages them to help maintain stability. If a Vault becomes too risky, it is automatically liquidated and sold in internal market-based auction mechanisms starting with collateral auctions. The aim is to cover the vault obligations plus a liquidation penalty fee pertaining to the collateral type. When all the debt and fees are covered via the auction proceeds, the system returns remaining collateral to the user. If the auction falls short of covering the Vault obligations, the deficit becomes Protocol debt, which the system tries to recover first through a buffer and then a debt auction if there is remaining debt. Additionally, there are other mechanisms and external actors that help maintain stability of Dai such as multiple trusted Oracle Feeds resembling a decentralized oracle infrastructure and keepers that participate in Maker auctions.

Maker Governance Community is responsible to govern the protocol by managing the platform and associated financial risks. Any user on the platform can propose a change or an update to the system while only MKR holders can vote. A user's MKR holdings determine their voting power in proposals. There are various incentives for governance to responsibly govern the protocol including the debt auction where MKR is minted and sold to recapitalize the system. Another highlight of the governance abilities is it can also protect the protocol from a malicious attack or long-lasting market irrationality by initiating an Emergency Shutdown as a last resort.

Dai (SAI) has shown resilience to fluctuations in ETH prior to 2020 [5]. Meanwhile, during the crypto market collapse in March 2020, Dai faced a near death situation where many vaults became under-collateralized resulting in a large number of auctions. Some of these auctions were won by zero-bidders who bid decimal amounts; consequently, there was a shortfall of more than 5.4 M DAI [24]. The system was recapitalized through debt auctions, where MKR was auctioned for Dai (reducing MKR value).

4.2 Synthetic Assets

Synthetic assets enable users to gain exposure to underlying assets without necessarily holding them [25, 26]. The leading example is the Synthetix protocol, which enables the issuance of synthetic assets called Synths on the Ethereum blockchain. The platform's native token SNX is used as collateral to mint Synths. Synthetic commodities that the platform supports range from cryptocurrencies, real-world assets such as gold, indexes and inverses [25]. These synthetic assets such as the synthetic USD (sUSD) or synthetic Ether (sETH) track the price of and hold a stable value with respect to the underlying asset (e.g., sUSD's price is around 1 USD) [5, 26]. Similarly, wrapped coins such as WBTC (which is a collateral type for DAI) can be considered as examples of synthetic assets. Wrapped coins are non-native coins on a blockchain tied to the value of another cryptocurrency that originates from a different blockchain [27]. This functionality of usability on a different blockchain is achieved by putting the backing coin in a type of digital vault called wrapper. Although synthetic assets are not necessarily always stablecoins, it is important to note them in this context as they can be similar to crypto-collateralized or crypto-backed stablecoins.

5 Algorithmically Stabilized Stablecoins

Algorithmic stablecoins do not essentially require the use of backing assets. Such coins typically solely depend on algorithmic stabilization, oracle price feeds and user participation (trading) to maintain their peg. Although a truly stable algorithmic coin remains to be achieved, there are an increasing number of projects in this area. In the following, we describe several types of algorithmic stabilization via examples.

5.1 Purely Algorithmic—Ampleforth

Ampleforth is an example of a purely algorithmic approach to reducing the volatility of cryptocurrencies. It is a synthetic commodity money based on algorithmically enforced elastic supply [28]. The Ampleforth platform has a single ERC20 token called AMPL [5]. It should be noted that Ampleforth does not claim to be a stablecoin but rather a low volatility coin that is designed to diversify risk [28, 29].

High correlation among cryptocurrencies results in a vulnerable ecosystem and introduces systemic risk. Ampleforth's elastic supply tackles this challenge by an algorithmic rebasing mechanism that applies countercyclical pressure against the fluctuation in the market. The rebasing mechanism helps maintain stability by incentivizing users to stabilize the system via arbitrage opportunities. If the market price of AMPL is above the Price Target plus the Price Threshold, then the algorithm expands the token supply, reducing the price. Whereas, if the market price of AMPL falls below the price target minus the price threshold, the algorithm contracts the supply by automatically and directly removing tokens in user accounts to increase the AMPL price. Moreover, the changes to algorithmically determined supply targets are graded over a defined time to distribute uniformly over this period. During the expansion phase, there is a limited sell opportunity for fast actors; while during a contraction phase there is limited to buy opportunity. This buy and sell opportunity incentivizes traders to correct the price and bring the system to equilibrium after expansion or contraction phases. As long as enough traders are willing to benefit from trading opportunities, the platform can be maintained theoretically.

5.2 Algorithmic Seigniorage—Basis

Despite the failed launch of the token, the algorithmic seigniorage design of the Basis token is noteworthy [30]. The protocol was designed to maintain stability by expanding and contracting the supply when the market price of the token deviates from the peg, while price information would be provided by oracles. It featured a three-token system, with Basis as the stablecoin pegged to the USD, bond and share tokens [31]. Basis and bond tokens would be issued, share tokens would have a fixed supply at genesis and return Basis to shareholders when the platform expanded. Basis failed to launch due to regulatory reasons associated with bond and share tokens and the funds were returned to investors [32].

If Basis would trade for less than one USD, then the protocol would issue and auction Bond tokens to users in order to remove coins out of circulation to increase token's price. The auctions would run continuously until enough Basis is destroyed. Bond tokens would be auctioned to contract Basis supply and then buyers would be able to redeem one Basis in the future for a price of less than one Basis at the time of the auction. Conversely, if Basis traded for more than one USD, new Basis would be issued to increase supply and lower the price towards the peg. During the expansion cycle, Bond holders would receive newly minted Basis tokens with oldest bonds first in queue order. However, bonds older than 5 years would be expired to prevent bonds from losing value. After the outstanding bonds are covered, remaining newly minted Basis would be evenly distributed across share tokens. Basis developers expected only small volatility as long as there is enough liquidity and speculators incentivized to participate in auctions to restore the peg.

5.3 Future Value Backed—MetaMUI

MetaMUI is the mainnet coin of MUI MetaBlockchain, a digital currency generation platform developed by Sovereign Wallet Network [33]. The value of MetaMUI coin (digital sovereign currency) is controlled and maintained algorithmically by a special

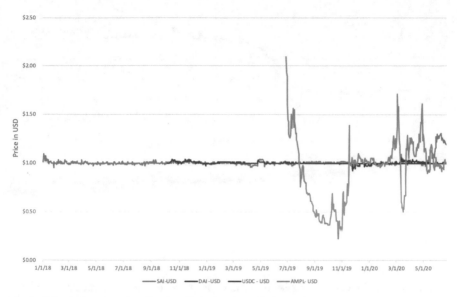

Fig. 2 Price (USD) chart for USDC, DAI, SAI, and AMPL (*source* https://www.coingecko.com/en)

AI-based algorithmic engine called ACB (Algorithmic central bank). In contrast to other stablecoins described in this survey that not only reduce volatility but also typically maintain a constant value with respect to a fiat currency, the MetaMUI coin is designed and developed to maintain low volatility while increasing value over long periods of time.

For publishing a digital currency, the central bank node of the target currency is required to deposit assets to MUI central bank node. These assets, called project funds, are collected and maintained by the MUI treasury. The project funds are used to buy MetaMUI coins from the market and increase its demand. Once the demand is high, the MUI ACB sells the coins at a higher value in the market and increases the circulation up to a capped limit.

The most important goals of the MetaMUI coins are to maintain an increasing value over time with respect to fiat currencies and maintain low volatility within any short window of time. To achieve this, MUI ACB calculates next target price based on two prices—previous market price of MetaMUI and leveraged market price of gold and other assets that are appreciating over time with respect to fiat currencies (Figs. 2 and 3).

6   Central Bank Digital Currencies (CBDCs)

Following Bitcoin, the rising popularity of stablecoins such as Libra intrigued central banks across the world to explore blockchain and another form of stable digital currency, namely, Central Bank Digital Currencies (CBDC). In fact, several central

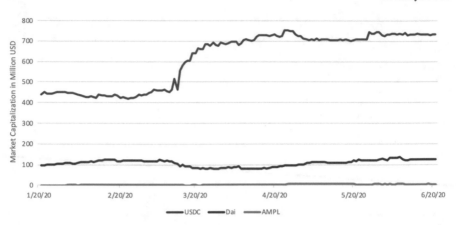

Fig. 3 Market Capitalization in USD for USDC, DAI, and AMPL (*source* https://coinmarketcap. com/)

banks across the world are exploring CBDC projects such as the Sveriges Riksbank (Sweden's e-krona), the Chinese Central Bank (digital yuan), the Eastern Caribbean Central Bank and the Central Bank of Brazil while other central banks such as the Bank of England and the Bank of Canada are considering CBDCs with ongoing research [34, 35]. CBDC can be thought of as a digital money equivalent to physical cash or reserves held at a central bank [36]. Depending on the particular scheme or use case, a CBDC design may or may not include a blockchain DLT. However, due to the benefits of DLT, blockchain-based CBDCs are widely considered. There is ongoing exploration of the role that commercial banks would play with respect to CBDC's [37].

Major benefits of DLT/blockchain-based CBDCs include faster and cheaper domestic or cross-border payments with respect to traditional payment methods, reduced friction associated with traditional banking, resilience against operational failures, physical disruptions, and cyberattacks that traditional systems are vulnerable to [35]. Although there are substantial perceived benefits to implementing CBDC, regulatory aspects, privacy concerns, challenges associated with the scalability of blockchain technology, energy consumption and negative impacts on the financial system (more specifically commercial banks and fractional banking) need further assessment and development [2, 35].

7   Challenges and Risks Regarding Stablecoins

Despite the advantages of stablecoins, there are legal, regulatory, and oversight risks and challenges associated with stablecoins. Firstly, from a legal perspective, the categorization of stablecoins is relatively ambiguous. Depending on the jurisdiction and the characteristics of a stablecoin, it may be considered an equivalent to money, a contractual claim, implicating a right against underlying assets, a security or a financial instrument [2]. It is hard to regulate stablecoins without legal certainty. This uncertainty also complicates consumer and investor protection where adequate information and disclosures including the risks and obligations should be available for customers or investors to make informed decisions. Additionally, due to the lack of proper supervision and effective regulations, stablecoins can be potentially used for illicit financial activities, money laundering, and financing terrorism, thereby compromising financial integrity [38]. To mitigate this problem, entities and issuers in a stablecoin system should comply with the highest international AML/KYC and countering the financing of the proliferation of weapons of mass destruction (CPF) standards [2]. Overall, the challenges and risks associated with each stablecoin depend on its design and structure as well as the jurisdiction that it is in. While some risks such as money laundering might be easier to address for certain types of stablecoins in certain jurisdictions, others might be more complicated. Additionally, the distributed nature of blockchain networks may make it difficult to enforce regulations such as tax compliance.

Depending on the backing of the stablecoin and how they are held, stablecoins might not be able to maintain stability and redeemability/convertibility with respect to the peg [39, 40]. Lack of transparency regarding collateral backing of some well-known stablecoins such as Tether has received scrutiny over the recent years [41]. This accelerated regulatory compliance efforts among the fiat backed stablecoin space. Likewise, large price fluctuations can pose financial or operational failure risks for the stablecoin as a payment system. Moreover, poorly designed or governed systems introduce systemic risk and pose disruptions to financial markets and the economy [2]. This risk is amplified for stablecoins that are adopted at a larger scale. Additionally, lack of interoperability among stablecoins and other payment systems can lead to inefficiencies and isolated financial silos [39]. Wrapped tokens and the use of atomic swaps can potentially improve interoperability between blockchains.

DLT benefits from eliminating risks from a single point of failure. They have resiliency benefits against various cyber and operational risks compared to centralized systems [42]. On the other hand, cyber security is still a significant concern for DLT systems [35, 43]. Operational resilience for a stablecoin is also essential as it can be compromised by black swan events, malicious attacks to the system or severe market downturns. Eventually, stablecoins might be subject to international standards such as ISO or IEC standards and regulations regarding operational and cyber risks [2]. Furthermore, holders may lose confidence in the stablecoin if the issuing organization or governance is not stable and accountable compared to central and commercial banks. As an emerging technology, DLT can also be susceptible to currently unknown risks [35].

Ensuring market integrity (fairness and transparency of the price information) is another challenge pertinent to stablecoins that needs to be addressed by maintaining fair and stable prices in primary and secondary markets [2]. Entities taking on multiple roles such as trading platform, market-maker, and custodial wallet might increase the risk of market misconduct due to conflicts of interest.

Stablecoins implemented on top of open, permissionless DLT platforms inherit some of the fundamental challenges associated with DLT protocols. These include high energy consumption with Proof of Work (which could be mitigated by alternative approaches under development such as Proof of Stake), interoperability as well as the problem of low transaction throughput [43]. On the privacy side, there are concerns about data collection and usage that might discourage users from using stablecoins [38]. Since blockchain serves as an immutable distributed ledger, it conflicts with a legal right in various jurisdictions, the "right to be forgotten" [43]. Thus, organizations need to carefully evaluate user's right to privacy especially with public blockchains in such jurisdictions. It should be mentioned that there are a number of efforts in the cryptocurrency space, e.g., Monero, Zcash, Aztec, and Nightfall focused on privacy using zero-knowledge proofs, which could be applied or extended to provide privacy guarantees for stablecoin transactions as well, though a challenge is that these schemes have to be compliant with AML laws as well [44].

In addition to the general risks and challenges discussed above, there are various distinct challenges that apply to the particular categories of stablecoins. Contrary to the idea of decentralization, fiat backed stablecoins are relatively centralized as they require a trusted institution or a consortium to issue, burn or hold assets. Due to backing requirements associated with handling assets, operations can be more costly compared to other stablecoins. Meanwhile, they may be less complex in design and less volatile than most crypto collateralized or algorithmic stablecoins.

In the case of crypto-collateralized stablecoins, loans may not be fully recovered in the case of default due to high fluctuations in the collateral's value. Additionally, tokens with multiple on chain collaterals incur a correlation risk, which implies the diversification benefit will be less if the collaterals' volatilities have high correlation [45]. Increasing exposure in one type of collateral can impose similar risks. Low-quality price feeds (often delivered through centralized oracles) are a significant source of risk relevant to crypto-collateralized and algorithmic stablecoins that can adversely affect stability and operational resilience. Crypto-collateralized stablecoins also need more careful design due to the possibility of liquidity issues and need to account for human factors with respect to incentives for opening/closing collateralized deposits.

Algorithmic stablecoins are highly complex; issuance and stability factors might not be fully understandable for users. Since they do not feature any collateral, pure-algorithmic stablecoins are most vulnerable to market crashes and "death spirals" [30]. Additionally, bond or share tokens in algorithmic seigniorage may be classified as securities in some jurisdictions as users can make profit through them. Algorithmic stablecoins depend more heavily on buy and sell activity of users with rational economic incentives to maintain stability but if the participants lose interest in buying and selling, the peg cannot be maintained.

The total market capitalization of stablecoins worldwide has recently reached approximately 10 Billion USD, which is still relatively extremely small compared to all the fiat money in the world [46]. Many challenges and risks as aforementioned need to be tackled for stablecoins to be adopted on a global scale. According to the Bank of England, efficiency and functionality benefits over current payment systems are needed for wider adoption of stablecoins [39]. If stablecoins achieve global scale, they could present risks and challenges to monetary policy, international monetary system, financial stability, and fair competition [2]. Future research is needed to determine at what level stablecoin usage could present a risk to implementation of monetary policy [47]. Many countries and well-known financial institutions are researching and/or developing CBDCs to reap the benefits of blockchain-based stable digital currencies while avoiding the potential risks and adverse impacts of stablecoins. CBDCs offer a more scalable, secure, and stable digital currency depending on their design and structure. The concept is still mostly in an experimental stage, and thus further research is required to assess the financial impacts of CBDCs.

8   Conclusions

We have presented a survey classifying and describing the many different kinds of stablecoins that are being researched and developed, including fiat/asset-backed stablecoins, crypto-collateralized stablecoins, and algorithmic stablecoins, as well as the closely related efforts on developing fiat-equivalent digital currencies. While some projects are further along in deployment, all projects are at a relatively early stage with a lot of open questions, particularly related to risk management. We believe that significant new research is needed to address these challenges.

References

1. Torpey, K.: BitPay Has No Current Plans for Bitcoin's Lightning Network, Seeing Growth in Stablecoin Use (2020). Retrieved July 2020 from https://cointelegraph.com/news/bitpay-shuns-lightning-and-liquid-says-actual-bitcoin-payments-still-dominate
2. G7 Working Group on Stablecoins: Investigating impact of global stablecoins (2019). Retrieved April 2020 from https://www.bis.org/cpmi/publ/d187.pdf
3. Antonine Le Calvez and Coin Metrics: The BitMEX Liquidation Spiral—Analyzing How Crypto's Nascent Market Structure Held Up During the Crash (2020). Retrieved May 2020 from https://coinmetrics.substack.com/p/coin-metrics-state-of-the-network-bf8
4. Mengqi Sun: OCC Says Banks Can Use Stablecoins in Payments (2021). Retrieved January 2021 from https://www.wsj.com/articles/occ-says-banks-can-use-stablecoins-in-payments-11610068515
5. CoinMarketCap: CoinMarketCap (2020). Retrieved June 2020 from https://coinmarketcap.com/
6. Coinbase: Introducing USD Coin (USDC)—Stablecoin by Coinbase (2020). Retrieved May 2020 from https://www.coinbase.com/usdc
7. Centre: Introducing USD Coin A stablecoin brought to you by Circle and Coinbase (2020). Retrieved May 2020 From https://www.centre.io/usdc
8. Diem: Libra Whitepaper (2020). Retrieved January 2021 from https://www.diem.com/en-us/white-paper/#the-libra-payment-system
9. Steinbeck, J.: First Look, LIBRA, Facebooks "Blockchain" Cryptocurrency (2019). Retrieved May 2020 from https://medium. com/@jimmiesteinbeck/first-look-libra-facebooks-blockchian-cryptocurrency-941abfc188f5

10. Hecker, R.: How Libra Failed, and How It Could Succeed in 2020 (2020) Retrieved May 2020 from https://www.coindesk.com/how-libra-failed-and-how-it-could-succeed-in-2020
11. Partz, H.: British Payment Firm Checkout.com Joins the Libra Association (2020). Retrieved May 2020 from https://cointelegraph.com/news/british-payment-firm-checkoutcom-joins-the-libra-association
12. Next Nature Network: The Eco Coin Whitepaper V1.0 (2018). Retrieved April 2020 from https://uploads-ssl.webflow.com/5c1b58255c613376879c2558/5c4970105b4d237571564f43_ECOcoin_white_paper_v1.0.pdf
13. Grijns, H.: Week of the Circular Economy #9: ECO coin (2020). Retrieved May 2020 from https://amsterdamsmartcity.com/posts/week-of-the-circular-economy-9-eco-coin
14. Next Nature: Eco Coin First Trial at DGTL 2017 (2017). Retrieved May 2020 from https://nextnature.net/2017/04/eco-coin-dgtl-report
15. Lipton, A.: Thomas Hardjono and Alex Pentland. 2018. Breaking the Bank. Scientific American (January 2018). Retrieved May 2020 from https://tradecoin.mit.edu/sites/default/files/documents/Tradecoin-Lipton-Pentland-ScientificAmerican-Jan-2018.pdf
16. Lipton, A.: Thomas Hardjono and Alex Pentland. 2018. Digital Trade Coin: Towards a More Stable Digital Currency. R. Soc. open sci. (2018). https://doi.org/10.1098/rsos.180155
17. Fnality International: Fnality (2020). Retrieved June 2020 from https://www.fnality.org/about-fnality
18. Fnality International: Fnality–The Catalyst for True Peer-to-Peer Financial Markets. Whitepaper V4.0 (2019). Retrieved July 2020 from https://www.fnality.org/the-catalyst-for-true-peer-to-peer-financial-markets?hsCtaTracking=744e4c75-c03f-4129-b2af-8cf17888c7c1%7C27fcc1d4-ab18-4d23-a119-6b8c3ae6be96
19. Fnality International: USC Continues to Evolve (2020). Retrieved June 2020 from https://www.fnality.org/news-views/usc-continues-to-evolve
20. Althouser, J.: Major Banks Join USC Project for Blockchain-based Cryptocurrency Banking (2017). Retrieved May 2020 from https://cointelegraph.com/news/major-banks-join-usc-project-for-blockchain-based-cryptocurrency-banking
21. MakerDAO: The Maker Protocol: MakerDAO's Multi-Collateral Dai(MCD) System (2020). Retrieved May 2020 from https://makerdao.com/en/whitepaper
22. MakerDAO: A Guide to Single-Collateral Dai (Sai) Shutdown (2020). Retrieved June 2020 from https://blog.makerdao.com/a-guide-to-single-collateral-dai-sai-shutdown/
23. Orcutt, M.: MakerDAO community greenlights first 'real-world' assets for use as collateral (2020). Retrieved June 2020 from https://sports.yahoo.com/makerdao-community-greenlights-first-real-182126086.html
24. MakerDAO: The Market Collapse of March 12–13, 2020: How It Impacted MakerDAO (2020). Retrieved June 2020 from https://blog.makerdao.com/the-market-collapse-of-march-12-2020-how-it-impacted-makerdao/
25. Warwick, K.: What Is Synthetix and How Does It Work?. (2020). Retrieved January 2021 from https://www.gemini.com/cryptopedia/synthetix#section-how-does-synthetix-work
26. Kuznetsov, N.: Synthetic dreams: Wrapped Crypto Assets Gain Traction Amid Surging Market (2021). Retrieved January 2021 from https://cointelegraph.com/news/synthetic-dreams-wrapped-crypto-assets-gain-traction-amid-surging-market
27. Binance Academy: What Are Wrapped Tokens?. (2021). Retrieved January 2021 from https://academy.binance.com/en/articles/what-are-wrapped-tokens
28. Kuo, E., Iles, B., Manny Rincon Cruz: Ampleforth: A New Synthetic Commodity (2019) Retrieved April 2020 from https://www.ampleforth.org/paper/
29. Hulliet, M.: Ampleforth Publishes Updated White Paper for Non-Correlated, Price-Stable Digital Asset (2019). Retrieved May 2020 from https://cointelegraph.com/news/ampleworth-publishes-updated-white-paper-for-non-correlated-price-stable-digital-asset
30. George Samman: The State of Stablecoins 2019 Hype vs. Reality in the Race for Stable, Global, Digital Money (2019). Retrieved April 2020 from https://reserve.org/stablecoin-report
31. Al-Naji, N., Chen, J., Diao, L.: Basis: A Price-Stable Cryptocurrency with an Algorithmic Central BankFormerly known as: Basecoin (2018). Retrieved May 2020 from http://www.basis.io/basis_whitepaper_en.pdf

32. del Castillo, M.: Crypto's Top Funded Startup Shutters Operations Following SEC Concerns (2018). Retrieved June 2020 from https://www.forbes.com/sites/michaeldelcastillo/2018/12/13/sec-rules-kill-cryptos-top-funded-startup/#2b9804542918
33. Yun, S., Kim, F., Jeong, J.: MUI-Metablockchain Whitepaper (2020). Retrieved January 2021 from https://sovereignwallet-network.github.io/whitepaper/MUI-MetaBlockchain-White-Paper.pdf
34. Bourgi, S.: China's Central Bank Plans Digital Yuan Pilot for Payments to Hong Kong (2020). Retrieved January 2021 from https://cointelegraph.com/news/china-s-central-bank-plans-digital-yuan-pilot-for-payments-to-hong-kong
35. World Economic Forum: Central Banks and Distributed Ledger Technology: How are Central Banks Exploring Blockchain Today? (2019). Retrieved April 2020 from http://www3.weforum.org/docs/WEF_Central_Bank_Activity_in_Blockchain_DLT.pdf
36. World Economic Forum: Central Bank Digital Currency Policy-Maker Toolkit (2020). Retrieved April 2020 from https://www.weforum.org/whitepapers/central-bank-digital-currency-policy-maker-toolkit
37. Francesca Carapella and Jean Flemming: Central Bank Digital Currency: A Literature Review (2020). (November 2020). Retrieved January 2021 from https://www.federalreserve.gov/econres/notes/feds-notes/central-bank-digital-currency-a-literature-review-20201109.htm
38. Adrian, T., Mancini-Griffoli, T.: Digital Currencies: The Rise of Stablecoins (2019). Retrieved May 2020 from https://blogs.imf.org/2019/09/19/digital-currencies-the-rise-of-stablecoins/
39. Bank of England: Central Bank Digital Currency Opportunities, challenges and design. (2020). Retrieved May 2020 from https://www.bankofengland.co.uk/-/media/boe/files/paper/2020/central-bank-digital-currency-opportunities-challenges-and-design.pdf
40. Bullmann, D., Klemm, J., Pinna, A.: In Search For Stability in Crypto-Assets: Are Stablecoins The Solution?. Occasional Paper Series No 230. (2019). Retrieved May 2020 from https://www.ecb.europa.eu/pub/pdf/scpops/ecb.op230~d57946be3b.en.pdf
41. Khatri, Y.: Tether Says Its USDT Stablecoin May Not Be Backed By Fiat Alone (2019). Retrieved January 2021 from https://www.coindesk.com/tether-says-its-usdt-stablecoin-may-not-be-backed-by-fiat-alone
42. Piscini, E., Dalton, D., Kehoe, L.: Blockchain & Cyber Security. Let's Discuss. Retrieved June 2020 from https://www2.deloitte.com/content/dam/Deloitte/ie/Documents/Technology/IE_C_BlockchainandCyberPOV_0417.pdf
43. World Economic Forum: Building Block(chain)s for a Better Planet (2018). Retrieved April 2020 from http://www3.weforum.org/docs/WEF_Building-Blockchains.pdf
44. Foxley, W.: Developers of Ethereum Privacy Tool Tornado Cash Smash Their Keys (2018). Retrieved June 2020 from https://www.coindesk.com/developers-of-ethereum-privacy-tool-tornado-cash-smash-their-keys
45. MakerDAO: MakerDAO Governance Risk Framework (Part 2) (2018). Retrieved May 2020 from https://blog.makerdao.com/makerdao-governance-risk-framework-part-2/
46. Voell, Z.: Stablecoin Supply Breaks $10B as Traders Demand Dollars Over Bitcoin (2020). Retrieved June 2020 from https://www.coindesk.com/stablecoin-supply-breaks-10b-as-traders-demand-dollars-over-bitcoin
47. Blockchain: The State of Stablecoins (2019). Retrieved April 2020 from https://www.blockchain.com/ru/static/pdf/StablecoinsReportFinal.pdf

Central Bank Digital Currencies

Nadia Pocher and Andreas Veneris

Abstract Today's societal digitization continues to advance at exponential speeds driven by technology trends. Billions of Internet of Things devices have made their way into our daily lives but also into healthcare, manufacturing, and supply chains. In contrast, the financial sector still largely operates on legacy infrastructures, where merchants receive their payments long after they released the digital/physical good to the consumer. In addition, the emergence of Decentralized Finance through blockchain technology, and the accumulation of data in private silos, has demonstrated a capacity to impact national sovereignty and monetary transmission channels. Against this backdrop, many central banks have recently started to research and test the issuance of digitally native fiat money—or Central Bank Digital Currencies (CBDCs)—in an effort to redesign the essence and use of physical cash. CBDCs present a broad variety of designs, which translate into manifold techno-legal and standardization policy questions. In this context, this chapter surveys the state-of-the art with specific focus on "retail" CBDCs. In doing so, it provides an overview of candidate architectures, heeds legal impacts and regulatory compliance issues, presents a set of case studies and touches upon cross-border CBDC challenges.

1 Introduction

The promise of an electronic version of cash, possibly grounded on blockchain and Distributed Ledger Technologies (DLTs), has electrified the world over the

N. Pocher (✉)
Faculty of Law, Universitat Autónoma de Barcelona, Bellaterra, Spain
e-mail: nadia.pocher@uab.cat

KU Leuven - CiTiP, Leuven, Belgium

Universitá di Bologna, Bologna, Italy

A. Veneris
Department of Electrical and Computer Engineering and Department of Computer Science, University of Toronto, Toronto, Canada
e-mail: veneris@eecg.toronto.edu

past decade. This prospect has created an excitement for technological disruption that reminds of the 1990s, when the Internet entered the mainstream. Indeed, cryptocurrency-related developments have been labeled to form an "Internet of Value(s)" [1] or an "Internet of Money" [2]. Their core premise lies in the basic functioning of blockchain systems: as they are not only secured by cryptography and economic incentives but also governed by decentralized consensus mechanisms, they enable value transfers that transcend the need to rely on a "central" authority. Accordingly, these setups have the potential to replace the legacy financial infrastructure, by eliminating multiple layers of intermediation and informing a new "hype" of direct participation of citizens and businesses to a new global economy [3–5].

Meanwhile, the prospect of a widespread adoption of decentralized "smart" (or "programmable") money has fascinated and unsettled both governments and the private sector. Not surprisingly, this exogenous and mainly privately-driven innovation has motivated monetary institutions to start rethinking payments, transmission channels, and even the very essence of "physical cash" [6–10], in a worldwide quest to adapt to a new reality. If the full potential of this value interconnection is fulfilled, the impact will not be limited to payments. They will have ripple effects on the most diverse fields such as privacy, national security, law and regulation, property rights. Besides cryptocurrencies and cryptoassets, in fact, in recent years billions of Internet of Things (IoT) devices have been deployed in our daily lives. These tools continuously collect valuable data related to large economic sectors, such as healthcare, manufacturing, supply-chains, infrastructures [11–14].

While this data is largely retained in privately-held and tightly-closed silos, often out of the reach of governments and local entities, their rightful owners are not in a position to profit from them [15]. Parallelly, domestic and international commercial micro-payment systems currently lack platforms and economic incentives that could underpin efficient public IoT/AI data marketplaces. Against this backdrop, it does not come as a surprise that also central banks have been investigating the deployment of innovative technologies to their own currencies. Their motivation partly lied in the possible disappearance of cash, which could deprive citizens and businesses of risk-free government-issued money. Further, as noted by an extensive literature [6–10, 16, 17], digital currencies can create novel payment channels, transactional communities, and novel safe networks-of-relations. Hence, they may potentially secure sovereign monetary identities, nourish past social investments but also safeguard geopolitical digital boundaries within the global economy [18].

For the sake of convenience, Tables 1 and 2 list the acronyms used in this chapter.

1.1   Central Bank Money

Following the footsteps of the rapid globalization and digitization of the economy, in the past decades, payment transmission systems have evolved significantly. This is related to infrastructural advancements in the institutional domain (e.g., real-time gross settlement/RTGS, fast retail payment systems, instant payments), but also to

Table 1 Technical terms

AI	Artificial Intelligence
CBDC	Central Bank Digital Currency
DeFi	Decentralized Finance
DCRI	Digital Currency Research Institute
DLT	Distributed Ledger Technology
IoT	Internet of Things
M2M	Machine-to-Machine
ML	Machine Learning
mCBDC	Multiple CBDC
NFC	Near Field Communication
P2P	Peer-to-Peer
PET	Privacy Enhancing Technology
PoC	Proof-of-Concept
RF	Radio Frequency
RCC	Range Controlled Communication
RTGS	Real-Time Gross Settlement System
TEE	Trusted Execution Environment

the activity of an emerging private sector (e.g., Big Techs, FinTech startups) [19]. As of today, the vast majority of efforts are pursued jointly, through mechanisms of public-private partnership (PPP). While those innovations have indeed improved the existing system, the advent of decentralized finance (DeFi) and IoT/5G/AI has brought along even more rapid developments. It is within this context that, in the wake of the release of the whitepapers of Bitcoin in 2008 [20], Ethereum in 2013 [21] and Libra (now Diem) in 2019 [22], legacy monetary institutions and central banks have started entertaining the idea of digitizing—more specifically, tokenizing (i.e., creating a digital representation of)—M0 sovereign money [6, 23, 24].

The literature offers various definitions of "sovereign currency". Namely, [25] assumes that it is one that is "*set as such by a sovereign law, issued by an authorized issuer, and whose value results from a statutory rule*". Traditionally, central banks and monetary authorities issue two types of "central bank money":

- "General purpose money" or "fiat money"—the official and sovereign currency, also known as physical money or cash, consisting of physical coins and banknotes. It is legal tender—i.e., it is legally recognized as a means to satisfactorily meet financial obligations –, which also means it must be accepted as such to extinguish a public or private debt, and it is available to the general public; and
- "Bank reserves" or "settlements accounts"—provided by central banks to authorized institutions that are participants in their RTGS systems—e.g., commercial banks and non-bank payment service providers (PSPs)—, through the opening of

Table 2 Monetary and regulatory terms

AML	Anti-Money Laundering
BIS	Bank for International Settlements
BoC	Bank of Canada
CBDL	Central Bank Digital Loonie
CBUAE	Central Bank of the United Arab Emirates
CDD	Customer Due Diligence
CPF	Counter-Proliferation Financing
CFT	Counter-Terrorist Financing
DCEP	Digital Currency Electronic Payment
ECB	European Central Bank
FATF	Financial Action Task Force
FI	Financial Institution
FINMA	Swiss Financial Market Supervisory Authority
HKMA	Hong Kong Monetary Authority
IMF	International Monetary Fund
KYC	Know-Your-Customer
MAS	Monetary Authority of Singapore
NB	Narrow Bank
PBoC	People's Bank of China
PoC	Proof-of-Concept
PPP	Public Private Partnership
PSP	Payment Service Provider
SDR	Special Drawing Right
STR	Suspicious Transaction Reporting

ad hoc reserves accounts. In practice, they are scriptural deposits recorded on a centralized ledger (i.e., database) held, settled and managed by the central bank.

Central bank money is a liability of the central bank. By extension, it can be considered a liability of the relevant sovereign government. By contrast, the majority of money that is in circulation belongs to the categories of "commercial bank money" or "electronic money (e-money)". Because it is issued by private stakeholders such as commercial banks, non-bank PSPs and e-money institutions (collectively, Financial Institutions or FIs), it essentially becomes a liability of those private entities to the public. When using commercial bank money, the end-user has a claim against an FI to receive central bank money (i.e., cash) upon request (i.e., the relevant monetary value can be redeemed at par). Since it is redeemable on demand, it extends central bank money. For articulate definitions and conceptual disambiguation, we refer the interested reader to [6, 25–29].

1.2 Typology of CBDCs

The idea of digitizing central bank money was originally focused on the mentioned category of "bank reserves" or "settlement accounts", thus limited to interbanking activities. Hence, ordinary public and private financial transactions were not the target of the first explorations. Only later, following the introduction of blockchain-based cryptocurrencies, institutions started to entertain the idea of issuing digital fiat money. Accordingly, as of today, there are two subsets of CBDCs, and they are developed in a parallel fashion because they respond to different payment needs.

On the one hand, a *wholesale-CBDC* is a RTGS-like settlement scheme between financial institutions. It is detached conceptually, but also practically, from the daily flows of physical cash. Although manifold designs have emerged over time, and different technologies have been deployed by both the public and the private sector, the goal behind this type of CBDC is to update or complement solutions in the area of central bank deposits [25]. In contrast, a *retail-CBDC* is offered to the public at large, and it is the most transformative subset of CBDCs. It embodies an evolution towards a more "democratic" public transmission channel to central bank monetary holdings/policies. In this case, a digital form of fiat money is offered in a legal tender fashion, to be used for everyday transactions. From this perspective, retail CBDCs seemingly draw from the features of cryptocurrencies, albeit minimizing related risks such as price volatility, the absence of regulatory compliance, and the limited/complex exchange mechanisms [30]. In other words, retail CBDCs not only expand the concept of central bank money as we have known it for the past centuries but also require central banks to safeguard monetary stability, efficiency and security when devising the issuance, use-case(s) and distribution of these instruments.

As the new concept of CBDCs lies at the crossroads between different disciplines—more notably economics, policy, technology, law, finance, and sociology—new definitions are necessary but also difficult. Illustratively, [31] provides a tech-oriented definition of a retail-CBDC as: "*A credit-based currency in terms of value, a crypto-currency from a technical perspective, an algorithm-based currency in terms of implementation, and a smart currency in application scenarios*". More broadly, [32] highlights that "*CBDC is not a well-defined term. It is used to refer to a number of concepts. However, it is envisioned by most to be a new form of central bank money. That is, a central bank liability, denominated in an existing unit of account, which serves both as a medium of exchange and a store of value*". Accordingly, [26] suggests that "*A CBDC is a digital form of central bank money that is different from balances in traditional reserve or settlement accounts*".

1.3 The Growing Interest in Issuing a CBDC

The discussion above illuminates the complex nature of CBDCs, in all terms of their definition, architecture, regulation, privacy and use-case. Likewise, over the past decade central banks, governments and monetary authorities have motivated a possible issuance in various ways. Indeed, the growing interest of central banks in CBDCs has had many drivers and opinions on their origin vary [7, 8]. However, three core factors seem to have sparked this interest.

First, the use of traditional cash by the general public has been decreasing, in favor of digital alternatives such as debit and credit card transactions and wire/electronic fund transfers. In some jurisdictions, like Sweden or Canada, the decline in the use of cash has arguably been particularly stark. The second factor relates to private altcoins and other tokenization initiatives that followed the advent of Bitcoin and later Ethereum. The latter also provides a Turing-complete smart contract language to build decentralized applications, as well as complex automated cost-effective and globally-reaching financial instruments coined as DeFi [33]. As of today, there are more than 5,000 blockchain-based cryptocurrencies in circulation. Cryptocurrencies trade at free-floating prices relative to fiat currencies and the majority of them feature volatile price histories, which in-effect limits their usability as "money". Attempts to limit their price volatility led to the development of stablecoins and, more recently, "mega-stablecoins" such as Facebook's Libra/Diem [22].

The development of digitally native finance applications outside of the legacy networks challenges the traditional bank-based payment and monetary policy transmission mechanisms [23]. This is because it poses the so-called risk of "currency substitution" [17, 34]. This fact prompted central banks to protect their raison d'être and financial stability by investigating their own tokenization of fiat currencies. Further, the growing interest in CBDCs mirrors an effort to leverage the programmability of "digital cash" technologies into a new functional form of M0 money. Evidently, this new form of money needs to have the proper technology characteristics to serve an ever-growing digital global economy that shapes a new perception, and relation, between the public and the central bank's monetary instruments [35, 36]. Finally, central banks are reportedly attracted to CBDCs to foster payment efficiency, create new monetary policy transmission channels, advance financial inclusion, safeguard safety/privacy and regulatory compliance [6, 23, 24].

2 Characteristics and Design Choices for CBDCs

General purpose *retail* CBDCs are system-critical technologies that millions of people will be using. Accordingly, far from being a small task, their issuance needs safeguard the local economies but also elicit in geopolitical trends. Reportedly, CBDC systems should namely demonstrate the following core characteristics:

- **Privacy**: maximized but complying with regulations such as Anti-Money Laundering and Counter-Terrorist Financing (AML/CFT);
- **Universal Access**: regardless of user's means, ability or geographical location;
- **Security**: resistant to the most sophisticated cyber-attacks;
- **Resilience**: operating continuously both online and offline; and,
- **Performance**: scaling for daily use within the jurisdiction but also cross-border.

By formulating the above objectives, CBDC systems should be layered so that third parties can build on top of the core platform. As such, they should rely on flexible, long-run sustainable architectures that separate the core system from the front-end user experience, but also one that is adaptable to new consumer trends, thus accommodating the ever-changing commercial use cases. In contrast to commercial systems that focus on a specific market(s), central bank digital money should guarantee universal access to all citizens irrespective of financial means or sight, dexterity or cognitive impairments, so as to ensure accessibility and financial inclusion. Further, this e-cash should also be usable in remote communities or places, even those without Internet access, and should also serve cross-border travelers.

Although user and transaction privacy should be protected, CBDCs must adhere to strict regulatory standards, in particular with regards to AML regulation, both domestically and internationally [37]. The underlying CBDC systems must also be resilient and robust without compromise to their fault-tolerance. They must be able to operate continuously and have low-latency while they remain scalable to serve large populations within their jurisdiction but also cross-border. Further, they should be able to communicate with existing retail payment systems and banking ecosystems, so to leverage past technology investments and established payment channels. This compatibility is also necessary to allow users to access their funds from accounts at commercial banks and merchants to accept CBDCs as a means of payment. Additionally, they need to employ architectural designs with service-quality metrics of the highest operational standards and exhibit low-cost efficiency. Finally, those designs should provide traditional seigniorage income to the underwriting central bank but also foster healthy competition in the payments market(s).

2.1   Core-Architecture Considerations

Traditionally, payment systems are classified as either *token-* or *account-based*. This taxonomy also applies to CBDCs, and it translates into how access is granted to the end-user and into the authentication/identification method used to conduct a transaction [29, 38]. On the one hand, access to a token-based means of CBDC-payment relies on the validity of the traded object (i.e., the validity of a token)—hence, in principle, it is an anonymous and a bearer-type instrument grounded solely on cryptographic principles. On the other hand, in an account-based CBDC, access depends on the identification and identity verification of the account holder. This reminds of traditional commercial bank or e-money accounts that require the public to undergo a Know-Your-Customer (KYC) process to use their payment systems [6, 19, 27, 39]. As argued by [19], "*in an account-based CBDC, ownership is tied to an identity, and transactions are authorized via identification. In a CBDC based on digital tokens, claims are honored based solely on demonstrated knowledge, such as a digital signature*". Hence, in account-based CBDCs the system comprises a bookkeeping ledger and a payment service, where the latter refers to how payments are initiated, verified, cleared, and settled [26, 40, 41].[1]

There are three different ways CBDC systems are currently envisioned in terms of their core layer architecture and method of distribution to the public. Traditionally, a "payment" refers to the transfer of the liability of the central bank as this is recorded on the ledger. From an architectural perspective, CBDCs have been classified according to their design choices as follows [7, 38, 40]:

1. *Direct*: the central bank holds the CBDC ledger and also handles the transactions. In case of account-based CBDCs this scheme requires the public to somehow hold reserve accounts with the central bank;
2. *Hybrid*: the central bank holds the CBDC ledger, but the payment service is provided by private actors such as FIs or Telcos. Some authors label these systems as *platform* CBDCs [36]; and,
3. *Synthetic*: the private sector updates the CBDC ledger—i.e., the ledger is held indirectly by the central bank by settling the reserve accounts through PPP schemes—, and also handles the transactions [7]. In these cases, FIs hold periodically-settled reserve accounts with the central bank, as it happens with electronic payments today. The three structures are depicted in Fig. 1.

The *direct* structure is usually described as "one-tier", as only the central bank is involved and the CBDC is a direct claim of the public. Evidently, this entails the central bank to initiate and continuously serve a relationship with all CBDC users, a move outside of most central banks' traditional and historic core competencies. On the contrary, *hybrid* and *synthetic* CBDC models are usually labelled as "two-tier" architectures, and their structures are less invasive than their "one-tier" counterpart. Similarly to traditional mechanisms, "two-tier" schemes require a cooperation between the government and private FIs [19, 42]. Notably, in *hybrid* structures, the CBDC remains a direct claim on the central bank, even if transactions are managed by private actors. By contrast, in *synthetic* CBDC schemes end-users interact with intermediaries, as with commercial bank money and e-money. In these cases, one can argue, the CBDC "emulates" a stablecoin offered by a private actor, and the stablecoin is essentially backed by its reserve account with the central bank. Hence, private intermediaries bear a responsibility to cover fully or in part—as provided by the respective jurisdiction—the liability of their stablecoins [29, 41, 43]. Reportedly, such a CBDC scheme resembles special-purpose licenses granted to non-bank FinTech firms in jurisdictions such as India, Hong Kong, China, and Switzerland [7].

In the world of CBDCs, the circumstance where end-users do not possess a direct claim on the central bank is seemingly relevant to the definition of the instrument as a CBDC. In more detail, the intricate nature of synthetic CBDCs can be leveraged to argue against their qualification as an actual "grass-roots" CBDC. This is because by definition it is assumed that a CBDC is a direct liability of the central bank [29].

---

[1] In this respect, [14] analyzes the repercussions of the distinction between account-based and token-based systems on integration scenarios between CBDC architectures and IoT developments in the context of Machine-to-Machine (M2M) transactions.

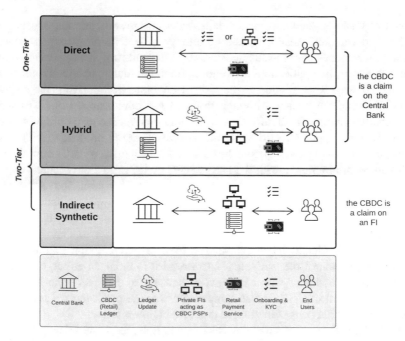

Fig. 1 *Source* Elaboration of the authors inspired by various publications by the Bank for International Settlements. Most recently, [17, 38]

Nonetheless, experts have also commented that if the stablecoin is pegged 1:1 to the sovereign currency by means of regulation, it is ostensibly as if users are holding central bank money—and this after all is the core essence of a CBDC [43].

2.2 The Offline-Usability Conundrum

A necessary requirement for CBDCs is to be usable even when users have (temporarily) no access to the Internet. Facilitating such offline transactions results in a *trade-off* between hardware/software security, costs, and convenience. Intuitively, this trade-off is balanced with the introduction of low-cost cards that can store only a small amount of money. The main security challenge is lost (or stolen) funds. Another equally important concern is an adversary that may attempt to double-spend offline, as they may have not yet been settled through the online system. Finally, offline transactions introduce new challenges when it comes to AML compliance.

One way to implement offline transactions is via tamper-proof hardware [6, 44, 45]. Many processor chips, including those in smartphones, have Trusted Execution Environment (TEE) enclaves/capability (e.g., SGX in Intel, TrustZone in ARM, KNOX in Samsung). With the use of TEE hardware capabilities, one can create appropriate hardware/software cryptographically-secured enclaves that store a small amount of CBDCs good enough for daily transactions and common expenditures (such as supermarket, restaurant, gasoline, and typical entertainment expenses) when access to a network is not available. Further, TEEs allow a smartphone to ensure third-party software applications are running on the hardware in an unmodified and untampered way. This eliminates the risk of adversaries modifying the software to double-spend the money. Although research has demonstrated that TEEs may occasionally exhibit vulnerability, they are widely used for secure transactions today.

An additional approach is to issue debit-like CBDC-cards, pre-loaded with a small number of CBDCs (e.g., $ 200) from the user's wallet when the wallet is online. These cards can be programmed, with the use of NFC or RCC, to store securely in their ROM chips items like a PIN number, or even biometric information. Afterwards, users can store CBDCs from their own smart device (smartphone, tablet or computer) when that device is online, thus crediting their online accounts. When the hardware of these CBDC-cards is activated by a nearby RF signal, they can perform sufficient power-efficient operations such as two-way cryptographic authentication and/or transmission of the encrypted data stored into them. In effect, external RF signals (like a merchant's terminal) powers them up so they can securely transmit offline the amount of CBDCs that compensates for the particular transaction—no different to what happens with modern credit/debit cards today.

Evidently, in the case of smart devices like smartphones, tablets, laptops, the process is even simpler—they already have their own power source and secured hardware to emulate the behavior of those RF-activated CBDC-cards. Moreover, these devices can act as terminals that can "activate" through RF other CBDC-cash-cards, provided their battery is not emptied. All these novel hardware designs and protocols call for new Design-for-Security embedded chip architectures—a semiconductor research area that demands a more holistic hardware design approach than just a traditional cryptographic implementation(s) [46]—but also global CBDC hardware/software co-design interoperability standards.

If a CBDC card is lost or stolen, the user will lose the funds stored in this card, just like with physical cash when a wallet is lost or stolen. As these cards require syncing with an online wallet to deposit/withdraw funds [44], and because the amount of e-fiat they can store is rather limited, this aids the AML process as well. In closing, these pre-loaded CBDC-cards act as "cold static storage" for small amounts of quasi-token CBDCs. Further, they can be used by international visitors and tourists, but also by those who don't have access to commercial bank accounts or smart devices, thus contributing to the promotion of financial inclusion.

2.3 The Public-Private Interplay Design Factor

In light of the foregoing, it is clear that different proposed CBDC architectures lead to diverging public-private dynamics from a monetary policy perspective. The topic is increasingly explored, as it relates to a broader discussion on the preferable degree of competition between public (e.g., central banks, government) and private actors (e.g., commercial banks and FIs, commercial corporations) in the deployment of digital currencies. With regard to CBDCs, the main controversy is whether society can best reap the opportunities of digital payments by central banks replacing private FIs/Fintech or by simply joining forces with them [41, 43, 47, 48].

The first policy option is mirrored by *direct* one-layered CBDCs, while the situation is more complex with regard to two-layered design approaches. Intuitively, the deployment of *hybrid* and *synthetic* schemes assumes that the relevant central bank is willing to waive a portion of its power [49]. Nonetheless, two-layered CBDCs enshrine a significant distinction with regard to the boundaries of involvement of private actors in the relevant value chain [50]. Most importantly, in *hybrid* structures central banks still hold the CBDC ledger and manage end-users accounts, while in both cases—*hybrid* and *synthetic*—payment services and relationships (along with the accompanied KYC/AML processes) with end-users are managed by the private sector no different to what broadly happens today.

The idea of outsourcing CBDC activities to private actors through PPP mechanisms has generated a lively academic and political debate. The pivotal aspects of the controversy revolve around how to guarantee payment innovation, efficiency, "fair" competition and financial inclusion against the risk this practice may entail to national monetary choices and financial stability—both traditional goals guaranteed by the central banks themselves [41]. Further, as outlined throughout this chapter, there are issues raised by the collection, use and dissemination of the associated user payments metadata. Clearly, the wobbling consumer confidence in the banking sector exert a significant influence on the debate [49]. More specifically, it was argued that public-private scenarios stimulate competition and disincentivize monopolies thanks to the participation of FIs. Likewise, experts maintain these mechanisms foster innovation, inclusion and credibility, while they ostensibly reduce risks and costs for central banks. By contrast, they may pose financial stability and liquidity risks in case of *synthetic* CBDCs, notably if the responsibility to maintain an adequate asset backing rests on private actors and associated regulation [17, 19, 35, 38, 41, 43, 47, 49, 51].

2.4   Cross-Border Perspectives (mCBDCs)

CBDCs are often examined as stand-alone projects, pursued by one central bank or another. This is especially true with regard to the *retail* subset, with the analysis often focusing on specific domestic projects, perhaps in comparison with similar ideas. Nevertheless, the cross-border feature of tokenized money is most relevant, and generates questions that are, for the most part, still to be answered. In the past months, the Bank for International Settlements (BIS) has addressed the interactions between CBDC systems, both *retail* and *wholesale*, by exploring these arrangements [52] and surveying current trends [53]. This sparked interest in academia as well [39, 54]. Two concepts emerge as crucial: "interoperability" and "standardization".

From the first perspective, the world of DLTs/blockchain is increasingly permeated by debates on interoperability—i.e., broadly speaking, the compound of *"any characteristics of systems that could help them exchange information"* [52]. In the CBDC realm, the notion is at least twofold. On the one hand, the systems devised by different jurisdictions ought to be able to communicate, also in terms of offering cross-currency capabilities. On the other hand, when CBDCs are developed through PPPs, it is crucial the various providers guarantee interoperability in the way they design the payment architecture, so not to generate closed payment silos and ensure users of different providers may transact with each other.

Secondly, interoperability relies on "standardization"—i.e., the development of industry-wide technical standards within the framework of international cooperation. In the words of [52], *"common technical standards, such as message formats, cryptographic techniques, data requirements and user interfaces can reduce the operational burden of participating in multiple systems. Aligned legal, regulatory and supervisory standards can simplify know-your-customer and transaction monitoring processes"*. Nonetheless, there are three different options to set up a cross-border and cross-currency CBDC mechanism: (i) developing compatible standards, (ii) interlinking different systems, (iii) creating a single multi-currency system. Only in the latter case the outcome is an integrated CBDC "payment system"—i.e., as outlined in [52, 53], a single set of participants, a single infrastructure, ledger, rulebook and governance. In the other cases, CBDC "payment arrangements" allow interoperability. For details on the pros and cons of these strategies, we refer to [52, 53].

In this context, the BIS argues through its CPMI working group for central banks to include cross-border and internationally-oriented considerations in their CBDC projects early on [52, 55]. Along these lines, the setup of "multi-CBDC" (or *mCBDC*) arrangements would deliver on the promise of improving cross-border payments efficiency against the backdrop of the increasing globalization. Arguably, the choice is between fostering communication between sovereign currencies (e.g., by handling settlement in different currencies) and witnessing the creation of a global private sector stablecoin, where the first option seems preferable [52]. It is against this backdrop that important joint CBDC sandbox initiatives have been put forward by major monetary institutions all over the world [52, 53].

3   History of CBDC Projects

Central bank interest in "digital money" started emerging in 2014. However, only the People's Bank of China (PBoC) initiated work for its e-CNY platform at the time—most other R&D pilots/reports on *retail* CBDCs gained notoriety over the last 2–3 years. As of today, central banks and governments continue to scrutinize both reasons and plans to issue a digital sovereign currency. Accordingly, extensive commentaries are published by a broad range of stakeholders on a regular basis, touching upon different aspects such as security, privacy, technology infrastructure, public opinion polls, regulation and cross-border challenges [7–10, 16, 26, 56, 57].

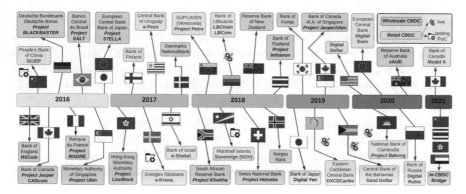

Fig. 2 Global roadmap on major *wholesale* and *retail* CBDC projects (figure taken from [58])

Indeed, central banks are no novices at the e-fiat expedition. The first pilots in wholesale interbanking CBDCs, DLT-based stock trading settlement and cross-border transfers started to emerge in 2015–16. The vast majority of those pioneers experimented with some form of blockchain technology. The work of [24] classifies CBDC projects as *early adopters*, *followers* and *new entrants*. Similarly, below we provide a historical summary, starting with blockchain-based settlement systems, and moving to CBDC products and other sandboxes today, as depicted in Fig. 2.

3.1 The Research Pioneers: 2015–16

In 2015–16, research pioneers started exploring CBDCs by addressing wholesale interbanking use-cases. Notable references are led by the PBoC as early as in 2014—*e-CNY*, also coined as the *Digital Yuan* or *Digital Currency Electronic Payment (DCEP)* system—and by the Bank of England (*RSCoin* [59]). Around the same time, the Bank of Canada (BoC) piloted the four-phased *Project Jasper*, one of the most comprehensive efforts up to date. As the Jasper series remains representative of sandboxing initiatives by other central banks, we provide reference to each phase:

- *Jasper I (2016)*: In this phase, the BoC experimented with DLT-based RTGS systems using the newly released permissionless platform of Ethereum.
- *Jasper II (2017)*: The BoC repeated the sandboxing from Phase 1 introducing additional liquidity requirements to the commercial banks for settlement. However, a main characteristic of that project was that the underlying network moved to the permissioned Corda one.
- *Jasper III (2018)*: In the third cycle, the Bank partnered with a set of commercial Canadian banks to extend the complexity/functionality of the Corda system from Phase 2. In particular, the new system allowed not only for RGTS settlement between commercial banks but also for settlement of stock trades from the Toronto Stock Exchange.

- *Jasper IV (2018–19)*: In this last phase, the BoC partnered with the Monetary Authority of Singapore (MAS)—that had just completed three phases of its own *Project Ubin*—to experiment on a cross-border, cross-currency, and cross-platform international payments system. Another interesting aspect of this joint expedition was that one Bank used the Corda network while the other utilized Quorum, so to test the interoperability of two foreign platforms.

During that same era, in Europe, the Deutsche Bundesbank and the Banque de France put forward projects *BLOCKBASTER* and *MADRE*, respectively. After the Banco Central do Brasil set up *Project SALT* and the US Federal Reserve started scouting the CBDC realm, two initiatives climaxed the first wholesale CBDC era in late 2016: the MAS launched *Project UBIN* and the four-phased *Project Stella* was piloted by the European Central Bank (ECB) and the Bank of Japan.

3.2 The Next Wave: 2017–19

While wholesale CBDCs remained in the limelight, with *Project LionRock* of the Monetary Authority of Hong Kong (HKMA) still addressing interbank settlements, the 2017–18 period saw the onset of *general purpose* CBDCs projects. Notably, central banks started exploring the relation between digital fiat money and cash, with one noteworthy example being the *e-Krona Project* initiated by the Sveriges Riksbank in Sweden, one of the trailblazers of the CBDC arena up to today. This is because cash usage in Sweden had dramatically declined in favor of e-payments.

The 2017–18 pilot initiatives are in both the retail and the wholesale domain, structured around CBDC concepts that are often diverse [9]. Wholesale plans were presented by the central banks of Denmark, South Africa with *Project Khokha*, Switzerland with *Project Helvetia*, New Zealand, Norway, and Thailand with *Project Inthanon*. Meanwhile, different understandings of retail use-cases were explored by the central banks of Finland (Project *E-hryvnia*), the National Bank of Ukraine, *Project Bakong* by the National Bank of Cambodia, Uruguay with *Project e-Peso*, Israel with *Project e-Shekel*, Venezuela with *Project Petro*, and the Marshall Islands.

In early 2019 around 70% of central banks responding to a BIS survey declared to be engaging in some CBDC-related activity [23]. Although only 30% voiced an intention to issue such instruments within the medium term, that year was arguably a breakthrough one in which research in CBDCs reached a new level of maturity, but also headlines. With little doubt, the watershed moment for this was the political and economic spark provided by Facebook's announcement of the Libra coin in late June 2019. In the same year, the ECB started to analyze the implications of cryptoassets on monetary policy [60] and in October 2020 a report [61] was issued on principles and configurations for a candidate retail *Digital Euro*. The goal was not to outline a specific design, but rather to gather insights from experts and the public at large. Following the reports of the Bank of Korea and the Bank of Japan, the first cross-border interbank settlement mechanism between two DLT-based currency platforms was concluded by the BoC and the MAS, noted earlier as *Project Jasper/Ubin IV*.

3.3 The Age of Maturity: 2020–21

At the beginning of 2020, central banks working on CBDCs had risen to 80% with nearly half of them at the PoC phase, and a smaller number with actual pilot projects [62]. Later in July, the Bank of Lithuania issued the first state-backed digital collector coin, LBCOIN, which can be transferred in a peer-to-peer fashion. LBCOIN is no legal tender (the Bank of Lithuania belongs to the Eurosystem) and can only be exchanged into a physical collector coin. The US that had remarkably been quite silent on its plans showed the first signs of life—in May 2020 the non-profit *Digital Dollar Project Initiative* released its whitepaper reasoning why the Fed should release a digital USD counterpart. Later, in June US congressional hearings took place in with regard to CBDCs that continued on April 15, 2021. Earlier that year, the Boston Fed had announced a collaboration with MIT's Media Lab on a digital dollar with an expected report to be released by the fourth quarter of 2021.

The month of October 2020 also saw the landmark launch of the first CBDC by the Central Bank of the Bahamas through the *Sand Dollar* platform. The Sand Dollar is pegged to the Bahamian dollar, which in turn is pegged to the US dollar on a 1:1 basis under currency board-like rules. This move also validates claims that smaller countries may want expedite implementation of their respective CBDCs due to risk of competition by CBDCs from larger foreign economies. That is, if foreign CBDCs are easier (or more "stable") to use, they may intermediate or present a risk of displacement to "local money" with whatever dramatic impact this may have on said domestic monetary/fiscal policies for those smaller economies. Meanwhile, the Eastern Caribbean Central Bank launched its CBDC labeled *DXCDCaribe*, in November 2020 Brazil's central bank launched the *PIX* instant-payment platform, and the Bank of Russia unveiled interest in a *Digital Ruble*. Also in 2020, the Reserve Bank of Australia started considering a wholesale CBDC system labeled *eAUD*.

Admittedly, the first half of 2021 testifies not only to the increasing interest in CBDCs but also to their growing maturity. Notably, 86% of central banks surveyed by BIS were exploring CBDCs: 60% of them at an advanced experimental or PoC stage and 14% at a pilot phase [63]. In January, the European Commission and the ECB announced a cooperation on a possible *Digital Euro* upon the conclusion of a public consultation. This report was published in April [64]. In February 2021, the *Digital Dollar* debate rekindled significantly in the US and the Swedish *e-Krona Pilot Project* was extended [65]. In the meantime, PBoC's testing of the *e-CNY* was widened to four cities and its launch was announced by the Winter Olympics at Beijing in early 2022. Concurrently, in February, the BoC unveiled three design proposals under their *Model X* challenge for a CBDC denominated in Canadian dollars (the *Digital Loonie*) by three universities [44]. In May 2021, the Bank of Korea issued an open competition for a PoC CBDC system to the private sector.

This era also demonstrates more mature projects in wholesale- and retail mCB-DCs. These projects examine the cross-border behavior of local RTGS CBDC systems by commercial and central banks. More notable is the 2019-20 *Project Aber* by the Saudi Arabian Monetary Authority and Central Bank of the UAE (CBUAE), and *Project Inthanon-LionRock* by the HKMA and the Bank of Thailand. It is certainly not a coincidence that in February 2021 the announcement by the HKMA, CBUAE, Bank of Thailand and PBoC for a major "mCBDC bridge" collaboration was not a surprise for those experienced players. Similarly, other projects address cross-border CBDC use in 2021—illustratively, *Project Dunbar* and *Project Jura* [53, 66].

3.4   Trends and Future Expectations

Along with the efforts by central banks to digitize fiat money, one cannot ignore the moves and associated geopolitical impact by commercial players. Most notably, Facebook's Diem consortium of more than 20 corporations, as viewed in terms of (i) strength in public cross-border reach and cross-border payments and (ii) data protection/surveillance policies. Facebook has more than 1.5B active daily users, trending to 2.4B active users per month. Upon launch, it becomes a corporation with an international reach large enough to compare to any central bank. For historical reference, in early 2020 Facebook renamed its Libra effort to Diem, and pursued a Swiss payment license by the Financial Market Supervisory Authority (FINMA). As the effort to attain such a license has not proved successful, in April 2021 Facebook announced Diem will focus only on the US public. In recent releases, they tap into their native coin as an "interim digital USD" backed 1:1 with assets to the US dollar.

PBoC's e-CNY launch by February 2022 and its aggressive moves to cross-border partnerships with regional players cannot also be underestimated. It has the potential to change the influence of the Remninbi, global payment systems and currencies, and the standardization of CBDCs. With no other major central bank having announced a CBDC launch, we should expect the next few years to be dominated by headlines and research from those two players—but also other independent actors. Further, one should expect the BIS, in its role of "*central bank to the world's central banks*" [34], to continue to lead the standardization playground for CBDCs, notably through its CPMI working group and newly introduced Innovation Hubs [67]. Indeed, in its June 2021 report the BIS has voiced the belief that, with more than 50 central banks entertaining the idea of issuing a digital currency, the time for the monetary system to reap the benefits of CBDC-related R&D has finally come [17]. All in all, CBDCs promise exciting new challenges and innovation over the next decade.

4   Regulatory and Compliance Issues

The socio-economic (r)evolution brought about by cryptocurrencies has raised legal and regulatory questions, many of which remain unanswered to this day. Indeed, these innovations do not only challenge most areas of the law, but they do this in an ever-evolving fashion. As such, experts have been pursuing the best approach to the transformations inspired by DLTs/blockchain, cryptoassets, tokenization and DeFi, among others. To this end, efforts were made to taxonomize policy options with regard to the interplay between law and technology. Accordingly, the following regulatory options were identified: (i) do nothing (i.e., a permissive "wait and see" approach), (ii) introduce tight restrictions (e.g., outlaw certain activities or the provision/acquisition of certain products/services), (iii) issue flexible "case by case" permissions, (iv) set up structured, albeit restricted, experiments (e.g., sandboxes), and (v) devise new regulatory frameworks [68–70].

When CBDCs started to emerge, it was clear their innovative techno-legal character was accompanied by a certain degree of traditionality in terms of the type of stakeholders involved (i.e., central banks, regulated/regulatable intermediaries). Thus, issues originated in the context of blockchain-driven developments are channeled into a more familiar structure of overseen and regulated environments. Nonetheless, CBDCs are far from being unfettered by regulatory questions. In this section, we outline a few outstanding dilemmas, with no attempt to offer a comprehensive account. Naturally, CBDCs raise manifold other issues, most of which belong to areas traditionally less harmonized across jurisdictions than the ones addressed here, as highlighted by [27, 29]. Illustratively, they relate to private and property law, contract law, tax law, insolvency law, private international law.

4.1   CBDCs and Monetary Law

Given the hype surrounding CBDC projects, it is interesting that almost no jurisdiction would currently allow their issuance without amending domestic laws. Indeed, a 2020 study by the International Monetary Fund (IMF) [29] highlighted how CBDC issuance itself poses several risks for the central banking community, burdening it with legal, financial and reputational questions. The two public law domains investigated by the report, "central bank law" and "monetary law", are crucial to warrant CBDCs a sound legal basis. The experts approached these domains separately, to conclude that while the first one could be rather addressed through legal reforms, the latter field poses structural policy challenges with a less straightforward solution.

First, if a CBDC is to be a liability of the central bank (i.e., in the *direct* and *hybrid* forms described above), its issuance must be regulated by "central bank laws", as defined by [29]. This is for the CBDC to be warranted a legal basis in compliance with the principle of attribution of powers and the central bank "mandate" (i.e., its "objective(s), functions and powers" [29]). Likewise, the qualification of a CBDC as "currency" must be regulated under "monetary law". If it is to be used as a mean of payment to extinguish monetary obligations, "monetary law" must treat is as such.[2]

Overall, according to [29], the legal treatment in both fields will largely depend on the specific design, from a technical and operational perspective. Namely, account vs. token-based, wholesale vs. retail, direct vs. indirect, centralized vs. decentralized, and the interrelations between these dichotomies. Hence, different reforms may be required to ensure the soundness of the underlying framework. Notably, controversies arise in relation to the lack of legal basis to issue (i) "token-based" instruments, and (ii) "account-based" CBDCs to the general public. Both aspects would require *ad hoc* amendments to the relevant "central bank law" and "monetary law" provisions.

4.2 Anti-Money Laundering and Counter Terrorist Financing

In the law and technology domain, DLT-related literature underlines how ubiquity and smart contract-driven opportunities have fuelled fears of cryptocurrencies being misused for illicit purposes. Due to their purported traits of anonymity and untraceability, they have been linked to transactions on the dark web, online gambling, money laundering, and to the financing of criminal activities and terrorism.[3] This extends into the regulatory frameworks to fight money laundering and combat the financing of terrorism and proliferation (AML/CFT/CPF), internationally overseen by the Financial Action Task Force (FATF).[4] These rules aim to protect the integrity of the financial system by preventing criminals from enjoying the profits of their deeds, and this compliance domain exerts a significant influence on CBDC projects.

Although most jurisdictions provide their specific provisions, the structure of AML measures is fairly harmonized. Usually, a set of regulated entities is required to give "active cooperation" to the authorities in light of their position as "gateways" with (perceived or actual) oversight capacity on monetary/value transactions. These entities range from commercial banks and financial institutions, to professionals (e.g., lawyers and notaries), to casinos and art galleries. In the crypto sphere, Virtual Asset Service Providers—i.e., a subset of providers of exchange and wallet services—were recently added to the list. In brief, AML duties revolve around licensing, Customer-Due-Diligence (CDD) obligations such as Know-Your-Customer (KYC) and ongo-

---

[2] In the words of [29], *"monetary law is the legislative and regulatory framework that provides the legal foundations for the use of monetary value in society, the economy and the legal system"* and *"the basic principle of monetary law provides that it is for a sovereign State"* (or monetary union) *"to determine and establish its own currency system"*.

[3] The Silk Road case, followed by the shutdown of Darknet markets (e.g., Alphabay, Valhalla, Wall Street Market), added to this skepticism and fear. For more information [58, 71, 72].

[4] The FATF is an intergovernmental, policy-making, monitoring and enforcement organization that sets standards and provides comprehensive guidance, e.g., its Recommendations. Its mandate was extended to combating the financing of terrorism in 2001 and of proliferation of weapons of mass destruction in 2020. In the remainder of the Chapter, AML refers to AML/CFT/CPF.

ing monitoring, record retention and Suspicious Transaction Reporting. The overall framework is informed by the risk-based approach, which means compliance duties are to be molded to preliminary risk assessments.[5] Ostensibly, the ultimate goal is for the competent authorities to be informed of suspicions of money laundering or financing of terrorism or proliferation.

Despite the fact that AML aspects of CBDCs are discussed extensively, these instruments are understandably not treated as cryptocurrencies in this regard, but as a form of fiat currency [8]. Nevertheless, and although CBDC-related AML considerations are detached from those for cryptocurrencies, several studies outline how different CBDC architectures may lead to various AML repercussions. A key question concerns the allocation of the responsibility for compliance duties, end-user account management, and related identity/transaction checks. As central banks do not traditionally interact with public end-users, two-layered CBDC structure may be favored. Indeed, two-tier models allow to outsource compliance aspects to PSPs and commercial banks, to be either managed directly or delegated. This intermediated access model is reportedly favored to leverage existing customer-facing services and avoid unnecessary duplication of resources.

4.3 Cash, Anonymity, and Identification

Even if the technology underpinning Bitcoin is largely acknowledged to inform a *pseudonymous* means of payment, rather than an *anonymous* one, a significant set of altcoins has increasingly evolved toward higher levels of anonymity and cryptographic complexities. Accordingly, the FATF emphasized growing money laundering concerns in terms of virtual-to-virtual "layering" mechanisms [73]. Concurrently, tech advancements in "privacy coins", such as Monero and ZCash, and pervasive transaction obfuscation mechanisms (e.g., mixers/tumblers) were complemented by the advent of decentralized exchanges, unhosted wallets and cross-chain atomic swaps [72, 74, 75]. In this context, the FATF identified several concrete examples of *anonymity* as "red flag indicators" of suspicious activities in the crypto sphere [76].

When it comes to electronic transactions, controversies on *anonymity* well preceded cryptocurrencies and CBDCs. Indeed, the debate dates back to the 90s, and targeted anonymous digital cash and e-cash [77–79]. To be more precise, the core issue had already flourished with regard to physical cash. As the trait of *anonymity* is inherent to latter, which is one of the purest examples of a fungible asset, the-break fight against financial crime has long faced the "anonymity problem", and has addressed it leveraging identification and traceability aspects. Indeed, (some form of) "identification" is argued to be necessary to safeguard the payment system. In a CBDC scenario, the issue is interlinked to the opportunities offered by digital identities (digital IDs) and digital identification, as recently underlined by [17].

---

[5] For instance, CDD must be "enhanced" in specific cases identified as posing noteworthy risks.

More specifically, [19] shows how AML and anti-fraud practices may imply a trade-off between access to the means of payment and traceability. If CBDCs are designed to replicate a situation that is similar to cash-like anonymity, but at the same time they overcome the material physical limitations of coins and banknotes, significant concerns may arise. In the words of [17], *"a token-based CBDC which comes with full anonymity could facilitate illegal activity, and is, therefore, unlikely to serve the public interest. Identification at some level is hence central in the design of CBDCs"*. What is interesting, however, is that cash being dangerous from an AML perspective was one of the reasons why e-money solutions, and the degree of control they can enable through their programmability, were sponsored in the first place [6, 47].

Indeed, monitoring and/or limiting the use of cash is a widespread means to counter criminal activities. Thresholds for customs declarations are provided and cash transactions above certain volumes trigger compliance duties and other measures. In the EU, CDD obligations arise for FIs upon the establishment of a business relationship or when the customer carries out transactions that amount to EUR 15,000 or more. In Canada and in the US, obliged entities must report transactions of CAD/USD 10,000 or more within 24-hours [80, 81]. The EU has considered to introduce restrictions to payments in cash [82], and the recent 2021 "AML Package" is proposing a EU-wide limit of 10,000 EUR to payments in cash, including bearer-negotiable instruments, for professional purposes [83, 84].[6] Meanwhile, some countries already limit its use between private individuals if no regulated intermediary is involved in the transaction [85]. Bearer's instruments, such as bearer's checks and passbooks, are often equated to cash. Illustratively, in Italy cash transactions that exceed EUR 1,000 are prohibited, but also in France (EUR 1,000), Portugal (EUR 1,000), Belgium (EUR 3,000), Slovakia (EUR 15,000), Spain (EUR 2,500), Bulgaria (EUR 5,000), and Greece (EUR 500). In those jurisdictions, transfers of higher values must be made through regulated intermediaries. Outside Europe, similar strategies are applied to specific types of transactions in Jamaica, Mexico, Uruguay, and India.

4.4 Privacy and Data Protection

A major driver behind the onset of cryptocurrencies has been the desire to exchange money privately, without the involvement of a third-party intermediary. Additionally, after the adoption of the EU General Data Protection Regulation (GDPR) in 2016, a wave of global-scale sensitivity to privacy and data protection concerns started to inform the law and technology domain. At times, AML frameworks and privacy/data protection may seem at odds. Scholars have focused on this possible contrast, especially when it comes to permissionless blockchains [86], and with reference to specific concepts (e.g., Privacy Enhancing Technologies (PETs),

---

[6] This is an example of the application of the risk-based approach to the threat posed by cash-intensive businesses. Meanwhile, EU Member States would still be able, if not encouraged, to maintain lower thresholds and/or adopt stricter provisions.

de-anonymization techniques). An extensive array of contributions addresses the interplay between blockchain, privacy and data protection [69, 86–89]. The topic appears as most relevant to the discussion on CBDCs, and it is at the heart of heated debates in the context of the initiatives put forward by central banks.[7]

Additionally, the public-private dynamics of different CBDC designs originate diverging questions, as private stakeholders may be made part of mechanisms of information exchange possibly detrimental to the individual privacy of end-users. Indeed, one of the reasons why AML aspects are discussed in CBDC projects is that they are seemingly opposed to privacy and data protection safeguards. The more information *is* or *can be* disclosed to obliged entities and law enforcement authorities, the more intrusive this may be with regard to financial aspects of end-users' lives.[8] By contrast, a system with full privacy would thwart compliance regimes. These considerations are mirrored by CBDC research, with manifold attempts to build anonymity oriented scenarios while ensuring a certain degree of oversight to avoid dangerous criminal repercussions. Relatedly, [91] puts forward a CBDC architecture that aims to combine privacy with regulatory oversight by holding CBDCs outside of custodial relationships, while [14] explores M2M scenarios.

The relevance of this debate is not exclusive to CBDCs, but to digital payments at large [90, 91]. Nonetheless, CBDCs have a significant potential to impact on the individual from a twofold perspective. As argued by [36], they may "*diminish individual privacy, whether defined as freedom from intrusion into private life or the ability of an individual to control her or his own personal information and protect against its misuse, or with reference to data protection, security, and safety, or even freedom from mass monitoring, profiling or surveillance*". Indeed, "*the combination of transaction, geolocation, social media and search data raises concerns about data abuse and even personal safety. As such, protecting an individual's privacy from both commercial providers and governments has the attributes of a basic right*" [17].

Relatedly, [36] highlights how the issues raised by CBDCs are informed by a broad conceptualization of "privacy".[9] Indeed, albeit often voiced as if they were a single concept, CBDC-related "privacy" concerns different stakeholders—e.g., the central bank, settlement and payment providers, retailers. In this sense, experts have focused on the governance of how network participants can access the CBDC system. This is crucial upon establishing the respective roles of public and private stakeholders in guarding identity and transaction data [19].

---

[7] The final report of the ECB public consultation on a candidate Digital Euro [64] is an example of the debate on the interplay between privacy, security and AML rules.

[8] As argued by [90], transaction privacy is severely hampered by user-level payment history datasets. The latter are increasingly generated by commercial payments platforms, while other dangers arise from subsequent monetization and/or clustering. Progress in AI/ML techniques amplifies the risks.

[9] On some of the privacy and data protection concerns raised by CBDCs, see also [35].

4.5   Privacy-Transparency Trade-Offs

CBDC-related AML issues diverge from those arising in cryptocurrencies. However, if e-fiat money is advertised as a "physical cash" substitute, any desire for a certain share of anonymity needs to avoid any detriment to the integrity of the financial system. Nonetheless, anonymity is not a binary zero-sum property, but rather *ranges* within a spectrum.[10] Further, online anonymity has a socio-technical nature [15, 93]: on the technical side, and within a DLT context, it is influenced by the deployment of specific privacy tools (e.g., PETs), governance considerations (e.g., centralized vs. decentralized systems), and the broader system architecture (e.g., relationship with other on/off-chain layers); on the social side, it refers to the actual possibility of identification and traceability and to the use of forensic techniques to "follow the (crypto) money", against the backdrop of the strategies to prevent this [58].

Although a tension between privacy and transparency seems to be inherent to CBDCs, at a closer look it appears as a *trade-off* [15]. Indeed, all means of payment provide varying degrees of privacy/anonymity, ranging from methods requiring the bank to monitor transaction/identity data (e.g., wire transfers), to anonymous transactions in physical cash. As opposed to the latter, digital cash allows to exert control, which means sensitive information may also be exposed [6]. Against this backdrop, not only CBDCs can be designed to embed various "privacy vs. transparency" trade-offs, but DLTs themselves are conducive to balancing the individual right to privacy against AML public interests. While a fully-transparent CBDC, with real-world identity transactions fully visible to law enforcement, may violate human rights, if privacy is provided without limitation (i.e., no information can be revealed about transactions) misuses for illicit purposes may not be averted. This option is not viable to regulated stakeholders, as it may generate dangerous societal impacts.[11]

Luckily, nuanced solutions are available, and most CBDCs position themselves in the *middle*, offering some privacy to end-users and some visibility, in terms of auditability, to authorities. The work in [58] addresses this trade-off and elaborates on the findings of [94] with regard to confidentiality and auditability. As outlined in Figure 3, different CBDC designs can be classified accordingly. While they entail different trade-offs, a correlation is to be noted between the latter and AML anonymity-related provisions. An interlink between technical and regulatory compliance assumes the latter can be embedded into technology. This concept informs *design-based* regulatory techniques and *regulation-by-design*, as a means to foster desirable outcomes by devising inherently compliant instruments.[12] In closing, research currently shows different data privacy preferences across the globe and CBDC initiatives embody context-specific inclinations, as shown in [19].

---

[10] de Koker [92] addresses the difference between anonymous, identified and pseudonymous clients and the AML impacts. "Crypto" digital payments enhance these complexities [58].

[11] Additionally, history shows that a regulated access of financial authorities to information on monetary/data flows resonates positively with citizens and businesses.

[12] Pocher and Zichichi [14], Pocher and Veneris [58] show how law and technology experts address this notion [47, 68, 95–97].

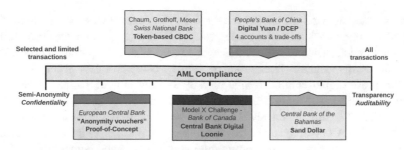

Fig. 3 *Source* Elaboration of the authors in [58]

5   A Deep Dive: Three CBDC Case Studies

The previous sections retraced the evolution of CBDCs from a techno-legal and historical perspective. Some specific projects, however, have played a particular technical and geopolitical role with regard to the trends and future development of the global CBDC ecosystem. In this section, we detail three of these instances, in a case-study fashion. First, we dive into the PBoC's *DCEP*—it is not only the first fully-operational CBDC system but also projects a major influence in the domestic and cross-border digital payment arenas, as also indicated by recent US Senate hearings. Next, we move to Facebook's *Diem*. Although one may argue Diem is not a CBDC—it is not offered by a central bank, but by a private consortium of corporations –, Diem holds most elements of a synthetic CBDC platform and, as noted earlier, it is now "advertised as such" by its founders. Finally, we outline an academic proposal to the BoC's February 2021 *Model X* challenge. At the time of this writing, we respectfully submit that the BoC has not publicly committed to issue a CBDC. Hence, the three *Model X* challenge proposals reflect only the opinions of their academic authors. As this published CBDC-related *Model X* challenge was the first of its kind by a central bank, but also due to its intricate design details, it warrants this third case study in a complementary position to the first two.

5.1   China's DCEP/e-CNY

The rapid rise of China's DCEP, also labeled e-CNY, as a CBDC leader is a natural outcome of the country's fast-paced mobile-based economy digitization in the past decade, even by the most competitive Western standards. According to a brief by Deloitte Digital [98], in 2018 more than 70% of China's 829 million net citizens use mobile devices to make payments, a swift 60% increase in just 3 years. In the first 9 months of 2020, mobile payments exceeded $48B in value—an 135x increase since 2012 [99]. The amount of data generated by China's commercial sector has already surpassed that of the US and is expected to grow to 48.6 ZB by 2025—in contrast to an expected 30.6 ZB projection for the US [98]. As another example, more than 96% of the revenue during China's Double 11 Festival in 2019 came from mobile payment systems. The maturity of this system now allows the public to utilize their personal IDs to essentially "individualize" their e-commerce experience.

With those digital cultural trends already spilling abroad, China's technology companies today claim more than 40% of their revenue sources from foreign actors. A main driver in this digital revolution has been the widespread adoption of the Alibaba and WeChat e-payment methods in the past decade. Today these platforms serve the vast majority of those commercial interactions/transactions.

Rationale and History for the DCEP: Until July 2021, the PBoC had issued no comprehensive published research paper that explained the technical architecture details behind the e-CNY and its underlying motives. Hence, over the past years, information has been mainly derived from public talks by Chinese officials, such as Mu Changchun (Director of the Digital Currency Research Institute (DCRI) of PBoC) and Qian Yao (former Head of the Institute of Digital Money at PBoC), or from newswires and Chinese/Western opinion articles. Under those conditions, the motives and drivers of DCEP seemed to include:

- The rapid digitization of the economy by private actors (WeChat, Alibaba, etc) and the risks generated by those companies as they silo the associated user data;
- The additional risks to China's monetary policy, capital flows and currency sovereignty by the emergence of alternative coins such as Bitcoin, Ethereum, Facebook's Libra/Diem—but also from other forthcoming CBDCs;
- The need for a SWIFT alternative to cross-border payments as the network has been claimed to be using its underlying data for US geopolitical interests [100];
- An add-on to China's recent Cross-Border Inter-Bank Payments System; and,
- The natural progression of China's efforts in the past 20 years to expand the internationalization and influence of its own currency, more notably to countries within the Road & Belt Initiative.

In a speech on December 25, 2020 at the Chinese Winter Olympics Group,[13] Mu Changchun said the PBoC's adventure into the world of digital currencies first started in 2014. Back then, the designated working group concluded there was no need to issue a digital currency, partly because 3G networks were not sufficient to support such a novel expedition. However, Mr. Changchun continued, due to the threat of Bitcoin to countries with closed capital controls like China, the Bank had formed the e-M0 group to further investigate the matter and to first build a prototype borrowing from Bitcoin's architecture. Later in that talk, he added that Facebook's 2019 Libra announcement had increased initial concerns. In 2016, the PBoC formed the DCRI working group. In the same year, the e-M0 group determined blockchain-based technology cannot serve the needs of a national digital currency. This is because the one-tier Bitcoin-based archetype does not prove adequate to the technical needs of a modern e-payment platform such as the one China's economy commands.

---

[13] https://www.youtube.com/watch?v=U6tUrUpDCW4&t=2126s&ab_channel=PlusToken.

Later, in 2017, the DCRI expanded its efforts by including more blockchain, legal and hardware-design expert staff. In the dawn of 2018, it announced the introduction of China's CDBC as a main priority. By mid-2019, the PBoC declared it was ready to launch DCEP, and by April 2020, pilot tests were conducted in four geographical regions: the Xiongan area in the Hebei Province, Suzhou in the Jiangsu Province, Chengdu in the Sichuan Province, and Guangdong's Shenzhen. This occurred by "airdropping" a limited number of e-CNYs to the public for use, and user-experience feedback, at a few merchant locations. Meanwhile, mCBDC *Project Inthanon/LionRock* was initiated by the HKMA and Bank of Thailand. In the following months, pilot tests were conducted in more targeted environments such as Shanghai's Tong Ren Hospital and Beijing's Metro Daxing Airport Express, while the state-owned Agricultural Bank of China launched the first e-CNY ATM machines. As the official launch of e-CNY is set for the Winter Olympics in Beijing in February 2022, a year earlier the UAECB, Bank of Thailand, HKMA and PBoC announced a cross-border DLT-based mCBDC project. On May 22, 2021, former PBoC governor Xhou Xiaochuan at a speech at the Tsinghua Wudaokou Global Financial Forum underlined how the DCEP is not built to displace existing payment systems, nor to replace the US dollar as a currency reserve.[14] The interested reader is referenced to [99] for an elaborate chronology of DCEP's history and deployment.

As anticipated, in July 2021, the e-CNY Working Group at the PBoC released its first R&D white paper [101]. Its main goals are to clarify the position of the PBoC and to explain its objectives and visions, as well as e-CNY's design frameworks and policy considerations, to the end of engaging into multi-stakeholder communication. Accordingly, the expert group highlights:

- the link between the rapid evolution of the digital economy and digital payments, and the need for new, safe, inclusive and adaptive retail payment infrastructures;
- the profound change in China's use of cash—according to a 2019 survey, *"the number and value of transactions via mobile payment accounted for 66% and 59% of the total, while those paid in cash accounted for 23% and 16%, and those paid by card 7% and 23%, respectively. Among those surveyed, 46% used no cash in any transaction during the survey period"* [101]—and the consequent need for digitalization to safeguard access to cash itself and financial inclusion;
- the rapid development of global stablecoins; and
- the attention paid by the international community to CBDCs and their different design options, as well as the importance of the internationalization of e-CNY and its role in cross-border payment programs.

Architectural DCEP Considerations: As defined in [101], the e-CNY *"is the digital version of fiat currency issued by the PBOC and operated by authorized operators. It is a value-based, quasi-account-based and account-based hybrid payment instrument, with legal tender status and loosely-coupled account linkage"*. The model features a centralized management and a two-tier operational system, where the

---

[14] https://mp.weixin.qq.com/s/OWkVaWw0-f2wSSFFH979rg.

PBoC is positioned at the center. The PBoC issues e-CNY—in parallel to the physical RMB—to "authorized operators" (i.e., commercial banks and licensed non-bank payment institutions) that, in turn, exchange and circulate it to end users.

Hence, the e-CNY is reportedly a direct cash-like claim on the central bank, with client onboarding and payment services managed by intermediaries. As Mu Changchun had added on December 25, 2020, DCEP is a "two-tier" architecture where the PBoC does not directly issue it to the public, but to a second tier of commercial players coined as "designated operating institutions", most likely in exchange for central bank reserves. Currently, the designated operating institutions are state-owned commercial banks, Alibaba (Ant Group), and WeChat (Tencent), together with the three major telcoms, namely China Unicom, China Mobile and Telecom. Later, Mr. Changchun commented, with the State Council's approval, Postal Savings Bank and Bank of Communications may be added. He further claimed a main reason for the system to be two-tiered is that there can be data breaches or hacking risks if it was built as an one-tier; the two layers prevent this with their diversification. In another report [42], it is claimed the infrastructure entails a mix of a conventional database and DLTs, where a copy of holding and transaction data is received and settled by the PBoC on a regular basis. To that end, it remains to be seen how China's President Xi Jinping's December 2019 promise for a national initiative to "seize blockchain opportunities" globally may materialize [102].

Interesting insights are provided by [101] with regard to the concept of "*controllable anonymity*" or "*managed anonymity*". Indeed, as commented by [58], the e-CNY is informed by the principle of "*anonymity for small value and traceable for high value*" and may offer four or five types of accounts/wallets. The decision on which account to assign to a given user rests on characteristics such as CBDC amounts, anticipated use, and other information provided during registration. Reportedly, the two most anonymous types of account—i.e., the "*least privileged wallets*" [101]—require few identifying information and no real-name identity. In these cases, risks of money laundering and other criminal abuses are mitigated by imposing strict balance and transaction limits—a daily transaction limit and a relatively low balance limit. On the contrary, depending on the provided information, the least anonymous types of wallets must be opened at a counter, can be linked to a bank account or even used as one. Further, the implemented restrictions (if any) vary, depending on the "*strength of customer personal information*", with regard to both types of transactions that can be performed and relevant amounts.

The e-CNY offers both software and hardware wallets [101]. Offline transactions are designed in a way that resembles the CBDL report [44] and the 2019's commentary by the Mount Union of Science and Technology.[15] Nonetheless, even in the most anonymous scenario among the account types, some identifying information is given when the account is opened. Hence, one may be expecting that the true identity of the user can always be retrieved. In any case, by implementing this multi-layered structure one can achieve a limited degree of user-to-user anonymity which is both controllable and tiered. Within this framework, commercial banks hold identifying

---

[15] https://www.mpaypass.com.cn/news/201912/06094420.html.

information and can de-anonymize suspicious transactions for AML purposes. Privacy and data protection issues raised by the e-CNY's two-layered structure are addressed by [35], although [101] argues e-CNY is expected to collect less transaction information than other e-payment systems, and to disclose information to third parties or other governmental agencies only if mandated by law. To this end, China's central bank plans to prohibit arbitrary use of e-CNY data and to set up an internal firewall, as well as to implement security and privacy protocols—e.g., separation of e-CNY from other business lines, tiered authorization system, internal audits.

Although on the surface the DCEP seems like a hybrid CBDC architecture, one should examine this statement under a prism of China's domestic policies/practices. Considering that major Chinese banks are state-owned/controlled, but also the history of authoritative power/actions by the Communist Party of China onto the domestic commercial sector, it becomes a belief that DCEP borrows many elements from a direct CDBC architecture that only borderlines to a typical hybrid model.

Domestic and Global Implications of the DCEP: Although denied in public speeches by China's government officials, the overwhelming rhetoric by news media from both the East and the West is that the DCEP presents a challenge to the US monetary system but also to the USD's currency reserve status. Some even take the view that DCEP's emergence will be used as a "digital weapon" against the US in economic, trade, and geopolitics as it will eventually allow China to obtain the data and track (or even block) international transactions just like the US has done with the SWIFT network in the past.[16] According to statistics by the World Bank, more than 1.7B adults around the world use cash because they don't have access to a bank account. Nevertheless, more than two-thirds of this population use mobile phones that can be eventually used to conduct mobile payments. Indeed, this is what happened in China (but also India) during the past decade: it is not uncommon in both large-population countries to see street merchants using QR codes to sell their products. Along with China's technology investment in the emerging Belt & Road initiative region, it becomes a realistic scenario for the e-CNY to enjoy distribution/adoption to those countries after it proved its maturity domestically.

The tremendous "early adopter" impact of the DCEP could likely go much further to establish novel e-commerce channels for China, as artfully articulated in [104]:

- *Business-to-Customer flows*: the e-CNY has the potential to massively level the operations between banks and big tech, while further squeezing merchant acquiring businesses. It also opens up new opportunities for licensed e-CNY providers looking to provide banking services to supply chains and end-consumers;
- *Cross-border Business-to-Business flows*: this relates to cross-border trade settlements, with China being already one of the larger exporters/importers in the global economy, but also a leader in global foreign direct investments; and
- *Consumer-to-FIs flows*: this relates to domestic and international e-service innovation due to the cost-competitive and tech-efficient nature of the DCEP.

---

[16] For example, see [103].

5.2 Libra/Diem by Facebook et al.

As widely acknowledged, the watershed moment for central banks was June 18, 2019 when Facebook and its associated consortium—the "Libra Association"—unveiled the forthcoming introduction of the Libra coin [22]. The announcement brought shock-waves across the globe to governments and the private sector alike. Within hours, the US Senate and Congress called Facebook testify on their plans. During those hearings, members from both chambers were critical of Facebook's past practices on data protection, but also of their plans to obtain regulatory clearance. The next day, both the EU and China made similar succinct commentaries.

It, therefore, comes as no coincidence that on June 23, 2021 the BIS in its Annual Economic Report [17] urged central banks to issue CBDCs as soon as possible, as *"the most significant recent development has been the entry of big techs into financial services. Their business model rests on the direct interactions of users, as well as the data that are an essential by-product of these interactions …the user data in their existing businesses in e-commerce, messaging, social media or search give them a competitive edge through strong network effects. The more users flock to a particular platform, the more attractive it is for a new user to join that same network, leading to a Data-Network-Activities or DNA loop"*. The report emphasizes additional concerns if central banks delay their CBDC introduction—notably, if digital currencies are introduced by the private sector first, the risk of "currency substitution" [17].

History and Economics of Libra/Diem: The Association's first Libra-coin revelation in June 2019 [22] intended to design it as a basket of the five sovereign currencies that compose the Special Drawing Right (SDR) by the IMF, no much different to what described in [105] a year earlier. The announcement displayed an Association of corporate and non-profit organizations—a list that included Visa, PayPal and Mastercard—that planned to support the ecosystem after an initial deposit of a minimum of $10M in return for Libra Investment Tokens. Following the backlash by domestic and foreign governments, by fall 2019 some members dropped out of the Association. In the spring of 2020, the project shifted to offering a set of stablecoins—USD, EUR, GBP and the SGD—and also abandoned its plans for a permissionless system. In December 2020, it rebranded itself as Diem. By April 2021, the Association petitioned for a payment service license from FINMA. A year later they dropped trying to obtain it, and focused on the US via a USD stablecoin.

For the sake of simplicity, in the coverage below we use the term "Diem" to indicate the project from infancy. Diem is the base currency in the system. At the time of writing, it appears to be pegged to the USD only. Each coin is backed by a reserve that contains mostly low-risk liquid assets (like highly-rated US government securities) but also cash accounts. This reserve protects the coin from the highly volatile price distributions of traditional cryptocurrencies. The Diem Association manages the currency reserves, with its members acting as liquidity providers during on-boarding and off-boarding periods. The Association mints and burns the Diem-coin based on the fiat deposits and withdrawals in its reserve. Frequent auditing provides continued public confidence into the ecosystem, while other designated dealers and regulated virtual asset providers are added as the network matures.

According to [99], the main use-cases of Diem include:

- *Local Payment & Commerce Systems*: bringing a unified experience in e-commerce— e.g. Facebook, Instragram, WhatsApp and other e-commerce platforms are powered by Diem to eliminate the costs and multiple layers of other existing and expensive payment mechanisms today;
- *A CBDC Sandbox*: this is the case where smaller countries choose the Diem ecosystem as a sandbox to build their own CBDCs, no different to typical open library-based software development practices today; and,
- *Cross-border Payments*: with recent shifts in US. markets,[17] this task may come into a jeopardy. With time though, Facebook's 2.5B reach is expected to promote system adoption, including audience in US "politically friendly" jurisdictions. As cross-border payments today remain expensive (it is estimated they cost up to 7% of the remitted amount—in less advanced economies this climbs above 12%), Diem has a potential to disrupt this sector economically but also geopolitically.

One cannot but only observe the stand-out parallels between Diem and the DCEP in their root motives, use-cases and objectives.

Diem's Baseline Architecture: At the outset, Libra was designed as a permissioned DLT, governed and operated by its consortium of private organizations. The DLT is maintained by the consortium members termed as *validators* in terms of the consensus protocol. Using a state replication paradigm designed on top of the Diem Byzantine Fault Tolerant (BFT) consensus mechanism, the validators preserve an identical database. Diem's BFT is a variant of the Hotstuff protocol [106]—also used in Ethereum's Casper and the Tendermint protocols: it guarantees safety and liveness in a partially synchronous system. Its conservative nature ensures that agreement over the state of the system is reachable by the validators at any point in time—even in the presence of byzantine faults. All the rules around validator management, governance, transaction processing, security policies, and incentives are implemented as smart contracts in Diem's programming language *Move* [107].

Indeed, the Move language—or, the "programming language of money" as it is advertised—is one of the core contributions of the Diem ecosystem. Designed by Facebook's Novi team, Move is a safe and flexible bytecode-based programming language with which one can create transaction scripts and smart contracts that can affect the system's state. A key feature of Move is the notion of "first-class resource types". Here, resource types have pre-defined semantics around their logic: they cannot be copied or discarded. This makes them secure and protected by definition. Move's other highlight is its inherent ability to prove the smart contracts' properties formally. In particular, along with the semantics of Move, a specification language and a formal prover have been provided by the Novi team to allow developers to add properties and formally verify that their contracts are functionally correct.

---

[17] Diem's announcement was posted shortly after the release of the said Citibank report.

Overall, Diem's open-source implementation and the completeness of Move are ingrained with features that are arguably essential to any CBDC "programmable money" infrastructure. With modularity as one fundamental design feature, it allows usability in other protocols as well. Given that a complete functional Move verification toolset/methodology is also provided, the language certainly stands out compared to other high-level smart-contract languages like Solidity and Vyper. As of today, the project is at a testnet stage, with the network set to go live by late 2021. Once it proves maturity, one should expect open access to third parties (i.e., regulated virtual asset providers) to submit Move-based decentralized apps—no different to what happens today with Google Play (Android) and Apple Store (iOS) apps.

Is Diem a CBDC? Diem is not a CBDC in the traditional sense of the definition, as it is not issued by a central bank. With no doubt, its goal is to serve the business interests of its private consortium members and its virtual asset provider partners. However, considering its recent partnership with the Silvergate Bank [108], but also the patronage by its leading economist Dr. Catalini as an "interim digital dollar" until the Fed "acts" [109], Diem is positioning itself with "proxy CBDC features". As synthetic CBDCs are usually compared to stablecoins, Diem's architecture and operation arguably bears strong similarity to synthetic CBDCs.

5.3   Model X: a Canadian Central Bank Digital Loonie

Soon after completing the four phases of *Project Jasper*, on February 25, 2020 the BoC published its *Contingency Planning for a Central Bank Digital Currency* [110]. In this plan, the BoC disclaimed it has no plans to launch a CBDC, but only wants to build the capacity to issue a general purpose, cash-like, CBDC should the need to implement one arises. It also noted that it will consider launching a CBDC if certain scenarios materialize, or appear to be likely triggered, such as:

- A continuous decline in the use of banknotes to the point where Canadians no longer can use them for a wide range of transactions; and/or,
- A situation where one or more alternative private sector digital currencies start to become widely used as an alternative to the Canadian dollar as a method of payment, store of value and unit of account.

Two months later, in April 2020, the Bank issued an academic competition-for-proposals under the *Model X* title, addressing the five policy objectives noted in Sect. 2—Privacy, Universal Access, Security, Resilience and Performance. The BoC also specifically requested a solution with an accompanied "business plan" that does not put it in direct contact with the end-users (e.g., services such as identity verification or account opening/servicing), although it remained open to providing a baseline service to them. Further, the solution should adhere to the highest service-quality metrics and foster healthy competition in the payments market.

The remainder of this subsection outlines a techno-legal economic proposal submitted by a team from the *University of Toronto* and *York University* [44] for a *Central Bank-issued Digital Loonie*, or *CBDL*. In brief, the proposal argues for a *two-phased* account-based KYC-backed approach. In the first phase, the BoC establishes a digital cash mechanism based on a *centralized platform* with an authentication protocol based on existing resources that safeguards users' privacy/data. In the second phase, the BoC expands this platform to a backbone that allows private enterprise to build *a decentralized messaging platform* under the auspices and supervision of the BoC and transforms CBDLs into "programmable e-money". Offline transactions are served through a quasi-token-like portable CBDL-card, similar to what described earlier. Finally, the proposal contains extensive reference to legal/regulatory considerations.

CBDL Principles: CBDLs have the following physical-cash characteristics: (i) they are a liability on the BoC's balance sheet where each CBDL is equivalent to one Canadian dollar, (ii) they are available to every registered Canadian resident and corporation, (iii) they transfer quasi-anonymously among verified e-wallets that require one-time e-KYC so they initially get set, (iv) transfers are in real-time with minimum fees, (v) they allow offline transactions, (vi) they generate seigniorage income for the BoC at creation, and (vii) they comply with AML regulations. Whether CBDLs bear interest or not, it is a viable system option yet a policy question.

Phase 1 Operation: In the first phase, the BoC establishes an entity that provides CBDL-accounts and processes all CDBL transactions within a tightly-closed centralized system. This phase also disrupts and establishes a new status-quo in cash-like payments by introducing CBDLs. In doing so, it requires an expansion of BoC activities by incorporating and overseeing an entity that provides CBDL-accounts to millions of residents and businesses, but it is also responsible for the processing of large numbers of transactions of BoC-issued CBDLs per day and conducts overnight AML—namely the *"Narrow Bank" (NB)*. The NB will have no physical location to serve end-users and its staff can reside within the BoC premises, for instance. Further, CBDL transaction messages in the first phase trigger push transactions providing immediate settlement by the NB. This is possible because those transactions are direct transfers between fully-funded CBDL-wallets that involve no credit.

An important proposal argument is that the CBDL platform should secure Canadians' privacy by default but also allow them to monetize their data. It is also suggested for AML to leverage existing public infrastructure (e.g., provincial service agencies, or Canada Post) and private sector solutions by Canadian-owned FIN-TRAC FI firms for KYC. Eligible Canadian residents and businesses obtain their wallet addresses after undergoing this e-KYC. Wallet addresses are represented by a quasi-anonymous identifier, built to not identify the user identity or the respective transaction data to other system parties. However, CBDL users are not anonymous when the homomorphic encrypted AML process triggers compliance flags, or to court orders that direct to reveal certain information. This onboarding and transaction processes bear similarities to India's Adhaar system [111] that provides each citizen with a digital biometric identity allowing them to transact without releasing identities or transaction-data between the parties. Finally, it is proposed user-wallets have upper limits (e.g., 10,000 CBDLs) sufficient for typical cash-like transactions, and special provisions, such as reduced functionality or with preset-expiration dates, for tourists or business visitors. It is also suggested e-KYC should not contract international parties to safeguard Canada's sovereignty and ensure data does not leave Canada.

Phase 2 Operation: The second phase introduces a permissioned quasi-decentralized payment messaging programmable layer on top of the Phase 1 infrastructure to improve scalability and promote digital and economic innovation. A select number of entities (such as major FIs) with experience in handling technology, AML and data will be invited to join the network as "validator nodes", to process CBDL-related transactions but also the execution of archetypal smart contracts. These private entities will bear the cost of this new phase while the NB will remain as a validator that ensures "everyone plays by the rules". The proposal goes at length to describe the lucrative opportunities at a global scale and respective incentives for FIs to participate. In this setup, the NB will transition to be one of the validator nodes but it will also be the single entity that performs overnight AML "housekeeping". Finally, the system could collapse back to a centralized platform in the rare case of a systemic crisis, exclusively operated by the NB under the basic operations of Phase 1.

The messaging layer in Phase 2 will be open-source, it will follow tight domestic/international standardization for interoperability, and it will continue releasing entry-level public APIs for third parties. This setup will enable the platform's core functionality to allow commercial parties that are non-validator nodes—such as other FIs, FinTechs/PayTechs, and service providers (non-FI corporations)—to build digital commerce services but also participate in the enhanced CBDL system. Evidently, to allow private entities offer technical services to increase and/or capture new markets, the NB will need to mandate *programmable-CBDC standardization* to allow third-parties to build network overlay fintech/data services, but also to "communicate" with other emerging foreign CBDC projects. Examples of these services include further data-protection/data-mining mechanisms, digital authorizations and e-signatures, asset-tokenization ecosystems, low-latency system processing/markets for IoT/AI operators, account and spending management tools, perks for users to exchange private data for services, and other overlay networks to permissionless/permissioned systems and/or foreign CBDCs.

The Business Rationale of CBDLs: CBDLs are a mix of *direct* CBDCs, with Phase 2 introducing "contained" elements from *hybrid* platforms, as the BoC (NB) still retains system control and distribution of CBDLs. The authors rationalize this architecture having a "carrot and stick" approach to positively disrupt established FI payment practices, and replace them for ones that benefit the public in a new global digital economy where one needs to remain innovative and relevant [112] while protecting their citizen's data. They also argue that current (outdated) payment systems are unreasonably expensive to the public acting as revenue "cash cow" streams for the FIs. Further, by concept and by architecture, CBDLs are intended as a digital complement for *cash* and it is only proper to be advertised as a competition to current cash payments. In contrast, commercial bank main service is to provide market liquidity through credit arrangements (e.g., loans, overdraft arrangements, lines of credit).

The authors urge against the use of synthetic CBDCs; they believe it does not balance the public's privacy interests, may dilute national sovereignty, and may not intrinsically promote healthy innovation in the private sector. They argue that, whatever contingency condition triggers BoC's plans, it is both necessary and sufficient to introduce CBDLs "stand alone", not to involve FIs in the distribution of Phase 1 and limit their operational jurisdiction in Phase 2 with close supervision. The reason is that FIs do not have incentives to cannibalize existing revenue streams by spearheading a new CBDC system. Following the authors' extensive analysis, there's a claim to be made that CBDLs resemble (within a geopolitical and policy regional context) the practices and implementation doctrines of the DCEP.

CBDL Legal Considerations: The CBDL report complements its techno-economic plan with an extensive set of legal recommendations. The latter are here summarized to the extent they mirror legal issues other central banks will likely face upon issuing a CBDC. At early CBDL design stages, the BOC should address the following issues:

1. The legal authority of the BoC to issue CBDLs;
2. Regulation and oversight of e-wallets and the exchange/settlement network;
3. Considerations relating to AML regulations.

The first question asks whether the BoC has explicit authority to issue digital currency under the current version of the BANK OF CANADA ACT. Any related legal or political challenges may result in reputational damages and implementation delays, which should be averted. The second question pertains to the appropriate regulatory body to oversee the network, including the establishment of the NB. Phase 1 presents the following two critical legal issues: (i) to support CBDL transactions, the model envisions the need for the BoC to issue CBDLs to the NB, or equivalently, a reserve account within the BoC and (ii) the legal environment in which the NB should be subject to regulatory oversight. Phase 2 involves expanding the network to BoC/NB-licensed private service providers that are permitted to develop innovative fintech/data services by creating proxy/service wallets that connect with the end user verified CBDL-wallets with the NB. These licensed service providers and network validators should still be brought into the regulatory framework.

Finally, the third question pertains to changes to AML requirements under the PROCEEDS OF CRIME (MONEY LAUNDERING) AND TERRORIST FINANCING ACT. This exploration should also include offline transactions through the quasi-token CBDL-cards that present additional issuance considerations as well as new AML concerns. The legal section of the CBDL proposal closes with additional aspects the BoC should be mindful in later stages of the design process, such as deposit insurance, consumer protection, privacy and tax implications.

6 Conclusions

Research in digital currencies and decentralization in this emerging digital world is a multi-disciplinary endeavor; technologists, regulators, economists, political scientists, and sociologists, among others, need to gather together and listen to each other to properly shape the "history of things to come". Even more, research for digital currencies by central banks is an exciting field that promises to occupy headline news stories and scientific practices in this drastically changing decade for our society. Along those lines, this chapter attempted to outline the key elements of central bank digital M0 money evolution, as mirrored by publications of leading institutions, private actors, and monetary authorities. As seen, the debate is heated and complex. Although many central banks declare they are not yet fully convinced that CBDC benefits outweigh their risks/costs, they still run PoCs and pilots as those words are typed here. From this angle, the case studies of the PBoC's *DCEP* and Facebook's *Diem* provide topical insights to assess the imminence of this worldwide shift in monetary policy, payment system modernization and geopolitical trends.

Section 1 set off by disambiguating "central bank money", to review the difference between *wholesale* and *retail* use-cases and the drivers underpinning their interest. Section 2 addressed different perspectives on candidate architectures for *retail* CBDCs, as emerging in a vast body of literature. In this context, dimensions such as public-private interplay, offline usage, and cross-border efforts were heeded. In Sect. 3, the reader could follow the history of CBDC projects, starting from pioneer efforts to existing initiatives and future trends. By pursuing a more specific avenue, Sect. 4 outlined a set of questions pertaining to the regulatory and compliance domains—i.e., monetary law considerations, AML scenarios and cash-like anonymity, privacy and data protection concerns, privacy-transparency trade-offs. Finally, Sect. 5 divided into the details of three major projects, pinpointed on the grounds of their key role within the global CBDC arena.

Even if the topic is subject to major developments on a daily basis, some conclusions may already be drawn at this stage. Evidently, CBDC systems are bound not only to serve millions of users but also to exert enormous influence on many aspects of the public's life from a techno-legal and socio-economic perspective. Further, they are strongly linked to risks of collected/siloed data and relevant publicly-available monetization practices. Likewise, their impact should be foreseen with regard to their economic/social influence from a domestic and international geopolitical viewpoint. Against this backdrop, it can be argued the deployment of e-fiat money involves a vast range of considerations that go way beyond the argument of "a new way of making purchases without using physical banknotes". It remains to be seen whether and how today's major economies will leverage CBDC-related innovations to capitalize on their position. Alternatively, it is to be expected that the strength of proactive private players and certain sovereign countries "over others" will further unfold.

Acknowledgements While this chapter is the result of joint research and editing effort carried out by both authors, Nadia Pocher is the author of Subsections 1.1, 1.2, 2.1, 2.3, 2.4, 3.2, 3.3, Sect. 4, and Andreas Veneris is the author of Subsections 1.3, 2.2, 3.1, 3.4., Sect. 5. The remainder is the result of joint drafting. The contribution of Nadia Pocher received funding from the European Union's Horizon 2020 research and innovation program under the Marie Skłodowska-Curie International Training Network European Joint Doctorate G. A. No 814177.

References

1. Tapscott, D., Euchner, J.: Blockchain and the internet of value: an interview with don tapscott. don tapscott talks with jim euchner about blockchain, the internet of value, and the next internet revolution. Res. Technol. Manag. **62**(1), 12–19 (2019). https://doi.org/10.1080/08956308.2019.1541711
2. Antonopoulos, A.M.: The Internet of Money-Volume Two. Merkle Boom LLC (2017)
3. Werbach, K.: The Siren song: algorithmic governance by blockchain. In: After the Digital Tornado: Networks. Algorithms, Humanity (2020)
4. Walch, A.: In Code(rs) we trust: software developers as fiduciaries in public blockchains. In: The Blockchain Revolution: Legal and Policy Challenges, pp. 1–27 (2018). https://papers.ssrn.com/sol3/papers.cfm?abstract_id=3203198
5. Casey, M., Crane, J., Gensler, G., Johnson, S., Narula, N.: The impact of blockchain technology on finance: a catalyst for change. Technical Report (2018)
6. Allen, S., Capkun, S., Eyal, I., Fanti, G., Ford, B., Grimmelmann, J., Juels, A., Kostiainen, K., Meiklejohn, S., Miller, A., Prasad, E., Wüst, K., Zhang, F.: Design choices for central bank digital currency: policy and technical considerations. Technical Report 13535 (Jul 2020)
7. Adrian, T., Mancini-Griffoli, T.: The rise of digital money. Technical Report (Jul 2019). https://www.imf.org/
8. Allen, J.G., Rauchs, M., Blandin, A., Bear, K.: Legal and regulatory considerations for digital assets. CCAF, Technical Report (Oct 2020). https://www.jbs.cam.ac.uk/faculty-research/centres/alternative-finance/publications/legal-and-regulatory-considerations-for-digital-assets
9. Auer, R., Cornelli, G., Frost, J.: Rise of the central bank digital currencies: drivers, approaches and technologies. Technical Report (Aug 2020). www.bis.org
10. Sandner, P., Gross, J., Grale, L., Schulden, P.: The Digital Programmable Euro. Implications for European Banks, Libra and CBDC (Jul 2020)
11. Al-Fuqaha, A., Guizani, M., Mohammadi, M., Aledhari, M., Ayyash, M.: Internet of things: a survey on enabling technologies, protocols, and applications. IEEE Commun. Surv. Tutor. **17**(4), 2347–2376 (2015)
12. Ahlgren, B., Hidell, M., Ngai, E.-H.: Internet of things for smart cities: Interoperability and open data. IEEE Internet Comput. **20**(6), 52–56 (2016)
13. Manyika, J., Chui, M., Bughin, J., Dobbs, R., Bisson, P., Marrs, A.: Disruptive Technologies: Advances that will Transform Life, Business, and the Global Economy, vol. 180. McKinsey Global Institute San Francisco, CA (2013)
14. Pocher, N., Zichichi, M.: Towards CBDC-based machine-to-machine payments in consumer IoT. In: Proceedings of the 37th ACM/SIGAPP Symposium on Applied computing (SAC 22). Association for computng machinery, New York, NY, USA, 308–31. (2022). https://doi.org/10.1145/347731
15. Rogaway, P.: The moral character of cryptographic work (2016)
16. European Central Bank: Tiered CBDC and the financial system, no. 2351, p. 42 (2020). https://www.ecb.europa.eu/
17. Bank for International Settlements: CBDCs: an opportunity for the monetary system. Technical Report (Jun 2021). https://www.bis.org/publ/arpdf/ar2021e3.pdf
18. Swartz, L.: New Money: How Payment Became Social Media. Yale University Press (Aug 2020)

19. Carstens, A.: Digital Currencies and the Future Monetary System. In: Hoover Institution Policy Seminar, vol. 89, no. 1, p. 17 (2021). https://www.bis.org/speeches/sp210127.pdf
20. Nakamoto, S.: Bitcoin: A Peer-to-Peer Electronic Cash System (2008). www.bitcoin.org
21. Buterin, V.: Ethereum whitepaper. Ethereum Foundation. Technical Report (2013). https://ethereum.org/en/whitepaper/
22. Amsden, Z., Arora, R., Bano, S., Baudet, M., Blackshear, S., Bothra, A., Cabrera, G.: The Libra Blockchain-White Paper, pp. 1–29 (2019). https://diem-developers-components.netlify.app/papers/the-diem-blockchain/2019-06-25.pdf
23. Barotini, C., Holden, H.: Proceeding with caution-a survey on central bank digital currency, vol. 101, no. (1682–7651), pp. 1–15. Bank for International Settlements (2019)
24. Opare, E.A., Kim, K.: A compendium of practices for central bank digital currencies for multinational financial infrastructures. IEEE Access **8**, 110 810–110 847 (2020)
25. ITU-T Focus Group on Digital Currency including Digital Fiat Currencies: Taxonomy and definition of terms for digital fiat currency, ITU-T. Technical Report (Jun 2019). https://www.itu.int/en/ITU-T/focusgroups/dfc/Documents/DFC-O-012_TaxonomyanddefinitionoftermsforDFC.pdf
26. Bank of International Settlements: Central bank digital currencies: foundational principles and core features. Technical Report 1 (2020). www.bis.org
27. Brummer, C.: Cryptoassets: Legal, Regulatory, and Monetary Perspectives. Oxford University Press (2019)
28. Geva, B.: Banking in the Digital Age-Who is Afraid of Payment Disintermediation? EBI Working Paper Series, no. 23 (2018)
29. Bossu, W., Itatani, M., Margulis, C., Rossi, A., Weenink, H., Yoshinaga, A.: Legal aspects of central bank digital currency: central bank and monetary law considerations by, Technical Report (Nov 2020)
30. Sidorenko, E.L., Sheveleva, S.V., Lykov, A.A.: Legal and economic implications of central bank digital currencies (CBDC). In: Economic Systems in the New Era: Stable Systems in an Unstable World, pp. 496–502. Springer International Publishing (2021)
31. Yao, Q.: A systematic framework to understand central bank digital currency. In: Science China Information Sciences, vol. 61, no. 3 (2018)
32. CPMI-MC: Central bank digital currencies. Technical Report (Mar 2018)
33. Amler, H., Eckey, L., Faust, S., Kaiser, M., Sandner, P., Schlosser, B.: DeFining the DeFi: challenges & pathway. Technical Report (2021). https://arxiv.org/abs/2101.05589
34. Golubova, A.: BIS backs central bank digital currencies: their time 'has come' (2021). https://www.kitco.com/news/2021-06-23/BIS-backs-central-bank-digital-currencies-Their-time-has-come.html
35. Neroni Rezende, I., Pocher, N.: Co-governing emerging socio-technical systems: investigating the implications of public-private partnerships in smart cities and central bank digital currencies [Unpublished]
36. Rennie, E., Steele, S.: Privacy and emergency payments in a pandemic: how to think about privacy and a central bank digital currency. Law Technol. Hum. **3**(1), 6–17 (2021)
37. Fanusie, Y.J.: Central bank digital currencies: the threat from money launderers and how to stop them. In: The Digital Social Contract: A Lawfare Paper Series, pp. 1–23, (Nov 2020). https://www.lawfareblog.com/
38. Auer, R., Böhme, R.: The technology of retail central bank digital currency. BIS Q. Rev. 85–100 (2020). (no. March)
39. Kochergin, D., Dostov, V.: Central Banks Digital Currency: Issuing and Integration Scenarios in the Monetary and Payment System. Lecture Notes in Business Information Processing **394**, 111–119 (2020)
40. Viñuela, C., Sapena, J., Wandosell, G.: The future of money and the central bank digital currency dilemma. Sustainability (Switzerland) **12**(22), 1–21 (2020)
41. Group of 30: Digital currencies and stablecoins: risks, opportunities, and challenges ahead (2020). https://group30.org/

42. Auer, R., Böhme, R.: Central bank digital currency: the quest for minimally invasive technology. Bank for International Settlements. Technical Report (Jun 2021). https://www.bis.org/publ/work948.pdf
43. Kriwoluzky, A., Kim, C.H.: Public or Private? The future of money. Technical Report (Dec 2019). https://www.europarl.europa.eu/RegData/etudes/IDAN/2019/642356/IPOL_IDA(2019)642356_EN.pdf
44. Veneris, A., Park, A., Long, F., Puri, P.: Central Bank Digital Loonie: Canadian Cash for a New Global Economy (2021). https://ssrn.com/abstract=3770024
45. Gang, X., Mount Union of Science and Technology: analysis of the central bank's digital currency DC/EP dual offline payment scenarios and solutions. Technical Report (2019). https://www.mpaypass.com.cn/news/201912/06094420.html
46. Yang, K., Blaauw, D., Sylvester, D.: Hardware designs for security in ultra-low-power IoT systems: an overview and survey. IEEE Micro **37**, 72–89 (2017)
47. Nabilou, H.: Testing the waters of the Rubicon: the European Central Bank and central bank digital currencies. J. Bank. Regul. **21**(4), 299–314 (2019)
48. Brunnermeier, M.K., Niepelt, D.: On the equivalence of private and public money. J. Monet. Econ. **106**, 27–41 (2019). https://doi.org/10.1016/j.jmoneco.2019.07.004
49. Jagati, S.: CBDCs with a twist: the public-private solutions needed for adoption (2020). https://cointelegraph.com/news/cbdcs-with-a-twist-the-public-private-solutions-needed-for-adoption
50. Adrian, T.: Evolving to work better together: public-private partnerships for digital payments (2020). https://www.imf.org/en/News/Articles/2020/07/22/sp072220-public-private-partnerships-for-digital-payments
51. Ojo, M.: Balancing public-private partnerships in a digital age: CBDCs, central banks and technology firms. In: Munich Personal RePEc Archive (2021). https://mpra.ub.uni-muenchen.de/107716/
52. Auer, R., Haene, P., Holden, H.: Multi-CBDC arrangements and the future of cross-border payments. Bank for International Settlements, Technical Report (Mar 2021)
53. Auer, R., Boar, C., Cornelli, G., Frost, J., Holden, H., Wehrli, A.: CBDCs beyond borders: results from a survey of central banks. Technical Report (Jun 2021)
54. Jung, H., Jeong, D.: Blockchain implementation method for interoperability between CBDCs. In: Future Internet (2021). https://doi.org/10.3390/fi13050133
55. Duffie, D.: Interoperable Payment Systems and the Role of Central Bank Digital Currencies. Finance and Insurance Reloaded, Institut Louis Bachelier Annual Report (2020)
56. Bank of Canada: Contingency planning for a central bank digital currency. Technical Report (2020). https://www.bankofcanada.ca
57. Khiaonarong, T., Humphrey, D.: Cash use across countries and the demand for central bank digital currency. Technical Report (Mar 2019)
58. Pocher, N., Veneris, A.: Privacy and transparency in cbdcs: a regulation-by-design aml/cft scheme. IEEE Trans. Netw. Serv. Manag. (2021)
59. Danezis, G., Meiklejohn, S.: Centrally banked cryptocurrencies. In: Network and Distributed System Security Symposium 2016, vol. 02 (2016)
60. ECB Crypto-Assets Task Force: Crypto-assets: implications for financial stability, monetary policy, and payments and market infrastructures. European Central Bank. Technical Report (May 2019)
61. European Central Bank: Report on a digital euro. European Central Bank. Technical Report (Oct 2020). https://www.ecb.europa.eu/
62. Boar, C., Holden, H., Wadsworth, A.: Impending arrival-a sequel to the survey on central bank digital currency. Technical Report 107 (Jan 2020)
63. Codruta, B., Wehrli, A.: Ready, steady, go?-Results of the third BIS survey on central bank digital currency. Technical Report 114 (Jan 2021). https://www.bis.org/publ/bppdf/bispap114.pdf
64. European Central Bank: Eurosystem report on the public consultation on a digital euro, no. (Apr 2021)

65. Sveriges Riksbank: E-krona pilot Phase 1. Technical Report (Apr 2021). https://www.riksbank.se/globalassets/media/rapporter/e-krona/2021/e-krona-pilot-phase-1.pdf
66. BIS Innovation Hub Hong Kong Centre, HKMA, Bank of Thailand, Digital Currency Institute PBoC, Central Bank UAE, Inthanon-LionRock to mBridge: Building a multi CBDC platform for international payments. Technical Report (Sept 2021). https://www.bis.org/publ/othp40.htm
67. PYMNTS: BIS to open four new innovation hubs over next two years (Jul 2020)
68. Zetzsche, D.A., Arner, D.W., Buckley, R.P.: Decentralized finance. J. Financ. Regul. **6**(2), 172–203 (2020)
69. Finck, M.: Blockchain regulation and Governance in Europe (2019)
70. Zetzsche, D.A., Buckley, R.P., Arner, D.W., Barberis, J.N.: Regulating a revolution: from regulatory sandboxes to smart regulation (2017)
71. Foley, S., Karlsen, J.R., Putnins, T.J.: Sex, Drugs, and Bitcoin: How much illegal activity is financed through cryptocurrencies? Rev. Finan. Stud. **32**(5), 1798–1853 (2019)
72. Pocher, N.: The open legal challenges of pursuing AML/CFT accountability within privacy-enhanced IoM ecosystems. In: CEUR Workshop Proceedings, vol. 2580 (2020)
73. FATF: Guidance for a risk-based approach: virtual assets and virtual asset service providers. FATF, Paris, Technical Report (June 2019). www.fatf-gafi.org/
74. Pocher, N.: Self-hosted wallets: the elephant in the crypto room? (2021). https://www.law.kuleuven.be/citip/blog/self-hosted-wallets/
75. Pocher, N.: Crypto-wallets and the new EU AML package: where are the battle lines drawn? (2021). https://www.law.kuleuven.be/citip/blog/crypto-wallets-and-the-new-eu-aml-package/
76. FATF: Virtual assets red flag indicators of money laundering and terrorist financing. Technical Report (Sept 2020). http://www.fatf-gafi.org/
77. Chaum, D.: Blind signatures for untraceable payments (1998)
78. Chaum, D., Grothoff, C., Moser, T.: How to issue a central bank digital currency. Swiss National Bank. Technical Report (2021). https://www.snb.ch/
79. Magnuson, W.: Blockchain democracy: technology, law and the rule of the crowd (2020). https://www.cambridge.org/core/books/blockchain-democracy/1E3D5E83BC932319E38BA622026C6239
80. FINTRAC: Canada's legislation: the proceeds of crime (Money Laundering) and terrorist financing act (2019)
81. 31 U.S.C. Title 31-Money and Finance, Subtitle IV-Money, Chapter 53-Monetary Transactions, Subchapter II-Records and Reports on Monetary Instruments Transactions, Sec. 5331-Reports relating to coins and currency received in nonfinancial trade or
82. Ecorys and Centre for European Policy Studies, Study on an EU initiative for a restriction on payments in cash. Technical Report (Dec 2017). https://ec.europa.eu/
83. European Commission: Anti-money laundering and countering the financing of terrorism legislative package (2021). https://ec.europa.eu/info/publications/210720-anti-money-laundering-countering-financing-terrorism_en
84. European Commission: Proposal for a Regulation of the European Parliament and of the Council on the prevention of the use of the financial system for the purposes of money laundering or terrorist financing (2021). https://ec.europa.eu/finance/docs/law/210720-proposal-aml-cft_en.pdf
85. Sands, P., Campbell, H., Keatinge, T., Weisman, B.: Limiting the use of cash for big purchases: assessing the case for uniform cash thresholds. Technical Report (Sept 2017)
86. Karasek-wojciechowicz, I.: Reconciliation of anti-money laundering instruments and European data protection requirements in permissionless blockchain spaces. J. Cybersecur. 1–28 (2021)
87. Salmensuu, C.: The general data protection regulation and blockchains (2018). https://papers.ssrn.com/sol3/papers.cfm?abstract_id=3143992
88. Berberich, M., Steiner, M.: Blockchain technology and the GDPR-how to reconcile privacy and distributed ledgers? I. Technical core features and use cases of the blockchain technology. Eur. Data Prot. Law Rev. **2**(3), 422–426 (2016)

89. Goodell, G., Aste, T.: Can cryptocurrencies preserve privacy and comply with regulations? Front. Blockchain **2**(4) (May 2019)
90. Garratt, R.J., van Oordt, M.R.: Privacy as a public good: a case for electronic cash. Bank of Canada Staff Working Paper, no. 2019-24 (2019)
91. Goodell, G., Al-Nakib, H.D., Tasca, P.: A digital currency architecture for privacy and owner-custodianship. Future Internet, no. 13 (2021). https://arxiv.org/abs/2101.05259
92. de Koker, L.: Anonymous clients, identified clients and the shades in between perspectives on the FATF AML/CFT standards and mobile banking. SSRN Electron. J. (no. d) (2015)
93. Sardá, T., Natale, S., Sotirakopoulos, N., Monaghan, M.: Understanding online anonymity. Media Cult. Soc. **41**(4), 557–564 (2019)
94. European Central Bank and Bank of Japan: Balancing confidentiality and auditability in a distributed ledger environment. Technical Report (Feb 2020). https://www.ecb.europa.eu/
95. Torra, V.: Data privacy: foundations, new developments and the big data challenge, vol. 28 (2017)
96. Casanovas, P., González-Conejero, J., De Koker, L.: Legal compliance by design (LCbD) and through design (LCtD): preliminary survey. In: CEUR Workshop Proceeding, vol, 2049, pp. 33–49 (2018)
97. Cavoukian, A.: Privacy by design. In: Office of the Information and Privacy Commissioner (2011)
98. Digital, D.: Technology trends: How do they translate in the Chinese market? Technical Report (Apr 2020)
99. Citi: Future of money. Technical Report (Apr 2021)
100. RT: Digital currencies may challenge SWIFT global payment network, says Russian central bank (2020). https://www.rt.com/business/510534-digital-currencies-swift-challenge/
101. Working Group on E-CNY People's Bank of China: progress of research & development of e-CNY in China. (Jul 2021). http://www.pbc.gov.cn/en/3688110/3688172/4157443/4293696/2021071614584691871.pdf
102. Kharpal, A.: With Xi's backing, China looks to become a world leader in blockchain as US policy is absent (2019). https://www.cnbc.com/2019/12/16/china-looks-to-become-blockchain-world-leader-with-xi-jinping-backing.html
103. U.S.-China Economic and Security Review Commission, An Assessment of the CCP's Economic Ambitions, Plans, and Metrics of Success (2021)
104. Ekberg, J., Ho, M.: A new dawn for digital currency: why China's eCNY will change the way money flows forever. Technical Report (Apr 2021)
105. Veneris, A., Park, A.: Special drawing rights in a new decentralized century. Technical Report (2019). https://arxiv.org/abs/1907.11057
106. Yin, M., Malkhi, D., Reiter, M.K., Gueta, G.G., Abraham, I.: HotStuff: BFT consensus in the lens of blockchain, pp. 1–23 (2018). http://arxiv.org/abs/1803.05069
107. Blackshear, S., Cheng, E., Dill, D.L., Gao, V., Maurer, B., Nowacki, T., Pott, A., Qadeer, S., Russi, D., Sezer, S., Zakian, T., Zhou, R.: Move: A language with programmable resources, pp. 1–26 (2020). https://diem-developers-components.netlify.app/papers/diem-move-a-language-with-programmable-resources/2020-05-26.pdf
108. Diem Association: Diem announces partnership with silvergate and strategic shift to the United States (May 2021). https://www.diem.com/en-us/updates/diem-silvergate-partnership/
109. Ledger Insights: Diem plans to replace USD stablecoin with gov digital dollar (May 2021)
110. Bank of Canada: Contingency planning for a central bank digital currency (Feb 2020). https://www.bankofcanada.ca/2020/02/contingency-planning-central-bank-digital-currency
111. Abraham, R., Bennett, E.S., Sen, N., Francis, N.B.S.S.: State of adhaar report 2016–17. ID Insight, Technical Report (May 2017)
112. Ricks, M., Crawford, J., Menand, L.: Fedaccounts: digital dollar. Vanderbilt Law, Research Paper 18-33, UC Hastings Research Paper No. 287 (2020). https://ssrn.com/abstract=3192162

Application and Policy

Ocean Protocol: Tools for the Web3 Data Economy

Trent McConaghy

Abstract This chapter describes a toolset to enable Web3 data economy. Ocean Protocol is an on-ramp for data services into crypto ecosystems, using datatokens. Each datatoken is a fungible ERC20 token to access a given data service. Ocean smart contracts and libraries make it easy to publish data services (deploy and mint datatokens) and consume data services (spend datatokens). Ocean contracts run on Ethereum mainnet to start, with other deployments to follow. Ethereum composability enables crypto wallets as data wallets, crypto exchanges as data marketplaces, data DAOs as data co-ops, and more. Ocean Market is an open-source community marketplace for data. It supports automatic determination of price using an "automated market maker" (AMM). Each datatoken has its own AMM pool. Anyone can add liquidity, aka stake (equivalent in AMMs). This is curation, as stake is a proxy to dataset quality. We envision thousands of data marketplaces, where Ocean Market is just one. In addition to Ocean Market being open-source (and therefore forkable), Ocean includes tools to help developers build their own marketplaces and other apps. Ocean's "Compute-to-Data" feature gives compute access on privately held data, which never leaves the data owner's premises. Ocean-based marketplaces enable monetization of private data while preserving privacy. These tools are part of a system designed for long-term growth of a permissionless Web3 Data Economy. The Ocean Data Farming program incentivizes a supply of data. The community-driven OceanDAO funds software development, outreach, and more.

T. McConaghy (✉)
Ocean Protocol, Berlin, Germany
e-mail: gtrent@gmail.com

© The Author(s), under exclusive license to Springer Nature Switzerland AG 2022
D. A. Tran et al. (eds.), *Handbook on Blockchain*, Springer Optimization
and Its Applications 194, https://doi.org/10.1007/978-3-031-07535-3_16

505

1 Overview and Introduction

Modern society runs on data [1]. Modern artificial intelligence (AI) extracts value from data. More data means more accurate AI models [2, 3], which in turn means more benefits to society and business. The greatest beneficiaries are companies that have *both* vast data and internal AI expertise, like Google and Facebook. In contrast, AI startups have excellent AI expertise but are starving for data; and typical enterprises are drowning in data but have less AI expertise. The power of both data and AI—and therefore society—is in the hands of few.

This chapter describes the Ocean Protocol, which aims to spread the benefits of AI by equalizing the opportunity to access and monetize data. We accomplish this by creating simple *tools* to publish data and consume data as decentralized *datatokens*. Datatokens interoperate with ERC20 wallets, exchanges, DAOs, and more. These data may be held on-premise to preserve privacy. Additionally, Ocean has tools for data marketplaces. We have implemented these tools as Solidity code running on Ethereum mainnet [4]; as Python and JavaScript/React libraries to ease higher-level integration; and as a community data marketplace web application. Over time, Ocean will also get deployed to other networks.

These tools are encapsulated in a broader system design for long-term growth of an open, permissionless data economy. The Data Farming program incentivizes a supply of data. OceanDAO funds software development, outreach, and more. OceanDAO will be funded by revenue from apps and services in the Ocean data ecosystem, Ocean network rewards, and Ocean Protocol Foundation. Everything described above has been deployed live as of November 30, 2020.

2 Ocean System

2.1 Goals

The top-level goal is to spread the benefits of AI by equalizing the opportunity to access and monetize data. We can make this more specific. The Ocean System has these aims:

- An overall system that is **sustainable and growing**, towards ubiquity.
- Basic design is **simple** to understand and communicate.
- In line with **Ocean Mission and Values** [5]. These include: unlock data, preserve privacy, spread of power, spread of wealth, stay within the law, censorship-resistant, and trustless. It should be permissionless, rent-free, and useful to the world [6]. It should be anti-fragile: get more resilient when "kicked", therefore also needing evolvability. It follows time scales of decades not months.

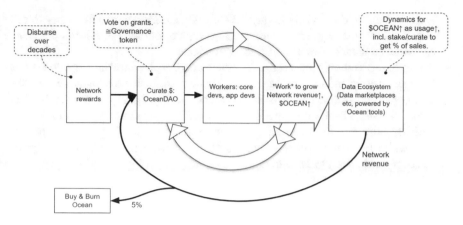

Fig. 1 Ocean system

2.2 The Design

Figure 1 shows the Ocean System design. At its heart is a **loop**, designed for "snowball" effect" growth of the ecosystem. The **Workers** (center) do ***work*** to help grow the **Data Ecosystem** (right). Marketplaces and other data ecosystem services generate revenue, using Ocean software tools. A tiny fraction of that revenue is **looped back** (arrow looping from right to left) as **Network revenue** to the Ocean community: to **Buy & Burn OCEAN** (bottom left) and back to workers curated by **OceanDAO** (center-left). To catalyze growth and ensure decent funding in early days, **Network rewards** (left) also feed to Workers via OceanDAO.

3 Data Ecosystem Powered by Ocean Tools

3.1 Introduction

The Data Ecosystem is a major subblock of the Ocean system. **Ocean tools** power the Data Ecosystem. This section describes Ocean tools.

3.2 USPs of Ocean Tools

Towards *building something that people want* [7], Ocean tools offer these unique selling propositions (USPs):

- **Earn by selling data, and staking on data**. Ocean Market makes it easy to sell your data, whether you are an individual, company, or city. Furthermore, anyone can stake on data to earn a % of transaction fees.
- **An on-ramp and off-ramp for data assets into crypto**, allowing: crypto wallets for data custody and data management, DEXes[1] for data exchanges, DAOs[2] for data co-ops, securitizing data assets, and more via DeFi composability. It's "data legos". The data itself do not need to be on-chain, just the access control.
- **Quickly launch a data marketplace, with many USPs**: Buy and sell private data while preserving privacy, non-custodial, censorship-resistant, auto price discovery, data audit trails, and more.
- **Decentralized data exchange platform**, enabling these characteristics: improve the visibility, transparency and flexibility in usage of data; share data while avoiding "data escapes"; needs little dev-ops support and maintenance; has high liveness; is non-custodial; and is censorship-resistant. E.g., Ocean as a traffic data management platform for smart cities. E.g., federated learning without having to trust the orchestration middleman [8].

3.3   Ocean Tools Foundation: Datatokens

Datatokens Introduction

Ocean is an on-ramp and off-ramp for data assets into crypto ecosystems, using *datatokens*. Datatokens are ERC20 tokens [9] to *access* data services. Each data service gets its own datatoken. Ocean smart contracts and libraries make it easy to publish data services (deploy and mint datatokens) and consume data services (spend datatokens).

Datatokens Goals

Here are the main goals.

- **Simple**. Complexity is *the* challenge in software design. Accordingly, we strive to make Ocean's overall design simple. This includes **simple developer experience**, **simple user experience**, and **simple code** (sufficiently simple to deploy to Ethereum mainnet, for additional benefits of being permissionless, stability, security, composability, and community).
- **Be a tool for existing workflows**; focus where Ocean adds value. "Tornado Cash and Uniswap … are successful in part because they are just tools that people can put into their existing workflows, and not ecosystems" [10]. We agree. Platforms imply asking you to *switch* from another platform. In contrast, tools humbly ask

---

[1] DEX = Decentralized Exchange.

[2] DAO = Decentralized Autonomous Organization.

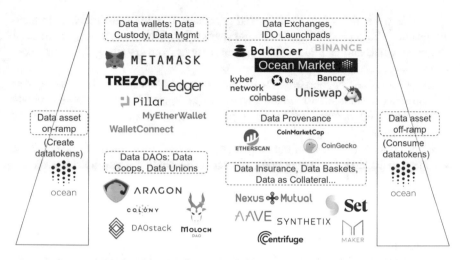

Fig. 2 Mental model. Datatokens are the interface to connect data assets with DeFi tools. Ocean is an on-ramp for data services into ERC20 datatoken data assets on Ethereum, and an off-ramp to consume data. [*Note* showing a logo does not imply a partnership]

you to try them out, and if you find them useful, to add them to your toolbox. Platforms are zero-sum; tools are positive-sum. This means Ocean should **leverage other infrastructure** wherever possible, and be maximally composable with other protocols and tools.

Ocean datatokens achieve these goals. They keep Ocean simple, composable, and make Ocean more a set of tools and less a platform.

Datatokens Mental Model

Figure 2 shows the mental model for Ocean datatokens. Ocean does the beginning (create datatokens) and the end (consume datatokens). In between are any ERC20-based applications, including Ocean-based marketplaces.

Datatokens are ERC20 Access Tokens

Traditional access tokens exist, such as OAuth 2.0 [11]. If you present the token, you can get access to the service. However, the "tokens" are simply a string of characters, and "transfer" is basically copying and pasting that string. This means they can easily be "double-spent": if one person gets access, they can share that access with innumerable others, even if that access was only meant for them. These tokens aren't the "tokens" we think of in blockchain.

How do we address the double-spend problem? This is where blockchain technology comes in. In short, there's a single shared global database that keeps track of who owns what, and can then easily prevent people from spending the same token twice. Footnote (^{3}) gives details.

This generalizes beyond Bitcoin tokens to other token assets. ERC20 [9] was developed as a standard for token ownership actions. It's been adopted widely in Ethereum and beyond. Its focus is *fungible* tokens, where tokens are fully interchangeable.

We can connect the idea of *access* with the ERC20 token standard. Specifically, consider an ERC20 token where **you can access the dataset if you hold 1.0 tokens**. To access the dataset, you send 1.0 datatokens to the data provider. You have custody of the data if you have at least 1.0 tokens. To give access to someone else, send them 1.0 datatokens. That's it! But now, the double-spend problem is solved for "access control", and by following a standard, there's a whole ecosystem around it to support that standard.

Datatokens are ERC20 tokens to access data services.[4] Each data service gets its own datatoken.

Datatoken Variants

At the smart contract level, datatokens don't differ. Variants emerge in the semantic interpretation by libraries run by the data provider, one level up. Here are some variants:

---

[3] Let's illustrate how the Bitcoin system prevents double-spending of Bitcoin tokens (bitcoin). In the Bitcoin system, you "control" an "address". An "address" is a place where bitcoin can be stored. You "control" the address if you're able to send bitcoin from that address to other addresses. You're able to do that if you hold the "private key" to that address. A private key is like a password—a string of text you keep hidden. In sending bitcoin, you're getting software to create a transaction (a message) that specifies how much bitcoin is being sent, and what address it's being sent to. You demonstrate it was you who created the transaction, by digitally signing the message with your private key associated with your address. The system records all such transactions on this single shared global database with thousands of copies shared worldwide.

[4] We could also use ERC721 "non-fungible tokens" (NFTs) [ERC721] for data access control, where you can access the dataset if you hold the token. Each data asset is its own "unique snowflake". However, datasets typically get shared among > 1 people. For this, we need fungibility, which is the realm of ERC20.

However, there is a more natural fit for NFTs: use an NFT to represent the base rights. The base rights are the ability to create access licenses (=mint ERC20 access tokens). The first base rights holder is the copyright holder, but they could transfer this to another entity as an exclusive deal (= transfer the NFT).

Ocean's V3 datatokens release uses just ERC20 tokens. There is a base rightsholder, it's just implicit: it's the entity that controls the ability to mint more ERC20 tokens (= "publisher"). This could conceptually be replaced with an explicit representation, which is NFT.

- Access could be **perpetual** (access as many times as you like), **time-bound** (e.g., access for just one day, or within specific date range), or **one-time** (after you access, the token is burned).
- Data access is always treated as a data service. This could be a service to access a **static** dataset (e.g., a single file), a **dynamic** dataset (stream), or for a compute service (e.g., "**bring compute to the data**"). For static data, we can tune variants based on the type of storage: Web2 cloud (e.g., AWS S3), Web3 non-permanent (e.g., Filecoin), Web3 permanent small-scale (e.g., Ethereum), Web3 permanent large-scale (e.g., Arweave), or go meta using IPFS but "pinned" (served up) by many places. For dynamic data, variants include Web2 streaming APIs (single-source), Web3 public data oracles (e.g., Chainlink), and Web3 private data oracles (e.g., DECO).
- **Read versus write etc. access**. This paper focuses on "read" access permissions. But there are variants: Unix-style (read, write, execute; for individual, group, all); database-style (CRUD: create, read, update, delete [12]), or blockchain database-style (CRAB: create, read, append, burn [13]).

The terms of access are specified in the metadata, which is on-chain (more on this later).

Datatokens and Rights

Having a token to physically access data implies the **right** to access the data. We can formalize this right: the datatoken would typically automatically have a **license** to use that data. Specifically: the data would be **copyrighted** (a form of intellectual property, or IP), as a manifestation of bits on a physical storage device. The license is a contract to use the IP in specific form. In most jurisdictions, copyright happens automatically on creation of the IP. Alternatively, encrypted data or data behind a firewall can be considered as a **trade secret**. Overall, enforcement in a given jurisdiction then falls under its existing IP framework.[5]

"Ownership" is a bundle of rights. "Owning" a token means you hold the private key to a token, which gives you the right to transfer that token to others. Andreas Antonopoulos has a saying: "Your keys, your Bitcoin. Not your keys, not your Bitcoin" [14]. That is, to truly own your Bitcoin, you need to have the keys to it. This crosses over to data: **"Your keys, your data. Not your keys, not your data"**. That is, to truly own your data, you need to have the keys to it.

---

[5] The World Intellectual Property Office (WIPO) is a United Nations (UN) agency dedicated to setting IP guidelines. Each jurisdiction can then choose how to implement the guidelines. For example, the United States Patent & Trademark Office (USPTO) implements WIPO guidelines on patents and trademarks in the US.

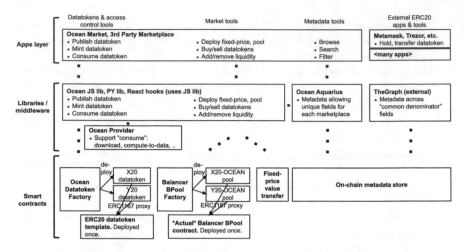

Fig. 3 Ocean tools architecture

Relation to Oracles

Oracles like Chainlink help get *data itself* on-chain [15]. Ocean is complementary, providing tools to on-ramp and off-ramp *data assets*. The data itself do not need to be on-chain, which allows wider opportunity for leveraging data in DeFi. Oracle datafeeds can be tokenized using Ocean.

Analogy to Shipping Containers

Here's an "intuition pump" to understand datatokens. Just as shipping containers are an *overlay protocol* that made physical supply chains more efficient, datatokens are an *overlay protocol* that makes data service supply chains more efficient. The essay [16] elaborates.

3.4 Ocean Tools Architecture

Overview

Figure 3 shows the Ocean tools architecture. The lowest level has the smart contracts, which are deployed on Ethereum mainnet.[6] Above that are libraries and middleware, which expose the contracts to higher-level languages and provide convenience utilities. The top layer is applications.

---

[6] Deployed to Ethereum mainnet to start, then to other networks in time. A later section elaborates.

Left to right are groupings of functionality: tools for datatokens, tools for markets (including pools), tools to consume data services and for metadata, and external ERC20 tools. The following subsections elaborate; more information yet is in the documentation [17] and the open-source code itself [18–22].

Datatokens and Access Control Tools

The publisher actor holds the dataset in Google Drive, Dropbox, AWS S3, on their phone, on their home server, etc. The dataset has a URL. The publisher can optionally use IPFS for a content-addressable URL. Or instead of a file, the publisher may run a compute-to-data service.

In the **publish** step, the publisher invokes **Ocean Datatoken Factory** to deploy a new datatoken to the chain. To save gas fees, it uses ERC1167 proxy approach on the **ERC20 datatoken template**. The publisher then mints datatokens.

The publisher runs **Ocean Provider**. In the **consume** step, Provider software needs to retrieve the data service URL given a datatoken address. One approach would be for the publisher to run a database; however, this adds another dependency. To avoid this, it stores the URL on-chain. So that others don't see that URL, it encrypts it.

To initiate the **consume** step, the data consumer sends 1.0 datatokens to the Provider wallet. Then they make a service request to the Provider. The Provider loads the encrypted URL, decrypts it, and provisions the requested service (send static data, or enable a compute-to-data job). The Appendix has details of this process.

Instead of running a Provider themselves, the publisher can have a third party like Ocean Market run it. While more convenient, it means that the third party has custody of the private encryption/decryption key (more centralized). Ocean will support more service types and URL custody options in the future.

Ocean JavaScript and Python libraries act as drivers for the lower-level contracts. Each library integrates with Ocean Provider to provision and consume data services, and Ocean Aquarius for metadata. **Ocean React hooks** use the JavaScript library, to help build webapps and React Native apps with Ocean.

Market Tools

Once someone has generated datatokens, they can be used in any ERC20 exchange, including AMMs. We elaborate on this later. In addition, Ocean provides **Ocean Market**. It's a vendor-neutral reference data marketplace for use by the Ocean community. It's decentralized (no single owner or controller), and non-custodial (only the data owner holds the keys for the datatokens).

Ocean Market supports fixed pricing and automatic price discovery. For fixed pricing, there's a simple contract for users to buy/sell datatokens for OCEAN, while avoiding custodianship during value transfer.

For automatic price discovery, Ocean Market uses Balancer pools [23]. Each pool is a datatoken—OCEAN pair. In the Ocean Market GUI, the user adds liquidity then invokes pool creation; the GUI's React code calls the Ocean JavaScript library, which calls **Balancer Factory** to deploy a **Balancer BPool** contract. (The Python library also does this.) Deploying a datatoken pool can be viewed as an "Initial Data Offering" (IDO).

Complementary to Ocean Market, Ocean has reference code to ease building **third-party data marketplaces**, such as for logistics (dexFreight data marketplace [24]) or mobility (Daimler [25]).

Metadata Tools

Metadata (name of dataset, dateCreated etc.) is used by marketplaces for data asset discovery. Each data asset can have a decentralized identifier (DID) [26] that resolves to a DID document (DDO) for associated metadata [27]. The DDO is essentially JSON [28] filling in metadata fields.

OEP8 [29] specifies the metadata schema, including fields that must be filled. It's based on the public DataSet schema from schema.org [30].

Ocean uses the Ethereum mainnet as an **on-chain metadata store**, i.e., to store both DID and DDO. This means that once the write fee is paid, there are no further expenses or dev-ops work needed to ensure metadata availability into the future, aiding in the discoverability of data assets. It also simplifies integration with the rest of the Ocean system, which is Ethereum-based. Storage cost on Ethereum mainnet is not negligible, but not prohibitive and the other benefits are currently worth the tradeoff compared to alternatives.

Due to the permissionless, decentralized nature of data on Ethereum mainnet, any last-mile tool can access metadata. **Ocean Aquarius** supports different metadata fields for each different Ocean-based marketplace. Third-party tool **TheGraph** sees metadata fields that are common across all marketplaces.

Third-Party ERC20 Apps and Tools

The ERC20 nature of datatokens eases composability with other Ethereum tools and apps, including **MetaMask** and **Trezor** as data wallets, DEXes as data exchanges, and more. The Applications section expands on this.

Actor Identities

Actors like data providers and consumers have Ethereum addresses, aka web3 accounts. These are managed by crypto wallets, as one would expect. For most use cases, this is all that's needed. There are cases where the Ocean community could layer on protocols like Verifiable Credentials [31] or tools like 3Box [32].

This subsection has described the Ocean tools architecture at a higher level [17, 33] and has further details.

3.5   Ocean Tools: Network Deployments

Toward a broadly open Data Economy, we aim for Ocean deployment across many chains as a thin layer for data assets and permissioning. It starts with a single deployment, and expands.

Initial Permissioned Deployment

"Decentralized" means no single point of failure. "Permissioned" means there are a set of gatekeepers that together control the entity. "Permissionless" means there are no gatekeepers; one needs no permission to have a hand in controlling an entity.

Ocean V1 and V2 have been decentralized and *permissioned*:

- Ocean smart contracts run on a permissioned Proof-of-Authority (POA) network [34].
- A small group of people can upgrade the smart contracts (in a multisig setting).

Permissionless via Ethereum Mainnet

Starting with Ocean V3.0 release, Ocean is decentralized and *permissionless.*

- **Ocean smart contracts are deployed to Ethereum mainnet**—a permissionless network.
- **The contracts will not have upgradeability built in**. Therefore, the only way to upgrade contracts is by *community consensus* to use a new set of smart contracts. (There's no longer a small handful of gatekeepers.)

This means that Ocean V3 also meets the "V5" target of being permissionless, specified in the Ocean roadmap [35, 36].

Despite using Ethereum mainnet, the "scale" aspect is now manageable [37]. This is possible because Ocean V3 contracts are simpler than the V1 and V2 contracts due to the datatokens architecture and more. This includes greatly reduced gas usage, and fewer wallet confirmations to purchase data. Finally, we have realized that data scientists are accustomed to some latency with Web2 payments; we don't need to be radically better than that.

Reduced Gas Costs via Sister Chain

The growth of DeFi has meant increased usage of Ethereum mainnet, and in turn a great increase in gas prices. To alleviate this cost, and for improved scalability, Ocean V3.x will have a "sister" deployment to another network. It may be Ocean V3.1 depending on other priorities. One possibility is xDai Chain [38], with an Arbitrary Message Bridge (AMB) [39] to connect OCEAN tokens *and* datatokens to Ethereum mainnet. Other strong possibilities include a Substrate + EVM deployment [40], or a rollup-based technology like OVM [41].

It's acceptable if the sister chain has a lower security, since users would have the choice between higher security (Ethereum mainnet) or lower gas costs (the sister, at potentially lower security). Then, data assets that start high value or need high security can continue to deploy directly to Ethereum mainnet. And, lower-value data assets can first get deployed to the sister network, then "graduate" to Ethereum mainnet if they increase sufficiently in value.[7]

Further Network Deployments

Over time, we envision every blockchain network to have an Ocean-powered data resource permissioning layer. We envision them being interconnected with datatokens and OCEAN flowing everywhere.

We envision Ocean as an add-on library in Parity Substrate [40] and [42] Cosmos SDK, making it only an "import" away for Polkadot and Cosmos blockchain families, respectively. Substrate 2.0 support for off-chain data integration is particularly promising.

We see potential integrations with layer 2 rollups such as Optimistic Virtual Machine (OVM) to improve privacy, throughput or cost [41].

We expect to see deployment to other EVM-based networks such as Binance Smart Chain, Matic Network, SKALE Network, NEAR Protocol, and Solana; each with their respective bridges. We envision deployment onto federated networks for consortia like Energy Web Chain. Finally, we eventually see deployment to non-EVM blockchains, especially ones that can hold a lot of data like arweave, or ones with built-in oracles like aeternity.

In all these deployments, OCEAN tokens will remain on Ethereum mainnet, and bridged to other networks. We will be encouraging the broader community to do each of these deployments, with funding from OceanDAO or the respective chain's grant mechanisms. In each of these deployments, with the interest of Ocean sustainability, there will be small transaction fees that go to the Ocean community via a bridge.

---

[7] Thanks to Simon de la Rouviere for this framing, which he in turn derived from Austin Griffith.

3.6 Ocean Tools: Data Marketplaces

Introduction

This section drills into details about Ocean-based data marketplaces.

Since each data service has its own ERC20 token, any ERC20 exchange can serve as a data marketplace. They can be AMM DEXes, order-book DEXes, order-book CEXes, and more.

But we can still make it easier for users. Specifically, marketplaces *tuned* for data can help users in the whole data flow, including publish data, set price, curate data, discover data, buy data, and consume data.

Marketplaces Architecture

Figure 4 shows the conceptual architecture. There are many data marketplace frontends; there is a common backend (for a given network).

The frontends include a community market (**Ocean Market**) [top middle] and independent third-party markets like those of dexFreight or Daimler [top left]. Each frontend runs client-side in the browser, using **Ocean React hooks**, which use the **Ocean JS library**, which interfaces with the backend.

The **Decentralized Backend** [bottom left] is Solidity code running on Ethereum mainnet and includes the datatoken and pool contracts, and the on-chain metadata store.

When a **Buyer** purchases a dataset on a frontend [far top, far left], most of the revenue goes to the **Data seller** [bottom right]. Some fees go to **LPs, the marketplace runner, and the broader community** [top right].

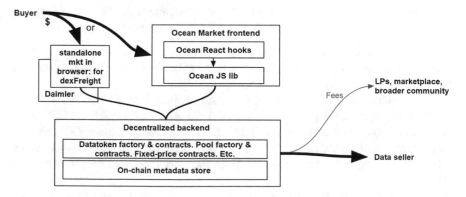

Fig. 4 Ocean Marketplaces share a decentralized backend

How Do You Price Data?

"How do I price the data?" is an oft-asked question, and rightly so. It's a real problem. Pricing data is hard [43, 44].

Fixed Pricing

Some data sellers will want to sell data for a fixed price (fixed # OCEAN). The challenge is to do it without intermediaries or complicated escrow contracts. Our solution is a simple smart contract that includes *transfer()* OCEAN one way, and *transfer()* datatoken the other way.

Automated Pricing via AMMs

If price can be discovered automatically, it would be of immense value. It's worth spending real effort on how to price data automatically. Order books, auctions, and AMMs are some possibilities. Let's review each.

- For a sale to occur, **order books** require bids and asks to match up in real time, aka a "double coincidence of wants". This is not feasible for newly created long tail assets like datatokens. (However, they are useful once a datatoken gets enough liquidity and traders.)
- **Auctions** occur over a time interval, such as an hour or a day. Auctions are useful for an initial pricing, but after that we still want automated price discovery for the rest of the lifetime of the assets. Auctions do not provide this.
- **AMMs** provide automated price discovery without the disadvantages of order books or auctions listed above. AMMs work for an initial asset offering and throughout the asset's lifetime. AMMs don't require a double coincidence of wants; they can be thought of as robots that are always ready to buy or sell.

Ocean's datatoken framing enables people to build data exchanges using any of the above approaches —order books, auctions, or AMMs. That said, it's worth spending effort to make the most promising approach easy to use.

From the analysis above, the most promising approach is *AMMs*. So, we focus on them. A given AMM pool would have (1) the specific datatoken and (2) some other more-established token such as ETH, DAI, or OCEAN. Having OCEAN as a convenient default (without forcing its use) helps drive demand for OCEAN, which in turns helps long-term sustainability [45]. In short, we focus on datatoken-OCEAN pools.

AMMs auto-discover price of data. In an Ocean AMM-based data market, a datatoken's price automatically goes up more datatokens are bought (as OCEAN is swapped for more datatokens). It goes down as datatokens are sold (as datatokens are swapped for OCEAN).

AMM options. The burgeoning Ethereum DeFi space offers many high-quality options. Bancor was first. Uniswap has low gas costs. Balancer allows non-equal weights among tokens in the pool (e.g., 90/10 versus 50/50). Many more have emerged recently, along with aggregators. One could build an Ocean-based data market with any of these.

The First Data Market AMM: Balancer

For Ocean Market, we chose to focus on one AMM tool to start: **Balancer**. There were a few reasons for this. First, Balancer allows to add liquidity through a single token, unlike most AMMs. Here's how this helps: when a new datatoken pool is published, at first only the publisher has datatokens. For others to add liquidity, they will need to add just OCEAN tokens. Balancer allows this.

Second, Balancer lightens the liquidity burden for publishers. Balancer uniquely allows non-equal weights among tokens in a pool. This allows a pool with 90% weight to the datatoken and 10% to OCEAN. Compared to a 50/50 pool, a data publisher only needs to provide 1/5 of the OCEAN liquidity for the same initial datatoken price. For the future, Balancer gives the possibility to *dynamically* change weights to bootstrap liquidity [46] and mitigate impermanent loss [47].

The default deployment of Balancer has high gas costs for deploying pools. We overcame this issue as follows. In Ocean, Balancer factory and pool contracts (BFactory, BPool) are tweaked to use the ERC1167 proxy pattern to reduce gas costs [48]. They can be viewed as an early version of Balancer V2.[8,9]

Benefits. Here are some benefits for Balancer-Ocean AMM data markets. The first is automatic price discovery, as discussed. Next, if the pool is the first market that this datatoken has been deployed to, then we can consider this the data asset's **"initial data offering"** (IDO) to kickstart liquidity. Finally, the AMM is decentralized and non-custodial. Later in this paper, we elaborate further on IDOs and marketplace benefits.

---

[8] The Ocean and Balancer teams have collaborated for years. Part of the collaboration between the Ocean and Balancer teams is discussion around Balancer V2. Ocean plans to use Balancer Lab's deployment by Balancer V2, if not sooner.

[9] The Ocean-tweaked Balancer pools are optimized for low gas cost, but because they do not come from Balancer's official factory, they aren't currently eligible for Balancer Liquidity Mining [Martinelli 2020]. If users choose, they can deploy datatoken pools using Balancer's official factory, which is more expensive but does receive BAL rewards. Balancer V2 will optimize gas costs, at which point we expect Ocean libraries to use Balancer official factory (and therefore be eligible for BAL rewards).

OCEAN Staking and Data Curation

AMMs require liquidity to be provisioned. Liquidity is the number of datatokens and OCEAN in the pool. Anyone can add liquidity. The higher the liquidity, the lower the slippage (change in price) when there is a purchase.

In providing liquidity, the price signal is authentic [49] or, equivalently, that the "market" is real.

A liquidity provider (LP) is staking because staking and liquidity provisioning are equivalent in AMMs [50].

This means:

OCEAN staking is the act of adding liquidity to a datatoken-OCEAN pool.

Furthermore, an LP is curating since the amount of liquidity is a proxy to quality of the data asset.

This means:

Ocean data curation is the act of adding/removing liquidity in a datatoken-OCEAN pool.

Liquidity providers (LPs) earn a cut of the transaction fee proportional to their stake. Since curators are LPs, it means that curators are incentivized to **curate towards the most valuable datasets** because it will earn them the most fees. It also means that to earn more fees, curators are incentivized to **refer** others to the data pools that they've staked on. In addition, this means that **curation in Ocean has authentic signals** of quality because it requires actual skin-in-the-game in the form of liquidity (stake).

Ocean Marketplace Actions

Functionality covered. A marketplace's core functionality is about connecting buyers to sellers for given assets: to make the assets discoverable, and buying/selling an asset of interest. For smoother user flow, Ocean Market supports adjacent functionality: publishing the asset in the first place, and consuming it. Each subsection will cover these, in the order that it would happen, with a focus on Ocean Market.

Action: publish dataset. When the user (publisher) clicks on "Publish", they end up here. They start to fill out metadata, including Title and Description. The publisher then provides the URL of where the data asset can be found. This URL gets stored encrypted and on-chain. When a buyer later consumes a datatoken, that URL will be decrypted. The publisher then fills out price information. It may be fixed price or dynamic (automatic). If automatic, they add liquidity as desired to be in line with their target price. Finally, they hit "publish" and Ocean Market will invoke blockchain transactions to deploy a datatoken contract, publish metadata on-chain, and (if automatic) do all AMM-related transactions.

Action: add/remove liquidity. If the datatoken has an AMM pool, *any* user can add/remove datatokens or OCEAN as liquidity. This is OCEAN staking, and curation, as discussed earlier.

Action: discovery. Ocean Market will have thousands of data assets. To help discovery, Ocean includes support for browsing, searching, and filtering data assets.

Action: buy/sell dataset. Here, a buyer comes to Ocean Market and connects their wallet. Their wallet has some OCEAN. The buyer clicks the "buy" button; then Metamask pops up and asks for the buyer to confirm a transaction to swap OCEAN tokens for 1.0 datatokens. The buyer confirms, and the swap happens on-chain. Now, the buyer now has 1.0 more datatokens in their wallet.

Action: Consume Dataset. Here, a datatoken owner comes to Ocean Market and connects their wallet. They go to the appropriate sub-page with the asset that they own. They consume the dataset by clicking the "use" button. They follow the prompts to end up with a downloaded dataset, or to get results of bringing compute to data.

Developer Tools and Third-Party Marketplaces

Ocean Market is just one data marketplace. We envision *many* data marketplaces. We can catalyze this, by making it easy for developers to create their own marketplaces. Using Ocean JavaScript or Python libraries, each of the following is 1–3 lines of JS or PY code:

- Create a data asset (provision data service, deploy datatoken contract, add metadata, mint datatokens).
- Create an AMM market (or fixed-price market).
- Add or remove liquidity.
- Swap OCEAN for datatokens, and vice versa (buy datatokens and sell datatokens).
- Submit a datatoken and consume a data asset.

The libraries interface to Ocean smart contracts. Ocean Provider is a support tool to provision data assets, and Ocean Aquarius and to help store data on-chain and to query metadata (with the help of a local cache).

With these tools in place, there are two main ways that a developer can build an Ocean-based data marketplace: (1) Fork Ocean Market, which uses Ocean tools, and (2) Build up their own marketplace using Ocean tools more directly (React hooks, Javascript library, etc.).

Group-Restricted Access in Marketplaces

Certain use cases need to restrict who can consume data. For example, only registered medical personnel can read sensitive patient data. Or, only bank employees can review a consumer's KYC application form.

Places to restrict access include:

1. Restrict inside the ERC20 contract itself.
2. Restrict access to the marketplace.
3. Restrict the ability to buy in the marketplace.
4. Restrict at the point of consumption.

(1) Involves modifying the ERC20 contract's transfer() and approve() or transferFrom(). Other projects have taken this route. However, this alters the spec of ERC20 contracts, which we are reluctant to do.

(2) Typically means the marketplace firewalling itself, and only allowing login with verified accounts. This is fairly common practice in the enterprise, and can work here.

(3) Means the marketplace maintaining a whitelist of Ethereum addresses that can purchase a given data asset.

(4) Means that the provider runs code that checks the person's credentials. Consumption is only allowed if the token is transferred and the credentials check out. This is akin to entering an R-rated movie: you need the ticket (token) and to show your ID (verifiable credentials).

We recommend (2), (3), and (4), depending on the scenario. (2) and (3) cannot fully prevent bad actors from acquiring a restricted datatoken outside of marketplaces, but they are easier to implement (and Ocean tools have affordances for this). (4) is the most secure but takes more effort by the provider.

There are a few blockchain-compatible ways to implement access at a group level. One is to use Verifiable Credentials (VCs), wherein an issuing authority digitally signs an attestation that a DID has a credential [31]. Another approach is to do a whitelist, via a Token-Curated Registry (TCR), or a custom smart contract implementation. A final way is to use a DAO with particular membership rules.

The Applications section details some use cases for group-restricted access.

Benefits of Ocean Data Marketplaces

Ocean marketplace tools make it easy to build and launch data marketplaces.
Ocean-based data marketplaces have these characteristics:

- *Interoperability*—data assets being bought and sold are ERC20 tokens on the Ethereum mainnet, which play well with the broader Ethereum ecosystem.
- *Don't need login*—users just connect their Web3 wallet (Metamask, etc.). Therefore to buy or sell datatokens, they're in and out in 2 min. This big UX improvement feels similar to DEXes, versus traditional CEXes.[10]
- *Non-custodial and decentralized*—no centralized middlemen controls the datatokens. No single point of failure.

---

[10] CEX = Centralized Exchange.

- *Censorship-resistant, with flexibility*—by default, everyone can transact with the marketplace on the same terms, regardless of their personal identifying characteristics. Or, to meet data regulations or KYC, there is the option of whitelists.
- *Buy and sell private data while preserving privacy*—using Ocean Compute-to-Data. Data won't leave the premises. This also gives sellers the option to make data *exclusive (in an economic sense)* which can give data a pricing premium.
- *Provenance*—sellers and buyers benefit from the auditability of purchase transactions (using, e.g., Etherscan).
- *Monetization*—marketplace has the option to take a commission on sales. This helps to ensure that data marketplace businesses can be built that can sustain themselves and grow over time.

AMM pools enable these additional characteristics:

- *Automated price discovery*—the pool holds OCEAN and datatoken as liquidity. Datatoken price goes up as more datatokens are sold.
- *Curation (= Staking = Provisioning liquidity)*—authentic signals for quality of a dataset (= amount of OCEAN staked).
- *Transaction fees for LPs/curators/stakers.*
- *Referrals*—LPs are incentivized to refer to pools that they get transaction fees from.

The essay [51] elaborates on more aspects of Ocean-powered marketplaces, especially Ocean Market. Also, [52, 53] elaborate on staking and selling data, respectively.

3.7 Ocean Tools: Compute-To-Data

Ocean Compute-to-Data provides a means to share or monetize one's data while preserving privacy. This section expands on this.

Motivation

Private data are data that people or organizations keep to themselves, or at least *want to* keep to themselves. It can mean any personal, personally identifiable, medical, lifestyle, financial, sensitive or regulated information.

Privacy tools are about asymmetric information sharing: get info to the people you want, for their benefit, while ensuring that others don't see the info.

Benefits of Private Data. Private data can help research, leading to life-altering innovations in science and technology. For example, more data improve the predictive accuracy of modern AI models. Private data are often considered the most valuable data because it's so hard to get, and using it can lead to potentially big payoffs. It's often considered as a competitive, or even decisive, advantage in their market by companies.

Risks of Private Data. Sharing or selling private data comes with risk. What if you don't get hired because of your private medical history? What if you are persecuted for private lifestyle choices? Large organizations that have massive datasets know their data is valuable—and potentially monetizable—but do not pursue the opportunity for risk of data escaping and the related liability.

Resolving the Tradeoff. There appears to be a tradeoff between benefits of using private data, and risks of exposing it. What if there was a way to get the benefits, while minimizing the risks? This is the idea behind Compute-to-Data: let the data stay on-premise, yet allow 3rd parties to run specific compute jobs on it to get useful analytics results like averaging or building an AI model. The analytics results help in science, technology, or business contexts; yet the compute is sufficiently "aggregating" or "anonymizing" that the privacy risk is minimized.

Conceptual Working

Figure 5 illustrates Compute-to-Data conceptually. Alice the data scientist goes to a **data marketplace** and purchases access to private data from seller Bob. She runs her AI **modeling algorithm** (which Bob has approved) on Bob's **private data** to **privately train a model**, which Bob also stores privately. She then runs the trained **private model** on new input data to get **model predictions**. Those predictions are the only data she sees. Everyone is satisfied: Alice gets predictions she wants, and Bob keeps his data private.

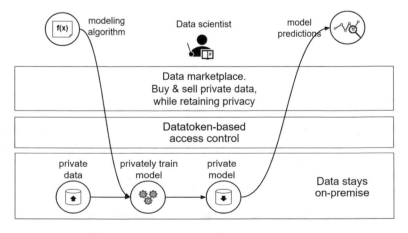

Fig. 5 Ocean compute-to-data

At the heart, there are **datatokens** which grant access to run compute next to the data. These datatokens can be used within a marketplace context (like described) or other contexts.

Compute-To-Data Flow Variants

This section describes variants of how Compute-to-Data may be used. For further detail yet, we refer the reader to [8].

In these variants, there is still model training next to the data:

- Alice is able to download the trained model, i.e., it can leave Bob's premises. This will happen if Bob believes the model has low risk of leaking personally identifiable information (PII), such as being a linear model or a small neural network.
- Alice learns a model across *many* data silos. This is "Federated Learning."

In the following, the compute that is run next to the data is *not* for training a model, but something else.

- A simpler "aggregating" function is run next to the data, such as an average, median, or simple 1-d density estimation. This means that Compute-to-Data is useful for simpler business intelligence (BI) use cases, in addition to more complex artificial intelligence (AI) use cases.
- An aggregating function is computed across *many* data silos. This is "Federated Analytics".
- A synthetic-data generation algorithm is run next to Bob's data. Alice downloads the synthetic data, which she then visualizes or trains models.
- A *hash* is computed for each {input variable, input value} combination of Bob's data, where the compute is done next to the data. Hashing naturally anonymizes the data. Alice downloads this hashed data and trains the model client-side. This is called "Decoupled Hashing".
- Random noise is added to Bob's dataset, sufficiently so for the data to be considered anonymized. Alice downloads this partly-randomized data and trains the model client-side. This is a variant of "Differential Privacy".

Share, or Monetize

Compute-to-Data is meant to be useful for data sharing in science or technology contexts. It's also meant to be useful for monetizing private data, while preserving privacy. This might look like a paradox at first glance but it's not! The private data isn't directly sold; rather, specific access to it is sold, access "for compute eyes only" rather than human eyes. So Compute-to-Data in data marketplaces is an opportunity for companies to monetize their data assets.

Compute-To-Data Architecture

New actors. Ocean Protocol has these actors: **Data Providers**, who want to sell their data; **Data Consumers**, who want to buy data; and **Marketplaces**, to facilitate data exchange. Compute-to-Data adds a new actor, the **Compute Provider**. The Compute Provider sells compute on data, instead of data itself.

New Components. Ocean technology has several components. Operator Service and Operator Engine were introduced for Compute-to-Data.

- **Operator Service**—a microservice in charge of managing the workflow and executing requests. It directly communicates and takes orders from Provider (the data provider's proxy server) and performs computation on data, provided by Provider.
- **Operator Engine**—a backend service in charge of orchestrating the compute infrastructure using Kubernetes as a backend. Typically, the Operator Engine retrieves the workflows created by the Operator Service in Kubernetes. It also manages the infrastructure necessary to complete the execution of the compute workflows.

New Asset Type. Before, datasets were the only asset type in metadata (DDO). Compute-to-Data introduces a new asset type—algorithm, which is a script that can be executed on datasets.

For further detail, [54] provides a worked example and further references on Compute-to-Data.

Marketplaces and Compute-To-Data

Marketplaces can allow their users to publish datasets with Compute-to-Data enabled. Some marketplaces may even require it.

Marketplaces choose what exact compute resources they want to make available to their end users within a K8s cluster, even having them choose from a selection of different images and resources.

Likewise, marketplaces can choose and restrict the kind of algorithm they want to allow their users to run on the datasets in a marketplace.

Trusting Algorithms

Only trust in a narrow facet is required: does the algorithm have negligible leakage of PII? For example, a simple averaging function aggregates data sufficiently to avoid leaking PII. AI algorithms also aggregate information.

The data owner typically chooses which algorithms to trust. It's their judgment call. They might inspect the code and perhaps run it in a sandbox to see what other dependencies it causes, communications it invokes, and resources it uses. Therefore, it's the same entity that risks private data getting exposed and chooses what algorithm to trust. It is their choice to make, based on their preference of risk vs. reward.

This also points to an opportunity for marketplaces of vetted algorithms: Ocean marketplaces themselves could be used, where liquidity provided is a proxy for quality and trust of the algorithm. Like with all pools, anyone can provide liquidity. This may be quite powerful, as it creates a "data science" side of the market.

Benefits of Compute-To-Data

Compute-to-Data has privacy benefits and other benefits as well:

- **Privacy**. Avoid data escapes, never leak personal or sensitive information.
- **Control**. Data owners retain control of their data, since the data never leaves the premises.
- **Huge datasets**. Data owners can share or sell data without having to move the data, which is ideal for very large datasets that are slow or expensive to move.
- **Compliance**. Having only one copy of the data and not moving it makes it easier to comply with data protection regulations like GDPR [55].
- **Auditability**. Compute-to-Data gives proof that algorithms were properly executed, so that AI practitioners can be confident in the results.

Relation to Other Privacy-Preserving Technologies

Compute-to-Data is complementary to other technologies such as encryption/decryption, Multi-Party Compute, Trusted Execution Environments, and Homomorphic Encryption. This compatibility and Ocean's "tools" framing helps make it easy to adopt Ref. [8] has details.

4   Ocean Applications

4.1   Decentralized Orchestration

Here's an example compute pipeline: input raw training data → clean the data → store cleaned data → build model → store model → input raw test data → run predictions → store predicted result y_test. Leveraging Ocean, a developer can write Solidity code [56] to define a pipeline and execute it, i.e., do "decentralized orchestration":

1. Write a smart contract that uses various data services. There's a datatoken for access control of each data service.
2. Deploy the smart contract to Ethereum mainnet.

3. For each different datatoken, do a tx that approves the required amount of datatokens to the SEA smart contract.
4. Finally, do a transaction that makes the actual call to the smart contract.

With a Set Protocol, ERC998 or similar to have a "basket" datatoken that holds all the necessary sub-datatokens in the compute pipeline, step 3 becomes simpler yet.

These Solidity scripts can be seen as "Service Execution Agreements" (SEAs) [57], a riff on "Service Level Agreements" for centralized orchestration.

A major sub-application is decentralized federated learning (decentralized FL). In traditional FL, a centralized entity (e.g., Google) must perform the orchestration of compute jobs across silos. So, PII can leak to this entity. OpenMined [58] could decentralize orchestration, using Ocean to manage computation at each silo in a more secure fashion [8].

4.2 Data Wallets: Data Custody and Data Management

Data custody is the act of holding access to the data, which in Ocean is simply holding datatokens in wallets. Data management also includes sharing access to data, which in Ocean is simply transferring datatokens to others.

With datatokens as ERC20 tokens, we can leverage existing ERC20 wallets. This includes browser wallets (e.g., Metamask), mobile wallets (e.g., Argent, Pillar), hardware wallets (e.g., Trezor, Ledger), multi-sig wallets (e.g., Gnosis Safe), institution-grade wallets (e.g., Riddle & Code), custodial wallets (e.g., Coinbase Custody), and more.

ERC20 wallets may get tuned specifically for datatokens as well, e.g., to visualize datasets, or long-tail token management (e.g., holding 10,000 different datatoken assets).

Existing software could be extended to include data wallets. For example, Brave browser has a built-in crypto wallet that could hold datatokens. There could be browser forks focused on datatokens, with direct connection to user browsing data. Integrated Development Environments (IDEs) for AI like Azure ML Studio [59] could have built-in wallets to hold and transfer datatokens for training data, models as data, and more. Non-graphical AI tools could integrate; such as scikit-learn or TensorFlow Python libraries using a Web3 wallet (mediated with Ocean's Python library).

As token custody continues to improve, data custody inherits these improvements.

4.3 Data Auditability

Data auditability and provenance is another goal in data management. Thanks to datatokens, blockchain explorers like Etherscan [60] now become data audit trail explorers.

Just as CoinGecko [61] or CoinMarketCap [62] provide services to discover new tokens and track key data like price or exchanges, we anticipate similar services to emerge for datatokens. CoinGecko and CoinMarketCap may even do this themselves, just as they've done for DeFi tokens.

4.4 Data DAOs: Data Co-Ops and More

Decentralized Autonomous Organizations (DAOs) [63] help people coordinate to manage resources. They can be seen as multi-sig wallets, but with significantly more people, and with more flexibility. DAO technology is maturing well, as we reference in the OceanDAO section. A "data DAO" would own or manage datatokens on behalf of its members. The DAO could have governance processes on what datatokens to create, acquire, hold, sell/license, and so on.

Here are some applications of data DAOs:

Co-ops and Unions (Collective Bargaining). Starting in the early 1900s, thousands of farmers in rural Canada grouped into the SWP [64] for clout in negotiating grain prices, marketing grain, and distributing it. Labor unions have done the same for factory workers, teachers, and many other professions. In [65], the authors suggest that data creators are currently getting a raw deal, and the solution is to make a labor union for data. A data DAO could be set up for collective bargaining, as a "data coop" or "data union". For example, there could be a data coop with thousands of members for location data, using FOAM proof-of-location service [66].

Manage a single data asset. There could be a DAO attached to a single data asset. One way is: create a Telegram channel dedicated to that dataset. You can only enter the Telegram channel if you have 1.0 of the corresponding datatokens (inspired by Karma DAO [67]). This can also be for Discord, Slack, or otherwise.

Datatoken pool management. There could be a data DAO to manage a datatoken pool's weights, transaction fees, and more, leveraging Balancer Configurable Rights Pools [68] (inspired by PieDAO which does this for a pool of DeFi assets [69]).

Index Funds for Data Investments. Using, e.g., Melon [70], an investment product can be constructed to allow people to buy a basket of data assets with the current plethora of mutual and index funds as a guide.

4.5   Permissioned Group-Restricted Access in Data Exchanges

In this operational model, "membership rules" apply for a group. These rules are governed by a verifiable credential, a TCR, a custom whitelist, a DAO, or otherwise. These membership rules enable the following applications for data sharing.

Contests, Hackathons, Impromptu Collaborations. A group of hackers or data scientists self-organize to try to solve an AI problem, such as a Kaggle competition or a hackathon. They want to be able to easily access each other" data and compute services as they progress, especially if they are working remotely from each other.

Regulatory Sandboxes. A government wants to give a means for organizations to run in a "monitored" regulatory environment that the government can observe. The organizations are vetted by the government and may have access to specially designated government compute services.

Enterprise data access. An enterprise might make some of its data available to only its employees, but want to be able to use Ocean services available in the broader network.

Sharing autonomous driving data. Individuals in each membership company of MOBI [71] need to access automotive data from any one of the MOBI member companies. It could be time-consuming and error-prone to specify permission for each member company individually. Furthermore, those permissions will fall out of date if MOBI members are added or removed; and updating the permissions one organization at a time could also be time-consuming or error-prone. This involves two levels of permissions: access of member companies into MOBI, and access of individuals in each member company (enterprise).

Sharing medical data. Researchers on European soil that wish to directly access German medical data need to demonstrate that they have been accredited by appropriate authorities. This will usually be through their hospital or university. There could be thousands of researchers accessing the data. As with automotive data, it will be time-consuming and error prone to specify and update permissions for each of these thousands of researchers. This may be two levels of permissions (hospital/university into EU authority; individual into hospital/university), or it may be among hospitals and universities in a more networked fashion.

Sharing financial data (while preserving privacy). Small and medium-sized credit unions in the US have a challenge: they don't have large enough datasets to justify using AI. Since the credit unions don't compete with each other, they would find great value to build AI models across their collective datasets.

4.6 Unlocking Latent Data of Individuals and Enterprises

Specialized apps could get built to tokenize and earn from latent data assets of individuals or organizations.

Individuals. For example, an app that you install on your phone, you give permissions to access data (Compute-to-Data for privacy), then it auto-creates an AMM market for your data. After that you earn money from the data on your phone. In a mashup with personal tokens [72], these would be your sovereign "personal datatokens" and you could launch your data as an "Personal Data Offering" (PDO).

One could even launch multiple PDOs. There would be one datatoken for each data type generated—smartphone data, smartwatch data, browser data, shopping data, and so on. Then, bundle the datatokens into a collection of personal datatokens (composable datatokens). One could even have a personal data marketplace.[11]

One could do the same with a browser plug-in, to sell the latent data on your browser (cookies, bookmarks, browsing history). New service firms could emerge to help enterprises tokenize and earn from their massive internal data troves.

These data assets might be sold one person or entity at a time. Or, they could be brought into a DAO to pool resources for more marketing and distribution muscle.

Enterprises. Large enterprises have massive datasets. They know their data are valuable: they spend millions annually to help protect it and insure it due to hacks. But what if rather than data being a *liability*, data was *assets* on enterprises' books? Ocean offers this possibility, by making it easy to turn the internal data into fungible assets (via datatokens), automatically discover the prices (via AMMs), while preserving privacy and control (via Compute-to-Data). Data not only become a new line of revenue, they become a *financial asset* that can be borrowed against to fund growth, and more.

4.7 Data Marketplaces

Earlier, we described Ocean Market, an out-of-the-box data marketplace that Ocean offers. Here are some ways that forks of Ocean Market could differentiate.

- **Focus on a given vertical**. For example, the Ocean-based dexFreight data marketplace focuses on the logistics vertical. Other verticals include health, mobility, and DeFi.
- **Focus on private data**. A marketplace focusing on Ocean compute-to-data, with features for curation of compute algorithms.
- **Different fee structures**. When a purchase happens in Ocean Market, the default 0.1% fee goes fully to LPs. Variants include: marketplace operator gets a cut, referrers get a cut, charge higher %, charge a flat fee.

---

[11] Thanks to Simon Mezgec for this suggestion.

- **Novel pricing mechanism**. Many price discovery mechanisms are possible [43, 44], including royalties (a % of sales), English / Dutch / Channel auctions [73], or income share agreements like Bowie or Dinwiddie bonds [74]. This may include a novel initial distribution mechanism, as the Initial Data Offerings section elaborates.
- **Different payment means**. Ocean Market takes OCEAN. Variants could take in fiat, DAI, ETH, etc. (Ocean Market may support these directly over time as well. PRs are welcome!)
- **Decentralized dispute resolution** using Aragon Court [45] or Kleros [75].
- Different Balancer weighting schemes. For example, make the 90/10 weights shift to 50/50 over time (Liquidity Bootstrapping). Or, surge pricing pools, which have higher fees when there is more demand for liquidity. These could be implemented with Configurable Rights Pools. PRs for Ocean Market are welcome!
- Different DEXes. Ocean Market currently uses a tweak of Balancer AMMs for lower gas costs. 3rd party marketplaces may use the original Balancer deployment, Uniswap, Bancor, Kyber, or other.

Beyond forks of Ocean Market, there is a larger variety of possible marketplaces. Here are some Web2 and Web3 variants.

- **AMM DEXes**. This could be a Uniswap or Balancer-like webapp to swap datatokens for DAI, ETH, or OCEAN. It could also have something like pools.balancer.exchange to browse across many datatoken pools.
- **Order-book DEXes**. It could use 0x, Binance DEX, Kyber, etc. It could leverage platform-specific features such as 0x's shared liquidity across marketplaces.
- **Order-book CEXes**. Centralized exchanges like Binance or Coinbase could readily create their own datatoken-based marketplaces, and to kickstart usage could sell datasets that they've generated internally.
- **Marketplaces in AI tools**. This could be an AI-oriented data marketplace app embedded directly in an AI platform or webapp like Azure ML Studio or Anaconda Cloud. It could also be an AI-oriented data marketplace as a Python library call, for usage in any AI flow (since most AI flows are in Python). In fact, this is already live in Ocean's Python library.
- **"Nocode" Data Marketplace builder**. Think Shopify [76] for data marketplaces, where people can deploy their own data marketplaces in just a few clicks.

4.8 Initial Data Offerings (IDOs)

In an IDO, people or organizations can launch data assets using the technology tools and marketing techniques to launch other tokens (e.g., for Initial Coin Offerings and Initial Exchange Offerings).

2017 brought a craze of Initial Coin Offerings (ICOs), for better and for worse. Great efforts were put into designing mechanics of token distributions, and marketing and legals. A lot was learned from that era, and the learning has continued.

Here are some innovations since then. Vitalik Buterin suggested **DAICOs**, which leverages DAO technology to decentralize fundraising effort [77]. Fabian Vogelsteller suggested **reversible ICOs** (rICOs), which gave investors the ability to pull their funds out, to minimize their risk [78].

Binance and other CEXes offer **Initial Exchange Offerings** (IEOs), which hold a fixed-price token sale, followed immediately by trading on the exchange. UMA and others have conducted **Initial DEX Offerings** using an AMM as the first market for their token [79]. Balancer's **Liquidity Bootstrapping Pools** (LBPs) [46] refine this by slowly releasing more of the token over several months while simultaneously increasing its weight in the pool (towards 50–50 liquidity).

Unisocks [80] and Karma DAO [67] each have a combination of **bonding curve + AMM**. The bonding curve [81] acts as a primary market, where price increases with each token minted. The AMM acts as a secondary market. YFI (and Bitcoin!) had no **"offering" at all**. Rather, tokens got distributed to users for doing "work" to add value to the system. The philosophy is "earned, not printed" [82]. Incidentally, this is how the majority of OCEAN are distributed too, curated by OceanDAO.

An Initial Data Offering (IDO) can use any of these techniques. A pragmatic starting point is **Initial DEX Offering** via an AMM. This is simple and well-suited to long-tail tokens like datatokens. Ideally, there's software to make launching such datatokens easy; call it an **IDO Launchpad**. Ocean Market makes it easy to publish a datatoken and create a Balancer AMM all at once; therefore **Ocean Market is the first IDO Launchpad**. We envision other IDO variants in the future, from the list above and new innovations. We hope to see more IDO Launchpads created by ambitious teams.

4.9 Data as an Asset Class for DeFi

The data economy is already 377B€ for Europe alone [83]. Tokenized data assets have great promise to grow the size of DeFi assets under management.

Data can be securitized and used as collateral [74]. An example is Bowie Bonds, where a fraction of David Bowie's IP (intellectual property) licensing revenue was paid to bondholders. Data is IP. To use it as a financial asset, one must price it. In Bowie's case, the value was established from previous years' licensing revenue. Alternatively, we can establish price by selling data assets in data marketplaces.

As such, data is an asset class. With datatokens, we can onboard more more data assets into each major DeFi service types:

- Data assets can be used as collateral in **stablecoins** and **loans**, therefore growing total collateral.
- Data assets bought and sold in **DEXes** and **CEXes** contribute to their $ volume and assets under management (AUM).
- There can be insurance on data assets. As described above, there can be data DAOs, data baskets, and more.

Data to Optimize DeFi Returns

We can close the loop with data helping DeFi, and vice versa. Specifically: **data can improve decision-making in DeFi to optimize returns**. This will catalyze the growth of DeFi further. Here are some examples:

- **Yield farming**. Data can improve the automated strategies to maximize APR in yield farming. Think yearn.finance robots, but optimized further.
- **Insurance**. Data to lower the risk models in insurance.
- **Loans**. Better prediction of default for under-collateralized loans.
- **Arb bots**. More data for higher-return arbitration bots.
- **Stablecoins**. Assessment of assets for inclusion in stablecoins.

Data-powered loops. DeFi *looping* techniques further boost returns. For each of the examples above, we envision *loops* of buying more data, to get better returns, to buy more data, and so on. To go even further, we could apply this to data *assets* themselves.

Data Management Platforms for Smart Cities and More

In 2001, the government of Estonia rolled out a data management platform called X-Road [84]. It then deployed an identity system on top; each citizen received an identity card with a digital signature. Since then, Estonia has rolled out apps for elections, health, taxes, parking, lawmaking, E-Residency and two dozen more government apps, plus third-party apps like banking [85].

Both Ocean and X-Road can be used as digital infrastructure for smart cities' data sharing. X-Road can be seen as a smart city example, since the majority of Estonians live in Tallinn. X-Road has a longer history, but has more centralized control and requires dev-ops effort. While Ocean is younger, by using a global permissionless infrastructure, it has lighter dev-ops requirements. We envision a future with both X-Road and Ocean-based data sharing in smart cities.

Ocean framed as a data management platform can be used not only for data sharing within a city (across citizens), but also within a province/state (across cities), within a nation (across provinces), within an international initiative (across nations, e.g., GAIA-X [86]), within a company (across employees), and within a multinational enterprise (across national offices).

Composable Datatokens

Datatokens can be composed into bundles, sets, or groups using ERC998 [87], Set Protocol [88], Melon Protocol [70], or others. This helps for the following use cases:

- **Group across time**. Package each 10-min chunk of data from the last 24 h into a single token.

- **Group across data sources**. Package 100 data streams from 100 unique Internet-of-Things (IoT) devices, as a single token.
- **Data baskets for asset management**. Group together 1000 datasets that each have individual (but small) value, to sell as a single asset to others wanting to hold data assets.
- **Data indexes**. Track the top 100 data assets and make it easy for others to invest in those as a single asset, similar to today's index funds.
- **On-chain annotations to metadata**. Use ERC998 in a bottom-up setting to "attach" tags or other information to the data asset. Uses include: reputation given by a marketplace's users, quality as computed by a marketplace's algorithms, input training data vs output vs a model, industry verticals, and more.

The essay [74] elaborates further.

5 Conclusion

This paper presented Ocean Protocol. Ocean tools help developers build marketplaces and other apps to privately and securely publish, exchange, and consume data. The tools offer an on-ramp and off-ramp for data assets into crypto ecosystems, using datatokens. Composability gives many application opportunities, including data wallets, data marketplaces, data DAOs, and more. Ocean Market is a live reference community marketplace that natively integrates Balancer AMMs, to facilitate "Initial Data Offerings".

Ocean tools are encapsulated in a broader system designed for long-term growth of an open, permissionless Web3 Data Economy. Ocean Data Farming incentivizes a supply of quality data. A key piece is OceanDAO to fund software development, outreach. OceanDAO is funded by revenue from Ocean apps and from Ocean's network rewards.

References

1. "The World's Most Valuable Resource is No Longer Oil, But Data", *The Economist*, May 6, 2017, https://www.economist.com/news/leaders/21721656-data-economy-demands-new-approach-antitrust-rules-worlds-most-valuable-resource
2. Banko, M., Brill, E.: Scaling to very very large corpora for natural language disambiguation. Proceeding Annual Meeting on Association for Computational Linguistics, July, 2001, http://www.aclweb.org/anthology/P01-1005
3. Halevy, A., Norvig, P., Pereira, F.: The unreasonable effectiveness of data. IEEE Intell. Syst. **24**(2) (2009), https://research.google.com/pubs/archive/35179.pdf
4. Buterin, V.: Ethereum: A Next-Generation Smart Contract and Decentralized Application Platform, Dec. 2013, https://ethereum.org/en/whitepaper/
5. McConaghy, T.: "Mission & Values for Ocean Protocol", July 23, 2018, https://blog.oceanprotocol.com/mission-values-for-ocean-protocol-aba998e95b8

6. Goldin, M.: "Mike's Cryptosystems Manifesto", Google Docs, 2017, https://docs.google.com/document/d/1TcceAsBlAoFLWSQWYyhjmTsZCp0XqRhNdGMb6JbASxc/edit
7. Graham, P.: "How to Start a Startup", Paul Graham Blog, March 2005, http://www.paulgraham.com/start.html?viewfullsite=1
8. McConaghy, T.: "How Does Ocean Compute-to-Data Related to Other Privacy-Preserving Approaches?", Ocean Protocol Blog, May 28, 2020, https://blog.oceanprotocol.com/how-ocean-compute-to-data-relates-to-other-privacy-preserving-technology-b4e1c330483
9. Vogelsteller, F., Buterin, V.: "ERC-20 Token Standard", Nov. 19, 2015, https://github.com/ethereum/EIPs/blob/master/EIPS/eip-20.md
10. Buterin, V.: Twitter, Feb. 18, 2020, https://twitter.com/VitalikButerin/status/1229819782230241280
11. Oauth Team, "Oauth 2.0 Spec", Oauth Website, last accessed Aug. 19, 2020, https://oauth.net/2/
12. "Create, read, update, and delete", Wikipedia, last accessed Aug. 19, 2020, https://en.wikipedia.org/wiki/Create,_read,_update_and_delete
13. Pregelj, J.: "CRAB—Create. Retrieve. Append. Burn", BigchainDB blog, Oct. 19, 2017, https://blog.bigchaindb.com/crab-create-retrieve-append-burn-b9f6d111f460?gi=209a8d51b780
14. Ogundeji, O.: "Antonopoulos: Your Keys, Your Bitcoin. Not Your Keys, Not Your Bitcoin", CoinTelegraph, Aug 10, 2016, https://cointelegraph.com/news/antonopoulos-your-keys-your-bitcoin-not-your-keys-not-your-bitcoin
15. "Chainlink", Homepage, last accessed Aug. 20, 2020, https://chain.link/
16. McConaghy, T.: "Ocean Datatokens: From Money Legos to Data Legos. How DeFi Helps Data, and Data Helps DeFi", Ocean Protocol Blog, Sep. 8, 2020, https://blog.oceanprotocol.com/ocean-datatokens-from-money-legos-to-data-legos-4f867cec1837
17. "Ocean Protocol Software Documentation", last accessed Aug 20, 2020, https://docs.oceanprotocol.com/
18. "Ocean Contracts", Ocean Protocol GitHub, last accessed Aug. 31, 2020, https://github.com/oceanprotocol/ocean-contracts
19. "Ocean-lib: JavaScript", Ocean Protocol GitHub, last accessed Aug. 31, 2020, https://github.com/oceanprotocol/ocean-lib-js
20. "Ocean-lib: Python", Ocean Protocol GitHub, last accessed Aug. 31, 2020, https://github.com/oceanprotocol/ocean-lib-py
21. "Ocean Market", Ocean Protocol GitHub, last accessed Aug. 31, 2020, https://github.com/oceanprotocol/market
22. "Ocean React Hooks", Ocean Protocol GitHub, last accessed Aug. 31, 2020, https://github.com/oceanprotocol/react
23. "Balancer", homepage, last accessed Aug. 21, 2020, https://balancer.finance
24. Ocean Protocol Team, "dexFreight and Ocean Protocol Partner to Enable Transportation and Logistics Companies To Monetize Data", Feb. 13, 2020, https://blog.oceanprotocol.com/dexfreight-ocean-protocol-partner-to-enable-transportation-logistics-companies-to-monetize-data-7aa839195ac
25. Ocean Protocol Team, "Ocean Protocol delivers Proof-of-Concept for Daimler AG in collaboration with Daimler South East Asia", July 7, 2020, https://blog.oceanprotocol.com/ocean-protocol-delivers-proof-of-concept-for-daimler-ag-in-collaboration-with-daimler-south-east-564aa7d959ca
26. Decentralized Identifiers(DIDs) v0.11: data model and syntaxes for decentralized identifiers, W3C Community Group, Draft Community Group Report 06 February 2019, last accessed Feb. 28, 2019, https://w3c-ccg.github.io/did-spec/
27. "Ocean Enhancement Proposal 7: Decentralized Identifiers," last accessed Feb. 11, 2019, https://github.com/oceanprotocol/OEPs/tree/master/7
28. Crockford, D.: "Introducing JSON", Webpage, last accessed Aug 19, 2020, https://www.json.org/
29. "Ocean Enhancement Proposal 8: Assets Metadata Ontology," last accessed Feb. 11, 2019, https://github.com/oceanprotocol/OEPs/tree/master/8

30. "Schema.org DataSet schema", last accessed Feb. 11, 2019, https://schema.org/Dataset
31. W3C, "Verifiable Credentials Data Model 1.0", W3C Recommendation, November 2019, https://www.w3.org/TR/vc-data-model/
32. 3Box Team, "3Box: User Data Cloud", 3Box Homepage, last accessed Aug 19, 2020, https://3box.io/
33. McConaghy, T.: "Ocean Protocol V3 Architecture Overview: Simplicity and Interoperability via a Datatokens Core," Ocean Protocol Blog, Oct. 12, 2020, https://blog.oceanprotocol.com/ocean-protocol-v3-architecture-overview-9f2fab60f9a7
34. McConaghy, T.: "Ocean on PoA vs. Ethereum Mainnet?", Ocean Protocol Blog, Feb. 12, 2019, https://blog.oceanprotocol.com/ocean-on-poa-vs-ethereum-mainnet-decd0ac72c97
35. "Ocean Roadmap ‖ Setting sail to 2021", Ocean Protocol Blog, Feb. 13, 2019, https://blog.oceanprotocol.com/ocean-roadmap-setting-sail-to-2021-70c2545547a7
36. "Ocean Product Update ‖ 2020", Ocean Protocol Blog, Mar .19, 2020, https://blog.oceanprotocol.com/ocean-product-update-2020-f3ae281806dc
37. McConaghy, T.: "The DCS Triangle: Decentralized, Consistent, Scalable", July 10, 2016, https://blog.bigchaindb.com/the-dcs-triangle-5ce0e9e0f1dc
38. Xdai Team, "Xdai Chain", homepage, https://www.xdaichain.com/
39. "About the AMB", TokenBridge homepage, last accessed Sept. 3, 2020, https://docs.tokenbridge.net/amb-bridge/about-amb-bridge
40. "Parity Substrate: Build your own blockchains", last accessed Aug. 19, 2020, https://www.parity.io/substrate/
41. Ethereum Optimism Team, "Optimistic Virtual Machine Alpha", Ethereum Optimism Blog, Feb. 12, 2020, https://medium.com/ethereum-optimism/optimistic-virtual-machine-alpha-cdf51f5d49e
42. "Cosmos SDK Documentation", homepage, last accessed Aug. 19, 2020, https://docs.cosmos.network/
43. Kuhn, E.: "Let's Talk About Data Pricing — Part I", Aug. 13, 2019, https://blog.oceanprotocol.com/lets-talk-about-data-pricing-part-i-bbc9cf781d9f
44. Kuhn, E.: "Let's Talk About Data Pricing — Part II", Aug. 21, 2019, https://blog.oceanprotocol.com/value-of-data-part-two-pricing-bc6c5127e338
45. "Aragon Court", homepage, last accessed Aug. 20, 2020, https://aragon.org/court
46. McDonald, M.: "Building Liquidity into Token Distribution", Mar. 4, 2020, https://medium.com/balancer-protocol/building-liquidity-into-token-distribution-a49d4286e0d4
47. Balancer Team, "Interest-Bearing Stablecoin Pools Without Impermanent Loss", Balancer blog, Oct. 30, 2019, https://balancer.finance/2019/10/30/interest-bearing-stablecoin-pools-without-impermanent-loss/
48. Murray, P., Welch, N., Messerman, J.: "EIP-1167: Minimal Proxy Contract", Ethereum Improvement Proposals, no. 1167, June 2018, https://eips.ethereum.org/EIPS/eip-1167
49. Simler, K., Hanson, R.: "The Elephant in the Brain: Hidden Motives in Everyday Life", Oxford University Press, Jan. 2018, https://www.amazon.com/Elephant-Brain-Hidden-Motives-Everyday/dp/0190495995
50. Monegro, J.: "Proof of Liquidity", Placeholder blog, May 22, 2020, https://www.placeholder.vc/blog/2020/5/22/proof-of-liquidity
51. McConaghy, T.: "Ocean Market: An Open-Source Community Marketplace for Data. Featuring OCEAN Staking, Automated Market Makers, and Initial Data Offerings," Ocean Protocol Blog, Sep. 24, 2020, https://blog.oceanprotocol.com/ocean-market-an-open-source-community-marketplace-for-data-4b99bedacdc3
52. McConaghy, T.: "On Selling Data in Ocean Market: Selling Your Proprietary Data, Others' Proprietary Data, and Value-Adds to Open Data," Ocean Protocol Blog, Oct. 19, 2020, https://blog.oceanprotocol.com/on-selling-data-in-ocean-market-9afcfa1e6e43
53. McConaghy, T.: "On Staking on Data in Ocean Market: Earning Opportunities — and Risks — in Ocean's Community Marketplace," Ocean Protocol Blog, Oct. 22, 2020, https://blog.oceanprotocol.com/on-staking-on-data-in-ocean-market-3d8e09eb0a13

54. Patel, M.: "Technical Guide to Ocean Compute-to-Data", Ocean Protocol Blog, May 19, 2020, https://blog.oceanprotocol.com/v2-ocean-compute-to-data-guide-9a3491034b64
55. "General Data Protection Regulations", Wikipedia, last accessed Aug. 21, 2020, https://en.wikipedia.org/wiki/General_Data_Protection_Regulation
56. "Solidity", Wikipedia, last accessed Aug. 21, 2020, https://en.wikipedia.org/wiki/Solidity
57. Dimitri de Jonghe, "Exploring the SEA: Service Execution Agreements", Ocean Protocol Blog, Nov. 30, 2018, https://blog.oceanprotocol.com/exploring-the-sea-service-execution-agr eements-65f7523d85e2
58. "OpenMined", homepage, last accessed Aug. 20, 2020, https://www.openmined.org/
59. Azure ML Team, "Microsoft Azure Machine Learning Studio", homepage, last accessed Aug. 19, 2020, https://studio.azureml.net/
60. "Etherscan", homepage, last accessed Aug. 21, 2020, https://etherscan.io
61. CoinGecko Team, CoinGecko website, last accessed Aug. 19, 2020 https://www.coingecko.com
62. CoinMarketCap Team, CoinMarketCap website, last accessed Aug. 19, 2020, https://www.coinmarketcap.com
63. "Decentralized Autonomous Organization", Wikipedia, last accessed Aug. 19, 2020, https://en.wikipedia.org/wiki/Decentralized_autonomous_organization
64. "Saskatchewan Wheat Pool", Wikipedia, last accessed Aug. 21, 2020, https://en.wikipedia.org/wiki/Saskatchewan_Wheat_Pool
65. Imanol Arrieta Ibarra et al.: "Should We Treat Data as Labor? Moving Beyond 'Free'", AEA Papers and Proceedings, **108**, 38–42. https://doi.org/10.1257/pandp.20181003, https://www.aeaweb.org/conference/2018/preliminary/paper/2Y7N88na
66. McConaghy, T.: "Radical Markets and the Data Economy", Ocean Protocol Blog, Mar. 5, 2020, https://blog.oceanprotocol.com/radical-markets-and-the-data-economy-4847c272f5
67. Lee, A.: "Announcing Karma DAO: First-Ever Token-Permissioned Networking Chat Group on Telegram", July 20, 2020, https://medium.com/@andrwlee/announcing-karma-dao-first-ever-token-permissioned-networking-chat-group-on-telegram-5feab7a54def
68. Balancer Team, "Configurable Rights Pool", GitHub Repository, published Aug, 2020, https://github.com/balancer-labs/configurable-rights-pool
69. Delmonti, A.: Introducing PieDAO, the asset allocation DAO, Mar. 3, 2020, https://medium.com/piedao/introducing-piedao-the-asset-allocation-dao-1af9eec5ee4
70. "Melon Protocol", homepage, last accessed Aug. 19, 2020, https://melonprotocol.com/
71. "MOBI—Mobility Open Blockchain Initiative". Homepage. Last accessed Aug. 20, 2020, https://dlt.mobi/
72. Chaturvedi, V.: The Man Who Tokenized Himself Gives Holders Power Over His Life, June 30, 2020, https://www.coindesk.com/man-who-sells-himself-now-wants-buyers-to-con trol-his-life
73. Azevedo, E.M., David, M.: Pennock, and E. Glen Weyl, "Channel Auctions", Aug 31, 2018, https://papers.ssrn.com/sol3/papers.cfm?abstract_id=3241744
74. McConaghy, T.: "Datatokens 3: Data and Decentralized Finance (Data * DeFi)", Ocean Protocol Blog, Dec. 3, 2019, https://blog.oceanprotocol.com/data-tokens-3-data-and-decentralized-fin ance-data-defi-d5c9a6e578b7
75. "Kleros", homepage, last accessed Aug. 20, 2020, https://kleros.io/
76. "Shopify", homepage, last accessed Aug. 21, 2020, https://shopify.com
77. Buterin, V.: "Explanation of DAICOs", Eth.Research Blog, Jan. 2018, https://ethresear.ch/t/explanation-of-daicos/465
78. Vogelsteller, F.: "rICO—The Reversible ICO", Lukso Blog, Apr. 13, 2020, https://medium.com/lukso/rico-the-reversible-ico-5392bf64318b
79. Uma Team, "UMA Initial Uniswap Listing", Apr. 22, 2020, https://medium.com/uma-project/umas-initial-uniswap-listing-afa7b6f6a330
80. "Unisocks", homepage, https://unisocks.exchange/
81. Simon de la Rouviere, "Introducing Curation Markets: Trade Popularity of Memes & Information (with code)!", Medium, May 22, 2017, https://medium.com/@simondlr/introducing-curation-markets-trade-popularity-of-memes-information-with-code-70bf6fed9881

82. @Bitcoin, Twitter, Apr. 4, 2020, https://twitter.com/Bitcoin/status/1246482664376414209
83. European Commission, "Building a Data Economy—Brochure", Sep. 2019, https://ec.europa.eu/digital-single-market/en/news/building-data-economy-brochure
84. "X-Road", Wikipedia, last accessed Aug. 19, 2020, https://en.wikipedia.org/wiki/X-Road
85. "E-Estonia", homepage, last accessed Aug. 19, 2020, https://e-estonia.com/
86. "GAIA-X", homepage, last accessed Aug. 21, 2020, https://www.data-infrastructure.eu/
87. Lockyer, M.: "ERC-998 Composable Non-Fungible Token Standard", Apr. 15, 2018, https://github.com/ethereum/eips/issues/998
88. "TokenSets", homepage, last accessed Aug. 19, 2020, https://www.tokensets.com/
89. "DashNexus", homepage, last accessed Aug. 20, 2020, https://app.dashnexus.org
90. Martinelli, F.: "Balancer Liquidity Mining Begins", Balancer Blog, May 29, 2020, https://medium.com/balancer-protocol/balancer-liquidity-mining-begins-6e65932eaea9

Blockchain in Supply Chain: Opportunities and Design Considerations

Gowri Sankar Ramachandran, Sidra Malik, Shantanu Pal, Ali Dorri, Volkan Dedeoglu, Salil Kanhere, and Raja Jurdak

Abstract Supply chain applications operate in a multi-stakeholder setting, demanding trust, provenance, and transparency. Blockchain technology provides mechanisms to establish a decentralized infrastructure involving multiple stakeholders. Such mechanisms make the blockchain technology ideal for multi-stakeholder supply chain applications. This chapter introduces the characteristics and requirements of the supply chain and explains how blockchain technology can meet the demands of supply chain applications. In particular, this chapter discusses how data and trust management can be established using blockchain technology. The importance of scalability and interoperability in a blockchain-based supply chain is highlighted to help the stakeholders make an informed decision. The chapter concludes by underscoring the design challenges and open opportunities in the blockchain-based supply chain domain.

G. S. Ramachandran (✉) · S. Malik · S. Pal · A. Dorri · R. Jurdak
Queensland University of Technology, Brisbane, Australia
e-mail: g.ramachandran@qut.edu.edu

S. Malik
e-mail: s.malik@qut.edu.edu

S. Pal
e-mail: shantanu.pal@qut.edu.edu

A. Dorri
e-mail: ali.dorri@qut.edu.edu

R. Jurdak
e-mail: r.jurdak@qut.edu.edu

V. Dedeoglu
CSIRO, Brisbane, Australia
e-mail: volkan.dedeoglu@data61.csiro.au

S. Kanhere
University of New South Wales, Sydney, Australia
e-mail: salil.kanhere@unsw.edu.au

1 Introduction

Supply chains drive many industries in multiple domains, including agriculture [34, 35], electronics [40, 80], and manufacturing [1, 38]. Enterprises procure materials and services from different stakeholders to build, store and deliver products to end consumers. Information about the product, referred to as "data", gets exchanged between manufacturers, shipment companies, auditors, regulators, and retailers in this product pipeline. Businesses rely on data to understand the status of the supply chain while making an informed decision about future demands and optimizing their operations. Besides, the data helps the companies manage the regulatory compliance and auditing processes.

To sum up, the information about the supply chain, in the form of digital data, can enhance the efficiency of the supply chain. However, the processes followed by the supply chain entities must be trustworthy to attain meaningful insights. Blockchain technology offers mechanisms to enhance trust.

Bitcoin [47], a blockchain platform, introduced an innovative architecture for creating, managing, and sharing a digital currency without involving a centralized intermediary such as a bank. It offers decentralization, immutability, and transparency through a clever combination of a consensus algorithm, cryptographic primitives, and a distributed ledger. After the arrival of Bitcoin, many blockchain platforms, including Ethereum [78], Tendermint [14], and EOS [25], entered the market with similar properties and support for cryptocurrencies. However, the properties of Bitcoin and other blockchain platforms that followed it seem extremely useful for applications beyond cryptocurrency, that require decentralization, immutability, and transparency.

Supply chain applications' natural characteristics include its involvement of multiple stakeholders in the form of producers, retailers, auditors, shipment companies, and in some cases, the end consumers. Having a transparent operational infrastructure driven by a decentralized architecture combined with an immutable ledger provide enormous business and practical advantages to the stakeholders. Here, the integration of blockchain technology into the supply chain requires IT infrastructure. Many of the supply chain companies already employ an IT infrastructure to manage their business processes digitally [79]. Integrating blockchain technology into the existing supply chain management infrastructure allows the supply chain participants to gain business advantages by improving traceability and transparency, automating processes for purchasing and payments, reducing conflicts and errors, improving regulatory compliance and cross-border transactions, and protecting the supply chain against counterfeiting through an immutable and tamper-proof distributed ledger. Therefore, the benefits of combining blockchain technology with the supply chain application are clear. Still, it is crucial to understand how blockchain technology can be merged reliably and seamlessly with supply chain networks to gain maximum advantages.

This chapter discusses blockchain applications in the supply chain and highlights the characteristics and requirements of supply chain applications. We also explain how supply chain stakeholders can share data and manage trust using blockchain technology. Besides, we also focus on the need for scalable blockchain technology to deal with the high transaction throughput and latency demands of real-world supply chain applications. The section on authentication and access control discusses methods to set up a trusted and secure supply chain infrastructure. The interoperability challenges of supply chain systems are also discussed to help companies understand the existing challenges and potential approaches. We conclude the chapter with an overview of design considerations and open challenges.

2 Characteristics and Requirements of Supply Chain Applications

A supply chain can be defined as the end-to-end process of producing and delivering goods and services from the acquisition of raw materials to the delivery of final products to the end consumers. As depicted in Fig. 1, a typical supply chain involves multiple stakeholders and has many characteristics and requirements. This section provides an overview of these characteristics and common requirements together with the main challenges facing today's supply chain.

2.1 Characteristics of Supply Chain Applications and Networks

In this section, we will review the characteristics of supply chain applications.

- **Collaborative**: Supply chain applications naturally involve multiple stakeholders in the form of producers, transporters, auditors, retailers, and regulators as shown in Fig. 1. Each of these stakeholders must collaborate with each other to operate the supply chain in the process of delivering goods and services to end consumers while gaining financial incentives. However, collaboration becomes challenging when the level of trust is low and there is no effective mechanism to manage and share supply chain data transparently among stakeholders.
- **International**: Globalization of production and trade has given rise to geographic dispersion of the supply chain, where the stakeholders may operate from different legal jurisdictions and countries. In a global supply chain, raw materials may be acquired from suppliers, processed by manufacturers and the final products may be delivered to consumers at geographically dispersed locations introducing new risks and challenges in the flow of goods and services. Note that the stakeholders may have to comply with the regulations of one or more jurisdictions in this setting.
- **Time-sensitive**: Supply chain processes are time-sensitive. Stakeholders rely on the timely flow of goods and information to make operational decisions. Any delay in the flow of goods and information may cause supply chain disruptions in the form of inefficiencies, quality degradation of products, lower service quality, market loss, and reduced gains.

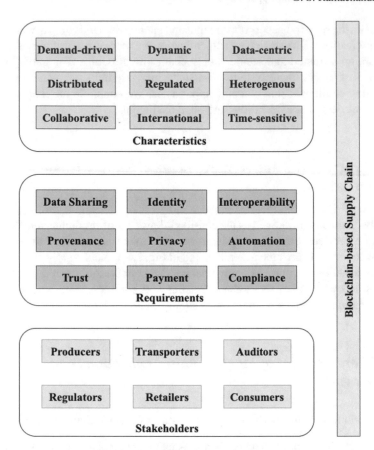

Fig. 1 Characteristics and requirements of blockchain-based supply chain

- **Distributed**: In a supply chain, the information technology (IT) and operation technology (OT) infrastructure that manages the flow of goods, documents, certifications, and other information is distributed among multiple stakeholders. These infrastructures are typically connected through enterprise systems and data sharing frameworks.
- **Regulated**: Supply chains are subject to national and international (import and export) regulations. Stakeholders must comply with the laws and regulations of their jurisdiction. In some cases, they may have to comply with jurisdictions associated with both the producer and consumer. Such complexities make regulatory compliance challenging, in particular for supply chains operating in multiple jurisdictions.
- **Heterogenous**: Supply chains involve multiple stakeholders using heterogeneous data standards and disparate systems. Note that different standards may introduce compatibility issues, which could make collaboration challenging and difficult in a supply chain application.

- **Demand-driven**: The dynamic nature of consumer behavior requires supply chains to be resilient against shifting consumer demand. Demand-driven supply chain management focuses on the demand signals and forecasts for the planning and operation of processes to optimize the delivery of goods and services to consumers. Building a supply chain that is able to respond to changing consumer demand may require supply chain stakeholders to invest in technology to collect, share, and react to real-time demand data.
- **Dynamic and Unpredictable**: Supply chains are dynamic systems due to ongoing changes in stakeholders, offered services and products, regulations, market dynamics, and technological advancements. For example, when a ship accidentally blocked the Suez Canal in 2021, many ships could not cross the Suez Canal for up to 6 days, which disrupted the supply chain massively. To respond to such unpredictable real-world events, flexible and agile supply chain mechanisms are required.
- **Data-centric**: For improved performance, supply chain operations should be based on data-driven decision making, which relies on real-time, accurate, and actionable data flow. Digitization of supply chains, data provenance and trust mechanisms, big data analytics, and effective data sharing systems help stakeholders make informed business decisions to enhance the efficiency of their supply chain.

2.2 Requirements of Supply Chain

In this section, we will review the common requirements of supply chain applications.

- **Trust**: Trust is a pivotal factor in enabling collaboration in supply chains. Since supply chains are complex constructs, establishing trust among the stakeholders and the systems used is a challenging requirement for the integrity of supply chains.
- **Payment**: In a supply chain transaction, a seller provides goods and services to a buyer and receives a payment in return. Payment systems enabling business-to-business payments without relying on third parties, accelerating the flow of funds, reducing the transaction costs, and lowering the risk of financial frauds improve the sustainability and efficiency of supply chains.
- **Compliance**: In a supply chain, stakeholders need to comply with various regulations, standards, and policies. Non-compliance may result in financial loss due to legal fines, potential loss of market, loss of non-compliant assets, and supply chain disruptions as well as brand and reputation damage for companies. Thus, supply chain stakeholders need to implement mechanisms to ensure regulatory compliance. Recently, regulatory technology (RegTech) has been proposed as a potential solution to address the complexities of compliance through digitization and automation of the compliance process.
- **Provenance**: Due to the rising consumer demand on the information related to the origin of products, the processes for manufacturing and production, transfer of custody, and ownership, provenance mechanisms become critical. In a supply

chain, provenance improves consumer confidence and trust in the authenticity of the products and accountability among stakeholders.

- **Privacy**: To improve the operational efficiency of supply chains, the stakeholders are required to share their data between stakeholders in a supply chain network. However, supply chain data includes commercially sensitive information. Companies may feel that revealing such data may cause a business to lose its competitive advantage. Thus, privacy preservation mechanisms are required for enabling data sharing while protecting the sensitive business data.

- **Automation**: In a typical supply chain, there are many time-consuming, error-prone, and repetitive processes that can be automated, such as processing orders and payments, inventory management, transportation arrangements, and manufacturing workflows. Using emerging technologies such as artificial intelligence, robotics, IoT, and big data, supply chain processes can be automated to enhance the efficiency by lowering operational costs, reducing manual labor, improving inventory management, and increasing productivity and accuracy while ensuring continuous compliance to regulations.

- **Data sharing**: Data sharing improves coordination and trust among stakeholders, efficiency of production, quality of products and services, while reducing operational costs and risk of non-compliance. Furthermore, establishing end-to-end provenance requires supply chain stakeholders to share their data with each other and the end consumers. Therefore, trusted and effective mechanisms are needed to enable data sharing among distributed and heterogeneous stakeholders. Lack of incentives may constitute a barrier for data sharing. Thus, incentive mechanisms can be utilized to encourage stakeholders to share their data.

- **Identity management**: Supply chains are distributed and complex networks, where the verification of the identities and credentials of interacting stakeholders and the digital identities for the products and devices involved is crucial for the cooperation among stakeholders and the operation of supply chains. However, stakeholders may not be willing to disclose or share the data related to supply chain identities due to business sensitivities and to uphold competitive advantage. As a consequence, supply chains may suffer from lack of transparency, which becomes a bigger problem as the supply chains grow. Traditional centralized and federated identity management approaches establish trust between stakeholders relying on centralized entities or platforms, which introduces data ownership and security problems. Thus, decentralized mechanisms are required to manage supply chain identities, where entities control their data shared with other stakeholders without relying on third parties while enabling authorities to verify the identities and credentials of the stakeholders.

- **Interoperability**: Due to the interconnected and interdependent nature of supply chain networks, interoperability is required to facilitate collaboration among the stakeholders. While interoperability can be achieved through standards, procedures, and inter-organizational protocols at the service and business levels, digital interoperability requires mechanisms and protocols for application interfaces, integration services, and data communications to enable reliable and seamless data sharing among the distributed and heterogeneous supply chain systems.

Supply chain applications could become trustworthy, transparent, and efficient if the above requirements are fulfilled. While existing supply chain systems fulfill these requirements through a combination of manual and automated processes, they broadly rely on systems managed by a single organization. Relying on a single organization introduces a single point of failure, wherein the entity may misbehave or tamper with the supply chain information. In the rest of this chapter, we will introduce blockchain technology and its application in supply chain to fulfill some of these requirements.

3 Overview of Blockchain Technology

The characteristics and requirements of the supply chain applications demand solutions that can operate in a distributed setting while offering trust, security, data sharing, payment, and transparency guarantees. Supply chain operators can set up an infrastructure to share information with other stakeholders using a custom-built and centrally managed platform. However, such solutions are susceptible to central points of failure, wherein the organization that runs the platform has complete control of the infrastructure. In a supply chain network, a centrally managed infrastructure could be compromised by a malicious stakeholder, resulting in incorrect dispersion of information. The stakeholders may not be willing to reveal essential traces and logs in the case of a dispute or unexpected events due to their error. Note that the organizations would like to ensure that their reputation is not damaged when they mismanage the supply chain. Therefore, the centrally managed infrastructure is not guaranteed to provide complete transparency to the stakeholders in the network.

3.1 *Introduction to BitCoin*

Blockchain technology provides support for distributed and decentralized infrastructure. At its core, it consists of a consensus algorithm, cryptographic protocols, and immutable storage. The first peer-to-peer electronic cash system, BitCoin [47], introduced a blockchain that cleverly distributes the data validation process to the computation nodes in the public network. Following the BitCoin protocol, any computationally capable node can create blocks that make up the blockchain using a consensus mechanism called Proof-of-Work (PoW). PoW lets the public BitCoin nodes solve a computationally intensive cryptographic puzzle, which demands significant computation resources. On average, the nodes in the network require 10 minutes to solve the puzzle. When a node solves the puzzle, it is selected as a winner for that round. The winning node gets to verify the new transactions and propose the new block. Subsequently, all the nodes in the network receive the newly proposed block, verify them, and append to their local copy of the blockchain ledger.

The ledger used by the blockchain technology offers immutability support, which means, once the data is written to the blockchain, it cannot be modified. When a malicious entity tries to alter the data on the blockchain, he gets to modify only his local copy of the ledger. Besides, other nodes in the network will not accept such modifications because the modified version will fail the integrity check when it is compared with the original blockchain ledger maintained by the majority of the nodes in the network. Note that hundreds to thousands of nodes in the network keep a copy of the ledger. It is nearly impossible to update the ledger on all of those nodes. Therefore, the write-once property of blockchain technology is one of the powerful features for applications that require trust and transparency.

Blockchain technology also enables the nodes in the network to maintain some level of anonymity through the use of public-key cryptography. Users submit transactions by using their public key and a signature, which forms the basis of their identity. When a transaction is submitted by the user, for example, to spend a Bit-Coin, the user digitally signs the transaction using her private key and then presents the signature and a public key to the network. Anyone in the BitCoin network can verify the authenticity of the signature using the public key. Note that the user need not share his private key.

In summary, the BitCoin platform, which started the blockchain revolution, uses PoW to create and manage blocks without a central intermediary securely. Besides, public-key cryptography protects users' privacy while allowing them to exchange digital currencies in a trusted manner. Lastly, the immutable ledger maintains a distributed and trusted record of all the transactions to provide complete transparency.

3.2 Other Blockchain Platforms

Following the introduction of BitCoin, many blockchain platforms entered the industry, offering immutability, transparency, and trust guarantees. Notably, Ethereum [78] is one of the most popular blockchain platforms, after BitCoin. While Ethereum's architecture closely resembles BitCoin, including its use of PoW[1], public-key cryptography, and immutable ledger, it introduced smart contracts as a new feature. Smart contracts allow the users to run computations within the blockchain platform to enable automation. For example, users can code up the supply chain events in the smart contract and let the smart contract automatically transfers payments to the relevant stakeholders when the event is associated with the product delivery event triggers the smart contract on the blockchain. This functionality of Ethereum resulted in many innovations in the finance, supply chain, and banking domains.

When BitCoin and Ethereum became popular blockchain technology for decentralized and trusted computing, a new blockchain deployment model entered the industry. Platforms such as BitCoin and Ethereum use an open and public network

---

[1] Note that the Ethereum community is developing a new Proof-of-Stake consensus protocol, which may soon replace the energy-inefficient PoW protocol.

model; wherein anyone can join the blockchain network to participate in the creation and management processes. A new permissioned deployment model was introduced to let enterprises set up a private blockchain ledger to manage multi-stakeholder transactions. Hyperledger Fabric [5] is one of the most popular permissioned blockchain platforms. It is a lightweight blockchain platform with a trusted and immutable ledger. In Hyperledger Fabric, the consensus process is managed by an order, which verifies the integrity of the transactions. And a set of nodes are assigned to manage the ledger for the entire network. While a permissioned blockchain network follows a private setup, it still offers trust, immutability, and transparency properties. However, users must prove their identity and follow an on-boarding process to become part of a permissioned blockchain network. In contrast, the nodes' identities are unknown in the public blockchain platforms such as BitCoin and Ethereum.

To sum up, blockchain technology provides a decentralized infrastructure with support for trust, transparency, immutability, payment, and computation (through smart contract). Supply chain applications could leverage these functionalities to fulfill the requirements listed in Sect. 2. The following section will review the popular blockchain-based supply chain applications.

4 Real-World Applications of Blockchain in Supply Chain

Many blockchain-based supply chain applications have been discussed in the literature [9, 30, 40, 72]. In this section, we will review two of the most popular and active real-world blockchain-based supply chain applications.

4.1 IBM FoodTrust

IBM developed a food supply chain system using blockchain technology to address the food inefficiency problems. Improper management of the food supply chain and the lack of visibility into the food production processes lead to high food waste, increased carbon footprint, health issues due to contamination, and high pricing of goods. To overcome these inefficiencies, IBM created FoodTrust [9], which connects industries, farmers, and other stakeholders in the food and agriculture industry through a blockchain-based system. The key features of FoodTrust include Insights, Trace, and Documents modules, which are discussed below:

- **Insights**: This module uses blockchain technology in combination with the Internet of Things (IoT) to provide visibility to the supply chain. With this module, the supply chain stakeholders can understand how fresh a food product is in near real time.

- **Trace**: This module provides end-to-end traceability for the supply chain stake-holders. It allows the members to trace the location and the status of the food products in a secure and trusted manner.
- **Documents**: This module helps the members in the supply chain network securely manage certificates and other digital documents. Regulators and auditors can ensure compliance through this module.

Through these modules, IBM FoodTrust allows the stakeholders to enhance their reputation in the market. IBM manages this project through the Platform-as-a-Service model. Organizations interested in providing transparency and provenance can join IBM's FoodTrust to manage their supply chain.

4.2 TradeLens

TradeLens [30] is a blockchain-based platform for a global supply chain. IBM and GTD Solution Inc jointly develop the platform. In particular, TradeLens focuses on shipping and transportation processes to cater to the demands of the logistics oper-ators. As of June 2021, the TradeLens platform handled more than 2 billion events, millions of documents, and more than 40 million containers. It is a permissioned platform with support for privacy preservation, immutable storage, enterprise-grade security, and standards-compatible.

TradeLens provides Open APIs to let organizations interact with a platform. The supply chain operators, including port authorities, shipping companies, and con-signment owners, can easily leverage the TradeLens platform through interopera-ble and standard-based APIs. Stakeholders with legacy enterprise and IT systems have to invoke the REST APIs, which securely register the supply chain data in the blockchain.

4.3 Other Applications

In addition to the platforms managed by IBM, a few other applications also employ blockchain technology in the supply chain. EverLedger Underwood [75] tracks the journey of diamond using blockchain technology. Another notable application is BeefLedger ?, which tracks the beef supplies using Ethereum and Proof-of-Authority consensus. Besides, the healthcare domain also explores blockchain technology to track medicines and manage patient's healthcare data Radanović and Likić [58], Garcia et al. [22].

5 Blockchain-Based Data sharing for Supply Chain

Supply chain applications constantly share data with other organizations in the application network. This information includes the product's status, invoices, regulatory requirements, demands, and customer feedback. A combination of digital and manual processes is needed to collate the information and then forward them to the necessary stakeholders through either their ERP systems or email messages.

In a multi-stakeholder environment, each organization may have a digital infrastructure and an operational procedure to exchange information relevant to their business. In some cases, a third-party data sharing platform is leveraged by all the stakeholders for data sharing. Businesses that rely on a third-party data sharing platform hand over the management responsibility to an organization that is not necessarily part of the supply chain network. In such circumstances, all the organizations in the supply chain networks must trust the third party. Such an architecture could fulfill the stakeholders' business and data sharing demands, but it assumes that the third party is honest and trustworthy.

Blockchain technology offers a decentralized, distributed, and transparent platform for application developers. Supply chain applications could leverage blockchain technology in a multi-stakeholder environment. Using blockchain technology, each organization that is part of the supply chain network could participate in the data sharing process while having access to the transparent and immutable ledger. In the rest of this section, we will review Trinity, a distributed publish-subscribe broker with blockchain-based immutability.

Multiple data sharing modalities have been considered for supply chain applications including publish-subscribe messaging model. Because of its lightweight design, resource efficiency, and scalability, the publish-subscribe messaging model is widely used in supply chain applications, which is evident from the list of customers including McDonald's that use PubNub, a publish-subscribe service provider. In a nutshell, the pub-sub messaging model connects the data providers (denoted as publishers) with the data consumers (denoted as subscribers) through a topic-based interaction model, which is shown in Fig. 2. For example, a shipment dispatched from a manufacturer's hub could be labeled as "stakeholderA/countryX/shipmentid6576" and the data associated with this shipment could be accessed by all the relevant stakeholders by simply subscribing to topic "stakeholderA/countryX/shipmentid6576". Following this design philosophy, a topic can be created for each shipment to let the stakeholders both send (publish) and receive (subscribe) data.

Despite its advantages, the publish-subscribe system follows a centralized architecture, wherein the data publishers and subscribers interact via a centralized broker. Here, the broker receives the data from publishers and routes it to the relevant subscribers. In this architecture, it is clear that the broker is critical for data sharing. Therefore, the organization that runs the broker must act honestly and it should not tamper with the data. Recall that the centralized solutions are prone to Byzantine failures. Trinity [63] is a Byzantine fault-tolerant and distributed publish-subscribe broker, that can operate in a multi-stakeholder setting while preventing Byzantine fail-

Fig. 2 Overview of
publish-subscribe broker

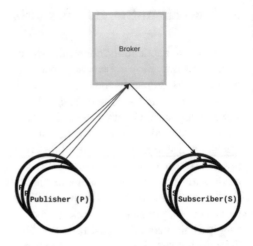

ures and without breaking the fundamental interaction behavior of publish-subscribe messaging model. Figure 3 shows the architecture of Trinity [63].

Trinity is a distributed Byzantine fault-tolerant pub-sub broker [63]. It is created for multi-stakeholder applications, in which each stakeholder is assumed to be operating a broker and a consensus node within their domain to serve its local publishers and subscribers. All the domains are connected through a consensus layer, which is responsible for validating published messages. Each message must be approved by more than one-third of the consensus nodes, before it is sent to the subscribers via a broker. Trinity ensures that when a subscriber in one of the domains receives a message, all the other subscribers in a non-faulty domain also receive the same message, provided more than one-thirds of the domains are non-faulty. A set of safety and liveness properties are presented here, if any reader is interested in understanding the trust and safety guarantees in detail [63].

Streaming Data Payment Protocol: Streaming data payment protocol (SDPP) is another blockchain-based data sharing platform, which incentivizes the data producers. In a supply chain setting, SDPP can be used to handle payments whenever the data associated with the supply chain is shared among stakeholders. It enables a data provider and data consumer to easily connect and transact with each other using micropayments for streaming data [59, 60]. SDPP is a peer-to-peer data sharing protocol. Its design carefully separates out three key components: the off-chain data communication channel (which is operated as a traditional Internet client-server application-layer protocol, atop TCP), a payment channel (implemented using a cryptocurrency protocol), and a records medium (implemented using a distributed ledger technology), as shown in Fig. 4.

Alternatively, applications can interface with third-party blockchain-as-a-service platforms such as IBM's FoodTrust and TradeLens, discussed in Sect. 4. Note that the supply chain systems may have to integrate with a blockchain through the APIs offered by the blockchain platform. When the protocols and hardware

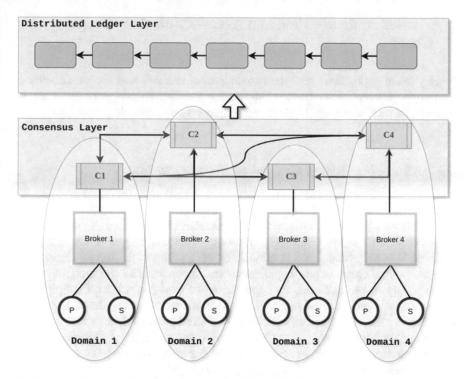

Fig. 3 The architecture of trinity: a distributed publish-subscribe broker [63]

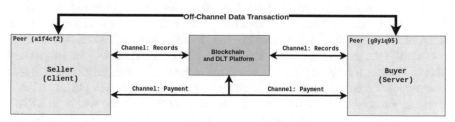

Fig. 4 Overview of streaming data payment protocol [60]

used by the existing supply chain systems are not compatible with the blockchain's API, it becomes challenging to interconnect a supply chain system with blockchain. While solutions such as Trinity, FoodTrust, and TradeLens provide a way to connect blockchain with the supply chain, it is still essential to develop interoperable APIs and protocols for the broader adoption of blockchain in the supply chain.

6   Trusted Authentication and Access Control for Supply Chain

Supply chain applications and their management demand appropriate security in terms of access control and trust. In this section, we discuss the importance of access control and trust in the supply chain, and explore some available proposals.

6.1   Managing Access Control in Supply Chains

Multiple organizations collectively provide and consume information in a supply chain network. Within each organization, a number of employees and devices may be responsible for registering supply chain data. Enabling access to the right entity with the right permission level is critical for the security of the supply chain. Note that a security breach at one organization could compromise the integrity of the entire supply chain. That said, the importance of access control in the supply chain is paramount. Access control is a security mechanism that assures the safe access of resources by the authorized objects. Access control places selective restrictions to access certain resources only by the trusted and authorized entities. These restrictions are governed by a set of access control policies that are controlled by an organization (or a number of organizations) that regulate access [52].

The access control mechanisms in the supply chain should take into consideration the various characteristics of a supply chain (e.g., distributed, time-sensitive, data-centric, etc) system from a multi-stakeholder point of view. Traditional access control mechanisms e.g., Role-Based Access Control (RBAC), Attribute-Based Access Contro (ABAC), Organization-Based Access Control (OrBAC), and Capability-Based Access Control (CapBAC) cannot, in isolation, provide a fine-grained and robust access control solution for the supply chain Hussein et al. [29], Pal [48]. In RBAC, access is enforced based on the specific roles of an entity. In other words, in RBAC, access is granted for a certain resource to a specific role, and the users who belong to this role can then access the resource. However, RBAC is highly centralized, therefore, given the distributed and collaborative nature of a supply chain, it is not an ideal solution at scale.

Alternatively, ABAC uses attributes (e.g., date, time, location, etc.) for enforcing access control policies, conditions and regulations. This approach provides much more flexibility in managing the identities of entities. However, ABAC does not use the concrete unique identity of an entity. This feature of ABAC may be promising for access control in the supply chain since it could provide some level of anonymity for the stakeholders. However, ABAC systems are also centralized in nature. Moreover, when the number of entities grows within the supply chain, ABAC by itself cannot handle the access control issues at scale Pal et al. [51].

OrBAC uses entities (e.g., subject, action, and object) to control access over the resources. It allows the policy designer to explicitly mention security policies independent of the implementation. However, access control enforcement in OrBAC is highly centralized in nature, thus reduces commercial flexibility. CapBAC provides access control solutions based on capabilities. A capability can be defined as a communicable, unforgeable token of authority. It contains access control policies (and associated conditions) for specific resources. These capability tokens can be evaluated at the edge nodes at the time of access. However, given the scale and requirements of a supply chain network, managing the number of capabilities and their distribution is difficult Pal et al. [50].

Access control for supply chain systems should be scalable, flexible, usable, trustworthy, and recognize the inherently decentralized nature of such systems. At the same time, the access control mechanism must be sufficient to protect the privacy, integrity, and confidentiality of the supply chain and its components Pal et al. [54]. The issues noted above around the existing approaches to access control in the supply chain mean that further consideration of access control for the area is needed. Furthermore, the traditional access control models are user-centric and therefore they do not consider the relationship among various users, services, companies, and organizations. As noted earlier (cf. Sect. 1), supply chain applications typically involve multiple stakeholders like producers, transporters, auditors, retailers, and regulators. Therefore, it is important to build trust between these entities for a scalable, flexible, and robust access control for the global supply chain network Zavolokina et al. [81].

We note that the use of blockchain in the supply chain attracted the interest of stakeholders. Blockchain technology is distributed and does not depend upon a centralized trusted authority—this minimizes the amount of trust required from a unit node in the blockchain ledger. To provide trust by commonly used access control mechanisms, in general, a centralized entity or a trusted third party acts as the intermediary to guarantee authenticity. In the case of blockchain, every node in the network is equal and ensures transaction authenticity by using consensus algorithms. So blockchain has the ability to provide a robust and flexible access control by providing distributed, secure, and trust-less features, which cannot be achieved using traditional access control solutions discussed above Song et al. [71].

From a supply chain point of view, the core business and the sub-processes must create a trusted environment for a long-term partnership. That is, to increase customer value and ensure sustainable collaboration between the supply chain parties. However, this collaborative relationship needs a foundation of trust among the various entities within the supply chain network Kidd et al. [36]. Commonly, supply chain managers are not able to foster a trusting relationship among the various partners over the supply chain. Therefore, it is significant to incorporate different aspects of access control as well as other contexts (e.g., social, behavioral, etc.) that help to integrate a number of perspectives for the development of trust in products, suppliers, and customer relationships Sahay [65]. Next, we discuss the significance of trust and its management in supply chain.

6.2   Trust Management in Supply Chain

There is no coherent and universal definition of trust Pal et al. [53]. It is a multi-dimensional and intangible concept without a clear understanding. This further creates difficulty in trust measurement, in particular, for large and complex supply chains. In different disciplines (e.g., social sciences and computing systems) the representation of trust is different.

- Trust in social sciences: It depends on social influences. This may come from characteristics and the individual behavior of an entity (e.g., a human) and are measured as honesty, cooperativeness as well as willingness to cooperate in a social setup. According to Samarati and de Vimercati [66], trust can be seen as "*the extent to which one party is willing to depend on somebody, or something, in a given situation with a feeling of relative security, even though negative consequences are possible*".
- Trust in computing systems: It is different in different areas. That said, in computing the definition of trust differs depending on the domain e.g., networking, security, artificial intelligence, e-commerce, etc. According to Cho et al. [17], it is a subjective belief, as follows: "*an agent's trust is a subjective belief about whether another entity will exhibit behavior reliably in a particular context with potential risks. The agent can make a decision based on learning from past experience to maximize its interest (or utility) and/or minimize risk*". In Jøsang et al. [32], trust is defined as "*the subjective probability by which an individual expects that another performs a given action on which its welfare depends*".

Overall, trust can be referred to as the honesty, truthfulness, or even the reliability of a trustee. Trust is always context-dependent and seen the way it is used. From a supply chain point of view trust can be seen as: "*at the heart of collaborative innovation capabilities. Without a foundation of trust, collaborative alliances can neither be built nor sustained. But, few companies are able to leverage trust for a sustainable competitive advantage; most companies lack high levels of trustworthy collaborations*" Fawcett et al. [21].

Recall, supply chain applications are comprised of stakeholders that may situate in multiple countries and typically function in a distributed infrastructure (with internal partners, supplier, and buyer). In such a mobile and dynamic operational infrastructure, businesses are making decisions based on the data shared by different entities. Therefore the propagation of trust in the supply chain plays an important role in the logistic integration and collaboration of various components in the network to uphold the integrity of the supply chain Su et al. [73]. The organizations should not only trust the data values coming from different entities, but they should also rely on a trusted infrastructure.

In Fig. 5, we illustrate an outline of propagation of trust in the supply chain and its integration in collaborative supply chain operation. On the left side of the figure, we show various antecedents of trust for the supply chain. They are: (1) *shared information* among the different entities that help to build up mutual expectations,

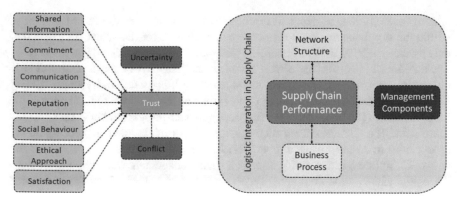

Fig. 5 Propagation of trust and its integration in collaborative supply chain operation

(2) *commitment* to the long-term orientation with both parties cooperating in supply chain values to maintain a trusted relationship, (3) *communication* must be efficient and effective and should allow for feedback, (4) *reputation* is an important factor to building trust by monitoring suppliers (and other components) to evaluate the performance, (5) *social behavior* effectively highlights the issues of cooperation trust for making awareness and sensitivity on social and environmental issues, (6) *ethical approach* represents corporate social responsibility to produce products and services, and (7) *satisfaction* helps to improve on-time service, production, and delivery to the supply chain members Kac et al. [33].

However, the integration of these trust "components" is highly influenced by the *uncertainty* and *conflict* present in the environment. Uncertainty may come from unpredictable events, for example, fluctuations in supply requirements, malicious activities, and frequent modifications to the parts of a supply chain management system. Similarly, conflicts may arise due to disagreements of opinions, business processes, and goals between the buyers and suppliers. Trust in the supply chain has a greater impact on the integration and management of the business process, management components, and network structure for improving operational performance Kolluru and Meredith [37]. Blockchain plays a significant role in this case–in the aggregation of different trust values (can be seen as the measurement of trust from various entities) that help the stakeholders validate the quality of information in supply chain Hou et al. [28].

Chen et al. [15] discuss a trust-based supply chain quality management framework supported by blockchain technology. The discussion is focused on the various use-cases to build a trust management framework within a supply chain network. With a similar view proposal, Agrawal et al. [3] discuss the blockchain-based traceability framework for traceability in multi-level supply chain in textile and clothing industries. Al-Rakhami and Al-Mashari [4], discuss the integration of blockchain technology in the supply chain to address trust issues and how to preserve data integrity between supply chain parties. The proposal also focuses on the efficient transmission of information between the partners in the supply chain. Blockchain is used to verify logged data and reproduces the actual observation of the entities within the supply chain. It helps in trust establishment for the actual (i.e., original) data by ensuring that the blockchain-logged data will be considered trustworthy.

Shahid et al. [70] discuss a trust management framework for agricultural supply chain supported by blockchain technology. In this model, every transaction is written inside the blockchain which finally uploads the data to a Interplanetary File Storage System (IPFS)—which is a protocol and peer-to-peer network that store and share data. The storage system then returns a hash of the data (stored inside the blockchain) to ensure trusted cooperation between the entities in the supply chain network. Similarly, Bai et al. [6] discuss a trust management framework to manage the equipment in agricultural supply where the trust values of sensors are stored in the blockchain. Malik et al. [44] present a trust management framework, called "TrustChain", for supply chain management using blockchain technology. The proposed framework uses a reputation model that evaluates the quality of commodities as well as the trustworthiness of entities based on multiple observations of supply chain events. The reputation score comes from various segments of a supply chain network–separate from participants and products. Blockchain helps in a transparent, efficient, secure, and automated calculation of reputation score to build trust in the supply chain. In Longo et al. [42], a software connector is developed to connect a blockchain with the enterprises' information systems. This proposal aims to allow various companies to share information with one another with different levels of visibility. This in turn, build the trust by checking data authenticity, integrity, and invariability over time through the blockchain.

6.3 Section Summary

In summary, access control and trust are two significant components for building a secure, safe and reliable supply chain infrastructure. Access control preserves confidentiality, integrity, and availability, and trust helps to maintain the integrity of information and relation among the various components. Access control, therefore, helps in trust-building in a supply chain infrastructure Pal et al. [49], Rabehaja et al. [57]. Typically, trust can be observed as a metric that is gathered by the interactions and observations based on the actors involved in a supply chain infrastructure. Finally, blockchain integration in the supply chain can provide secure, scalable, and interoperable communication of information with their trading partners with better access control management and addressing trust issues.

7 Interoperability

Supply chain networks are increasingly decentralized and global. The distributed, heterogeneous and regulated characteristics of supply chain demand blockchain systems to have inherent features for interoperability. Interoperability can be defined as the ability to communicate and access information across various blockchain systems. As there may be multiple small-scale solutions and use-cases within supply chains, interoperability of these solutions is important for compliance and data sharing purposes.

Interoperability is needed as blockchain solutions within supply chain co-exist on enterprise level. These solutions differ depending upon the mutual interest of stakeholders, for example, technology and platform choices, commercial sensitivity of the data, governance and access control, scalability, etc. Furthermore, multiple blockchain ecosystems can be utilized together in a supply chain for various services and functionalities, e.g., a public blockchain can be used for digital payments, whereas a separate consortium blockchain can be used to record supply chain transactions and provide identity and trust management services as in Malik et al. [46]. These blockchain solutions must be able to *communicate* seamlessly without having to worry about the technical and design differences within each ecosystem. To enable these independent blockchain solutions to effectively communicate, careful categorization of the interoperability challenges is required.

7.1 Interoperability Challenges

In this section, we will review the main challenges with respect to interoperability of blockchain enabled supply chain systems.

- **Governance and Data Privacy**: As discussed in Sect. 2, one of the significant requirements of supply chain applications is the regulatory compliance that varies according to the country and different legal jurisdictions. More importantly, auditing and regulation heavily rely on the stakeholders and their willingness to share data across global boundaries. Due to the obvious competitive advantage, data privacy remains of utmost importance which may lead to interoperability barriers. Thus, lack of standardization and governance is a challenge for traceability as well as interoperability.
- **Platform and Data Heterogeneity**: Blockchain platforms are heterogeneous in nature even within the use-case of supply chain systems. There must exist some technical compatibility among the two blockchain platforms to communicate with each other in terms of consensus mechanisms, smart contract operations, and data format. Furthermore, blockchain systems can store various types of data, for example, raw data, hashed event data, and encrypted data. Thus, in addition to blockchain platform heterogeneity, data heterogeneity also introduces challenges for interoperability.
- **Blockchain Infrastructure**: The components that enable blockchain services may include the network infrastructure, back-end oracles, network nodes, cloud servers, etc. Enabling disparate blockchain systems with propriety and legacy enterprise systems to communicate and work seamlessly with each other is another interoperability challenge.

7.2 Existing Approaches for Interoperability

Blockchain interoperability has been a topic of recent interest in literature. Most of these approaches target the interoperability challenge in general, without focusing on the application specific blockchains such as in supply chain or healthcare. In this section, we first discuss some of the literature in interoperability of blockchains and then categorize these approaches toward the end.

Lafourcade and Lombard-Platet [39] discuss the possibility of two blockchains to be interoperable. According to their analysis, two public blockchains are only interoperable if they follow the same transaction, consensus, and block structure which is conceptually equal to having a single blockchain. Similarly, Hardjono et al. [26] present a design philosophy for interoperability of blockchains. Blockchain interoperability can be categorized into mechanical and value levels. Mechanical level involves protocols, encryption standards, consensus mechanisms, transaction structure, etc. whereas value level corresponds to real assets, fiat currencies, etc. which can be associated with coins or tokens. The authors highlight that interoperability at the mechanical level is necessary for interoperability at the value level but it may not guarantee it. In their subsequent work, Hardjono et al. [27] present blockchain gateways as a key notion of interoperability between two blockchains. Dedicated gateways within each blockchain system interact with each other for an asset transfer from one blockchain to another.

A distributed publish/subscribe model, Trinity, with blockchain-based immutability is presented by Ramachandran et al. [63]. Instead of using a centralized broker, the authors propose a blockchain enabled pub-sub broker system. The data to be exchanged is distributed among all the brokers in the network through a consensus mechanism, validated through smart contracts, and then stored on the ledger. Trinity evaluations exhibit that it consumes minimal resources, and the data management processes can be automated using the smart contracts.

Ghaemi et al. [23] present an interoperability solution for permissioned blockchains using a similar publish/subscribe architecture. The proposed solution uses a broker blockchain network as opposed to any third-party. Broker blockchain uses its connector and topic smart contracts to provide connectivity among two other types of blockchains: a *publisher* blockchain network and a *subscriber* blockchain network. Publisher blockchain is the source from which the information is required by a subscriber blockchain. The broker blockchain stores the *topic*, a copy of the information that needs to be shared between the two blockchain networks. The publisher blockchain not only creates a topic on broker blockchain but is also responsible to update any change in the information constituting to the topic. A PoC was developed using Hyperledger Besu, an Ethereum client, and two different versions of Hyperledger Fabric.

An interoperability architecture between private and public blockchain platforms is proposed by Ghosh et al. [24]. The proposed solution, *CollabFed*, leverages on decentralized gateways and smart contracts. A consumer request for information is generated from an open network and logged on the public blockchain network.

The consortium members of the private blockchain also take part in endorsing these user requests by a two-third majority vote. The logged requests trigger the smart contracts in private blockchains which propagate and schedule the requests based on a predefined business logic. The results are then transferred back to the consumers. PoC implementation of CollabFed was developed using Ethereum as the public blockchain platform and Hyperledger Fabric, and Burrow as the private blockchain platforms.

HyperService is a platform proposed by Liu et al. [41] that delivers interoperability and programmability across heterogeneous blockchains. HyperService contributes a programming framework that allows developers to build cross-chain applications in a unified programming model. In addition, it also provides the cryptography protocol that enables realizing these cross-chain applications on blockchains.

An interoperability API implementation, called Bifrost is presented by Scheid et al. [69]. Bifrost allows users to store and retrieve arbitrary data on multiple blockchain systems. Bifrost consists of three components: APIs, blockchain adapters, and a database. The API consists of store, migrate, and retrieve function calls. The blockchain adapters covert user data into a transaction or a query based on store or retrieve function calls. Each blockchain must possess an adapter for communicating with other blockchains. Finally, the database stores the transaction hashes and necessary credentials required to retrieve data.

A token-based cross-blockchain platform for blockchain interoperability is presented by Borkowski et al. [12, 13]. A cross-blockchain asset transfer token, PAN, is introduced using claim-first transactions. The authors describe a method of generating a cryptographic Proof of Intent (PoI). PoI certifies that the sender on source blockchain system is willing to transfer a given amount of assets to a wallet address on a destination blockchain. This PoI then can be used on the destination blockchain to claim the transferred assets. To update the asset records on source blockchain, observing parties (called witnesses) are incentivized for their role.

Various other literature outline the need and requirements of blockchain interoperability. A good survey on blockchain interoperability is provided by Johnson et al. [31] and Belchior et al. [7]. To summarize, the interoperability approaches can be categorized into the following broad categories:

- **Side chains, Management chains, and Relays**: Side chains are considered to be the blockchains which are typically the extension of the main chain. A cross-chain communication protocol is then used for asset exchange between the two chains. Sometimes they are also used as relays to incorporate offline payment or query mechanisms from a financial chain to the main chain. A few examples of such interoperability mechanisms are Broker blockchain by Ghaemi et al. [23], Wood [77], Ethereum 2.0 Sharding with beacon chain, AION, Blocknet, etc.
- **Notaries**: A notary is a trusted entity or an organization who acts as an intermediary between the two blockchain systems and monitors transactions and triggers. It is responsible for exchange of information from the source to destination blockchain. These schemes are employed as centralized or decentralized exchanges, such as Uniswap by Adams et al. [2], and liquid by blockstream [10].

- **APIs/Gateways**: An application programming interface (API) is a piece of code that governs the access point to a server and the rules developers must follow to interact with a database, library, software tool, or programming language. APIs or gateways act as a translator, taking requests from source blockchain and converting them to language or format understandable by the destination blockchain. Application of such approaches are discussed above in Liu et al. [41] and Bifrost by Scheid et al. [69]. A very well-known application is Hyperledger Cactus, a blockchain integration framework which validates cross-chain transactions using a validator network.
- **Hashed/Time locked contracts (HTLCs)**: These approaches use hashlocks and time locks to enforce operational atomicity between the two parties, typically on disparate blockchain systems. A trader provides a cryptographic proof of committing to a transaction before a time-out. HTLCs enable cross-chain atomic operations or atomic swaps, such as for conditional payments from Bitcoin to Ethereum. Few other examples include Bitcoin lightening network and BTC relay.
- **Pub-Sub Models and Tokens**: Pub-Sub models employ the concept of message exchange between publisher and subscriber blockchain networks. Publisher blockchains are often termed as the source of information whereas the subscriber is the requester. The message exchange can be done through either tokens, APIs or side-chains such as systems proposed by Ramachandran et al. [63], Ghaemi et al. [23] and Borkowski et al. [12, 13].

7.3 Suitability of Interoperability Approaches in Supply Chains

The approaches discussed in the previous section can widely be adopted depending upon the interoperability requirement. In supply chains, governance and regulatory compliance is very important for monitoring and auditing purposes, which can be achieved by adopting an interoperability approach based on side chains combined with notaries. A consortium of notaries may play an important role in defining access policies for privacy sensitive data. For integrating financial transactions with physical asset exchanges, tokenization and Pub-Sub models are best suited. Similarly, the APIs and gateways may help to reduce the infrastructure challenges by translating data read/write requests. Although HTLCs and smart contract approaches are complex in terms of implementation and may not be generalized in their design, these approaches are well suited for supply chain systems to enforce standardization based on legal jurisdictions across various chains.

8 Importance of Scalable Blockchains for Supply Chain

Supply chain involves multiple stakeholders that communicate through transactions and smart contracts leading to a huge volume of transactions in blockchain. Conventional blockchains are not directly applicable in supply chain as they suffer from low scalability which is rooted in blockchain resource consumption, latency, efficiency and throughput which are discussed in greater details below.

- **Resource Consumption**: Recall that blockchains are managed distributively by all participating nodes. The conventional methods that facilitate the distributed management of blockchain incur significant computation, memory, and bandwidth overheads which limits its scalability. The computational overhead of the blockchains is rooted by resource consuming consensus algorithms. Additionally, Once a new transaction/block is generated, all participants shall verify the same. The verification of transactions also involves verifying the history of transactions which requires the verification to store blockchain database. However, the size of the blockchain database will significantly increase as due to its immutability, removing or modifying previously stored data is not possible and will compromise the blockchain consistency. Blockchain broadcast the blocks/transactions which consumes significant bandwidth due to large number of participants in supply chain applications.
- **Latency**: There is a non-trivial delay associated with committing a new transaction in the blockchain and receiving confirmation. This delay involves the delay in committing the transaction (i.e., following the consensus algorithm) and the delay in receiving confirmation. The latter involves waiting for a particular number of blocks to be appended to the block in which a particular transaction is stored which in turn protects against double spending and ensures the transaction will not be placed in forked blocks. However, as shown in Fig. 1, supply chain involves transactions that require real-time transaction settlement.
- **Efficiency**: In most of the existing consensus algorithms, multiple validators, i.e., miners, work simultaneously to commit the same block in the blockchain and only the one that follows the consensus algorithm first is permitted to commit the block and thus receive incentive while the resources of other nodes will be wasted. This in turn limits the blockchain efficiency.
- **Throughput**: Conventional blockchains suffer from limited throughput, i.e., total number of transactions that can be committed in blockchain per second. However, supply chain applications demand high throughput due to high transaction generation rate. The number of transactions is continuously increasing as new applications and users join the system which highlights the demand for a throughput management algorithm. The latter ensures that the blockchain throughput can accommodate the load in the network.

Blockchain optimization for applications beyond cryptocurrency, including supply chain, has received significant attention in recent years. The authors in Thakur and Breslin [74] employed off-chain transactions to increase the blockchain scalability. The participating nodes that are involved in the life time of a product, jointly create a channel where all transactions related to the product are stored. This in turn reduces the number of transactions that needs to be committed in the blockchain as for each channel only two transactions will be committed in the blockchain. The rest of the communication happens in the channel.

To increase the blockchain scalability, the authors in Malik et al. [43] employed the concept of *sharding*. In general, sharding refers to dividing the network into smaller groups, i.e., shards where the transactions/blocks generated in each shard are broadcast and verified inside the shard. Thus, sharding increases the blockchain scalability by limiting the scope of the ledger. Sharding improves the blockchain throughput and reduces delay linearly as the improvement rate is directly impacted by the number of shards that exist in the network.

The authors in Dorri et al. [20] introduced a verification and communication model for blockchain, known as *Vericom*, to the blockchain bandwidth consumption by introducing a dynamic multicasting and traffic routing algorithms. To route traffic, i.e., blocks and transactions, the authors employed a PK-based routing algorithm where the traffic is routed based on the PK of the destination. A group of high resource available nodes in the network jointly form a backbone network. Each backbone node is allocated to a particular *routing character* which refers to the most significant characters of the hash function output. Depending to the hash of their PK, the participating nodes join a backbone node to receive transactions sent for them. The backbone nodes use conventional IP-based routing algorithms to route traffic in backbone. Unlike conventional blockchains where transactions are broadcast, Vericom multicasts the transactions to a randomly selected group of nodes known as verifier sets. The latter is unique per transaction/block and is selected based on the hash function output of the transaction/block content.

The consensus algorithm is fundamental in blockchain scalability. Most of the conventional consensus algorithms consume significant resources, limit throughput, lack efficiency, and involve significant delay in committing and confirming transactions. Various optimized consensus algorithms have been introduced in recent years Sankar et al. [67]. Intel introduced Proof of Elapsed Time (POET) Sankar et al. [67] consensus algorithm which relies on the Trusted Execution Environment (TEE) in Intel CPUs. When a validator aims to commit a new block to the blockchain, it needs to wait for a random period of time identified by TEE. In case during the waiting period, it receives a block consisting the same transactions, it shall drop the block and start committing a new block.

In Dorri and Jurdak [19] the authors introduced a scalable fast consensus algorithm with near real-time transaction settlement which is known as *Tree-chain*. Tree-chain is a leader-selection consensus algorithm where a leader is selected for a particular period of time and commit transactions, with specific features, to the blockchain. Tree-chain relies on the hash function output to achieve randomization in two levels: blockchain level and transaction level. In blockchain level, a *Consensus Code Range (CCR)* is distributed randomly and in an unpredictable way between the validators. CCR refers to the most significant characters of the hash function output. Each potential character in the hash function output is associated with a weight defined in a weight dictionary. The validators calculate a Key Weight Metric (KWM) which is essentially the sum of the weight of all the characters in the hash of a PK. The KWMs are then stored in a list in descending order and the node with the highest value of KWM is dedicated to the first CCR range.

In the transaction level, the validator of each transaction is randomly selected based on the hash of the transaction content. The validator whose CCR matches with the most significant characters of the hash of a particular transaction is the validator in charge for committing that particular transaction in the blockchain. Each validator creates a unique ledger chained to the genesis block for committing its corresponding blocks which ensures fast transaction settlement. Thus, Tree-chain, by design, embraces the concept of forking.

9   Design Considerations and Open Challenges

This section reviews the design considerations and open challenges to help stake-holders carefully adopt blockchain technology for their supply chain applications.

9.1   Public Versus Permissioned Blockchains

A blockchain-based supply chain application can be designed using either a permissioned or a public blockchain. When selecting a blockchain platform, the stakeholders must carefully weigh the pros and cons of these approaches.

A public blockchain primarily relies on nodes maintained by the community members. In the case of BitCoin [47] and Ethereum [78], any community member with a computationally capable node can join the network and participate in the PoW consensus process. As discussed in Sect. 3, the node that successfully solves the cryptographic puzzle gets rewarded with a cryptocurrency. Here, the reward comes from the users that submitting transactions to the network. It is important to note that the user must pay a transaction fee whenever a transaction is submitted to the blockchain network. Therefore, applications that rely on public blockchain platforms must consider the transaction fees. Note that a supply chain application that produces hundreds of transactions per day must spend tens to hundreds of dollars per day for leveraging the services of a public blockchain.

Alternatively, applications can use a permissioned or private blockchain for a supply chain application. Following this model, the stakeholders that are part of a supply chain network would share their computing nodes for managing and maintaining the blockchain in a distributed and trusted fashion. Technically, a private blockchain network can be established with platforms such as Hyperledger Fabric [5], Tendermint [5, 14], or a customized private version of Ethereum. Such platforms typically employ a lightweight consensus algorithm, which differs from a computationally expensive PoW consensus model. Since the blockchain network is established among the stakeholders that are part of a supply chain network relevant to their business following a lightweight consensus process, users are not required to pay a transaction fee. It is essential to note the operational cost of running a private blockchain network are covered by businesses that leverage them. Besides, the cost would be marginal since the companies are using the Internet and computing code without relying on computationally intensive consensus processes. And, the blockchain itself does not demand users to pay a transaction fee, as seen in public blockchains. Therefore, private blockchain platforms are cost-effective while allowing organizations to leverage trust and immutability in a distributed and multi-stakeholder settings.

In summary, the stakeholders must weigh in these pros and cons when selecting a blockchain platform as the operational cost of a blockchain may discourage some stakeholders from submitting transactions to a public network, potentially reducing the effectiveness of a blockchain-based supply chain application [18].

9.2   Preventing Garbage-In-Garbage-Out Problem

Blockchain technology allows the stakeholders to track and trace their products throughout the supply chain reliably. When products and goods move through a supply chain, each stakeholder must gather and log data associated with them. For example, the arrival time, temperature, humidity, and expiry date of a product must be logged at each supply chain endpoint to ensure that the food products are in edible condition. Under this operational setting, blockchain technology can receive information about the products and store them in an immutable ledger following a consensus process involving multiple stakeholders in a distributed network. It is essential to underscore how blockchain can also receive incorrect information from the stakeholders and store them in a blockchain. Note that blockchain technology does not offer any mechanism to verify the correctness of the data submitted by a user. This issue is referred to as *Garbage-In-Garbage-Out* problem [83], which means the blockchain technology can also store inaccurate information in the ledger and present them to the users in the future. Stakeholders may think that blockchain technology is helping them provide transparency to the end-users. On the contrary, inaccurate provenance information may mislead the end consumers while ruining the reputation of the organizations that are part of the supply chain network. Therefore, the stakeholders must ensure that the blockchain ledger's information is accurate for a genuinely dependable supply chain ecosystem. Figure 6 highlights the importance of end-to-end trust.

Garbage-In-Garbage-Out problems require solutions outside of the blockchain. In particular, the interface between the physical and the digital world must be built using robust methods to ensure end-to-end trust and transparency. Therefore, supply chain applications must include approaches involving humans, sensing devices, trusted third parties including regulatory bodies and certification laboratories to weed out false claims in a multi-stakeholder supply chain system.

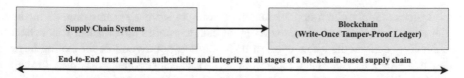

Fig. 6 Garbage-In-Garbage-Out problem: storing data in a blockchain ledger alone does not guarantee authenticity and trust. All the layers of a blockchain-based supply chain require trusted methods

9.3 Automated Verification of Compliance

A supply chain network is established involving multiple stakeholders operating from numerous jurisdictions. Within each jurisdiction, stakeholders must comply with the regulations enforced by the local government and other regulatory bodies. In a supply chain network, examples of regulations include export controls and technical regulations. Contemporary supply chain applications rely on paperwork, and many human operators at various jurisdictions ensure regulatory compliance. When leveraging blockchain technology for the supply chain, it may be easier to automate the compliance verification process through a smart contract [11]. Recall that smart contracts allow the stakeholders to track and manage the supply chain autonomously. Authorities could translate the regulatory standards into one or more smart contracts to speed up the compliance verification process.

It is important to note that the automated verification of regulatory compliance requires accurate data from the stakeholders. For regulatory authorities to adopt automation in this context, the Garbage-In-Garbage-Out problem discussed in Sect. 9.3 must be solved. Ramachandran et al. [61] introduces an *Assisted Autonomy* framework to automate the compliance process in cross-border supply chains gradually. Such approaches are promising, but data trust remains a significant problem.

9.4 Lack of Common Data Standard

When establishing a supply chain network involving multiple stakeholders from multiple jurisdictions, each stakeholder may follow a custom data standard local to their country. An absence of a common data standard in a distributed and multi-stakeholder setting would result in interoperability issues, wherein each stakeholder may have to develop new extensions to their existing infrastructure, which would hamper adoption [64]. It is important to note that an organization responsible for shipping products may be dealing with several products belonging to various stakeholders. For such an organization, it is almost impossible to develop new extensions to interface with each organization seamlessly.

Besides, supply chain applications may want to store a combination of media files such as images and videos and digital data to enable provenance. Blockchain technology can handle a few bytes of data, but it is not a good fit for storing large data items such as videos and images. In such cases, stakeholders may have to keep the data either locally and store the hash on the blockchain or in decentralized file storage such as IPFS [8] and store the file pointer on the blockchain, as discussed here [62].

Given these issues around the absence of a common data standard and large data files, supply chain applications must develop and adopt a universal and application-agnostic data standard.

9.5 Privacy Concerns

The information about the products helps the stakeholders understand the status of the supply chain. Depends on the product and the supply chain network, the stakeholder may have to share sensitive details, including the ingredients used (in the case of a food producer) and location (in the case of shipping companies). Stakeholders may not be willing to openly share sensitive details since it may leak their intellectual property or other business secrets. In transportation and shipping companies, people's location information gets exposed to several stakeholders. Stenberg et al. [72] reports that the truck drivers are not comfortable with live tracking of their location in a real-world supply chain. Besides, the data protection regulations [55] such as General Data Protection Regulation (GDPR), California Consumer Privacy Act (CCPA), and Australian Privacy Act (APA) provide guidelines about the use of data associated with products and companies operating from a particular jurisdiction. Such regulations and the privacy concerns of the stakeholders must be considered when designing a blockchain-based supply chain application. Malik et al. introduce PrivChain [45], a blockchain-based solution for protecting the privacy of the supply chain stakeholders using Zero-knowledge proof and Pederson commitments.

9.6 Lack of Interoperability with Legacy IT Systems

Many supply chain organizations have been using enterprise systems to manage their supply chain processes. Such systems enable stakeholders to share information about their products, handle payments, and generate paperwork for regulatory compliance. Note that organizations may have made significant investments to build and deploy these systems [72]. Connecting these legacy systems with a blockchain-based supply chain may require a considerable overhaul, making it more expensive and disruptive. Therefore, blockchain-based supply chain applications must support protocols and services leveraged by legacy and enterprise systems.

9.7 Operational Costs

Supply chain systems need a set of new processes and methods to establish a trusted blockchain-based supply chain. Such extensions may require personnel with specialized skills. Besides, organizations may also have to invest in new hardware and software infrastructure. Such issues may make organizations hesitate to adopt a new blockchain-based supply chain system. Allowing organizations to leverage legacy systems and processes would reduce the operational cost to a large extent.

In addition, organizations have to deploy consensus nodes and manage them if they are part of a private blockchain network. The deployment and management of consensus nodes introduce a significant overhead both in terms of hardware and networking. IBM's TradeLens and FoodTrust offer blockchain-as-a-service to organizations. Following this model, companies need not deploy new hardware for leveraging blockchain for traceability and provenance. However, organizations must trust the third-party service provider since the blockchain infrastructure is not owned and managed by multiple organizations in a distributed and transparent setting.

9.8 Lack of Engagement from Field Operators

A supply chain application involves field operators responsible for driving trucks or manually recording information at a supply chain terminal. The information entered by these field operators is highly critical for tracking the flow of goods in a supply chain. The integration of blockchain to a supply chain network may expose some of the data entered by these field operators to other entities in the supply chain in a raw format. Note that current systems may record the raw data in a server belonging to the field operator's organization. As part of the information sharing process, each organization may only share a summary or high-level statistics with other organizations, filtering sensitive information recorded by the field operators. With the introduction of blockchain, sensitive information may get stored directly on the blockchain, which may infringe the privacy of the field operators. Due to the perceived risk of privacy violation, field operators may hesitate to record data, which would reduce the effectiveness of blockchain-based supply chain applications. The integration of privacy-preserving features may provide confidence to the field operators, which would increase their engagement with the system.

In addition, blockchain-based supply chain systems should not include cumbersome processes demanding additional effort from field workers. For example, a truck driver may not favor logging into a terminal every few minutes to register the status of the supply chain [72]. Therefore, it is essential to consider how the new processes affect the day-to-day work of the field operators.

9.9 Payment Processing Challenges

In a supply chain application, multiple stakeholders provide products and services to other stakeholders. When a physical product moves in a forward direction toward the end consumers, the payment associated with the products flows in the reverse direction. The literature on supply chain argues the need for the inclusion of a digital infrastructure and automation processes to prevent payment delays [76]. Having a cumbersome manual process with paper-based documents would be detrimental to supply chain efficiency. Therefore, digitization and the automation of the payment process would be massively beneficial to the supply chains.

Blockchain technology includes smart contracts, an innovative and decentralized computation engine. Using smart contracts, businesses can code up the payment process, which supply chain events could trigger. However, the events that drive the smart contracts must carry accurate information since the organizations may not want to make the payment without having confidence in the data. Therefore, the *garbage-in-garbage-out* problem discussed in Sect. 9.3 must be resolved to confidently automate payments.

9.10 Sustainability Demands

Global warming and ever-growing carbon emissions force authorities to impose stricter regulations around sustainability. Supply chain companies are asked to incorporate sustainable standards and practices in their operations. Many of the existing blockchain-based supply chain systems predominantly track the flow of products toward the end consumers and process payments in the reverse direction. Recycling of products is not given much importance. The emerging sustainability standards emphasize the importance of recycling. Many blockchain-based solutions have been created to incentivize sustainable behavior in the food supply chain [16, 82]. Following these initiatives, the supply chain stakeholders must introduce mechanisms to manage the recycling and decommissioning processes. Note that the blockchain technology offers support for tokens, which can be used to incentivize end consumers for their sustainable actions [56, 68].

10 Decision Tree for Supply Chain

Figure 7 provides a guideline for supply chain stakeholders. It helps the stakeholders choose a suitable solution for their supply chain based on the design and operational constraints. In a nutshell, applications producing hundreds to thousands of events have to opt for a private blockchain to minimize the operational costs. From the cost perspective, the number of transactions is a good indicator, but other con-

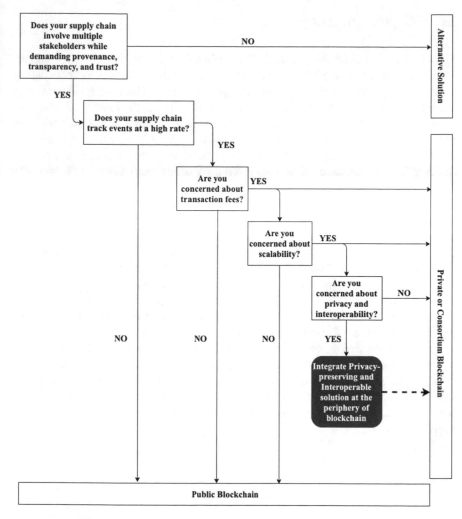

Fig. 7 A Decision-tree for selecting the suitable solution for supply chain applications

strains including privacy and the need for interoperability warrant new solutions at the periphery of the blockchain. While we acknowledge that this decision tree is not comprehensive for a supply chain domain, but it highlights the importance of minimizing the operational cost at the early design phases to establish an effective and sustainable supply chain.

11   Chapter Summary

Supply chain applications operate in a distributed and collaborative environment involving multiple stakeholders from various jurisdictions. Stakeholders including producers, regulatory bodies, logistics companies, and end consumers increasingly demand transparency and provenance to fully understand the history of products. Centralized digital and enterprise systems offer a viable solution for tracking products in a supply chain, but they suffer from single point of failure, wherein the organization that runs and manages the infrastructure may misbehave compromising the integrity of the entire supply chain. Blockchain technology offers features such as immutable ledger, decentralization, and smart contracts, which are tailor-made for the creation of a trusted and decentralized supply chain networks.

This chapter introduced the challenges and the requirements of a supply chain application. In particular, it introduced how blockchain technology can be used to share data in a multi-stakeholder supply chain environment. The discussions on trust and access control underscores the importance of security and authentication requirements. To help the supply chain stakeholders decide on the choice of blockchain platforms, this chapter discussed the scalability challenges of blockchain-based systems. The section on interoperability highlighted the need to support legacy systems and multi-blockchain frameworks to enable broad adoption.

Lastly, the section on open challenges and opportunities provided guidelines to supply chain stakeholders as they start to adopt blockchain technology for their supply chain. This chapter showed that blockchain technology offers several advantages to supply chain applications, but the issues around scalability, interoperability, privacy, regulations, and data trust must be solved to establish a dependable multi-stakeholder supply chain.

References

1. Abeyratne, S.A., Monfared, R.P.: Blockchain ready manufacturing supply chain using distributed ledger. Int. J. Res. Eng. Technol. **5**(9), 1–10 (2016)
2. Adams, H., Zinsmeister, N., Salem, M., Keefer, R., Robinson, D.: Uniswap v3 Core. Technical Report, Uniswap. Technical report (2021)
3. Agrawal, T.K., Kumar, V., Pal, R., Wang, L., Chen, Y.: Blockchain-based framework for supply chain traceability: a case example of textile and clothing industry. Comput. Ind. Eng. **154**, 107130 (2021)
4. Al-Rakhami, M.S., Al-Mashari, M.: A blockchain-based trust model for the internet of things supply chain management. Sensors **21**(5), 1759 (2021)
5. Androulaki, E., Barger, A., Bortnikov, V., Cachin, C., Christidis, K., De Caro, A., Enyeart, D., Ferris, C., Laventman, G., Manevich, Y., et al.: Hyperledger fabric: a distributed operating system for permissioned blockchains. In: Proceedings of the Thirteenth EuroSys Conference, pp. 1–15 (2018)
6. Bai, Y., Fan, K., Zhang, K., Cheng, X., Li, H., Yang, Y.: Blockchain-based trust management for agricultural green supply: a game theoretic approach. J. Clean. Prod. 127407 (2021)
7. Belchior, R., Vasconcelos, A., Guerreiro, S., Correia, M.: A survey on blockchain interoperability: past, present, and future trends (2020). arXiv:2005.14282

8. Benet, J.: Ipfs-content addressed, versioned, p2p file system (2014). arXiv:1407.3561
9. Blockchain, I.: Ibm food trust. a new era for the world's food supply (2019). Accessed 15 Oct 2019
10. Blockstream. Liquid network: faster, more confidential bitcoin transactions (2021). https://blockstream.com/liquid/
11. Bocek, T., Rodrigues, B.B., Strasser, T., Stiller, B.: Blockchains everywhere-a use-case of blockchains in the pharma supply-chain. In: 2017 IFIP/IEEE Symposium on Integrated Network and Service Management (IM), pp. 772–777 (2017). https://doi.org/10.23919/INM.2017.7987376
12. Borkowski, M., Ritzer, C., McDonald, D., Schulte, S.: Caught in chains: claim-first transactions for cross-blockchain asset transfers. Technische Universität Wien. Technical Report, TU Wien (2018)
13. Borkowski, M., Ritzer, C., Schulte, S.: Deterministic witnesses for claim-first transactions. Technische Universität Wien. Technical Report, TU Wien (2018)
14. Buchman, E.: Tendermint: Byzantine fault tolerance in the age of blockchains. Ph.D. thesis (2016)
15. Chen, S., Shi, R., Ren, Z., Yan, J., Shi, Y., Zhang, J.: A blockchain-based supply chain quality management framework. In: 2017 IEEE 14th International Conference on e-Business Engineering (ICEBE), pp. 172–176. IEEE (2017)
16. Chidepatil, A., Bindra, P., Kulkarni, D., Qazi, M., Kshirsagar, M., Sankaran, K.: From trash to cash: how blockchain and multi-sensor-driven artificial intelligence can transform circular economy of plastic waste? Adm. Sci. **10**(2), 23 (2020)
17. Cho, J.-H., Swami, A., Chen, R.: A survey on trust management for mobile ad hoc networks. IEEE Commun. Surv. Tutor. **13**(4), 562–583 (2010)
18. De Giovanni, P.: Blockchain and smart contracts in supply chain management: a game theoretic model. Int. J. Prod. Econ. **228**, 107855 (2020). ISSN 0925-5273. https://doi.org/10.1016/j.ijpe.2020.107855. https://www.sciencedirect.com/science/article/pii/S0925527320302188
19. Dorri, A., Jurdak, R.: Tree-chain: a fast lightweight consensus algorithm for iot applications. In: 2020 IEEE 45th Conference on Local Computer Networks (LCN), pp. 369–372. IEEE (2020)
20. Dorri, A., Mishra, S., Jurdak, R.: Vericom: a verification and communication architecture for iot-based blockchain (2021). arXiv:2105.12279
21. Fawcett, S.E., Jones, S.L., Fawcett, A.M.: Supply chain trust: the catalyst for collaborative innovation. Bus. Horiz. **55**(2), 163–178 (2012)
22. Garcia, R.D., Zutião, G.A., Ramachandran, G., Ueyama, J.: Towards a decentralized e-prescription system using smart contracts. In: 2021 IEEE 34th International Symposium on Computer-Based Medical Systems (CBMS), pp. 556–561 (2021). https://doi.org/10.1109/CBMS52027.2021.00037
23. Ghaemi, S., Rouhani, S., Belchior, R., Cruz, R.S., Khazaei, H., Musilek, P.: A pub-sub architecture to promote blockchain interoperability (2021). arXiv:2101.12331
24. Ghosh, B.C., Bhartia, T., Addya, S.K., Chakraborty, S.: Leveraging public-private blockchain interoperability for closed consortium interfacing (2021). arXiv:2104.09801
25. Grigg, I.: Eos-an introduction. White paper (2017). https://whitepaperdatabase.com/eos-whitepaper
26. Hardjono, T., Lipton, A., Pentland, A.: Towards a design philosophy for interoperable blockchain systems (2018). arXiv:1805.05934
27. Hardjono, T., Lipton, A., Pentland, A.: Toward an interoperability architecture for blockchain autonomous systems. IEEE Trans. Eng. Manag. **67**(4), 1298–1309 (2019)
28. Hou, Y., Xiong, Y., Wang, X., Liang, X.: The effects of a trust mechanism on a dynamic supply chain network. Expert Syst. Appl. **41**(6), 3060–3068 (2014)
29. Hussein, D., Bertin, E., Frey, V.: A community-driven access control approach in distributed iot environments. IEEE Commun. Mag. **55**(3), 146–153 (2017)
30. Jensen, T., Hedman, J., Henningsson, S.: How tradelens delivers business value with blockchain technology. MIS Q. Exec. **18**(4) (2019)

31. Johnson, S., Robinson, P., Brainard, J.: Sidechains and interoperability (2019). arXiv:1903.04077
32. Jøsang, A., Ismail, R., Boyd, C.: A survey of trust and reputation systems for online service provision. Decis. Support Syst. **43**(2), 618–644 (2007)
33. Kac, S.M., Gorenak, I., Potocan, V.: The influence of trust on collaborative relationships in supply chains. E+M Ekonomie a Manag. **19**(2), 120–131 (2016)
34. Kamble, S.S., Gunasekaran, A., Sharma, R.: Modeling the blockchain enabled traceability in agriculture supply chain. Int. J. Inf. Manag. **52**, 101967 (2020)
35. Kamilaris, A., Fonts, A., Prenafeta-Boldύ, F.X.: The rise of blockchain technology in agriculture and food supply chains. Trends in Food Sci. Technol. **91**, 640–652 (2019)
36. Kidd, J., Richter, F.-J., Li, X.: Learning and trust in supply chain management. Manag. Dec. (2003)
37. Kolluru, R., Meredith, P.H.: Security and trust management in supply chains. Inf. Manag. Comput. Secur. (2001)
38. Kurpjuweit, S., Schmidt, C.G., Klöckner, M., Wagner, S.M.: Blockchain in additive manufacturing and its impact on supply chains. J. Bus. Logist. **42**(1), 46–70 (2021)
39. Lafourcade, P., Lombard-Platet, M.: About blockchain interoperability. Inf. Process. Lett. **161**, 105976 (2020)
40. Lee, J.-H., Pilkington, M.: How the blockchain revolution will reshape the consumer electronics industry [future directions]. IEEE Consum. Electron. Mag. **6**(3), 19–23 (2017)
41. Liu, Z., Xiang, Y., Shi, J., Gao, P., Wang, H., Xiao, X., Wen, B., Hu, Y.-C.: Hyperservice: interoperability and programmability across heterogeneous blockchains. In: Proceedings of the 2019 ACM SIGSAC Conference on Computer and Communications Security, pp. 549–566 (2019)
42. Longo, F., Nicoletti, L., Padovano, A., d'Atri, G., Forte, M.: Blockchain-enabled supply chain: an experimental study. Comput. Ind. Eng. **136**, 57–69 (2019)
43. Malik, S., Kanhere, S.S., Jurdak, R.: Productchain: scalable blockchain framework to support provenance in supply chains. In: 2018 IEEE 17th international symposium on network computing and applications (NCA), pp. 1–10. IEEE (2018)
44. Malik, S., Dedeoglu, V., Kanhere, S.S., Jurdak., R.: Trustchain: trust management in blockchain and iot supported supply chains. In: 2019 IEEE International Conference on Blockchain (Blockchain), pp. 184–193. IEEE (2019)
45. Malik, S., Dedeoglu, V., Kanhere, S., Jurdak, R.: Privchain: provenance and privacy preservation in blockchain enabled supply chains (2021). arXiv:2104.13964
46. Malik, S., Gupta, N., Dedeoglu, V., Kanhere, S.S., Jurdak, R.: Tradechain: decoupling traceability and identity in blockchain enabled supply chains (2021). arXiv:2105.11217
47. Nakamoto, S.: Bitcoin: A peer-to-peer electronic cash system. Technical Report, Manubot (2019)
48. Pal, S.: Internet of things and access control: sensing, monitoring and controlling access in IoT-enabled healthcare systems, vol. 37. Springer Nature (2021)
49. Pal, S., Hitchens, M., Varadharajan, V.: On the design of security mechanisms for the internet of things. In: 2017 Eleventh International Conference on Sensing Technology (ICST), pp. 1–6. IEEE (2017)
50. Pal, S., Hitchens, M., Varadharajan, V.: Towards a secure access control architecture for the internet of things. In: 2017 IEEE 42nd Conference on Local Computer Networks (LCN), pp. 219–222. IEEE (2017)
51. Pal, S., Hitchens, M., Varadharajan, V., Rabehaja T.: Fine-grained access control for smart healthcare systems in the internet of things. EAI Endorsed Trans. Ind. Netw. Intell. Syst. **4**(13) (2018)
52. Pal, S., Hitchens, M., Varadharajan, V., Rabehaja, T.: Policy-based access control for constrained healthcare resources. In: 2018 IEEE 19th International Symposium on A World of Wireless, Mobile and Multimedia Networks (WoWMoM), pp. 588–599. IEEE (2018)
53. Pal, S., Hitchens, M., Varadharajan, V.: Towards the design of a trust management framework for the internet of things. In: 2019 13th International Conference on Sensing Technology (ICST), pp. 1–7. IEEE (2019)

54. Pal, S., Rabehaja, T., Hill, A., Hitchens, M., Varadharajan, V.: On the integration of blockchain to the internet of things for enabling access right delegation. IEEE Internet Things J. **7**(4), 2630–2639 (2019)
55. Pantlin, N., Wiseman, C., Everett, M.: Supply chain arrangements: the abc to gdpr compliance- a spotlight on emerging market practice in supplier contracts in light of the gdpr. Comput. Law Secur. Rev. **34**(4), 881–885 (2018)
56. Patel, D., Britto, B., Sharma, S., Gaikwad, K., Dusing, Y., Gupta, M.: Carbon credits on blockchain. In: 2020 International Conference on Innovative Trends in Information Technology (ICITIIT), pp. 1–5. IEEE (2020)
57. Rabehaja, T., Pal, S., Hitchens, M.: Design and implementation of a secure and flexible access-right delegation for resource constrained environments. Futur. Gener. Comput. Syst. **99**, 593–608 (2019)
58. Radanović, I., Likić, R.: Opportunities for use of blockchain technology in medicine. Appl. Health Econ. Health Policy **16**(5), 583–590 (2018)
59. Radhakrishnan, R., Krishnamachari, B.: Streaming data payment protocol for the internet of things. In: The 1st International Workshop on Blockchain for the Internet of Things, held in conjunction with IEEE Blockchain (2018)
60. Radhakrishnan, R., Ramachandran, G.S., Krishnamachari, B.: Sdpp: streaming data payment protocol for data economy. In: 2019 IEEE International Conference on Blockchain and Cryptocurrency (ICBC), pp. 17–18. IEEE (2019)
61. Ramachandran, G., Deane, F., Malik, S., Dorri, A., Jurdak, R.: Towards assisted autonomy for supply chain compliance management. In: Proceedings of the 2021 Third IEEE International Conference on Trust, Privacy and Security in Intelligent Systems and Applications (TPS-ISA). Institute of Electrical and Electronics Engineers Inc. (2021). https://eprints.qut.edu.au/226429/. eCF Paper Id: 1637720482626
62. Ramachandran, G.S., Radhakrishnan, R., Krishnamachari, B.: Towards a decentralized data marketplace for smart cities. In: 2018 IEEE International Smart Cities Conference (ISC2), pp. 1–8 (2018). https://doi.org/10.1109/ISC2.2018.8656952
63. Ramachandran, G.S., Wright, K.-L., Zheng, L., Navaney, P., Naveed, M., Krishnamachari, B., Dhaliwal, J.: Trinity: a byzantine fault-tolerant distributed publish-subscribe system with immutable blockchain-based persistence. In: 2019 IEEE International Conference on Blockchain and Cryptocurrency (ICBC), pp. 227–235, (2019). https://doi.org/10.1109/BLOC.2019.8751388
64. Renner, S.A., Rosenthal, A.S., Scarano, J.G.: Data interoperability: standardization or mediation. In: 1st IEEE Metadata Conference. Citeseer (1996)
65. Sahay, B.S.: Understanding trust in supply chain relationships. Ind. Manag. Data Syst. (2003)
66. Samarati, P., de Vimercati, S.C.: Access control: policies, models, and mechanisms. In: International School on Foundations of Security Analysis and Design, pp. 137–196. Springer (2000)
67. Sankar, L.S., Sindhu, M., Sethumadhavan, M.: Survey of consensus protocols on blockchain applications. In: 2017 4th International Conference on Advanced Computing and Communication Systems (ICACCS), pp. 1–5. IEEE (2017)
68. Saraji, S., Borowczak, M.: A blockchain-based carbon credit ecosystem (2021). arXiv:2107.00185
69. Scheid, E.J., Hegnauer, T., Rodrigues, B., Stiller, B.: Bifröst: a modular blockchain interoperability api. In: 2019 IEEE 44th Conference on Local Computer Networks (LCN), pp. 332–339. IEEE (2019)
70. Shahid, A., Almogren, A., Javaid, N., Al-Zahrani, F.A., Zuair, M., Alam, M.: Blockchain-based agri-food supply chain: a complete solution. IEEE Access **8**, 69230–69243 (2020)
71. Song, Q., Chen, Y., Zhong, Y., Lan, K., Fong, S., Tang, R.: A supply-chain system framework based on internet of things using blockchain technology. ACM Trans. Internet Technol. (TOIT) **21**(1), 1–24 (2021)
72. Sternberg, H.S., Hofmann, E., Roeck, D.: Newblock the struggle is real: insights from a supply chain blockchain case. J. Bus. Logist. **42**(1), 71–87 (2021). https://doi.org/10.1111/jbl.12240. https://onlinelibrary.wiley.com/doi/abs/10.1111/jbl.12240

73. Su, Z., Li, M., Fan, X., Jin, X., Wang, Z.: Research on trust propagation models in reputation management systems. In: Mathematical Problems in Engineering (2014)
74. Thakur, S., Breslin, J.G.: Scalable and secure product serialization for multi-party perishable good supply chains using blockchain. Internet of Things **11**, 100253 (2020)
75. Underwood, S.: Blockchain beyond bitcoin. Commun. ACM **59**(11), 15–17 (2016)
76. Viriyasitavat, W., Hoonsopon, D., Bi, Z.: Augmenting cryptocurrency in smart supply chain. J. Ind. Inf. Integr. **21**, 100188 (2021)
77. Wood, G.: Polkadot: vision for a heterogeneous multi-chain framework. White Paper (2016)
78. Wood, G., et al.: Ethereum: A secure decentralised generalised transaction ledger. Ethereum Project Yellow Paper **151**(2014), 1–32 (2014)
79. Wu, F., Yeniyurt, S., Kim, D., Cavusgil, S.T.: The impact of information technology on supply chain capabilities and firm performance: a resource-based view. Ind. Mark. Manag. **35**(4), 493–504 (2006)
80. Xu, X., Rahman, F., Shakya, B., Vassilev, A., Forte, D., Tehranipoor, M.: Electronics supply chain integrity enabled by blockchain. ACM Trans. Des. Autom. Electron. Syst. (TODAES) **24**(3), 1–25 (2019)
81. Zavolokina, L., Zani, N., Schwabe, G.: Designing for trust in blockchain platforms. IEEE Trans. Eng. Manag. (2020)
82. Zhang, Y., Guin, U.: End-to-end traceability of ics in component supply chain for fighting against recycling. IEEE Trans. Inf. Forensics Secur. **15**, 767–775 (2019)
83. Ziolkowski, R., Miscione, G., Schwabe, G.: Decision problems in blockchain governance: old wine in new bottles or walking in someone else's shoes? J. Manag. Inf. Syst. **37**(2), 316–348 (2020)

Tokenization of Assets

Raghu Bala

Abstract An asset is classified as any resource owned that can be used to produce positive economic value, and that economic value is usually recorded on a ledger. Typically, these ledgers have been the balance sheet of enterprises. With the advent of distributed ledgers, in several use cases, it has been found that in certain asset classes it is better to register these assets on a decentralized ledger.

In this paper, we will examine how assets are recorded on distributed ledgers, and how such ownership can be encapsulated in the form of a ***token***. The tokenization of assets refers to the process of issuing a blockchain token that digitally represents a real tradable asset—in many ways similar to the traditional process of securitization [14]. We discuss the various factors that influence the value of a token, the different forms of tokens, and how tokens can be traded on exchanges.

The assets we will tackle include digital, physical, dynamic, and financial assets. For instance, a song is an asset in the music industry, and tokenization enables one to create liquidity for stakeholders. These stakeholders may include artistes, distributors, producers, music labels and more—who can now go directly to consumers using tokens thereby disintermediating the process. This typifies the opportunities that asset tokenization brings about.

1 What Are Assets?

Assets are resources that hold value, value which is typically converted to cash or some other class of asset by the owner of the asset. For instance, one may wish to purchase an automobile and assuming one purchases the car for cash, then the transaction is simply one that involves two parts: (a) The transfer of assets between two parties and (b) the transfer of title between the parties.

R. Bala (✉)
NetObjex, Irvine, CA, USA
e-mail: raghu@netobjex.com

Typically for small everyday items, e.g., the purchase of a bag of apples at supermarket does not involve the transfer of title, and simply is the exchange of payment (cash or cash equivalent) for the goods (in this case, apples). For larger ticket items, for instance, the purchase of a car or home, or stocks in the equity markets, then both steps are involved.

In financial accounting, assets are found on the balance sheet and further stratified into current assets and fixed assets. Current assets are short-term assets which are cash or cash-equivalents, receivables, inventory, prepaid expenses, and short-term investments. In other words, current assets are *liquid assets*. Liquid assets are assets that can be easily converted to cash and can be used to fund the operations of an entity. By comparison, fixed assets would include furniture and fixtures, buildings and equipment—resources that an entity relies on helping to generate future revenues. These assets are very *often illiquid* and take time to convert to cash. There are also other asset types such as intangible assets such as goodwill, trademarks, patents or the brand equity of a company.

2   Asset Ownership

The concept of asset ownership dates back many centuries and has a checkered history. Many revolutions started between peasants and landowners; wars waged between countries over territorial disputes; lawsuits between companies over patent infringements; legal disputes in domestic divorce cases involve the ownership of assets.

Despite this history littered with disputes, many jurisdictions to this day, do not maintain clear data over the titles of ownership. In more developed countries, such as the United States of America, there are strict laws governing the ownership of major asset classes, e.g., real estate, automobiles, financial assets such as stocks and bonds. In the case of home ownership, the process typically involves a title, insurance, and escrow (if one is borrowing funds to purchase a home).

However, there is more room for improvement, over other asset classes with significant value, e.g., jewelry, art, and memorabilia. In many developing economies, one still hears about usurpation of land by some from their rightful owners, in a process called "squatting".

3   Determining Asset Value

The valuation of assets is a complex topic in that the price for an asset can be set in a number of different ways.

(1) **Standard Global Market Value**: For some asset classes such as gold, silver, or crude oil, there are global markets that trade these commodities and set a price for them. So in any corner of the world, that price would be the benchmark that is used.

(2) **Local Market Value**: For some assets classes such as the price of Gasoline, the prices tend to have a localized range, e.g., California gas prices range from $3.50 to $4.50 for a gallon, while Texas gas prices range from $2.50 to 3.50 a gallon. These localized pricing is due to taxation, holding costs, labor costs and other factors. Many multinational corporations sell their products at prices which are attuned to the local economies, e.g., a laptop for a given brand may cost slight more or less in one country versus the next.

(3) **Industry Pricing**: Many products fall within this category and prices can fluctuate significantly. For instance, in luxury goods, manufacturers set their own prices based on their brand equity. Similarly, works of art trade for values set by art connoisseurs. In some cases, there are companies which taken on the mantle of being the standard bearer for a given industry to help guide consumers on pricing. For instance, in the automotive industry, Kelly Blue Book, acts as a yardstick for prices of used cars.

4 Asset Valuation Factors

In the last section, we examined the concept of value of an asset. This value is obviously impacted by a number of other factors:

(1) **Condition**: Something that is in mint condition versus that is in a state of disrepair obviously would command a higher valuation.

(2) **Provenance**: The provenance reflects the origin of the asset. For mined assets such as diamonds or rubies, then it would indicate the place from which it was unearthed. In the case of manufactured goods, it would be the place where it was manufactured, e.g., garments. The provenance of assets is a hot topic given news items surrounding the use of sweatshop labor in various garment factories around the world; or the sale of blood/conflict diamonds emanating from countries with civil wars, genocide and other atrocities being committed.

(3) **Authenticity**: An important factor in determining the value of an asset is whether it is the genuine article or a knock off. The global market for fake or counterfeit goods is US$5 T. Major targets of knock-off makers are auto parts and luxury items.

(4) **Ownership History**: Ownership history is another important factor in determining the value of an asset. In some industries, used products are passed off as new, e.g., Printer Ink cartridges, Diamonds and more. In the diamond industry, about 40–50% of all diamonds sold in retail as new, are actually used diamonds. This is allowed to flourish, due to the weak governance around ownership history.

(5) **Grade**: In some asset classes, quality or grade is assessed in terms of its purity, e.g., octane levels in gasoline (89,91,93), gold (18 K, 22 K, 24 K), diamonds (color, cut, clarity and carats—the 4Cs). In other assets it may be the location, e.g., real estate, where certain zipcodes command a higher value.

These are five factors, among several more, that determine how asset values are determined. It important to recognize that these five factors are orthogonal in nature. Let's use diamonds as a use case. One can have an authentic diamond in mint condition, with the finest 4C attributes, from a conflict zone, but has changed hands multiple times. Or, an auto part that is in mint condition, but actually produced in an unauthorized manner at the same factory at which the genuine part is produced.

5   Distributed Ledgers

Thus far, we have learnt that assets were tracked on financial statements, and that there are several factors that influence the value of an asset. These methods of tracking assets are known as centralized forms of tracking assets where one party maintains and tracks this data. Quite commonly used accounting packages such as Quickbooks or Peachtree are ledgers and track assets for enterprises, non-profit organizations and more.

Quite often these centralized ledgers have to be inspected or audited. This may be for tax purposes, investment due diligence, or as part of governance and compliance in public entities. Such audits are conducted by third-party auditors, e.g., PriceWaterhouseCoopers or Deloitte are among the "Big-4" firms with audit practices often used by large public entities.

Despite third-party audits, each year we see many cases of accounting fraud or malfeasance in the news. This is primarily due to the fact that the ledgers are not transparent and management can easily alter the ledgers without adequate oversight.

This brings us to the topic of distributed ledgers. In this section, we will begin to explore the concept of distributed ledgers and its features and functionality that make it suitable for the purposes of asset tokenization.

5.1   Types of Distributed Ledgers

A distributed ledger is defined as a database that is synchronized and accessible across different sites and geographies by multiple participants [25, 33]. The following illustration compares and contrasts how Centralized ledgers stack up to distributed ledgers (Fig. 1).

In a centralized ledger, the control over the data is held by a single entity whereas in a distributed ledger there is no single trusted authority. Instead there is a set of protocols and supporting infrastructure that allows computers at different locations to

Distributed ledger system

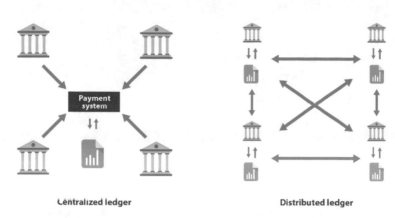

Fig. 1 Centralized versus Decentralized Ledgers

propose and validate transactions and update records in a synchronized way across a network. In distributed ledger networks, transactions are conducted in a peer-to-peer fashion and broadcast to the entire set of participants who work to validate them in batches.

Distributed ledgers can be implemented using a number of underlying technologies. Blockchain is one implementation of distributed ledger technology or DLT. There are other implementations of DLT including Directed Acyclic Graphs (DAG) [26], and Distributed Hash Table (DHT). So one can think of Blockchains as using Linked Lists, DAGs using Tree data structures, and DHT using Hash tables. The diagram below depicts Blockchain (a) versus DAG (b) (Fig. 2).

5.2 Blockchain Types and Architectures

Blockchains are interesting data structures because they embody three key capabilities:

- **Data**: Blockchain nodes carry data of a distributed ledger.
- **Network**: Blockchain nodes work together in a network to reach consensus on transactions.
- **Logic**: Blockchain nodes also embody logic in the form of *smart contracts* [31].

There are a number of different types of Blockchains as shown in [27, 28, 30]:

- **Permissionless/Public**: Permissionless blockchains are DLTs that are available to anyone who wishes to validate blocks, without requiring permission from any central authority. Often, permissionless blockchains are implemented as open source software, and freely downloadable.

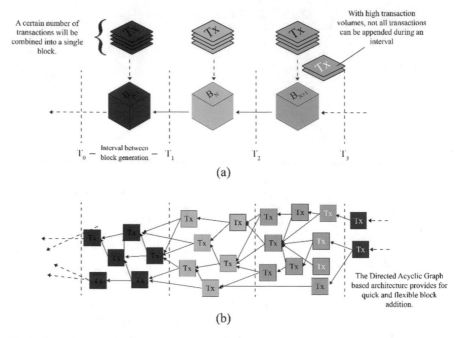

Fig. 2 Blockchain versus Directed acyclic graphs

- **Permissioned/Private**: Permissioned blockchains are DLTs where a single entity holds influence over a blockchain and invites other entities to participate in the network. With only authorized nodes maintaining the distributed ledger, it is possible to restrict access privileges such as who can read information, or who can issue transactions.
- **Hybrid**: Hybrid blockchains combine the privacy benefits of a permissioned blockchain, while maintaining the security and transparency benefits of a permissionless network. This enables operators of a hybrid blockchain the flexibility to decide on what data to make public and transparent and what data to remain private.
- **Consortium (or Federated)**: A Federated blockchain network is similar to permissioned blockchain, but in this case multiple entities jointly operate the network.

The diagram below illustrates these different architectures:

The relevance of these blockchain architectures to the concept of managing assets comes into play when one wishes to develop an asset marketplace. These asset marketplaces could be for any class of assets. When deploying an asset marketplace, one can select from one of the above architectures depending on their business model. For instance, in a B2C type business model, a permissionless architecture may make sense as the participants would be end-consumers and the system would be open to the public. In a B2B type business model, one may elect to go with a permissioned architecture as participants may be limited to a select few invitees (Fig. 3).

TYPES OF **BLOCKCHAIN**

Public Blockchain

- Anyone can join and participate in the consensus on the public blockchain.
- A ledger system that is completely decentralized, secure, and immutable.
- Anonymous, but transparent transactions.

Private Blockchain

- A single company will have control over the network under a Private blockchain.
- Improved production, energy efficiency, and privacy.
- The data handling process has been simplified, however, it is not available to everyone.

Federated Blockchain

- The blockchain network is influenced by multiple organizations.
- System is decentralized, incredibly quick, and scalable.
- Network security and privacy are protected by network regulations.

Hybrid Blockchain

- Decentralized, controlled, and highly scalable system.
- Authoritative access, only certain aspects are private.
- Flexible, control over what data is kept public and private.

Fig. 3 Types of blockchain

5.3 Smart Contracts

As we have seen above, blockchains are the embodiment of data, network, and logic. This logic is codified in the form of *smart contracts*.

A smart contract is an agreement between two or more entities codified in the form of computer code and runs on a blockchain. In the case of the Ethereum blockchain, for instance, smart contracts are written in the solidity language whereas in the case of Algorand blockchain, smart contracts are written in Teal. Smart contracts are autonomous in the sense that they run on the blockchain without having to be executed by any entity. The smart contracts execute when certain conditions are met.

In the context of this chapter, let us assume a soccer federation wishes to issue NFTs in the form of digital playing cards for the players playing in its league. Furthermore, the federation wishes to receive a 2% transaction fee for every transaction involving a sale of these NFTs. In this example, the smart contract would be codified in a manner such that 2% fee is levied on the buyer and funds would transfer from the buyer's wallet to the soccer federation's wallet.

6   Asset Tokenization

A quick recap of the issues relating to asset valuation factors before we delve further.

- **Ownership**: Across many asset classes, there is a need for better governance around asset ownership. There is a need for a single source of truth to unambiguously determine ownership.
- **Authenticity**: Authenticity is currently established independently by manufacturers without an authoritative registrar enabling counterfeiters to easily circumvent current measures.
- **Provenance**: Provenance records for many assets are unavailable, or, lacks the infrastructure

Blockchain technology has a few characteristics that make it suitable to help solve problems related to ownership, authenticity, and provenance.

(1) **Decentralized**: One of the problems when asset records are not decentralized and centralized, there is a lack of oversight from industry watchdogs and/or other interested parties.
(2) **Immutable**: In many ecosystems, while data is captured and stored, it is open to mutation or manipulation. Such is the case with land and property title deeds, in many jurisdictions, where corrupt officials can be bribed to modify ownership of such assets without the knowledge of the real owner. This type of issue can be stamped out in permissionless ledger, or, a permissioned ledger with sufficient oversight.
(3) **Security**: Closely related to Immutability, is the fact that the data stored in a ledger is secure. In distributed ledger technology, the data is cryptographically hashed to ensure transactions are stored securely [1].

6.1   Types of Assets

Assets can be classified in several ways. Earlier on in this paper, we saw how financial accountants classified assets as current, fixed, and intangible assets. In the blockchain context, an often quoted term is a *digital asset*. This term refers to *anything that exists in binary data which is self-contained, uniquely identifiable, and has a value or ability to use* [4]. We would like to lend more color to this definition, and decompose digital assets into four sub-categories:

(1) **Digital**: Assets that are innately digital, e.g., music, movies, avatars, etc.
(2) **Physical**: The digital twins of physical assets, e.g., the digital representation of a physical object, e.g., automobile, diamond, etc.
(3) **Dynamic**: Dynamic assets which generate data, e.g., energy consumption, water consumption, carbon emissions, etc.

(4) **Financial**: Financial assets could be assets that represent money, e.g., tethered coins to the US Dollar such as USDC, USDT, and Tether; assets tied to the value of a stock or commodity, e.g., using Kwenta.io one can trade a number of derivatives of financial assets.

When we represent assets on a blockchain, they will usually carry a unique identifier and in some asset classes, e.g., cars and diamonds, they may relate to a physical world identity number such as a VIN or GIA certificate number.

6.2 What is Tokenization?

The tokenization of assets refers to the process by which a token is issued to make that asset tradeable [14]. It brings about several distinct benefits by offering the potential for a more efficient and fair financial process through the reduction of friction in the creation, buying, and selling of assets. We will explore four key advantages that tokenization provides for both buyers and sellers:

Liquidity

The tokenization of assets for typically illiquid assets such as fine art, diamonds, real estate, and more increases liquidity by providing access to a broader base of traders. Liquidity benefits sellers by creating a "liquidity premium", thereby capturing greater value from the underlying asset.

Lower Cost

The transaction of tokens between buyers and sellers is executed using smart contracts. As smart contracts are software algorithms integrated into a blockchain with trigger actions based on pre-defined parameters, these transactions are to a large extent automated. This results in reduced administrative costs with fewer intermediaries needed, leading to speedier deal execution and lower transaction fees.

Transparency

When transactions are conducted using tokens, several ground rules can be established and codified into the token and/or the smart contracts. For instance, the blockchain can have a running record of the ownership history of a token and thereby the asset with which it is associated. Quite often, most token transactions require the transacting parties to go through a KYC/AML (Know Your Customer/Anti Money Laundering) process. This adds transparency to transactions, enabling parties to know who they are dealing with and who had previously owned this token.

Accessibility

Tokenization has opened investment in assets to a broader audience primarily because tokens are highly divisible, meaning investors can purchase tokens that represent incredibly small percentages of the underlying assets. In addition, since the processing cost of transactions is lower, it reduces the amount of capital required. This makes the process more inclusive thereby enabling small investors to participate. Furthermore, the higher liquidity of tokens and the availability of global and 24/7 markets make trading far more accessible to the masses.

6.3   Types of Tokens

In general, tokens can be classified in a number of different dimensions.

Fungibility

Fungibility is one way of classifying tokens. In this dimension, there are at least 3 categorizations. As defined by Investopedia [13]:

   Fungibility is the ability of a good or asset to be readily interchanged for another of like kind. Like goods and assets that are not interchangeable, such as owned cars and houses, are **non-fungible**.

1. **Fungible Tokens:** Financial assets typically can be represented using fungible tokens. Several central banks are considering the use of Central Bank Digital Currencies or CBDCs. These are a digital representation of regular fiat currency. Cryptocurrencies are also often represented as Fungible tokens. For instance, Tom and Joe each had one Bitcoin (BTC) token. There is no Identifier for each token and the only piece of information stored would be the value and the wallet that value belongs to. If Tom sends Joe 1 BTC, and Joe sends Tom 1 BTC—it is fungible because they each replaced their BTC with an identical item. Ethereum offers the ERC-20 token standard upon which many fungible tokens have been developed.
2. **Non Fungible Tokens:** Non-fungible tokens, or NFTs, can be used to capture the value in digital, physical, and dynamic assets. Each token is unique and bears its own unique identifier. Ethereum, EOS, Algorand, and others offer the ability for one to mint their own NFTs. ERC-721 is a popular token standard upon which many NFTs have been launched.

   Typical assets that can be "NFT-field" include:

- Digital assets, e.g., music files, video clips, digital avatars, digital art.
- Physical assets, e.g., memorabilia—signed basketballs, author signed first edition copies of books, etc.
- Dynamic assets, e.g., energy—in the form of renewable energy certificates, carbon credits, and more.

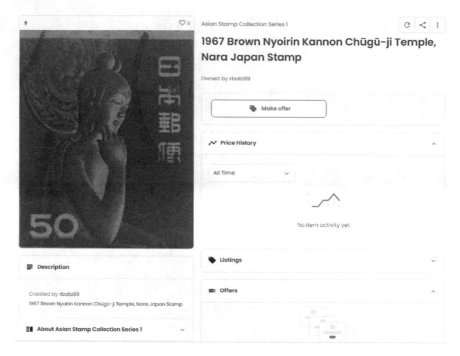

Fig. 4 Example of an NFT listing

An example of an NFT is shown in the diagram below [29] (Fig. 4):

3. **Semi-Fungible Tokens:** SFTs, or semi-fungible tokens, can be thought of as hybrid tokens. This token could represent a "10 dollar Amazon coupon" in which case it is fungible as each coupon is the same as the other. However, once a coupon is redeemed it becomes non-fungible, it no longer can be traded as a normal token. An example of as semi-fungible token is the ERC-1155 standard that operates on the Ethereum network (https://boxmining.com/erc-1155/) and the following diagram shows the ability of the token to be swapped for different sets of items during the conversion or redemption phase (Fig. 5).

Security, Utility, or Payment

In the last section, we examined various types of tokens in terms of their fungibility attribute. Another way to classify tokens is in terms of their utility quotient. In the USA, the security and exchange commission has used the "Howey Test", to determine if a token is a utility or a security. It is important to point out that this is an area where the case law is actively evolving and each jurisdiction around the world may have different definitions over time.

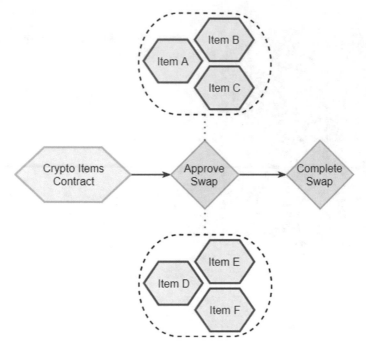

Fig. 5 Semi fungible tokens

Security

A token is deemed a security token if it represents an investment of money in a common enterprise with a reasonable expectation of profit derived from the efforts of others [16]. A security token provides rights and obligations similar to securities or investments like shares or debt instruments [17].

Utility

The German financial regulator BaFin (Bundesanstalt für Finanzdienstleistungsaufsicht, i.e., Federal Financial Supervisory Authority) defines a utility token as tokens that grant holders access to a current or prospective product or service but do not grant holders rights that are the same as those granted by specified investments [17].

Payment

Payment tokens are defined as tokens that are used as an alternative form of payment and exchange. Unlike fiat currencies such as the US Dollar, Euro, or the Japanese Yen, payment tokens such as Bitcoin are not legal tender, and not backed by a government. Instead, their main objective is to be a decentralized tool for buying and selling goods and services without the traditional intermediaries [15, 17].

6.4   Digital Asset Marketplaces and Blockchain Architectures

Earlier, we studied several blockchain architectures and in this section we will relate these architectures to digital asset marketplaces. Digital asset marketplaces act as a trusted intermediary between issuers and investors and facilitate transactions between the different stakeholders [34].

The relevance of these blockchain architectures to asset marketplaces is the fact that one can employ permissionless, permissioned, hybrid, or consortium architectures depending on the business model.

Let us examine two popular marketplaces.

- **NBA Top Shots**: This marketplace sells NFTs associated with the National Basketball Association (NBA). This marketplace uses a "walled-garden" approach and can be thought of as a permissioned ledger. The benefit to operating in this manner is the fact that the smart contracts can be codified to provide recurring revenues to the operators on every transaction. Such benefits only accrue when one operates the network. If the NFTs are sold in open marketplaces such as Nifty or Rarible, these recurring fees cannot be accrued as one does not control the smart contracts of these networks.
- **Synaptic Health Alliance**: A good example of Federated/Consortium Blockchain is the Synaptic Health Alliance whose members include Aetna, Centene, Humana, United Healthcare, and several other large players in the healthcare arena. This marketplace is an example of an information exchange where the asset is the information about healthcare provider data.

6.5   Valuation of Tokens

The value of a Token and its exchange rate are determined by the following factors: velocity, utility, volatility, and liquidity [11, 12, 35].

To find the price of a token, we will use Vitalik Buterin's definition below.

Token Price

$$MC = TH$$

where: M = total money supply (or total number of coins), C = price of the currency (or $1/P$, with P being price level), T = transaction volume (the economic value of transactions per time), $H = 1/V$ (the time that a user holds a coin before using it to make a transaction), V = velocity of the asset (the number of times that an average coin changes hands every day).

Using this definition, to solve for the token price, one must solve for C:

$$C = TH/M$$

Replacing $H = 1/V$ in the equation we get

$$C = T/MV$$

One can see that the velocity of the coin is inversely proportional to the value of the token, i.e., the longer people hold the token, the higher the price of each token. This is intuitive, because if the transactional activity of an economy is \$100 billion (for the year) and coins circulate 10 times each over the course of the year, then the collective value of the coins is \$10 billion. If they circulate 100 times, then the collective coins are worth \$1 billion. On the other hand, if transactions are absent, then the token lacks liquidity and its velocity equals zero. Consequently, the asset will trade at a discounted rate. There must be some minimal velocity for a token to reach its full value. Thus, understanding and calculating the velocity in any token economy is extremely important.

Given the above, it is important to understand that token markets are no different from any other market. They are driven to a large extent by supply and demand—two levers of Economics. How much is one willing to pay for something?—ultimately this is the major driving force of any asset's price.

6.6 Use Cases for the Adoption of Tokens

In this section, we identify a non-exhaustive list of use cases for the use of tokens. For each use case, we will highlight its impact on velocity, transaction volume, and hold time—concepts we covered in the previous section. Velocity is defined as the number of times that an average coin changes hands every day; hold time is defined as the time that a user holds a coin before using it to make a transaction.

Loyalty

A token can be used to engender loyalty among its users both businesses and consumers. Loyalty itself is a concept that can be decomposed into rewards and redemption. Let's assume our token is called RWD token.

Impact: Increases velocity and transaction volume

6.7   Rewards

In a typical B2C scenario, consumers will receive RWD tokens as rewards for purchases made through the RWD marketplace using the RWD App. One can also launch the RWD token in a third-party marketplace willing to adopt the token. The RWD marketplace will list 1000 s of items that consumers can purchase through their app. Consumers will receive tokens during retail promotions. For example, a manufacturer on the RWD marketplace sets a promotion that states that a consumer will receive a 5% rebate for every 50$ spent. If a customer spends $100, he/she would receive 5$ worth of RWD tokens at checkout.

6.8   Redemption

Consumers can use the RWD tokens when shopping on the RWD marketplace as a replacement for fiat currency to reduce their payment for goods.

Payment

Every business typically engages with multiple stakeholders in the ecosystem including:

- Distributors,
- Wholesalers,
- Consumers,
- Logistics companies,
- and more.

In all transactions with the above entities, there is an exchange of value, i.e., Payment for Goods/Services. The use of the RWD token is an attractive option when it comes to micropayments where the transfer of funds to/from fiat is prohibitively expensive. Transaction fees levied by banks and others make transferring funds to fiat incur at a minimum 2–3% if not more in transaction fees.

Having the RWD token as an accepted form of payment within the RWD ecosystem will result in stakeholders opting to leave the funds in the form of RWD tokens and only transfer to fiat periodically.

> > Increases velocity and transaction volume

Staking

Staking is defined as the process of actively participating in transaction validation (similar to mining) on a proof-of-stake (PoS) blockchain. On these blockchains, anyone with a minimum-required balance of a specific cryptocurrency can validate transactions and earn Staking rewards [18].

> > Decreases velocity and increases hold time

Governance

Governance tokens are cryptocurrencies that represent voting power on a blockchain project [19]. With these tokens, one can create and vote on governance proposals. By doing so, the user directly influences the direction and characteristics of a protocol, e.g., Vote on the change of a user interface, how fees are distributed, how development funds are used, and other vital questions.

> > Decreases velocity and increases hold time

Yield Farming

Yield farming [32], also known as liquidity mining, is a way to generate rewards by locking up cryptocurrencies and getting rewards.

In decentralized finance, DeFi for short, many markets are governed by Automated Market Making (AMM) algorithms codified in smart contracts. These AMM algorithms need a pool of funds known as a Liquidity Pool to perform its function of making markets—i.e.,providing the optics of a perfect market by buying from sellers and selling to buyers any particular asset class.

These liquidity pools need funds and those who provide funds to them are known as Liquidity Providers (LP). These LPs provide funds to the liquidity pool in return for rewards generated from fees charged for transactions by the underlying DeFi platform. Yields are typically paid in tokens and these tokens can then be reinvested in other LPs to derive even more yield. Yield farms usually shift their allocation of funds among various LPs to maximize their yield.

> > Decreases velocity and increases hold time

7 The Cyber-Physical Divide

As we have seen, blockchain has a number of desirable attributes such as immutability, being cryptographically secure and decentralized. Despite these positive attributes, there are still some gaps in the use of blockchain. At some level, blockchains are disconnected from the physical world. They are ledgers maintained in the cloud for a given asset class, e.g., an automobile. So, how does one corroborate the digital record with the physical object? Therein lies the Last Mile Problem as highlighted in [3].

Bridging the cyber-physical divide is relevant and important to the problem of asset tokenization because in many cases, the value of an asset is determined by data that may reside external to the blockchain. For instance, in the oil and gas industry, when crude is sold from the upstream oil wells to the mid stream refineries and flows long distances through oil pipelines, the price of oil may have changed drastically due to volatility in the markets.

7.1 The Last Mile Problem

The last mile problem is the fact that we need some mechanisms in the physical world (as opposed to the cyber world of the blockchain) to ensure:

(1) Data is acquired and transmitted accurately—otherwise blockchain is susceptible to the Garbage-In-Garbage-Out problem that afflicts many systems that have been devised before it. Some techniques include encrypting data during transmission to avoid attacks such as the Man-In-The-Middle attack (https://www.csoonline.com/article/3340117/what-is-a-man-in-the-middle-attack-how-mitm-attacks-work-and-how-to-prevent-them.html) [5].
(2) When managing physical assets, there needs to be a method of ensuring that the physical object and the record on the digital ledger match. In the case of Automobiles, there is a Vehicle Identification Number or VIN number which is stamped on the windshield, engine block, and on driver's side door.

Similarly, diamonds have laser inscriptions that match the Gemological Institute of America (GIA) Certificate. In the GemIdentity blockchain [8], ownership records link the identity of the owner to that of the diamond in the blockchain (Figs. 6 and 7).

7.2 Digital Twins

Following the last section, Digital Twins are defined as *a virtual representation that serves as the real-time digital counterpart of a physical object or process* [9].

Digital twins are important in the context of assets when we have to manage physical and dynamic assets. And in these cases, quite often we have to collect data

Fig. 6 Locations where a VIN number is etched on a motor vehicle [6]

Fig. 7 How a GIA Certificate matches a Laser inscription on a diamond [7]

Fig. 8 A smart meter

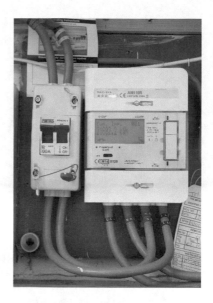

from these assets and that is where Internet of Things technology can play a vital role in bridging the Cyber-Physical divide. In other words, linking the assets in the physical or real world to the asset definitions in the cyber world of blockchain.

Let us consider the example of a dynamic asset such as energy. Energy may be harvested from a number of different sources including wind, hydro, and solar. This energy produced is recorded in smart meters such as the one shown below [10] (Fig. 8).

This energy data can be tokenized in the form of NFTs called renewable energy certificates and is currently taking place in several parts of the world.

7.3 *Blockchain Oracles*

The Cyber-Physical divide can be overcome through the use of Blockchain Oracles. Blockchain oracles are defined as third-party services that provide smart contracts with external information. They serve as bridges between blockchains and the outside world [20].

Typically data that is available on the blockchain is called "on-chain" and data which is acquired or stored outside the blockchain is called "off-chain". From time to time, smart contracts, operating on the blockchain, need to retrieve from off-chain resources to complete their task. This is when the smart contract leverages a blockchain oracle.

Connect to Any API   Send Payments Anywhere

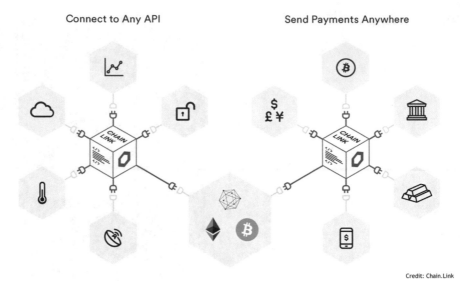

Credit: Chain.Link

Fig. 9 How ChainLink oracles can be used

To an oracle, the "outside world" is anything that off-chain data, i.e., data that is outside of the blockchain network. This outside world can mean physical objects such as the retrieval of temperature data from a sensor, or it could mean getting the latest price of crude oil from the New York Mercantile Exchange.

One of the leading blockchain projects in bridging the Cyber-Physical divide is Chainlink, and the following diagram illustrates at a high level the role of a decentralized oracle service [21] (Fig. 9).

As the diagram above illustrates, oracles can serve many purposes, and as such we can classify them into the following categories [22, 23]:

- **Software Oracles**: These oracles handle information data that originates from online sources such as the prices of commodities and goods, flight or train arrival times, and so on.
- **Hardware Oracles**: In some cases, smart contracts require information directly from the physical world, for example, a reading from a Smart Electricity Meter, or odometer reading from an automobile.
- **Consensus-based Oracles**: Primarily used in human consensus and prediction markets like Augur and Gnosis, these oracles aggregate data from several oracles using proprietary methods for determining their authenticity and accuracy to prevent market manipulation.
- **Outbound Oracles**: These oracles enable smart contracts with the ability to send data to the outside world. An example would be a smart lock in the physical world, which receives payment on its blockchain address that triggers an action to unlock automatically.

As we have seen, blockchain oracles, provide a link between off-chain and on-chain data. They are a vital part of the blockchain ecosystem because they broaden the scope in which smart contracts can operate. Without blockchain oracles, smart contracts would be limited to only accessing data from within their networks.

Coming back to our oil and gas example, with the advent of smart contracts and oracles, more real-time information on the value of assets in transit (in pipelines or tankers) can be gathered easily. So, if one were tokenizing a commodity such as crude oil, then oracles would play a very pivotal role in the valuation of such assets.

7.4 How Do Oracles Work?

Here is a step-by-step breakdown of how Oracles work using Chainlink [24]—a popular decentralized Oracle network.

1. Assume that a client smart contract requires external data, i.e., data residing outside the blockchain. This client smart contract is invoked by a transaction to execute a function. To retrieve the external data, this function prepares a request using a Job ID along with parameter values. It then executes the Oracle contract with the request.
2. Next, the Oracle contract publishes an event with the Job ID, parameter values, and payment (in LINK tokens) by the client smart contract
3. At this time, the event notifies all Chainlink nodes attached to the blockchain network of the job request
4. The Chainlink node that had the job ID deployed is assigned the task of executing the request. It forms the execution context for the job by combining the job specification corresponding to the job ID along with the parameters contained in the event.
5. Once the job is executed, the desired data is acquired. The Chainlink node signals the fulfillment of the Job request by submitting a transaction to the Oracle contract. This transaction payload contains the job execution results.
6. The Oracle contract looks up the corresponding requestor using the request ID and issues a call back to the requestor contract with the Oracle result data
7. At this point, the client smart contract would have successfully obtained the data from the outside world, and proceeds to execute the original transaction request by the client application.

The diagram below illustrates how Chainlink works under the hood [24] (Fig. 10):

8 Trading in Tokens of Assets

In this section, let us explore the practical aspects of trading in the tokens of assets.

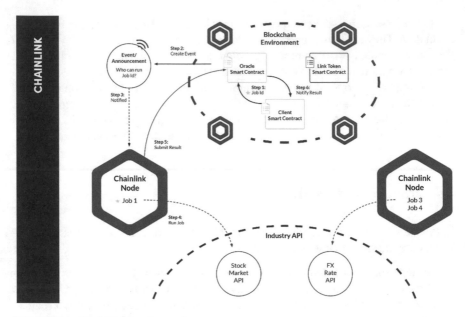

Fig. 10 How ChainLink oracles work

8.1 Wallets

Before one can participate in the trading of tokens, one would require a wallet. This wallet will hold the tokens one plans to buy and sell. Wallets come in two flavors [2]:

- **Custodial**: These can be thought of as centralized wallets or accounts, similar to your bank account where the bank holds your funds.
- **Non-Custodial Wallets**: These can be thought of as wallets or purses that you hold your funds in. They are in your possession and not under the control of any institution. When using non-custodial wallets, one will have to download or install some software on your PC or mobile phone.

Each wallet is secured by private and public key cryptography. In order to send and receive funds, one would share their public address to their wallet with the party they are transacting with. But one should never reveal their private key much like one would never reveal their bank account password.

8.2 Exchanges

There are many ways in which tokens can be exchanged between two parties.

- **Direct**: Two parties can move tokens between their wallets directly. This is common in private party transactions.
- **Centralized Exchanges**: There are many centralized exchanges that one can join, e.g., Binance, Kraken, etc. to trade tokens.
- **Decentralized Exchanges**: Unlike centralized exchanges where the transactions are handled by a central authority, in a decentralized exchange, automated market makers and liquidity pools codified in the form of smart contracts operate the exchange.
- **Storefronts**: Some organizations which sell NFTs may wish to control the shopping experience, and also impose certain royalties on the sale of NFTs. For this reason, there are walled-garden storefronts and marketplaces such as NBA TopShots.
- **Marketplaces**: For NFTs, there are marketplaces such as Nifty, OpenSea, Rarible where one can trade in NFTs.

9 Methods of Pricing Tokens

Pricing of tokens—both regular tokens and NFTs—has many variations. Let us examine a non-exhaustive list of pricing models.

9.1 Fixed Pricing

In the case of NFTs, at a given point in time, prices are fixed. For instance, the price requested by the seller for a trading card may be $5. When we say fixed price, we are saying it is not fluctuating in any way.

9.2 Time Based

In many Initial Coin Offerings (ICOs), the price of tokens are adjusted on a time-based schedule. For example,

- January 1–Feb 28: 10c per token
- March 1–April 30: 12c per token
- May 1–June 30: 15c per token
- July 1: ICO price 20c per token

The reason for setting this schedule is that the rounds of sale prior to the ICO are considered the pre-sale period. During this pre-sale period, early investors are attracted to the ICO via steep discounts with the promise to lock-in profits at the time of the ICO. These early round monies are also used to promote the token and pay for various administrative and legal obligations.

9.3   Auction

Auction-based pricing is becoming common in NFTs especially where a given asset is scarce. During the auction, bids are placed and the NFT goes to the highest bidder. Often times, there is a *reserve price*-a price below which the seller will not be willing to sell the asset.

9.4   Dynamic Pricing

Prices can also be dynamic and there are many forms of dynamic pricing. Let us describe one such model. Imagine there are ten limited edition NFTs of a Basketball trading card of a famous player. The pricing for such cards could be set up in the following way:

- 8–10 NFTs left: $100
- 5–7 NFTs left: $250
- 3–4 NFTs left: $500
- 2 NFTs left: $1000
- Last NFT left: $5000

So the item's price is not fixed and tied to its scarcity. There may also be a lock-up period for buyers before they can sell it on a secondary market for the above model to work.

9.5   Market Pricing

Most tokens sold on exchanges (centralized or DEX) are subject to the laws of demand and supply. Hence, the markets determine the price of the token much like financial instruments such as stocks, bonds, commodities, and more.

9.6   Tethered Pricing

Some tokens are known as tethered tokens or stable coins, e.g., USDT, USDC where their value is linked to a fixed amount of fiat currency or other assets, e.g., gold. For instance, 1 USDC represents 1 US Dollar. Central Banks around the world are also preparing to issue Central Bank Digital Currencies (CBDCs) and these would be tethered to the respective fiat currencies, e.g., Yen, Yuan, Rupee, etc.

10   Opportunities Created by Asset Tokenization

A song is an asset in the music industry, and tokenization enables one to create liquidity for stakeholders. These stakeholders may include artists, distributors, producers, music labels, and more. The music industry is one where all the stakeholders claim that there are massive gaps in revenue attribution and that they are not receiving their fair share of the rewards for their works.

With asset tokenization, these stakeholders can now go directly to consumers using tokens thereby disintermediating the process.

In the energy industry, previously the production of electricity which we all need was the domain of the energy grid operators. With the availability of solar panels, anyone around the world can literally become an energy producer and live "off-the-grid" if they chose to. Asset tokenization provides those with solar panels to generate energy, tokenize it and sell it to others. It democratizes energy.

Asset tokenization is a powerful concept. It disintermediates, democratizes, and decentralizes traditional centralized infrastructures to help make markets more open and transparent.

References

1. https://academy.binance.com/en/articles/what-makes-a-blockchain-secure
2. https://www.gemini.com/cryptopedia/crypto-wallets-custodial-vs-noncustodial
3. https://hbr.org/2018/06/what-blockchain-cant-do
4. https://www.securities.io/what-are-digital-assets/
5. https://www.csoonline.com/article/3340117/what-is-a-man-in-the-middle-attack-how-mitm-attacks-work-and-how-to-prevent-them.html
6. https://www.yourmechanic.com/article/how-to-decode-a-vin-vehicle-identification-number-by-jason-unrau
7. https://belgiumdiamonds.net/laser-inscribed-diamonds/
8. https://www.gemidentity.com
9. https://en.wikipedia.org/wiki/Digital_twin
10. https://www.cse.org.uk/advice/advice-and-support/smart-meters
11. https://medium.com/newtown-partners/velocity-of-tokens-26b313303b77
12. https://medium.com/blockchannel/on-value-velocity-and-monetary-theory-a-new-approach-to-cryptoasset-valuations-32c9b22e3b6f
13. https://www.investopedia.com/terms/f/fungibility.asp#:~:text=Key%20Takeaways-,Fungibility%20is%20the%20ability%20of%20a%20good%20or%20asset%20to,houses%2C%20are%20non%2Dfungible.
14. https://www2.deloitte.com/content/dam/Deloitte/lu/Documents/financial-services/lu-tokenization-of-assets-disrupting-financial-industry.pdf
15. https://boxmining.com/erc-1155/
16. https://www.startupblog.com/sec-chair-defends-agency-action-on-icos/
17. https://www.planetcompliance.com/what-is-the-difference-between-utility-security-and-payment-tokens/
18. https://help.coinbase.com/en/coinbase/trading-and-funding/staking-rewards/staking-inflation#:~:text=What%20is%20staking%3F,transactions%20and%20earn%20Staking%20rewards.

19. https://academy.shrimpy.io/post/what-are-governance-tokens#:~:text=Governance%20t okens%20are%20cryptocurrencies%20that,in%20order%20to%20remain%20decentrali zed.&text=With%20these%20tokens%2C%20one%20can%20create%20and%20vote% 20on%20governance%20proposals.
20. https://academy.binance.com/en/articles/blockchain-oracles-explained
21. https://coincentral.com/what-is-chainlink-a-beginners-guide-to-decentralized-oracles/
22. https://blockchainhub.net/blockchain-oracles/
23. https://medium.com/@teexofficial/what-are-oracles-smart-contracts-the-oracle-problem-911f16821b53
24. https://www.kaleido.io/blockchain-blog/how-chainlink-works-under-the-covers
25. https://www.investopedia.com/terms/d/distributed-ledgers.asp#:~:text=A%20distributed%20l edger%20is%20a,to%20have%20public%20%22witnesses%22.&text=Blockchain%20is% 20a%20type%20of%20distributed%20ledger%20used%20by%20bitcoin.
26. https://medium.com/@kotsbtechcdac/dag-will-overcome-blockchain-problems-dag-vs-blockchain-9ca302651122
27. https://searchcio.techtarget.com/feature/What-are-the-4-different-types-of-blockchain-techno logy
28. https://www.itu.int/en/ITU-T/focusgroups/dlt/Documents/d12.pdf
29. https://opensea.io/assets/0x495f947276749ce646f68ac8c248420045cb7b5e/107437774021 06773030368817456720297121761225507412828130218384382837846441985
30. https://xord.com/publications/hybrid-and-federated-blockchain-networks/
31. https://www.bitdegree.org/crypto/tutorials/what-is-a-smart-contract
32. https://academy.binance.com/en/articles/what-is-yield-farming-in-decentralized-finance-defi
33. https://www.bis.org/publ/qtrpdf/r_qt1709y.htm
34. https://tokeny.com/whats-the-purpose-of-a-digital-asset-marketplace/
35. https://investmentbank.com/token-velocity/

The New Economy of Movement

Tram Vo and Chris Ballinger

Abstract The convergence of a number of emerging technologies—including AI, IoT, and Blockchain—permits any entity, whether a vehicle, smartphone, sensor, road, or another piece of transportation infrastructure, to have a trusted identity, be intelligent, communicate, and autonomously participate as an independent economic agent in transactions. These transactions will become a large part of the new, pay-as-you-go, mobility services economy at the "edge". The potentially large number of independent agents, combined with the frequency and near real-time latency requirements of these transactions, will require edge connectivity, processing, execution, settlement, and new types of digital identifiers. For a roaming, connected entity—such as a person, vehicle, smartphone, electric vehicle (EV) battery, or package—one of the most important and valuable attributes is its location in time and space. Combining secure identity with trusted time-stamped locations creates a "Trusted Trip" and, for the first time, enables marginal cost pricing for many new classes of mobility transactions such as urban road tolling, meter-free parking, congestion management, carbon and pollution taxing, usage-based insurance, and many other usage-based Mobility as a Service (MaaS) applications. Together, these new transactions will comprise a multi-trillion-dollar ecosystem that we call the New Economy of Movement.

1 The Role of Industry Federations

Many organizations, including most MOBI members, have experimented to varying degrees with blockchain and related technologies. Hundreds of blockchain Proof of Concepts (PoCs) have been conducted by vehicle manufacturers and mobility services companies. While these PoCs are generally successful in that the blockchain

T. Vo (✉) · C. Ballinger
MOBI, Los Angeles, CA, USA
e-mail: tram@mobi.world

C. Ballinger
e-mail: chris@dlt.mobi

© The Author(s), under exclusive license to Springer Nature Switzerland AG 2022
D. A. Tran et al. (eds.), *Handbook on Blockchain*, Springer Optimization
and Its Applications 194, https://doi.org/10.1007/978-3-031-07535-3_19

603

technology itself operates as intended, they have not reached enterprise-scale adoption and revenue generation. This is also seen in other industries. Due to this reason, there is increasing interest in blockchain industry federations creating standards and building trusted shared digital infrastructures, the foundations for new economies at the edge. Consortia-built edge networks provide necessary core services of governance, authority, identity, and assurance (GAIA) for their industries.

The reason for the lack of enterprise-scale adoption is obvious in hindsight: blockchain is an inherently collaborative technology designed for multi-party networks, yet previous PoCs have been conducted by organizations largely within their own walls. Companies found that putting a vehicle, a device, an identity, or any other asset on a chain was relatively easy; however, commercial applications only emerge when mutually trusted data about these assets is shared across a business network or value chain. Simply put, blockchains work best for transactions in large and complicated networks where the frictional cost of trust is high. This is unlikely to be the case within a single or small group of organizations, where simpler and cheaper means are available to establish trust, data provenance, and transaction integrity. The result is that these PoCs, while successfully demonstrating the technology, lack a clear rationale to move ahead and advance to a consumer-facing application. The hard problem turns out not to be the technology itself, but the path to scale. Summing up his dozen years of experience in the blockchain space, cryptographer and blockchain pioneer W. Scott Stornetta observed, "Successful blockchain efforts do not begin with technology… they begin with a community". Or as Brian Behlendorf, open-source guru and Executive Director of Hyperledger/Linux even more succinctly quipped, "Blockchain is a team sport". In blockchain, it is more about the "Minimum Viable Community" than the minimum viable product.

Consortia are necessary to develop a shared language and business logic to speed the technology's adoption. Likewise, a shared digital infrastructure is a necessary condition for building trust across a business ecosystem. As a general rule, while the number of business partners in a value chain increases linearly, frictional costs increase geometrically, quickly overwhelming gains from digitizing business networks for industries like mobility where manufacturing supply, service, and sales value chains are especially complex.

Blockchain consortia and business networks—such as MOBI, Global Battery Alliance (GBA), TradeLens, RiskStream, and OpenIDL—are opening up new business models for their members and seeing strong membership growth. In particular, there is excitement about possible new applications, use cases, and services in these networks and a clearer understanding of the cost reductions that come with business automation. These efforts have generally garnered a positive reception from global government, finance, and regulatory authorities.[1] Within MOBI, much of this interest involves ledger-based business models for mobility, permissioned data, the "Trusted Trip", and a member-owned identity services network.

---

[1] The European Commission, for example, joined both MOBI and the GBA.

2 Roaming, Connected Devices, and the New Economy of Movement

Over the last hundred years, transportation has gotten much more congested; however, the fundamental modes of urban mobility have not changed much. The personal vehicle began to dominate mobility soon after Henry Ford's assembly line and mass production made it affordable. It was powered by an internal combustion engine and fueled with cheap gas. Privately owned and driven, it seated four to seven people and fit the average family with room for groceries and a spare tire. Cities and homes were redesigned around it. Car-based commuting opened new options for urban workers and newly accessible suburbias emerged outside cities. Large merchants replaced neighborhood retailers as driving distance replaced walking distance as the feasible shopping radius. Existing communities were erased to accommodate the roads, parking structures, traffic systems, and other infrastructure of the age of automobiles.

These fundamental modes are beginning to change. We're seeing the emergence of a new age of mobility services such as on-demand ride hailing, delivery, peer-to-peer car sharing, and micro-mobility rentals. The changes so far are just the tip of the iceberg to come. The new mobility paradigm would not fully be realized until a few more foundational technologies mature.

2.1 The BASICs

We have identified five technologies that are poised to change the mobility ecosystem for the better. We call these the **BASICs**.

Blockchain

A tamper-evident distributed ledger that records transactions and enables entities—be they individuals, organizations, vehicles, connected infrastructure, or objects—to directly exchange value and coordinate behavior. Blockchain technology is poised to enable new services and automate transactions by means of a radical, decentralized approach to business data and accounting.

Artificial Intelligence

Artificial intelligence allows machines to solve complex problems that would otherwise require human input. Cars, buses, subways, and other vehicles will become increasingly autonomous, not only in terms of moving from point A to point B but also in their ability to initiate and execute vehicle-to-everything (V2X) transactions. Autonomous vehicles and AI more generally will radically change not only mobility but the entire economy.

Services

Digital technologies are turning products into services. We are seeing a rapid acceleration of MaaS and usage-based consumption models. Automakers will sell fewer cars to private owners while expanding their mobility and fleet management services. Insurance companies increasingly see their future in usage-based mobility insurance.

Internet of Things (IoT)

Improvements in internet connectivity, speed, sensors, and computing power are turning vehicles and mobility infrastructure (roads, signals, tolls, charging stations, etc.) into nodes on the IoT. Connected vehicles and devices produce and manage real-time data, which is used to unlock a vast array of smart applications and new revenue opportunities. This data offers powerful insights about usage-based mobility services and the movement of goods around the globe, enabling the automation of more efficient and sustainable business processes.

Connectivity

People, in addition to vehicles, are becoming increasingly connected as well. Nearly half of the world's population has a smartphone. This means billions of people are constantly connected to the internet and to ubiquitous internet commerce platforms. Connected vehicles are the "fourth screen" and an additional access point to these commerce platforms.

The BASICs will disrupt mobility. Artificial intelligence makes machines intelligent, autonomous agents. Connectivity and the Internet of Things allow these agents to communicate and exchange information. Blockchains give vehicles secure identities that let them share value and coordinate behavior in edge networks. All of these come together to enable rich new services and usage-based pricing models. Promising blockchain PoCs for smart cities and mobility have been demonstrated; however, few can currently be deployed at commercial scale. This is because of the lack of a connected mobility ecosystem and standards upon which these technologies can be built. MOBI tackles this challenge by bringing together industry partners, subject matter experts, and innovators in working groups (WG) to validate use cases, set blockchain-based standards, and coordinate multi-stakeholder pilots. While our standards and use cases are motivated by all of the BASICs, we believe that blockchain is the key to scaling applications because it enables low friction automation and lowers the cost of trust for multi-stakeholder collaboration within extended ecosystems.

2.2   The New Economy of Movement

The New Economy of Movement is defined by smart and connected mobility services enabled by the BASICs. It will be greener, safer, and more affordable than the last hundred years of the "old" economy. The "new" economy will unbundle mobility from individual vehicle ownership and data from platforms while unleashing capital, creating a less centralized ecosystem, and enabling new usage-based payment business models. Blockchains play a critical role in the new economy of movement by creating trust, coordinating decentralized transactions, and incentivizing sustainable behaviors.

Trusted Identity in Decentralized Ecosystems

The BASICs permit any connected device to autonomously participate as an independent agent in decentralized economic transactions using **World Wide Web Consortium (W3C) Verifiable Credentials (VCs) standard**. These transactions require a new type of digital identifier—one which is machine-readable and which anchors any subject (e.g., a person, organization, thing, data model, etc.) to relevant attributes, characteristics, and capabilities—known as **Decentralized Identifiers (DIDs)**, as defined by W3C in their DIDs standard. Their design enables the controller of the DID to claim ownership and/or authority of their identity without requiring confirmation or permission from any other party.

Think of a DID as a globally unique identifier. Rather than having to manage several distinct forms of identification, entities can enjoy secure, hassle-free transactions with any number of ecosystem participants while having control over their own data.

In a decentralized system, DIDs are embedded in VCs. When an entity issues a VC, they attach their DID (digital signature) to that credential for future verification. An entity can create as many different DIDs as they wish, using separate DIDs for different digital relationships and contexts to prevent data correlation. The DIDs are registered on a decentralized network (for MOBI's community, this is the Integrated Trust Network (see Sect. 5.2)). DIDs do not contain/store any personally (or organizational) identifiable information (PII).

DIDs and VCs are the crucial building blocks for connected mobility and IoT commerce, enabling countless privacy-preserving multiparty applications and yielding increased transparency, coordination, and transaction automation between stakeholders.[2]

In 2019, MOBI released its first standard, MOBI VID, which leverages the internationally-accepted VIN standard and W3C's DID standard to define a vehicle's Self-Sovereign Digital Twin™ (SSDT™). The vehicle's SSDT™ combines the vehicle's unique identifiers with key life events to give the vehicle a trusted,

---

[2] https://www.w3.org/TR/did-core/#introduction.

machine-readable, tamper-evident, decentralized Self-Sovereign Identity. The physical vehicle and its SSDT™ are inextricably linked by the Vehicle Identification Number (VIN). This SSDT™ can then be used to store data for trusted IoT transactions, enabling interoperability and business automation across value chains. In a decentralized ecosystem, SSDTs™ are linked to DIDs anchored in a trusted network.

Blockchain and distributed ledger technologies (DLTs) open up business networks and other multi-party ecosystems for broader and more efficient collaboration. Stakeholders with relevant read and/or write permissions will be able to interact with the data stored on the tamper-evident decentralized ledger.

Ubiety

One of the most important and valuable attributes of any roaming entity is its *ubiety*. Ubiety, defined as an entity's unique position in space and time, is an uncommon word but a critical concept for IoT commerce. In the New Economy of Movement, an entity's trusted ubiety is what establishes the potential to provide a service or data at that location. A vehicle cannot, for example, provide data for real-time mapping or offer itself for hire at a location where it is not present. A vehicle cannot negotiate a right of way with another unless both vehicles are in close proximity and reliably know each other's position.

Today, the position of vehicles is typically established by triangulating signals from high orbit GPS satellites. In the best case, GPS can reliably establish vehicle position within a few meters. In difficult and congested environments like urban streets, GPS location accuracy deteriorates markedly or fails entirely. In addition, GPS itself can be spoofed.[3] In the near future, the addition of many new sources of triangulation signals—such as low and mid orbit satellites, fixed infrastructure and cellular stations, and even the vehicle network itself—will improve location accuracy by at least one and perhaps two orders of magnitude, even in difficult and congested environments.[4]

Combining Vehicle Identity and Ubiety

Combining a secure identity with a time-stamped location and recording the sequence of identity/ubiety pairs through time creates a **Trusted Trip**. Trip information is

---

[3] GPS spoofing is cheap, common, and readily available. It is used in everything from masking military movements, to cheating at Pokemon Go, to watching sporting events for free outside one's "local market." See, for example, https://helpdeskgeek.com/reviews/7-apps-to-fake-your-gps-location-on-android/.

[4] https://www.rewiresecurity.co.uk/blog/gps-and-telematics-new-trends-2021.

extremely valuable as a source of monetizable information about consumer preferences and social/commercial behavior. The trip is the basic unit of information for monetizing mobility in a services economy for consumers, producers, and infrastructure owners.

For consumers, the pairing of ubiety and identity extends the efficiencies of digital economics to the physical world. For the private sector, trips can be directly monetized through mobility services like rideshare, rental, and delivery; or indirectly monetized by extracting valuable data for real-time mapping, autonomous driving, usage-based insurance, and many other data-dependent applications. For manufacturers and shippers, it allows precise control of components, supply chains, and production. For infrastructure owners, the trip contains information needed to charge usage fees, including urban road tolling, congestion pricing, and taxation of carbon emissions. In short, it enables digital routing and tracking of the physical world.

Hence the enormous investment in developing map applications and centralized mobility platforms geared toward harvesting, aggregating, and profiting from users' mobility data. The strong network effects of these platforms give the companies an almost unassailable advantage in data-intensive mobility products, including such staples as rideshare, rental, and multi-modal trip coordination; as well as in developing algorithms for the emerging killer apps of the future, such as autonomous driving.[5]

MOBI Trusted Trip™

For a trip to be useful to economic agents in a transaction, it must be trusted.

MOBI Trusted Trip™ (MTT) links an entity's self-sovereign identity with its time-stamped location in the form of digital credentials (MTT Verifiable Credentials, or MTTVC) and enables the linkage to be certified throughout a trip in a trusted network. MOBI released the MTT Standard in October 2021.

The MTTVC is based on W3C's DID and VC Specifications. Using DIDs allows holders, on their own terms, to share trusted data with another verified entity without disclosing identity-related information and other selective attributes. MTT allows a holder of such credentials to prove to verifiers that it was present, completed a trip, used a resource, provided a service, or performed other relevant activities on the trip while safeguarding its (PII). This significantly reduces the risk of data being erroneous, tampered, or spoofed, ensuring the value of the shared network is preserved and enabling marginal cost pricing for countless types of decentralized mobility transactions. MTT is the key primitive for the New Economy of Movement.

How can third parties verify that a roaming entity started or completed a trip in a decentralized network?

At a minimum, the verification requirements include:

---

[5] https://www.investopedia.com/terms/n/network-effect.asp.

- **DIDs**: Verifier ID, user ID, vehicle ID, device ID, etc. all properly registered in a network.
- **Physical Presence**: Verification of physical presence at A and B, where A and B may be geospatial locations, transponder reads, charging/fueling stations, or time/date stamps.
- **Physical Movement**: Verification of physical movement on particular travel segments between A and B.
- **Documents**: Verification of documents created during transport, especially in logistics (e.g., bill of lading, manifests, receipts).

The basic interoperability requirements for a verifier to issue a standardized MTTVC within an automated business network include:

- Minimum dataset (proofs) with the essential information included in a Trusted Trip
- Unique trip identifier, referring to a trip that is globally unique and verifiable
- A trust anchor, including digital infrastructure, for establishing the authenticity and validity of certificates presented by certificate holders.

Pay-As-You-Go Mobility and MOBI Trusted Trip™

MTT unlocks almost every imaginable use case for decentralized, usage-based pricing for mobility services. For example, consider the case of greenhouse gasses (GHGs) and carbon footprint. In many countries, transportation is the largest source of GHG emissions. Yet there are limited tools to measure, monitor, and manage these emissions at their source—the vehicle's tailpipe. Combining identity and ubiety, along with other data verified by a minimum set of industry-accepted "proofs" within a decentralized shared network, enables a multitude of applications that advance smart, green mobility and decarbonization. Figure 1 illustrates the basic model for MTT.

Marginal Cost Pricing for Services and Data—Unbundling Mobility

In general, the improvements and efficiencies permitted by blockchains, ledgers, and automated business networks have the potential to unbundle mobility services and reduce the market power of mobility data aggregators. Unbundling services reduces the monopoly tendencies of data platforms and allows more providers to compete. For example, today there exist mobility services platforms that bundle trip planning, matchmaking, contracting, various reputation scores for drivers and riders, payments, and more. A decentralized ecosystem with better data rights management and open access will produce more competition, enhanced services, and a better customer experience. It might produce a system where a driver (or autonomous vehicle) hires a platform, rather than the other way around. Consumers and businesses benefit from open competition, lower or no monopoly rents, greater efficiency, lower consumer prices, and more choices.

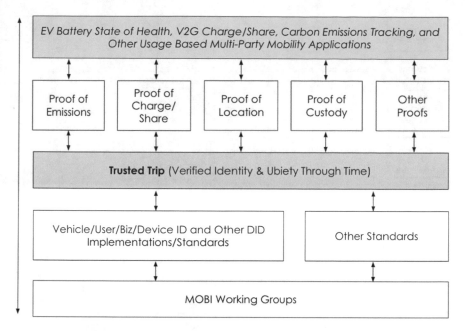

Fig. 1 The New Economy of Movement revolves around MOBI Trusted Trip™

Marginal Cost Pricing for Infrastructure—Fixing the Mobility "Commons"

In 1968, ecologist Garret Hardin wrote an article that called attention to the overuse of public resources by innocent but self-interested resource users.[6] This idea—that without property rights rational individuals have strong incentives to deplete a resource before it is depleted by others—can be traced back almost 250 years to British economist William Forster Lloyd, [4] who wrote about the effects of unregulated grazing on public land known as a "common." The concept became widely known as the "tragedy of the commons" and is a root cause of environmental damage and overuse of public resources. In mobility, it manifests in congestion, air pollution, crumbling infrastructure, and many other places where modern modes of transportation are most lacking. Blockchains and Trusted Trips give smart cities and governments new tools to manage their assets, fund their projects, balance supply with demand, incentivize green behavior, and solve the tragedy of the commons.

Supply Chain and MOBI Trusted Trip™

MTT is also a key primitive for improving efficiency in the manufacturing and mobility services supply chain. Modern vehicles contain thousands of parts from

---

[6] Hardin, G (1968). "The Tragedy of the Commons". *Science.* **162** (3859): 1243–1248. Bibcode:1968Sci...162.1243H. https://doi.org/10.1126/science.162.3859.1243. PMID 5699198.

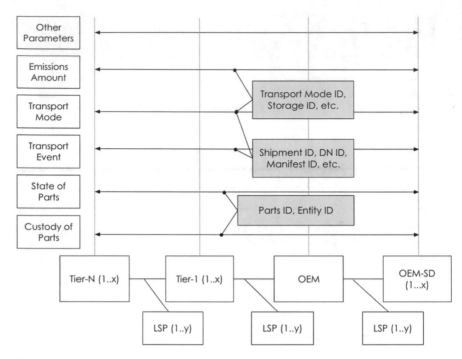

Fig. 2 MOBI Trusted Trip™ will increase transparency within the mobility supply chain, enabling more seamless parts tracking and traceability

hundreds of possible suppliers. Current supply chains use many siloed and paper-based processes, which means parts in transit are often hidden from the rest of the parties in the chain, making the system slow, opaque, and prone to error.

Tracking the location and trusted identity of a supply chain asset as it travels introduces many efficiencies and cost savings into the manufacturing process that ultimately result in safer, greener, and more affordable vehicles. That said, supply chain benefits don't end when the vehicle leaves the factory—even after manufacture, parts traceability reduces counterfeiting and improves the specificity, efficiency, and regulatory compliance of vehicle and part recalls, in addition to increasing visibility in the recycling process and unlocking an array of second and third-life use cases (particularly for electric vehicle batteries). Figure 2 illustrates MTT for parts traceability up and down the supply chain using a minimum set of industry-accepted proofs. This will enable scaling for a multitude of use cases such as efficient maintenance/recalls, authenticity verification, and ethical/sustainable labor practices.

Blockchains for ADAS and Autonomous Driving—Data, Algorithms, and Market Power in the New Economy of Movement

A vehicle, its owner, and various other mobility services providers will each create valuable content stemming from the movement and behavior of the vehicle, its parts, occupants, customers, and more. Onboard sensors of modern connected vehicles create massive amounts of data, and vehicles are increasingly using this data to plan routes, avoid accidents, and improve road safety. Advanced Driver Assistance Systems (ADAS) are active safety systems controlled by onboard computers that can override driver input and, in fully autonomous vehicles (AVs), drive without any driver input. The competition over better algorithms for ADAS and AVs is essentially a war for better and more driving data. Today, this war is being won by Silicon Valley companies running their machine learning on huge datasets scraped and aggregated from their applications' users.

The concentration of market power and wealth in the hands of data aggregators is a long trend, but digitization and the internet put the trend on steroids. Most of the biggest and fastest-growing global companies are digital natives that owe their dominance to controlling a chokepoint of the data economy. Centralization in the digital age is a core problem and the "original sin of the Internet", in the words of Michael Casey, author, futurist, and Chief Content Officer at Coindesk. It is, he writes, "a pet topic for those of us who believe the ideas behind blockchain technology can point us toward a better economic model".[7]

Blockchains can improve control over personal data and the rights of content creators in three ways. First, blockchains can strengthen digital rights management. Second, blockchains can improve data provenance and chain of custody, allowing data to be traced to a trusted source and preventing malicious or accidental changes. Third, blockchains can make it easier for individual users to monetize their data and for aggregation services to fairly compensate content creators.

3 Introduction to MOBI

Mobility Open Blockchain Initiative (MOBI) is a global nonprofit smart mobility consortium. MOBI and our members are creating blockchain-based standards to identify vehicles, people, businesses, and MOBI Trusted Trips. MOBI and its members are building the Web3 digital infrastructure for connected vehicle and IoT commerce with the goal of making transportation more efficient, equitable, decentralized, and sustainable, all while preserving the data privacy of users and providers alike. MOBI is incorporated as a US 501c6 nonprofit corporation funded and directed by its members. MOBI provides a neutral platform where members openly innovate and develop standards for blockchain in the mobility services industry. MOBI is technology and vendor agnostic.

---

[7] https://www.coindesk.com/can-blockchain-save-us-from-the-internets-original-sin.

3.1 Community

The MOBI community is made up of many of the world's largest vehicle manufacturers, along with startups, governments, non-governmental organizations (NGOs), transit agencies, insurers, toll road providers, smart city leaders, financial institutions, and technology companies. MOBI is a global organization comprising many large and small organizations distributed across Asia, Europe, and the Americas. MOBI is entirely supported by member contributions.

3.2 History and Motivation

The economic and environmental cost of population growth and urbanization demands a redesign of our cities and transportation systems. Novel technologies in development along with emerging market trends have the potential to solve our most difficult urban mobility challenges. These are now disrupting or poised to disrupt industries across many mobility verticals. MOBI was established to bridge smart mobility use cases to real-world implementation through the creation of standards and a shared, protocol-agnostic digital infrastructure.

Prior to MOBI's launch, many companies from the mobility and tech communities were experimenting with blockchains, building PoCs to demonstrate the technology. These companies found that putting a vehicle, data, or service on a chain was easy, but scaling was hard. Without common, agreed-upon standards of identifying things, sharing data, and transacting within a business network, the technology itself had little use. MOBI was officially launched on May 2, 2018 after several years of trials and discussions between major vehicle manufacturers and tech companies which underscored the need for a consortium approach.

3.3 Use Cases

MOBI's use cases span the entire mobility value chain. Use cases include, but are not limited to, vehicle identity, battery identity and state of health, multiparty supply chain track and trace, autonomous machine payments, connected mobility and IoT commerce, EV charging, decentralized energy storage, data markets, emissions tracking, vehicle and ride-sharing, usage-based mobility and insurance, fleet mangement (e.g., dealer floorplan automation), and congestion management. MOBI has active projects in several of these areas and continues to launch more each year.

3.4 Goals and Vision

Before the adoption of any revolutionary technology, industries need standards and specifications to build the foundational infrastructure enabling the creation of products and services that can communicate and work together. But standards, while necessary, are not sufficient for the adoption of truly new and disruptive technology. Standards also need to be adopted and implemented by a minimum viable community to enable use case applications to scale. To this end, MOBI is building the Web3 infrastructure for connected mobility and IoT commerce: the Integrated Trust Network (ITN) and Citopia. Together with the MOBI consortium, the new initiatives form what we refer to as the MOBI Technology Stack (MTS). The MTS comprises the foundational technologies needed to verify decentralized transactions between connected entities. Each layer provides a different architecture and function, together forming a holistic approach to Web3 applications for the connected ecosystem.

4 Overview of MOBI Working Groups and Standards

A MOBI working group is created when there is sufficient interest among members in a particular vertical. MOBI conducts regular surveys of its members to confirm areas of greatest interest. Once a vertical is identified, MOBI invites at least two established members who have deep subject matter expertise and are interested in co-chairing the working group. The co-chairs collaborate with the MOBI team to create a working group charter explaining the possible subordinate and derivative use cases. Additional members with both subject matter expertise and interest are assembled to review the charter and launch the working group. New working groups are established depending on demand from members, resource availability, and industry trends.

Each working group identifies the most representative and valued use cases in the respective vertical and creates blockchain-based standards outlining data architectures, security, permissions, and governance. The purpose of MOBI standards is to facilitate business network collaboration for each use case in the vertical. The typical time from inception to release of a MOBI standard is about 1 year.

4.1 MOBI Working Groups

As of January 2022, MOBI has launched seven working groups: Vehicle Identity I; Vehicle Identity II; Usage-Based Mobility and Insurance; Electric Vehicle Grid Integration; Connected Mobility Data Marketplace; Finance, Securitization, and Smart Contracts; and Supply Chain. Each working group is discussed in more detail below.

Vehicle Identity (VID)

MOBI's two VID working groups aim to define a digital document that is a verifiable link to a specific vehicle, a minimum representation of that vehicle's Self-Sovereign Digital Twin™, which can be used to establish existence, manage access control, confirm ownership history, and contain key events in the life of a vehicle. The vehicle is immutably anchored to MOBI VID by the physical Vehicle Identification Number (VIN), thus enabling a secure way for stakeholders with relevant read and/or write permissions to interact with the data stored through the tamper-evident blockchain. MOBI VID I establishes a machine-readable Vehicle Birth Certificate (VBC), capturing the vehicle's characteristics when it leaves the factory. The VBC is the first "link" in the full lifecycle of a vehicle. This gives each vehicle its own unique digital identity, which is able to capture features like the vehicle's make, model, production year, fuel efficiency, classification, and more. It is similar to a human birth certificate in that it begins at the vehicle's manufacture.

VID I was co-chaired by Groupe Renault and Ford, with support from Accenture, AIOI USA, BMW, Car Vertical, Cerebri AI, Cognizant, ConsenSys, CPChain, DMX, DLT Labs, GM, Honda, Hyperledger, IBM, IOTA Foundation, KAR, Luxoft, MintBit, Netsol Technologies, Oaken Innovations, On The Road Lending, Quantstamp, Trusted IoT Alliance, and Xapix.

VID II was co-chaired by BMW and Ford, with support from Accenture, AWS, AutoData Group, Bosch, Car IQ, CEVT, DENSO, DMX, Hitachi America, Ltd., Honda, IBM, KAR Auction Services, Luxoft, Nara Institute, Quantstamp, Ownum, and USAA.

The VID I working group released its technical standard on the machine-readable Vehicle Birth Certificate in July 2019. Following the release of the VID I Standard, MOBI started developing VID II in October 2019. VID II Standards were released in January 2021 to continue defining vehicle identity and lifetime events such as registration and maintenance traceability.

Usage-Based Mobility and Insurance (UBMI)

The UBMI working group aims to define the general framework that would allow people to plug and play a myriad of UBMI data sources, identities, and applications. The first standards are set to define the system design, multi-party processes, and data structures with appropriate identity, data, and permissioning proceeds to enable the access, sharing, and consumption of all data—real-time or static—generated within the mobility ecosystem to price risk and create usage-based mobility and insurance products.

UBMI is co-chaired by Achmea and AIOI USA, with support from Accenture, Cerebri AI, Cognizant, ConsenSys, DENSO, Deon Digital, DMX, Ford, GM, Honda, IBM, Luxoft, Netsol Technologies, Ocean Protocol, On the Road Lending, Quantstamp, R3, Renault, Reply, RouteOne, Streamr, Swiss Re, Tezos Foundation, USAA, USC, Volkswagen, and ZF.

In the evolving landscape of traditional business models toward service-based offerings, UBMI is the new model for modern mobility. For example, within UBMI, usage-based insurance (UBI) premiums are determined not by the vehicle type and various self-reported factors when the insurance contract is purchased, but by the actual trip mileage, driver history and behavior, local conditions, how the vehicle is being used, location, and more, all calculated in real time. This can help address challenges for multiple stakeholders within the insurance market. Since premiums rise for high-risk activity, drivers are incentivized to adopt safer driving behaviors. With real-time access to a rich universe of risk data, insurance providers can determine and price the risk of a policy with high accuracy, and therefore benefit from paying out fewer claims. The first UBMI standard aims to identify and facilitate the collection of all the different pieces of data needed by insurance providers to calculate UBI premiums.

Electric Vehicle Grid Integration (EVGI)

The EVGI working groups aims to support the adoption of electric vehicles by creating interoperable systems for governments, utilities, and the mobility industry alike, focusing on systems and data requirements for three core use case areas: Vehicle-to-Grid (V2G), Peer-to-Peer (P2P), and Tokenized Carbon Credits (TCC).

EVGI I is chaired by Honda and GM with support from Accenture, AWS, Cognizant, CPChain, DENSO, DOVU, Hitachi America, Ltd., IBM, IOTA Foundation, KAR, KoinEarth, Pacific Gas & Electric (PG&E), Politecnico di Torino, R3, Sphericity, and Swedish Blockchain Association. EVGI II is co-led by Accenture, Anritsu, ASJade, AWS, DENSO, Ford, Henshin Group, Hitachi, Honda, ITOCHU, peaq, and Stellantis.

The increasing adoption of electric vehicles accompanies a rise in innovative smart mobility applications. However, without standards, it is difficult to scale them and link different applications. To address this challenge, the EVGI working groups are developing interoperable systems by creating system specifications and standard data schemas to facilitate the increased adoption of electric vehicles. EVGI's focus on V2G, TCC, and P2P energy trading applications will enable a more efficient and resilient way to manage grid loads, anticipate demand, generate carbon offsets, and implement P2P services.

In September 2020, the EVGI I working group released its first technical specification focused on V2G, TCC, and P2P. The EVGI I Standard establishes a foundation on which a wide range of use cases can be built. This means that rather than describing a single application, the standard ensures that all of the functionalities and relevant data attributes for each use case are available for organizations to use in building their own applications.

Connected Mobility Data Marketplace (CMDM)

The CMDM working group aims to enable a blockchain-based permissioned-data marketplace for all stakeholders of the mobility ecosystem—including vehicle manufacturers, advertisers, insurance providers, and others—to effectively share and exchange data. CMDM focused on the exchange of vehicle, infrastructure, and user data for use cases that include vehicle coordination and safety, V2X payments, monetizing mobility data, targeted content delivery, and creating better driving algorithms through federated machine learning.

CMDM is co-chaired by GM and DENSO, with support from Accenture, AMO Labs, CEVT, Cognizant, Constellation Labs, Continental, CPChain, DMX, Fifth-9, Filament, Ford, IBM, NuCypher, Ocean Protocol, Reply, RouteOne, ShareRing, Swedish Blockchain Association, and Toyota Insurance Management Solution (TIMS).

The CMDM working group released its first standards in March 2021 covering V2X and infrastructure-to-infrastructure (I2I) use cases. In each of these use cases, connected devices enable secure data sharing, ID authentication among parties, and secure transaction recording. CMDM standards define a robust system for interoperable data sharing among connected devices and provide a new certificate for a connected sensor identity—in short, they create a framework for identity, authority, and assurance within decentralized networks.

Finance, Securitization, and Smart Contracts (FSSC)

The FSSC working group assesses the potential value proposition of blockchain and interoperability standards that stakeholders of the mobility finance ecosystem—such as original equipment manufacturers (OEMs), auto financiers, and dealerships—can implement to reduce the cost of vehicle ownership and improve customer satisfaction with use cases such as credit on the blockchain, securitization, tokenization of mobility assets, and fractional ownership of mobility assets.

FSSC is chaired by Orrick, Herrington and Sutcliffe, and RouteOne with support from Accenture, Altaventure, BMW Bank, CEVT, Connections Insights, CO-OP Financial Services, ConsenSys, D.E. Consulting, Ford Credit, Global Debt Registry, IOTA Foundation, On the Road Lending, Quant Network, Quantstamp, Reply, Tezos Foundation, and USAA.

Launched in March 2020, the FSSC working group strives to improve accuracy and transparency, create operational efficiencies, minimize fraud risks, and save on costs and time in the execution of financings, including securitizations, for all entities in the mobility value chain. Such entities include consumer loan originators, credit facility providers to dealers, securitization sponsors/issuers, servicers, investors, rating agencies, trustees, and regulators.

The FSSC working group released standards in June 2021 that prescribe a set of core services, as well as a set of logical schemas, which capture all pertinent data attributes utilized throughout the vehicle finance lifecycle. Those data attributes are redundantly stored within a distributed network for use within financial applications and new business solutions. The standard specifies a rich digital infrastructure, providing frameworks for identity, permissioning, and more. These frameworks are intended to achieve interoperability between the "walled garden" financial ecosystems that hinder collaboration between ecosystem stakeholders, raise costs, and decrease transparency. Integrating these systems allows for solutions that provide operational efficiencies, superior insights, new revenue streams, and much more.

Supply Chain (SC)

Blockchain applications have proven very successful in improving the transparency and traceability of supply chains. The SC working group was created to explore blockchain's value proposition for mobility supply chain management and establish standards to set a common framework. The group aims to create interoperability standards to bring efficiencies and increased visibility through the N-tiers of supply chains; enable provenance, tracking, and authenticity of parts and vehicles; and improve conflict resolution and settlement with DLT.

SC is chaired by BMW and Ford with support from Accenture, AIOI USA, Arxum, AutoData Group, AWS, CEVT, DENSO, DLT Labs, DMX, Fifth-9, Hitachi America, Ltd., Honda, IBM, IOTA Foundation, ITOCHU, Marelli, Nara Institute, Politecnico Di Torino, Quantstamp, R3, Reply, SyncFab, Thirdware, and Vinturas.

The working group released interoperability standards in June 2021 to improve provenance, tracking, and authenticity of parts and components up and down the supply chain and address auditability and settlement through blockchain-based systems.

In addition to parts traceability, the working group also considered several use cases including authenticity of components, component traceability, mineral provenance, fair labor practices, automation of payments via smart contracts, organization IDs, master data management, 3D printing workflow traceability, supplier site data collection, and trade agreement origin certification.

4.2 MOBI Standards

There are four types of MOBI standards: business white papers, use cases and business requirements, technical specifications, and reference implementation architectures. Their full descriptions are below.

Business White Papers (WP)

MOBI business white papers are high-level business reviews that discuss issues and propose solutions to the world's most pressing transportation challenges with consideration to ecosystem stakeholders, new strategies, emerging technologies, and global policies.

Use Cases and Business Requirements (UC)

MOBI use cases and business requirements documents describe pain points, stakeholder responsibilities, and the high-level business requirements that potential solutions must meet in order to resolve stakeholder needs. UCs also detail workflows for particular applications and are technology-agnostic.

Technical Specifications (TS)

MOBI technical specifications define recommended minimum interfaces between systems/modules and data specification exchanged in the process leading up to a reference implementation. This process enables interoperability in independently developed systems.

Reference Implementation Architectures (RI)

MOBI reference implementation architectures prescribe and recommend a solution architecture that stakeholders can refer to when they deploy solutions, ensuring that stakeholder requirements described in TS and UC are met in the process. RIs are vendor agnostic.

5   MOBI Web3 Technology Stack (MTS)

In the New Economy of Movement, Web3 federated networks and decentralized platforms are needed to manage DIDs and VCs to enable trusted data sharing and business automation between connected entities. MOBI's Web3 infrastructure consists of three member-owned and operated layers—collectively the MOBI Technology Stack (MTS)—needed to verify these transactions. Each layer provides a different architecture and function, together forming a holistic approach to Web3 applications for decentralized connected ecosystems.

The foundational layer is the *MOBI consortium*, which creates standards to identify connected entities and shared business processes.

The middle layer is *the Integrated Trust Network (ITN)*, a layer-two, protocol-agnostic digital infrastructure to provide trusted decentralized identity services.

The top layer is *Citopia*, a member-owned and operated trustless decentralized marketplace to onboard Self-Sovereign Digital Twins™ and enable VCs issuance for business automation and trusted track and trace.

5.1   MOBI Consortium

MOBI is a global nonprofit smart mobility consortium consisting of public and private stakeholders from around the world. MOBI and its members are creating blockchain-based standards to identify vehicles, people, businesses, and *MOBI Trusted Trips*, with the goal of making transportation more efficient, equitable, decentralized, and sustainable. MOBI officially launched in May 2018 and released its first standard, MOBI VID, the following year. As of July 2022, MOBI has formed seven working groups, released 15 standards, and launched the DRIVES (Distributed Registry for Intelligent Vehicle Ecosystem and Sustainability) Program in early 2021 to ideate and demonstrate MOBI Standards to extend multi-party value chains in a decentralized ecosystem.

5.2   Integrated Trust Network (ITN)

Autonomous IoT transactions will become an increasingly large part of the multi-trillion-dollar, pay-as-you-go, services economy. In order to execute these transactions, a trusted identity and settlement network is needed. The ITN offers cross-industry stakeholders around the globe an open and inclusive core services infrastructure for decentralized transactions at the edge.

The ITN is a member-owned and operated technology-agnostic network designed for multi-party vertical applications enabled by a secure joint core services infrastructure. The ITN's goal is to provide a Federated Trust Layer of Core Services and Business Automation Interoperability Infrastructure for User Agents, bringing organizations together within a trusted network while protecting their intellectual property (IP) rights, customers, value chains, and brands. MOBI is working with other consortia to co-build this edge network for connected mobility and IoT commerce. Figure 3 illustrates ITN Solution Architecture; Fig. 4 illustrates ITN architecture for User Agents.

IoT devices worldwide are forecast to reach 31 billion by 2025 and are doubling every 3 years.[8] The number of possible automated device transaction pairs for digital services grows as the square of devices, and the trust problem is further complicated by the fact that many of these devices are roaming. The goal of the ITN is to unlock monetization opportunities across usage-based services by automating application interoperability and multi-party data sharing. Consider it the infrastructure for a "Point of Sale" network for the IoT device economy.[9]

---

[8] https://www.statista.com/statistics/1101442/iot-number-of-connected-devices-worldwide/.

[9] The reference to the point of sale infrastructure created by the BankAmericard (later rebranded as VISA) network is intentional. The BankAmericard network was created to provide shared services of governance, authority, identity, and assurance to its members, who needed a shared infrastructure to bank the new merchant point of sale transactions occurring outside the bank's premises. In today's parlance, these were the original edge transactions, and the member owned network was required

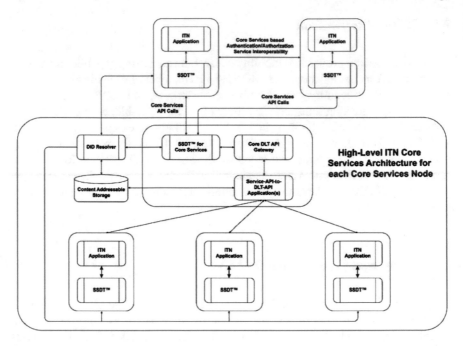

Fig. 3. ITN solution architecture of core services

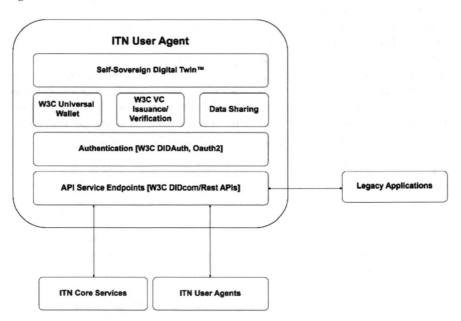

Fig. 4. ITN architecture for user agents

5.3   Citopia

At the top of the MTS is Citopia, a member-owned and operated federated Web3 marketplace and tokenized ecosystem for connected mobility and IoT commerce where ecosystem stakeholders can securely transact. Citopia facilitates the onboarding of Self-Sovereign Digital Twins™ and issuance of VCs for automating multiparty transactions in the Web3 economy.

Citopia SSDTs™ are linked to DIDs anchored in the ITN to create unique, portable, tamper-evident self-sovereign identities for all entities in the Citopia marketplace and permit them to participate in trusted, decentralized transactions. The federated Web3 marketplace consists of Citopia vinTRAK, Citopia partsTRAK, and Citopia MaaS.

Citopia vinTRAK enables privacy-preserving and secure vehicle track-and-trace applications, including fleet management (e.g., dealer floorplan automation), vehicle maintenance and repair, emissions tracking, road usage charging, and usage-based insurance. Citopia partsTRAK enables trusted multiparty track-and-trace for assets in the mobility value chain. partsTRAK applications include battery passport, EV charge/pay/share and SOH Tracking, maintenance/recall traceability, lifecycle decarbonization, and ethical/sustainable sourcing. Citopia MaaS offers secure, seamless, customizable travel experiences by allowing end users to plan, reserve, and pay for multimodal trips all in one place.

Citopia allows users to interact and transact with a diverse network of providers from a single marketplace, eliminating the need for multiple logins, user cards, apps, and payment methods. Users can personalize their trips by specifying trip preferences and choosing from route options such as fastest, cheapest, least transfers, and most sustainable.

Urban populations are on the rise and cities need solutions to manage infrastructure demands. At the same time, the way people like to move around is also evolving. They want a seamless and efficient experience of integrated modals of transit. Current mobility platforms rely on a centralized operator. In centralized platforms, integration costs are high for providers; at the same time, retaining users' and providers' data controllability and privacy, along with fair business practices is not possible. For these reasons, centralized platforms are not scalable. Figure 5 illustrates how Citopia works with the ITN to enable the execution of trusted decentralized transactions.

Citopia uses Zero Knowledge Proofs and other advanced cryptographic methods to verify transactions without exposing users' and organizations' sensitive data. As a result, users and providers on Citopia maintain full control over who sees their data, how much data is shared, and how that data is used. Federated learning also offers potential cooperation benefits such as better customer experience (e.g., better matchmaking results). Users will also benefit from better services at a lower cost, as Citopia enables marginal cost pricing for a variety of services by eliminating the need for expensive third-party intermediaries and promoting competitive pricing among

---

to deliver KYC at the edge. MOBI and the MTS are required to deliver the vehicle and device equivalents of KYC for edge mobility.

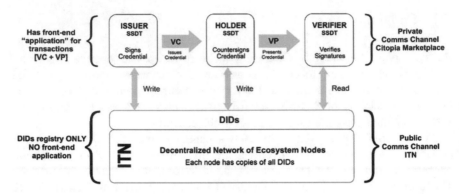

Fig. 5 Separation for decentralization: MOBI Web3 Infrastructure

providers in the ecosystem. Figure 6 illustrates Citopia's network effect. Citopia's track-and-trace functionalities allow providers to streamline operations, adapt to user trends, and respond to service gaps in real time, resulting in increased service quality and higher user satisfaction.

Citopia has designed a native stable token, backed by transportation assets, with several desirable features for mobility applications. However, the native token isn't required for Citopia payments. Given the regulatory uncertainties surrounding stable tokens and digital payments, we expect that the initial iterations of Citopia payments will be the familiar fiat currency instruments that are widely used today. As private or central bank digital currencies gain approval and acceptance, Citopia can expand its settlement choices. To the extent that these new digital currencies reduce transaction fees, more small-value transactions and new mobility applications will become feasible.

The MOBI community is currently working on several completed and ongoing pilots to demonstrate the capabilities of

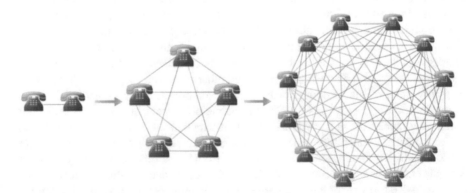

Fig. 6 The network effect on Citopia

- Citopia and the ITN: Citopia for EVs | Use Cases: EV Charging, Reservation, Payment, State of Health, and Emissions Reporting
- Citopia partsTRAK | Use Case: EV Battery Track and Trace for Vehicle Manufacturers, Suppliers, Dealers, and Vehicle Owners
- European Commission Pilot with Citopia & ITN on CO2 Emissions Monitoring
- Citopia vinTRAK — Use Case: Dealer Floorplan Audit
- Citopia MaaS — Transit IDEA Award

6 DRIVES (Distributed Registry for Intelligent Vehicles Ecosystem Sustainability) Program

MOBI launched the DRIVES (Distributed Registry for Intelligent Vehicle Ecosystem and Sustainability) Program in early 2021 to ideate, incubate, and demonstrate multi-party use cases and to accelerate the implementation of MOBI standards. The roadmap after a standard's release is as follows:

- Use case selection by working group members.
- Proof of value discussions by the WG.
- Minimum value ecosystem (MVE)—demonstration of key primitives for use cases and ITN core services.
- Pilots and prototypes—broaden the MVE for the WG to perform pilots.
- Enterprise-scale commercial applications—MOBI's role is to provide trusted identity services via the ITN. Members build and scale individual applications.

6.1 The What and Why of DRIVES

DRIVES is a structured agile development environment to:

- Test the MOBI Technology Stack and allow for community collaboration.
- Demonstrate MTT as a key primitive for smart mobility use cases such as battery state of health, usage-based fees, and rewards/incentives.
- Demonstrate MTT as a key primitive for Supply Chain use cases such as parts traceability, emissions tracking, recycling, and safe disposal.
- Provide a testnet for connected mobility providers to collaborate and scale multi-party use cases built on blockchain/DLT.
- Enable new business models, monetization, and incentive mechanisms for smart and low-carbon mobility.

For participants, DRIVES provides a neutral laboratory to incubate pilots, analyze the value chain for innovative business models, and understand the benefits, costs, and feasibility of multi-party applications. It leverages MOBI's standards, infrastructure, and network of peers to reduce the cost of onboarding participants, share resources, and lower risks.

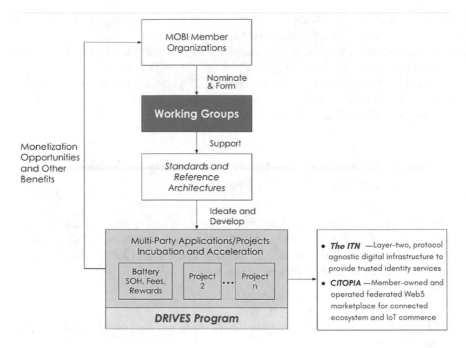

Fig. 7 Overview of DRIVES program and other MOBI initiatives

6.2 DRIVES Program in the MTS

The DRIVES Program plays a key role in incubating and demonstrating the MOBI Technology Stack. MOBI completed its first Trusted Trip demo in October 2021. Our ongoing and future demos and pilots will accelerate scaling for the MOBI Technology Stack and unlock countless opportunities for connected mobility and IoT commerce. Figure 7 illustrates where and how DRIVES delivers value for MOBI members; Fig. 8 illustrates the implementation architecture.

7 Conclusion

Since the first Model T rolled off Henry Ford's assembly line in 1908, mobility has been dominated by private ownership and the internal combustion engine. This mobility model led to improvements in living standards, lifestyles, and opportunities unimaginable to the average person at the dawn of the twentieth century; it also led to pollution, long commutes with hours stuck in traffic, global warming, and other ills that challenge us at the dawn of the twenty-first. The old mobility paradigm is not sustainable. Our roads cannot fit a private car for every person, our air cannot absorb more carbon, and our days cannot accommodate more hours in traffic.

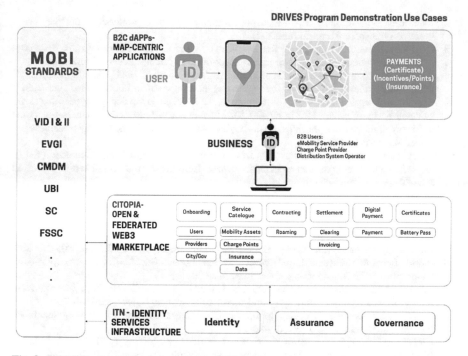

Fig. 8 DRIVES program implementation architecture

As machines take on more jobs now done by people, we may have more leisure, but fewer ways to earn and afford the benefits of that leisure. Our personal data, our movements, our preferences, and our content creation—where we choose to go and what we choose to do when we arrive—cannot be fully replicated by machines and may become the biggest source of human value creation. It will be critical that new modes of digital rights evolve to protect this value, and that we monetize this value ourselves rather than cede it to others.

Blockchains, Self-Sovereign Digital Twins™, and a more decentralized mobility services economy, along with better batteries, decarbonization, IoT sensors, and faster connections, enable a better mobility future. As for the buyers of the Model T, it is a future that is brighter and more affordable in ways unimaginable to us today. But ultimately, the most important legacy of blockchains and the New Economy of Movement may be to help people choose when and how to monetize their movement and all that their movement reveals about their preferences, identities, and behaviors.

References

1. Accenture: Mobility as a Service (2019). https://www.accenture.com/_acnmedia/PDF-71/Accenture-Mobility-Service.pdf. Accessed June 2021

2. Asian Development Bank. Urban Transport. https://www.adb.org/sectors/transport/key-priori
   ties/urban-transport. Accessed June 2021
3. Auto Remarketing: Takata Airbag Recall Effect on Used Pricing (2015). https://www.autore
   marketing.com/trends/takata-airbag-recall-effect-on-used-pricing. Accessed June 2021
4. Stockton, B.: 7 Apps to Fake Your GPS Location on Android, HelpDeskGeek (2020). https://hel
   pdeskgeek.com/reviews/7-apps-to-fake-your-gps-location-on-android/. Accessed June 2021
5. Carson, B., Romanelli, G., Walsh, P., Zhumaev, A.: Blockchain beyond the hype: what is the
   strategic business value? McKinsey (2018). https://www.mckinsey.com/business-functions/
   mckinsey-digital/our-insights/blockchain-beyond-the-hype-what-is-the-strategic-business-
   value. Accessed June 2021
6. Casey, M.J.: Can Blockchain Save Us from the Internet's Original Sin? Coindesk (2017).
   https://www.coindesk.com/can-blockchain-save-us-from-the-internets-original-sin. Accessed
   June 2021
7. Denton, J.: Forget Nio and XPeng. This company and Tesla will be the top two electric-vehicle
   plays by 2025, says UBS. Market Watch (2021). https://www.marketwatch.com/story/for
   get-nio-and-xpeng-this-company-and-tesla-will-be-the-top-2-electric-vehicle-plays-by-2025-
   says-ubs-11615306959. Accessed June 2021
8. Elks, S.: Green transport set to overtake cars in world's major cities by 2030 (2020)
9. World Economic Forum. https://www.weforum.org/agenda/2020/02/green-transport-cars-
   major-cities-2030-climate-change/. Accessed June 2021
10. Ellen McArthur Foundation: Cities And The Circular Economy (2021). https://www.ellenm
    acarthurfoundation.org/explore/cities-and-the-circular-economy. Accessed June 2021
11. Environmental Defense Fund. Health impacts of air pollution. https://www.edf.org/health/hea
    lth-impacts-air-pollution. Accessed June 2021
12. Global Market Insights: Usage-based insurance market to hit $107bn by 2024: Global Market
    Insights, Inc. (2018). https://www.globenewswire.com/news-release/2018/12/03/1660531/0/
    en/Usage-based-Insurance-Market-to-hit-107bn-by-2024-Global-Market-Insights-Inc.html.
    Accessed June 2021
13. Goodall, W., Fishman, T.D., Bornstein, J., Bonthron, B.: The rise of Mobility as a Service.
    Deloitte (2017). https://www2.deloitte.com/nl/nl/pages/consumer-industrial-products/articles/
    the-rise-of-mobility-as-a-service.html. Accessed June 2021
14. Hannon, E., Nauclér, T., Suneson, A., & Yüksel, F.: The zero-carbon car: abating material
    emissions is next on the agenda. McKinsey (2020). https://www.mckinsey.com/business-fun
    ctions/sustainability/our-insights/the-zero-carbon-car-abating-material-emissions-is-next-on-
    the-agenda. Accessed June 2021
15. Hardin, G: The tragedy of the commons. Science **162**(3859), 1243–1248 (1968). https://sci
    ence.sciencemag.org/content/162/3859/1243. Accessed June 2021
16. Herweijer, C., Combes, B., Swanborough, J., Davies, M.: Building block(chain)s for a
    better planet. World Economic Forum (2018). https://www.weforum.org/reports/building-
    block-chain-for-a-better-planet. Accessed June 2021
17. Intergovernmental Panel Climate Change: Special report: global warming of 1.5 °C summary
    for policymakers (2018). https://www.ipcc.ch/sr15/chapter/spm/. Accessed June 2021
18. International Energy Agency. Data and Statistics. https://www.iea.org/data-and-statistics?cou
    ntry=WORLD&fuel=CO2%20emissions&indicatorCO2BySector. Accessed 24 March 2021
19. Investopedia. Network Effects. https://www.investopedia.com/terms/n/network-effect.asp
    Accessed June 2021
20. Jibrell, A.: Automotive recall bill grew 26% to $22 billion in 2016, study says. Automotive
    News (2018). https://www.autonews.com/article/20180130/RETAIL05/180139974/Auto-rec
    all-bill-grew-26-to-22-billion-in-2016-study-says. Accessed June 2021
21. MaaS Alliance. What is MaaS? https://maas-alliance.eu/homepage/what-is-maas/. Accessed
    18 March 2021
22. McKinsey & Company: What's driving the connected car? (2014). https://www.mckinsey.com/
    industries/automotive-and-assembly/our-insights/whats-driving-the-connected-car. Accessed
    June 2021

23. Moore, S.: How to Stop Data Quality Undermining Your Business. Gartner (2019). https://www.gartner.com/smarterwithgartner/how-to-stop-data-quality-undermining-your-business/. Accessed June 2021
24. National Highway Traffic Safety Administration (2019) Odometer fraud. https://www.nhtsa.gov/equipment/odometer-fraud. Accessed June 2021
25. Pastori, E., Vergnani, R.: Odometer tampering: measures to prevent it. Research for TRAN Committee, European Parliament (2019). https://www.europarl.europa.eu/RegData/etudes/STUD/2017/602012/IPOL_STU(2017)602012_EN.pdf. Accessed June 2021
26. Plungis, J.: Who owns the data your car collects? Consumer Reports (2018). https://www.consumerreports.org/automotive-technology/who-owns-the-data-your-car-collect/. Accessed June 2021
27. Rewire Security: New Trends in GPS & Telematics in 2021 and Beyond, Rewire Security Blog (2021). https://www.rewiresecurity.co.uk/blog/gps-and-telematics-new-trends-2021. Accessed June 2021
28. Saint John, J.: General Motors pledges a zero-emissions light-duty vehicle fleet by 2035. Green Tech Media (2021). https://www.greentechmedia.com/articles/read/general-motors-pledges-a-zero-emissions-light-duty-vehicle-fleet-by-2035. Accessed June 2021
29. Schmahl, A., Burchardi, K., Egloff, C., Govers, J., Chan, T., Giakoumelos, M.: Resolving the blockchain paradox in transportation & logistics. Boston Consulting Group (2019). https://www.bcg.com/en-gb/publications/2019/resolving-blockchain-paradox-transportation-logistics.aspx. Accessed June 2021
30. Stamford, C.: (2019) Gartner predicts 90% of current enterprise blockchain platform implementations will require replacement by 2020. Gartner. https://www.gartner.com/en/newsroom/press-releases/2019-07-03-gartner-predicts-90--of-current-enterprise-blockchain. Accessed June 2021
31. Statistica: IoT and non-IoT connections worldwide 2010–2025 (2021). https://www.statista.com/statistics/1101442/iot-number-of-connected-devices-worldwide/. Accessed June 2021
32. Tillemann, L., Wolff, C., Ben Dror, M.: The road ahead: a policy research agenda for automotive circularity. World Economic Forum (2020). http://www3.weforum.org/docs/WEF_A_policy_research_agenda_for_automotive_circularity_2020.pdf. Accessed June 2021
33. United Nations: 68% of the world population projected to live in urban areas by 2050, says UN (2018). https://www.un.org/development/desa/en/news/population/2018-revision-of-world-urbanization-prospects.html. Accessed June 2021
34. US Environmental Protection Agency: Sources of greenhouse gas emissions (2021). https://www.epa.gov/ghgemissions/sources-greenhouse-gas-emissions. Accessed June 2021
35. World Wide Web Consortium: Decentralized Identifiers (DIDs) v1.0 Core architecture, data model, and representations (2020). https://www.w3.org/TR/did-core/#introduction. Accessed June 2021
36. World Economic Forum: Building value with blockchain technology: how to evaluate blockchain's benefits (2019). http://www3.weforum.org/docs/WEF_Building_Value_with_Blockchain.pdf. Accessed June 2021
37. World Economic Forum: Driving the sustainability of production systems with fourth industrial revolution innovation (2018). https://www.weforum.org/whitepapers/driving-the-sustainability-of-production-systems-with-fourth-industrial-revolution-innovation. Accessed June 2021

Blockchain-Based Data Management for Smart Transportation

Mirko Zichichi, Stefano Ferretti, and Gabriele D'Angelo

Abstract Smart services for Intelligent Transportation Systems (ITS) are currently deployed over centralized system solutions. Conversely, the use of decentralized systems to support these applications enables the distribution of data, only to those entities that have the authorization to access them, while at the same time guaranteeing data sovereignty to the data creators. This approach not only allows sharing information without the intervention of a "trusted" data silo, but promotes data verifiability and accountability. We discuss a possible framework based on decentralized systems, with a focus on four requirements, namely, data integrity, confidentiality, access control, and persistence. We also describe a prototype implementation and related performance results, showing the viability of the chosen approach.

1 Introduction

In the last decade, Intelligent Transportation Systems (ITS) have emerged as a way to efficiently improve mobility, travel security and increase the options for travelers. As defined in the European Union directive 2010/40/EU [12], ITS are advanced applications for the provision of innovative transport and traffic management services, with the ultimate purpose of aiding individuals within the infrastructure to make safe and timely decisions. The general idea is usually that of devising a sort of data management middleware to build advanced applications for the provision of innovative transport and traffic management services, with the aim of enabling users "to be better informed and make safer, more coordinated and 'smarter' use

M. Zichichi (✉)
Ontology Engineering Group, Universidad Politécnica de Madrid, Madrid, Spain
e-mail: mirko.zichichi@upm.es

S. Ferretti
Department of Pure and Applied Sciences, University of Urbino "Carlo Bo", Urbino, Italy
e-mail: stefano.ferretti@uniurb.it

G. D'Angelo
Department of Computer Science and Engineering, University of Bologna, Bologna, Italy
e-mail: g.dangelo@unibo.it

of transport networks" [12]. Vehicles and transportation infrastructures are becoming increasingly "smarter", which means that they are equipped with sensors that track and process a huge amount of different types of information, e.g., data sensed by the interior of the vehicle, the surrounding environment, road conditions, etc. This enables the creation of applications "without embodying intelligence as such", which brings out the real essence of an infrastructure of this kind. The interaction processes between two individuals, or an individual and a vehicle, or an individual and the infrastructure, within the ITS, should include the least possible presence of a human intermediary. All of this constitutes a network of user-owned and infrastructure devices that is usually referred as VANET (Vehicular Ad-hoc NETwork) [31]. In this vision, the intelligence shifts from that of a human third-party to that of an artificial intelligence that has been optimized for this use case. This artificial intervention leads to the creation of "innovative services relating to different modes of transport and traffic management" [12], that take advantage of faster processing and better performances. When there are no human intermediaries, indeed, traditional processes become faster to execute.

In addition, the growth of smartphones and Internet-of-Things devices enables individuals' ubiquitous connectivity and the ability to collect environmental and personal information or crowd-sensed data [42]. Thus, users become an active part of the infrastructure itself. The entirety of such crowd-sensed information is essential for building sophisticated smart services that aim at improving traffic management, transportation efficiency and safety, raising awareness about the environment, and thus improving the liveability and health status of the community of a given territory [11, 18, 37].

A variety of applications and protocols can be enforced altogether to obtain advanced and improved transportation systems. However, to fully exploit their potential and promote the development of smart mobility applications and services for social good, several novel challenges must be faced, that require substantial changes in transportation system models. The "desiderata" for such novel applications and systems revolve around data management: more in particular, the mentioned data gathering, communication, analysis and distribution among individuals' vehicles, infrastructures, and services. Data sharing is placed on a middle ground between devices that produce data and the systems that process data to create new smart services. Data-driven innovation will bring enormous benefits for ITS users and not only [19]. The generation and sharing of such an amount of data create the need for trading mechanisms that are at the basis of the productivity and competitive markets for smart service providers, but also fundamentals in health, environment, transparent governance, and convenient public services. In turn, this creates the need for evaluating data, in terms of interoperability and quality. These two features, together with data structure, authenticity, and integrity are key elements for an effective data exploitation [19].

In the context of ITS, one of the main issues is the unreliability of the exchanged information [11, 42, 54]. This problem is typically due to the physical errors of the sensors, malfunctions, poor network and GPS coverage. Such noisy data lead to inaccurate information. Another problem is due to the fact that some users might be

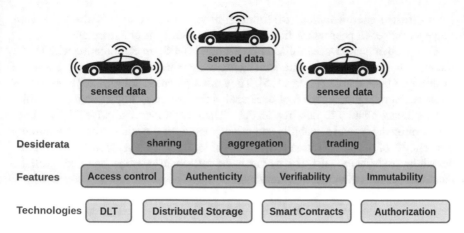

Fig. 1 Intelligent transportation system data management schema

interested in deliberately transferring forged information. Examples are insurance frauds, as well as free-riders that decide to share false data, randomly generated without using their sensors, in order to gain some revenues/credits for such fake data sharing. Thus, one of the main goals to pursue is the identification of strategies for the generation and distribution of secure and trustable crowd-sensed information.

The need for trustful data trading and sharing leads to the rationale that anyone should be allowed to verify the authenticity and the immutability of shared information (see Fig. 1). From the point of view of the data sharer, on the other hand, the features of verifiability and access control are needed to be combined: making data completely public would make them more verifiable but would also lower their value; but on the other hand, completely closing the access to the data would lower its verifiability.

This is where a (relatively) new kind of technology can come to aid. Distributed Ledger Technologies (DLTs) are thought to provide a trusted and decentralized ledger of data. DLTs are a novel keyword that extends the famous "blockchain" buzzword, to include those technological solutions that do not organize the data ledger as a linked list of blocks. Currently, DLTs are widely utilized in scenarios where: i) multiple parties concur in handling some shared data, ii) there is no complete trust among these parties, and often iii) parties compete for the access/ownership of such data [8, 14]. This is a typical scenario of smart transportation services that exploit data sensed from multiple sources, i.e., vehicles and infrastructure.

DLTs provide the technological guarantees for trusted data management and sharing, as they can offer a fully auditable decentralized access control policy management and evaluation [33]. Indeed, these decentralized architectures are an ideal choice for data management and sharing [50] because of their features: (i) Transparency: auditability of access permissions by authorized third-parties; (ii) Security: shifting trust towards the consensus mechanism allows mitigating the vulnerabilities of (semi-)trusted intermediaries; (iii) Immutability: on-chain data will always be veri-fiable; (iv) Peer-to-peer interactions: potential of user-to-user agreements.

On the other hand, Decentralized File Storages (DFS), in combination with DLTs, increase the possibilities in data management and sharing as they provide a range of different but suitable features [55]. They are a potential solution for storing files while maintaining the benefits of decentralization, offering higher data availability and resilience thanks to data replication. Their combined use with DLTs allows overcoming the typical scalability and privacy issues of the latter, while maintaining the benefits of decentralization [38]. In practice, DFS are leveraged for storing the actual data outside the DLT, i.e., by means of "off-chain" storage, and tracing all the data references in the DLT, i.e., "on-chain".

Finally, smart contracts, built upon some DLT implementations, allow checking the terms of an agreement without requiring the presence of a trusted human third-party validator. These may enable auditability of access permissions by authorized third-parties and mitigate privacy vulnerabilities of (semi-)trusted intermediaries when accompanied by off-chain security mechanisms [53].

To sum up, various technologies enable the deployment of viable and scalable systems for the support of smart services in the ITS domain. This work aims to sur-vey the possible ways to handle data management and governance, showing their strengths and limitations, in order to provide a framework. Furthermore, we investi-gate the feasibility of this framework by offering an implementation based on current DLT and DFS solutions and discussing its performance in comparison to ITS needs.

2   Data Management Strategies

2.1   The Classic Centralized Approach for Crowd-Sourced Data Aggregation

The most straightforward approach for managing data and services in ITS resorts to a cloud computing infrastructure (Fig. 2) [44]. Vehicles and smartphones collect data and transmit them to the cloud, in platforms where it is possible to extract information and utilize it for models, visualizations, and/or decision-making [32]. In this scenario, cloud computing enables ubiquitous, cost-effective, on-demand network access to a shared pool of configurable computing resources that can be rapidly provisioned and deployed with minimal management effort. Large online platforms, backed up by data centers and centralized computing facilities, provide the advantage of efficient data aggregation, data mining, analysis optimization, storage, batch processing, and computation, i.e., the cloud can compute the gigantic amount of data and complex computations in a very short time [44].

There is a trade-off, however, that leads to imbalances in market power as large online platforms, where a small number of players may accumulate large amounts of data, gather important insights and competitive advantages from the richness and variety of the data they hold [19]. The current practice of data controllers, i.e.,

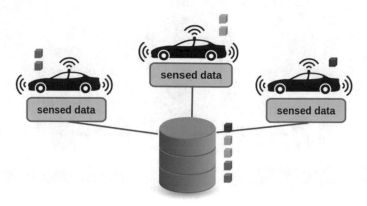

Fig. 2 Crowd-sensed data aggregated in a storage maintained by a single central entity

entities that collect and manage data coming from users' devices and infrastructure, is to centralize resources in "silos". These controllers usually have a data-driven business model that gives them no incentive to freely share data among each other and to other entities, nor to provide users transparency of their data usage. About the users' location and activities, this information relates to the personal sphere of the individual and composes a part of the dataset called personal data, i.e., any piece of information that can identify or be identifiable to a natural person. Thus, when such a kind of system model is used, it becomes difficult for users to maintain control over their own personal data. That is, individual control, in particular with regard to one's person, has been described as a reflection of fundamental values such as autonomy, privacy, and human dignity [30]. This can indeed represent a problem. It is not by chance that regulations, such as the European Union's General Data Protection Regulation (GDPR) [13] and California Consumer Privacy Act (CCPA) [9], are being implemented, to protect the right that "natural persons should have control of their own personal data" (GDPR p. 2).

2.2 Pure P2P: Keep Data Locally and Distribute Upon Request

At the opposite corner, with respect to the centralized solution, there is a pure peer-to-peer (P2P) approach.

It has been years since P2P technology attracted attention for making it possible to share and exchange resources such as text files, music, videos, uncensored between one user and another, i.e., peer-to-peer. Initially, it seemed to be expected to become the main component of the Internet, although interest in this technology had waned due to the growth of cloud computing. However, a new wave of interest has recently emerged with the advent of blockchain and cryptocurrencies.

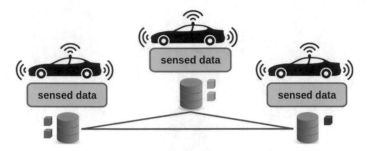

Fig. 3 Pure P2P data aggregation

In general, P2P applications usually run on top of an existing network, such as the Internet. This overlay network can support different P2P architectures, usually depending on the type of application that needs to be served. In a P2P environment, a node is not connected to all the other peers in the network, but instead has a limited number of connections to peers that are defined as "neighbors". Consequently, the fact that each node is only connected to a certain number of other nodes makes it necessary to relay multiple messages between peers in order to disseminate information to the whole network. Furthermore, there is an aspect to consider in the structure of a P2P network, namely the dynamism of peers that can (freely) join and leave the network. This often requires the use of some protocol to keep the network healthy and connected.

To summarize, there are two important aspects related to the functioning of a P2P system: (i) how messages are exchanged and relayed between peers; (ii) how the overlay is constructed and maintained to cope with churns (i.e., nodes dynamically coming and going in the system) [21, 45].

In the case of a Vehicular network, the idea here is very basic. As in classic P2P systems, each user's vehicle, or IoT device, or smartphone, maintains locally its generated data (Fig. 3). Upon request, it is free to decide if sharing such data with someone else or not. At a first sight, such a solution might seem quite simple to implement. Moreover, it solves a lot of issues concerned with data sovereignty. In fact, each node maintains its data and makes decisions about sharing them.

But clearly enough, this is not a practical solution. In fact, in order to provide sharing capabilities, each user's device should be always connected, i.e., there should be a mechanism cope with churns in the vehicular network. Users' devices should provide some guarantees related to storage, computation, and communication capabilities, in order to maintain, handle and transmit their data.

These issues can be solved by switching to an edge-computing like solution. Basically, each user's device has a sort of delegated agent representing it, which is located on the Internet. The device stores its data at this edge node, which is thus in charge of handling such data. Therefore, storage, computation, and communication requirements are shifted to an Internet node, which might more easily provide higher availability guarantees, rather than a user's device. Yet, the availability and reliability of a device's data is as available and reliable as its delegate. Moreover, while this solution is certainly viable, from a distributed system point of view, it still requires some additional protocols to manage the data sharing in ITS. Finally, it does not offer guarantees concerned with traceability, verifiability, and immutability of data.

2.3 A Distributed Ledger Technology to Register Data

As we pointed out in the last example, user's vehicles, IoT devices, and smartphones equipped with sensors can transfer data to the network, by interacting with a gateway. In the last example, this role was played by a delegated agent. However, we argue that such sensed data can be stored and managed in a DLT network (Fig. 4). Thus, each device interacts with a DLT node, transmitting sensed data on a periodical basis. In order to provide a level of traceability, verifiability, and immutability of the generated data, the data itself, or a related digest (when data consist of a large file or sensitive information) is added to a DLT [54]. According to this approach, for instance, a vehicle's on-board computing unit is able to issue messages to a DLT node, thanks to authentication. These messages are then converted to transactions added to the ledger. In general, all public DLTs provide such functionalities by exposing APIs that allow entities, external to the DLT, to send novel transactions. The main point here is that these transactions must be registered in the DLT in a fast way. Second, a good level of scalability must be guaranteed. Third, since a high amount of data is produced, the DLT should offer low fees (or no costs at all). Finally, we need to treat all these transactions as a data-stream, easy to retrieve. These main requirements make not all the existing DLTs eligible in this context.

DLTs can be distinguished for their level of scalability and responsiveness. For instance, Ethereum [8] provides a distributed virtual machine able to process any kind of computation through smart contracts. However, it is well known such a blockchain technology has some scalability issues [5]. Conversely, DLTs such as the

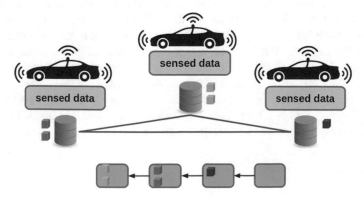

Fig. 4 DLT for data registration

IOTA DLT [41] provide features thought to guarantee scalability, but they lack the support for smart contracts. By design, IOTA is recognized as a responsive, scalable, feeless DLT, with tools for supporting data streams [54]. Among other solutions, it is worth mentioning the implementation of sharding techniques in DLTs. In a few words, sharding consists in breaking the ledger into smaller, more manageable chunks, and distributing those chunks across multiple nodes, in order to spread the load and maintain a high throughput. Currently, however, these technologies are still in their infancy, e.g., Radix [43], or being developed, e.g., Ethereum 2.0 [23].

2.4 A Decentralized File System for Crowd-Sensed Data

In order to overcome the typical DLTs' scalability and cloud services' privacy issues, Decentralized File Storages (DFS) are a potential solution for storing files while maintaining the benefits of decentralization. They offer higher data availability and resilience thanks to data replication. DFSs are crucial for DLTs, as they can be leveraged to store data outside the DLT, i.e., off-chain, when the consensus mechanism discourages on-chain storage. To guarantee data integrity and verifiability, encrypted sensed data could be stored directly on the DLT, i.e., on-chain. However, preventing the on-chain storage is a preferable solution, not only for retaining high data reads availability and better performances for data writes [55], but also because on-chain personal data are generally incompatible with data protection requirements [22].

A principal example of DFS is the InterPlanetary File System (IPFS) [3], a protocol that builds a distributed file system over a P2P network. IPFS creates a resilient file storage and sharing system, with no single point of failure and without requiring mutual trust between nodes. IPFS [3] is a DFS and a protocol thought for distributed environments with a focus on data resilience. The IPFS P2P network stores and shares files and directories in the form of IPFS objects that are identified by a CID (Content IDentifier).

This technology is useful to store data that is not convenient to put on DLTs, and where, in order to retrieve an object, only the file digest is needed, i.e., the result of a hash function applied on the data. The CID is the result of the application of a hash function to a file and it is used to retrieve the referenced IPFS object in the network. Put in other words, the file digest is the identifier of the IPFS object. Users that want to locate that object use this identifier as a handle. When an IPFS object is shared in the network it will be identified by the CID retrieved from the object hash, for instance, a directory with CID equal to *Qmb-WqxBEKC3P8tqsKc98xmWNzrzDtRLMiMPL8wBuTGsMnR*. If any other node in the network tries to share the same exact directory, the CID will be always the same.

IPFS can be used together with the InterPlanetary Linked Data (IPLD) [27] to ensure that a logical object always map to the same physical digital object. IPLD consists of a set of standards and technologies leveraged to create universally addressable data structures, where the CID itself contains the hash and data decoding information. IPLD enables to link resources identified by hashes that can refer to diverse resources.

3 A Framework for Data Sharing and Management Based on DLTs and DFS

Based on the possible approaches described in the previous section, in this section, we provide a framework for the management and sharing of data in ITS. The main pillar of this proposed framework is the concept of moving the processes for the management of ITS data close to the individual that enacted their production, or at least making them completely transparent to this one. This means, for instance, that a user of a smart vehicle should have the last say on the processing of his own sensed/personal data (e.g., the geo-location while driving) and that both the sensing device manufacturer company and the user should (proportionately) benefit from the value of that data. In our vision, technologies such as DLTs and DFS can help to reach this objective.

DLTs, indeed, allow avoiding all the typical drawbacks of centralized server based approaches (censorship, single point of failure, see Sect. 2.1), or those of pure P2P applications (no data verifiability and traceability, see Sect. 2.2). The use of DLTs to represent and transact with data would also grant data validation and access control.

Crucial here is the use of smart contracts, since they provide a new paradigm where unmodifiable instructions are executed in an unambiguous manner during a transaction between two parts. Without the presence of a third-party, smart contract instructions can make sure that the constraints on how and when data are accessed are always respected. Every process is completely traced and permanently stored in the smart contract enabled DLT.

All these properties are necessary in order to create digital data spaces managed both by users and organizations. DFS can help in this sense, since they compensate for certain deficiencies in DLTs. In fact, large-sized data can be better handled off-chain, as well as data that are not meant to be stored forever in a distributed ledger. In these cases, DFS is more suitable for data storing; still, this approach can be combined with the use of DLTs, as we will see in this section. Moreover, it is possible to use DFS for maintaining continuous data availability. To sum up, the framework we need must answer three main functional requirements: (i) ensure data integrity, (ii) ensure data confidentiality, (iii) control who has access to data, and (iv) ensure data persistence. A solution for each one of them will be detailed in the next subsections and will consist of depicting the same framework from different points of view.

3.1 Data Integrity

We already mentioned that crowd-sensed data, coming from users' smartphones and IoT devices, allow building sophisticated smart services (e.g., to improve traffic management, transportation efficiency, and safety) [54]. One requirement, however, is crucial for the creation of secure services and for giving a real value to the sensed data, that is data integrity. To be valuable, indeed, data sensed in an ITS must be reliable in its entirety, and this property should be easily verifiable. DLTs ensure the verification of data integrity in a simple and straightforward way, since the ledger is immutable. Of course, this does not completely assure reliability, as data integrity does not coincide with data security or quality. Indeed, incorrect information about an assertion can be introduced into the DLT, i.e., the GIGO problem [2]. However, the ledger maintains a trace that makes it possible to investigate the insertion process of data. Thus, DLTs can be leveraged to ensure data integrity (Fig. 5). However, this does not necessarily mean that data is stored on-chain. This consideration stems from two observations:

- Storing data into a DFS usually requires lower latencies with respect to DLTs, which typically require some time-consuming consensus mechanism, e.g., Proof-of-Work.
- On-chain data cannot be deleted or modified, becoming an issue when user intentions or regulations require the opposite. For instance, due to the GDPR right to be forgotten or to the right for rectification [13], personal data must not be stored directly in the DLT, even when encrypted [22].

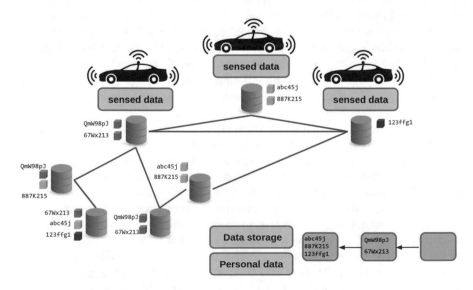

Fig. 5 Data integrity diagram

With this in view, the framework considers DFS for data storage, while adopting the mechanism of storing hash pointers in DLTs for content addressed data. In content addressing, data are identified by their content "fingerprint" instead of their location (such as in the HTTP protocol). A cryptographic hash function is used to identify the content and its result, i.e., a digest, which can be disseminated in a distributed environment to easily refer to the same piece of data. The advantages of content addressing with respect to location addressing are that: (i) links are permanent, i.e., hash pointers; (ii) the link itself does not reveal any of the content, but the content can be used to derive the link; and (iii) it increases the integrity of data since altering the content would produce a new link.

It is worth noticing that storing data off-chain (i.e., in a DFS) and the hash pointers on-chain offer the same levels of data integrity in respect to storing data completely on-chain. Having access to the off-chain stored data, indeed, enables the possibility to compute the hash function over the data and compare the result with the hash pointer that has been immutably stored on-chain.

Moreover, this mechanism enables data deletion [38] and privacy [55], since data in DFS are not immutable and not always public.

3.2 Data Confidentiality

In the previous subsection, we referred to the process of storing data in DFS, but there is one aspect that needs to be pointed out. Since DFS protocols can be executed in public networks [3], data needs confidentiality before any sharing and/or storing. Indeed, the value of a piece of data also depends on who can access it, e.g., a private information becoming public may lose its value in certain use cases. For this reason, personal data is pre-processed by an encryption algorithm before publishing it to the DFS. We refer to the result of this operation as the "encrypted data". The encryption algorithm can assume any form, but it should be implemented in a way that it does not break data integrity, i.e., it must be possible to verify that the hash pointer and the (encrypted) data correspond. The encryption is a critical part for approaching Privacy by Design [10] and crosses vertically all the other parts of a framework for data sharing. For the sake of simplicity, here we refer to a symmetric encryption algorithm that encrypts a piece of data with a new randomly created symmetric key (but more on this can be found in [53]).

In a simple and generic approach, the personal data generated from a data source (smartphone or IoT device) or held by a data controller is encrypted using a symmetric content key (possibly using an efficient symmetric key cryptography algorithm). It is important to differentiate between two instances of personal data that are needed to protect: (i) types of data that can be defined "static", e.g., personal information regarding the name of a driver rarely changes. In this case, each datum can be protected using a content key with no particular relations to other data and that can be created in a pseudo-random way and then kept in safe. (i) types of data that more frequently update the property of a person, e.g., the location of a subject can be updated each second in a stream of data. In this case, we mostly deal with time series data that may be more useful when aggregated. Hence it might be more useful to have content keys related between each other. For instance, employing a symmetric key derivation that exploits relations.

Up to this point, the specified framework includes a device within the ITS, e.g., vehicle, IoT device, or smartphone, that:

1. fetches a piece of data from a sensor (found in the device or belonging to another device with which it has a direct or indirect communication);
2. encrypts the data with a symmetric key;
3. stores the result, i.e., the encrypted data, on a DFS (making a request to a DFS node);
4. stores the hash pointer of such encrypted data on a DLT (making a request to a DLT node).

It is important to note that steps 1–4 can be directly instantiated in a user's personal device, allowing him/her to directly control its data. Alternatively, only the first step can be instantiated in the personal device and the other three can be instantiated in one (or more) devices managed by another entity (that is then considered as a data controller). However, at this point, the framework still misses a mechanism for accessing shared data.

3.3 Data Access Control

A second part of the framework, which is complementary to data confidentiality, sees the data consumer as the main actor who is willing to have access to some shared data. Practically, this actor needs the symmetric key for the decryption of some data and, thus, an authorization service should be placed between the key and this actor. There are several methods to design this service, but here we distinguish between:

- **Centralized**—the most feasible solution, that is where only one service provider is involved in the authorization service and that holds the entire set of secret keys to access data. The data consumer contacts the server directly to retrieve the keys he is eligible to get. This design implies that users trust the server, since this entity has complete access to user data. It also covers the case in which the data provider or controller directly implements this service. The drawback of this approach is that it does not cope with the possibility that the authorization server is honest-but-curious, i.e., it follows the protocol correctly but, if curious, can decrypt and thus access data.
- **Decentralized**—the vision to decentralize the service would help to shift the trust from one entity to a protocol [29]. In this case, indeed, nodes in a decentralized network may be considered semi- or un-trusted, but a consensus mechanism together with a dedicated cryptographic mechanism, would enable the user to be more confident in the protocol [6, 16, 20, 47].

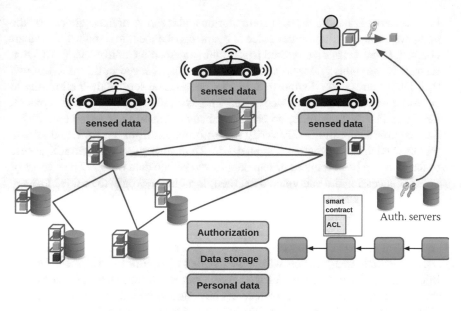

Fig. 6 Data access control diagram

In the framework that we present here (Fig. 6), the authentication service leverages a decentralized environment to provide authorizations to consumers for different reasons: (i) to avoid, also in this case, a single point of failure; the failure here includes both service interruption and privacy leakages; (ii) to release the data provider device from the burden of completely handling keys distribution; in fact, this service may become very expensive in terms of communication in case of fine-grained access; (iii) to exploit smart contract distributed computation for implementing a "fair" and automatic access control mechanism; (iv) to exploit DLT's transparency for the auditability of access permissions.

The protocol for the decentralized authentication service used in the framework includes two parts, an on-chain and an off-chain part [53]. On-chain smart contracts are exploited for the management of an Access Control List and for the distributed computation of the access mechanism. However, this is complemented with an off-chain keys distribution mechanism. In particular:

- **On-chain Access Control List (ACL)**— The access to the encrypted data, stored in a DFS, is managed through smart contracts, that regulate access rights to data. In practice, each piece of data can be referenced in a specific smart contract and bundles of data can be referenced through Merkle Trees [34]. The form of this data is the one we have seen in the previous sections. In addition, with regard to the data, the contract can also maintain data schemes and indications to the data kind, in order to have interoperability when interacting with such data [24]. The best way to exploit the data indeed is to provide these schemes in a machine-readable format for specifying what to expect from them. This is needed by a possible data consumer in order to better handle the data computation and it can be defined directly by the entity who manages the source device or by means of a specific standard. The smart contract, however, is mostly used to maintain an Access Control List (ACL) that represents the rights to access some data. Consumers, listed in the ACL through their DLT address, can demonstrate to an authorization service their eligibility to access some particular data. Once a Consumer is eligible to obtain certain content, i.e., he is listed in the ACL, he can access such content through a content key. Service providers can directly verify this information from the ACL and release the key needed to decrypt the encrypted data. Thus, through smart contracts, access to the data can be purchased or can be allowed by the data owner. The release of keys for accessing the encrypted data, then, is authorized only to entitled users.

- **Off-chain Keys Distribution**—When this service of keys distribution is operated by a single central provider, trust must be given to this one, since the keys are kept in one place only. Assuming that this provider is honest-but-curious, such that it follows the protocol correctly (e.g., an online social network sharing a user's vehicle geo-location with a user's friends, if curious, can access to this information), decentralization of the service can be put in place in order to shift the trust to the protocol itself, instead of the single provider. In the decentralized case, nodes in a network can be considered semi- or un-trusted, but a data protection/cryptographical mechanism, built into their execution protocol, allows the whole system to be trusted [51, 56]. When a data consumer is entitled to access some data on-chain (i.e., in a smart contract ACL), it can then request the release of the associated distributed keys to the nodes of such a network. Since it is not possible to store secret keys or decrypt messages on-chain, due to its public execution, an off-chain keys distribution mechanism is needed. However, distributed computation should be used to maintain decentralization in the key distribution process, e.g., MultiParty Computation (MPC) [35, 51]. Two cryptographical schemes can be used in this case for the content keys distribution:

  - **Secret Sharing (SS)**—This scheme splits the content key in n shares, but only t shares are enough to reconstruct the key. A (t, n)-threshold scheme is employed to share a secret between a set of n participants, with the possibility to reconstruct the secret using any subset of t (with $t \leq n$) or more shares, but no subset of less than t. In a network where multiple nodes store secret shares, a consensus can be reached by t nodes to provide the shares to a data consumer, allowing him to know the secret. This can be employed to provide privacy to a user that is sharing a secret, since none of the nodes can obtain the whole secret without the help of other $t - 1$ nodes. Thus, single nodes alone are unable to reconstruct the content key because they only save a portion of this key.
  - **Threshold Proxy Re-Encryption (PRE)**—The content key can be re-encrypted by a proxy node using the proxy re-encryption (PRE) scheme. PRE [1], is a type of encryption where a proxy entity transforms a ciphertext encrypted with a content key k_1, into a ciphertext decryptable with a key k_2, without learning anything about the underlying plaintext. This is possible using a re-encryption key rk_{1-2} generated by the data owner who has initially encrypted the plaintext

with k_1. PRE is a scheme that usually involves only one semi-trusted proxy node. However, it can be the case that this node decides to re-encrypt data immediately, rather than to apply conditional policies as instructed, or it may collude with the Consumer to attack the data owner's private key. A threshold proxy re-encryption scheme can be used to solve this problem. Instead of using a single re-encryption key, a (t, n)-threshold scheme is used to produce "re-encryption shares", in such a way that these can be combined client-side by the data consumer.

Both these techniques can be supported in a decentralized data access control and come with different advantages and disadvantages. SS relieves the user from any interaction during each key distribution, but at the same time if t nodes are malicious then the user cannot intervene to stop the keys from getting leaked. On the other side, PRE has the drawback of requiring the user to generate a re-encryption key for each new consumer, however, he has the option to stop producing new re-keys if some nodes are malicious.

3.4 Data Persistence

In P2P systems a general issue affects the availability of data when the network nodes have no incentive to keep their storage occupied. For instance, it is well known that in file-sharing systems, such as BitTorrent, the data availability of some popular content may become poor quickly, and eventually it is hard to locate and download it [25]. Similarly, when there is no incentive to maintain files, also DFS cannot offer guarantees on the persistence of data. Indeed, data is usually stored in the DFS as long as some node has some disk space to maintain a replica. The more the nodes that maintain a copy of a given file, the higher the reliability and the higher the guarantees that the file can be properly retrieved. To cope with this issue, incentivization mechanisms can be employed to obtain that the distributed system permanently stores files, i.e. users can reward nodes that maintain a copy of their data (Fig. 7).

In order to provide incentives to nodes for maintaining data, some DFS integrate DLTs, bringing together clients' requests with storage nodes' offers. In practice, participants are rewarded with cryptocurrencies for serving and hosting content on their storage. This strategy does not alter the protocol on how nodes exchange data in the DFS but, "simply", network nodes are paid to store and not erase them. This payment is generally based on a proof that these nodes publish in the integrated DLT, e.g., Proof-of-Spacetime [4].

Some DFS [4, 49] also integrate smart contracts in order to reward hosts for keeping files. These contracts are usually referred to as File Contracts, i.e., a particular kind of smart contract employed to arrange an agreement between a storage provider and their clients.

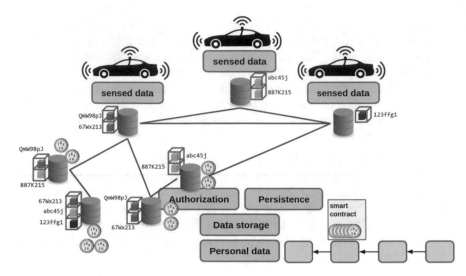

Fig. 7 Data persistence incentives diagram

In Filecoin [4], for instance, file storage is treated as an algorithmic market thanks to File Contracts. Some nodes provide the storage and the prices of the service are not controlled by a single enterprise, but rather depend on a supply-demand market model. In this model, the user pays miners to store and retrieve files:

- A storage agreement is an agreement between a user and a storage miner in which the former pays in advance and the latter periodically demonstrates that it is still storing the file until the expiry time. In addition, the user can auction the storage to miners who meet certain requirements. The storage miner stores files for users and extracts additional blockchain coins by performing storage tests. The user is guaranteed storage by storing evidence on the blockchain.
- A retrieval deal is an agreement between user and retrieval miner in which the latter retrieves files from storage for the former. This means that the retrieval miner can be a different entity from a storage miner. Unlike storage deals, retrieval deals are managed off-chain and their value may depend on the speed of retrieval. Thus, payments are made off-chain, are incremental, and are typically based on reliability or speed of recovery.

In the Filecoin blockchain, storage miners require powerful hardware in addition to storage, as they are also tasked with creating new blocks (every 30 s) and running proofs. The blockchain acts as a ledger for proofs, agreements, and coin transfers, therefore storage proofs are public. An important concept in generating proofs for Filecoin is the sector, i.e., the basic storage unit. The size and time increments for commitments are standardized and are chosen as a trade-off between security and usability, while the duration of a sector depends on the storage market. A sector is fully occupied by commitments made through File Contracts and must be sealed before proof. Unused space is considered committed by auto-deal, i.e., by the miner to itself.

The sector and its content ID (CID) are sealed in a replica, i.e., encoded in a way that can be used for proofs. Then the Proof of Replication (PoRep) is performed on the replica. This proof is based on the fact that an honest storage miner has already sealed the blocks and can respond quickly before a timeout. The sectors are sealed in a time-consuming process (due to the fact that the sectors are 32Gb or 64Gb) using a key that depends on the node hardware. If a user wants their files stored as multiple copies on different miners, each copy will have a different seal and an attacker would have to download each block of the file and reseal it.

Over time, randomly selected storage miners have random sectors questioned, from which the data is read for verification and compressed into a Proof of SpaceTime (PoSt). Both users (on demand) and miners (periodically) check that blocks are valid and storage miners are "punished" if blocks are not available before an agreement expires. Storage miners, therefore, are required to provide public (i.e., stored on-chain) proof that a given data encoding has existed in physical storage continuously for a period of time. This means that all miners are asked to perform a PoRep on a random sector of their storage when mining blocks, i.e., a WinningPoSt, or every day, i.e., WindowPoSt.

4 Implementation and Evaluation

In this section, we will discuss the feasibility of the implementation of the presented framework. Our implementation features several technologies that fall into the realm of DLTs and DFS. Moreover, we will provide an assessment of the performance of such implementation considering a specific ITS use case and its general constraints.

4.1 Implementation

IPFS as DFS and Filecoin for Incentivization

Within the framework, the DFS plays the role of common data space, where all the system components store data. In our implementation, such a role is played by IPFS [3]. Thus, when a piece of data is uploaded to the network, the data digest is returned as a reference. This reference can be stored in the DLT and then employed to retrieve the specific data. Thus, the piece of data is published as an IPFS object and then (asynchronously) referenced through its hash into a DLT transaction. The digest, as explained before, allows verifying the integrity of the IPFS object.

To upload files on IPFS, a node running the IPFS protocol is necessary. Due to the fact that it is (still) not feasible to run an IPFS node on constrained devices (such as smartphones or vehicles on board units), other solutions must be explored. In our implementation, we resort to an IPFS service provider (i.e., Infura [26]) to let users disseminate files in the IPFS network.

Data persistence is implemented through incentivization, thanks to the use of Filecoin smart contracts [4]. Filecoin is the typical incentive used on top of IPFS, where participants are rewarded with Filecoin tokens for serving and hosting content on their storage.

IOTA as DLT

For what regards the DLT implementation, we refer to IOTA. The IOTA ledger is not structured as a blockchain, but instead as a Direct Acyclic Graph (DAG) called the Tangle [41]. Such public data ledger claims to be particularly targeted toward the IoT industry. The IOTA transactions validation approach, indeed, is thought to address two major pain points that are associated with traditional blockchain-based DLTs, i.e., latency and fees. IOTA has been designed to offer fast validation, and no fees are required to add a transaction to the Tangle [7]. This makes IOTA a candidate choice to support smart services built through crowd-sensed data.

An important feature offered by IOTA is the Masked Authenticated Messaging (MAM). MAM is a second-layer data communication protocol that adds functionality to emit and access an encrypted data stream over the Tangle. Data streams assume the form of channels, formed by a linked list of transactions in chronological order. Once a channel is created, only the channel owner can publish encrypted messages. Data consumers that possess the MAM channel encryption key are enabled to decode the message. Messages are pushed on the channel in chronological order, and each message has a link to the next message to be created. This allows access from one message onward, while restricting access to prior messages in the channel [7].

The data access to new data may be revoked simply by using a new encryption key. We consider that the hash pointer of each piece of data stored in IPFS is stored in a MAM message, and the symmetric content key to be shared with data consumers is the MAM encryption key of the message.

Decentralized Access Control Based on Ethereum Smart Contracts and Cryptographic Threshold Schemes

The access control is fully managed by several predetermined Authorization Servers. These nodes perform on-chain tasks related to the smart contracts execution, but also off-chain tasks such as the keys distribution. As to on-chain tasks, a set of Ethereum smart contracts [8] are used to implement the access control. These smart contracts refer to the information that is written into IOTA MAM Channels, i.e., hash pointers, using the MAM Channels addresses called "roots". The software that has been used for managing the ACL is the OpenEthereum client [36], a popular Ethereum blockchain client. In particular, it offers the implementation of a "SecretStore" where nodes can distribute keys on the basis of smart contracts extracted information. Regarding off-chain keys distribution, we refer to two cryptographic schemes:

- **OpenEthereum Secret Sharing (SS)**—To implement this scheme, we resorted to the Secret Store provided by the OpenEthereum client [36], in which the content key k is split in n shares and $t < n$ are enough to reconstruct it. Considering the key to decrypt a data as a secret, the secret sharing among a network of nodes enables to be in a situation where, potentially, up to $t - 1$ nodes can be malicious. Indeed a consensus is reached from t nodes upward when a data consumer asks to receive a content key. Moreover, any node in the network created through the OpenEthereum cannot access the data on its own, as it would need the help of other $t - 1$ nodes.
- **Umbral Threshold Proxy Re-Encryption (TPRE)**—In Umbral [16], the content key k is initially encrypted in a public key encryption scheme using the public key of the data provider. The result of such an operation is then used in a (t, n)-threshold Proxy Re-Encryption schema, i.e., it can be re-encrypted using t "re-encryption shares". This process will produce a re-encrypted key that can be decrypted using the private key of the data consumer to obtain the initial content key k. Among many schemes, Umbral uses a single-use uni-directional proxy re-encryption where the re-encryption function is one way.

4.2 Performance Evaluation

Our experimental scenario was based on a hypothetical real ITS application. In particular, we conducted a trace-driven experimental evaluation where traces were generated using a real dataset of mobility traces of buses in Rio de Janeiro (Brazil) [15]. Based on these traces, we simulated several users' devices on board of buses that, during their path, periodically generate sensed data. We considered one user for each bus. These data may represent temperatures, air pollution values, etc. In this case, we focused on two different types of data: (i) small-sized data, such as hash pointers or geodata, i.e., latitude and longitude (100 bytes); (ii) large-sized data, such as photos (1 MB). Here, we present a summary of results which are discussed in detail in [52, 54, 55].

IOTA MAM Channels

One user per bus was emulated by a single process issuing messages containing a hash pointer to a MAM Channel, based on the data-trace (a MAM channel was associated with each user). Based on the bus paths, each user was set to generate approximately 45 messages/h. This resulted in a message to be issued to the DLT every 80 s, which is a reasonable time interval to sense data in an urban scenario. For each test configuration, we replicated the experiment 12 times for 1 h. We considered different heuristics for the selection of an IOTA full node from a (dynamic) pool of public nodes to pair to each user ($\sim$60 active nodes). The rationale behind this choice was based on the assumption that users' smartphones or buses' computing units may not have computation capabilities to behave as full nodes [17]. One of the heuristics, called Fixed Random, requires each user to be assigned to a random IOTA full node from the pool for the whole duration of the test. Another heuristic was the Adaptive RTT: each user keeps a trace of past interactions with full nodes and creates a ranking based on the experienced Round Trip Time (RTT) [28]; then for each message to be uploaded on IOTA a full node is chosen based on ranking. For each MAM message, we recorded the outcome of the upload request, i.e. successful or unsuccessful, as well as the latency between the transmission of the message to a node and the confirmation of its insertion in the ledger. This interval of time is characterized mainly by two operations that are needed for storing the transaction in the IOTA ledger: (i) the tips selection consists of selecting from the Tangle two random transactions that do not have a successor yet, i.e., tips; (ii) the Proof-of-Work that requires computation to obtain a piece of data difficult (costly and time-consuming) to produce but easy to verify [41].

Table 1 shows a summary of the results obtained for different repetitions of a specific test [52, 54]. Two main measures experienced during a series of tests are reported: (i) the average latencies including both the tips selection and PoW phases, and (ii) the percentage of errors, that is the number of messages that failed to be added to the Tangle, due to full nodes' errors. Results show that, on one side, the measured latencies are relevant. Indeed, the random selection of a full node for issuing a transaction does not lead to good results, since the amount of errors is quite high, as well as the measured latencies. On the other hand, the good news is that if we carefully select the full node to issue a transaction, the performances definitely improve. In fact, the use of the "Adaptive RTT" heuristic has a low amount of errors, on average around 0.8% and average latency amounts to 23 s. However, this is still far from a real-time update of the DLT and the level of acceptability of latency values truly depends on the application scenario. In terms of scalability, results in Table 1 show that in all cases, average latencies increase significantly with the number of users. There is an important difference between the 60 and 240 users scenario. In the case of 240 users, we have a message generation rate of about ∼3 msg/sec to be issued to the IOTA DLT. If we assume that the workload is evenly distributed among all the nodes in the pool, then, each node receives on average a new message request every ∼20 s. Bearing in mind that, at best, it takes 23 s for a full node to completely

Table 1 Results on IOTA, with 60, 120, 240 users [52, 54]

# users	Heuristic	Avg latency (s)	Conf. Int. (95%) (s)	Errors (%)
60	Fixed random	72.68	[70.43, 74.94]	15.37
	Adaptive RTT	22.99	[22.69, 23.29]	0.81
120	Fixed random	87.75	[85.38, 90.12]	29.49
	Adaptive RTT	27.35	[27.11, 27.58]	1.1
240	Fixed random	177.62	[174.25, 181.0]	42.81
	Adaptive RTT	73.26	[72.68, 73.85]	7.55

process a message, then we see that an initial overhead of a few seconds leads to a huge increase at the end of the test, i.e., ~73 s in the "Adaptive RTT" heuristic. This means that further improvements are needed to solve scalability issues in this scenario.

IPFS

In this assessment, we used a single DFS node with two different implementations, while varying the number of users, i.e., we tested different cases with a specific number of users associated with a single DFS node. We assessed two different scenarios: (i) we set up an IPFS node, i.e., Proprietary, on a dedicated device connected to other nodes in the main IPFS network and devoted it to only handle requests coming from our application; (ii) we tested a public IPFS node Service, i.e., the Infura service provider [26], that offers free access to IPFS. Tests were conducted in order of dimension (small files first, then larger ones) and users number (from 10 to 100). The performance evaluation has been designed as a stress test in which each simulation sends requests to the two different types of IPFS nodes following the buses real traces. A simulation lasted 15 min and sent exactly 15 messages for each user.

Table 2 shows the latency recorded during the tests. In general, we noticed better performance when a dedicated node is employed. The results show that the IPFS Service has a similar behavior with both small and large files. This is due to the fact that it has more resources than the Proprietary node, which is then unable to cope with larger files. In the stress test that we implemented, the IPFS Proprietary performances get worse when increasing the number of users. Furthermore, there is a turning point with 80 users where, overall, performance degrades in the presence of large files, while latencies with small files remain stable (or even decrease). In general, IPFS Proprietary always works better except for over 80 users in the case of large files. This means that a dedicated node is always preferable, but must be limited to a rate of 60–70 users requests per minute.

Decentralized Access Control

In a second experiment we measured the amount of time required to perform access control operations using the implementation of the keys distribution, i.e., OpenEthereum client [36], called SS, and Umbral [16], called TPRE. The tests were performed using a network of 25 interconnected nodes with the aim to emulate the real DLTs and the distributed systems use cases.

Table 2 Latencies and errors when sending messages to IPFS nodes [55]

# users	IPFS node	Data size	Avg latency (s)	Conf. Int. (95%) (s)	Errors (%)
10	Proprietary	Small	0.19	[0.18, 0.2]	0.0
		Large	1.22	[1.17, 1.28]	0.0
	Service	Small	9.49	[9.09, 9.9]	0.0
		Large	6.16	[5.75, 6.57]	0.0
40	Proprietary	Small	0.59	[0.57, 0.62]	0.0
		Large	3.42	[3.31, 3.54]	0.0
	Service	Small	7.5	[7.18, 7.83]	0.0
		Large	11.3	[11.01, 11.58]	0.0
70	Proprietary	Small	2.65	[2.56, 2.74]	0.0
		Large	7.48	[7.3, 7.66]	0.0
	Service	Small	6.22	[6.09, 6.34]	0.0
		Large	8.58	[8.42, 8.74]	0.0
100	Proprietary	Small	1.53	[1.48, 1.58]	0.0
		Large	20.27	[19.71, 20.83]	83.33
	Service	Small	6.81	[6.69, 6.92]	0.0
		Large	12.91	[12.68, 13.14]	0.21

We emulated from 10 to 100 data consumers asking for access to the user's bus data after they have been added to the ACL, then we averaged the result. The results of the test carried out allow us to evaluate the goodness system in terms of performances:

- **Threshold variation**: involves the variation of t from 5 to 25, with number of nodes $n = 25$ and message size set to 30 Bytes. As the Table 3 shows, the encryption latency time remains mostly constant, ~180 ms for TPRE and ~1045 ms for SS, while the decryption time increases linearly with t. The biggest time difference between the two schemes comes from the actual generation and distribution of the key shares (in the encryption phase), i.e., a surplus of ~792 ms from TPRE to SS.

Table 3 Threshold latencies (mean) when encrypting (+ distributing) and decrypting messages for a Decentralized Access Control [53]

Threshold	SS encryption (ms)	TPRE encryption (ms)	SS decryption (ms)	TPRE decryption (ms)
5	1024	176	192	130
10	1017	182	233	189
15	1030	183	265	245
20	1045	185	349	309
25	1069	190	371	376

Table 4 Nodes number latencies (mean) when encrypting (+ distributing) and decrypting messages for a Decentralized Access Control [53]

# nodes	SS encryption (ms)	TPRE encryption (ms)	SS decryption (ms)	TPRE decryption (ms)
5	397	120	148	104
10	549	135	148	101
15	666	147	175	108
20	843	163	178	108
25	952	175	188	110

Table 5 Message size latencies (mean) when encrypting (+ distributing) and decrypting messages for a Decentralized Access Control [53]

Size	SS encryption (ms)	TPRE encryption (ms)	SS decryption (ms)	TPRE decryption (ms)
10 B	1026	174	126	111
50 B	1022	174	126	109
100 B	1025	177	125	110
500 B	1025	175	125	109
1 KB	1027	176	126	108
5 KB	1027	185	129	109
10 KB	1031	178	135	109
50 KB	1071	177	178	110
100 KB	1127	178	231	116
500 KB	1541	196	642	127
1 MB	2054	220	1150	151
5 MB	6214	394	5278	305
10 MB	11456	608	10452	502

- **Number of nodes variation**: threshold value t was set to 2 and the message size was set to 30 KB. Generally, as expected, the time costs of operations increase with the number of nodes n. However, we must note the fact that the results in Table 4 grow much faster for SS rather than for TPRE. This makes the TPRE method more scalable.
- **Size of messages variation**: n was set to 25 and $t = 2$, while the size of the message varied. Results reported in Table 5 suggest that the TPRE scheme scales better than SS. From 10 Bytes to 1 MB TPRE latency raises slightly, while SS has a clear inflection point when the message size is set to 100 KB and then skyrockets from 1 MB onward.

5 Discussion

To fully exploit their potential and to promote the development of ITS, several novel challenges must be faced. In ITS, data gathering, communication, analysis, and distribution among individuals vehicles, infrastructures and services can be built in a "centralized" way or in a "pure P2P", as shown respectively in Sects. 2.1 and 2.2. However, in both cases it may become difficult for users to maintain control over their own data and guarantees for data traceability, verifiability, and immutability are not always met. For this reason, we argue that the integration of DLTs, DFS, and smart contracts can help with the creation of a framework providing data integrity, confidentiality, access control, and persistence. An important and critical outcome of this work is concerned with the implementation and experimental assessment we performed, showing the related results of the technologies available at the moment. It is well known that decentralized and secure DLTs that enable the distributed execution of smart contracts, such as Ethereum in our implementation, still have scalability issues [5]. Here, thus we focused on testing out DLTs and DFS where data is uploaded.

Latencies measured to store data into the considered DFS, i.e., IPFS, can be considered acceptable for general ITS scenarios. In this case, as a measure of scalability, the best performances were obtained when the number of dedicated IPFS nodes followed the equation $\#nodes = \frac{\#requests}{sec}$, where $\#requests$ is the number of IPFS upload requests generated by the users in our scenario. On the other hand, for what concerns the employed DLT, i.e., IOTA, we conclude that at the moment of the test execution the results were not viable for real-time ITS applications, but acceptable for less demanding services. Tests show a latency between 23 and 27 s for 0.75–1.5 MAM messages insert requests per second, with, at best, an experienced tps, i.e., transactions per second, of 0.13 (considering 1 MAM message roughly equal to 3 IOTA transactions). This means that, for a latency on average of ∼25 s, with a configuration similar to ours during tests, available IOTA nodes in such a scenario should scale following $\#nodes \geq \frac{k \times \#requests}{sec}$, with $k = 53$ and where $\#requests$ is the number of MAM messages insert requests generated by the users in our scenario. Hypothetically, having a DLT protocol that allows $tps = 1$, would require having available IOTA nodes in such a scenario that scale following the same formula but with $k = 2$.

We have focused on data protection through encryption, using For what does concern decentralized access control we employed two different schemes, i.e., SS and TPRE. At first we discussed their qualitative differences, then we compared them in terms of execution time. Our performance evaluation shows that, in respect to SS, TPRE is: (i) faster when increasing the size of the messages; (ii) more scalable, as it better manages the increase in the number of nodes executing the protocol; (iii) more efficient when increasing the threshold value, due to its shares generation method. On the other hand, TPRE has the drawback of requiring the data owner to generate a re-encryption key for each new data consumer.

Clearly enough, an adequate ITS infrastructure must be set in all the cases, in order to build a scalable architecture able to properly handle a possibly high data generation rate from multiple vehicles. In other words, we think that the issue is concerned more with the system deployment rather than on the DLT/DFS protocol. For instance, an edge computing architecture can be merged with the framework and used to geographically place node gateways, which receive data from vehicles and insert them into DLT/DFS.

5.1 New DLT Proposals

To overcome the scalability issues described in the above discussion, several new approaches are emerging in the DLTs scenario. New proposals include solutions where improvements are made directly on-chain, i.e., at layer one, and solutions that build on top of that layer and are executed off-chain, i.e., layer two.

Off-chain solutions are implemented separately from the layer one DLT protocol and require no changes, in order to derive their security directly from the layer one consensus. Optimistic Rollups, for instance, are a layer two scaling solution built for the Ethereum blockchain where computation is partly executed off-chain and data is maintained on-chain [46]. Its aim is to increase the blockchain transactions per second by a factor of a hundred, or even a thousand, and then also to decrease transaction fees. Another layer two scaling solution is the use of side chains. In particular, it consists of using another blockchain, i.e., the sidechain, that runs a faster or lighter protocol, in order to manage assets in the original blockchain, i.e., mainchain. For instance, the sidechain can ensure asset security using the Plasma framework and a decentralized network of Proof-of-Stake validators [40]. Furthermore, there is currently a rise of technologies that operate using the same protocol for executing Ethereum Smart Contracts but with fewer latencies and reduced gas price. For instance, Polygon [39] has achieved up to 10 000 transactions per seconds on a single sidechain through internal tests. In general, it has been shown that implementing an Ethereum private network using Proof of Authority consensus mechanism, with optimal configuration, can reach up to 1000 transactions per second [48].

On-chain solutions usually involve the use of different forms of ledgers, for instance, a DAG such as in IOTA or a sharded ledger. Given that we already presented and discussed on the IOTA DLT latencies, furthermore we conducted preliminary tests with other possible DLTs that implement novel techniques to improve responsiveness and scalability, i.e., Radix, which is specifically based on sharding, obtaining interesting results [43]. In a few words, sharding consists in breaking the ledger into smaller, more manageable chunks, and distributing those chunks across multiple nodes, in order to spread the load and maintain a high throughput. At the time of writing, the Radix technology is still in its infancy. Nevertheless, we exploited the test network to issue transactions on the ledger. Thus, obtained results cannot be considered accurate and it is too early to give an overall judgment on this DLT. However, we obtained very low latencies (below 1 s), with a non-negligible (but low) error rate. We stress the fact that these results cannot be compared with those obtained for IOTA. In fact, in IOTA we exploited the main network, while in Radix we had to employ a preliminary testnet, with few nodes involved in the ledger management (~6 nodes) and basically no additional workload, apart from our tests. As a matter of fact, comparable results can be obtained if tests are executed on the IOTA test network, where the PoW is faster (we obtained average latencies around ~2 s). Moreover, two possible problems must be faced in this case. The first one is related to the fact that Radix requires fees to add a transaction to its ledger. This might make the costs, for supporting a smart transportation system application, prohibitive. Moreover, an open security question arises, i.e. if we decrease the number of nodes that validate transactions (as the sharding does), then does the risk of a security breach increase?

6 Conclusions

The framework we presented in this work shows a possible specification for taking advantage of decentralized architectures to build reliable and modern services for Intelligent Transportation Systems (ITS), on the basis of data sharing and management. The resulting framework integrates DLTs, DFS, smart contracts, and authorization protocols. This was a response to the need and importance of being able to optimize the use of resources and data in ITS, but not limited to that use case. The solution we have shown optimizes data sharing from four points of view, namely data integrity, confidentiality, access control, and persistence. Our experimental evaluation of the implementation of such a system using currently available technologies shows an acceptable feasibility for less demanding ITS services, i.e., non-real-time services. However, many issues are left open and pave the way for new studies on the optimization of such a system and the integration of new technologies: (i) the optimization of the use of algorithms for managing distributed ledgers (many solutions are particularly CPU and network intensive), (ii) the optimization of data placement, and (iii) the optimization of the infrastructure supporting such a distributed system by an adhoc deployment of nodes.

Acknowledgements This work has received funding from the European Union H2020 research and innovation program under the MSCA ITN European Joint Doctorate grant agreement No 814177 Law, Science and Technology Joint Doctorate—Rights of Internet of Everything.

References

1. Ateniese, G., Fu, K., Green, M., Hohenberger, S.: Improved proxy re-encryption schemes with applications to secure distributed storage. ACM Trans. Inf. Syst. Secur. (TISSEC) **9**(1), 1–30 (2006)
2. Babich, V., Hilary, G.: Om forum-distributed ledgers and operations: what operations management researchers should know about blockchain technology. Manufact. Serv. Oper. Manag. **22**(2), 223–240 (2020)
3. Benet, J.: Ipfs-content addressed, versioned, p2p file system (2014). arXiv preprint arXiv:1407.3561

4. Benet, J., Greco, N.: Filecoin: a decentralized storage network. Protoc Labs (2018)
5. Bez, M., Fornari, G., Vardanega, T.: The scalability challenge of ethereum: an initial quantitative analysis. In: 2019 IEEE International Conference on Service-Oriented System Engineering (SOSE), pp. 167–176. IEEE (2019)
6. Blakley, G.R.: Safeguarding cryptographic keys. In: 1979 International Workshop on Managing Requirements Knowledge (MARK), pp. 313–318. IEEE (1979)
7. Brogan, J., Baskaran, I., Ramachandran, N.: Authenticating health activity data using distributed ledger technologies. Comput Struct. Biotechnol. J. **16** (2018)
8. Buterin, V., et al.: Ethereum white paper (2013). https://github.com/ethereum/wiki/wiki/White-Paper
9. California State Legislature: California consumer privacy act (2020). https://leginfo.legislature.ca.gov/faces/billTextClient.xhtml?bill_id=201720180AB375
10. Cavoukian, A.: Privacy by design. In: Take the Challenge. Information and Privacy Commissioner of Ontario, Canada (2009)
11. Chiasserini, C.F., Giaccone, P., Malnati, G., Macagno, M., Sviridov, G.: Blockchain-based mobility verification of connected cars. In: 2020 IEEE 17th Annual Consumer Communications Networking Conference (CCNC), pp. 1–6 (2020)
12. Council of European Union: Directive 2010/40/eu on the framework for the deployment of intelligent transport systems in the field of road transport and for interfaces with other modes of transport
13. Council of European Union: Regulation (eu) 2016/679 - directive 95/46
14. D'Angelo, G., Ferretti, S., Marzolla, M.: A blockchain-based flight data recorder for cloud accountability. In: Proceedings of the 1st Workshop on Cryptocurrencies and Blockchains for Distributed Systems (CryBlock) (2018). https://doi.org/10.1145/3211933.3211950
15. Dias, D., Costa, L.H.M.K.: CRAWDAD dataset coppe-ufrj/riobuses (v. 2018-03-19). Downloaded from https://crawdad.org/coppe-ufrj/RioBuses/20180319 (2018). https://doi.org/10.15783/C7B64B
16. Egorov, M., Wilkison, M., Nuñez, D.: Nucypher kms: decentralized key management system (2017). arXiv preprint arXiv:1707.06140
17. Elsts, A., Mitskas, E., Oikonomou, G.: Distributed ledger technology and the internet of things: a feasibility study. In: Proceedings of the 1st Workshop on Blockchain-enabled Networked Sensor Systems (BlockSys) (2018)
18. ETSI: Etsi en 302 637-2 v1.4.1 (2014). https://www.etsi.org
19. European Commission: A European strategy for data (2020)
20. European Union Agency for Cybersecurity: Data Pseudonymisation: Advanced Techniques and Use Cases. Technical report, European Union Agency for Cybersecurity (2021). https://www.enisa.europa.eu/publications/data-pseudonymisation-advanced-techniques-and-use-cases
21. Ferretti, S.: Shaping opportunistic networks. Comput. Commun. **36**(5), 481–503 (2013). https://doi.org/10.1016/j.comcom.2012.12.006
22. Finck, M.: Blockchain and the General Data Protection Regulation: Can Distributed Ledgers be Squared with European Data Protection Law?: Study. European Parliament (2019)
23. Foundation, E., et al.: Ethereum 2.0 specifications (2021). https://github.com/ethereum/eth2.0-specs
24. Foundation, T.S.: Innovation meets compliance: Data privacy regulation and distributed ledger technology. Technical report, The Sovrin Foundation (2020)
25. Guo, L., Chen, S., Xiao, Z., Tan, E., Ding, X., Zhang, X.: A performance study of bittorrent-like peer-to-peer systems. IEEE J. Sel. Areas Commun. **25**(1), 155–169 (2007)
26. Infura Inc: Infura: Secure and scalable access to ethereum apis and ipfs gateways. (2020). https://infura.io/
27. IPLD Team: Interplanetary linked data (ipld) (2016). https://specs.ipld.io/
28. Jacobson, V.: Congestion avoidance and control. In: ACM SIGCOMM Computer Communication Review, vol. 18. ACM (1988)
29. Jemel, M., Serhrouchni, A.: Decentralized access control mechanism with temporal dimension based on blockchain. In: 2017 IEEE 14th International Conference on e-Business Engineering (ICEBE), pp. 177–182. IEEE (2017)

30. Koops, B.J., Newell, B.C., Timan, T., Skorvanek, I., Chokrevski, T., Galic, M.: A typology of privacy. U. Pa. J. Int'l L. **38**, 483 (2016)
31. Leiding, B., Memarmoshrefi, P., Hogrefe, D.: Self-managed and blockchain-based vehicular ad-hoc networks. In: Proceedings of the International Joint Conference on Pervasive and Ubiquitous Computing: Adjunct (2016)
32. Lucic, M.C., Wan, X., Ghazzai, H., Massoud, Y.: Leveraging intelligent transportation systems and smart vehicles using crowdsourcing: an overview. Smart Cities **3**(2), 341–361 (2020)
33. Maesa, D.D.F., Mori, P., Ricci, L.: A blockchain based approach for the definition of auditable access control systems. Comput. Secur. **84**, 93–119 (2019)
34. Merkle, R.C.: A digital signature based on a conventional encryption function. In: Pomerance, C. (ed.) Advances in Cryptology – CRYPTO '87, pp. 369–378. Springer, Berlin (1988)
35. Micali, S., Goldreich, O., Wigderson, A.: How to play any mental game. In: Proceedings of the Nineteenth ACM Symposium on Theory of Computing, STOC, pp. 218–229. ACM (1987)
36. OpenEthereum: OperEthereum Secret Store (2020). https://openethereum.github.io/SecretStore
37. Palazzi, C.E., Roccetti, M., Ferretti, S.: An intervehicular communication architecture for safety and entertainment. IEEE Trans. Intell. Transp. Syst. **11**(1), 90–99 (2010). https://doi.org/10.1109/TITS.2009.2029078
38. Politou, E., Alepis, E., Patsakis, C., Casino, F., Alazab, M.: Delegated content erasure in IPFS. Futur. Gener. Comput. Syst. **112**, 956–964 (2020). https://doi.org/10.1016/j.future.2020.06.037
39. Polygon: Polygon - Ethereum's internet of blockchains (2021). https://polygon.technology/papers/
40. Poon, J., Buterin, V.: Plasma: scalable autonomous smart contracts. White paper, pp. 1–47 (2017)
41. Popov, S.: The tangle (2016). https://iota.org/IOTA_Whitepaper.pdf
42. Prandi, C., Mirri, S., Ferretti, S., Salomoni, P.: On the need of trustworthy sensing and crowdsourcing for urban accessibility in smart city. ACM Trans. Internet Technol. **18**(1) (2017). https://doi.org/10.1145/3133327
43. RadixDLT: Radix knowledge base (2019). https://docs.radixdlt.com/kb/
44. Raza, S., Wang, S., Ahmed, M., Anwar, M.R.: A survey on vehicular edge computing: architecture, applications, technical issues, and future directions. Wirel. Commun. Mob. Comput. **2019** (2019)
45. Serena, L., Zichichi, M., D'Angelo, G., Ferretti, S.: Simulation of dissemination strategies on temporal networks. In: Proceedings of the 2021 Annual Modeling and Simulation Conference (ANNSIM), pp. 1–12. Society for Modeling and Simulation International (SCS) (2021)
46. Sguanci, C., Spatafora, R., Vergani, A.M.: Layer 2 blockchain scaling: a survey (2021). arXiv preprint arXiv:2107.10881
47. Shamir, A.: How to share a secret. Commun. ACM **22**(11), 612–613 (1979)
48. Toyoda, K., Machi, K., Ohtake, Y., Zhang, A.N.: Function-level bottleneck analysis of private proof-of-authority ethereum blockchain. IEEE Access **8**, 141611–141621 (2020). https://doi.org/10.1109/ACCESS.2020.3011876
49. Vorick, D., Champine, L.: Sia: Simple Decentralized Storage. Nebulous Inc (2014)
50. Xie, J., Tang, H., Huang, T., Yu, F.R., Xie, R., Liu, J., Liu, Y.: A survey of blockchain technology applied to smart cities: research issues and challenges. IEEE Commun. Surv. Tutor. **21**(3), 2794–2830 (2019). https://doi.org/10.1109/comst.2019.2899617
51. Yao, A.C.C.: How to generate and exchange secrets. In: 27th Annual Symposium on Foundations of Computer Science (sfcs 1986), pp. 162–167. IEEE (1986)
52. Zichichi, M., Ferretti, S., D'Angelo, G.: Are distributed ledger technologies ready for intelligent transportation systems? In: Proceedings of the 3rd Workshop on Cryptocurrencies and Blockchains for Distributed Systems (CryBlock 2020), co-located with the 26th Annual International Conference on Mobile Computing and Networking (MobiCom 2020), ACM, pp. 1–6. ACM (2020)

53. Zichichi, M., Ferretti, S., D'Angelo, G., Rodríguez-Doncel, V.: Personal data access control through distributed authorization. In: 2020 IEEE 19th International Symposium on Network Computing and Applications (NCA), pp. 1–4. IEEE (2020)
54. Zichichi, M., Ferretti, S., D'Angelo, G.: A framework based on distributed ledger technologies for data management and services in intelligent transportation systems. In: IEEE Access, pp. 100384–100402 (2020)
55. Zichichi, M., Ferretti, S., D'Angelo, G.: On the efficiency of decentralized file storage for personal information management systems. In: Proceedings of the 2nd International Workshop on Social (Media) Sensing, Co-Located with 25th IEEE Symposium on Computers and Communications 2020 (ISCC2020), pp. 1–6. IEEE (2020)
56. Zyskind, G., Nathan, O., et al.: Decentralizing privacy: using blockchain to protect personal data. In: 2015 IEEE Security and Privacy Workshops, pp. 180–184. IEEE (2015)

Crypto Regulation and the Case for Europe

Philipp G. Sandner, Agata Ferreira, and Thomas Dunser

Abstract The blockchain phenomenon has seen an extraordinary rise to prominence. The technology has grown at a revolutionary speed across many sectors and within little over a decade. It is no longer a niche technology for geeks, but a formidable innovation capable of triggering a paradigm shift, not only in finance, but within the society as a whole. While the blockchain industry has experienced an unprecedented growth, regulators struggled to keep pace with this innovation, both in terms of understanding this phenomenon and adapting or providing adequate legal and regulatory frameworks. As a result, legal and regulatory uncertainties are some of the main obstacles for blockchain innovation. This paper seeks to analyze regulators' and policymakers' efforts to understand and develop an adequate regulatory approach to crypto assets, tokens, and the distributed ledger technology (DLT) in Europe and illustrates the evolution of regulatory perception and recognition of this innovation. As the EU regulator remained passive for some time toward blockchain innovation except for a few inconsequential statements or reports, the EU countries tried to address this innovation individually and mostly attempted to apply existing legal framework to blockchain, with limited success. This paper gives an example of Liechtenstein as a jurisdiction that developed a comprehensive, bespoke and unique law that creates an entirely new legal architecture and principles to enable the token economy. It also outlines the EU latest initiative to create unique and bespoke regulation to govern markets in crypto assets and highlights the challenge of regulating the dynamically developing blockchain technology for the entire European region.

P. G. Sandner (✉)
Frankfurt School of Finance and Management, 60322 Frankfurt am Main, Germany
e-mail: email@philippsandner.de

A. Ferreira (✉)
Warsaw University of Technology, 00-661 Warsaw, Poland
e-mail: agata.ferreira@pw.edu.pl

T. Dunser
Office for Financial Market Innovation and Digitalisation, L9490 Vaduz, Liechtenstein

1   Introduction

With Bitcoin, a new type of technology was born in 2008 when Satoshi Nakamoto released the white paper for a new cash payment system (Nakamoto 2008), which effectively invented blockchain technology. By 2015 the technology already gained a lot of interest among startups, financial institutions, and industrial enterprises. Besides Bitcoin, many other crypto assets emerged with various design approaches such as stablecoins, utility tokens, security tokens, decentralized finance (DeFi), and non-fungible tokens (NFTs). Many of these tokens have an identifiable issuer to whom existing regulatory frameworks could potentially apply. However, other types of assets that are based on fully decentralized protocols are governed entirely by technology and either do not have an issuer (like in the case of Bitcoin) or the initiators designed the technology in an "issuerless" way—and have no relation to any "real-world asset". It is the latter class of assets that are truly new and that have recently attracted increasing attention from regulatory authorities, international organizations, standard-setting bodies, and the like.

On the part of regulators and policymakers, interest in and the activity surrounding cryptocurrencies, crypto assets, and stablecoins peaked in 2019 so far. Of the several key regulators and policymakers at the supra-national level, nearly all issued a report, warning, study, or recommendations on some aspect of blockchain technology in financial markets.[1] This spike in interest is related to the increasing business activity in this area and growing interest of investors and consumers. The exponential rise in the price of Bitcoin also attracted the interest of a wider audience [1]. The increasing business activity always preceded the actions of regulators and policymakers, thus rendering the activities of the latter a "reaction" to the market developments.

According to the Financial Stability Board (FSB), crypto assets reached an estimated total market capitalization of $830 billion on January 8, 2018, before falling sharply in subsequent months [2]. While the global value of the crypto assets market is still relatively small compared to the entire financial system, its absolute value and daily transaction volume are substantial, and its rapid development continues, gaining increasing market acceptance [3].

This paper seeks to analyze regulators' and policymakers' efforts to understand and develop an adequate regulatory approach to crypto assets, tokens, and the distributed ledger technology (DLT) in general. After several years of innovation in the space of decentralized technologies, several principles became clear on how to treat both issuer-based tokens and issuerless tokens. However, when regulators and policymakers tried at first to understand these new decentralized technologies and the assets they enable, it was not clear to them from the beginning how to treat assets based on this new technology. Only recently has it been possible to identify best regulatory practices and to disentangle good approaches to regulation from the "noise"

---

[1] These include: European Central Bank, European Banking Authority, European Securities and Markets Authority, Bank for International Settlements, Financial Stability Board, Organisation for Economic Co-operation and Development, International Monetary Fund, Financial Action Task Force, International Organisation of Securities Commission and G7.

of warnings, recommendations, or studies. Liechtenstein has adopted a remarkable perspective on and vision for crypto assets and tokens by creating a set of abstract definitions and models and applying them in their bespoke regulatory approach. The Liechtenstein Token Act has therefore inspired other policymakers and subsequent regulatory actions.

The remainder of this paper is structured as follows. First, we seek to present the history of "opinions" on behalf of regulatory bodies and policymakers over the last years. These opinions often lacked clear definitions, understanding, and models but also included valuable contributions. In the next section, we present key definitions and models of the Liechtenstein Token Act and describe how these have been included in Liechtenstein's national framework to build a solid basis for the emerging token economy. Thereafter, we describe how the European Union's approach to regulate crypto assets—the Markets in Crypto Assets Regulation (**MiCA**)—tackles crypto assets and tokens, and how it relates to the Liechtenstein Token Act. In the subsequent section, we review a variety of regulatory approaches and strategies. Finally, we offer concluding remarks.

2   Evolution of Regulatory Views on Blockchain

2.1   First Institutional Statements Before 2016—Cryptocurrencies in Focus

The first official statements and analysis focused on virtual currencies. The European Banking Authority (**EBA**) first issued a public warning against risky and unregulated virtual currencies in 2013. The role of the EBA is to monitor new and existing financial activities and adopt guidelines and recommendations to promote the safety and soundness of markets and convergence of regulatory practice. The EBA followed with an opinion on virtual currencies in 2014. It identified more than 70 risks arising from virtual currencies across several categories including risks to users, non-user market participants, financial integrity, existing payment systems, regulatory authorities, and the risk of money laundering and other financial crime [4]. In 2014, the EBA did not recommend a comprehensive regulatory approach addressing all identified risks, but did suggest immediate fragmented measures including governance requirements, capital requirements, and the segregation of client accounts. It also discouraged credit institutions, payment institutions, and e-money institutions from buying, holding, or selling virtual currency to shield regulated financial services from virtual currency schemes. The first assessment of cryptocurrencies has been one of mistrust.

2.2   Year 2016—Cryptocurrencies and First Analysis of DLT

In 2016, the European Central Bank (**ECB**) issued an analysis of virtual currency schemes. It reiterated and confirmed its earlier considerations and reaffirmed that risks from virtual currency schemes are actually low and not material in terms of monetary policy, price stability, financial stability, and the operation of payment systems [5]. The ECB also acknowledged potential advantages of virtual currencies for users, including challenging existing payment solutions regarding costs, global reach, payer anonymity, and speed of settlement. Furthermore, the ECB noted that virtual currency schemes could potentially become more successful than those incumbent, specifically in virtual communities, closed-loop environments, and cross-border payments. This was a more positive and encouraging stance on cryptocurrencies than the earlier EBA opinion.

After the first wave of official statements and reports on virtual currencies, the ECB issued a paper in 2016 analyzing DLT in securities post-trading [6]. In the paper, the ECB speculated that DLTs might enter securities markets. At the time (in 2016), the technology was still in the early development stage, and it was uncertain whether it would be "widely adopted in the securities market, and whether its adoption will address current market inefficiencies" ([6], p. 3). In parallel with the ECB, in early 2017, the European Securities and Markets Authority (**ESMA**) issued a report on the DLT applied to securities markets [7]. ESMA is an independent EU authority that contributes to safeguarding the stability of the EU's financial system by enhancing the protection of investors and promoting stable and orderly financial markets. It has full accountability toward the European Parliament, the Council of the European Union, and the European Commission. In its 2017 report, the ESMA identified several challenges of DLT to be addressed before its benefits can materialize. It noted interoperability issues, lack of common standards, and potential privacy and scalability problems. While the ESMA emphasized that the existing regulatory framework could apply to blockchain, it also acknowledged that some regulatory requirements could become less relevant and additional regulations might be needed to mitigate emerging risks. In 2017, the ESMA only considered potential regulatory impediments for the emergence of blockchain technology, as it was premature to fully appreciate the impact of the technology and resulting regulatory needs. The ESMA has not found any impediments in the EU regulatory framework to prevent blockchain technology from developing and fully emerging. Although the ESMA focused on securities markets, it also highlighted the need to clarify broader legal issues beyond financial regulations, including legal certainty and issues pertaining to corporate, contract, competition law, and DLT. Although the technology has now developed beyond speculations about whether it would be adopted in financial markets, most issues persisted, and many problems remained unresolved, including interoperability and the lack of common standards.

2.3 Year 2017—ICOs Controversies and First Acknowledgements of Crypto Assets

Not long ago, DLT was just starting to be noticed and scrutinized by regulators and supervisory bodies. Blockchain was still considered an immature technology and any dedicated regulation precipitate. The potential impact on financial markets and uptake of the technology in financial services was also unclear. However, the blockchain industry has rapidly grown, and 2017 was marked by a meteoric rise in Initial Coin Offerings (ICOs) and a massive increase in the value of various cryptocurrencies [8].

Also in 2017, first, the US Securities and Exchange Commission (SEC) warned investors about ICOs [9], and second, China and South Korea banned ICOs [10, 11], calling them "illegal fundraising". Third, in Europe, the ESMA issued two statements on ICOs, one on risks for investors and another on the rules applicable to firms involved in these offerings [12]. Regulators began realizing both the potential of this technology for financial markets and the magnitude of the associated risks. In its Fintech Action Plan of 2018, the **European Commission** acknowledged that crypto assets had become a worldwide phenomenon and a promising new type of financial asset; however, their high volatility, fraud, operational weaknesses, and vulnerabilities posed many risks. It also admitted for the first time that it was necessary to assess the suitability of the EU regulatory framework regarding crypto assets. The European Commission decided to continue monitoring the development of crypto assets and work together with supervisors, regulators, industry, civil society, and international partners to determine any further course of action [24]. It has also mandated the EBA and ESMA to assess the applicability and suitability of the existing EU financial services regulatory framework to crypto assets.

2.4 Year 2018—Cryptocurrencies and Crypto Assets—Focus on Risks and Concerns

In 2018, two reports commissioned by the **European Parliament** were produced. The first report on virtual currencies and central banks' monetary policy acknowledged that financial regulators may dislike virtual currencies because of their anonymity or cross-border circulation, money laundering risks, financing of illegal activities, tax avoidance, circumvention of capital controls, and fraudulent financial practices [13]. However, the report recommended that regulators treat virtual currencies as any other financial transaction or instrument proportionally to their market importance, complexity, and associated risks. The report also suggested the cross-border harmonization of regulations. The borderless and disintermediated character of the technology was becoming an issue confronting regulators, and only international cooperation could provide comprehensive regulatory solutions to this new phenomenon. The second commissioned report, on cryptocurrencies and blockchain, focused on the use of cryptocurrencies in financial crime, money laundering, and tax evasion [14]. It recommended that the fight against these activities should focus on cases of the illicit use of cryptocurrencies, while leaving blockchain untouched from the perspective of money laundering, terrorist financing, and tax evasion. The EU also amended its Fourth Anti-Money Laundering Directive to include virtual currency trading platforms and hosting wallets as entities subject to AML and combating the financing of terrorism (CFT) requirements.[2]

Finally, 2018 concluded with a Financial Stability Board (**FSB**) report on the crypto assets market and potential channels for future financial stability implications [2]. The FSB is an international body established to coordinate the work of national financial authorities and international standard-setting bodies to develop and promote the implementation of effective regulatory, supervisory, and other financial sector policies. In its report, the FSB concluded that although crypto assets did not pose a material risk to global financial stability, they raised several broader policy issues. It recommended vigilant monitoring. The primary identified risks that could have future implications for financial stability are related to market liquidity, volatility, leverage, technology, and operations. By 2018, several national regulatory bodies in the EU were already actively monitoring the regulatory implications of crypto assets, increasing their oversight and supervision; and issuing guidance, warnings, and clarifications on the applicability of the legal framework.

2.5 Year 2019—Peak of Interest in ICOs and Crypto Assets and the Impact of Libra

Pre-Libra Institutional Activity

After a busy 2018, during which the interest and activity of regulatory bodies in crypto assets, blockchain, and virtual currencies increased, 2019 witnessed an explosion of reports, statements, and recommendations issued by several EU and international regulatory and supervisory bodies. As such, crypto assets firmly entered the regulatory agenda. In the meantime, however, the market for token sales and new ICOs collapsed in 2018 and stalled in 2019 [15].

In January 2019, the **ESMA** issued advice on ICOs and crypto assets [16], and the **EBA** issued a report on crypto assets [31]. The ESMA recognized that the main challenge from the increasing presence of crypto assets in the market is the lack of clarity on the applicability of the existing regulatory framework to these new types of assets. It noted that while the current regulatory framework might apply to some crypto assets, it might need to be clarified and reconsidered for new types

---

[2] Directive (EU) 2018/843 of the European Parliament and of the Council of 30 May 2018 amending Directive (EU) 2015/849 on the prevention of the use of the financial system for the purposes of money laundering or terrorist financing, and amending Directives 2009/138/EC and 2013/36/EU, PE/72/2017/REV/1, OJ L 156, 19.6.2018, p. 43–74.

of assets. However, the ESMA also emphasized considering whether the regulations should be expanded to cover crypto assets and related activities that remain outside the regulatory governance framework. In these considerations, the ESMA advocated a technology-neutral approach to ensure that similar activities are subject to the same standards regardless of their form. It identified and made recommendations regarding regulatory gaps, for when crypto assets qualify as transferable securities or other types of financial instruments and are subject to the relevant financial regulations,[3] and when they do not fall within an existing regulatory framework (when they do not qualify as financial instruments or other regulations relating to non-financial instruments like the E-Money Directive,[4] for example). Notably, as some EU member states initiated regulatory efforts to establish national rules, the ESMA highlighted a concern over the divergent national approaches to crypto assets in the EU and the emerging bespoke regulations at the national level, which given the cross-border nature of crypto assets, could hamper regulatory harmonization across the EU. Similarly, the EBA expressed concern about the proliferation of uncoordinated legislative and supervisory actions at the national level, which can give rise to many risks to consumer protection, operational resilience, a level playing field, and market integrity. The EBA also concluded that activities related to crypto assets in the EU are thus far limited and pose no risks to financial stability overall. In addition, it recommended that the European Commission undertake a cost-benefit analysis to decide whether EU-level action is appropriate and feasible ([4], p. 4).

The Bank of International Settlements (**BIS**), which serves central banks in their pursuit of monetary and financial stability and fosters international cooperation, acknowledged in a statement in March 2019 that while the crypto assets market is still relatively small, the continued growth of these products and trading platforms can increase concerns related to financial stability and the risks faced by banks [17]. BIS considers crypto assets an immature asset class in constant evolution and lacking agreed standards. It highlighted many risks for banks, such as those related to liquidity, markets, operations, money laundering and terrorist financing, and legal and reputational aspects. It also noted wider implications and risks from the future growth of crypto assets, including implications for monetary policy, payment systems, consumer protection, market integrity, deposit insurance and guarantee schemes, and data privacy, and taxation. As such, BIS issued a discussion paper seeking stakeholders' views on designing a prudential treatment of crypto assets [18].

---

[3] Including: Markets and Financial Instruments Directive (MFID II) 2014/65/EU and Regulation (EU) 600/2014; The Prospectus Regulation (EU) 2017/1129 of the European Parliament and of the Council of 14 June 2017; The Prospectus Directive 2010/73/EU; Market Abuse Regulation (EU) 596/2014; Transparency Directive 2013/50/EU of the European Parliament and of the Council of 22 October 2013; Regulation (EU) No 909/2014 of the European Parliament and of the Council of 23 July 2014 on improving securities settlement in the European Union and on central securities depositories; Directive 98/26/EC of the European Parliament and of the Council of 19 May 1998 on settlement finality in payment and securities settlement systems.

[4] Directive 2009/110/EC of the European Parliament and of the Council of 16 September 2009 on the taking up, pursuit and prudential supervision of the business of electronic money institutions.

The **OECD** has also issued a report on ICOs in which it highlights the regulatory vacuum in the crypto assets market [19]. It considers legal and regulatory uncertainties to be the main impediments for the development of ICOs as a form of financing small and medium-sized enterprises. The OECD cites the lack of a clear regulatory framework applicable to an ICO offering, unclear legal rights and obligations of token issuers and holders, and a poor understanding by the investment community of potential legal and regulatory requirements of token issuances as the main limitations of ICO offerings. The lack of regulatory clarity also applies to the underlying DLT and related legal issues of enforceability, liability, and recourse in the use of smart contracts. The OECD emphasized the risk of regulatory arbitrage and risks to investors stemming from the lack of transparency in the absence of disclosure requirements. Furthermore, it considers the clarification of regulatory and supervisory frameworks applicable to ICOs, as well as international cooperation, as stepping stones to overcome current limitations and risks, prevent regulatory arbitrage, and realize the potential of ICOs for the financing of blockchain-based enterprises while protecting investors ([19], p. 43).

In addition, the **FSB** prepared a report to update G20 Finance Ministers and Central Bank Governors on the global outlook and work underway on regulatory and supervisory approaches to crypto assets and potential gaps [20]. The report recommended that G20 keeps the topic of regulatory approaches and potential gaps under review and adopts a forward-looking risk assessment in the rapidly evolving crypto asset ecosystem. The FSB recognized that a regulatory response needs to balance the need for a coherent multilateral approach with inherent jurisdictional differences, resulting in regulatory asymmetries. Furthermore, the FSB determined that crypto assets are at the nascent stage, do not present material risks to global financial stability, and that most issues can be addressed within existing regulatory frameworks.

The International Organization of Securities Commissions (**IOSCO**), an international body and global standard setter for the securities sector, also contributed and published a report in 2019 on issues and risks associated with the trading of crypto assets on crypto asset trading platforms. The purpose of the report is to assist regulatory authorities and provide a toolkit of measures regulatory authorities can use in policymaking to govern crypto asset trading platforms. Recommended considerations for regulators include rules on access and on-boarding, safekeeping of participant assets, transparency of operations, market integrity and trading rules, price discovery mechanisms, and the resiliency and security of the technology [21].

In May 2019, the **ECB** noted that while crypto assets do not pose an immediate threat to financial stability in Europe because of their small relative value and limited links to the financial sector, diverse and unconnected national regulatory initiatives could be ineffective, facilitating regulatory arbitrage and ultimately inhibiting the resilience of the financial system as a whole. The ECB recommended a broader and balanced approach to the regulation of crypto assets, particularly with regard to risks arising from unregulated entities, including "gatekeeping" services (like custody, trading, and exchange services). In addition, the ECB noted that regulatory intervention could be complicated because of the distributed architecture of crypto assets

[22]. Thus, it distinguished two possible regulatory approaches. First, if centralized service providers carry out crypto asset activities, the existing regulatory framework may be applicable. For decentralized activities, the ECB suggested a principle-based approach to regulations coupled with an additional formal validation mechanism. Shortly after the ECB report, the **FSB** produced another report, on financial stability and the regulatory and governance implications of decentralized financial technologies [23]. The FSB noted the challenges stemming from decentralized financial technologies such as blockchain for financial regulatory and supervisory frameworks, which were designed for a centralized financial system. Such decentralized technologies could be used to avoid regulations, compromise regulatory enforcement, and increase jurisdictional uncertainty. To combat these risks, the FSB suggests considering the appropriateness, applicability, and effectiveness of current financial regulations and potential regulatory gaps. New methods of regulatory enforcement and potential gaps in supervisory systems should also be considered. The FSB recommends that any regulatory action should involve multi-stakeholders, be proportional to the risks, and technology neutral.

Based on the myriad of reports, statements, and opinions, a set of firm regulatory recommendations started to emerge, highlighting the risks, acknowledging regulatory gaps, and recommending specific regulatory approaches. In parallel with European and other international organizations and bodies, the Financial Action Task Force (**FATF**) is actively considering the implications of virtual assets for international financial systems. FATF is an inter-governmental standard-setting body that promotes the effective implementation of legal, regulatory, and operational measures for combating money laundering, terrorist financing, and other threats to the international financial system. In October 2018, FATF adopted changes to its Recommendation 15 to clarify that it applies to virtual assets and virtual asset service providers [24]. The amended FATF Recommendation 15 requires that virtual asset service providers are regulated for AML/CFT purposes, licensed or registered, and subject to monitoring and supervision. In June 2019, FATF adopted an Interpretative Note to Recommendation 15 [25] that requires a risk-based approach to virtual asset financial activities and virtual asset service providers. It introduces licensing and registration obligations, and the monitoring and supervision of virtual asset service providers by competent authorities rather than self-regulatory bodies. It also extends the application of a range of sanctions for non-compliance. In addition, FATF recommends the application of all relevant preventive measures including customer due diligence, recordkeeping, and suspicious transaction monitoring.

The Impact of Libra Announcement

The second half of 2019 was dominated by the controversies and consternation surrounding stablecoins, a new type of crypto asset that seeks to stabilize its price by linking its value to an asset or a pool of assets. The debate on crypto assets underlying DLT, stablecoins, and their potential impact on the financial ecosystem accelerated after Facebook announced its project to issue Libra, a global stablecoin.

The Libra project announcement had an extraordinary impact and provoked immediate and firm official reactions worldwide. Promptly, several authorities, including the FSB, Bundesbank, the Bank of England, and the US Federal Reserve issued statements addressing Libra [26–29]. The Governor of the Bank of England and The U.S. House of Representatives' Committee on Financial Services have also issued a statement each, highlighting that Libra has not been received with "an open door" and requesting that Libra meets the "highest standards of prudential regulation and consumer protection" [30, 31]. The U.S. House of Representatives' Committee on Financial Services went as far as to request that Facebook and its partners immediately cease implementation plans "until regulators and Congress have an opportunity to examine these issues and take action" and requested a moratorium on any movement forward on Libra. The overall sentiment expressed through those first statements was akin to panic and the statements were dominated by concerns over serious potential risks and challenges of such global stablecoins arrangements. Uncertainties related to the lack of a clear regulatory framework, scrutiny, and recognition of global stablecoins were potentially hampering the actual issuance of Libra.

The **G7** meeting that took place in July 2019 was dominated by concerns over the Libra project. The Chair of the Committee on Payments and Market Infrastructures (CPMI) and member of the ECB Executive Board highlighted several serious risks posed by global stablecoin projects in his speech to the G7 in July 2019, including anti-money laundering/combatting the financing of terrorism (AML/CFT), consumer and data protection, cyber resilience, fair competition, tax compliance, issues related to monetary policy transmission, financial stability, and the smooth functioning of and public trust in the global payment systems [32]. At the same time, the need to improve access to payment services to ensure faster and cheaper payments and cross-border remittances has also been acknowledged as well as other benefits of stablecoins including greater competition in payment services and greater financial inclusion. Nevertheless, proposed recommendations illustrated a firm and skeptical approach to global stablecoins projects. The need to ensure public trust by meeting the highest regulatory standards, prudent supervision and oversight, and globally consistent regulatory approaches has been emphasized. Legal compliance of stablecoins projects across jurisdictions was also considered essential, including adequate governance and a risk management framework to ensure operational and cyber resilience and safe, prudent, transparent, and consistent management of the underlying assets. The G7 meeting official closing statement acknowledged that "projects such as Libra may affect monetary sovereignty and the functioning of the international monetary system" and "raise serious regulatory and systemic concerns, as well as wider policy issues" [33]. G7 strongly concluded that any stablecoin projects would need to meet the highest standards of financial regulation, especially with regard to AML/CFT, to guarantee they do not affect the financial system's stability or undermine consumer protection.

Several other bodies and organizations have issued statements and assessments of stablecoins, including the ECB, G7 Working Group, and FSB. The **ECB** published a report analyzing the taxonomy of stablecoins and assessing their macroeconomic impact on financial stability and monetary policy, noting a strong correlation between the type of stablecoin and its price volatility [21]. The report acknowledged significant uncertainties regarding the governance and regulatory treatment of stablecoins, which might hamper their uptake. Less innovative stablecoins were considered to be less volatile than more innovative ones. The ECB also acknowledged the possibility that stablecoins can be made redundant if financial institutions use the same underlying technology for traditional assets.

A **G7** working group investigated the impact of global stablecoins, identifying a long list of risks stemming from stablecoins of any size [34]. The risks are related to legal certainty and governance issues, investment rules of the stability mechanism, illicit finance, safety, the efficiency and integrity of payment systems, cybersecurity, operational resilience, and market integrity. Stablecoins are thought to pose challenges to data privacy and data protection, consumer and investor protection, and tax compliance. The biggest risks of global stablecoins can be attributed to their scale, which could affect monetary policy, monetary sovereignty, financial stability, fair competition, and the international monetary system overall. G7 strongly contends that no global stablecoin project should go ahead without adequately addressing all these risks. Regulations should be appropriately adjusted to address the specifics of global stablecoins. The recommended regulatory approach should be technology neutral, functional, mindful of the risks of regulatory arbitrage, and ensure a level playing field that encourages competition. The report also acknowledged the weaknesses of existing cross-border payments systems and the need to improve access to financial services and cross-border retail payments. However, instead of acknowledging stablecoins´ potential in addressing these issues, the report was skeptical given the uncertainty created by the significant legal, regulatory, supervisory, and operational challenges posed by stablecoins. Instead, the G7 Working Group recommended focusing on improving the efficiency and inclusiveness of existing, established financial systems and financial services.

FINMA, the Swiss financial authority with which the Libra project has been submitted for an assessment of its project under Swiss law, published a supplement to its ICO guidelines outlining the treatment of stablecoins [35]. FINMA adopted a technology-neutral approach and "same risks, same rules" principle focusing on "substance over form" and looking at tokens´ economic function and purpose. The supplement concluded that stablecoins vary, and therefore the laws that apply to them may include money laundering, securities laws, banking, and fund management regulations. FINMA emphasized legal uncertainties regarding transferability and enforceability under civil law of claims linked to tokens.

At the same time, elsewhere in Europe, fear and rejection of the Libra project dominated. Just two days after the FINMA guidance was published, Germany and France issued a joint statement addressing Libra and declaring that the project had failed to convince that risks would be properly addressed and reiterating that "no private entity can claim monetary power, which is inherent to the sovereignty of nations". The statement emphasized the risks including financial security, investor protection, AML/CFT, data protection, and financial and monetary sovereignty [36].

Furthermore, the **FSB** emphasized in its report delivered to a G20 meeting in October 2019 the need to assess any regulatory gaps in existing regulatory and supervisory frameworks in the context of stablecoins at the national level and in the cross-border and cross-authority context to minimize the risk of regulatory arbitrage. At the international level, stablecoins could affect existing international regulatory and supervisory standards. The FSB noted that global stablecoin projects could, in fact, alter the then assessment that crypto assets do not pose a material risk to financial stability and acknowledged that stablecoins can indeed pose systemic risks due to their large user base and potential to become of systemic importance, particularly in individual jurisdictions where they could replace domestic currencies. The FSB recognized that global stablecoin could disrupt banks´ funding and have implications for financial stability, market integrity, competition, and data protection. Thus, the FSB recommends strengthening international cooperation and coordination to address potential concerns of global financial stability and systemic risk. However, in spite of a long list of risks and challenges and some high-level recommendations, no specific regulatory steps have been suggested leaving stablecoins´ issues with much uncertainty [37].

The **ECB** also issued a more comprehensive study on stablecoins, acknowledging benefits that global stablecoin projects could make international payments cheaper and faster, and facilitate financial inclusion while also highlighting previously recognized risks, including potential impacts on operational robustness, safety and soundness of payment systems, customer protection, risks to financial stability and monetary sovereignty, and AML/CFT compliance [38].

The **EU Council** and the **Commission** officially joined the trend with their joint statement on stablecoins, which was rather repetitive and similar to other statements and reports issued in the aftermath of the Libra announcement [39]. It acknowledged the benefits of financial innovation in promoting competition and financial inclusion, broadening consumer choice, increasing efficiency, and delivering cost savings and the benefits of cheap and fast payments. However, the statement mainly highlighted challenges to consumer protection, privacy, taxation, cybersecurity and operational resilience, AML/CFT, market integrity, governance, and legal certainty. It emphasized risks to monetary sovereignty, monetary policy, the safety and efficiency of payment systems, financial stability, and fair competition. The Council and the European Commission committed to providing a framework for stablecoins and ensuring appropriate consumer protection standards and orderly monetary and financial conditions. In a follow-up step, the EU public consultation on an EU framework for markets in crypto assets, issued in December 2019, included questions seeking stakeholders' views concerning stablecoins.

To wrap up the year, the International Monetary Fund (**IMF**) issued a note in December 2019 in which it identifies selected elements of regulation and supervision to assist policymakers in framing the discussion on the regulation of crypto assets [40]. Note that the IMF considers crypto assets at the core of the Fintech revolution, and any regulation should not stifle innovation but build trust ([40], p. 17). The IMF provides several high-level recommendations for regulators, including a sequential, risk-based, and proportional approach to developing regulatory frameworks based on priorities and resources. Furthermore, it recommends a continuous comprehensive assessment of the risks and strategies. It emphasizes cross-sector and international cooperation and coordination as key elements in enhancing investor protection and minimizing the potential for regulatory arbitrage while maintaining regulatory flexibility to adapt to technological progress. In its advice, the IMF focused on the main aspects of crypto assets: offering, trading, custody, and exposure. In addition, it acknowledged relatively low societal financial and technology literacy and the need to ensure that participants, investors, and customers are adequately informed about the particularities and risks of crypto assets. Therefore, appropriate disclosure requirements at the time of the initial offer and thereafter are essential in protecting investors. Regarding the trading of crypto assets, the IMF follows IOSCO's report [27], recommending robust governance requirements for platform operators, onboarding compliance requirements for access to the platform, and resilient and safe operating systems and controls. Regulators should also consider the applicability of market abuse and transparency rules. The IMF suggests that a regulatory determination be made regarding the types of assets to be permitted for trading and safe custodial services. Clarifying the legal position of crypto asset ownership is also important in ensuring the effective clearing and settlement of crypto asset trading. The IMF highlighted its concern over the lack of a global standard for the prudential treatment of exposure to crypto assets for banks or other regulated entities ([40], p.16). The ongoing BIS consultation in this regard should address this concern [18]. It recommends a conservative approach such as capital deductions or the imposition of high-risk weights and robust assets segregation and separation. This relates to both direct and indirect exposure to crypto assets from derivatives, financial instruments linked to crypto assets, cyber insurance to wallet providers, or loans to crypto investors. Finally, the IMF acknowledges that formulating an adequate regulatory framework for crypto assets involves intense monitoring, a flexible approach, and international cooperation [18].

At the EU level, 2019 finished with the final report of the Expert Group on Regulatory Obstacles to Financial Innovation (**ROFIEG**) to the European Commission on recommendations for regulation, innovation, and finance [41]. Essentially, in relation to crypto assets, ROFIEG recommends accelerating the work to assess the existing regulatory framework and develop solutions to fill potential regulatory gaps. This should include addressing the lack of a common taxonomy and the resulting fragmented national approaches to crypto assets. The main risks to be addressed include money laundering, terrorist financing, tax evasion, governance and operational resilience, client asset protection, disclosure requirements, consumer protection, and the prudential treatment of exposure to crypto assets ([41], p. 16). The commercial law aspects of crypto assets, including the conflict of laws rule might also need to be addressed at the EU level.

2.6   Year 2020—Stablecoins and MiCA

Interest in crypto assets further intensified in 2020. At the request of the **European Parliament**'s Committee on Economic and Monetary Affairs a study on key developments, regulatory concerns, and responses on crypto assets was published in May 2020, a week before the announcement of the second version of the Libra project [42]. Interestingly, the study reiterated that stablecoins remain a marginal phenomenon among crypto assets and their impact remains local. It recognized that stablecoins pose challenges and risks to financial stability and monetary policy and that AMLD5 lags behind and should be enhanced and the current EU financial regulatory framework is not sufficiently tailored to crypto assets resulting in legal uncertainty. The risks stem also from financial institutions gaining exposure to highly volatile crypto assets.

The **ECB** issued more comprehensive official report on stablecoins and their implications for monetary policy, financial stability, market infrastructure and payments, and banking supervision in the euro area, in which it characterizes stablecoin arrangements, emphasizes the role of technology-neutral regulation in preventing arbitrage, and the importance of comprehensive Eurosystem oversight, irrespective of stablecoins' regulatory status [43]. The report goes further in its analysis of stablecoins than previous official documents and analyzes various scenarios for the uptake of stablecoins and the associated public policy, regulation, and supervision implications. The ECB estimates that the uptake of stablecoins collateralized with euro-denominated assets is a more likely scenario in the eurozone. It emphasizes potential implications of such a scenario for Eurosystem's monetary policy transmission and concludes that stablecoins could become a new payment method and could reach a scale, giving rise to financial stability risks due to fragilities of stablecoin arrangements and their links with the financial system. Again, the need for adequate, internationally coordinated regulation, and cooperative oversight has been recognized, as well as the importance of "same business, same risks, same rules" principle to ensure a level playing field and prevent regulatory arbitrage. The same principle for stablecoins´ regulation has been recognized by the **FSB** in its report on regulation, supervision, and oversight of global stablecoins, in which the FSB calls for completion of international standard-setting work, establishment of cooperation arrangements among authorities and adjustment of regulatory, supervisory and oversight frameworks [44]. The FSB also acknowledged the need for a holistic regulatory approach that addresses any potential regulatory gaps and clarifies regulatory powers, including internationally coordinated regulatory efforts to help achieve common regulatory outcomes across jurisdictions and reduce opportunities for cross-sectoral and cross-border regulatory arbitrage. In terms of cross jurisdictional analysis, the FSB identified regulatory gaps that include incomplete or non-existent implementation of the revised FATF standards, lack of capacity to provide regulatory supervision of global stablecoin arrangements, lack of adequate competition policies, and inadequate consumer protection measures. The FSB formulated a number of recommendations that include application of international standards to global stablecoins on a functional basis and proportionate to their risks and comprehensive governance frameworks with clearly allocated accountability, effective risk management frameworks, operational resilience, and AML/CFT measures.

Finally, the latest step in the recognition of crypto assets is the proposal of the **European Commission** of a regulation on Markets in Crypto-Assets (MiCA) [45]. The European Commission differentiates between crypto assets that are already governed by EU legislation and which will remain subject to existing legislation including MiFID II [46], and other crypto assets. For crypto assets that will remain subject to existing legislation, the European Commission proposes a pilot regime [47] for market infrastructures that wish to try to trade and settle transactions in financial instruments in crypto asset form to enable market participants and regulators to gain experience with the use of DLT exchanges that would trade or record shares or bonds on the digital ledger. MiCA forms a part of a digital finance package adopted by the European Commission on September 24, 2020 [48], which also includes a digital finance strategy. MiCA sets out a bespoke regime for previously unregulated crypto assets, including "stablecoins" and it has four main objectives: legal certainty, innovation support, consumer and investor protection and financial stability. MiCA introduces compliance requirements for issuers and crypto asset service providers wishing to apply for an authorization to provide their services in the single market. The requirements include capital requirements, custody of assets, a mandatory complaint holder procedure available to investors, and rights of the investor against the issuer. In addition, issuers of significant asset-backed crypto assets will be subject to more stringent capital requirements, liquidity management, and interoperability requirements (see Sect. 4 for more details).

2.7 Regulatory Uncertainties

The blockchain ecosystem has been evolving rapidly in the last decade and it outpaced regulators, authorities, and policymakers. As illustrated by our earlier analysis, multiple authorities and institutions analyzed cryptocurrencies, DLT and eventually crypto assets and either issued a statement, a report (or multiple reports), or participated in the debate by undertaking another form of analysis of these new phenomena.

However, for quite some time, this activity has not led to clear regulatory guidelines, set of principles, or proactive regulatory steps. Therefore, in the early phases of blockchain development, market participants faced high regulatory uncertainty. At first, some regulators tried to apply or formulate regulations to govern blockchain application within existing legal frameworks and normative principles. Only a few countries viewed crypto assets—in particular decentralized protocols—as a novel technology that commanded new principles for a regulatory framework and a bespoke regulatory approach.

One of these countries is Liechtenstein which sought to create an all-encompassing framework on how to treat tokens from a regulatory perspective. In Liechtenstein, a vision for the future token-based economy emerged. This vision guided the regulator to formulate the Liechtenstein Token Act which rests on multiple normative models and principles on how crypto assets and tokens should be viewed and regulated.

3 The Liechtenstein Token Act

3.1 Background

On January 1, 2020, the first comprehensive regulation of the so-called "token economy" came into force in the Principality of Liechtenstein with the Law on Tokens and TT Service Providers (Token Law or TVTG). The Government of Liechtenstein had explicitly developed a very broad regulation approach to create legal certainty for all applications of blockchain in the economy. This approach therefore is fundamentally different from other regulations that focus on the virtual assets, stable coins, digital securities, and related financial services.

3.2 The Vision of the Token Economy

The vision of the token economy (see also Duenser [13]) is based on the token's property to create digital information which cannot be manipulated or copied. This property is not only relevant for digital money or digital securities, but for many other assets and rights of the existing legal system. The token economy refers to the possibility to tokenize any kind of assets, such as a physical item like a car or a house, by representing a right corresponding to a physical item in a token. Such a right can be the property rights or usage rights of an item and they usually derive from official registers (like in the case of real estate), civil law (like in the case of physical items), or from contracts in all possible forms. To tokenize such rights means to create a unique object representing this right, which can then be owned and transferred like a physical item. This innovation is similar to the concept of creating physical security by representing an investor's rights relating to a company on a piece of paper. The invention of physical security was one of the drivers of the modern economy. The token is expected to trigger a similar development, but in a much broader sense. As the creation and transfer of a token is very efficient, there are almost no limitations for potential applications. As such, every purchase contract could be concluded and settled with tokens, for example, The purchase of a bicycle is equivalent to the transfer of the property right token versus digital money tokens. By using a token, the buyer would instantly receive a proof of ownership of the bicycle. By storing such a token in a personal wallet, the owner can show this digital proof of ownership to everyone in the world. On the other hand, the seller would instantly receive the digital money.

Therefore, the concept of the objectivization of rights via a token therefore has a similar effect on the legal system as the invention of physical security on the modern economy. By using digital, programmable contracts (e.g., smart contracts), tokens can now be used to transfer the rights described in a contract. Moreover, the same right can then be transferred to another person. Consequently, an additional layer of unique and objectivized rights will evolve, which will help to prove true legal ownerships [13]. This innovation is therefore expected to bring an unprecedented level of legal certainty to the digital economy.

The concept of tokenization can be applied to all processes and transactions: In supply chain management or in international trade, tokens can be used to prove the transfer of a good. In e-commerce, tokens can prove the successful purchase of a good or a right. Tokens can be used to secure the intellectual property of music, books, or movies. On a festival, tokens can help to simplify the order process of drinks and snacks.

Even if the vision of the token economy includes the application of digital money and securities, it covers a much broader field of applications. This has a significant impact on regulation. The regulatory approach of Liechtenstein is based on this broad vision of a token economy.

3.3 Classification of Tokens in Liechtenstein Token Act

As the legal classification of tokens triggers legal consequences, many countries have tried to fit the current applications of blockchain within existing legal classifications, such as currencies, security tokens, and the new forms as utility coins. Liechtenstein deliberately did not rely on existing classifications but introduced the (general) token as a new legal element (Liechtenstein Token Model). In the Token Act, the token is defined as "a piece of information on a TT System [i.e., a DLT Transaction System] which can represent claims or rights of memberships against a person, rights to property or other absolute or relative rights; and is assigned to one or more TT Identifiers [i.e., a Wallet-Address]". This step has wide-reaching consequences.

First, it provides the legal fundament for all possible applications of blockchain technology, including the current and future forms. Virtual currencies, like Bitcoin, are tokens which do not represent any rights and have no reference to real-world values. Utility coins are—for example—tokens representing usage rights of a DLT system. Security tokens, like share tokens, represent voting and/or dividend rights regarding companies, while bond tokens might represent the right of interest payments and redemptions. In this token model, stablecoins are tokens that represent, for example, the right to receive fiat money or gold. But more important is the fact that with the general token definition, all rights regarding physical items can be tokenized, such as the property right of a painting, the usage right of a car, or the right to receive a drink. It is also possible to tokenize license rights of intellectual property, such as the right to listen to music, and rights to use a patent. With the approach Liechtenstein has chosen, many more forms of tokens are covered with a legal fundament, which enables the secure use of tokens for almost every application in the economy. Hence, the "Liechtenstein approach" is intended to be a legal fundament of the token economy. As such, this token model itself is a revolution, since it enables the bridge between the existing legal framework and a digital transaction infrastructure. It supports the objectivization of any rights of the Liechtenstein legal system (which means to create objects which represent a right), so that they can be digitally possessed and transferred like a physical item. With this step, Liechtenstein has seen the potential to increase the legal certainty of any economic (and by that: legal) transaction of the digital and analogue economy.

Second, it solves the central problem of unsuited legal consequences appearing when using existing classifications for tokens. For example, if a country generally classifies tokens as securities, all laws on securities and financial instruments, especially financial market laws, and related tax rules would apply, making it impossible in practice to use a token for applications other than investment, such as a means of payment. The classification of tokens as a currency would trigger the application of other laws, so that the use of such tokens in or by a decentralized network would not be possible in certain jurisdictions. Therefore, focusing on existing classifications bears the risk of hindering innovation in the context of a fundamental technology like DLT, which can be used for almost every application. With the Liechtenstein Token Model, the legal consequences depend on the right which is represented by the token: If a security is represented, security laws shall apply, whereas in the case a token represents intellectual property rights, intellectual property laws should be applied, etc. With this approach, Liechtenstein is relying on the principle "substance over form". The sole act of creating a token has no legal implications in Liechtenstein. In particular, only the fact that a token is transferable does not trigger the application of security laws. This treatment of a token is crucial for the broad application of blockchain technology in the economy outside of financial markets.

Third, by introducing the token as a new element into the existing legal system, it is possible and also necessary to clarify all civil law questions relating to tokens: Can a token be owned, can it be stolen? How can a token be legally transferred? (see next section).

Fourth, Liechtenstein's approach offers a solution to potential conflicts between tokens and real-world assets. From the perspective of the token economy, it becomes clear that most tokens will have a reference to the real-world rights or assets. Pure virtual currencies or virtual assets without reference to the real world, such as Bitcoin, will rather be an exemption. Tokens with reference to the real-world face the challenge that the real-world asset or right is not synchronized with the token representing this right. For example, if a token represents the property right of a car, a conflict can arise if the owner of the property right token is not the same person as the holder of the car. This can happen, if the car is stolen, or sold to another person not knowing that the property right is tokenized and sold to another person. For the functioning of the Token Economy, the synchronization of online and offline rights is essential. The Liechtenstein Token Model is offering the legal fundament for the clarification of such conflicts within the legal system.

3.4   Civil Law of the Token in Liechtenstein Token Act

By introducing the token as a new legal element, Liechtenstein had to consider several fundamental legal questions: As the properties of a token are similar to those of a physical item, in theory the property law could be used to clarify the open questions about possession and ownership of a token. But as many legal rules and the jurisprudence of property law is based on corporeality, this option has proven to raise other significant legal issues and—by that—increased the legal uncertainty. One aspect of these considerations is the potential confusion between a physical object and its digital twin, if property law applies for both. Since a true digital object like a token is new, it became apparent that it would be better to introduce a new legal fundament for tokens in order to avoid confusion and the interference of corporeality.

To put the concept of a digital item into effect, Liechtenstein has developed new concepts of possession and ownership of tokens. The DLT has special properties which have to be respected. Tokens themselves cannot be owned or possessed because they are always assigned to some kind of address. Both, the token and the address are part of the transaction ledger (i.e., the blockchain) and cannot be owned in a traditional sense. In terms of ownership, the key with which a person can sign new transactions is particularly relevant. For DLT with asymmetric encryption, the key is often referred to as a "private key", but the Liechtenstein law is intended to be technology neutral, so that the law defines the "TT-key" in an abstract manner as "a key that allows for disposal over Tokens" (Article 2 TVTG).

Therefore, the Liechtenstein Token Act defines the holder of the key who is able to initiate transactions as the person who is possessing a token. To avoid confusion with the terms of property law, Liechtenstein introduced the term "a person with the power of disposal over the token" as corresponding with "a person possessing a token".

Another fundamentally relevant decision is the differentiation between possession and legal ownership of tokens. This is particularly interesting because of the discussions about the "code is law" principle among blockchain pioneers, which implies that the legal ownership is identical with the possession of a token. Liechtenstein acknowledged that the legal ownership and the possession can diverge in practice, such as when a token is stolen, or if the token or the key is transferred to a delegate, like a custodian. Legal owners of tokens might face difficulties when seeking to rely on a legal system where neither the token is legally clearly defined, nor is there a legal construction for clarifying token theft, i.e., by hacking a wallet. Liechtenstein decided that introducing the concept of legal ownership is crucial to clarify the integration of DLT in the legal system, so that both the service providers in their terms and conditions as well as the authorities and the courts are able to manage all circumstances properly. Therefore, the term "person with the right to dispose of a token" is introduced as an equivalent of legal ownership. This two-layered approach is especially important in the common use of custodial service providers, as they often either have access to the key or are assigning the token to their own respective wallet´s address. The Token Act, therefore, also provides the legal fundament for clarifying problems that can arise between service providers and their clients.

Even though this differentiation is crucial for legal certainty for token holders, it brings up additional questions that have to be clarified in the legal system. How can a person seeking to buy a token be sure that the seller is the legal owner? It would hinder the efficiency of the token economy, if the buyer had to verify the legal ownership of a token before each transaction. In order to protect users and to increase the efficiency of the token economy, Liechtenstein introduced the legal assumption that the person possessing the power of disposal over a token also has the right to dispose over the token. In addition, it is regulated that "those who receive tokens in good faith, … for the purpose of acquiring the right of disposal … are protected in [their] acquisition, even if the transferring party was not entitled to the disposal over the Token unless the recipient party had been aware of the lack of right of disposal or should have been aware of such upon the exercise of due diligence" (Article 9 TVTG). By these rules, Liechtenstein has introduced a civil law concept to protect both the buyer and the legal owner of a stolen token.

Similar to the fact that tokens cannot be owned directly, it is also not possible to transfer tokens directly. Technically spoken, a token is transferred by changing its assignment to another address. Therefore, a transfer transaction changes the power of disposal to the person which is holding the key of the new address. Thus, Liechtenstein legally defined that the disposal over tokens is equivalent to the transfer of the right of disposal over the token (Article 6 TVTG). With that legal definition, buyers of tokens can now be sure that after a successful technical transfer the legal transfer is also ensured. These elements build the fundament of legal certainty for tokens, i.e., in the digital layer.

As another pillar of legal certainty of token transfer, Liechtenstein had to clarify the requirements for the disposal over tokens. It is important to clearly define at what point of time a transfer is legally fulfilled. Article 6 of the Token Act therefore states three conditions: First, the (technical) conclusion of the transfer according to the rules of the DLT system, second, the declaration of both parties about the will to transfer the token, and third, the legal ownership of the transferring party. Only if all three conditions are met, the token is legally transferred.

The Token Act is consequently oriented to the token economy, acknowledging that most tokens have a reference to the analogue world. In addition to the legal clarification of the token transfer, Liechtenstein also had to regulate the consequences of a token transfer regarding the rights or assets in the analogue world. The synchronization of "online" and "offline" dimensions is crucial for the legal certainty of token owners. Therefore, the TVTG defines in Article 7: "(1) Disposal over the token results in the disposal over the right represented by the token". Because it is possible for tokens to represent any kind of rights on a DLT system, this rule clarifies that with the transfer of the token, the receiving party also gets the represented right. By that, it is possible to create a transferable object of every right in the existing legal system. This is the key for enabling the token economy. These rules have to be accompanied with collision rules: "If the legal effect under (1) does not come into force by law, the person obliged as a result of the disposal over the Token must ensure, through suitable measures, that [...] the disposal over a Token directly or indirectly results in the disposal over the represented right, and [...] a competing disposal over the represented right is excluded" (Article 7/2). The TVTG even contemplates the possibility of enforcement proceedings: "The disposal over a token is also legally binding in the event of enforcement proceedings against the transferor and effective vis-à-vis third parties, if the transfer: (a) was activated in the TT system prior to the commencement of the legal proceedings, or (b) was activated in the TT the system after the initiation of the legal proceedings and was executed on the day of the proceeding's openings, provided that the accepting party proves that he was without knowledge of the proceedings openings or would have remained without knowledge upon the exercise of due diligence".

By considering that more applications of tokens represent rights within the legal system and the economy and are, in particular, not purely virtual, like Bitcoin, it becomes important that a token can be cancelled. This is not an option for virtual currencies or many forms of utility coins as the original applications, but for every other tokenized asset, this is crucial for the legal certainty: If, for instance, the property right of a house is tokenized, and the token is lost or stolen, or becomes nonfunctional, it is necessary that there exists a legal procedure to cancel the token and create a new one. This is also a relevant feature if a multi-DLT-environment is considered: If a token owner decides to move a token to another DLT system, the cancellation procedure is also necessary to create legal certainty for all participants.

Tokens can also represent securities, as the shareholder rights to an equity in a company or debt rights. For both parties, the obligor and the obligee, it is necessary that tokens can be used to fulfill the legal part of such arrangements. Therefore, the TVTG defines in Article 8 the legitimacy and exemption: "(1) The person possessing the right of disposal reported by the TT System is considered the lawful holder of the right represented in the token in respect of the Obligor. (2) By payment, the Obligor is withdrawn from his obligation against the person who has the power of disposal as reported by the TT system, unless he knew, or should have known with due care, that he is not the lawful owner of the right". By this rule, an obligor of a security represented in a token can be sure that his or her obligations are fulfilled if the payments (interests or dividends) are transferred to the token holder. This rule also enhances the legal certainty of token holders.

Because of the special features of securities, the Liechtenstein law defined specific rules for security tokens. It is possible to create so-called uncertificated rights or book-entry securities, where it is explicitly stated that the book-entry register can be implemented by using DLT systems. This means that companies can directly create such book-entry securities by generating a security token without extra efforts. This way to create digital securities is very efficient and is intended to support innovation in this sector while having a high level of legal certainty.

3.5 Regulation of Service Providers in Liechtenstein Token Act

With the Token Act, Liechtenstein has also introduced a regulation of specific service providers. As the law was intended to be open for innovation, the service provider regulation is formulated in a role- and principles-based manner. This means that no existing business models are regulated as a whole, such as crypto exchanges, but only functions or roles. For example, if a company offers custodian services, it must comply with the relating obligations, no matter if this is the only service or if this is part of a comprehensive business, such as the provision of a trading facility. For all single roles, specific duties are introduced to address the relating specific risks. These duties are formulated in an abstract and principles-based manner, so a company is free to choose how to implement its service as long as the principles are achieved.

The Token Act covers 10 roles in total:

(1) "Token Issuer": a person who publicly offers the tokens in their own name or in the name of a client;
(2) "Token Generator": a person who generates one or more tokens;
(3) "TT Key Depositary": a person who safeguards TT Keys for clients;
(4) "TT Token Depositary": a person who safeguards token in the name and on account of others;
(5) "TT Protector": a person who holds tokens on TT Systems in their own name on account for a third party;
(6) "Physical Validator": a person who ensures the enforcement of rights in accordance with the agreement, in terms of property law, represented in Tokens on TT systems;
(7) "TT Exchange Service Provider": a person, who exchanges legal tender against Tokens and vice versa and Tokens for Tokens;
(8) "TT Verifying Authority": a person who verifies the legal capacity and the requirements for disposal over a Token;
(9) "TT Price Service Provider": a person who provides TT System users with aggregated price information on the basis of purchase and sale offers or completed transactions;
(10) "TT Identity Service Provider": a person who identifies the person in possession of the right of disposal related to a token and records it in a directory.

As Liechtenstein is a member of the European Economic Area (**EEA**), the financial market regulations are derived from EU laws. Therefore, the Token Act does not cover financial market functions, such as operating a trading facility or providing investment advice, because this would collide with the harmonized European single market. The Token Act is therefore applicable to all non-financial market applications, but—as long as the European regulation is not adopted—should also cover those financial market applications which currently are not in the scope. In this sense, Liechtenstein regulated such services for all tokens regardless how they are classified.

3.6 User Protection Regulation in Liechtenstein Token Act

From the perspective of users of blockchain systems, it is of utmost relevance that neither the tokens themselves nor the possession of tokens can be manipulated. Therefore, the technical quality of the "blockchain technology" is relevant for the level of user's security. On a technical level, the integrity of a DLT-transaction-database depends on several features: Number of nodes, distribution of nodes, consensus mechanism, level of cryptographic security, and many more. This means that the technical quality of a specific DLT system is derived from static design features, but also from dynamic aspects, like the distribution of nodes and miners. If, for example, one group of persons comes in the position of dominating the mining process by providing an extraordinary amount of computing power, it might also have the power to change the transaction ledger in certain DLT systems. Both aspects make it difficult for average token holders to assess the risks. To protect token holders, some governments might consider regulating the quality of the technology itself to define the minimum standard of DLT systems accessible in a jurisdiction.

To ensure user protection, Liechtenstein has decided to choose a fundamentally different approach. The Token Act does not regulate the quality of a DLT system, but obliges service providers to ensure that the chosen DLT system is appropriate for specific use. Consequently, the ten service providers regulated in the Token Act (TT service providers) have to fulfil the legal obligations (see last section).

The Government of Liechtenstein deliberately has refrained from regulating the quality of the technology in order to not hinder innovation. Even after a decade, DLT is still a relatively new technology with a high pace of development. The Government of Liechtenstein argued that a technology-based regulation would only be able to cover the currently known forms of DLT, so that not only new forms would be restricted in their implementation, but also the legal certainty for users of new forms would be undermined, since the legal protection would only cover the old forms. Especially with regard to civil law applicable to tokens, a restricted legal scope would cause severe risks to holders of tokenized assets. Liechtenstein therefore established the legal definitions very carefully and in a technology-neutral manner so that all current and future forms of DLT are provided for. Instead of the common term "DLT" which might focus only on the current forms, the Token Act defines the term "TT Systems" as "transaction system which allows for the secure transfer and storage of Tokens and the rendering of services based on this by means of trustworthy technology" And "trustworthy technology" is defined as "Technologies through which the integrity of tokens, the clear assignment of tokens to TT Identifiers and the disposal over tokens is ensured". TT identifiers are defined as "an identifier that allows for the clear assignment of tokens". Even though Liechtenstein refrained from using common terms such as blockchain and DLT, this set of definitions is intended to cover all similar kinds of such transaction systems. In particular, it highlights that neither the distribution nor the encryption is relevant for being considered as TT systems, but the fact that a digital information (the Token) can be owned and possessed as a physical item without any reference to a (central) intermediary or other counterparties. This means that every technology allowing such features falls under the scope of the law, regardless of its technological implementation. As a consequence, the use of DLT by one or several central intermediaries to provide a transaction system does not fall under the scope of the Token Act. For example, a private, permissioned blockchain used for the supply chain in industry is not covered by the Token Act. Also a bank using a private DLT system for its core banking database is not covered by this definition. If such applications want to benefit from the civil law fundament, it is possible to declare that civil law provisions expressly apply (Article 3 TVTG).

Therefore, the quality of the DLT system is not relevant for the application of the Token Act. The definition of a TT system is based on principal characteristics, and not on quality criteria. The scope of the definition of TT system is not only relevant for the civil law implications of tokens, but also for the obligation to register as a TT service provider according to the Token Act.

Natural and legal persons exercising one or more services based on a TT system need a registration prior to market entry and are obliged to fulfill the legal requirements. Consequently, both aspects have to be considered: The use of a TT system and the provision of a service. For example, a person providing custody services for tokens in a private DLT system which is not open to the public is not obliged to register as a TT service provider, since in such a situation, the need for client protection additional to standard consumer protection does not seem to be necessary only because DLT is used. On the opposite, if such a service is brought on an open TT system, it does not matter which instrument is tokenized for licensing obligation to apply. Consequently, a person holding a token representing a book (for example, the right to access and read a book) as a service for a third-party user, falls also under the scope of regulation like a custodian for cryptocurrencies or security tokens. Users of DLT systems which do not rely on the service of a registered service provider must check the quality of the blockchain systems by themselves and be informed of the current developments.

4 MiCA in Comparison to the TVTG

In September 2020, the European Commission has published the digital finance package [48], which also contains MiCA and a proposal for a pilot regime for multilateral trading facilities using DLT [47].

MiCA comprises a regulation of issuers of e-money tokens (title IV), of so-called asset-referenced tokens (title III), of crypto asset service providers (title V), and issuers of other crypto assets which are not regulated under title III and IV (title II). The regulation of issuers of e-money tokens and asset-referenced tokens is distinguished between significant and non-significant tokens.

The scope of MiCA applies "to persons that are engaged in the issuance of crypto assets or provide services related to crypto assets in the Union" ([45], Article 2.1). It is not applicable to crypto assets that qualify as financial instruments, electronic money, deposits, structured deposits, or securitization.

The term crypto asset is defined as "a digital representation of value or rights which may be transferred and stored electronically, using distributed ledger technology or similar technology". Comparing this definition to the Liechtenstein Token Act, the terms "crypto" and "distributed" refer to the current state of development. In case a new technology is introduced without using distribution or cryptography, this might lead to legal uncertainty whether the law should be applied or not.

With the definition of crypto asset, the EU Commission decided to use a very broad definition of a crypto asset, but made it clear that crypto assets can appear in different forms, including as financial instruments. Instead of declaring every crypto asset a financial instrument, the European Commission has chosen an approach similar to the Liechtenstein Token Model. This step is important, since it clarifies that financial market laws are basically applicable to tokenized financial instruments, but, in parallel, it also opens the possibility to draft new regulation for other forms of crypto assets. Consequently, the European Commission had to solve problems in the existing financial market framework caused by the fundamentally new technology used. In particular, the secondary markets for financial instruments require to use a central securities depository (CSD), which hindered many projects of security token trading, because almost no existing CSD had been able to register tokens. In addition, DLT has the potential to dispense with the CSD function to avoid unnecessary costs.

With the regulation of crypto asset service providers (title V), the European Commission developed a new regulatory framework for financial services with crypto assets that do not qualify as financial instruments. Therefore, the crypto asset services regulated in title V are very close to the definitions of investment services and activities regulated in MiFID II [46]:

- custody and administration of crypto assets on behalf of third parties,
- the operation of trading platform for crypto assets,
- the exchange of crypto assets for fiat currency that is legal tender,
- the exchange of crypto assets for other crypto assets,
- the execution of orders for crypto assets on behalf of third parties,
- placing of crypto assets,
- the reception and transmission of orders for crypto assets on behalf of third parties,
- providing advice on crypto assets.

Similar to Liechtenstein's TVTG, the European Commission decided to introduce a more role-based and principles-based regulation, so that companies with certain focused activities are regulated adequately, whereas the regulation still is open for innovation.

By comparing the TT service providers regulated in the TVTG, it becomes clear that both laws are addressing different actors: MiCA intends to regulate financial services with crypto assets, and the Token Act regulates the fundamental services which are relevant for the whole token economy. As Liechtenstein is a member of the European Economic Area (EEA), the European financial market regulation is also applicable in Liechtenstein. Therefore, the TVTG is designed as a complementary regulation to financial market laws: If a service using tokens is considered to fall under financial market regulation, the service provider must comply with both, the TVTG and the particular financial market law. This is introduced to ensure that financial service providers who want to use DLT have sufficient knowledge and well-defined processes for creating a sufficient level of client protection. In cases where no special regulation is applicable to token service providers, providers only have to comply with the TVTG. Therefore, the TVTG is a "catch all regulation" for token services. By enhancing the financial market laws to other crypto assets, another special regulation for token service providers is introduced. In particular, crypto asset service providers which operate a trading platform, execute orders, place crypto assets, receive and transmit orders or provide advice are not regulated in the TVTG, while the other services, such as custody and exchange services, are also regulated in Liechtenstein. This means that—in the case the government of Liechtenstein will not adopt the law—some CASP might have to comply with both laws, and some only with one of the laws. But as the level of regulation is quite similar, this is not expected to raise any additional burden to service providers.

Considering that MiCA intends to introduce financial market regulation for crypto assets and the fact that tokens can be used for almost every activity in the economy, also for non-financial-market services, MiCA potentially can expand the application of financial market regulation to many real-economy activities, which have not been covered with such regulation up to now. So, a precise legal definition of "crypto asset" is of utmost importance to clearly separate the tokenized financial market from the rest of the token economy.

As the civil law fundament, if crypto assets lies not within the competence of the EU-commission, MiCA is lacking a similar legal fundament for tokens as the Token Act. In order to get legal certainty for the European market, each country will have to face the challenge to adopt the civil law. Even if many countries have a civil law for physical or digital securities, the special features of security tokens often make an adjustment necessary. Considering the further applications of tokens for the objectivization of any kind of rights, most countries are lacking clear and profound civil law fundament. This might hinder the further development of the token economy on cross-border activities.

5   Review of Regulatory Approaches and Strategies

Exponential speed of technological developments, growing awareness and knowledge gap, and the novelty and complexity of new technological advancements such as DLT make it difficult to find an appropriate and balanced regulatory response. Regulators struggle to keep up and often focus more on risks and challenges and less on the opportunities DLT offers. With the controversies of the developments like Libra project, regulators and lawmakers can also become unduly biased in their regulatory approach to DLT technology, aiming at capturing and controlling such developments with stringent compliance and regulatory burdens rather than providing innovation conducive environment supporting innovation and promoting entrepreneurship in blockchain industry. Vigilance for risks and their mitigation is justified and within the mandate of most regulators within the world. However, a balanced regulatory approach requires weighting out levels or risks with short- and long-term opportunities and needs.

It is not possible to define a unified regulatory solution or approach to DLT given the diversity of legal systems, regulatory parameters and mandates, levels of economic development, and political environment across the globe. However, well-recognized regulatory principles, such as "same risks, same activity, same rules, and same supervision", can assist the regulators in their efforts to respond to blockchain innovation and ultimately could help global harmonization of laws and regulations applicable to blockchain. The latter is particularly important given the inherently borderless nature of blockchain innovation. Adopting a risk-based approach allows assessing the levels of risk presented by various blockchain innovation and enables providing regulatory measures that are proportionate to such risks. DLT innovation should not be considered as raising the same risks every time it is being used. Sometimes, there is no real risk elevation from the use of blockchain technology, like its use for loyalty cards which already exist today without controversies. Other blockchain-based innovations present entirely new sets of risks, like global stablecoins that trigger concerns over financial stability, market integrity, and monetary policy. Risk-based approach to regulation helps differentiate between those higher-risk applications of blockchain that perhaps warrant a more stringent regulatory approach, and lower-risk or even risk-neutral blockchain innovation that may not need regulatory intervention at all.

Too much or too stringent regulation out of fear for the "worst-case scenario" could be very damaging for the industry, stifle innovation, and also deprive consumers of the benefits of this technological innovation. Ultimately, such disproportionate and out-of-sync regulation would be damaging to the regulators themselves as a positive impact of regulation would be effectively diminished, the market would get distorted and negative externalities of technology amplified, necessitating even more invasive regulatory action and creating self-perpetuating regulatory spiral widening the gap between law and technology. Given the "novelty" of blockchain innovation, an activity-based regulatory approach could also allow regulators to identify and focus on activities that require regulatory intervention. This approach, coupled with some entity-based requirements for big players, like big tech companies, for example, could help address the risks arising out of new blockchain innovation and new significant market players entering the financial sector [49]. Further, a principle-based approach with the focus on an outcome rather than detailed rules would also be suitable for blockchain innovation, particularly for the type for which there is no regulatory precedent. To that end, principles of consumer protection, prevention of ML/FT, or level playing field, would be the examples of desired outcomes, which could be assessed against particular blockchain innovation. Flexibility of such approach allows more fluid adjustments to regulatory framework to keep at pace with technological developments unlike prescriptive detailed rules, which may need frequent amendments to close any potential gaps and address new developments.

These regulatory principles should be helpful in formulating an appropriate regulatory strategy toward blockchain-based innovations, taking into account needs, goals, and priorities of a particular jurisdiction. It is not uncommon that certain blockchain developments trigger robust prohibitive regulatory response. For example, China and South Korea banned ICOs and several countries introduced some kind of ban on cryptocurrencies. Such prohibitive approaches contribute to creating regulatory arbitrage opportunities and have several other negative effects, like reducing financial inclusion or criminalizing certain innovation. It is also usually most damaging to the country introducing such outright ban, as the innovation and capital simply moves elsewhere. The only justified use of a prohibitive regulatory strategy toward blockchain innovation could be to grant the authorities additional time for research and assessment of new innovation in order to enable it later in a controlled and informed fashion.

However, the most common current regulatory approach is that of a "wait-and-see", due to the novelty, lack of urgency on the part of the regulators and lack of established regulatory precedents. Many regulators also lack expertise, capacity, or resources to formulate adequate and timely responses to blockchain innovation. While regulators wait, innovation can freely develop and mature in such jurisdictions. However, lack of regulatory interest could also deter innovation due to lack of regulatory clarity, or, in the worst-case scenario, compromise consumer protection or even enable fraud if there are no regulatory boundaries. Where existing regulations apply to blockchain, regulators may also choose to be passive, observe and study technological developments in the meantime or issue guidelines to assist the industry, like FINMA did when issuing guidelines for ICOs and then the stablecoins [35]. Among the range of other possible approaches could be the introduction of accelerators, sandboxes, or similar collaborative measures that provide a safe and controlled environment for the innovation to develop under scrutiny but also with the support of authorities. Examples of sandbox regulatory approaches include the

EU PILOT regime or a recent initiative of the Central Bank of Brazil that recently announced a regulatory sandbox allowing stablecoin development under regulatory supervision [50].

Finally, regulators could opt for a bespoke regulatory framework for blockchain innovation. The examples include MiCA, which deals with financial market applications of crypto assets and the Liechtenstein Token Act. MiCA creates a bespoke regulatory framework for crypto assets not covered by existing financial regulation and many of the rules mirror financial regulation. Considering that the regulatory discussions have focussed almost exclusively on financial market applications of DLT, MiCA is a consequent step to address the lack of legal certainty, user protection, and AML/CFT rules. In contrast, the Liechtenstein Token Act is aiming at a much broader scope of application of DLT and therefore offers a completely new, abstract and neutral approach to tokens, those blockchain based and others. Even though MiCA and the Token Act appear as competing approaches to regulation, they are in fact based on similar regulatory concepts and build a complementary set of regulations. If the token economy in this broad sense should be enabled throughout Europe, all countries would potentially need a similar civil law fundament as Liechtenstein, so that all DLT users can benefit from a high level of legal certainty.

Regardless of a particular choice for blockchain regulation, regulators should not approach this topic through the narrow lens of underlying blockchain technology but should remain technology-neutral in their approach and any regulatory efforts should be collaborative and include all stakeholders. Regulation should not hinder innovation and entrepreneurship or impair competition and all market participants should be subject to general principles of transparency, prudence, integrity, and consumer protection.

6   Conclusions

The blockchain phenomenon has seen an extraordinary rise to prominence. The journey of Bitcoin from the publication of the white paper in 2008 to a market capitalization of $1 trillion in February 2021 (Chavez-Dreyfuss and Wilson [30]) illustrates market acceptance. Blockchain has demonstrated a high degree of resilience and the potential to not only transform existing capital markets but also to create new asset classes. The technology has also grown beyond financial applications and is being adopted in other sectors, from logistics to healthcare. All these developments took place at a revolutionary speed, within little over a decade.

However, laws and regulations tend to develop more incrementally, at a much slower pace and through a cumulative step-by-step process built around geographically divided legal jurisdictions, doctrines, jurisprudence, and legal practice. Also, while startups and IT developers produce new blockchain-based innovation literally every month, the education—and therefore the understanding—among regulators and policymakers progresses at a much slower speed. It is therefore not surprising that regulators struggled to keep pace with blockchain innovation, which has not only developed rapidly, but also in a decentralized and borderless fashion. Legal and regulatory uncertainties are some of the main obstacles for blockchain innovation, as market participants were often left without clear regulatory guidance how to specifically apply existing laws and regulations—created to cater to centralized and intermediated market design—to blockchain, build around decentralization and disintermediation principles. The rise of DLT is also indicative of a broader economic and societal transformation toward decentralization and peer-to-peer connections.

Slowly, these innovations have attracted the attention of authorities, but their first reactions, views and statements were those of mistrust, caution, and even dismissal. Even though several key regulators and policymakers at the supra-national level issued a report, warning, study, or recommendations on some aspect of blockchain technology, those actions were not only often out of sync with market developments (like in the case of ICOs), but also mostly lacked clear regulatory solutions, regulatory answers, specific regulatory steps, and recommendations. The blockchain potential has not been fully recognized by authorities until only very recently, triggered by the Libra project.

This paper illustrates the road to regulatory recognition of DLT, including cryptocurrencies and crypto assets, in the EU. For quite some time the EU regulator remained passive toward blockchain innovation except for a few inconsequential statements or reports. EU countries tried to address this innovation individually. In the absence of an off-the-shelf regulatory framework model or high-quality regulatory architecture for this new phenomenon, countries mostly attempted to apply existing legal framework to blockchain, with limited success.

Liechtenstein, however, developed a comprehensive, bespoke and unique law that creates an entirely new legal architecture and principles to enable the token economy. The Liechtenstein Token Act grants legal recognition and protection to a token, provides a bridge between tokens and the existing laws, addresses civil law issues around tokens, defines service providers roles and responsibilities to ensure the seamless connection between the digital and physical world and is flexible enough to cater to future technological developments.

Eventually, the EU has also embarked on a path of a unique and bespoke regulation, MiCA. MiCA is of a momentous importance for the entire blockchain ecosystem in Europe and beyond. It can either benefit or prejudice Europe and it will certainly influence and shape regulatory approaches in other countries, possibly setting global standards. MiCA is an applaudable effort by the EU regulator, which not so long ago paid little attention to blockchain innovation. However, such a bespoke, prescriptive, and detailed pan-EU regulation aimed to govern a dynamically developing blockchain technology will shape the future of the entire region and has to be carefully considered and meticulously calibrated.

References

1. European Banking Authority (2014). EBA opinion on "virtual currencies". EBA/Op/2014/08. https://eba.europa.eu/sites/default/documents/files/documents/10180/657547/81409b94-4222-45d7-ba3b-7deb5863ab57/EBA-Op-2014-08%20Opinion%20on%20Virtual%20Curr encies.pdf?retry=1. Last accessed 29 Jan 2020
2. Financial Stability Board (2018). Crypto-asset markets. Potential Channels for Future Financial Stability Implications
3. BBC (2017). China bans initial coin offerings calling them 'illegal fundraising'. https://www.bbc.com/news/business-41157249. Last accessed 30 Jan 2020
4. European Banking Authority (2019). Report with advice for the European Commission on crypto-assets
5. European Central Bank (2019a). Crypto-assets–trends and implications. https://www.ecb.eur opa.eu/paym/intro/mip-online/2019/html/1906_crypto_assets.en.html. Last accessed 29 Jan 2020
6. Pinna, A., Ruttenberg, W.: Distributed ledger technologies in securities post-trading. Revolution or evolution? Occasional Paper Series No 172 (2016). https://www.ecb.europa.eu/pub/pdf/scp ops/ecbop172.en.pdf. Last accessed 29 Jan 2021
7. European Securities and Markets Authority (2017a). Report: the distributed ledger technology applied to securities markets. ESMA50-1121423017-285. https://www.esma.europa.eu/sites/default/files/library/dlt_report_-_esma50-1121423017-285.pdf. Last accessed 29 Jan 2020
8. Basel Committee on Banking Supervision (2019). Discussion paper: designing a prudential treatment for crypto assets. https://www.bis.org/bcbs/publ/d490.pdf. Last accessed 29 Jan 2020
9. US Securities and Exchanges Commission (2017). SEC issues investigative report concluding DAO tokens, a digital asset, were securities. Press Release. https://www.sec.gov/news/press-release/2017-131. Last accessed 30 Jan 2021
10. Binham, C., Giles, C., Keohane, D.: Facebook's libra currency draws instant response from regulators'. Financial Times (18 June) (2019). https://www.ft.com/content/5535fb3a-91ea-11e9-b7ea-60e35ef678d2. Last accessed 12 Feb 2021
11. Kim, C.: South Korea bans raising money through initial coin offerings. Reuters (2017). https://www.reuters.com/article/us-southkorea-bitcoin/south-korea-bans-raising-money-thr ough-initial-coin-offerings-idUSKCN1C408N. Last accessed 30 Jan 2021
12. European Securities and Markets Authority (2017b). ESMA highlights ICOs risks for investors and firms. Press Release. https://www.esma.europa.eu/press-news/esma-news/esma-highlights-ico-risks-investors-and-firms. Last accessed 30 Jan 2020
13. Dünser, Th.: Legalize Blockchain. How States Should Deal with Today's Most Promising Technology to Foster Prosperity. BoD Verlag (2020)
14. Houben, R., Snyers, A.: Cryptocurrencies and blockchain legal context and implications for financial crime, money laundering and tax evasion. European Parliament, Policy Department for Economic, Scientific and Quality of Life Policies Directorate General for Internal Poli-cies, PE 619.024 (2018). http://www.europarl.europa.eu/cmsdata/150761/TAX3%20Study%20on%20cryptocurrencies%20and%20blockchain.pdf. Last accessed 30 Jan 2021
15. Fromberger, M., Haffke, L.: ICO market report 2018/2019–performance analysis of 2018's Initial Coin Offerings (2019). https://doi.org/10.2139/ssrn.3512125
16. European Securities and Markets Authority (2019a). Advice, initial coin offerings and crypto-assets. ESMA50-157-1391. https://www.esma.europa.eu/sites/default/files/library/esm a50-157-1391_crypto_advice.pdf. Last accessed 30 Jan 2020
17. Bank of International Settlements (2019a). Statements on crypto assets. https://www.bis.org/publ/bcbs_nl21.htm. Last accessed 31 Jan 2020
18. Bank of International Settlements (2019b). Designing a prudential treatment for crypto assets'. Basel Committee on Banking Supervision. Discussion Paper. https://www.bis.org/bcbs/publ/d490.pdf. Last accessed 31 Jan 2020
19. OECD (2019). Initial Coin Offerings (ICOs) for SME financing. www.oecd.org/finance/initial-coin-offerings-for-sme-financing.htm. Last accessed 31 Jan 2021

20. Financial Stability Board (2019a). Crypto-assets work underway, regulatory approaches and potential gaps. https://www.fsb.org/wp-content/uploads/P310519.pdf. Last accessed 31 Jan 2020
21. Bullmann, D., Klemm, J., Pinna, A.: In search for stability in crypto-assets: are stablecoins the solution? European Central Bank. Occasional Paper Series, No 230 (2019). https://www.ecb.europa.eu/pub/pdf/scpops/ecb.op230~d57946be3b.en.pdf. Last accessed 31 Jan 2020
22. European Central Bank (2019b). Occasional paper series crypto-assets: implications for financial stability, monetary policy, and payments and market infrastructures. ECB Crypto-Assets Task Force, No 223. https://www.ecb.europa.eu/pub/pdf/scpops/ecb.op223~3ce14e986c.en.pdf. Last accessed 31 Jan 2020
23. Financial Stability Board (2019b). Decentralised financial technologies report on financial stability, regulatory and governance implications. https://www.fsb.org/wp-content/uploads/P06 0619.pdf. Last accessed 31 Jan 2020
24. Financial Action Task Force (2018). Regulation of virtual assets. https://www.fatf-gafi.org/pub lications/fatfrecommendations/documents/regulation-virtual-assets.html. Last accessed 31 Jan 2020
25. Financial Action Task Force (2019). Public statement on virtual assets and related providers. https://www.fatf-gafi.org/publications/fatfrecommendations/documents/public-sta tement-virtual-assets.html. Last accessed 31 Jan 2020
26. Alois, J.D.: Federal Reserve Chairman Jerome Powell comments on Libra crypto at FOMC press briefing. Crowdfund Insider (19 June) (2019). https://www.crowdfundinsider.com/2019/06/148598-federal-reservechairman-jerome-powell-comments-on-libra-crypto-at-fomc-press-briefing/. Last acessed 16 Mar 2021
27. Board of the International Organization of Securities Commissions (2019). Issues, risks and regulatory considerations relating to crypto-asset trading platforms. Consultation Report. CR02/2019. https://www.iosco.org/library/pubdocs/pdf/IOSCOPD627.pdf. Last accessed 31 Jan 2020
28. Stacey, K., Binham, C.: Global regulators deal blow to facebook's Libra currency plan. Financial Times (25 June) (2019). https://www.ft.com/content/0c1f3832-96b1-11e9-9573-ee5cbb 98ed36. Last accessed 16 Mar 2021
29. The Forex Review (2019). Bundesbank warns against Libra (22 July). https://theforexreview.com/2019/07/22/bundesbank-warns-againstlibra/. Last accessed 16 Mar 2021
30. Chavez-Dreyfuss, G., Wilson, T.: Bitcoin hits $1 trillion market cap, surges to fresh all-time peak. Reuters (19 February) (2021). https://www.reuters.com/article/us-crypto-currency-bit coin-idUSKBN2AJ0GC. Last accessed 9 Apr 2021
31. The U.S. House of Representatives' Committee on Financial Services (2019). Committee democrats call on facebook to halt cryptocurrency plans (2 July). https://financialservices.house.gov/news/documentsingle.aspx?DocumentID=404009. Last accessed 16 Mar 2021
32. Cuervo, C., Morozova, A., Sugimoto, N.: Regulation of crypto assets. International Monetary Fund. FinTech Note 19/03 (2019)
33. G7 (2019) Chair's summary: G7 Finance Ministers and Central Bank Governors' meeting (17–18 July). https://minefi.hosting.augure.com/Augure_Minefi/r/ContenuEnLigne/Dow nload?id=7C00115F-99CD-4FC1-A520-1EF0126E1A7C&filename=G7%20Chair%27s% 20summary.pdf. Last accessed 16 Mar 2021
34. G7 Working Group on Stablecoins (2019). Investigating the impact of global stablecoins. https://www.bis.org/cpmi/publ/d187.pdf. Last accessed 1 Feb 2021
35. Swiss Financial Market Supervisory Authority FINMA (2019). Supplement to the guide-lines for enquiries regarding the regulatory framework for Initial Coin Offerings (ICOs)' (11 September)
36. France and Germany (2019). Joint statement on Libra (13 September). https://www.gouver nement.fr/sites/default/files/locale/piece-jointe/2019/09/1417_-_joint_statement_on_libra_ final.pdf. Last accessed 16 Mar 2019
37. Financial Stability Board (2019c). Regulatory issues of stablecoins. https://www.fsb.org/wp-content/uploads/P181019.pdf. Last accessed 1 Feb 2020

38. European Central Bank (2019c). Stablecoins–no coins, but are they stable? 3 In Focus
39. The Council and the European Commission (2019). Joint statement by the council and the commission on "stablecoins" (5 December). https://www.consilium.europa.eu/pt/press/press-releases/2019/12/05/joint-statement-by-the-council-and-the-commission-on-stablecoins/. Last accessed 16 Mar 2021
40. Dabrowski, M., Janikowski, L.: Virtual currencies and central banks monetary policy: challenges ahead. European Parliament, Policy Department for Economic, Scientific and Quality of Life Policies, Directorate General for Internal Policies, PE 619.009 (2018). https://www.europarl.europa.eu/cmsdata/149900/CASE_FINAL%20publication.pdf. Last accessed 30 Jan 2020
41. Expert Group on Regulatory Obstacles to Financial Innovation (ROFIEG) (2019). Thirty recommendations on regulation, innovation and finance. Final Report to the European Commission. https://ec.europa.eu/info/sites/info/files/business_economy_euro/banking_and_finance/documents/191113-report-expert-group-regulatory-obstacles-financial-innovation_en.pdf. Last accessed 1 Feb 2019
42. Houben, R., Snyers, A.: Crypto-assets–key developments, regulatory concerns and responses; study for the committee on economic and monetary affairs. Eur. Parliam. Policy Dep. Econ. Sci. Qual. Life Policies PE **648**, 779 (2020)
43. European Central Bank (2020). Occasional paper series. Stablecoins: implications for monetary policy, financial stability, market infrastructure and payments, and banking supervision in the Euro area. ECB Crypto-Assets Task Force, No 247
44. Financial Stability Board (2020). Regulation, supervision and oversight of "Global stablecoin" arrangements final report and high-level recommendations
45. European Commission (2020b). Proposal for a regulation of the European Parliament and of the council on markets in crypto-assets, and amending. Directive (EU) 2019/1937. COM(2020) 593 Final
46. European Parliament and Council (2014). Directive 2014/65/EU of the European Parliament and of the council of 15 May 2014 on markets in financial instruments and amending. Directive 2002/92/EC and Directive 2011/61/EU
47. European Commission (2020c). Proposal for a regulation of the European Parliament and of the council on a pilot regime for market infrastructures based on distributed ledger technology. COM(2020) 594 Final
48. European Commission (2020a). Digital finance package. https://ec.europa.eu/info/publications/200924-digital-finance-proposals_en. Last accessed 16 Apr 2021
49. Restoy, F.: Fintech regulation: how to achieve a level playing field'. Bank for International Settlements, Financial Stability Institute, Occasional Paper Series No 17 (2021)
50. Gusson, C.: Banco Central do Brasil abre inscrição para Sandbox que pode permitir emissão de stablecoins e criptomoedas. Cointelegraph (22 February) (2021). https://cointelegraph.com.br/news/brazilian-central-bank-opens-registration-for-sandbox-that-may-allow-issuing-of-stablecoins-and-cryptocurrencies. Last accessed 30 Mar 2021

Economic Perspectives
on the Governance of Blockchains

Ilia Murtazashvili and Martin Weiss

Abstract The structure and operation of blockchains are dynamic, which means that mechanisms must exist for implementing changes. The New Institutional Economics (NIE), with its emphasis on how rules govern the performance of any complex organization or network, provides an especially useful framework to consider governance of blockchains. We consider how NIE has been applied to blockchain and future applications. Our analysis is divided into consideration of blockchain networks as institutional technologies, blockchain networks as knowledge commons and polycentric enterprises, and the ways to empirically research blockchain networks. The Institutional Analysis and Design (IAD) framework developed by Elinor Ostrom is particularly useful to develop an empirical research agenda for comparing the institutional features of blockchains and, ultimately, to comparing their performance.

1 Introduction

Blockchain-based systems are useful when they create value for their users. Thus, it is natural to consider the economic aspects of these systems. Blockchains are also artificial constructs whose performance depends on rules internal to a given network and their relationship in the larger system. It is therefore natural to consider their governance as well.

While there are many economic analyses that can be performed related to blockchains [25], we focus on those aspects that relate to governance. In general, governance is needed to enable an economic system of rules (an *institution*) to adapt dynamically to changing preferences, circumstances, environmental factors, etc. Our discussion is based on the New Institutional Economics (NIE) framework in the tradition of (Nobel-prize winning economists) Ronald Coase, Oliver Williamson, Elinor

I. Murtazashvili (✉) · M. Weiss
Center for Governance and Markets, University of Pittsburgh, Pittsburgh, PA, USA
e-mail: ilia.murtazashvili@pitt.edu

M. Weiss
e-mail: mbw@pitt.edu

© The Author(s), under exclusive license to Springer Nature Switzerland AG 2022
D. A. Tran et al. (eds.), *Handbook on Blockchain*, Springer Optimization
and Its Applications 194, https://doi.org/10.1007/978-3-031-07535-3_22

Ostrom, Douglass North, and their associates. The focus of NIE is on problems of information, transaction costs, property rights, and other formal and informal economic institutions (that is, laws, norms, and rules). In short, NIE is concerned with *governance*—the study of good order and working relations [73]. NIE provides a useful dynamic framework for thinking about governance, since the institutions must adapt to changes in the technical and social environment.

NIE is a useful framework for blockchain systems because blockchains themselves are implemented in software and use well-defined protocols (i.e., rules) to communicate. These protocols require design choices, the results of which constitute aspects of blockchain governance. As well, blockchain systems support human exchange and operate in the context of existing social institutions, so it is highly relevant to understand how these external institutions interact with a given blockchain network to understand how they function [8]. How blockchains respond to uncertainty [43], including their rules and enforcement [34], is becoming increasingly central to economic analysis of blockchains.

From all of this, we suggest the polycentricity framework of Elinor Ostrom, along with Ostrom's Institutional Analysis and Design (IAD) framework, offers an especially useful framework to consider blockchain. The polycentricity framework serves as a descriptive framework to analyze how blockchains are governed and as a normative guide for policy related to blockchain, especially opportunities for establishing new blockchain communities (such as a smart city on a blockchain). The IAD framework provides additional guidance for how to analyze blockchains in terms of their specific institutional features, as well as a framework to empirically research blockchains.

We will focus on permissionless (or public) blockchains in this chapter for two reasons. First, much of the literature has focused on these, and second, from an economics perspective, permissioned blockchains function much as a corporation does, in that certain rights are established and maintained, and a central organization determines who can use the network and for what purposes. Governance remains significant in permissioned blockchains, as the choice of rules is akin to constitutional design [7], though permissionless blockchains are unlike corporations and permissioned blockchains in that they allow potentially anyone to participate in them. Thus, analysis of the public blockchains is an especially fruitful are for institutionally oriented research.

2   A Brief Introduction to New Institutional Economics

From its birth in the late eighteenth century, economics was concerned with institutions. Adam Smith—who initiated modern economics with his publication of *The Wealth of Nations* in 1776—recognized that successful markets required an appropriate constitutional framework [19]. Despite Smith's appreciation for institutions, starting in the middle of the twentieth century, economics became increasingly mathematical, focusing on behavior of utility-maximizing individuals. The mathematical turn in economics largely left institutions aside in focusing on modeling utility-maximizing individual behavior [52]. The maximizing individual represented a shift from the conventional economics' focus on organization of economic activities, and away from the concern with families, households, or firms [21]. The push for formalization resulted in a discipline became much more focused on atomistic individuals rather than consideration of the origins or consequences of the rules themselves [39].

Institutionalists recovered an insight of Adam Smith regarding the role of institutions in understanding how economies function. Though there are a diversity of approaches in institutional economics, our focus is on the NIE, as that approach to date has received more emphasis in the blockchain literature.

Central to the NIE is the concept of institutions. In economic terms, institutions are the formal and informal "rules of the game". Ostrom distinguished between "rules in form" and "rules in use", focusing on the latter in much of her work, especially in *Governing the Commons* (1990). The governance of institutions refers to the way in which these rules are applied (usage) and made (collective action). These include laws and constitutions (formal rules) and norms and conventions (informal rules). Together, institutions and culture (values transmitted from one generation to the next) are thought to explain economic phenomena, including the wealth of nations [6]. Working rules, or rules in use, are central to much of the analysis, as the formal rules often leave much to the discretion of individuals working within organizations.

For institutionalists, the structure of rules reflects transaction costs. Transaction cost economics, proposed by Coase and elaborated by scholars such as Williamson, North, Demsetz, and others, emphasizes the economic cost of engaging in exchange. These costs include search costs, the costs of writing contracts, the cost of enforcing contracts, etc. Trust reduces transaction costs because the contracting and enforcement costs are lower. The general level of transaction costs can influence economic organization as shown by Williamson [74]: high transaction costs can be reduced by integrating contracting entities into the same firm and low transaction cost environments are more likely to be organized as markets. Governments and firms offer different solutions to transaction costs, and the extent of hierarchy is determined by them.

Institutionalists in what came to be known as the Bloomington School (as its primary founders, Elinor and Vincent Ostrom, were at IU-Bloomington for most of their careers) emphasize polycentricity as an important feature of any given social, economic, or political arrangement [1]. The defining feature of a polycentric enterprise is multiple levels of authority in an enterprise, each with shared autonomy. Polycentricity is both a description for organizational performance and a normative perspective on how an organization ought to be organized [69].

Though many of the institutionalists conceptualize of government as a key organization, or firms (a legal entity sanctioned by government), [54] recognized that informal associations or organizations can provide sources of order. A large literature considers how order arises outside government under conditions of anarchy [48]. A central conclusion here is that self-governance often works well, including when individuals establish their own rules [48]. Conventionally, these rules have focused on smaller-scale organizations, such as a prison gang [67], though increasingly, it has been used to analyze communities on the Internet, such as digital piracy [39], as well as electromagnetic spectrum management, which also has features of a commons and which opens robust opportunities for cooperation, including spectrum sharing [22].

3 Economic Conceptualizations of Blockchains

NIE is being applied to blockchain in several ways. These include (1) through conceptualization of blockchains as an institutional technology (as in the approach of institutional cryptoeconomics), (2) by describing blockchains as a knowledge commons and polycentric enterprise, and (3) through the IAD framework, NIE offers an empirical approach to analyze the structure of blockchains and their performance. Though there has been progress in each area, much of this research agenda has just begun.

A. Institutional Cryptoeconomics

As summarized by Davidson et al. [30], blockchain can be viewed in several different ways. First, it might be considered a general-purpose technology that can have a transformative effect across many sectors of the economy. Second, it might be considered a technology that reduces transaction costs and thus leads to a re-organization of firms, markets, and the functions they perform. Finally, it might be considered as an institutional technology, which views blockchains as a new way of organizing economic activity. As North [54] shows, these effects are linked with each other (i.e., good institutional technologies reduce transaction costs).

Through distinguishing blockchains from cryptocurrencies, Davidson et al [30] describe blockchain as a new institutional technology. Berg et al. [16] coin the approach institutional cryptoeconomics, given the view that blockchains is a new institutional building block for economies. They argue that the trustless nature of these distributed ledgers is what make blockchain a new institutional technology, especially when contrasted with the more common centralized ledgers that underpin the modern economy and property rights systems. These centralized ledgers, which capture the economic "ground truth" about the distribution of assets and wealth at any moment in time, require an organization or entity that must be trusted by the transacting parties. This entity may be a government, corporation, etc. Establishing and maintaining this trust is costly both in real economic terms and in the time it takes to establish that trust.

This perspective is how blockchains are defined as a new building block of the economy: they can substitute for some government and market functions. They are a non-market contracting arrangement. Legal institutionalists view law as having a central role in enabling market economies, especially as capitalism matures into industrial capitalism and later to managerial capitalism, where legal relations are central to the behavior of firms ([21], Chap. 3). Rather than rely on firms, blockchain enables contracting without any hierarchies. Smart contracts and especially distributed autonomous organizations (DAOs), discussed below, are a manifestation of this view. Since DAOs enable contracting with other DAOs and with people, they come close to the realization of Coase [26] and Jensen and Meckling [44] view of firms as a nexus of contracts—but they can do so without requiring any legal relationship, as conventionally understood as enforced by third parties (though if disputes arise about smart contracts, parties may appeal to the law to enforce them).

B. Blockchains as Knowledge Commons and Polycentric Enterprises

Ostrom's analysis provides a framework to analyze not only specific blockchains, but any blockchain and its relationship to government. The idea of polycentrism in governance was developed to describe the overlapping jurisdictions that often characterize metropolitan government [27]. In their work the Ostroms and their colleagues showed that these apparently confusing overlaps resulted in more efficient outcomes across almost all measures than a more structured approach to metropolitan governance. In their later work, this concept was applied to the governance of common pool resource systems, where they showed that what [38] called the tragedy of the commons was not a necessary outcome of commons governance, and that "enclosure" or privatization was not always the most efficient response to commons. The Ostroms, together with their colleagues in the Bloomington Workshop on Political Economy, showed that resource commons (such as fisheries, forests, and irrigation systems) could be sustainably managed through appropriate governance. Polycentrism applies to commons governance as well because a resource commons may exist in the context of multiple local, regional, or national government jurisdictions. In the body of her work, Ostrom showed that a factor in successful commons governance was delegation of authority to manage the commons by governmental authorities [56].

The idea of "commons", applied initially to natural resources, has subsequently been applied to domains of knowledge [41] and to innovation and technological commons [62], such as electromagnetic spectrum [70]. Rather than privatization, open commons is explored as something that can work with the commons [15, 72]. Blockchain systems exhibit many characteristics of a commons: the state of the ledger is the shared resource that must be managed, and the code base determines the agreed upon rules for this management.

Legal researchers interested in knowledge commons refined the concept of knowledge commons in developing the Governing Knowledge Commons (GKC) framework. As Frischmann et al. [33] emphasize, the GKC framework focuses on an approach to the governing and management of objects of the human mind, such as knowledge, information, culture, etc. As such, it distinguishes itself from Ostrom's earlier work, which focused on natural resources as the object of governance and management. The GKC approach, and the related innovations commons perspective, recognizes that new technologies often depend for their success on their openness [3].

The open-source aspects of permissionless blockchain constitute a forum for adaptation and change in response to new opportunities and problems as they arise. In this regard, it parallels defenses of peer production in the Internet [13]. Allen et al. [4] explain blockchain as a way to govern a knowledge commons through shared knowledge of who owns what, including relevant rules for ownership of property. Blockchain is especially significant as a technology to verify facts—who owns what, what is owned, etc.

The GKC approach offers novel insights into blockchains and their governance. It recognizes that while some aspects of blockchains may be private property, such as individuals owning tokens, or smart contracts being written over real-world property, blockchain networks depend on shared technology [18]. And in the case of permissionless blockchains, the open character of participation leads to governance dilemmas similar to other knowledge commons, such as the Internet and peer production communities [53].

C. An Empirical Research Agenda

Blockchain is also a polycentric enterprise in which the performance of any given network depends on rules internal to blockchains (protocols, collective choice rules) rules external to the blockchain (laws and regulations). This provides an important opportunity to apply the Institutional Analysis and Design (IAD) framework of Elinor [58].

The IAD framework generalizes the earlier conclusions about self-governance of resource commons, which prioritized clearly defined boundaries to limit open competition over resources, inclusive decision-making rules, appropriate and inclusive dispute resolution rules, and forums to resolve disputes, along with some semblance of external recognition of community rights to organize. The IAD framework is meant to explain and predict outcomes as a consequence of features of an action arena and the rules governing them as a way to understand the governance of common pool resource systems. The action arena can essentially be anything, as it is applied to any domain of collective decision-making; several authors have considered its applicability to blockchains [8].

There are seven types of rules in the IAD framework. Position rules are the number of possible positions for actors in the action situation (which may include a formal role just as a job or an informal one such as a social role). Some of these rules in a blockchain include founders, investors, and token holders. Boundary rules are the characteristics or requirements to participate in a position role. For a blockchain, these include rules such as transaction fees and requirements to vote based on staked tokens. Choice rules specify capacity for action for individuals in a position (voting yes or no, proposing upgrades, selling votes, etc.) Aggregation rules relate how interactions between participants in the action situation are accumulated into final outcomes (specific voting rules, such as quadratic voting). Information rules specify the types of information and information channels available to participants in their respective positions. Pay-off rules are the likely rewards for participation in the action situation. Scope rules are any criteria for requirements that exist for the final outcomes from the action situation.

Combining insights from the IAD and GKC perspectives, the empirical process for analyzing a blockchain would then look something like this:

1. Description of the origin, history, and operation of a given blockchain network.
2. Description of the code-based rules and informal norms that are used for governance of the blockchain.
3. Clarify the ways in which legal rules and regulations influence the performance and autonomy of a given blockchain network, including the changes in these rules over time.
4. Analysis of the relevant contract, property, and other regimes relevant to the blockchain.
5. How blockchain users relate to users outside the networks (such as Bitcoin users versus users of conventional currency), as well as disputes and their resolution among users and non-users.
6. Explication of dispute resolution both within the blockchain and external to it.

Blockchains provide a significant opportunity for comparative institutional analysis because there are many way to organize a blockchain network. Within those constitutional rules, essentially any voting system can be experimented with. One possibility is quadratic voting—individuals pay for as many votes as they wish using a number of "voice credits" in the votes they buy [46]—which can provide a better way than majority voting. This view is an application of radical markets ideas of Posner and Weyl [61], which conceptualize of new ways of property ownership and elections to address inequities arising from power concentrations in both markets and politics. Blockchain, as a decentralized forum, is a radical organizational form that opens up ideas for experimentation with new voting rules.

An Ostromian analysis offers insight into how governance institutions arise in response to threats from concentration of power; the extent to which blockchains are fully democratic, as envisioned by Nakamoto's white paper on permissionless blockchains; how exit and voice are exercised in blockchains, and how that exit and voice is influenced by rules in the blockchain network. Disputes arise in blockchain; the only question is what institutions emerge to resolve them, and how well they work.

4 Permissionless Blockchains and Their Governance

One of the distinguishing features of NIE is that it is useful for considering systems as dynamically responding to change [55]. As stakeholder preferences, environmental context, etc. change, so must the formal and informal institutions that govern economic behavior. For example, external pressure to reduce the energy consumption of Proof-of-Work consensus mechanism has stimulated changes (e.g., Ethereum is moving to Proof-of-Stake as of this writing), as well, the concentration of mining activity and the creation of mining pools was not anticipated by Nakamoto, which may lead to changes in the consensus mechanism. The manner in which institutions change is through their governance.

In blockchains, there are several clear governance situations. Some examples of these are:

- Updating the ledger is a governance action because recorded the state of the world changes. Because blockchain is distributed, it requires agreement among stakeholders to make this kind of change. In blockchain systems, this governance action is automated and is encoded into software.
- It may occur that the software governing the adding of transactions must change. This may occur because of the discovery of software bugs, improvements in the execution of code, etc. but may also be because of the need to change operations, such as Ethereum's migration from Proof-of-Work to Proof-of-Stake.
- Blockchain-based systems, especially those executing "smart contracts" exist within the existing legal frameworks. Thus, blockchain systems must respond to changes in this institutional context. As well, blockchain systems may seek to influence this institutional framework.

Thus, the idea of polycentrism as a framework for understanding blockchain governance is useful. The governance functions, including those described above, are "nested" within each other. Each operates in a different way. Appropriate delegation of authority to the different governance layers contributes to the efficient dynamic operation and sustainability of each blockchain's (eco)system. Polycentric organization allows for independent innovation at each governance layer.

A. Institutional governance within blockchains

In blockchain systems, usage rules are built into the system and its protocols and are thus fixed in software. As described above, these software systems may require a change from time to time for several reasons. Executing governance is codified in collective action rules, which vary across blockchains [64]. For example, Bitcoin's governance is through consensus among the stakeholders. Because there is no central organization that controls this, these protocol changes are often made through voting and by stakeholders adopting the updated software. Other blockchains may choose other methods; for example, issuing governance tokens that determine the right to vote on protocols changes.

When governance efforts fail to reach consensus, an outcome may be "protocol forks", in which the underlying ledgers may become incompatible with each other over time. This is particularly the case with significant changes (e.g., changes in the block size) that are not backwards compatible. With a "hard fork", identical blockchains with a similar history emerge from a schism. One of the most significant hard forks, which also illustrate governance dilemmas on blockchain networks, is The DAO hack (which resulted in Ethereum splitting into Ethereum and Ethereum Classic). Importantly, the fork was resolved in autonomous fashion, but through coalitions of developers and programmers. Another was the disputes over transaction ordering that gave rise to the schism on Bitcoin and its division into Bitcoin and Bitcoin cash. As Alston et al. [5] explain, governance dilemmas illustrate that the machines cannot govern themselves, thus requiring consideration of leadership in blockchain networks.

B. Consensus

Since (permissionless) blockchains do not rely on a central institution, they must build consensus about the correct state of the world (i.e., state of the ledger) through a distributed mechanism. An effective consensus mechanism must be designed to provide technical or economic incentives to discourage or eliminate "cheating" (i.e., recording incorrect information on the consensus ledger). Consensus approaches employed by blockchain systems make entering illegitimate transactions costly (in the case of Proof-of-Work) or risky (in the case of Proof-of-Stake). If the cost is sufficiently high or the risk sufficiently great, individuals will be deterred from making false entries.

An important part of the question of trust is the reliability of the consensus mechanism. While some systems may be initially reliable, this can change. For example, as discussed above, the concentration of mining power in the Proof-of-Work mechanisms that we are witnessing today was not anticipated in Nakamoto's white paper. This leads to the real possibility that large mining pools could confederate to successfully conduct a "51% attack", in which control over blockchain entries lapses to a small group of miners, resulting in a significant possibility of fraud.

C. Governance External To Blockchains

There are many avenues to consider the nested aspects of blockchain. The issue posed by technology is not one of eliminating the need for law, but for law to adapt to new conditions. The issue is that many relations previously defined by vertical relations are increasingly horizontal. Those horizontal aspects constitute the "flat" world [36]. Blockchain, as a new form of trust, has done much to flatten the world, so much so that some still contend that it reduces the need for government and law.

For some applications, blockchains compete with the government. For example, the use of cryptocurrencies for illegal transactions can undermine government. Ransomware paid in Bitcoins is another example where privacy provided by cryptocurrencies can undermine governments.

Despite such Blockchain is polycentric enterprise. On one hand, governments can and do regulate blockchain. People who sign smart contracts want them to have legal force. States such as Wyoming have recognized blockchain smart contracts, as well as regulated banks. El Salvador passed a law that required accepting Bitcoin alongside the US dollar. It was highly controversial: the government thought it would provide lower fees, and savings opportunities, but institutions such as the World Bank criticized the move, as did the IMF.

A wealth of legal studies have also shown that blockchain depends on the law. Smart contracts, discussed above, are an example. The evolving legal consensus is that law can adapt, and that law for blockchain—what de Primavera and Wright [32] call lex cryptographia—is good for blockchain. Wyoming wants to be the cryptocapital of the world.[1] It is doing so by enacting dozens of blockchain-friendly regulations. Whether these are successful is an open question for future research. What is clear is that they illustrate blockchain, rather than an island, as a nested enterprise.

---

[1] https://slate.com/technology/2021/06/wyoming-cryptocurrency-laws.html.

Oracles are an example. An oracle is necessary to translate real-world data into a blockchain. In this regard, oracles are the interface between the real world and the blockchain network [2]. Hence, they are subjected to disputes and require negotiation. Indeed, a body of such negotiation law has emerged to address disputes, including those involving oracles, and a market for oracle services has also emerged [60].

5   Smart Contracts and Their Governance

Some permissionless blockchain systems support a scripting language that allows for automated program execution. These programs are referred to as smart contracts because they can support a transaction between parties when certain pre-determined conditions are met.[2] These smart contracts have been hailed by some as having the potential to replace traditional contracts. Werbach [71], Alston et al. [8], and others have argued that these smart contracts still exist within the traditional legal system, so that they do not avoid traditional forms of governance (e.g., arbitration, legal proceedings). This occurs for several reasons, including the observation that no contract can anticipate all future states of the world (principle of incomplete contracts) and because, in many cases, the smart contracts are between humans (maybe after several layers of indirection) and that humans may use the legal system to adjudicate disputes.[3] This "nesting" of governance regimes, from the legal framework to the smart contracts, is very much in line with the notions of polycentric governance that was developed by Vincent and Elinor Ostrom [59]. This framework allows for local management of resources and resolution of disputes where possible, which is consistent with the emergent regimes in blockchain-based contract governance.

Howell and Potgieter [42] examine smart contracts in the context of the NIE framework. They observe that smart contracts are not contracts in the economic sense, but rather they are a static automating of an agreement reached independently between human actors. Given this, and the fact that they are indelible, they are not easily adaptable to traditional contractual responses to uncertainty, including renegotiation, relational contracting, arbitration, etc. In this context, they anticipate that smart contracts will emerge more slowly and in well-defined contexts for some time. As well, they anticipate interest in "self-driving contracts", or contracts that are able to adapt over time using Artificial Intelligence (AI) technologies [24]. That said, as shown by the experiment presented in Shay et al. [66], automated enforcement of laws and rules is far from simple and obvious.

---

[2] A simple example of a smart contract is a vending machine. When one party enters the correct payment, the counterparty (the vending machine) automatically dispenses an item.

[3] Following on with the vending machine example, most vending machines clearly provide a telephone number or other contact information in case of malfunction. For example, if the machine does not fulfill its part of the smart contract, the human initiator (purchaser) may seek recourse with the (human) vendor.

Nor is the implementation of a law governing blockchain contracts as simple as it may seem. Lemieux [49] considers some additional challenges to smart contracts. For example, Arizona's smart contract law defines a smart contract as "An event-driven program, with state, that runs on a distributed, decentralized, shared and replicated ledger and that can take custody over and instruct transfer of assets on that ledger". Following the usual laws of contract, Arizona's law recognizes a contract in full force when it is digitally signed, witnessed, validated, confirmed, and entered into a blockchain ledger by a pre-determined number of notes, in which case it can no longer be repudiated. With blockchains, non-repudiation is confirmation of a transaction. However, Arizona's blockchain law does not definitively answer how many nodes must be updated before a transaction is confirmed, an answer which depends on the design of the blockchain, and there is ambiguity in the phrase "that runs", rather than "has run", since the former suggests that draft contracts have full legal status.

This example also illustrates the polycentric aspects of blockchains: regardless of the challenges, there is something to be gained from formalization. Individuals using these smart contracts petition governments to enforce them, and in many cases, the government is attempting to adapt law to a ledger-centric contracting environment.

6 Blockchain and Government

One of the features of NIE is that it can be used to compare alternative institutions, especially governments and markets. Markets tend to be better at identifying what and how to produce goods and services. Governments tend to be better at enforcing agreements. Since blockchains are self-enforcing, to an extent, they eliminate a role for the government in enforcing agreements.

The institutional technology view sees blockchain as an alternative to government, as it replaces government. But as noted above, that is not the only way to see blockchain. Here, we suggest that for some applications, blockchain it indeed a new technology of freedom. At the same time, its database features suggest it will be a way to revolutionize public sector governance.

A. Blockchains as a New Technology of Freedom

Part of the emphasis in the economics of blockchain is that it provides freedom from government, including government regulation [10]. In some ways, blockchain is the ultimate source of consumer sovereignty, as some cryptocurrencies provide for nearly entirely anonymous transactions. Suppose one is concerned about what one purchases online (online pills, fetishes, etc.). Cryptocurrencies like z-cash provide great privacy in transaction, even compared to Bitcoin.[4] Through crypto-secession, individuals may even be able to choose their own rules to govern themselves politically [2].

---

[4] As Bitcoin relies on a public address, enough transactions can make it easier to identify the owner of that address. Z-cash emerges to meet the demands of people who demand even greater privacy. See https://z.cash/.

Central to this approach is the idea that blockchain provides an exit option—in this case from government. One of the central insights of economic approaches is that people who are engaged in otherwise safe behavior ought to be able to do so without worrying too much about government, save for the occasional and reasonable tax on activities to address social costs of activities. Cowen [28] considers the implications of DAOs for governance. For any given firm, the government may wish to assert control over activities. Surveillance of small, informal networks is costly. But a handful of firms served by a few telecommunications companies and a small number of content providers is the target, state actors can more easily target the network and, ultimately, collect information on people's public and private lives. In the best case scenario, a bilateral bargaining situation emerges where firms can challenge the government, but in most situations, the government has a preponderance of bargaining power and can make things difficult for firms and, more importantly, for customers.

Cowen's example is attempted by the government to regulate sex work, including the US government prohibition and seizure of Backpage, a website for sex workers to meet and vet clients. Suppressing sex work through the harassment of participants (usually female sex workers) is costly, but the government can more easily prohibit platforms that advertise sex work and remotely coordinate sex activities. Private governance becomes a challenge given the usual platform technology.

Blockchain changes the game. Previously, sex work platforms had personal or commercial owners easily identified. DAOs, in contrast, have no owners and remain as long as people sustain them. Enter SpankChain, a blockchain startup built on Ethereum for sex workers that promises a safe and secure framework to regulate sex activities of consulting adults outside the reach of state regulation.

SpankChain is a blockchain-based payment processing service for adult live streaming. The SpankChain gives people a way to accept crypto and "monetize yourself".[5] It promises low fees, no chargebacks (as the system is built on self-executing smart contracts), anonymity for performers and views, and safety and security that comes with a decentralized, distributed ledger technology. Members of the community-owned bank, SpankBank, stake claims in SPANK, and receive BOOTY, which is used for fees, tips, and transactions. SpankChain has also gotten in on NFTs and offers digital porn.

Even though SpankChain is decentralized, it is not fully so. The company still requires users to trust SpankChain since they provide the only channel to pay into. SpankChain also has the responsibility to send the correct amount of cryptocurrency payments to each performer.

The sex work examples are an opportunity for voluntary contracting parties to avoid government, and arguably improve the economic welfare of participants. Government does retain some interest in this application because there is a real risk of non-voluntary contracting, as might result from human trafficking, which violates

---

[5] https://spankchain.com/.

many other laws that governments may have in place. Government may have a more challenging time tracking down participants, given anonymity of blockchains, so we might anticipate some accommodation from SpankChain to support these laws and social norms. Still, there remain the challenges of power in the blockchain—it is not fully peer to peer—and governments could in principle prohibit such uses of blockchain, though success in regulating something such as this may be questionable.

Significantly, Rozas et al. [65] are developing a novel Ostromian approach to understand how blockchains relate to commons-based peer production (CBPP) communities. This research agenda marries insights from Benkler with Ostrom in considering the ways in which blockchains can unleash prospects for self-governance in CBPP communities. Here, as with the example above, blockchains enable prospects for self-governance. The difference is that [65] focus on peer production as self-governance, while Cowen's [29] example of SpankChain recognizes that blockchains have the potential to liberate businesses that have been subjected to predatory regulations or regulatory indifference. In each case—CBPP and industries that may confront pernicious regulation—blockchains are seen as enabling self-governance.

B. Blockchain as Smart Public Sector Governance

There is also the fact that blockchain is just a better database for many applications [23]. Blockchains can be used to create more secure digital identities. These digital identities can be linked to government services, such as payments for healthcare. Government healthcare records, which are already centralized, and hence subject to vulnerabilities from criminals or simply mistakes, could be put on a blockchain to reduce the risks from fraud and abuse. Paperwork, one of the great features of government, could be greatly reduced. Though blockchain has only begun to realize its promise for real estate [9], governments could facilitate moving property transactions to a blockchain. This would be of great benefit to anyone seeking to reduce reliance on third parties (brokers, mortgage lenders).

Still, there remains uncertainty about how effective blockchains will be in recording government information. There remains the challenge of first digitalizing records, which is a non-trivial administrative task in many political jurisdictions which currently rely on substantial paperwork. As Lemieux [49] points out, blockchain record keeping concentrates power in the hands of a few social actors—a techno-plutocracy—without the usual guarantees of rule of law.

These issues suggest further comparisons, including to Internet governance. Benkler [14] described a similar problem in Internet governance, as power concentrations undermined prospects for a fully open Internet (and even more obvious challenge given the prominence of Big Tech and their ability to assert substantial influence by de-platforming). This market power undermines prospects for peer mutualism, as well as peer anarchy [12]. Writing laws is also tricky, as the smart contract example illustrates. For example, Vermont's blockchain law which recognizes digital records recorded in a blockchain as legally binding requires that a person certifies it, but without clarifying what constitutes a "qualified person", opening the possibility for disputes [49].

C. Polycentrism as a Normative Goal

The concept of polycentrism is both descriptive and normative. As a descriptive matter, blockchains are generally polycentric: they generally depend on their effectiveness in the interrelationship between multiple levels of governance, both internal to and external to a blockchain.

Polycentrism is also a normative goal. Based on the previous discussion, governments arguably should recognize more opportunities for blockchain. Individuals may wish to set up cities on a blockchain. Since cities in general receive their authority from higher levels of government, the extent of autonomy reflects a political choice. Blockchains open up realms for experimentation with government. Thus, according to normative principles of self-governance, it may be desirable to allow such experimentation. But this also means that there are few reasons to expect blockchains to fully compete with a government. People may choose to put their groups on a blockchain, but as long as they have to live somewhere, they require that the government provides them with autonomy to do so. Indeed, one of Elinor Ostrom's most significant design principles for self-governance was that higher levels respect community autonomy. For many applications of blockchains that promise to provide for more choice, and more local governance, blockchains offer opportunities for experimentation.

In this regard, blockchain creates opportunities for new types of what [68] calls foot voting. Somin's examples were moving from jurisdictions or seeking out associations, such as a condo association. Blockchain allows more of that, provided the state is on board. People may foot vote by forming a blockchain city, or a blockchain neighborhood, with transactions on blockchains and taxes paid in cryptocurrency. That might realize Vincent [57] view that polycentrism is necessary for self-governance. In this case, self-governance depends on higher level political benediction, but the prospects for such self-governance are expanded because of blockchain.

7 Summary and Future Directions

The growing literature on the crypto-economics as well as the increasing public attention being paid to blockchains suggest that a systematic approach to understanding blockchains as an element of human and economic systems is increasingly important. To that end, Allen et al. [4] have proposed a language for describing and analyzing blockchain systems that is rooted in the IAD and GKC frameworks. The proliferation of blockchain-based systems increases the usefulness of such a language to allow for the longitudinal analysis of such systems. This ability to perform comparative analysis is useful for engineers and computer scientists who are building blockchain-based systems, investors seeking to understand the underlying value of their investments and economists who are seeking deeper understanding of this emerging phenomenon. This would be a potentially productive domain of collaboration between institutional economists, engineers, and computer scientists due to the many dimensions that can be explored.

Engineers and computer scientists who are engaged in system development often face the choice of which blockchain-based system is most suitable for the application they are building. If none is suitable, it may be necessary to build a new blockchain system with the desired characteristics. While "suitability" is often thought of in terms of system performance measures, it is equally important that the economic features of the blockchain in the system be understood so that the goals of the system's users are aligned with the function of the system. While the ideas of polycentrism and governance were introduced in this paper for the purpose of analyzing blockchain-based systems, they are also useful more broadly for systems (whether or not they use blockchains).

Digital currencies based on blockchain (such as Bitcoin, Litecoin, Dogecoin, etc.) have been the object of financial speculation in recent years and have garnered significant public attention due to the significant appreciation of the value of these (and other) digital currencies. For people wishing to engage in this activity, the clarity of understanding offered by the IAD/KC framework(s) can provide insight into the relative characteristics of the digital currency and thus provide some understanding of the potential (future) value of the underlying system.

Although the economic literature on crypto-economics is growing, it is still very much in the formative stages, with perhaps a dozen active research groups worldwide. As described above, the early work has proven to be helpful in the development of an analytical framework, but there is much more that could be done. For one, the current body of literature lacks an empirical foundation, so the development of a consistent body of data to support the longitudinal analysis of blockchain systems would benefit both the technological and economic communities. Ostrom [56] showed the value of this approach in her work on common pool resource systems, as her work contributed to economics, political science, and anthropology (and more). As with common pool resource systems, an empirically oriented longitudinal analysis could contribute to the understanding of the essential characteristics of successful and sustainable blockchain-based systems.

Blockchains are slowly making their way into government, and, as of this writing, there is much more to be understood about blockchain and government. When governmental leaders acknowledge and accept (perhaps even embrace) blockchain systems, the idea of polycentrism contributes usefully to the framework for this coexistence and collaboration. There is some work in this tradition [34]. There are many potential roles for blockchains in government or for use in governmental systems. For example, blockchains can be used to record land rights, including boundaries and uses. Blockchains can also be used to support elections and voting systems, protecting against ex post-vote tampering. Another application of blockchains is public procurement and delivery of services, including cash payments from governments. As government transactions can be subject to fraud or errors, an immutable record of transactions can reduce these risks.

Though not our focus, much of the debate on blockchain centers on its role as a currency as it relates to "digital currency", and how this relates to government (see, e.g., [40, 50]. Since these currencies offer an alternative to fiat currencies, governments around the world are evaluating different relationships with these digital currencies. As of this writing, one country (El Salvador) will begin accepting Bitcoin as a legitimate medium of exchange [45]. Other countries[6] are proposing to develop a government sponsored "official" cryptocurrency that would enable them to maintain a level of control and monitoring not possible with "public" cryptocurrencies (such as Bitcoin). Other governments seek to ban cryptocurrencies outright while many others seek an "appropriate" regulatory relationship.[7] There is every reason to believe that the relationship between governments and cryptocurrencies will be dynamic and continue to evolve as regulators gain a richer understanding of what can be regulated and what should be regulated (and to what end). A significant question is how governance perspectives in the NIE tradition relate to formalism using microeconomic theory more extensively. Halaburda, Haeringer, Gans, and Gandal (forthcoming), consider the dynamics of supply, demand, trading price, and competition among cryptocurrencies. They consider economic questions—with disagreements (usually termed a "fork"), how are those disagreements resolved? Bakos and Halaburda [11] develop a model of transaction safety in permissioned and permissionless blockchains to study this tradeoff and find that in several settings there may be no tradeoff at all. With a minimal level of trust in the blockchain operators and the supporting institutions, well-designed permissioned blockchains can offer both higher operational efficiency and higher transaction security. Models of platform competition—microeconomics of cryptocurrency considers the dynamics of competition (Gandal and Halaburda [35]).

Institutionalists offer a complementary view of blockchains. An institutional approach might consider the ways in which competition among blockchain platforms influences the evolution qualities and features of blockchains. Another complementary feature is to recognize that there is a diversity of qualities and features of blockchains. This suggests that binaries of public and private might be less useful than considering publicness and privateness as a continuum. In addition, Ostromian perspectives would suggest that while microeconomics offers precise analysis of the tradeoffs that come with differences in blockchain networks, the real world will include a diversity of institutions that cannot be predicted by any theory.

References

1. Aligica, P., Boettke, P.: Challenging Institutional Analysis and Development: The Bloomington School. Routledge (2009)
2. Allen, D.W., Lane, A.M., Poblet, M.: The governance of blockchain dispute resolution. Harv. Negot. Law Rev. **24**, 75–101 (2019)
3. Allen, D.W., Potts, J.: How innovation commons contribute to discovering and developing new technologies. Int. J. Commons **10**(2), 1035–1054 (2016)

---

[6] Venezuela attempted to develop a cryptocurrency, an attempt that was ultimately unsuccessful [17]. China is developing a "digital yuan" based on blockchain [20].

[7] For an overview of the legal status of blockchain across the world, see https://en.wikipedia.org/wiki/Legality_of_bitcoin_by_country_or_territory (retrieved 28 June 2021).

4. Allen, D., Berg, C., Davidson, S., MacDonald, T., Potts, J.: Building a grammar of blockchain governance. Medium (2021). Retrieved from https://medium.com/cryptoeconomics-australia/building-a-grammar-of-blockchain-governance-c2cb4b70f915, 23 June 2021
5. Alston, E., Law, W., Murtazashvili, I., Weiss, M.: Blockchains networks as constitutional and competitive polycentric orders. J. Inst. Econ. (2022). https://doi.org/10.1017/S174413742100093X
6. Alston, E., Alston, L.J., Mueller, B., Nonnenmacher, T.: Institutional and Organizational Analysis: Concepts and Applications. Cambridge University Press (2018)
7. Alston, E.: Constitutions and blockchains: competitive governance of fundamental rule sets. Case West. J. Law Technol. Internet (2020)
8. Alston, E., Law, W., Murtazashvili, I., Weiss, M.B.: Can permissionless blockchains avoid governance and the law? Notre Dame J. Emerg. Tech 2(1). https://ndlsjet.com/wp-content/uploads/2021/04/E.-Alston-Permissionless-Blockchains.pdf (2021)
9. Arruñada, B.: Blockchain's struggle to deliver impersonal exchange. Minn. J. Law Sci. Technol. 19, 55–106 (2018)
10. Atzori, M.: Blockchain technology and decentralized governance: is the state still necessary? SSRN 2709713 (2015)
11. Bakos, Y., Halaburda, H.: Tradeoffs in Permissioned vs Permissionless Blockchains: Trust and Performance. NYU Stern School of Business working paper. https://papers.ssrn.com/sol3/papers.cfm?abstract_id=3789425 (2020)
12. Benkler, Y.: Practical anarchism: peer mutualism, market power, and the fallible state. Polit. Soc. 41(2), 213–251 (2013)
13. Benkler, Y.: Peer Production and Cooperation. Edward Elgar Publishing, In Handbook on the Economics of the Internet (2016)
14. Benkler, Y.: Degrees of freedom, dimensions of power. Daedalus 145(1), 18–32 (2016)
15. Benkler, Y.: The political economy of commons. Upgrade Eur. J. Inform. Prof. 4(3), 6–9 (2003)
16. Berg, C., Davidson, S., Potts, J.: Understanding the Blockchain Economy: An Introduction to Institutional Cryptoeconomics. Edward Elgar Publishing (2019)
17. Berning, J.: The strange story of venezuela's failed cryptocurrency. Freethink 12 Feb 2021 (2021). Retrieved from https://www.freethink.com/shows/coded/season-3/venezuela-cryptocurrency, 28 June 2021
18. Bodon, H., Bustamante, P., Gomez, M., Krishnamurthy, P., Madison, M. J., Murtazashvili, I., Murtazashvili, J.B., Mylovanov, T., Weiss, M.B.: Ostrom amongst the machines: blockchain as a knowledge commons. Cosmos + Taxis 10:3+4
19. Brennan, G., Buchanan, J.M.: The Reason of Rules: Constitutional Political Economy. Cambridge Univeristy Press, New York (1985)
20. Broby, D.: China's digital currency could be the future of money – but does it threaten global stability? The Conversation 10 May 2021 (2021). Retrieved from https://theconversation.com/chinas-digital-currency-could-be-the-future-of-money-but-does-it-threaten-global-stability-160560, 28 June 2021
21. Bromley, D.W.: Possessive Individualism: A Crisis of Capitalism. Oxford University Press, New York (2019)
22. Bustamante, P., Gomez, M.M., Murtazashvili, I., Weiss, M.B.: Spectrum anarchy: why self-governance of the radio spectrum works better than we think. J. Inst. Econ. (2020)
23. Bustamante, P., Cai, M., Gomez, M., Harris, C., Krishnamurthy, P., Law, W., Madison, M.J., Murtazashvili I., Murtazashvili J.B., Mylovanov, T., Shapoval, N., Vee, A., Weiss, M.: Government by Code? Blockchain applications to public sector governance Frontiers in Blockchain. vol 5 (2022). https://doi.org/10.3389/fbloc.2022.869665 ISSN=2624-7852
24. Casey, A.J., Niblett, A.: Self-driving contracts. J. Corp. Law 43(1), 1–33 (2017)
25. Catalani, C., Gans, J.S.: Some simple economics of the blockchain. National Bureau of Economic Research (2016)
26. Coase, R.H.: The nature of the firm. Economica 4(16), 386–405 (1937)
27. Cole, D.H., McGinnis, M.D. (eds.): Elinor ostrom and the bloomington school of political economy: polycentricity in public administration and political science. Lexington Books (2015). ISBN 978-0-7391-9101-9

28. Cowen, N.: Markets for rules: the promise and peril of blockchain distributed governance. J. Entrep. Public Policy **9**(2), 213–226 (2020)
29. Cowen, N.: Markets for rules: the promise and peril of blockchain distributed governance. J. Entrep. Public Policy. vol. **9**, no. 2, pp. 213–226 (2019)
30. Davidson, S., De Filippia, P., Potts, J.: Blockchains and the economic institutions of capitalism. J. Inst. Econ. **14**(4), 639–658 (2018)
31. De Filippi, P., Wright, A.: Blockchain and the Law: The Rule of Code. Harvard University Press (2018)
32. Filippi, P.D., Wright, A.: Blockchain and the law: The rule of code. Harvard University Press (2018)
33. Frischmann, B.M., Madison, M.J., Strandburg, K.J.: Governing knowledge commons. Oxford (2014). ISBN: 9780199972036
34. Frolov, D.: Blockchain and institutional complexity: an extended institutional approach. J. Inst. Econ. 1–16 (2020)
35. Gandal, N., Halaburda, H.: Can we predict the winner in a market with network effects? Competition in cryptocurrency market. Games **7**(16), (2016). https://doi.org/10.3390/g70 30016
36. Hadfield, G.: Rules for a Flat World: Why Humans Invented Law and How to Reinvent it for a Complex Global Economy. Oxford University Press, New York (2016)
37. Halaburda, H., Haeringer, G., Gans, J., Gandal, N.: The Microeconomics of cryptocurrencies. J. Econ. Lit. (forthcoming)
38. Hardin, G.: The tragedy of the commons. Science **162**, 1243–1248 (1968). https://doi.org/10.1126/science.162.3859.1243
39. Harris, C., Cai, M., Murtazashvili, I., Murtazashvili, J.B.: The Origins and Consequences of Property Rights: Austrian, Public Choice, and Institutional Economics Perspectives. Cambridge University Press
40. Hendrickson, J.R., Hogan, T.L., Luther, W.J.: The political economy of bitcoin. Econ. Inq. **54**(2), 925–939 (2016)
41. Hess, C., Ostrom, E.: Understanding knowledge as a commons. Cambridge, MIT Press (2007)
42. Howell, B.E., Potgieter, P.H.: Uncertainty and dispute resolution for blockchain and smart contract institutions. J. Inst. Econ. 1–15 (2021)
43. Howell, B.E., Potgieter, P.H.: Uncertainty and dispute resolution for blockchain and smart contract institutions. J. Institutional Econ. vol. **17**, no. 4, pp. 545–559 (2021)
44. Jensen, M.C., Meckling, W.H.: Theory of the firm: managerial behavior, agency costs and ownership structure. J. Financ. Econ. **3**(4), 305–360 (1976)
45. Jones, S., Avelar, B.: El Salvador becomes first country to adopt bitcoin as legal tender. The Guardian (2021)
46. Lalley, S.P., Weyl, E.G.: Quadratic voting: how mechanism design can radicalize democracy. In AEA Papers and Proceedings, vol. 108, pp. 33–37 (2018)
47. Leeson, P.T.: Logic is a harsh mistress. J. Inst. Econ. (2020)
48. Leeson, P.T.: Anarchy Unbound: Why Self-Governance Works Better than you Think. Cambridge University Press (2014)
49. Lemieux, V.L.: Blockchain and Public Record Keeping: Of Temples, Prisons, and the (Re)Configuration of Power, vol. 2. Frontiers in Blockchain (2019). https://doi.org/10.3389/fbloc.2019.00005
50. Luther, W.J.: Getting off the ground: the case of bitcoin. J. Inst. Econ. **15**(2), 189–205 (2019)
51. Madison, M.J., Frischman, B.M., Strandburg, K.J.: Constructing commons in the cultural environment. Cornell Law Rev. **95**, 657–710 (2010)
52. McCloskey, D.: Max U versus Humanomics: a critique of neo-institutionalism. J. Inst. Econ. **12**(1), 1–27 (2016)
53. Murtazashvili, I., Murtazashvili, J.B., Weiss, M.B.H., & Madison, M. J. Blockchain Networks as Knowledge Commons. Int. J. Commons. **16**(1), 108–119 (2022). https://doi.org/10.5334/ijc.1146

54. North, D.: Institutions, Institutional Change, and Economic Performance. Cambridge University Press, Cambridge, MA (1990)
55. North, D.: Understanding the Process of Institutional Change. Cambridge University Press, New York (2005)
56. Ostrom, E.: Governing the Commons: The Evolution of Institutions for Collective Action. Cambridge University Press, New York (1990)
57. Ostrom, V.: The Meaning of American Federalism: constituting a Self-Governing Society. Institute for Contemporary Studies, San Francisco (1994)
58. Ostrom, E.: Understanding Institutional Diversity. Princeton University Press, Princeton (2005)
59. Ostrom, E.: Beyond markets and states: polycentric governance of complex economic systems. Am. Econ. Rev. (2010)
60. Poblet, M., Allen, D.W., Konashevych, O., Lane, A.M., Diaz Valdivia, C.A.: From Athens to the Blockchain: Oracles for Digital Democracy. Frontiers in Blockchain (2020)
61. Posner, E.A., Weyl, E.G.: Radical Markets: Uprooting Capitalism and Democracy for a Just Society. Princeton University Press, Princeton (2018)
62. Potts, J.: Governing the innovation commons. J. Institutional Econ. vol. 14, no. 6, pp. 1025–1047 (2018)
63. Rajagopalan, S.: Blockchain and Buchanan: code as a constitution. In: Buchanan, J.M , Wagner, R.E. (eds.) A Theorist of Political Economy and Social Philosophy, pp. 359–381. Palgrave Macmillan (2016)
64. Rajagopalan, S.: Blockchain and Buchanan: Code as Constitution. In: James, M.B., Wagner, R.E. (Eds.) A Theorist of Political Economy and Social Philosophy, pp. 359–381. Palgrave Macmillan (2019)
65. Rozas, D., Tenorio-Fornés, A., Díaz-Molina, S., Hassan, S.: When ostrom meets blockchain: exploring the potentials of blockchain for commons governance. SAGE Open 11(1), 21582440211002530 (2021)
66. Shay, L.A., Hartzog, W., Nelson, J., Conti, G.: Do robots dream of electric laws? An experiment in the law as algorithm. In: Calo, R., Froomkin, A.M., Kerr, I. (eds.) Robot Law, pp. 274–305. Edward Elgar Publishing, Cheltenham, UK (2016)
67. Skarbek, D.: The Puzzle of Prison Order: Why Life Behind Bars Varies Around the World. Oxford University Press (2020)
68. Somin, I.: Free to Move: Foot Voting, Migration, and Political Freedom. Oxford University Press, New York (2020)
69. Tarko, V.: Polycentricity. In: Melenovsky, C. (ed.) The Routledge Handbook of Philsophy, Politics, and Economics (forthcoming)
70. Weiss, M.B.H., Lehr, W.H., Acker, A., Gomez, M.M.: Socio-technical considerations for Spectrum Access System (SAS) design. IEEE Int. Symp. Dyn. Spectr. Access Netw. (DySPAN) 2015, 35–46 (2015). https://doi.org/10.1109/DySPAN.2015.7343848
71. Werbach, K.: The Blockchain and the New Architecture of Trust. MIT Press (2018)
72. Werbach, K.: Supercommons: toward a unified theory of wireless communications. Tex. Law Rev. 82, 863–973 (2004)
73. Williamson, O.: The economics of goverannce. Am. Econ. Rev. 95(2), 1–18 (2005)
74. Williamson, O.E.: Markets and hierarchies: analysis and antitrust implications: a study in the economics of internal organization. New York, Free Press (1975)

Printed in the United States
by Baker & Taylor Publisher Services